AutoCAD® 2002: Complete

AutoCAD® 2002: Complete

BILL BURCHARD
ART LIDDLE
DAVID PITZER

autodesk® Press

Australia • Canada • Mexico • Singapore • Spain • United Kingdom • United States

autodesk® Press

AutoCAD® 2002: Complete

Bill Burchard Art Liddle David Pitzer

AutoDesk Press Staff

Business Unit Director:
Alar Elken

Executive Editor:
Sandy Clark

Acquisitions Editor:
James DeVoe

Development Editor:
John Fisher

Editorial Assistant:
Jasmine Hartman

Executive Marketing Manager:
Maura Theriault

Channel Manager:
Mary Johnson

Marketing Coordinator:
Karen Smith

Executive Production Manager:
Mary Ellen Black

Production Manager:
Larry Main

Production Editor:
Tom Stover

Art/Design Coordinator:
Mary Beth Vought

Production Services:
TIPS Technical Publishing

Printed in Canada
1 2 3 4 5 XXX 05 04 03 02 01

For more information contact Autodesk Press, 3 Columbia Circle, PO Box 15015, Albany, NY 12212-5015.

Or find us on the World Wide Web at
http://www.autodeskpress.com

Library of Congress Cataloging-in-Publication Data
Burchard, Bill.
AutoCAD 2002 complete/
Bill Burchard, Dave Pitzer, Art Liddle.
p. cm.
ISBN 0-7668-4253-3
1. Computer graphics. 2. AutoCAD.
I. Pitzer, David. II. Liddle, Art. III. Title

T385 .B853 2002
620'.0042'02855369--dc21
2001047173

Notice To The Reader

Publisher and author do not warrant or guarantee any of the products described herein or perform any independent analysis in connection with any of the product information contained herein. Publisher an author do not assume and expressly disclaims any obligation to obtain and include information other than that provided to it by the manufacturer.

The reader is expressly warned to consider and adopt all safety precautions that might be indicated by the activities herein and to avoid all potential hazards. By following the instructions contained herein, the reader willingly assumes all risks in connection with such instructions.

Publisher and author make no representation or warranties of any kind, including but not limited to, the warranties of fitness for particular purpose or merchantability, nor are any such representations implied with respect to the material set forth herein, and the publisher takes no responsibility with respect to such material. Publisher and author shall not be liable for any special, consequential, or exemplary damages resulting, in whole or part, from the readers' use of, or reliance upon, this material.

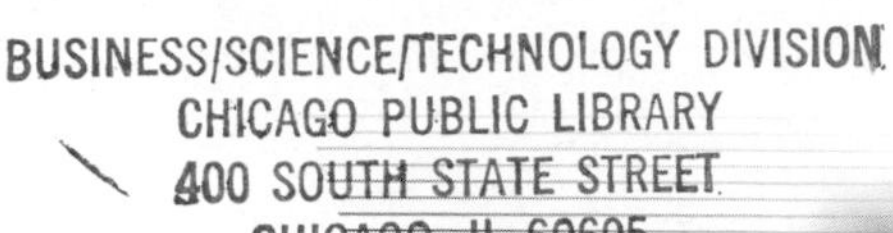

CONTENTS

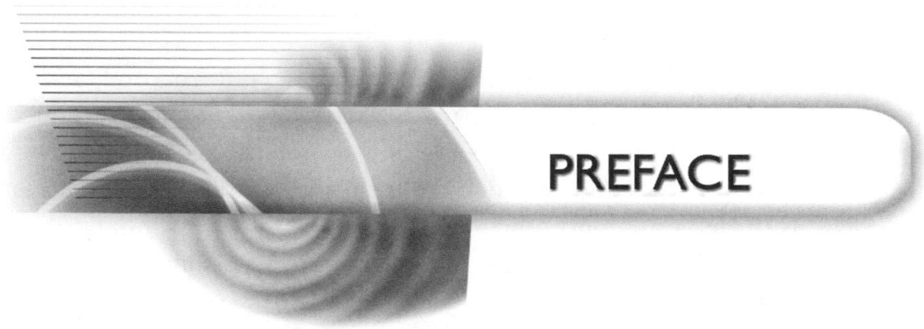

PREFACE

AutoCAD 2002 Complete is concisely written for those new to AutoCAD, as well as for experienced AutoCAD users, who need a clear, concise, easy-to-understand guide to using AutoCAD professionally. Because the book starts at the beginning level and progresses through more advanced topics chapter-by-chapter, those new to AutoCAD will easily learn the ins-and-outs of using this powerful application, while those already familiar with AutoCAD can easily zero-in on the advanced tools and techniques they need to master. Whether you are new to AutoCAD or a veteran AutoCAD user, AutoCAD 2002 Complete provides the insight you need to master AutoCAD's hidden power to quickly become an AutoCAD expert.

WHY YOU NEED THIS BOOK

AutoCAD 2002 Complete is written for both beginners and experienced AutoCAD users who need a clear, concise, easy-to-understand guide to using AutoCAD in today's professional marketplace, and who want to master the tools and techniques needed to use AutoCAD 2002 to create 2D drawings as well as create and render professional-level 3D models.

AutoCAD 2002 Complete is written by experts who incorporate their knowledge of using AutoCAD in today's professional marketplace. You start by learning AutoCAD's simple tools, and progress through step-by-step examples of creating 2D drawings, into creating and rendering professional-level 3D models. This book begins by laying down a solid knowledge foundation on how to start new projects with AutoCAD, and progresses into creating and editing drawings, as well as annotating, hatching, dimensioning, and plotting 2D drawings. It then builds upon this foundation by clearly explaining AutoCAD's 3D modeling tools and concepts, including photo-realistic rendering, and finishes by covering advanced concepts, such as using Object Linking and Embedding (OLE) and working with external databases.

WHAT YOU WILL FIND IN THIS BOOK

AutoCAD 2002 Complete details using AutoCAD's linetypes and lineweights, as well as mastering AutoCAD's user-defined coordinates systems (UCSs) to accurately draw with unprecedented precision. This book describes creating and editing drawings, where you learn how to create elementary objects such as lines and arcs, as well as more complex objects such as polylines and splines. Then the book walks you through using AutoCAD's object query and editing tools, and explains how to apply these tools to quickly locate and extract object geometry information, and to manipulate and edit objects, covering both basic and advanced editing techniques. Next, you build upon this solid foundation and enter more advanced subjects, where you learn to work with blocks and externally referenced (xref) drawings. You learn how to annotate your drawings, and how to use AutoCAD's newly enhanced associative dimension and hatch pattern tools to quickly create professional level 2D drawings. AutoCAD 2002 Complete then walks you through the complex yet powerful features for working with model space and paper space, as well as using AutoCAD's Layouts, which are the basis for plotting your drawings. Then, detailed examples show you how to use features such as AutoCAD's Plotter Manager, Plot Styles Manager, and Page Setups to easily produce professional-level hardcopy output of your designs. Finally, this book shows you how to use AutoCAD's 3D modeling and rendering tools by first introducing you to 3D concepts, then building upon these concepts to work in 3D space and build complex 3D surfaces, and to work with AutoCAD's 3D solids tools for creating impressive 3D models. The discussions and examples covering AutoCAD's 3D tools and features is comprehensive and finishes by showing you how to render your 3D designs to produce photo-realistic images that make your computer models look real. As an added bonus, AutoCAD 2002 Complete provides concise, step-by-step explanations on how to use AutoCAD's advanced tools that let you work with OLE objects and external databases. You master the secrets of interconnecting files and documents from multiple applications to and from AutoCAD drawings. You learn about the power of connecting to external databases, and linking AutoCAD objects to records in tables, a capability that lets you automatically label objects in drawings using records from databases, which is a capability that few AutoCAD users know about or understand how to use.

HOW TO USE THIS BOOK

AutoCAD 2002 Complete is organized to help you get up-to-speed as quickly as possible by focusing on the information you need. The book is organized into five parts, with each part focusing on a specific AutoCAD theme. Part 1, Getting Started, presents the basic elements of AutoCAD's drawings. Part 2, Creating, Editing and Manipulating Objects, presents AutoCAD's fundamental tools and objects used for working with drawings. Part 3, Adding the Finishing Touches, explains the advanced tools and techniques for completing and plotting drawings. Part 4, Working in 3D, presents AutoCAD's 3D tools and objects, where you learn to build and render 3D models. Part 5, Advanced Concepts, presents AutoCAD's tools and features for working with external files, including Object Linking and Embedding (OLE) and external databases.

CONVENTIONS USED IN THIS BOOK

Most topics covered in this book are detailed explanations of new features. Where needed for clarity, examples are presented that step you through the process of using a particular feature. Many of the examples use files located on the accompanying CD.

USING THE CD

All drawings, databases, and other files used in this book's examples are included on the accompanying CD. Typically, if an example uses a file from the CD, the example walks you through the process of copying the file to a new folder on you system, and then clearing the file's Read Only attribute.

ABOUT THE AUTHORS

Bill Burchard is Corporate CADD Manager for Psomas, a California-based land surveying and civil engineering firm. Mr. Burchard has served the AEC industry for 25 years. His range of experience includes surveying and engineering design, computer modeling and applications development, plan preparation and technical publications, as well as Web development, Geographic Information Systems (GIS), 3D modeling, 3D photo-realistic renderings, and 3D animations. Additionally, Mr. Burchard is a registered Autodesk author/publisher, and along with co-author David Pitzer, has written numerous books, including Inside AutoCAD 2000 Limited Edition, Inside AutoCAD 2000, Inside AutoCAD 14 Limited Edition, and Inside AutoCAD 14. Mr. Burchard also writes regularly featured articles for CADalyst magazine, published by Advanstar Communications, for Inside AutoCAD Journal, published by Element K Content, and for Autodesk's Toplines eNewsletter, which is accessed from Autodesk's Point A Web site (www.autodesk.com/pointa). He writes both feature articles and a bimonthly column for CADalyst and for Toplines, and writes monthly articles for Inside AutoCAD Journal. Additionally, Mr. Burchard sits on the Advisory Committee Board for Computer Sciences at Riverside Community College, and lectures on the subject of GIS for the University of California, Irvine.

David Pitzer has been using AutoCAD since 1987. He was formerly CADD Manager for an engineering firm in northern California where he lives. He has written numerous articles for CADENCE Magazine and CADalyst Magazine for whom he is a Contributing Editor. He is a member of the faculty at Santa Rosa Junior College where he teaches advanced AutoCAD courses. He has co-authored five AutoCAD books for New Riders Publishing. He has, in addition, written technical material for Autodesk.

Art Liddle is a registered architect and has more than thirteen years of hands-on experience with AutoCAD. He started writing reviews and technology stories about AutoCAD in 1990. He has been an editor with CADalyst magazine for the past nine years.

ACKNOWLEDGMENTS

The authors wish to thank and acknowledge the many professionals who reviewed the manuscript, helping us publish this book, including the Autodesk Press Team: Jim DeVoe, Acquisitions Editor, John Fisher, Developmental Editor, Mary Beth Vought, Art & Design Coordinator, Stacy Masucci, Production Editor, Karen Smith, Marketing Coordinator, and Jasmine Hartman, Editorial Assistant.

Special thanks is due to Michael Todd Peterson for writing the two dimensioning chapters in Part 3, Adding the Finishing Touches.

Bill Burchard wishes to thank his family and friends for their support during the last several months while writing this book. Additionally, Bill thanks his colleagues at Psomas for their support, suggestions, guidance, and help, with special thanks to Joanne Diaz, Sherri Bayer, and Amy McLaughlin, as well as his friends at Autodesk, including Glenn Cooper and Neill Vickers. Finally, Bill thanks his fellow C3 Group members for tolerating his absence during the last several months while writing this book, with special thanks to Gene Redmon, President, and Devon Diaz, past President.

DEDICATIONS

Bill Burchard:

I dedicate this book to my two moms, Mimi Houghtaling and Millie Black. I thank you for your support through these trying times, and love you both deeply.

Dave Pitzer:

I would like to dedicate this book to my students—past, present and future. Their excitement with AutoCAD is a constant source of enjoyment.

Getting Started

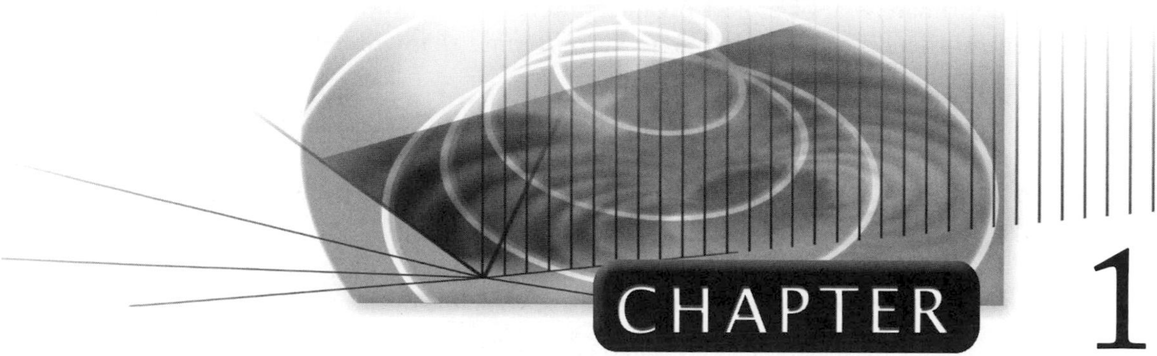

Before You Start Drawing

AutoCAD is designed to save you time by allowing you to draw efficiently. However, drawing efficiently with AutoCAD is only one way to save time on your projects. By analyzing the overall process required to complete a project, you can identify elements of your project that are not necessarily drawing based but have a lot to do with saving time. Then, through proper planning and organization methods, you can maximize efficiency and complete your project on time. In addition, AutoCAD 2002's new CAD Standards tools will help you maintain consistency in your drawings.

This chapter covers the following topics:

- Project organization
- Initial drawing setup
- Configuring Active Assistance
- Leveraging AutoCAD's features
- DesignCenter
- Managing CAD Standards

PROJECT ORGANIZATION

The ultimate goal of project organization is to save time by working efficiently. In general there is one simple rule to keep in mind that will help you work efficiently and save time:

> Reduce repetition: Do not repeat what you have already done.

Before you can reduce repetition, you must first identify the organizational elements of your project. The following sections discuss various organizational elements that you should consider prior to beginning your project:

- How many drawings are needed to complete the project?
- How much detail is required on each drawing?
- What is an efficient method of workflow management?
- How many different ways will you display the drawing?
- What project elements can be used more than once?

DETERMINING HOW MANY DRAWINGS ARE NECESSARY

When you sit down to start drawing a project, one of your earliest tasks is to determine just how many and what type of drawings you will need. You should consider two important issues when you establish the number of drawings to create.

First and foremost, you should address the computing power-versus-drawing size factor. Computing power consists of storage space, both long-term (such as a hard disk, tape, or network server) and short-term (such as RAM), coupled with the speed of your graphics card and CPU. Autodesk and other experts toss about many factors relating to how much memory and available hard disk space you need to handle each megabyte of drawing. Unfortunately, no hard and fast rule exists. If you use xrefs, extended entity data, attributes, and 3D solids, and if you will need to open multiple drawings sessions in AutoCAD at once, then your needs vary substantially from those of someone who stores 2D vector information in a drawing without any bells and whistles. Therefore, you must ascertain your ideal drawing size through experience and simple trial and error.

Tip: A good set of guidelines to use when estimating the ideal maximum drawing size for your system is to allow 32 times your drawing size in memory and 64 times your drawing size in free disk space.

When you have a good feel for how large your drawings can be before the performance of your system hinders your work, you can determine how you want to organize your drawing data into individual drawings. The following examples discuss various options for organizing a project into drawings with AutoCAD:

- Your model exists in one drawing, and you can view the model through various Layout pages (with paper space viewports). Keep in mind that a model can be 3D or 2D. The model can then be viewed and annotated differently in each Layout page.
- Each of your drawings comprises a very detailed portion of a much larger product. You can then assemble the various portions into a single model using external references (xrefs). The xrefs help you organize your drawings by listing the location of and relationship between each drawing.
- You can develop a new drawing by cutting and pasting elements from existing drawings. You can use AutoCAD's Multiple Document Environment (MDE) and open multiple drawings in a single session of AutoCAD, making copying elements between drawings very simple.
- Your drawings are only a small portion of your project documentation, and the drawings are linked to text documents, images, and other files using Object Linking and Embedding (OLE).

You will learn more about the use of xrefs in Chapter 13, Working with Large Drawings and External References; about the use of paper space in Chapter 19, Layouts and Paper Space; and about OLE in Chapter 27, Taking Advantage of OLE Objects in AutoCAD.

It is important that you consider a variety of approaches when you create your drawings and that you develop a tried-and-true method that delivers exactly what you need for a majority of the time. After you develop your approach to building your drawings, you can begin to predict how many drawings you will need, what their contents will be, and how they will relate to each other. This will be critical to estimating how long it will take to complete a given project.

Developing Drawings and Task Lists

Once you have developed the approach you will use to build your drawings, you should then develop a list of your project's drawings and a task list that goes with each drawing. This list of drawings and their associated task lists should be flexible and expandable. Spreadsheet applications, such as Microsoft Excel, are excellent tools for creating such lists.

A convenient way to associate a task list with a drawing is to link a spreadsheet into the drawing's paper space layout. When you link the task list in the drawing, you create a drawing-specific storehouse for your task list that is visible in each drawing. You also can plot this task list with your drawing so that you can easily measure your progress when you create progress prints of your project. Finally, you can update this task list as you work in each drawing. By using a linked spreadsheet, you can quickly add and edit tasks in your list while working in the drawing.

DETERMINING HOW MUCH DETAIL IS REQUIRED

Another important factor in the organization of your project is determining how much detail is required for each drawing. Because AutoCAD is so accurate, you can easily fall into the trap of creating minuscule details that might not have anything to do with the actual production of your project from your drawings. It is important that you implement only what the final user of your drawing(s) will need in order to accomplish the specific job. In addition, you should keep in mind that the person actually creating your project might have better, more effective methods of building the product in the field than you do on the drawing. The following list of questions will be discussed in this section to help you determine the level of detail included in each drawing and in your project as a whole:

- What should be documented?
- When should it be implemented?
- How should it be approached?
- Who should explain it?
- Where should it be presented?

The answers to each of these questions might seem obvious at first, but after you take a look at how they affect your project's organization, you may realize that the questions are not as easy to answer as they may seem.

Answering the Questions

In answering the first question, "What should be explained?," you should not be concerned with only the parts of your project that you will draw. You should also consider how to organize your drawings in a way that describes your project in the most efficient and effective manner. For example, you might need to define a continuous chamfer that surrounds a face plate, as well as the location of the holes in the face plate, and the specific location of tooling on the face plate. To adequately describe these conditions, you would need two sections—one horizontal and one vertical through the face plate—as well as a front view of the face plate. Two isometric views can accomplish the same thing with one less drawing required.

Your answer to the second question, "When should it be explained?," might mean that you will not need to draw some details of your project at all because you will get back from a vendor a set of shop drawings that determines how the detailing will be done. Answering this second question also might determine the order of the drawings within a set of project drawings.

It might be obvious to you that the third question, "How should it be explained?," concerns the type of drawing (such as isometric drawings, or plan and section drawings) that you need. However, you should also think about using photographs, annotation, and shaded 3D models as part of the explanation.

Your answer to the fourth question, "Who should explain it?," might determine that you have a vendor finish your work for you in the form of shop drawings, but may also involve finding a person within the project team to work on the design. A conceptual designer, for example, might not have the CAD skills that are needed to complete a detailed 3D model. Therefore, you might have to alter your preferred choice of drawing technique based upon who is doing the design and who is doing the drawing.

After you have answered the first four questions, you are prepared to ask the fifth and final question, "Where should it be explained?," which ultimately determines the organization of your drawings. For example, you must determine whether you have so many details that they must be displayed on their own detail sheets, or whether these details will fit as blow-ups on the same sheets that display your working model. You also must establish a systematized approach to the order of your drawings that works for every project. Furthermore, you should address whether you can readily insert new drawings into the drawing sequence if you discover that one is needed when you are well into your project. Two key concepts that can help you manage this portion of your effort are as follows:

- Developing an efficient numbering system used within the drawings
- Developing a file-naming/folder-naming system that corresponds to the drawing numbering system

Developing an Efficient Numbering System

You need a way to name or number your drawings that works the same way every time and allows flexibility in the order of your drawings and the number of each type of drawing. A perfect example of a system that works in this way is a library catalog, better known as the Dewey Decimal System. If you organize your drawings into categories and then sequence the drawings within a category, this type of system might work for you. For example, in the AEC industry, drawings have categories such as Cover Sheet and General Information Sheets, Floor Plans, Elevations, and Sections. Each category might contain a sequence of drawings, such as the First Floor Plan and the Second Floor Plan. Following this format you could define a drawing numbering system that meets the goals of predictability, flexibility, and expandability. The following list details a numbering system you could implement:

0.00, 0.01, 0.02, and so on—Cover Sheets and General Information

1.00, 1.01, 1.02, and so on—Floor Plans

2.00, 2.01, 2.02, and so on—Elevations

3.00, 3.01, 3.02, and so on—Sections

etc.

Note: You can add a new floor plan drawing to the end of the Floor Plans category at any time without adversely affecting the drawing numbers in any category that follows, such as the Elevations category. The limitation of the numbering system as shown is that a maximum of nine drawing categories, with a maximum of 99 drawings in each category, can be created. If you need more categories or drawings, simply add more numbers to your file names, as in, for example, 01.000.

Developing a File- and Folder-Naming System

You can carry the numbering system shown in the previous section a bit further and use it to name the drawing files while you also add project numbers and a drawing description. If you design your drawing-numbering and file-naming conventions with some forethought, the file names will sort themselves in the Open or Save File dialog boxes in the desired order so that you will know what a drawing contains without opening the drawing. The only limitation in Windows 95/98 and Windows NT is that the file name cannot contain more than 255 characters, and it cannot contain any of the following characters: \ / : * ? " < > |. Of course, if you make your file names 255 characters long, you cannot have the files in a folder, you will spend a lot of time typing the file names, and many file dialog boxes will not display the entire name within them. Therefore, a responsible file-naming system might include a project or work order number, a drawing number, and a predictable maximum number of characters for a description. The following example illustrates the format of a filename using this system:

<project number>-<drawing number>-<revision number><drawing owner>-<sheet title>.dwg

9701-1.00-01cws-OVERALL BUILDING PLAN.dwg

Note that Windows 95/98 and Windows NT also remember the upper-case letters separately from lower-case letters. Using upper- and lower-case letters can help with the legibility of the drawing file name. Also note that the final extension for the file name is always .dwg, because AutoCAD must see the .dwg file extension to load the file. Even if you can see only the first fifteen characters of this file name, you would know that the file belongs to project number 9701, that it is a plan of some type, that it is revision number 1, and that someone with the initials "cws" originated the drawing. This file-naming convention also sorts the drawing files first by project number, then by drawing number, then by revision, and then by author.

Note: If you create new drawings on a regular basis, an involved file name might become counterproductive. In other words, the amount of information that you include in your file name is partly based upon the shelf life of your drawings.

If you organize your drawings into folders that have some logical hierarchy to them, the file names can also become simpler.

It is always a good idea to keep the project number or work order number as part of the drawing file name so that drawings that might be accidentally misplaced can be tied to their project with relative ease. You also must consider how many versions of any one portion of the file name you might have. For example, the project number 9701 enables you to have ninety-nine projects in the year 1997. If you think you will have more than ninety-nine projects, you should use a number with five digits, as in 97001. Likewise, 1.00 means that only ninety-nine floors can exist in the building (1.01 through 1.99), and using 01 for a revision marker means that up to ninety-nine revisions can exist. Additionally, with the advent of the year 2000, it may be useful to include the year's full numeric value for project numbers, such as in 1997001. Otherwise, you may have difficulty displaying drawings in proper numeric order if you continue with a numbering system that displays only the last two digits of the year—a unique version of the Y2K problem.

Make sure that you do not limit yourself so much that you have to change your file-naming scheme in a year or two. Take the time to study the types and number of events that you need to record in your file name by looking over past projects that you or other members of your company have completed. Do not hesitate to run your ideas by others to see if they can spot any shortcomings.

DETERMINING HOW YOUR DRAWINGS ARE RELATED

This topic can become very complex because not only can AutoCAD drawings be linked to other documents, and other documents be linked to your drawings, complex relationships can also exist between the drawings themselves. You can organize your thoughts about the interrelationships between your drawings and other drawings or documents by asking a few simple questions:

- What information must be shared between drawings?
- What other documents will be included in the drawings?
- What other documents will the drawings be included within?

At this point it is usually a good idea to create a mock-up set of drawings using a standard form. If you have a number of drawings and documents that make up your project and your sheets are standard and uniformly sized, then a mock-up is both easy to do and beneficial. If your projects are small in terms of document count or widely varied in document format and size, you may not find a mock-up of a form very helpful.

You can create this mock-up as a basis for your drawing set in AutoCAD, or you can draw a free-hand mock-up using preprinted sheets and a pencil. Creating electronic mock-ups of your project will be helpful in the long run because the effort will contribute to the creation of the final documents.

However you choose to create a mock-up, it should organize key information to answer the three questions just reviewed. Additionally, useful information such as drawing number, drawing name, date, project name, and author could be included.

If you want to create a mock-up electronically, you can place the mock-up right on the drawing to aid in its creation. You can set this up ahead of time by using a form in place of a title block in a template drawing. (This process will be briefly discussed later in this chapter.) After the mock-up has served its purpose, you can substitute a real title block for the mock-up form. If you keep your mock-ups separate from your drawings, then it is a good idea to create your mock-up as an 8½-by-11-inch sheet so that you can plot "mini" sets of your entire project until you complete the production drawing set.

Using Object Linking and Embedding (OLE)

When you insert a document within your AutoCAD drawing, you can use Object Linking and Embedding (OLE). When you use OLE, you can either link the inserted object to the original document or embed it as a duplicate of the original document. If you link the inserted object to the original document, every time the drawing is opened, AutoCAD automatically locates the original document and then updates the inserted object based on the latest version of the original document. Linking the inserted object to the original document ensures that the latest version of the original document is always displayed in your drawing, which is a very useful feature if the original document is being edited on a regular basis.

Embedding an object, on the other hand, inserts a duplicate of the original document but does not link the inserted object to the original document. Consequently, any edits made to the original document are not reflected when the drawing is reopened. This feature is useful if you want to insert a copy of the original document without fear of the copy being overwritten by an updated version.

Note: For a more complete review of using OLE, refer to Chapter 27, Taking Advantage of OLE Objects in AutoCAD.

DEVELOPING EFFICIENT WORKFLOW MANAGEMENT

If more than one person will work on a drawing, you must determine how each person will know which drawing is the current drawing. This aspect of project delivery is known as workflow management. Consider a scenario, for example, in which outside consultants or contractors work on your drawings, and you want to make changes to these drawings. How will you know if the drawing you have contains the most recent information? If you do not address this concern, you might edit an outdated drawing.

Another problem you may encounter is one in which two people need to work on the same drawing at the same time. By applying workflow management to your project, you can avoid conflicts and limit mistakes.

Proper Network File Control

Your organization of the project and its documents and your management of file access are the key to avoiding disaster with multiple document users or authors. If you work on a network that provides multiple access to a drawing, then your network software must support file locking. Most contemporary network products enable you to open files in read-only mode and will not let you overwrite the file on which someone else is working. Proper network file control means that the first person to open a drawing is the only editor that can actually save changes to the drawing. All subsequent users can open the drawing only in read-only mode. If someone absolutely must record changes to a drawing that is currently open by someone else, they must save their changes as a new drawing. Recording these changes on a unique layer name and saving the changes as a new drawing enables the changes to be merged with the original drawing later.

Using Redlines

Users also can record redlines on a unique layer that can be merged with a drawing. Redlines list comments, questions, and editing instructions for a drawing that are acted upon by someone else later.

Tip: Because redlines can be created using basic AutoCAD commands, you could create your own toolbar that contains basic redlining tools such as text, lines, arcs, circles, and leaders. In this way, individuals who are not sophisticated AutoCAD users can still contribute to your drawings electronically, saving time and paper.

An alternative to creating a new redline layer in the current drawing is to xref the drawing into a new drawing and create the redlines in the new drawing. In either event, you should name the drawing file and the layers used for redlines based upon a standard. For example, the file name, or the layer name, could include the redliner's initials, a date, and the term "redlines" in its name. This makes identifying the redlines easy.

DETERMINING WHICH ELEMENTS CAN BE USED MORE THAN ONCE

As mentioned earlier, you can use xrefs to create more than one drawing from the same drawing. Although the use of xrefs can help ensure that numerous drawings contain exactly the same information and reduce total project drawing storage needs, xrefs can also help you avoid creating drawing elements more than once.

If you use a drawing template, as demonstrated in the previous exercise, you might not want to include a drawing border in the template drawing. Instead, you could devise a project border sheet that contains the project name, issue date, project number, project address, your firm's logo and address, and other information, and then insert it into drawings as an xref. Therefore, each new project drawing would include the border as an xref, and changes to the border would need to be made only once in the original xref.

Note: If you use a project border sheet, you should design the xref so that it contains only the information common to all project drawings. For example, elements like issue date and sheet revision data would stay unique to the drawing file and would not appear in the xref file.

Using an element more than once can mean more than creating an exact duplicate of objects. You might want to use different guides in all of your project drawings, such as the format of text, a key plan with different portions hatched in each drawing, or a sheet grid that does not plot. For example, you can create a block that contains only attributes that fill in your title block. The drawing author, checker, sheet number, sheet name, and date can be filled in separately for each drawing, but such elements as the text style, height, and layer will always be the same. (You will learn more about blocks in Chapter 12, Making the Most of Blocks.) If your projects usually involve a large number of drawings, by combining the use of block attributes inside blocks and xref drawings, you can increase efficiency by creating a template drawing that contains the border as an xref, the sheet specific text as attributes, and a sheet grid on a non-plottable layer.

An important factor to consider is that at the start of a project, you must map out as many multiple-use opportunities as you can, including organizing portions of drawings that you can reuse from earlier projects or from standard drawings that you have developed over numerous projects. Using the mock-up set discussed earlier, you should also map out the use of viewports to display the necessary views of your model.

DETERMINING HOW A DRAWING WILL BE DISPLAYED

The final and possibly most complex issue to consider when laying out a project is the number of ways that a drawing will be displayed. For instance, you might need to plot the drawing on a number of different sheet sizes, or you might need to create both drawings and renderings of the drawings for the project. Also, you might need to publish your drawings in a shop manual or technical publication. As you can imagine, large drawings with a significant amount of detail are not displayed very well on computer screens. (If they were, the AutoCAD Zoom command would not exist.)

If you create documents that will be used for marketing, published in a technical publication, and plotted on a sheet of paper, something will have to give. You might need to create completely different drawings for each type of media due to one single factor: your text and symbols will not work for each and every possible publishing method. Although a perfect solution does not exist for these broad publishing requirements, planning for the project's needs from the start can save a lot of time and headaches. If your publication needs vary from project to project, it might be a good idea to obtain from a previous project a drawing that is similar to the one you will use for the current project. You could try to publish this drawing under all the conditions that you must meet. This process would help you uncover any problems that you might encounter.

Note: Use AutoCAD's Publish to Web wizard to publish your drawings to the Internet. It quickly takes you step by step through the process of creating files that are ready to be posted on your Web site.

INITIAL DRAWING SETUP

Once you have pondered each of the issues discussed so far, it is time to set up your drawings. You must follow a sequential set of steps to determine how you can accomplish this task. If you follow these steps, you will avoid having to revise text sizes and drawing configurations as well as a number of other complications later in a project:

1. Determine the paper size.
2. Determine the drawing scale.
3. Develop title blocks.
4. Determine units and angles.

DETERMINING PAPER SIZE

The first and foremost determination you must make is the final plot size of your drawings. If you plot your drawings to paper, the paper sizes that your plotter handles define your options for the drawing size. The first step in determining paper size is using a mock-up process—whether a formal one, such as the process discussed earlier, or simply figuring out how much paper area is required for an appropriate scale of your drawings. Paper comes in an extensive set of sizes, and each industry generally settles upon a set of standard sizes. One important factor to consider about paper is whether you can create a modular approach to your paper sizes. An ideal progression of paper sizes will let you expand or contract between sheet sizes while maintaining the same aspect ratio between sheets. Maintaining the same aspect ratio means that you can enlarge or minimize your drawings without concern for whether the drawing will fit the same way on the larger or smaller sheet.

Table 1-1 lists standard paper sizes that are available.

DETERMINING DRAWING SCALE

After the paper sizes have been established, the next step is to determine the appropriate scale for your drawings. The scale of a drawing is a deceptively simple concept, but it involves more than simply figuring out what size your drawing must be to fit on the paper. The real issue about drawing scale is that the information con-

Table 1–1: *Standard Paper Sizes*

Paper Size	Standard	MM	In
Eight Crown	IMP	14611060	57 1Ú241 34
Antiquarian	IMP	1346533	5321
Quad Demy	IMP	1118826	4432 12
Double Princess	IMP	1118711	4428
Quad Crown	IMP	1016762	4030
Double Elephant	IMP	1016686	4027
B0	ISO	10001414	39.3755.67
Arch-E	USA	9141,219	3648
Double Demy	IMP	889572	3522 12
E	ANSI	8641118	3444
A0	ISO	8411189	33.1146.81
Imperial	IMP	762559	3022
Princess	IMP	711546	2821 12
B1	ISO	7071000	27.8339.37
Arch-D	USA	610914	2436
A1	ISO	594841	23.3933.11
Demy	IMP	584470	23´1812
D	ANSI	559864	2234
B2	ISO	500707	19.6827.83
Arch-C	USA	457610	1824
C	ANSI	432559	1722
A2	ISO	420594	16.5423.39
B3	ISO	353500	13.9019.68
Brief	IMP	333470	13 1Ú818 12
Foolscap Folio	IMP	333210	13 1Ú88 14
Arch-B	USA	305457	1218

Table 1–1: *Standard Paper Sizes (Continued)*

Paper Size	Standard	MM	In
A3	ISO	297420	11.6916.54
B	ANSI	279432	1117
Demy Quarto	IMP	273216	10 3Ú48 12
B4	ISO	250353	9.8413.90
Crown Quarto	IMP	241184	9 1Ú27 14
Royal Octavo	IMP	241152	9 1Ú26
Arch-A	USA	229305	912
Demy Octavo	IMP	222137	8 3Ú45 38
A	ANSI	216279	8.511
Legal	USA	216356	8.514
A4	ISO	210297	8.2711.69
Foolscap Quarto	IMP	206165	8 1Ú86 12
Crown Octavo	IMP	181121	7 1Ú84 14
B5	ISO	176250	6.939.84
A5	ISO	148210	5.838.27
	USA	140216	5.58.5
	USA	127178	57
A6	ISO	105148	4.135.83
	USA	102127	45
	USA	76102	35
A7	ISO	74105	2.914.13
A8	ISO	5274	2.052.91
A9	ISO	3752	1.462.05
A10	ISO	2637	1.021.46

tained on the drawing must be legible, yet the drawing scale must be standard in your industry, and the sheet size must be as convenient to handle as possible. Your drawing must place the model, notes, dimensions, hatching, and symbols in their most favorable and legible light. If you have to cram a drawing full of symbols and text, the lines that represent the object of your drawing may become difficult to discern. Creating the drawing at the appropriate scale allows for space between text, dimensions, and symbols both within and around your drawing.

Of course, if you always create drawings that are full-scale, then the only option you are faced with is the selection of the sheet size for your paper. Using the full-scale size might require you to cut your drawing up into sections rather than display the entire model on one sheet of paper. You might think that a drawing spanning more than one sheet seems inconvenient, but legibility is more important in this case.

Using Paper Space Viewports to Scale a Model

When plotting, you can set the scale of your model in two ways. You can either provide a scale factor at plot time and plot from model space, or you can pre-scale your model through paper space viewports. Both result in a drawing that is plotted at the appropriate scale, so which one is the best choice?

One reason users avoided paper space was because zooming and panning operations forced a regeneration, which made working in paper space very time-consuming. However, since Release 14, panning and zooming in paper space no longer causes a regeneration. Consequently, the chief argument against using paper space viewports to scale models is gone. A major advantage to using paper space viewports is that you get instant feedback from your drawing regarding what drawing scale fits on your sheet, because you immediately see what your plotted sheet will look like in your layout. Therefore, scaling views of your model in paper space viewports is the best choice.

To use a paper space viewport for plotting to scale, you must zoom in on your model at a predetermined scale factor. The following example shows how to determine and set the proper zoom-scale factor.

1. First you must calculate the required scale factor. If your drawings use a decimal scale, this is a relatively simple feat. For example, a drawing created at a scale of 1:10 uses a scale factor of 0.10. The AEC industry uses non-decimal scale factors, however, and the calculation requires a few more steps. A "1/8 inch = 1 foot" scale drawing requires a scale factor of 1/96. To convert AEC scales, simply multiply your drawing scale (in this case, 1/8) by 1/12 (12.0/0.125). Therefore, the scale factor for a "¼ inch = 1 foot" drawing is 1/48, and the scale factor for a "3 inches = 1 foot" drawing is ¼.
2. Click inside the paper space viewport to make it the focus.

3. Perform a Zoom Extents to see all of your model.
4. Perform a Zoom Center and select on the model the point that you want to be in the middle of the viewport.
5. Finally, when you are prompted for magnification or height, enter your scale factor, followed by **XP**.

The area of the drawing you selected is centered in the viewport at the desired scale.

Using this technique you can set up numerous plot scales of your drawing for any specific needs that you might have. Note also that any text, symbols, or other elements placed in model space will be scaled as well. If you want to display the same model at different scales, you must create symbols, dimensions, and text on different layers and at appropriate sizes for each scale, or you must draw them all in paper space. The great thing about drawing symbols, text, or other elements in paper space is that you can create them full-sized without having to convert to scale. This means, for example, that text that is 1/8 of an inch high is the same height in all drawings, no matter what scale is used to plot the model. Be mindful, however, of the fact that if the objects in model space move at all, the relative position of the annotation objects in paper space and in the model will be changed and may no longer align properly.

DEVELOPING TITLE BLOCKS

Almost any drawing, whether it is a work order, a maintenance drawing, or a sophisticated manufacturing document, should have a title block. A title block provides informational—and often legally required—verification of what the drawing represents in terms of the object of the drawing, the time of day the drawing was created, and the origin of thc drawing. If you only publish your drawings electronically, then the title block might differ considerably from the title block of a drawing that eventually will be used for plotting or printing. For now it is assumed that a paper plot is the ultimate goal of an AutoCAD drawing. The following list serves as a guide for the elements your title block should include:

- The name, address, and phone numbers of the firm originating the drawing
- The name, address, and phone numbers of any consultants working on the drawing
- The date that the drawing was originally created and approved for use
- A revision history, including who performed the revision, what the revision was, and when the revision occurred
- A drawing title
- A project name or work order title
- A location for seals, stamps, and/or approval signatures

- A drawing number
- A project or work order number
- The author of the drawing and the name of individual(s) who checked the drawing, if required
- The name of the AutoCAD drawing file
- The date that the drawing was printed or plotted
- A copyright notice, if required
- Additional general information, such as a project address, plant name, and owner's name
- Line work that organizes the title block information and its relationship to the drawing

The design of title blocks is often a source of great debate within a company. No perfect title block design exists, and your needs might include items not listed here. Generally, the more information (either critical or organizational in nature) that you can place in the lower-right corner of the sheet, the easier it will be for others to quickly find the desired drawing. The title block should provide the information legibly for all size plots but not dominate the sheet. You also might need to develop a title block for multiple sheet sizes. Most likely you will not be able to use the same title block for an 8½-by-11-inch sheet and a 34-by-44-inch sheet. You will need to experiment with different designs until you have a set of title blocks that works for all content possibilities and sizes.

Additionally, the title block can contain a grid design that promotes the modular development of your drawings. For example, if you typically develop details that can be printed on 8½-by-11-inch paper, then you could develop a drawing module that would enable you to piece together a number of small modular drawings into a larger drawing. In this case, you should be concerned with the drawing area within the title block for the module size, not for the size of the sheet of paper. This is because you will transfer the drawing area from one sheet to the next.

DETERMINING UNITS AND ANGLES

The discipline and country in which you work determines whether you will use fractional inches, feet and fractional inches, decimal feet, decimal inches, meters, or centimeters in the creation of drawings. Additionally, you must determine how accurately to display the dimensions. AutoCAD does not understand any specific system of the division of distance—the program simply draws using units. As a result, you must tell AutoCAD how you want those units displayed. You change the display of units and angles in AutoCAD from the Drawing Units dialog box, which is accessed by choosing Units from the Format pull-down menu. The Drawing Units dialog box is shown in Figure 1–1.

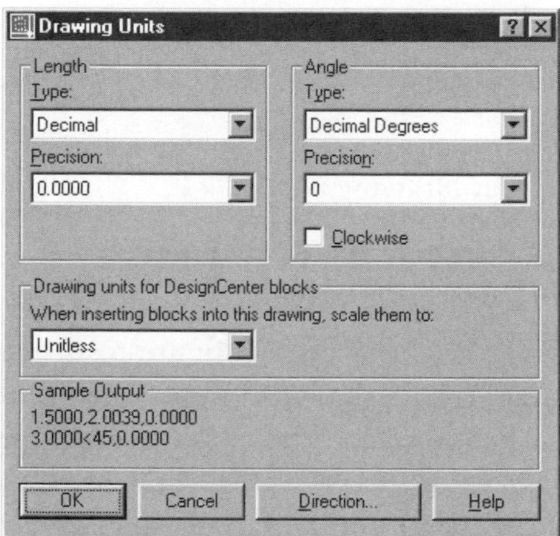

Figure 1–1 *The Drawing Units dialog box is used to set AutoCAD units and angles.*

You should set up the default units that you will typically use in your template drawings. When you do so, all new drawings that use the template will automatically use the default units and angles settings. You should also note that you can select the precision of the display of your units and angles.

Tip: Do not confuse the precision setting for units and angles with the precision setting for dimensions. You set the dimension precision independently when you define dimensioning defaults. Chapter 17, Dimensioning Basics, discusses dimensioning in greater detail.

When selecting the units and angles precision, your primary concern should be how much precision you need to see when you create your AutoCAD drawings. High-precision settings often cause AutoCAD to display the drawing coordinates using scientific notation, such as 1.07E+10, which usually is not much help. On the other hand, if you are trying to track down a drafting error, high-precision settings can tell you that a line has been drawn at an angle of 179.91846 degrees instead of at 180 degrees. The simple process of trial and error can help you determine the best settings for your needs.

Converting between Units

If you need to convert your drawings from feet and inches to metric units, the units in which you create your drawing will not automatically convert. This is because you are drawing in units, not in real-world sizes. For example, when you create a drawing in feet and inches, one unit is an inch. When you convert to a metric drawing, you must change units to decimal units and then convert the drawing to a metric drawing by scaling the drawing by the proper conversion factor. As a result of this component of AutoCAD's units architecture, you must determine what the drawing should represent before you start creating lines, circles, and arcs. You can instruct AutoCAD to dimension objects by scaling them from one unit of measure to another (accomplished by setting a linear scale factor for dimensioning), but the model will not be drawn true to size in the converted units. For more information, refer to Chapter 6, Precision Drawing.

CONFIGURING ACTIVE ASSISTANCE

Introduced in AutoCAD 2000i, Active Assistance is an automatic or on-demand context-sensitive Help tool that you can configure to display automatically whenever you start AutoCAD or, alternatively, only when you specifically invoke it through the new ASSIST command. You can also configure Active Assistance so that it activates and displays context-sensitive information anytime a command is invoked. If you choose this latter configuration, whenever you invoke a command supported by Active Assistance (not all commands are), the Active Assistance window displays basic information about the command. For example, when you invoke the ZOOM command, the Active Assistance window displays the information shown in Figure 1–2. As you can see, the information is not extensive and is really only useful to novice AutoCAD users. If you are seeking guidance for, say, the Chain option when filleting a solid, Active Assistance provides no assistance. In fact, the help window for the FILLET command does not mention solids at all. Nor should it; Active Assistance provides rudimentary assistance only and is ideal for those new to AutoCAD.

ACTIVE ASSISTANCE SETTINGS

Settings for Active Assistance are accessed by right-clicking in either the Active Assistance window or on the Active Assistance icon displayed in the Windows system tray located on your Windows taskbar, and then selecting Settings. When selected, AutoCAD displays the Active Assistance Settings dialog box as shown in Figure 1–3.

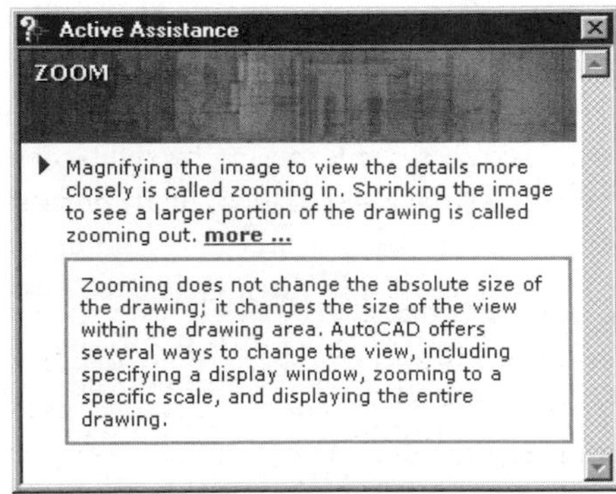

Figure 1–2 *The Active Assistance window displays basic information about commands.*

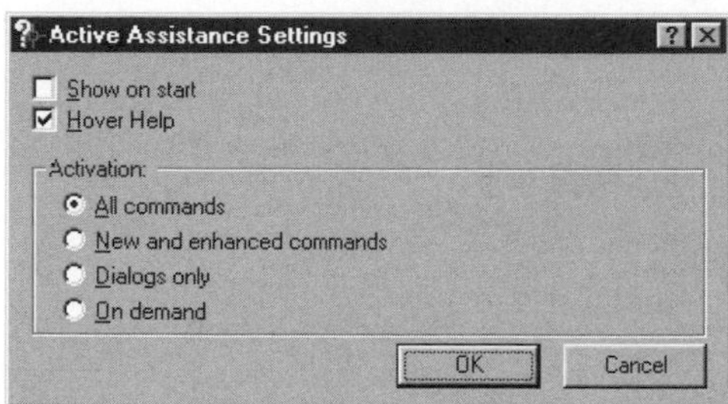

Figure 1–3 *The Active Assistance Settings dialog box.*

The Active Assistance Settings dialog box provides the following controls and activation modes:

- **Show on Start—**When this box is checked, AutoCAD will automatically display the Active Assistance window when you start AutoCAD.
- **Hover Help**—When this box is checked and a dialog box is active, Active Assistance displays context-sensitive information whenever the cursor is placed over a dialog box option or control. Figure 1–4 shows this feature when the cursor is over the Angle of Array option. The Active Assistance window displays a brief explanation of this option.
- **Activation section**—Only one of the following methods may be chosen at a time:
  - **All Commands**—When this method is selected, the Active Assistance window is opened whenever any command is invoked.
  - **New and Enhanced Commands**—When this method is selected, the Active Assistance window is displayed when new or enhanced AutoCAD commands are invoked.
  - **Dialogs Only**—When this method is selected, the Active Assistance window is opened whenever a dialog box is displayed.
  - **On Demand**—When this method is selected, the Active Assistance window is displayed only by double-clicking on the Active Assistance icon in the Windows desktop system tray, or by right-clicking on the same icon and choosing Show Active Assistance, or by entering ASSIST at the command prompt.

Active Assistance is obviously intended to primarily help beginning AutoCAD users as they learn the various commands and options. Its utility lies in providing brief but immediate help as a command is being executed. It is not intended as a substitute for AutoCAD's more extensive Help facility. Also, you can completely disable it by right-clicking on the Active Assistance icon in the Windows system tray and choosing Exit.

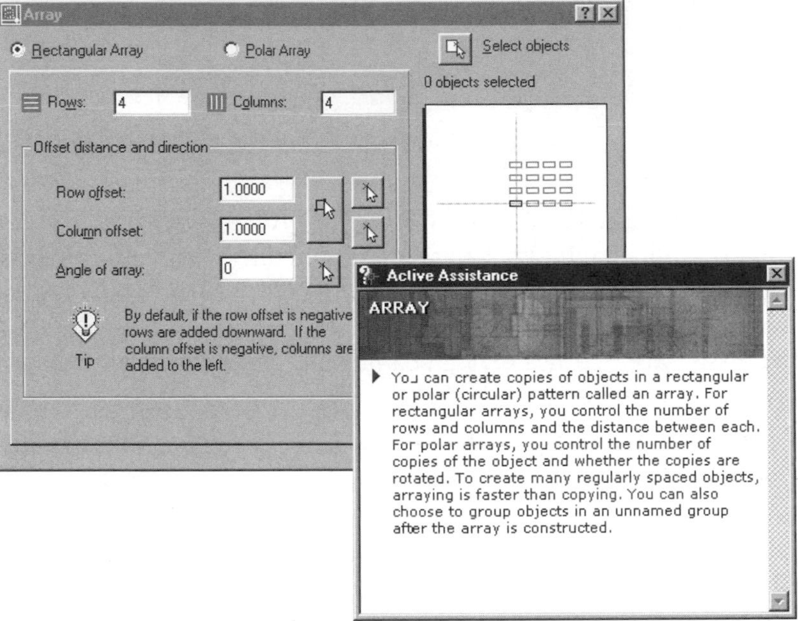

Figure 1–4 *The Active Assistance window with Hover Help active.*

Tip: While experienced AutoCAD users will not likely find the Active Assistance very useful, they may, however, wish to enable either the New and Enhanced Commands or the On Demand activation modes as they transition into AutoCAD 2002.

LEVERAGING AUTOCAD'S FEATURES

In this book you will learn how to use many AutoCAD features that help you work on a project more effectively. When you set up your projects, you must keep the capability of certain AutoCAD features or commands in mind as you develop your approach to a project or to project standards.

While this chapter discusses the use of these commands, its focus is on how these commands relate to project setup. To understand how to use these commands to their fullest, you should read this book's more detailed discussions in the appropriate chapters. For now, concentrate on the concepts that are being presented instead of concerning yourself with the detailed use of a command or concept.

When you learn how to use the commands mentioned in this chapter, think about how they relate to project delivery. After you have learned how to use these commands, you can return to this chapter to review their use in terms of project delivery.

DRAWING LAYERS

One of the most powerful features offered by AutoCAD—and a feature that makes project delivery more efficient—is the use of drawing layers. Layers allow you to organize different objects in your drawing. Layers also can minimize the number of drawings needed for a project by reducing replication of objects. By using layers effectively, you can increase efficiency and reduce errors.

Drawing layers are created using AutoCAD's LAYER command and are used to set up drawing data in hierarchical groups that can be turned on and off or that can be locked from editing. In this way you can, for example, create text that defines how a shop manufactures a part on one layer, and create text that helps a salesman explain the product on a different layer. By turning layers off and on, you can plot from the same drawing two drawings that serve different purposes.

You also can use AutoCAD's capability to freeze layers (a condition in which the layer information is not displayed or loaded into memory) to save drawing load-time, as well as to display various objects of the same model. You can freeze layers of drawings that are xref-ed into the current drawing, or you can freeze layers within individual paper space viewports. This means that one AEC drawing or viewport could contain a floor plan, another drawing or viewport could contain a reflected ceiling plan, and you could use the same model for both drawings. In addition, you can develop nonprinting layers for information that you do not want to be plotted but will use with the drawing.

Using AutoCAD's DEFPOINTS Layer and Non-Plotting Feature

AutoCAD has a special layer called DEFPOINTS that is never plotted. Objects on the DEFPOINTS layer appear on the screen but are not plotted. You can therefore draw information for reference purposes—such as floor plan areas, thread counts, or volumes and weights—on the DEFPOINTS layer. The DEFPOINTS layer automatically is added to your drawing when you create an associative dimension, or you can add it manually by creating a new layer named DEFPOINTS.

You can also create non-plotting layers. To do so, simply create a new layer and then choose the Printer icon in the Layer Properties Manager to turn printing off. The layer remains visible on the screen but will not be plotted.

Non-plotting layers, as well as the DEFFPOINTS layer, help you avoid replotting drawings because you do not have to remember to freeze or turn off layers that you do not want to plot.

Other Uses for Layers

As discussed earlier in this chapter, you can use layers for redlining purposes. Additionally, you can store project data and design information on layers that are needed by other software. Many third-party applications require that certain data be stored on layers with specific names. For example, point data from a site survey must be on a specific layer so a third-party application can locate the data and use it to generate site contours. You must take these types of specialized layers into account if you use third-party products when you develop your prototype drawings.

Furthermore, you can organize different objects in your model by placing them on different layers. By using different layers, you can quickly apply a new color, linetype, or lineweight for all objects on a layer. This method is useful for controlling the display of hidden lines, for example. By turning hidden lines off on the screen, you remove clutter and make viewing the model easier. Then you can turn hidden lines back on for plotting purposes.

Meeting Industry Standards

Finally, your industry might have organizations that have developed layering standards for use in electronic drawings. If your drawings are required to meet certain industry standards, you must set up your layers accordingly. Other industries, such as the AEC industry, might have guidelines only for layer names and use, but these guidelines will help you in developing your own layering standards. The point is that some research is required before you define your use of layers in your projects. After you have developed layering standards, you can easily store them in template drawings. You can also create custom pull-down menus and create, set, and reset layers from a quick menu.

Layers in AutoCAD are created and managed in the Layer Properties Manager shown in Figure 1–5. You will learn more about layer creation and management in Chapter 4, Managing Layers.

DEFINING LINETYPES

Linetypes represent different things for different industries. AutoCAD comes with a variety of predefined linetypes and also enables you to define a wide variety of your own linetypes. You can define linetypes that contain symbols within them or that have varying spacing between line components. Linetypes can also be defined as an industry drafting standard. As such, you should create a set of linetypes that meet your industry's requirements and include them in the linetype file. You can also load linetypes into your template drawing so that they are readily available. It is not absolutely necessary to preload linetypes, however, because they are relatively easy to retrieve from the AutoCAD linetype file. You will learn more about creating and loading linetypes in Chapter 5, Applying Linetypes and Lineweights.

DEFINING LINEWEIGHTS

Lineweights allow you to control the width of a line. Lineweights may be displayed on the screen and be plotted on paper. By using lineweights you can more easily identify different objects in your drawing. As with linetypes, discussed previously, varying lineweights can be used to differentiate objects that are hidden or lie

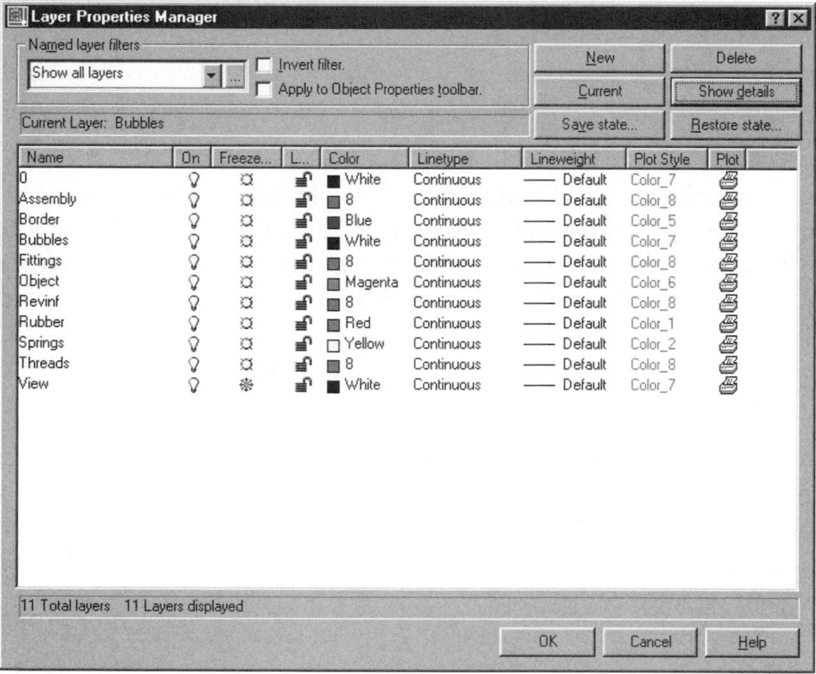

Figure 1–5 *The Layer Properties Manager dialog box manages layers in AutoCAD.*

beneath other objects. You will learn more about assigning lineweights in Chapter 5, Applying Linetypes and Lineweights. Figure 1–6 shows the Lineweight Settings dialog box.

SELECTING TEXT STYLES

AutoCAD enables you to use any font that comes with Windows (.ttf), as well as text fonts (.shx) that are supplied with AutoCAD. You must select the font or fonts

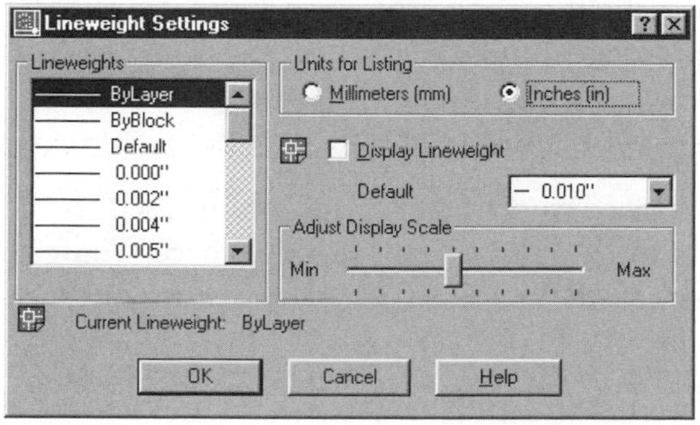

Figure 1–6 *The Lineweight Settings dialog box.*

that work best with your drawing size and plotter or printer. It is equally important that the fonts you use are legible in a wide variety of reproduction sizes. For example, many drawings are placed on microfilm as a matter of storage. To ensure that the lettering is visible on the microfilm, you should use text styles that are a minimum of 1/8 of an inch in plotted height.

After you have experimented with a variety of plotted output and text styles, you will want to select a few styles for general text, sheet title block information, drawing titles, and emphasized text. Everyone has distinct tastes in text appearance, so AutoCAD supports a host of variations in the appearance of text. Therefore, it might be best for you to find examples of text that you and your firm find acceptable, and then find text settings and fonts in AutoCAD that best approximate your desired results.

Text can have different angles, line spacing, heights, weights, effects (upside down, backwards, and/or vertical), and width factors. You will learn more about the use of text in Chapter 15, Annotating with Text.

SAVING VIEWS

AutoCAD enables you to save views of your drawings to which you can return time and time again. The reuse of views can result in considerable time savings if you have a predictable set of views upon which all users can rely. For example, everyone will need a view of the entire drawing, including its title block. You could save this view as a view named Overall by using the AutoCAD VIEW command. Views can be saved in either paper space or model space. You might want to save standard views that are ¼ portions of the drawing and call them UL, UR, LL, and LR for upper-left, upper-right, lower-left, and lower-right. Saving these views in a template drawing ensures that they are available in all of your drawings.

PAPER SPACE VERSUS MODEL SPACE VIEWPORTS

AutoCAD uses paper space layouts for real-world paper sizes and model space for real-world model sizes. The proper use of these two spaces means that you never need to be concerned about scaling a drawing up or down for a plot. In pre-paper space days, users had to remember the scale of each drawing so that when the drawings were plotted, the users could enter the correct scale factor in response to plot setup questions presented by AutoCAD.

Proper use of paper space layouts takes some getting used to because you view your model through ports from paper space to model space. You create your model and work on the model in model space, which allows you to view the model as close to full-screen as possible.

Unless your drawings are very predictable, you will not be able to create multiple viewports that will work every time. However, using the overall viewport saves some

Tip: You might want to consider creating a viewport that is as big as the drawing area within your title block to store with your template drawing. This viewport can provide a good starting place for any drawing and it can be resized, copied, or turned off as the drawing develops.

time when you create a drawing. You will learn more about the use of paper space in Chapter 19, Layouts and Paper Space.

SETTING DIMENSION STYLES

Each discipline or industry has its own standards and preferences for dimensioning. AutoCAD enables you to store dimensioning standards as dimension styles. You can set the color of individual portions of a dimension, the text style the dimension uses, the arrow style, the way the dimension extension lines work, and the format of the dimension annotation. Each dimension feature, however, must be scaled appropriately for the scale that you use for the drawing. You can set dimensions to be scaled based upon paper space viewport scaling, and AutoCAD will adjust the dimension features for you.

You should set up a generic dimension style for each type of dimension that you use (such as radial, leader, and linear) and save them in your template drawing. Figure 1–7 shows the Dimension Style Manager dialog box, which grants access to custom dimension styles and sets the dimension scale factor.

Note that you must set up your text style standards in your drawing before you set the dimension text style. You also can create custom blocks for use as arrows.

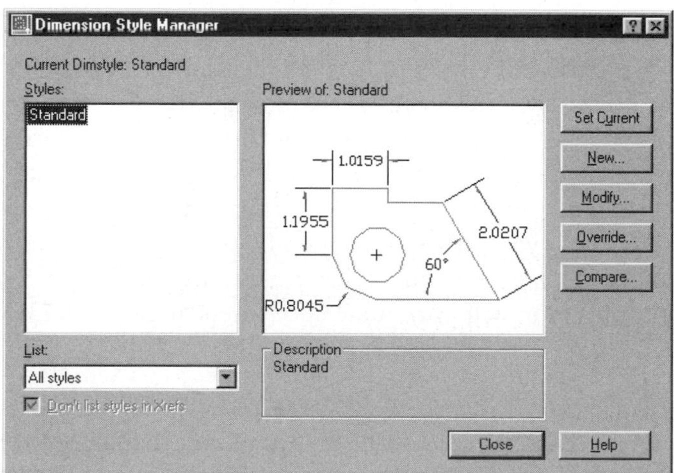

Figure 1–7 *Using the Dimension Style Manager dialog box enables you to set up custom dimension styles.*

Dimensions are made up from a complex set of options, and it will take some time and study on your part to tailor them to your needs and tastes. You will learn more about using dimension styles and the DDIM command in Chapter 17, Dimensioning Basics, and in Chapter 18, Managing Dimensions.

USING XREFS

This chapter has already briefly discussed a variety of situations in which xrefs can be helpful in the delivery of a project. You should be aware of a few things when using xrefs. First, if you intend to share your drawings with consultants or clients, you must make them aware of your folder structure. While organizing your drawings and xrefs with folders is good practice, if those you share your drawings with are not aware of your structure, AutoCAD may not find xrefs attached to an opened drawing. While you can avoid this problem by locating all drawings, including xrefs, in the same folder so that AutoCAD can find all pertinent files, it is better practice to use a proper folder structure and pass this structure on to those with whom you share your drawings.

A second issue arises when problems are found in an xref drawing. In many situations you will notice that something must be fixed in a drawing while you are working on an entirely different aspect of the drawing. For example, you could be dimensioning walls in a floor plan that is xref-ed into the current drawing when you discover that a wall is drawn wrong. In-place Reference Editing allows you to edit blocks and xrefs and save changes back to the original object or drawing file. More importantly, it allows you to do this from within the current drawing that contains the questioned xref.

A third point to keep in mind is that if you need to use only part of an xref-ed drawing, you can insert and clip the xref so that the desired portion of the drawing remains. You still have the advantages of current updates to the xref-ed portion of the drawing, but you will not have the entire xref attached to the current drawing. If you do not anticipate changes to the xref-ed drawing, you can attach the xref to make it a permanent part of the drawing. In this case, the xref-ed drawing becomes an AutoCAD block that you can explode, modify, and clip as you desire. This chapter discusses blocks in a later section, which also points out the differences between using blocks and xrefs. You will learn more about xrefs in Chapter 13, Working with Large Drawings and External References.

Tip: With Viewport Clipping, you can create non-rectangular viewports in paper space.

CREATING MULTILINE STYLES

AutoCAD enables you to draw multiple lines, which comprise a set of lines offset from a centerline path, by using the Multiline Styles dialog box. If you draw streets with curbs, multiple data lines, cavity walls, or other multiple-line objects, you might find it helpful to create standard multiline styles and save them for use in your projects.

While multilines are very convenient, their usefulness is somewhat limited. Multiline components must be edited via the MLEDIT command, which has editing limitations. For example, it is difficult to insert a door or window into a wall created by a multiline because you cannot easily break the wall. Also, if you grip-edit a multiline, you can edit only the outer boundary of the multiline.

Despite their editing limitations, you may find multilines a useful feature in certain situations. Because multilines behave as a single object, you can save considerable time creating and editing drawings if you plan for their use. For example, if you must change a room configuration, editing a single multiline changes all wall lines instantly.

Tip: In general, any time that you have basic objects that are made up of multiple lines, you can make good use of multilines. If, however, multiple variations exist in the width, composition, and interruptions of the multilines, then you should carefully consider their use. Also, multilines cannot be converted into 3D lines, nor can their intersecting lines have a radius applied.

USING BLOCKS

Blocks are primary components that contribute to huge time savings in project delivery because they can be used in many ways. The importance of blocks cannot be overstated. Blocks can be used for symbols, components, details, standard text notation, and many other situations. The use of blocks also saves drawing disk space if multiple instances of their use exist, because one definition of the block is saved in the drawing, and that definition is copied throughout the drawing without repeating all of the block's components for each insertion. You should build a library of blocks that are used for repeating objects in your projects and keep them readily available for use. Or, you could purchase block libraries from third-party vendors.

Blocks can be created on layer 0 so that they inherit the characteristics of the layer in which they are inserted, or they can contain multiple layer, color, linetype, and lineweight definitions so that they maintain their appearance regardless of the layer on which they are inserted. Blocks also can be used with the ARRAY, MINSERT, DIVIDE, and MEASURE commands so that multiple copies can be created easily for such things as stair treads, elevations of fences and grilles, and flooring patterns. As with multilines, the use of blocks requires careful planning and repetitive standards for layer names, linetypes, block names, and component design. A sophisticated use of blocks also enables the assembly of projects from a kit of parts. For example, many national companies use a kit of parts for the assembly of their new branch offices throughout the country. You will learn more about blocks in Chapter 12, Making the Most of Blocks.

The use of xrefs has also been discussed in this chapter. As a general rule of thumb, you should address two prime considerations when deciding whether to use xrefs or blocks. First, if the need for continual updating is not required, then blocks are more convenient to use than xrefs. Second, and more importantly, if you want to freeze your drawings in time so that the electronic copy on your disk matches the last version plotted, you should not use xrefs, because they will be updated each time you open the drawing.

DESIGNCENTER

AutoCAD DesignCenter (ADC) enables you to easily locate and insert blocks using a Find feature. Figure 1–8 shows a sample window of the DesignCenter. When you enter a block's name or description, ADC will search all drawing files in the specified folders for blocks matching the description. Then icons of the blocks are displayed, along with detailed descriptions. ADC also allows you to drag and drop blocks into a drawing; if a block has drawing units associated with it, AutoCAD automatically scales the block based on the current drawing's units.

ADC also allows you to drag and drop symbol table items. This feature allows you to quickly copy symbol table items from existing drawings and gives you the ability to restore the current drawing's original settings. You can quickly insert layers, dimension styles, text styles, and linetype styles from other drawings. You will learn more about ADC in Chapter 11, Applying AutoCAD DesignCenter.

MANAGING CAD STANDARDS

Maintaining CAD drawing standards is becoming increasingly important as drawings are shared and exchanged across Intranets, local networks, and the Internet. Whether you need to communicate about drawing files with team members, colleagues, contractors, or clients, the advantages of establishing and then conforming to standards such as layer conventions can save time and reduce the costly errors caused by poor communications. To address this issue, a new set of tools called CAD Standards has been incorporated into AutoCAD 2002.

CAD Standards defines a set of common properties for named objects in a drawing—layers and text styles, for example. You can create, apply, monitor, and audit compliance to the standards using any drawing file to enforce consistency. This ability is especially helpful in an environment where many individuals may contribute drawings to a project or group effort. The overall scheme for using CAD Standards is simple and straightforward:

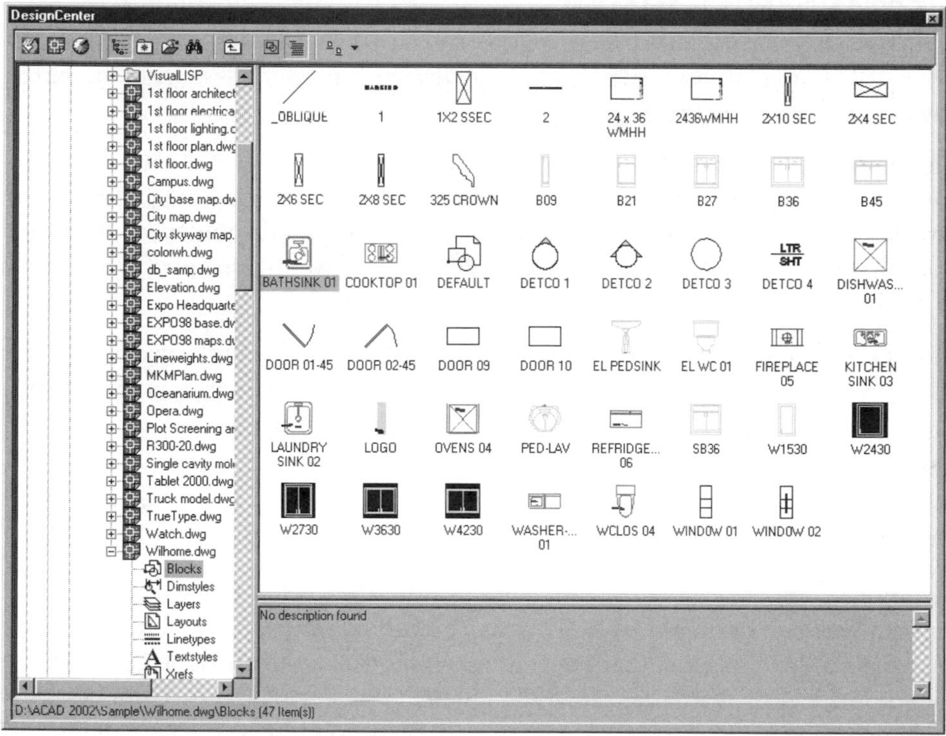

Figure 1–8 *The AutoCAD DesignCenter quickly locates blocks in other drawings.*

1. **Create a Standards drawing file**, defining the various standards you want to enforce. AutoCAD's CAD Standards allow you to define standards for layers, dimension styles, linetypes, and text styles.

   You may be able to use a template drawing with these named objects already defined to establish your standards drawing.

2. **Save the standards drawing** to the new .dws drawing file extension. The Save Drawing As dialog box lists the new .dws extension as an option.
3. **Associate a standards** .dws file with the drawing(s) you want to check for standards compliance. Use the Configure Standards dialog box for this step.
4. **Check for standards compliance** using the CHECKSTANDARDS command available on the CAD Standards toolbar or from the Tools>CAD Standards>Check menu. Use the Check Standards dialog box for this step.
5. **Interactively review standards** for deviations and either ignore or correct each deviation, much as you do with a standard spelling checker.
6. **Check a batch of drawing files** (optional) and generate a report of the number and type of standards deviations in the group of files.

THE STANDARDS MANAGER

The Standards Manager (the Configure Standards dialog box shown in Figures 1-9 and 1-10) is used to associate a standards drawing file, or files, with the current drawing. It consists of two folders.

The Standards folder displays any standards (.dws) files associated with the current drawing. If more than one standards file is associated with a drawing, the topmost-listed standard takes precedence. Standards files can be added, removed, or reprioritized. Figure 1–9 shows only one standards file—Site.dws—associated with the current drawing.

The Plug-ins folder shows all standards plug-ins that are currently available. Figure 1–10 shows standards plug-ins for dimension styles, layers, linetypes, and text styles. The data displayed in the Plug-ins folder is informative only.

INTERACTIVE STANDARDS AUDITING

Once a standards **.dws** file is associated with a drawing, the new CHECKSTANDARDS command displays the Check Standards dialog box (see Figure 1–11).

Deviations from the standards are reported in the Problem window. In Figure 1–11 the contours layer is reported as having a nonstandard property. The Preview of Changes window indicates that the contours layer is currently assigned a continuous linetype instead of the dashed linetype required by the standards. Controls in the dialog box allow you to ignore or fix the deviation or to move to the next found deviation. If you elect to fix the deviation, CAD Standards will automatically perform the needed change. If you choose to ignore the deviation, it may still appear in subsequent audits, depending on the user's setting.

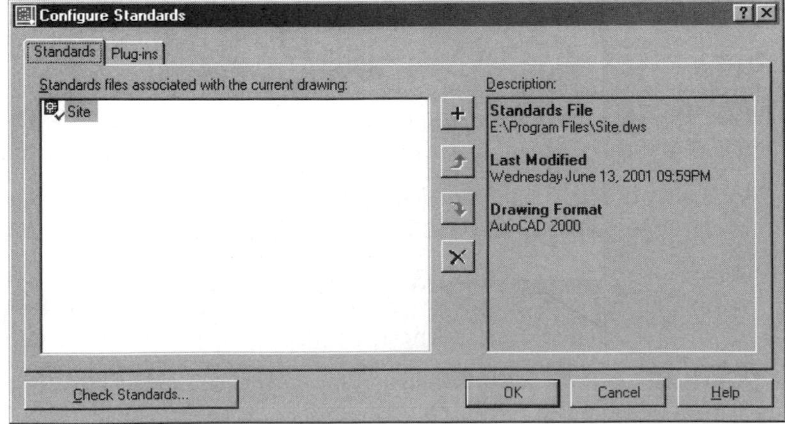

Figure 1–9 *The Configure Standards dialog box: Standards folder.*

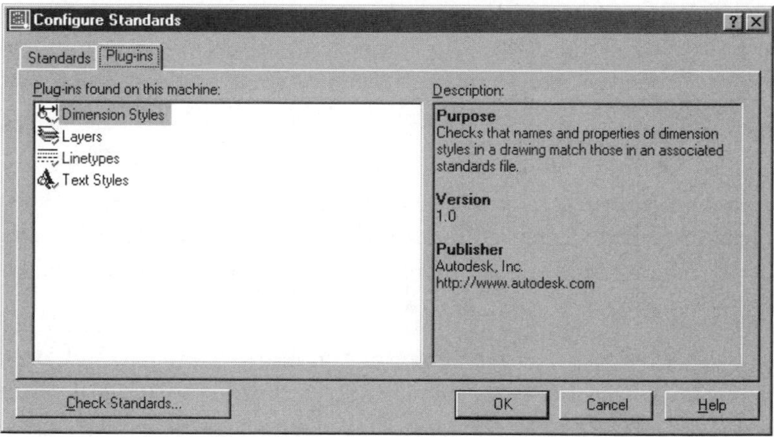

Figure 1–10 *The Configure Standards dialog box: Plug-ins folder.*

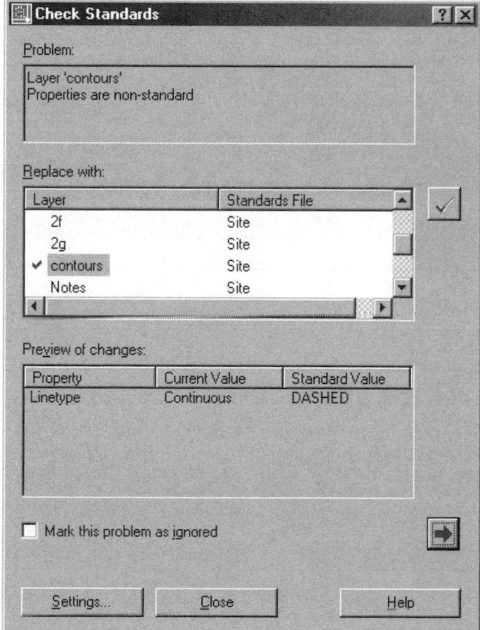

Figure 1–11 *The Check Standards dialog box.*

BATCH AUDITING

The Batch Standards Checker audits a series of drawings for standards violations and creates an XML-based summary report detailing all violations. To use the Batch Standards Checker, you must first create a standards check file that specifies what drawings to audit and the standards files that will be used to verify them. By default each drawing is checked against the standards file(s) associated with it. You can, however, override the default and choose another set of standards files. An XML report to which notes can be added can be generated upon completion of a batch audit. This report can be distributed to the owners of the drawing files for corrective action.

SUMMARY

In this chapter you learned about the components of an effective project-delivery system. You also learned how to use AutoCAD features to help you in the management of project delivery. You learned how to get organized before creating a drawing, as well as how to use OLE to create a task list that is embedded in each drawing. You also learned how to use a mock-up process to view your entire project as a whole for planning purposes based upon priorities, how much detail is necessary, how your drawings relate to other drawings and documents, and how to account for the people who work on the drawings.

Other sections discussed the use of repetitive elements and considerations for the different ways drawings can be displayed or published. You also reviewed the basic factors used in the initial drawing setup, such as paper, scale, units, angles, and precision. Finally, you learned about the CAD Standards tools that help you to perform Quality Control (QC) checks on drawings.

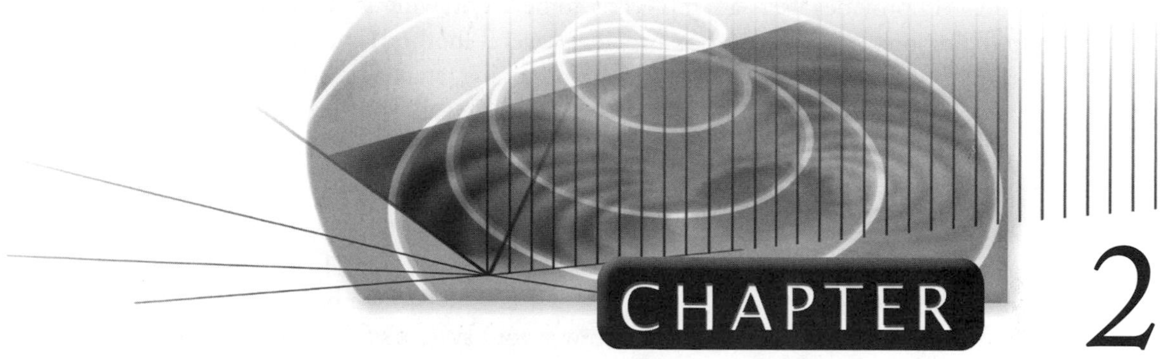

Beginning a Drawing

AutoCAD is a versatile drawing application that is used by many industries to draft and design a multitude of products. It is used by architects to create buildings; by civil engineers to design streets, highways, and utilities; and by cartographers to map the world. Mechanical engineers use AutoCAD to develop three-dimensional working models of a wide array of useful tools used in the day-to-day activities of working professionals. From designing heavy construction equipment to the most delicate medical instruments, AutoCAD provides state-of-the-art tools for turning dreams into reality.

With such a wide range of industries using AutoCAD, the program is designed to allow users to customize AutoCAD's drafting settings easily to suit their unique needs. In this chapter we explain AutoCAD's drafting settings and look at how AutoCAD simplifies setting up your drawing environment with the Startup dialog box.

This chapter covers the following topics:

- AutoCAD's default values
- Using wizards to set up a drawing
- Using templates to start a drawing
- Revisiting drawing limits
- Controlling drawing units
- Configuring drafting settings

THE STARTUP DIALOG BOX

AutoCAD provides users a simple way to begin drawing with AutoCAD: the Startup dialog box. This dialog box provides several methods for starting your drawing session that allow you to open an existing drawing, start a drawing from scratch, use a predefined drawing template, or use a wizard to walk you through the steps necessary to define your drawing's configuration.

The Open a Drawing button allows you to select an existing drawing. When you choose this button, the Startup dialog box displays a list of the most recently opened drawing files, as shown in Figure 2–1. You may select one of the files in the list or choose the Browse button to locate other files. The Browse button opens the Select File dialog box, which allows you to browse your system's folders for the drawing file you wish to open.

The other three buttons allow you to start new drawings using different approaches. Depending on the button you choose, the Startup dialog box presents you with choices that automate the process of setting up your new drawing's default values.

The following sections describe in detail the various ways you can use the Startup dialog box to set your drawing's default values, and they provide examples and exercises that demonstrate its ease of use.

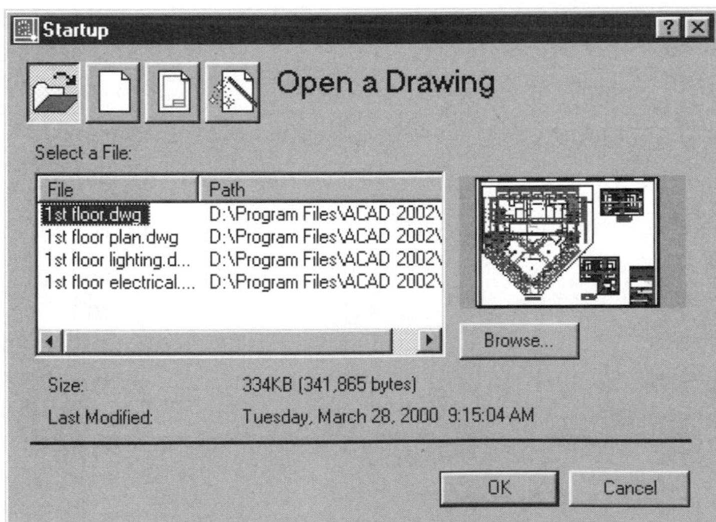

Figure 2–1 *The Open a Drawing button displays the most recently opened files in the Startup dialog box.*

STARTING FROM SCRATCH: UNDERSTANDING AUTOCAD'S DEFAULT VALUES

The Start from Scratch button is the simplest to use when starting a new drawing. When you choose this option, it presents you with only two choices in the Default Settings area. You can choose English or metric units of measurement, as shown in Figure 2–2. If you choose English units, AutoCAD creates a new drawing based on the Imperial measurement system, which uses the acad.dwt template file, and sets the drawing boundaries, called limits, to 12 by 9 inches. If you choose metric units, AutoCAD creates a new drawing based on the metric measurement system, which uses the acadiso.dwt template file, and sets the drawing limits to 420 by 297 millimeters.

When you choose either English or metric units, you are actually setting two system variables, MEASUREMENT and MEASUREINIT. The MEASUREMENT system variable sets the drawing units as either English or metric for the current drawing, whereas the MEASUREINIT system variable sets the drawing units as either English or metric for new drawings when they are created.

Specifically, the two system variables control which hatch pattern and linetype files an existing drawing or a new drawing uses when it is opened. When the system variables are set to zero, the English units are set and AutoCAD uses the hatch pattern file and linetype file designated by the ANSIHatch and ANSILinetype registry settings. When the system variables are set to 1, the metric units are set and AutoCAD uses the hatch pattern file and linetype file designated by the ISOHatch and ISOLinetype registry settings.

Figure 2–2 *The Start from Scratch button displays only two options for default settings.*

Note: Each time you launch AutoCAD and the Startup dialog box is displayed, you may choose the Cancel button. The Cancel button dismisses the Startup dialog box and creates a new drawing using the current values for MEASUREMENT and MEASUREINIT, including the associated drawing template and drawing limits. This occurs because AutoCAD stores these values in its system registry AutoCAD.

It is important that you determine which system of measurement you will use before you start your drawing because the system of units you select influences how objects appear in your drawing. For example, Figure 2–3 shows two rectangles that contain a hatch pattern. Both hatch patterns are the same, ANSI31, and their rotation and scale are identical, zero and 1.0000, respectively. However, it is obvious that the hatch pattern on the left displays lines much closer together than does the hatch pattern on the right.

The difference in the two hatch patterns occurs because the hatch pattern on the left was drawn first with the MEASUREMENT system variable set to zero (in English units). Next, the MEASUREMENT system variable was set to 1 (in metric units). Then the same hatch pattern was drawn on the right. This example demonstrates that AutoCAD uses two different hatch pattern files based on the current value of MEASUREMENT. Therefore, to ensure that AutoCAD inserts the proper hatch patterns and linetypes into your drawing, be sure to choose the correct units of measurement.

Tip: To correct a hatch pattern inserted using the wrong units of measurement, after changing the MEASUREMENT system variable, select the hatch pattern, right-click, and then choose Hatch Edit. Once AutoCAD displays the Hatch Edit dialog box, choose the same hatch pattern from the list, and then choose OK. AutoCAD updates the hatch pattern based on the current units of measurement.

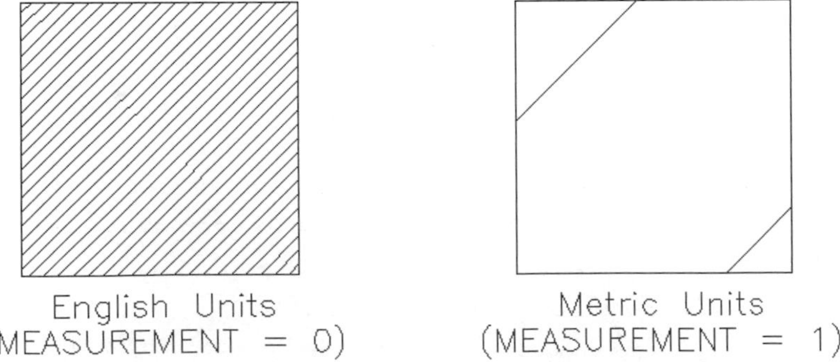

Figure 2–3 *The MEASUREMENT system variable affects the appearance of hatch patterns when they are inserted in a drawing.*

USING WIZARDS TO SET UP A DRAWING

AutoCAD provides wizards to help you set up a new drawing. As you complete setting one value, the wizard takes you to the next, using a standard Windows wizard interface that allows you to go ahead to the next default setting or back to edit previous settings. By using AutoCAD's wizards, you can quickly set certain default values and change them again if needed.

Understanding the Wizard Selections

When you choose the Use a Wizard button, the Startup dialog box offers two wizard choices—Advanced Setup and Quick Setup—as shown in Figure 2–4. The Advanced Setup wizard takes you through five dialog boxes, allowing you to set values for units, angle, angle measure, angle direction, and area. The Quick Setup wizard offers a subset of these values, removing those options for setting the various values related to angles. Specifically, the Quick Setup wizard allows you to set values for units and area only.

You can set the following values with the Startup wizard:

- **Units**—The Units dialog box allows you set the way linear units are displayed in your drawing.
- **Angle**—The Angle dialog box allows you set the way angles are displayed in your drawing.
- **Angle Measure**—The Angle Measure dialog box allows you to set the base angle from which all angles are measured. (It represents zero degrees.)
- **Angle Direction**—The Angle Direction dialog box allows you to set the direction in which all angles are measured from the base angle (clockwise or counterclockwise).
- **Area**—The Area dialog box allows you to set the drawing's limits, which represent the area within which your model should be drawn and contained.

The Units dialog box allows you to set how AutoCAD displays measured values, such as the length of a line, the radius of a circle, or the area of a polygon. It also controls how AutoCAD displays coordinate values. For example, if the units are set to Decimal, and you list the values of a circle object in a drawing, AutoCAD displays values as follows:

```
center point, X= 5.8796  Y= 4.2286  Z= 0.0000
      radius  1.1396
circumference  7.1601
        area  4.0797
```

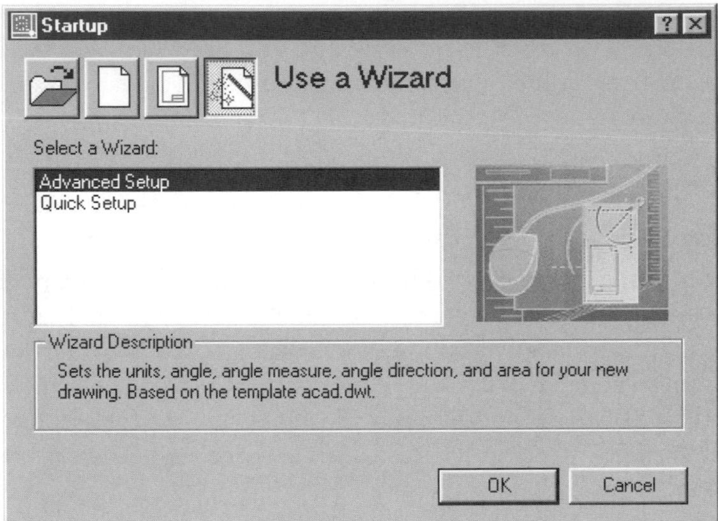

Figure 2–4 *The Use a Wizard feature offers two wizards to choose from.*

Note: In the discussion of the Start from Scratch option, you learned how to select between English and metric measuring systems, which control the hatch pattern and linetype files AutoCAD uses in a drawing. However, the Units dialog box discussed here controls how units in the selected measurement system are displayed.

However, if the units are set to Fractional, and you list the values of the same circle, AutoCAD displays values as follows:

center point, X= 5 7/8 Y= 4 1/4 Z= 0

radius 1 1/8

circumference 7 3/16

area 4 1/16

Notice the difference between the two formats. The decimal format displays values less than one unit as decimals. In contrast, the fractional format displays values less than one unit as fractions. This is true even though values for exactly the same object are listed.

Tip: If you use the fraction-based unit settings, measurements will be rounded to the nearest fractional unit. This can lead to drafting inaccuracies if these values are assumed to be exact values. Therefore, if you require precise values, switch to using decimal-based units instead.

Note: The selected units display mode does not affect AutoCAD's dimensions, which are controlled by their own system variables.

The Units dialog box also allows you to control the precision of the measured values. For decimal mode, this represents the number of decimal places displayed for a value. For fractional mode, it represents the smallest fractional increment displayed for a value (for example, 1/32 or 1/64).

When you select a units value, you are actually setting the LUNITS system variable. The LUNITS system variable is based on integer values, with each integer value representing a units display mode, as shown in the Table 2-1.

Additionally, another system variable, LUPREC, controls precision for units. Like the LUNITS system variable, LUPREC uses integer values to determine the number of decimal places displayed with a value. The range of acceptable values is zero through 8. In the case of values displayed in fractional mode, the integer value represents a fractional increment. Table 2-2 shows examples of the precision display for each value.

The Angle dialog box allows you to set how AutoCAD displays angle values, such as the angle of an arc or the angle between two lines. It also allows you to set the precision of displayed angles.

Like the Units dialog box, the Angle dialog box controls two system variables, AUNITS and AUPREC. When you select an angle value, you are setting the AUNITS system variable. The AUNITS system variable is based on integer values, with each integer value representing a units display mode, as shown in thc Table 2-3.

The AUPREC system variable controls the precision of displayed angles and functions like the LUPREC system variable, using integer values to determine the num-

Table 2–1: *Units Display Mode Examples for the LUNITS Variable*

Integer	Display Mode	Example
1	Scientific	1.1356E+00
2	Decimal	1.1356
3	Engineering	0'-1.1356"
4	Architectural	0'-1 1/8"
5	Fractional	1 1/8

Table 2–2: *Units Precision Display Examples for the LUPREC Variable*

Integer	Decimal Mode	Fractional Mode
0	0	0
1	0.0	0 1/2
2	0.00	0 1/4
3	0.000	0 1/8
4	0.0000	0 1/16
5	0.00000	0 1/32
6	0.000000	0 1/64
7	0.0000000	0 1/128
8	0.00000000	0 1/256

Table 2–3: *Angular Display Mode Examples for the AUNITS Variable*

Integer	Display Mode	Example
0	Decimal degrees	23.9493
1	Degrees/minutes/seconds	23d56'57"
2	Gradians	26.6103g
3	Radians	0.4180r
4	Surveyor's units	N 66d3'3" E

ber of decimal places displayed with a value. The range of acceptable integer values is zero through 8. In the case of values displayed in degrees/minutes/seconds or surveyor's units, the integer values control the display of minutes and seconds. If the precision value defines the precision beyond minutes and seconds, the seconds are displayed in decimal format. Table 2-4 shows examples of the precision display for each value.

Tip: You can adjust the value of a system variable by typing its name at the command prompt and then entering the desired value.

Table 2–4: *Angle Precision Display Examples for the AUPREC Variable*

Integer	Decimal	Surveyor's Units
0	8	N 82d E
1	8.2	N 81d46' E
2	8.23	N 81d46' E
3	8.235	N 81d45'55" E
4	8.2347	N 81d45'55" E
5	8.23469	N 81d45'55.1" E
6	8.234689	N 81d45'55.12" E
7	8.2346894	N 81d45'55.118" E
8	8.23468944	N 81d45'55.1180" E

Note: The selected precision display mode does not affect the accuracy AutoCAD uses in its calculations. All calculations are double precision and carried out to 16 decimal places.

Notice that the surveyor's units for integer values 1 and 2, and for values 3 and 4, respectively, display the same level of precision. This occurs because minutes and seconds must be displayed as full values. This is true because there are 60 minutes of angular arc in one degree, and there are 60 seconds of angular arc in one minute.

Note: When displaying angles in degrees/minutes/seconds or surveyor's units, the ' symbol stands for minutes, not feet, and the " symbol stands for seconds, not inches.

The Angle Measure dialog box allows you to set the base angle from which all angles are measured. AutoCAD uses the default value of zero degrees as due east. The Angle Measure dialog box allows you to choose from four preset base angles (east, north, west, or south), and enter your own user-defined base angle. The value for the base angle is stored in the ANGBASE system variable and can represent any angle value.

Note: The base angle is associated with the X-axis of the current UCS. If you rotate the UCS, the base angle also rotates.

The Angle Direction dialog box works in conjunction with the Angle Measure dialog box and allows you to set the direction in which all angles are measured. The only possible directions are counterclockwise and clockwise. The current direction value is held in the ANGDIR system variable, with zero indicating a counterclockwise direction, and 1 indicating a clockwise direction.

The Area dialog box allows you to set the drawing's limits, which generally represents the area in which your model is contained or the area within which you will be creating your drawing. The limits are intended to help you manage your drawing.

When you set the width and length of the limits, AutoCAD defines the limits as a rectangle, with its lower-left corner located at coordinates (0,0), and its upper-right corner based on the values you enter for width and length. The width and length values are stored in the LIMMIN and LIMMAX system variables, respectively.

You can instruct AutoCAD not to allow you to draw accidentally outside the drawing limits, thereby adding a level of error checking to your drawing session. This is called Limits Checking, which is controlled with the LIMCHECK system variable. Setting the variable to zero turns Limits Checking off, and setting it to 1 turns Limits Checking on. The LIMCHECK variable will only protect you against acquiring points outside the limits. You can copy objects outside limits and position other items as well.

The drawing area within the limits is also the area in which AutoCAD will display its grid. The grid is made up of a series of dots and is used as a frame of reference when creating objects. The grid is useful for quickly drawing objects based on grid locations. The grid's visibility is controlled from the GRID button located on the status bar at the bottom of AutoCAD's screen. You may also press the function key F7 to toggle the grid on and off.

Finally, you can use the limits with the Zoom command when plotting. When zooming, you can select the All option to immediately display the entire limits area on AutoCAD's screen. When plotting from model space, you can select the Limits option to plot the entire area defined by the limits setting.

You have learned how AutoCAD makes starting a drawing easy when you use its wizards to walk you through the setup process. Next you will learn how to take advantage of previous drawing setups by using templates.

USING TEMPLATES TO START A DRAWING

A template is a drawing file that contains predefined drawing settings and/or geometry, such as a title block, and is used to begin your drawing. Templates are intended as a quick way to take advantage of an existing drawing that contains the proper drawing settings and base geometry you use in every drawing.

For example, suppose you are creating ten drawings, and those drawings will use the same units, angle settings, and title block. To minimize your work effort, you can create the drawing once, defining the proper units and angle settings and inserting the desired title block, and then save the drawing as a template. As you begin each new drawing, you select the template file you previously created, which contains the correct units, angle settings, and title block. The template acts as a base for your new drawing and thereby eliminates a lot of repetitious preparation and setup work.

When you choose the Use a Template button, AutoCAD displays the list of available templates, as shown in Figure 2–5. AutoCAD comes with more than sixty predefined template files, and you can add your own. You create a template file by defining the proper units and angle settings, inserting any desired geometry, and saving the drawing as a .dwt file.

Creating a Template File

Although Autodesk has made a considerable effort to provide a broad array of templates, chances are you will probably want to customize your own templates. You can do so by modifying one of the template files provided with AutoCAD or by creating a drawing from scratch.

To save any drawing as a template file, choose the AutoCAD Drawing Template File (*.dwt) option from the Save as Type drop-down list, as shown in Figure 2–6.

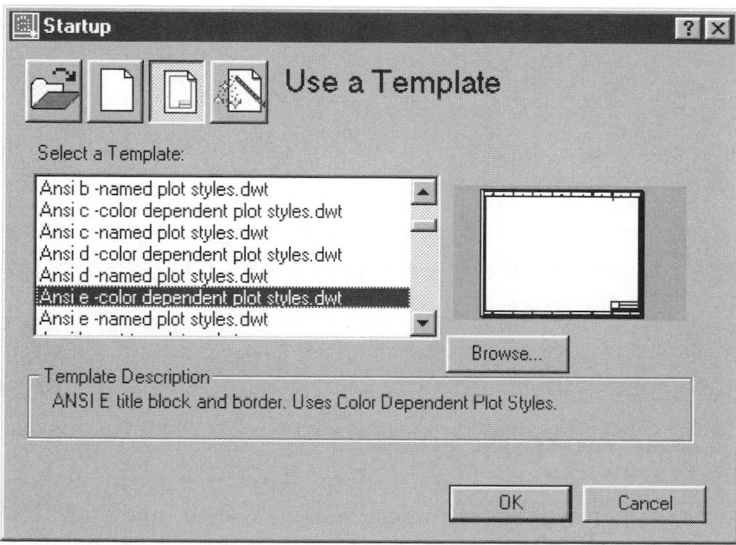

Figure 2–5 *The Use a Template feature offers more than sixty predefined templates.*

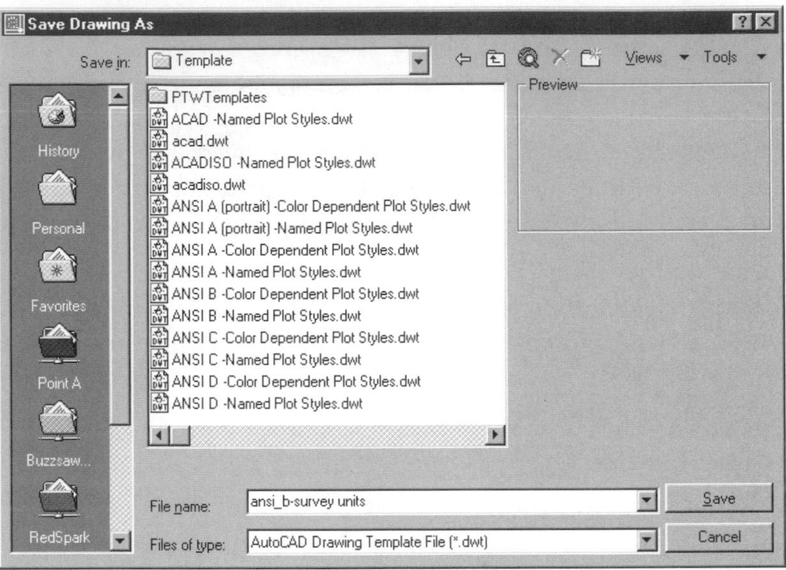

Figure 2–6 *To save a drawing as a template file, choose the AutoCAD Drawing Template File (*.dwt) option from the Save as Type drop-down list.*

When you do this, AutoCAD automatically switches to the Template folder. While you do not have to save your template in this folder, it is the default folder that the Startup dialog box looks in for template files. Therefore, it is a good place to store your custom template files unless the files are project specific.

After you type in a name for your template file, AutoCAD displays the Template Description dialog box, which lets you add a custom description for your template drawing, as shown in Figure 2–7.

The ability to create your own custom templates is very powerful. You can minimize repetitive tasks and reduce errors by creating custom template files and using the same templates over and over.

So far you have seen how to control drawing settings using the Startup dialog box. Next you will learn how to control drawing settings during the current drawing session.

SETTING DRAWING DEFAULTS

AutoCAD provides the ability to modify drawing settings during the current drawing session. By changing these settings, you can control how AutoCAD behaves. Therefore, you can modify the current session to optimize your productivity.

In the next few sections you will learn how to control drawing settings for the current drawing session. Some of the settings are similar to previous versions of AutoCAD, and some are brand new to AutoCAD 2002.

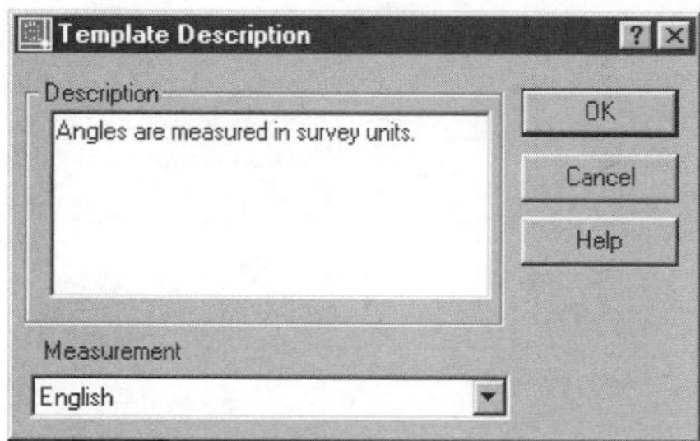

Figure 2–7 *The Template Description dialog box allows you to enter a description for your template file.*

Note: The LUPREC, AUPREC, and AUNITS system variables can be set using the Drawing Units dialog box.

REVISITING DRAWING LIMITS

Previously you learned about setting the limits for a drawing using a wizard. While this information is useful, it does not provide an easy method to modify the limits during the current drawing session.

So how do you easily change the limits during the current drawing session? The process is simple. To change the current drawing's limits, from the Format menu, choose Drawing Limits. When you choose Drawing Limits, AutoCAD starts the LIMITS command, which prompts you to enter new values for the lower-left and upper-right corners of the limits rectangle. You can enter values by picking them on the screen using your pointing device or by entering the coordinates explicitly using your keyboard. Once you enter the new limits values, AutoCAD resets the drawing limits to the new values.

CONTROLLING DRAWING UNITS

In the previous discussion about using AutoCAD's Advanced Setup wizard, you learned how to set various unit and angle values and how those values affect AutoCAD's display. You also learned how to set those values using the appropriate system variables. AutoCAD's Drawing Units dialog box offers another means of controlling these settings (see Figure 2–8). Notice that the counterclockwise/clockwise direction for angles is controlled by toggling the Clockwise feature on or off.

In addition to controlling the previously discussed settings, the Drawing Units dialog box works in conjunction with the DesignCenter to control the unit of measure used for block insertions. This feature automatically adjusts the size of blocks as they are inserted. If a block created in different units is inserted into the drawing, it is automatically scaled and inserted in the specified units of the current drawing.

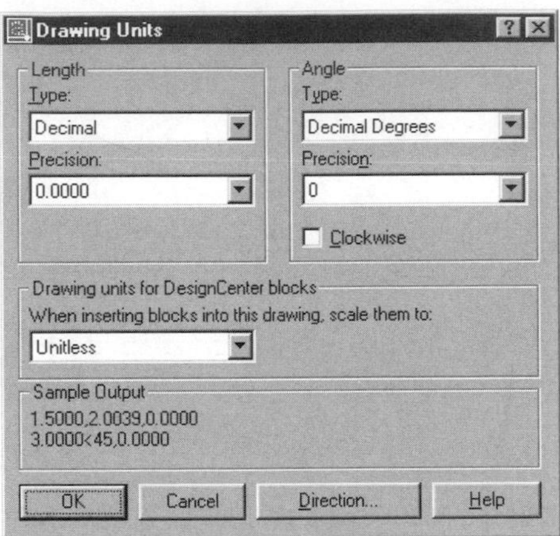

Figure 2–8 *The Drawing Units dialog box lets you control drawing units and angle direction.*

This is a powerful feature, making the insertion of blocks with predefined units very simple. On the other hand, if you do not want to adjust the block automatically, select the unitless mode to insert the drawing as a block and not scale the block to match the specified units. You will learn more about how to use this feature in Chapter 11, Applying AutoCAD DesignCenter.

While most of the features and functions of the Drawing Units dialog box were covered in the discussion about AutoCAD's Advanced Setup wizard, one handy feature was left out. This feature is accessed from the Direction Control dialog box and allows you to define a new base angle by picking points on the screen. By choosing Other and then the Angle button, shown in Figure 2–9, you can select two points on the screen using your pointing device. AutoCAD calculates the angle and uses it as the new base angle.

CONFIGURING DRAFTING SETTINGS

Drafting settings are useful tools you can use as an aid when drawing in AutoCAD. These features can increase accuracy, ease object editing through on-screen visual enhancements, and automate object creation and editing by providing a mouse-only interface. By controlling AutoCAD's Drafting Settings, you can make working with AutoCAD easier, quicker, and more accurate.

The Drafting Settings dialog box is accessed from the Tools menu by choosing Drafting Settings. Once the dialog box is displayed, you will see three folders, as shown in Figure 2–10. These folders control snap and grid, polar tracking, and object snap.

For detailed information on the functions and features of the Drafting Settings dialog box, see Chapter 6, Precision Drawing.

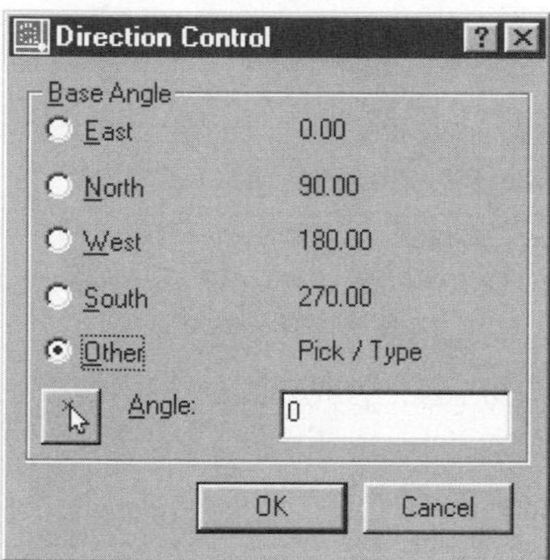

Figure 2–9 *The Angle button allows you to set the base angle by picking two points on the screen.*

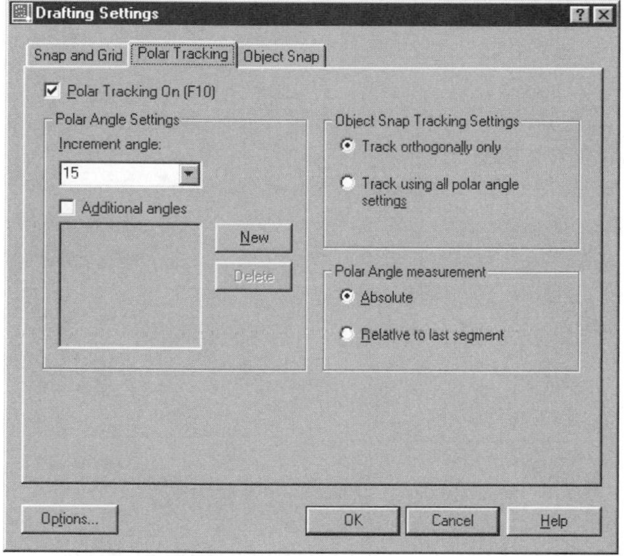

Figure 2–10 *The Drafting Settings dialog box controls settings for snap and grid, polar tracking, and object snap.*

SUMMARY

In this chapter you have learned how to control various AutoCAD settings through the Startup dialog box when you create a drawing, and through different menu commands and dialog boxes during the current drawing session. By controlling AutoCAD's drafting settings, you can help develop an environment that is appropriate for your needs.

The insight you gained in this chapter is expanded upon in the next chapter, where you will learn about controlling AutoCAD's behavior with the core system control—the Options dialog box.

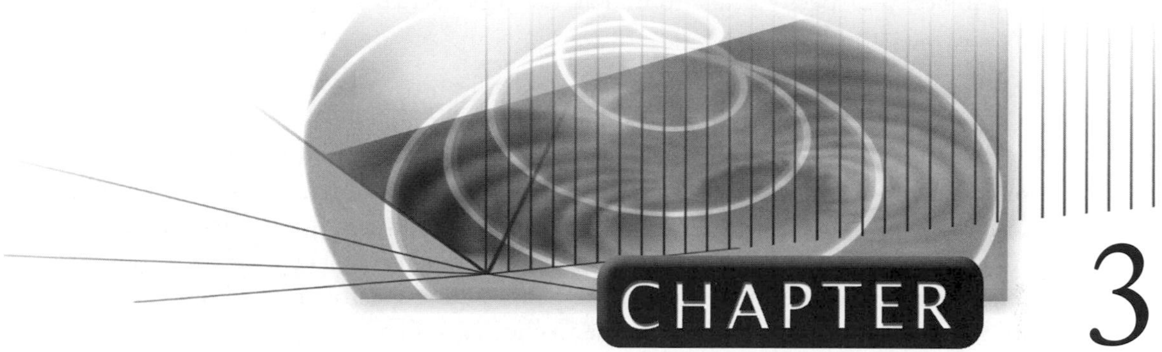

Fine-tuning the Drawing Environment

In Chapter 2 you learned how to control drawing settings using the Startup dialog box, the Drawing Units dialog box, and the Drafting Settings dialog box. AutoCAD also allows you to control many other features through its Options dialog box. Through the Options dialog box you can control where AutoCAD searches for and saves files, how its display appears, as well as drafting features such as AutoSnap, AutoTracking, and object selection methods.

This chapter covers the following topics:

- Default support paths
- Display options
- Open and Save folder
- Setting drafting options
- Controlling selection methods
- Creating profiles

DEFAULT SEARCH PATHS AND FILE NAMES

When you start AutoCAD, it determines where certain files that may be used during the drawing session are. These items include support files and device drivers, which are located in various folders on your computer system. AutoCAD also determines where to store certain file information, like temporary and backup files. In the Files folder, you specify where to find all the files needed and where to save temporary and backup files.

The Files folder is located in the Options dialog box, which is accessed from the Tools menu by choosing Options, and is shown in Figure 3–1. The Files folder identifies all necessary files and their locations in several folders. The folders are used to organize and display the information in a list of logical groups. As you select each folder in the list, a description is displayed in the description field under the list. The yellow folder icons in the list specify where AutoCAD searches for support, driver, menu, and other files. The white papers icons specify optional, user-defined settings such as which dictionary to use to check spelling.

By modifying the paths and files in the folders and adding new information, you can control the files AutoCAD uses during its drawing session.

To add a new search path, choose Options from the Tools menu and click on the Files tab if the Files folder is not already displayed. Double-click on the Support File Search Path folder to display its contents and then choose the Add button. You can type in a path, or select the Browse button to display the Browse for Folder dialog box and then choose OK when you have selected the correct folder. AutoCAD adds the

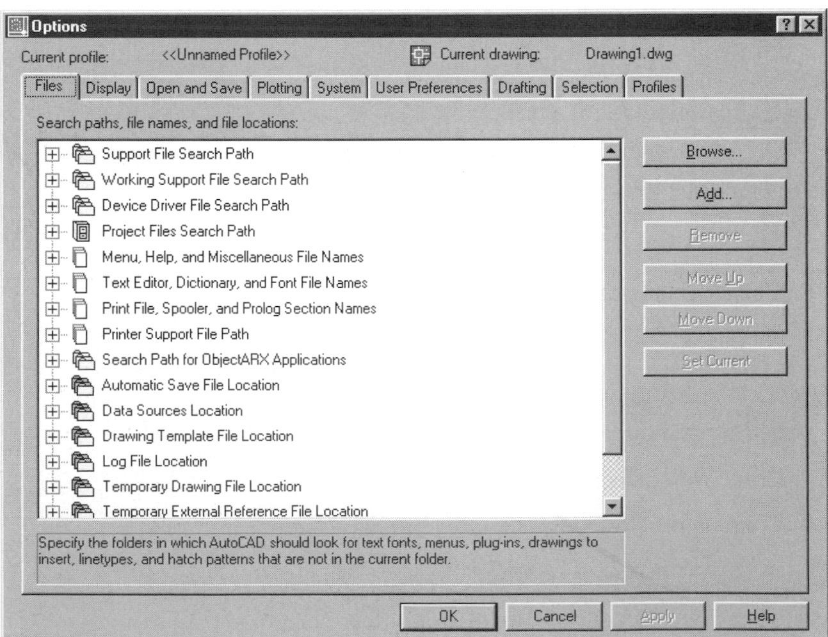

Figure 3–1 *The Options dialog box controls many of AutoCAD's features.*

Note: AutoCAD has default locations for all search paths. While you can change these locations, caution should be used. If you modify a location, AutoCAD may not be able to find files it needs in order to run properly.

Note: Third-party products will frequently place their own menus and LISP routines in separate directories created during installation. If you find that your third-party software is not performing properly, make sure the Support File Search Path and Device Driver File Search Path in the Files folder contain the proper path references.

new path to the Support File Search Path folder, as shown in Figure 3–2. You can remove a search path by highlighting it in the list and choosing the Remove button.

As you just learned, adding and removing paths is very simple. It is also very simple to redefine the files AutoCAD uses during the editing session. For instance, to replace the default alternate font file setting, Simplex, with Arial, choose Options from the Tools menu and then click on the Files tab. Next, choose the plus sign (+) in front of the Text Editor, Dictionary, and Font File Names item to expand its list, and then choose the plus sign (+) in front of the Alternate Font File item. This displays the current setting of Simplex.shx. After highlighting Simplex.shx, choose the Browse button to display the Alternate Font dialog box. From here, scroll to the top of the Font Name list, choose the Arial font, and then choose OK.

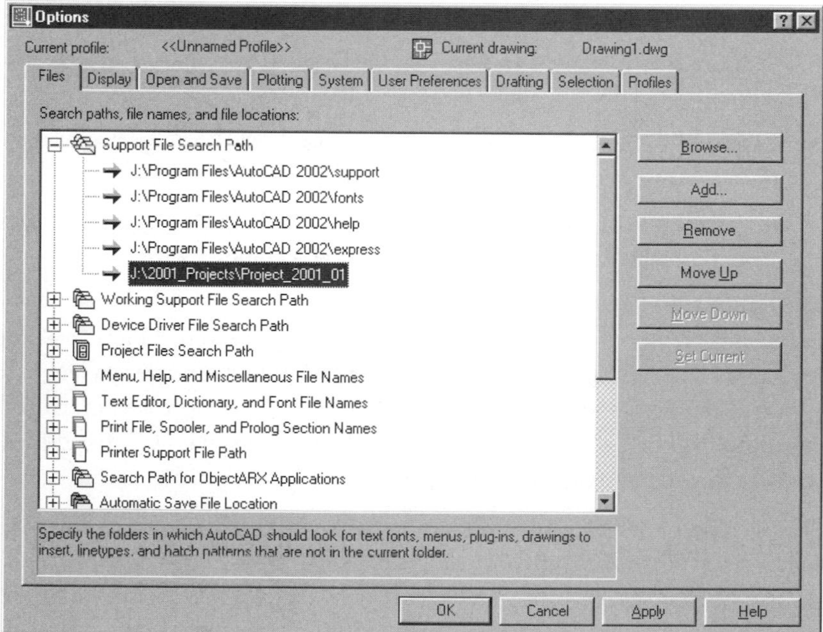

Figure 3–2 *A new path is added to the Support File Search Path folder.*

AutoCAD replaces the Simplex Alternate Font File with Arial, as shown in Figure 3–3. If you are satisfied with the changes to the paths and files, you can choose the Apply button. However, if you do not want to apply the modified settings, choose the Cancel button.

As you just learned, modifying the default paths and files that AutoCAD uses is easy. Next you will learn about customizing AutoCAD's display.

DISPLAY OPTIONS

AutoCAD's display represents the look of your drawing session. By editing the settings found in the Display folder, you can control how AutoCAD looks and how layouts appear, and even increase performance. By understanding the options in the Display folder, you can make your drawings easier to view, increase display performance, and increase your productivity.

The Display folder is organized into six areas, as shown in Figure 3–4. The six areas are as follows:

- **Window Elements**—This area controls different display settings in the AutoCAD window, including the number of lines in the command line window, and the screen's background color.
- **Layout Elements**—This area controls the appearance of paper space layouts.
- **Crosshair Size**—This area controls the size of the cursor's crosshairs.
- **Display Resolution**—This area controls the appearance of objects on the screen.
- **Display Performance**—This area controls display settings that affect AutoCAD's performance.
- **Reference Edit Fading Intensity**—This area specifies how much background objects fade during in-place reference editing.

WINDOW ELEMENTS

In the Window Elements area you can turn on or off the scroll bars that appear in each drawing window. The scroll bars allow you to pan the current view by sliding the scroll bar buttons along each scroll bar. You can also depress the arrow keys at either end of a scroll bar, or click in the bar itself to pan the view. A handy feature of scroll bars is that you can pan to a new view during a command.

The Window Elements area also allows you to control the display of the screen menu, which is a leftover relic from much earlier AutoCAD versions. The screen menu was used before pull-down menus were available and well before toolbars were developed. The screen menu appears on the right side of the screen, and it is still available for those who have been using AutoCAD for many years and prefer it as the main interface to AutoCAD commands. While it is perfectly acceptable to exe-

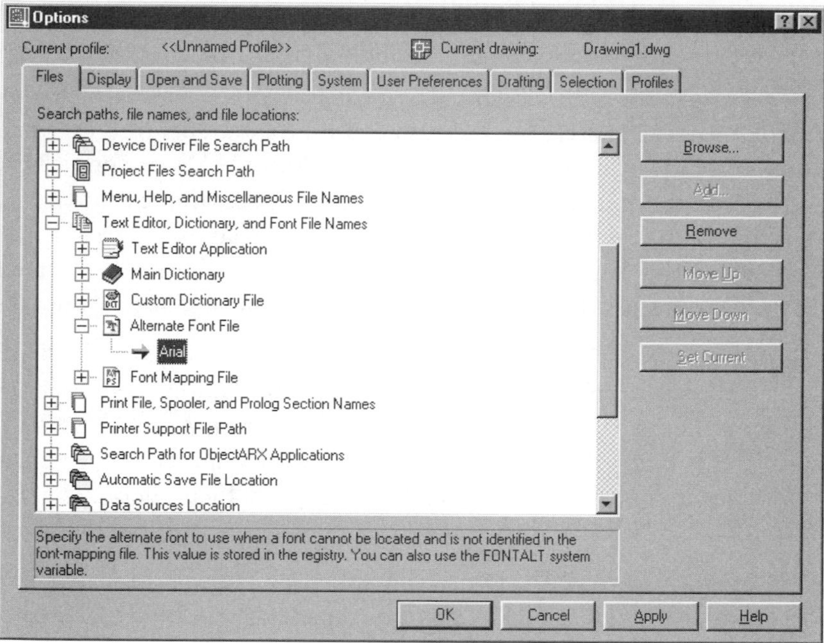

Figure 3–3 *The Alternate Font File is changed to Arial.*

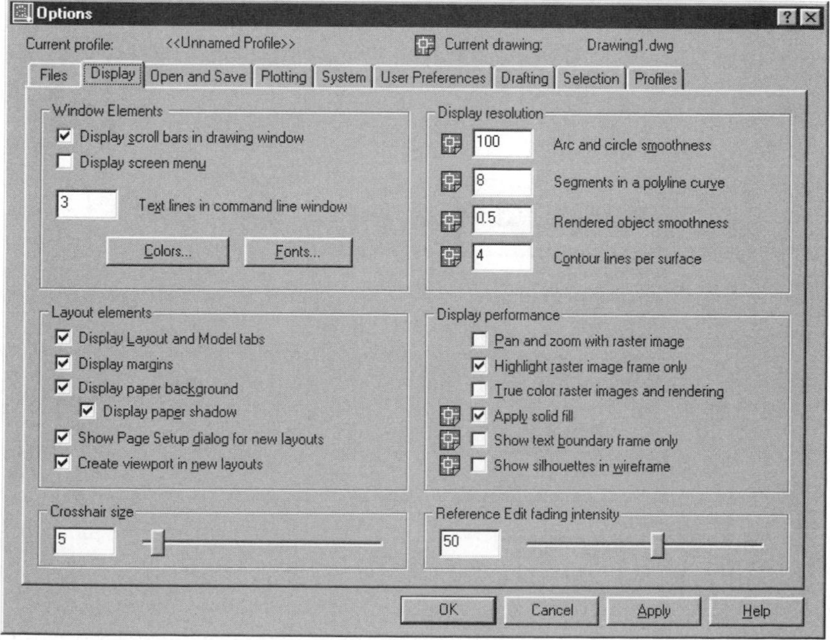

Figure 3–4 *The Display folder in the Options dialog box controls AutoCAD's screen appearance.*

cute commands in AutoCAD using the screen menu, if you are new to AutoCAD, you should avoid using it because it is not as intuitive as the pull-down menus nor as easy to use as the toolbars.

In the Window Elements area, you can also control the number of lines that appear in the command line window at the bottom of AutoCAD's screen. The default number of lines is three, and you can enter a new value in the text box. A value of three will show the previous two prompts plus provide another line for the active command line prompt. You can also modify the number of lines on the screen without using the Options dialog box by dragging up or down the top of the command line window.

The Window Elements area also allows you to control the font that appears in the command line window. When you choose the Font button, AutoCAD displays the Command Line Window Font dialog box, as shown in Figure 3–5. To change the font, choose the desired font, font style, and size, and then choose the Apply & Close button.

Finally, you can control AutoCAD's screen colors by choosing the Color button, which displays the Color Options dialog box, as shown in Figure 3–6. From this dialog box you independently control the background screen color of model space, paper space layouts, and the command line window. You can also change the color of the crosshairs and AutoCAD's AutoTracking vectors.

By modifying the colors of the various elements, you can adjust the display of your drawings to make viewing more comfortable and place less strain on your eyes, which is a real advantage if you spend eight hours or more a day working with AutoCAD.

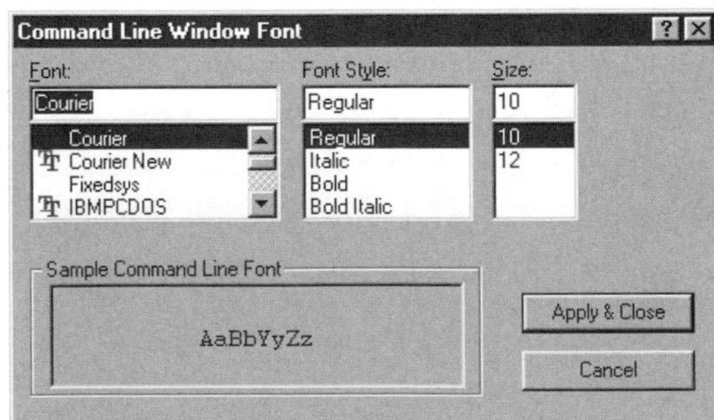

Figure 3–5 *The Command Line Window Font dialog box controls the font that appears in the command line window.*

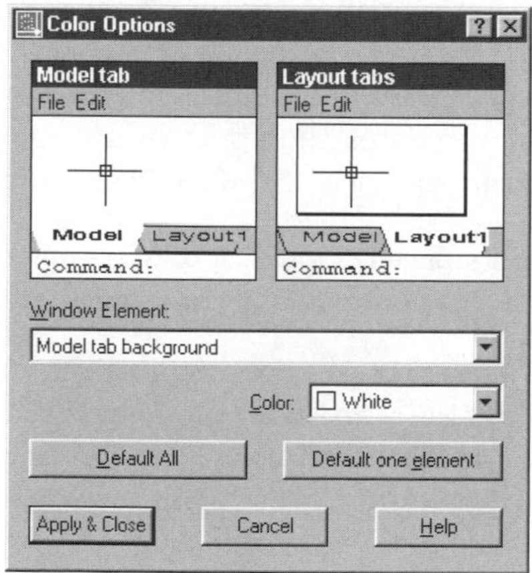

Figure 3–6 *The Color Options dialog box controls the color of AutoCAD's screen.*

After you have modified the screen's colors, you can easily set them back to their original color scheme. When you choose the Default All button, you change all color settings back to their original mode. When you choose the Default One Element button, you change the currently selected item back to its original color.

LAYOUT ELEMENTS

In the Layout Elements area you control options for existing and new layouts. A layout is an individual paper space environment in which you set up drawings for plotting. The various options toggle features off or on. Paper space layouts are discussed in detail in Chapter 19, Layouts and Paper Space.

CROSSHAIR SIZE

The Crosshair Size area controls the size of AutoCAD's crosshairs. The crosshairs appear on the cursor when you move the cursor into AutoCAD's drawing area. The valid range is from 1 to 100 percent of the total screen. At 100 percent the ends of the crosshairs extend to the edges of the drawing window. When the size is decreased to 99 percent or below, the crosshairs have a finite size, and the ends of the crosshairs are visible when the cursor is situated at the edge of the drawing area. The default size is five percent. Some find setting this to 100 percent helpful for visually aligning points using the cursor display.

DISPLAY RESOLUTION

The Display Resolution area controls the number of segments AutoCAD uses when displaying curved objects or curved areas. When AutoCAD draws curved objects, it actually simulates the shape of the curve by drawing short, straight line segments. By using a high number of line segments, AutoCAD enhances the appearance of the curved object. The smaller the number, the fewer line segments AutoCAD uses, and the poorer the simulation of the curve. The advantage of using smaller numbers for the values is that regeneration times are shorter and AutoCAD displays objects more quickly. For example, Figure 3–7 shows the effect of changing the Arc and Circle Smoothness value. The circle on the left simulates the effect of a value of 100, and the circle on the right simulates the effect of a value of 4. When you set the value to 100, the on-screen appearance of the circle is smooth.

Note: Due to the high speed of today's processors, you may not experience much degradation in AutoCAD's performance using high smoothness values.

Note: The number of line segments used by AutoCAD to display objects on the screen does not affect the accuracy AutoCAD uses when creating or plotting the objects. The values in the Display Resolution area only affect how objects appear on the screen.

DISPLAY PERFORMANCE

The Display Performance options control how AutoCAD deals with raster images (bitmaps) when panning or zooming, when editing them, and when displaying them in true color. By toggling these options off or on, you can dramatically affect you system's performance. For example, when you toggle the Pan and Zoom with Raster Image option off, AutoCAD displays only an outline of the raster image when you pan or zoom, thereby improving display performance. Otherwise, when this option is toggled on, as you pan or zoom, AutoCAD continuously redraws the image.

Similarly, when you toggle the Highlight Raster Image Frame Only option on, you increase your system's performance because it only highlights the image's frame when you select it for editing. Otherwise, when this option is off, AutoCAD highlights the image's frame and the entire area of the image.

You also affect display performance by toggling off the True Color Raster Images and Rendering option. When this option is off, you increase performance because AutoCAD does not use 16.7 million colors to display the image; it uses 256 colors instead. While the number of colors is significantly smaller with this option toggled off, it is usually adequate for display purposes.

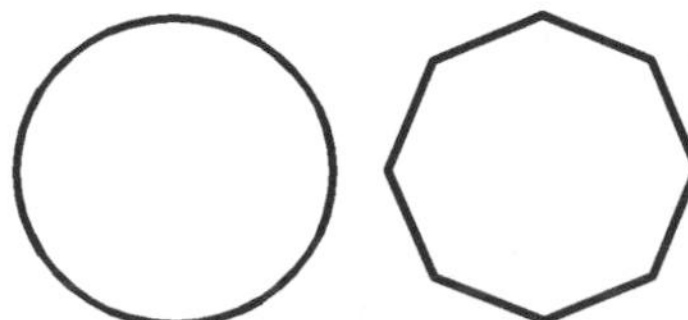

Figure 3–7 *The circle on the left uses a high smoothness value, while the circle on the right uses a low smoothness value.*

Other options include the Apply Solid Fill option and the Show Silhouettes in Wireframe option. The Apply Solid Fill option controls whether solid fills are displayed in objects like arrowheads on dimension lines, and in polylines, where a width greater than zero is assigned. When you toggle this feature off, only a wireframe representation of the object is displayed. The Show Silhouettes in Wireframe option controls whether silhouette curves of 3D solid objects are displayed as wireframes. It also controls whether the wire mesh is drawn or suppressed when a 3D solid object is hidden. To increase performance, toggle off these two options. Silhouettes are discussed in Chapter 24, Creating 3D Surfaces.

One final option is the Show Text Boundary Frame Only option, which displays a box representing the location of text objects instead of displaying the text. If you have a drawing with numerous text objects and you notice your system is performing slowly, toggle this option on. AutoCAD will then replace all text objects with rectangular outlines representing the limits of the text objects. When this feature is toggled off, the rectangular outlines are removed, and the original text reappears.

REFERENCE EDIT FADING INTENSITY

In-place reference editing allows you to edit blocks and xrefs from the drawing they are inserted in and save the changes back to their original location. The Reference Edit Fading Intensity value controls the visibility of objects that are not being edited and causes them to be displayed at a lesser display intensity than objects being edited. This makes focusing on the object(s) being edited much easier. The valid range is zero through 90 percent. The default setting is 50 percent. This feature is discussed in detail in Chapter 13, Working with Large Drawings and External References.

OPEN AND SAVE FOLDER

The Open and Save folder controls a variety of features associated with opening and saving files. By editing the features, you can control the format in which AutoCAD saves files, whether AutoCAD automatically saves files, and whether AutoCAD makes a backup copy when it saves a file. These and other related features are discussed in this section.

The Open and Save folder is organized into four areas, as shown in Figure 3–8. The four areas are described as follows:

- **File Save**—Controls various features associated with saving files
- **File Safety Precautions**—Controls elements dealing with automatically saving backup files
- **External References (Xrefs)**—Controls several features associated with external references
- **ObjectARX Applications**—Controls several features associated with AutoCAD's Runtime Extension (ARX) files

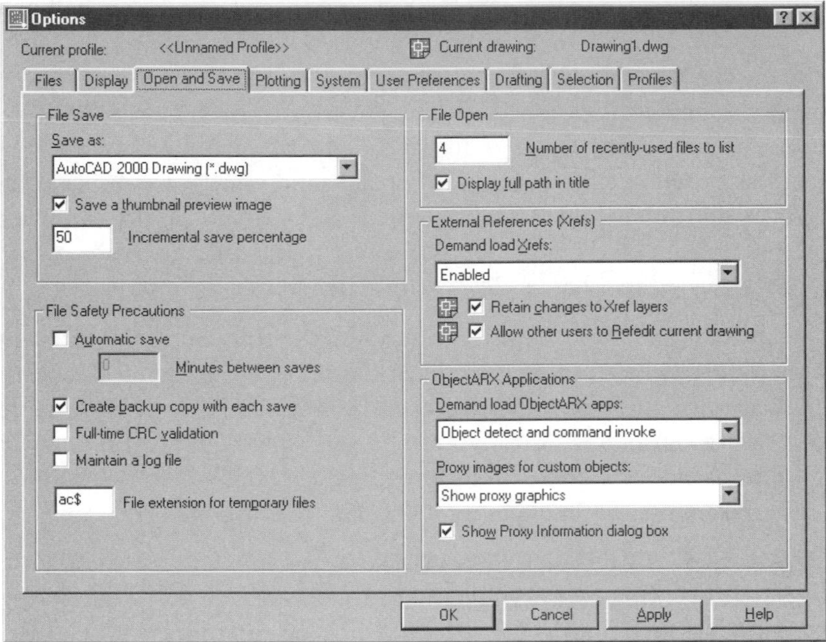

Figure 3–8 *The Open and Save folder in the Options dialog box controls how AutoCAD handles files.*

FILE SAVE

In the File Save area you control the type of file AutoCAD saves the current drawing as, whether a thumbnail image is saved with the drawing, and AutoCAD's incremental save percentage.

AutoCAD provides several different file types to save the current drawing as. These file types represent the default file types and can be overridden using the Save As command from the File menu. The file types include DWG formats for Releases 2000, 14, and 13; AutoCAD's Drawing Template Format (DWT); and DXF formats for Releases 2000, 14, 13, and 12.

You also control whether AutoCAD saves a thumbnail image with the current drawing. When you toggle this option on, AutoCAD snaps an image of the current drawing's display and saves it with the drawing. When you open an existing drawing, the image appears in the Select File dialog box when you select the drawing's file name. This feature is useful for visually identifying a drawing without opening it. This thumbnail image is not saved in DXF format files.

Finally, you can set the percent value for AutoCAD's Incremental Save Percent feature. This feature controls the frequency with which AutoCAD performs a full save when saving a file. Performing a full save removes wasted space from the drawing's database, which reduces a drawing's file size. The percent value refers to the amount of wasted space that is allowed in the drawing's database. Once the percentage of allowable wasted space is reached, AutoCAD performs a full save, removing the wasted space.

The advantage of this feature is that if you make small modifications to your drawing and frequently save the drawing, you do not spend a lot of time waiting for AutoCAD to finish a full save. The disadvantage is that because AutoCAD does not perform a full save each time it saves a drawing, the drawing's file size is larger than necessary because the wasted space is not removed. Generally speaking, the default value of 50 percent is adequate for most users' needs. However, if your drawing's file size is too large and is consuming too much disk space, reduce the number to 25 percent to make files sizes smaller. Reducing the number to below 20 percent optimizes drawing file sizes but degrades performance when saving drawings, because full saves are performed much more frequently.

Note: The Incremental Save Percent feature does not affect how much data is saved. AutoCAD always saves all data, including edits, during an incremental save. What AutoCAD does not do during an incremental save is remove wasted space, which makes the DWG file as small as possible. Therefore, if AutoCAD crashes, even though your last save was an incremental save, all data is retained. The only data lost is the new edits performed since your last save.

FILE SAFETY PRECAUTIONS

In the File Safety Precautions area, AutoCAD provides tools to help you avoid data loss and detect errors in drawings. The following list gives you an overview of these features:

- **Automatic Save**—Controls whether AutoCAD saves the current drawing automatically and how frequently it does so. Enter the save frequency value in minutes and AutoCAD will automatically save the current drawing when the time limit is reached. Once the time limit is reached, AutoCAD starts tracking the time from zero.
- **Create Backup Copy with Each Save**—When this feature is on, AutoCAD saves the current drawing and then makes a backup copy (BAK file). Generally it is a good idea to let AutoCAD save a backup copy of your drawing. However, bear in mind that the backup file is just as large as the drawing file, which can consume a lot of disk space.
- **Full-time CRC Validation**—This option specifies whether a cyclic redundancy check (CRC) should be performed each time an object is read into the drawing. CRC is an error-checking mechanism. If your drawings are being corrupted and you suspect a hardware problem or AutoCAD error, turn this option on.

- **Maintain a Log File**—Controls whether AutoCAD saves the contents of the text window to a log file. This feature is useful if you need to retrace the command performed during an editing session. This file can also get very large.
- **File Extension for Temporary Files**—This feature controls the file extension AutoCAD uses for the temporary files it creates during an editing session. The default extension is .ac$, which generally is adequate. When an AutoCAD editing session is ended improperly, AutoCAD does not have the opportunity to remove temporary files. Consequently, you must search your system for files ending with the designated extension and remove them.

Note: You define the location at which AutoCAD stores temporary files in the Files\Temporary Drawing File Location folder.

EXTERNAL REFERENCES (XREFS)

In the External Reference (Xrefs) area, AutoCAD allows you to control demand loading, retain modifications to xref-dependent layers, and allow the current drawing to be edited from another drawing.

The Demand Load Xrefs feature allows you to disable demand loading, enable demand loading, or enable demand loading with copy. Demand loading is discussed in detail in Chapter 13, Working with Large Drawings and External References.

The Retain Changes to Xref Layers option allows you to save the current state of xref-dependent layers in the current drawing. Layer states such as freeze/thaw and on/off can be changed in the current drawing for xref-dependent layers. When you toggle this option on, changes made to the layer states are saved and reinstated when the drawing is opened again.

The Allow Other Users to Refedit Current Drawing option controls whether the current drawing can be edited when it is attached as an xref. This toggles the in-line reference editing feature off or on, and is discussed in detail in Chapter 13, Working with Large Drawings and External References.

OBJECTARX APPLICATIONS

The ObjectARX Applications area allows you to control settings that relate to AutoCAD Runtime Extension (ARX) applications and proxy graphics.

The Demand Load ARX Apps feature specifies if and when AutoCAD demand-loads a third-party application if a drawing contains custom objects created in that application. Demand loading means AutoCAD loads the application in the current drawing session in order to display the custom object(s). This feature has four options:

- **Disable Load on Demand**—Turns off demand loading.
- **Custom Object Detect**—Demand-loads the source application when you open a drawing that contains custom objects. This setting does not demand-load the application when you invoke one of the application's commands.
- **Command Invoke**—Demand-loads the source application when you invoke one of the application's commands. This setting does not demand-load the application when you open a drawing that contains custom objects.
- **Object Detect and Command Invoke**—Demand-loads the source application when you open a drawing that contains custom objects or when you invoke one of the application's commands.

The Proxy Images for Custom Objects feature controls the display of custom objects in drawings. Proxy images are created when you open a drawing that contains custom objects but you do not have access to the application that created the objects. Because AutoCAD cannot properly display the objects with the application, it creates a place holder called a proxy object. This feature has three options:

- **Do Not Show Proxy Graphics**—Does not display proxy objects in drawings
- **Show Proxy Graphics**—Displays proxy objects in drawings
- **Show Proxy Bounding Box**—Shows a box in place of custom objects in drawings

The Show Proxy Information Dialog Box option specifies whether AutoCAD displays a warning when you open a drawing that contains custom objects. The dialog box tells you the total number of proxy objects in the drawing (both graphical and non-graphical), tells you the name of the missing application, and provides additional information about the proxy object type and display state.

SETTING PLOTTING OPTIONS

The Plotting folder controls a variety of features associated with plotting. By editing the features, you can control default plotting settings, the general plotting environment, plot style behavior in all drawings, and options related to script files.

The Plotting folder is organized into three areas, as shown in Figure 3–9. The three areas are described as follows:

- **Default Plot Settings for New Drawings**—Controls settings that relate to the default plotting settings
- **General Plot Options**—Controls options that relate to the general plotting environment
- **Default Plot Style Behavior for New Drawings**—Controls options related to plot style behavior in all drawings

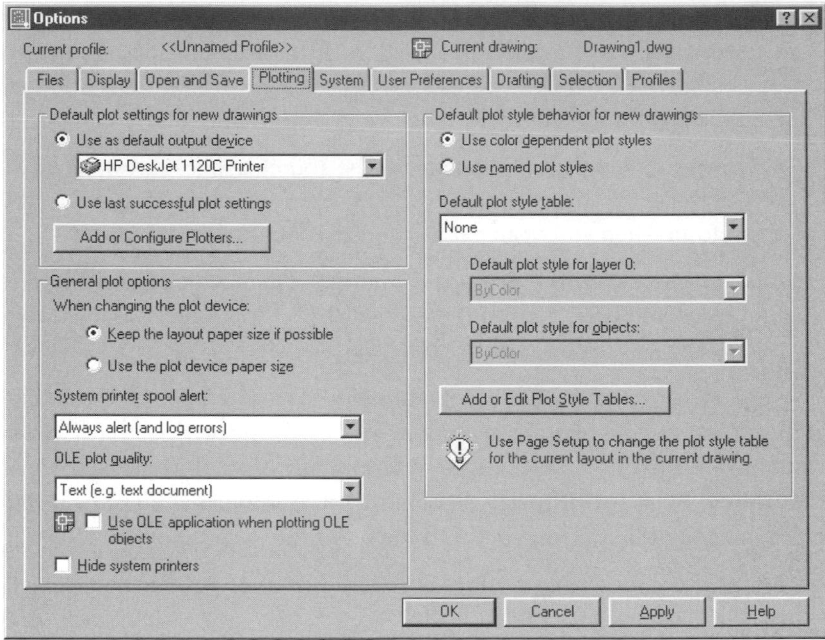

Figure 3–9 *The Plotting folder in the Options dialog box controls how AutoCAD handles plotting features.*

DEFAULT PLOT SETTINGS FOR NEW DRAWINGS

The Default Plot Settings for New Drawings area allows you to control various settings that relate to the default plotting settings. These include settings such as the default output device used for new layouts and model space, the plotting settings based on the settings of the last successful plot, and the Autodesk Plotter Manager (a Windows system window). These features are discussed in detail in Chapter 20, Basic Plotting.

GENERAL PLOT OPTIONS

The General Plot Options area allows you to control options that relate to the general plotting environment, including paper size settings, system printer alert behavior, and OLE objects in an AutoCAD drawing. These features are discussed in detail in Chapter 20, Basic Plotting.

DEFAULT PLOT STYLE BEHAVIOR FOR NEW DRAWINGS

The Default Plot Style Behavior for New Drawings area controls options related to plot style behavior in all drawings. A plot style is a collection of property settings defined in a plot style table and applied when the drawing is plotted. These features are discussed in detail in Chapter 20, Basic Plotting.

CONFIGURING THE SYSTEM

The System folder controls AutoCAD's system settings. These settings control the current 3D graphics display, options relating to the current pointing device, options relating to database connectivity, and other general options.

The System folder is organized into six areas, as shown in Figure 3–10. The six areas are described as follows:

- **Current 3D Graphics Display**—Controls settings that relate to system properties and configuration of the 3D graphics display system
- **Current Pointing Device**—Controls options that relate to the pointing device
- **Layout Regen Options**—Controls how the display list is updated in the model and layout folders
- **dbConnect Options**—Controls options that relate to database connectivity
- **General Options**—Controls general options that relate to system settings
- **Live Enabler Options**—Specifies how AutoCAD checks for object enablers

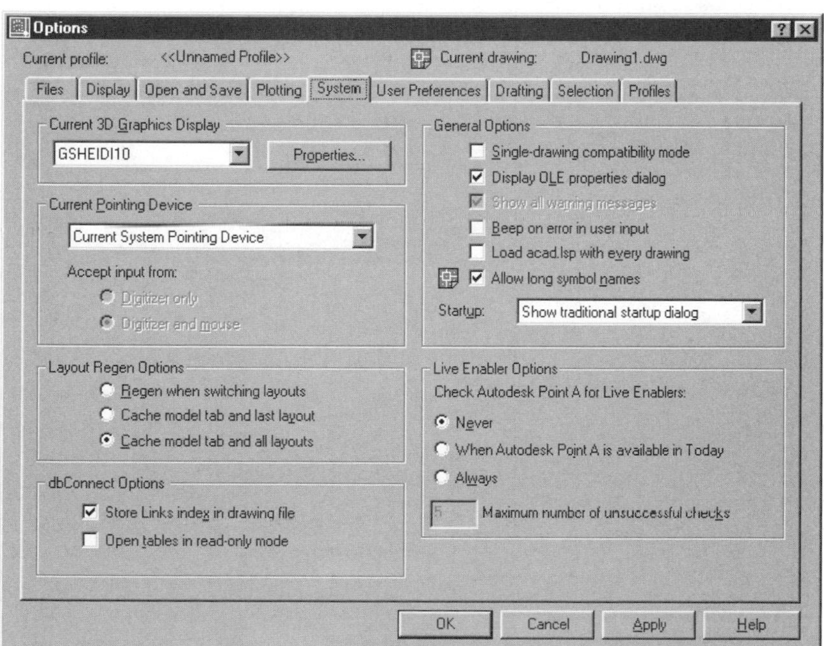

Figure 3–10 *The System folder in the Options dialog box controls AutoCAD's system settings.*

CURRENT 3D GRAPHICS DISPLAY

The Current 3D Graphics Display area has two features. The first allows you to select from the list of available 3D graphics display systems. The second displays the 3D Graphics System Configuration dialog box.

These two features work in conjunction with each other. For example, when you select a 3D graphics display system from the drop-down list, choosing the Properties button displays the 3D Graphics System Configuration dialog box set for the current 3D graphics display system.

The default 3D graphics display system is the Heidi 3D graphics display system. When this system is selected, the 3D Graphics System Configuration dialog box display appears as shown in Figure 3–11. If you are using a different graphics display system, the options on the 3D Graphics System Configuration dialog box will be different from the ones shown here. You will need to refer to AutoCAD's documentation for more information.

CURRENT POINTING DEVICE

The Current Pointing Device area controls options that relate to the pointing device. In this area you can select the desired pointing device from a drop-down list that displays the available pointing devices.

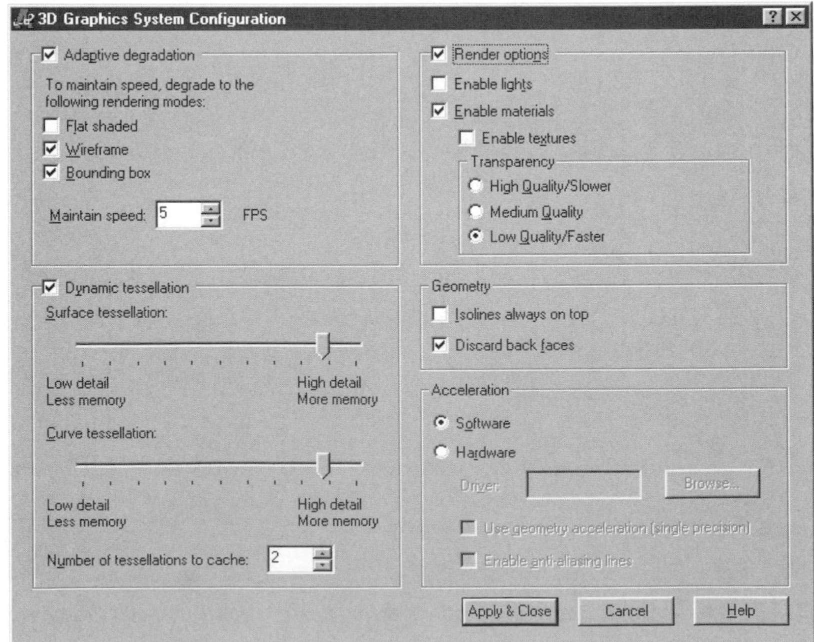

Figure 3–11 *3D Graphics System Configuration dialog box.*

AutoCAD comes installed with two pointing device options. The first is the Current System Pointing Device option, which sets the current Windows pointing device (typically your mouse) as the AutoCAD pointing device. The second is the Wintab Compatible Digitizer option, which sets the Wintab-compatible digitizer current. (This option is only available if you have a Wintab-compatible digitizer installed on your system.) If you choose the Digitizer option, you can specify whether AutoCAD accepts input from both a mouse and a digitizer or ignores mouse input and accepts only digitizer input.

LAYOUT REGEN OPTIONS

The Layout Regen Options area offers three options for controlling how the display list is updated when you switch between model and layout folders. The options are described as follows:

- **Regen When Switching Layouts**—The display list is always updated with a regen when you switch between folders.
- **Cache Model Tab and Last Layout**—The display list for the model folder and most recently viewed layout folder are cached. No regeneration occurs (except for modified objects) when you switch between these two folders. A regeneration does occur when you switch to any other folder.
- **Cache Model Tab and All Layouts**—The display lists for the model folder and all layout folders are cached. A regeneration occurs the first time you switch to a folder, but thereafter regenerations are suppressed (except for modified objects).

The default value is Cache Model Tab and All Layouts. Changing this setting can affect performance, depending on the size and type of drawing and available memory.

DBCONNECT OPTIONS

The dbConnect Options area allows you to control two features that deal with connecting to external databases. The features are described as follows:

- **Store Links Index in Drawing File**—This feature stores the database index within the AutoCAD drawing file. Select this option to enhance performance during SQL queries. Clear this option to decrease drawing file size and to enhance the opening process for drawings with database information.
- **Open Tables in Read-Only Mode**—Specifies whether to open database tables in read-only mode within the AutoCAD drawing file. Select this option to open tables in read-only mode, preventing unwanted edits. Clear this option to allow the table to be edited.

The dbConnect Options are discussed in detail in Chapter 28, Understanding External Databases.

GENERAL OPTIONS

The General Options area allows you to control such features as whether you can open multiple drawings in a single AutoCAD session, whether the Startup dialog box is displayed when you launch AutoCAD, or whether you can use long symbol names. There are seven features you can control in this area:

- **Single-Drawing Compatibility Mode**—This toggle specifies whether a Single-Drawing Interface (SDI) or a Multi-Drawing Interface (MDI) is enabled in AutoCAD. If you select this option, AutoCAD opens only one drawing at a time. If you clear this option, AutoCAD can open multiple drawings.
- **Display OLE Properties Dialog**—This toggle controls the display of the OLE Properties dialog box when you insert OLE objects into AutoCAD drawings. OLE objects are discussed in detail in Chapter 27, Taking Advantage of OLE Objects in AutoCAD.
- **Show All Warning Messages**—This toggle controls whether all dialog boxes that include a Don't Display This Warning Again option will be displayed. If you select this option, AutoCAD displays all dialog boxes with warning options, regardless of previous settings specific to each dialog box. If you clear this option, dialog boxes with the Don't Display This Warning Again option toggled on will not be displayed.
- **Beep on Error in User Input**—This toggle controls whether AutoCAD sounds an alarm beep when it detects an invalid entry.
- **Load acad.lsp with Every Drawing**—This toggle specifies whether AutoCAD loads your acad.lsp file into every drawing. If this option is cleared, only your acaddoc.lsp file is loaded into all drawing files. Clear this option if you do not want to run certain LISP routines in specific drawing files.
- **Allow Long Symbol Names**—Named objects can include up to 255 characters. Names can include letters, numbers, blank spaces, and any special character not used by Windows and AutoCAD for other purposes. When this option is enabled, long names can be used for layers, dimension styles, blocks, linetypes, text styles, layouts, UCS names, views, and viewport configurations. This option is saved in the drawing.
- **Startup**—This toggle controls whether the Startup dialog box is displayed when you start AutoCAD or when you start a new drawing in the current AutoCAD session.

LIVE ENABLER OPTIONS

The Live Enabler Options area specifies when AutoCAD checks online for object enablers, which let you display and use custom objects in drawings without the custom ObjectARX application. The options are described as follows:

- **Never**—AutoCAD never checks for object enablers.
- **When Autodesk Point A is Available in Today**—AutoCAD checks for object enablers only if Autodesk Point A (Autodesk's portal for designers) is enabled in the Today window and if a live Internet connection exists.
- **Always**—AutoCAD always checks for object enablers, regardless of how the Today window is configured.
- **Maximum Number of Unsuccessful Checks**—Determines the maximum number of times AutoCAD will attempt to find object enablers.

SETTING USER PREFERENCES

The User Preferences folder controls various features within AutoCAD. These features include the behavior of your pointing device when you right-click, how AutoCAD responds to input of coordinate data, how AutoCAD sorts objects during certain functions, as well as other features.

The User Preferences folder is organized into six areas, as shown in Figure 3–12. The six areas are:

- **Windows Standard Behavior**—This area controls whether Windows behavior is applied when you work in AutoCAD.
- **AutoCAD DesignCenter**—This area allows you to control settings that relate to the new AutoCAD DesignCenter (ADC).
- **Hyperlink**—This area controls settings that relate to the display properties of hyperlinks.
- **Priority for Coordinate Data Entry**—This area controls how AutoCAD responds to coordinate data input.
- **Object Sorting Methods**—This area provides options that determine the sort order of objects during specified AutoCAD functions.
- **Associative Dimensioning**—This toggle turns on and off the associative dimensioning for new dimensions.

WINDOWS STANDARD BEHAVIOR

The Windows Standard Behavior area allows you to control such features as whether AutoCAD accelerator keys adhere to Windows standards, and how your pointing device functions when you right-click. There are three features you can control in this area:

- **Windows Standard Accelerator Keys**—This option controls whether AutoCAD follows Windows standards in interpreting keyboard accelerators (for example, CTRL+C equals COPYCLIP). If this option is cleared, AutoCAD interprets keyboard accelerators by using AutoCAD standards rather than Windows standards (for example, CTRL+C equals Cancel, and CTRL+V toggles among the viewports).

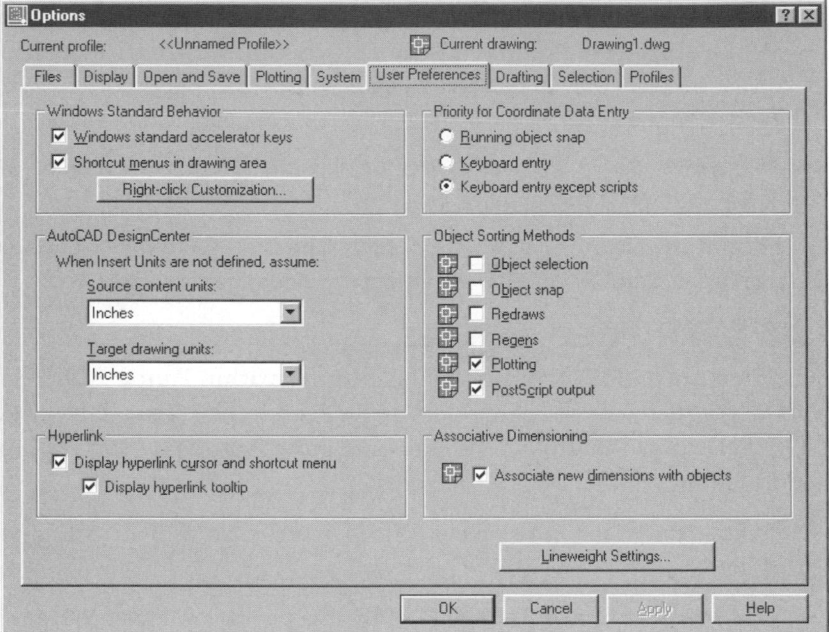

Figure 3–12 *The User Preferences folder in the Options dialog box controls certain aspects of AutoCAD's behavior.*

- **Shortcut Menus in Drawing Area**—This option controls various things such as whether your pointing device displays a shortcut menu or issues Enter when you right-click in the drawing area.
- **Right-Click Customization**—This button displays the Right-Click Customization dialog box shown in Figure 3–13. This dialog box allows you to control how the right-click feature functions under certain conditions. You can determine whether the right-click repeats the last command, is the same as pressing Enter, or displays the shortcut menu. You can also specify the behavior when you right-click under different circumstances, such as when no objects are selected, when one or more objects are selected, or when a command is in progress.

AUTOCAD DESIGNCENTER

The AutoCAD DesignCenter area allows you to control how blocks are scaled when they are inserted in the current drawing. To scale blocks properly when they are inserted from another drawing, you must consider two units: the units used in the source drawing, and the units used in the target (current) drawing. There are two features that allow you to control how AutoCAD deals with the units of the source drawing and the target drawing when no units are specified:

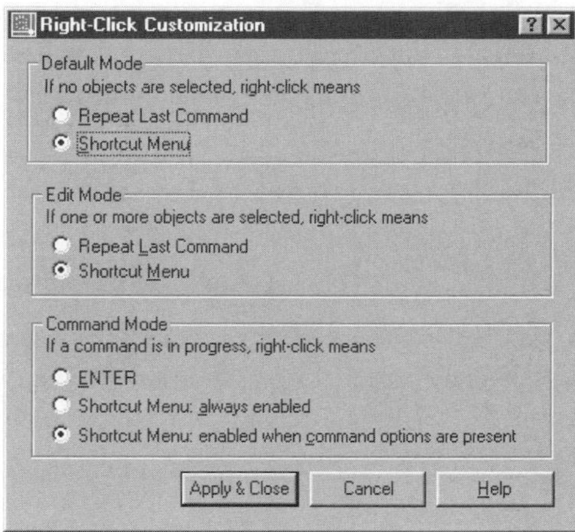

Figure 3–13 *The Right-Click Customization dialog box controls how your pointing device behaves when you right-click.*

- **Source Content Units**—Sets which units should be automatically used for an object being inserted from another drawing into the current drawing when no insert units are specified with the INSUNITS system variable
- **Target Drawing Units**—Sets which units should be automatically used in the current drawing when no insert units are specified with the INSUNITS system variable

The available units settings for both options include Unspecified-Unitless, Inches, Feet, Miles, Millimeters, Centimeters, Meters, Kilometers, Microinches, Mils, Yards, Angstroms, Nanometers, Microns, Decimeters, Decameters, Hectometers, Gigameters, Astronomical Units, Light Years, and Parsecs. If Unspecified-Unitless is selected, the object is not scaled when it is inserted. AutoCAD DesignCenter (ADC) is discussed in detail in Chapter 11, Applying AutoCAD DesignCenter.

HYPERLINK

The Hyperlink area allows you to control two options that deal with how your cursor reacts when it moves over a hyperlink in a drawing:

- **Display Hyperlink Cursor and Shortcut Menu**—This option controls the display of the hyperlink cursor and shortcut menu. Select this option to have the hyperlink cursor appear whenever the pointing device moves over a hyperlink, and to make the shortcut menu available. The hyperlink shortcut menu provides additional options when you right-click over a hyperlink in a drawing. If this option is cleared, the hyperlink cursor is never displayed, and the shortcut menu is not available.

- **Display Hyperlink Tooltip**—This option controls the display of the Hyperlink ToolTip. Select this option to display a Hyperlink ToolTip when the pointing device moves over an object that contains a hyperlink.

PRIORITY FOR COORDINATE DATA ENTRY

The Priority for Coordinate Data Entry area allows you to set one of three options that control whether running object snaps take precedence over coordinates entered from the keyboard:

- **Running Object Snap**—Select this option to have running object snaps override keyboard-entered coordinates at all times. Running object snaps are discussed in Chapter 6, Precision Drawing.
- **Keyboard Entry**—Select this option to have keyboard-entered coordinates override running object snaps at all times.
- **Keyboard Entry Except Scripts**—Select this option to have keyboard-entered coordinates override running object snaps at all times except when you run scripts.

Running object snaps are discussed in Chapter 6, Precision Drawing.

OBJECT SORTING METHODS

When you create an object, AutoCAD adds it to the current drawing's database. As subsequent objects are created, they are added to the end of the database, in the order in which they are created. Therefore, objects are stored in the drawing's database in the order in which they are created.

When AutoCAD sorts through a drawing's database, it can do so by sorting in the order in which the objects are created, or by randomly selecting objects. The advantage of sorting in the order in which the objects are created is you have predictability on how an object will be drawn or selected. For example, when you plot objects in the order in which they are created, you know that objects created first will lie under objects created last. The disadvantage is that it usually takes AutoCAD longer to sort through objects in the order in which they are created than by sorting randomly.

The Object Sorting Methods area allows you to control how AutoCAD sorts through objects in certain situations. By toggling options off or on, you can control the order in which AutoCAD deals with an object. If an option is selected (toggled on), AutoCAD sorts objects in the order of those created first to those created last. If an option is cleared (toggled off), AutoCAD sorts objects randomly. There are six options you can set to control object sort methods:

- **Object Selection**—If this option is selected, the order in which AutoCAD selects objects is affected. For example, if two overlapping objects are chosen during object selection, AutoCAD recognizes the newest object as the selected object.

- **Object Snap**—If this option is selected, the order in which AutoCAD selects an object to snap to is affected. For example, if two overlapping objects are chosen when using object snap, AutoCAD recognizes the newest object as the object to to which to snap.
- **Redraws**—If this option is selected, when you use either the REDRAW or REDRAWALL command, AutoCAD redraws objects on the screen in the order in which they were created.
- **Regens**—If this option is selected, when you use either the REGEN or REGENALL command, AutoCAD regenerates objects on screen in the order in which they were created.
- **Plotting**—If this option is selected, when you plot a drawing, AutoCAD plots objects in the order in which they were created.
- **PostScript Output**—If this option is selected, AutoCAD exports objects in the order in which they were created.

The two rectangles shown in Figure 3–14 demonstrate the effect of the options in the Object Sorting Methods area. The dashed rectangle on the right was drawn first, and the solid rectangle on the left was drawn second. When a regeneration is executed, with the Regens option in the Object Sorting Methods area toggled off, AutoCAD randomly selects objects. In this particular case, AutoCAD draws the solid rectangle on the left first and then draws the dashed rectangle on top of the solid rectangle. However, with the Regens option in the Object Sorting Methods area toggled on, AutoCAD is forced to sort objects in the order in which they were created. Consequently, AutoCAD draws the dashed rectangle first, and then draws the solid rectangle on top of the dashed rectangle, as shown in Figure 3–15.

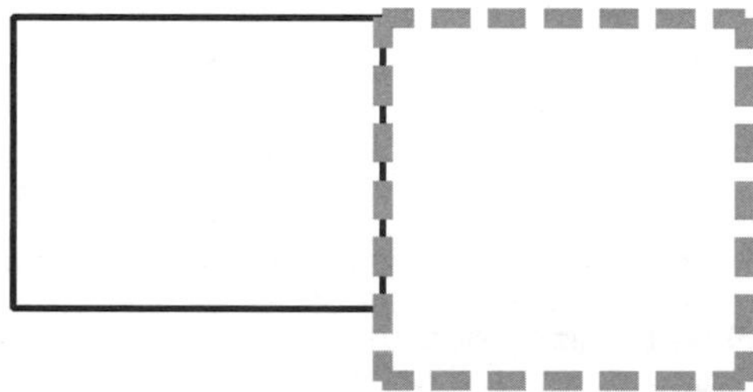

Figure 3–14 *Even though the dashed rectangle was created first, with the object sorting option Regens turned off, AutoCAD randomly selects objects and regenerates the dashed rectangle second, on top of the solid rectangle.*

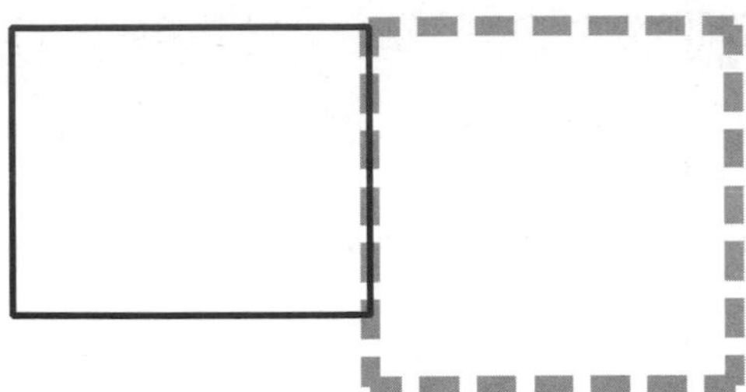

Figure 3–15 *With the object sorting option Regens turned on, AutoCAD selects objects in the order in which they were created and regenerates the solid rectangle second, on top of the dashed rectangle.*

ASSOCIATIVE DIMENSIONING

The Associate New Dimensions with Objects toggle in this area lets you switch between true associative dimensions (toggled on) and legacy dimensions (toggled off) for all new dimensions in the drawing. It does not affect existing dimensions. AutoCAD's new associative dimensions are discussed in detail in Chapter 17, Dimensioning Basics.

LINEWEIGHT SETTINGS

The User Preferences folder also includes a button labeled Lineweight Settings. This button displays the Lineweight Settings dialog box, which sets lineweight options, such as display properties and defaults, and the current lineweight. You will learn how to use lineweights in Chapter 5, Applying Linetypes and Lineweights.

SETTING DRAFTING OPTIONS

The Drafting folder controls settings that relate to object snaps, AutoTracking, and the AutoSnap marker. You can also control how AutoCAD displays alignment vectors, and the display size for the aperture (the square in the center of the crosshairs).

The Drafting folder is organized into five areas, as shown in Figure 3–16. The five areas are described as follows:

- **AutoSnap Settings**—Controls settings that relate to object snaps.
- **AutoSnap Marker Size**—Sets the display size for the AutoSnap marker. The marker is a geometric symbol that displays the object snap location when the crosshairs move over a snap point on an object. When you drag the button left or right, you increase or decrease the size of the AutoSnap marker.
- **AutoTrack Settings**—Controls the settings that relate to AutoTracking behavior.
- **Alignment Point Acquisition**—Controls the method of displaying alignment vectors in a drawing.

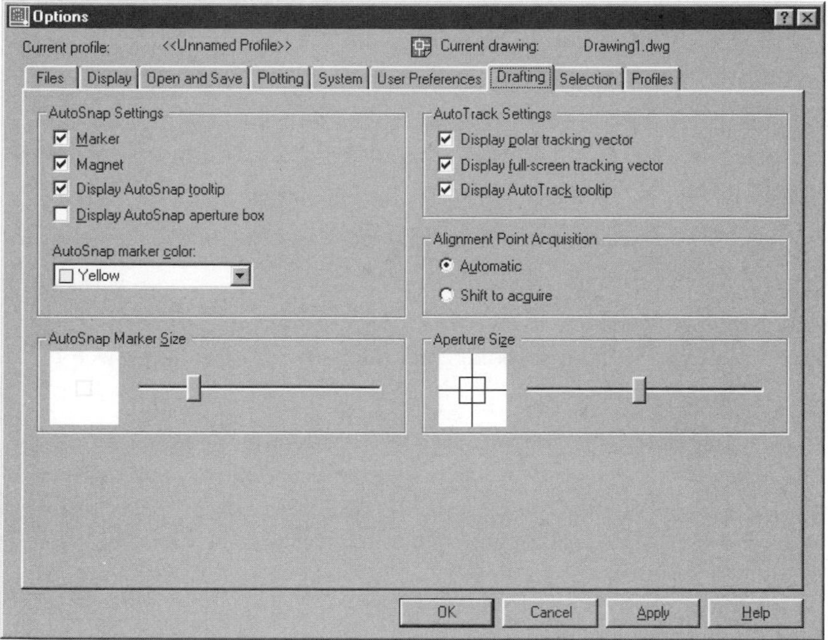

Figure 3–16 *The Drafting folder in the Options dialog box controls various snapping features.*

- **Aperture Size**—Sets the display size for the aperture. The aperture box is the box that appears inside the crosshairs when you select an object snap. When you drag the button left or right, you increase or decrease the size of the aperture.

AUTOSNAP SETTINGS

The AutoSnap Settings area allows you to control settings that affect the display and behavior of the crosshairs when using Object Snaps. There are five features you can control, described as follows:

- **Marker**—Controls the display of the AutoSnap Marker. When you move over an object, a geometric symbol displays, indicating the type of snap.
- **Magnet**—This option turns the AutoSnap magnet on or off. The magnet is an automatic movement of the crosshairs that locks the crosshairs onto the nearest snap point. The magnet is affected by the size of the AutoSnap Marker. With this option selected, when AutoCAD displays a marker, and the center of the crosshairs enter the marker symbol, AutoCAD snaps the crosshairs to the center of the Marker.
- **Display AutoSnap Tooltip**—This option controls the display of the AutoSnap tooltip. The tooltip is a text flag that describes which object snap is active.

- **Display AutoSnap Aperture Box**—This option controls the display of the AutoSnap aperture box. The aperture box is a box that appears inside the crosshairs when you select an object snap. When AutoSnap is activated, the AutoSnap markers only appear on objects that cross or lie within the aperture box.
- **AutoSnap Marker Color**—This option allows you to specify the color of the AutoSnap marker to make viewing the AutoSnap marker easier.

AutoSnaps and object snaps are discussed in detail in Chapter 6, Precision Drawing.

AUTOTRACK SETTINGS

The AutoTrack Settings area allows you to control settings that affect the display and behavior of AutoCAD's new polar tracking feature. There are three features you can control from this area:

- **Display Polar Tracking Vector**—This option turns polar tracking on or off. When selected, this feature allows you to draw lines based on predefined angles. These predefined angles are known as tracking vectors. Enabling polar tracking causes AutoCAD to snap to the tracking vector as you move the cursor during a drawing command.
- **Display Full-Screen Tracking Vector**—This option controls the display of tracking vectors. When you select this option, AutoCAD displays tracking vectors as infinite construction lines, extending through the snap point and the cursor, crossing the width of the screen. When you clear this option, the tracking vectors extend as rays from the snap point through the cursor to the edge of the screen.
- **Display AutoTracking Tooltip**—This option controls the display of the AutoTracking ToolTip. The ToolTip is a text flag that displays the tracking coordinates. When you select this option, the AutoTracking ToolTip is displayed.

The polar tracking feature is discussed in detail in Chapter 7, Basic Drawing Commands.

ALIGNMENT POINT ACQUISITION

The Alignment Point Acquisition area allows you to set the method used to display polar tracking alignment vectors in a drawing, either automatically or manually, as follows:

- **Automatic**—If this option is selected, AutoCAD displays tracking vectors automatically when the Note: aperture pauses over an object snap.
- **Shift to Acquire**—If this option is selected, AutoCAD displays tracking vectors only when you press the Shift key and move the crosshairs over an object snap.

When a tracking vector is set, a small "X" appears in the center of the object snap. The polar tracking feature is discussed in detail in Chapter 7, Basic Drawing Commands.

CONTROLLING SELECTION METHODS

The Selection folder controls settings that relate to object selection methods, display size for the pickbox, and AutoCAD's grips. The folder is organized into four areas, as shown in Figure 3–17. The four areas are described as follows:

- **Selection Modes**—Controls settings that relate to object selection methods.
- **Pickbox Size**—Controls the display size of the AutoCAD pickbox. The pickbox is the box that appears inside the crosshairs and is used to select objects. When you drag the control left or right, you increase or decrease the size of the pickbox.
- **Grips**—Controls the settings that relate to grips.
- **Grip Size**—Controls the display size of AutoCAD grips. Grips are small squares displayed on an object after it has been selected. When you drag the control left or right, you increase or decrease the size of the grips.

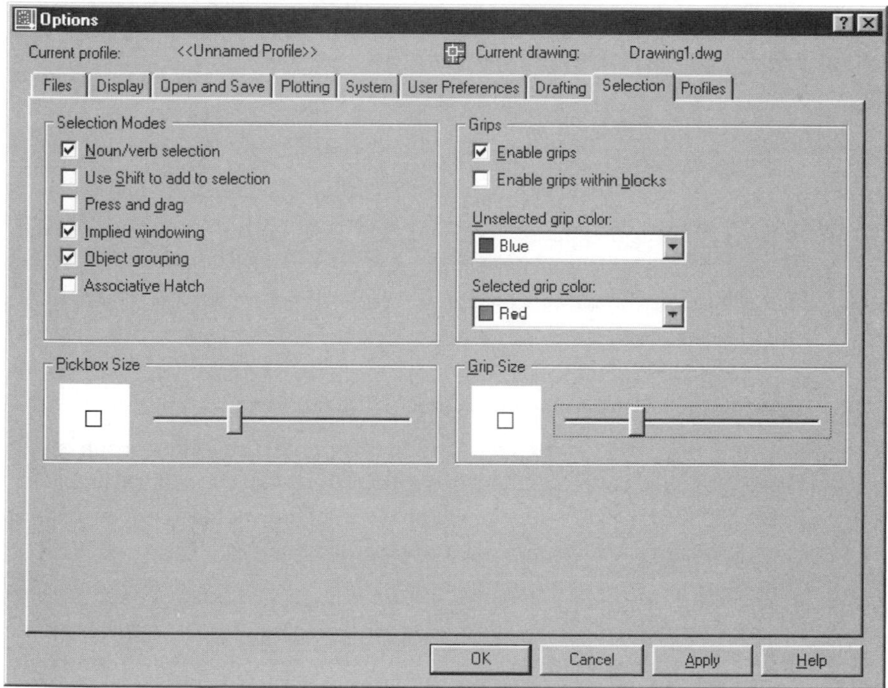

Figure 3–17 *The Selection folder in the Options dialog box controls object selection methods and various grip features.*

SELECTION MODES

The Selection Modes area allows you to control settings that affect AutoCAD's behavior when selecting objects. There are five features you can control:

- **Noun/Verb Selection**—When selected, this option allows you to select an object to edit before invoking a command. You can also select multiple objects to edit before invoking a command. When this option is cleared, you can still select multiple objects, but invoking a command clears the current selection set and prompts you to create a new selection set.
- **Use Shift to Add to Selection**—This option controls how multiple objects are added to a selection set during the selection process. When it is selected, you must press the Shift key to add selected objects to the selection set. When the option is cleared, AutoCAD automatically adds multiple objects to the selection set as you select them.
- **Press and Drag**—When this feature is on, this option lets you draw a selection window by clicking a start point and dragging the pointing device to the end point. If this option is cleared, the dragging feature is disabled, and you must draw a selection window by clicking a start point and then clicking the end point.
- **Implied Windowing**—When this feature is on, you can automatically initiate a selection window by avoiding objects when picking on screen and picking a blank area instead. Once a blank area is picked, the selection window is invoked, and dragging the selection window from left to right initiates a window selection, which only selects objects within the window's boundaries. Dragging from right to left initiates a crossing window selection, which selects objects within and crossing the window's boundaries. When this option is cleared, you must invoke a selection window during a command by typing **W** to initiate a window selection or **C** to initiate a crossing window selection.
- **Object Grouping**—When this feature is on, an entire object group is selected when you select one object in that group. When it is cleared, selecting an object that is in a group selects only that object. For detailed information about object grouping, see Chapter 10, Advanced Editing.
- **Associative Hatch**—When this feature is on, an associative hatch's boundary objects are selected along with the associative hatch. When it is cleared, selecting an associative hatch selects only the associative hatch, not its boundary elements. For detailed information about associative hatches, see Chapter 16, Using Hatch Patterns.

GRIPS

The Grips area allows you to control settings that affect AutoCAD's grips. Grips are small squares displayed on an object after it has been selected. There are four features you can control:

- **Enable Grips**—When selected, this option controls whether grips are displayed on an object after you select it. You can edit an object with grips by selecting a grip to make it "hot." When the grip is hot, you can move the grip, or you can invoke the shortcut menu to select a command. When you clear this option, you disable grips.
- **Enable Grips within Blocks**—This option controls how grips are displayed on a block after you select it. If this option is selected, AutoCAD displays all grips for each object in the block. You can edit an object with grips by selecting a grip to make it hot. When the grip is hot, you can move the object by the grip, or you can invoke the shortcut menu to select a command. If this option is cleared, AutoCAD displays one grip located at the insertion point of the block.
- **Unselected Grip Color**—You determine the color of an unselected grip from this drop-down list. AutoCAD displays an unselected grip as the outline of a small square.
- **Selected Grip Color**—You determine the color of a selected grip from this drop-down list. AutoCAD displays a selected grip as a small, filled square.

For detailed information about using grips, see Chapter 9, Basic Editing, and Chapter 10, Advanced Editing.

CREATING PROFILES

So far you have dedicated quite a bit of time to reviewing the numerous features in the Options dialog box. As you may have noticed, there are well over 100 different settings you can control. While having over 100 different settings provides you with the ability to set up your drawing environment exactly the way you want, it does not necessarily help if you must work on a different computer temporarily, one that does not have your custom setup. A worse situation occurs when someone else works on your computer while you are gone and wipes out your custom settings with their own custom settings. Not only is it frustrating to be forced to redefine your custom settings, it may be impossible to completely restore them if you cannot remember what your settings were.

Fortunately, the Options dialog box provides a simple method for saving and restoring your custom settings. By saving your custom settings in a profile, you not only can restore settings if they are accidentally lost, you can even copy your custom settings to another computer.

In the following exercise you will create two profiles and use them to restore default and custom settings in the Options dialog box.

EXERCISE: CREATING PROFILES

1. Launch AutoCAD and start a new drawing from scratch.
2. From the Tools menu, choose Options, and then choose the Profiles tab.

   The Profiles folder is displayed in the Options dialog box, as shown in Figure 3–18. If no one has added any profiles, the only profile that is displayed in the Available Profiles list is <<Unnamed Profile>>, which is AutoCAD's default profile.

   Notice that the current profile listed at the top of the Options dialog box in Figure 3–18 is <<Unnamed Profile>>. When you make changes in the Options dialog box, it is very important to note which profile is current because those changes are immediately saved to the current profile.

   Next you will create two new profiles that contain the current default settings.
3. Choose the Add to List button to display the Add Profile dialog box.
4. In the Profile Name text box, enter **Default Profile**.
5. In the Description text box, enter **This is AutoCAD's Default Profile** as shown in Figure 3–19.

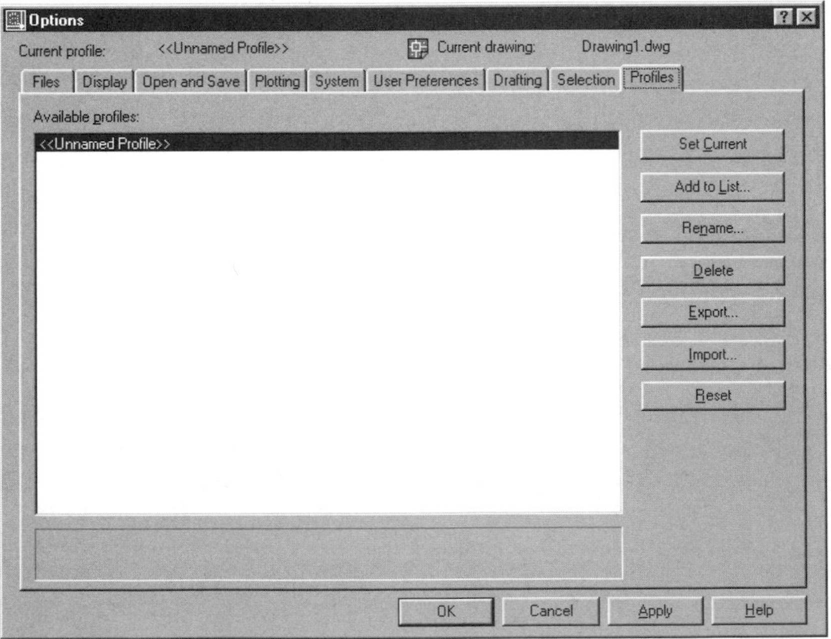

Figure 3–18 *AutoCAD's default profile, <<Unnamed Profile>>, is the current profile.*

6. Choose the Apply & Close button.
7. Choose the Add to List button again to create another profile.
8. In the Profile Name text box, enter **My Profile**.
9. In the Description text box, enter **This is my custom Profile** as shown in Figure 3–20.
10. Choose the Apply & Close button.

    Two new profiles appear in the Available Profiles list, as shown in Figure 3–21. The two new profiles you created currently have the same settings as the current profile. For this exercise you will only modify settings in the newly created My Profile.

    Modifying the settings in a profile is simple. To do so, make the profile you wish to modify the current profile. Then make the desired changes to the settings.
11. From the Available Profiles list, choose My Profile, and then choose the Set Current button. From this point on, any changes that you make to settings in the Options dialog box are saved to My Profile.
12. Choose the Display tab.

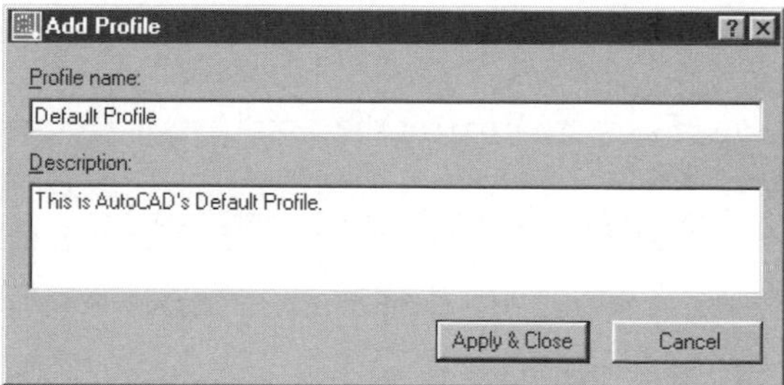

Figure 3–19 *The Add Profile dialog box with the proper settings for Default Profile.*

Figure 3–20 *The Add Profile dialog box with the proper settings for My Profile.*

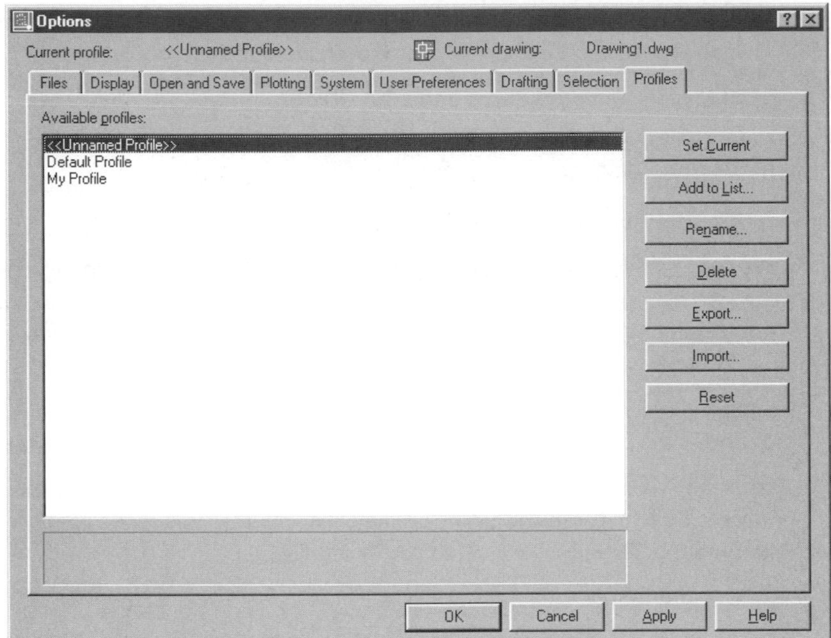

Figure 3–21 *Two new profiles appear in the Available Profiles list.*

13. In the Window Elements area, set the Text Lines in Command Line Window value to 6.
14. In the Crosshair Size area, set the value to 35, as shown in Figure 3–22.
15. Choose the Apply button and then choose OK.

The new values are saved to the current profile, and AutoCAD resets the drawing environment based on the new values, as shown in Figure 3–23. Notice the crosshairs are much larger than they were originally, and the command line window can now display six lines of text.

In addition to storing settings of the Options dialog box, profiles can also save the display and position of toolbars. In the next exercise you will add a toolbar to My Profile.

EXERCISE: ADDING TOOLBARS TO THE CURRENT PROFILE

1. Continuing from the previous exercise, move your pointer over any existing toolbar and then right-click.
2. From the shortcut menu, choose Customize to display the Toolbars dialog box.
3. When you select an empty box next to a toolbar name, AutoCAD adds the toolbar to the screen. For this exercise, be sure to select the ACAD menu from the Menu Group drop-down list.

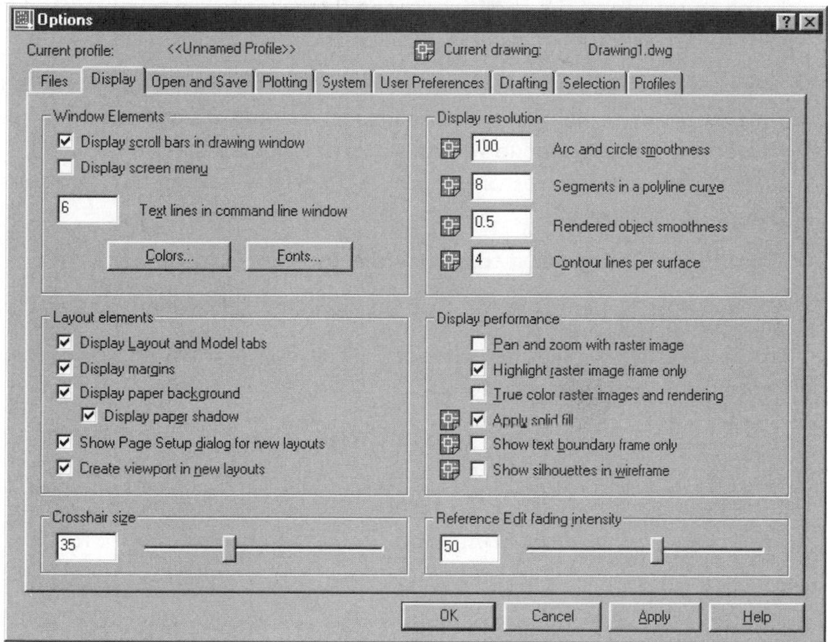

Figure 3–22 *The new values are entered in the Display folder.*

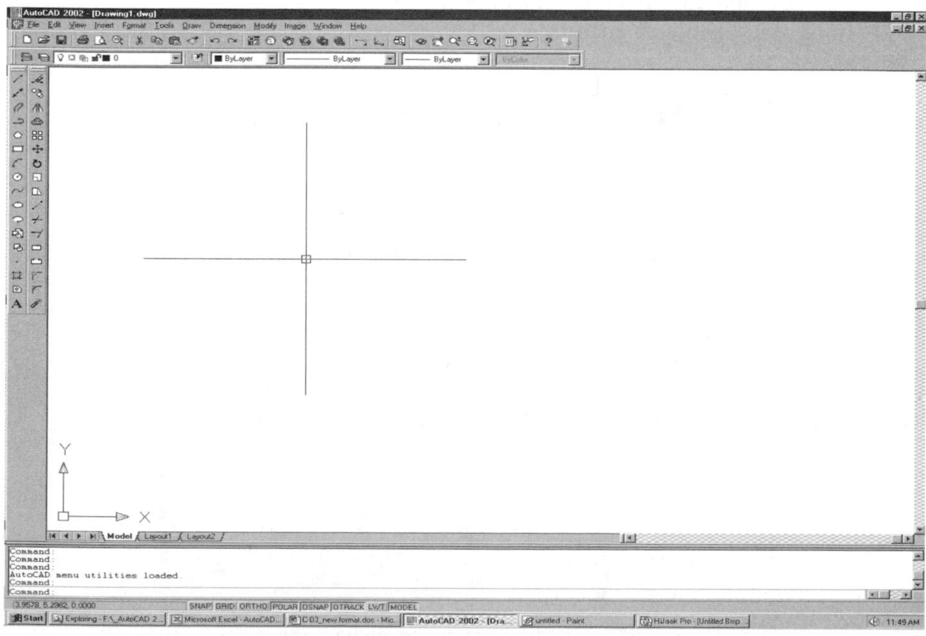

Figure 3–23 *The new values are saved to the current profile, and AutoCAD implements the changes.*

Scroll to the top of the toolbars list and then choose the box next to Dimension to display the Dimension toolbar as shown in Figure 3–24.

The profile named My Profile now contains the information necessary to display the Dimension toolbar and restore it to its current position on the screen. To demonstrate this, continue with this exercise and change the current profile to Default Profile, which restores AutoCAD's original settings and removes the Dimension toolbar.

4. Choose the Close button to dismiss the Toolbars dialog box.
5. From the Tools menu, choose Options, and then choose the Profiles tab.
6. In the Available Profiles list, choose Default Profile, then choose the Set Current button, and then choose OK.

When you set Default Profile as the current profile, AutoCAD restores its settings, removing the Dimension toolbar and setting the crosshairs and the command line window back to their original sizes. To restore the Dimension toolbar and increase the sizes of the crosshairs and the command line window, make My Profile the current profile.

The Profile folder has several commands that allow you to manage profiles. You can rename a profile by highlighting it in the Available Profiles list and choosing the Rename button. Similarly, you can delete unwanted profiles by highlighting them in the Available Profiles list and choosing the Delete button. (AutoCAD allows you to highlight only one profile at a time.)

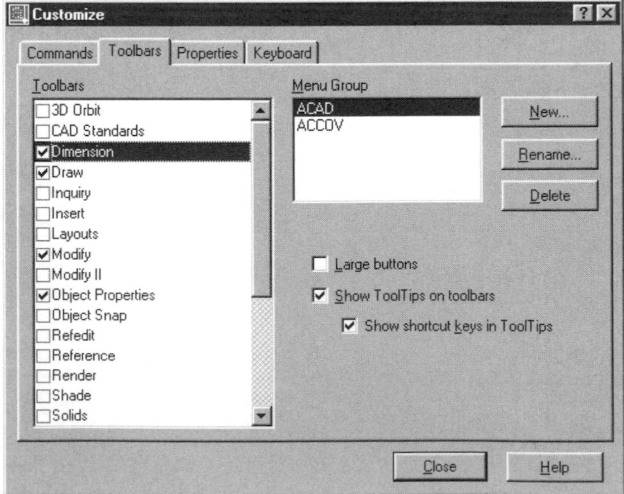

Figure 3–24 *Toolbar settings can be saved with the current profile.*

You can also import and export profiles. To export a profile, highlight it in the Available Profiles list and choose the Export button. When you export a profile, AutoCAD saves it as an ARG file to the folder you select. Once a profile is saved as an ARG file, you may import it by choosing the Import button.

Finally, if you want to set a profile back to AutoCAD's original default settings, highlight the profile in the Available Profiles list and choose the Reset button.

SUMMARY

This chapter covered a great deal of information regarding AutoCAD's drawing environment. You learned about the Options dialog box and its more than 100 settings. Among the many features you learned about covered were how to define support paths and how to control AutoCAD's display. You saw how to configure plotters and how to set your own user preferences.

By mastering the topics discussed in this chapter, not only will you increase your knowledge on how to work with AutoCAD efficiently, you will gain a fundamental understanding of AutoCAD at a level few users possess.

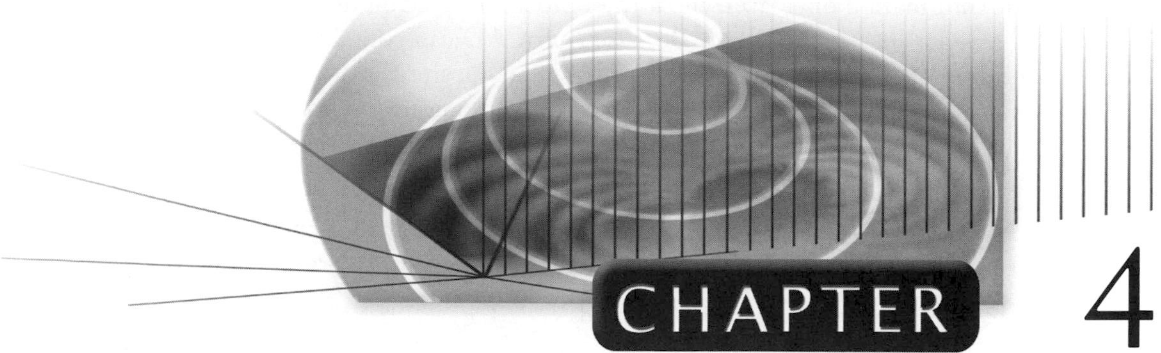

Managing Layers

Prior to the use of Computer-Aided Design (CAD), projects consisted of dozens of mylar and vellum drawings that were ultimately printed as bluelines and then taken into the field or the shop, where they were used in the construction of buildings or the fabrication of parts. While drawings are still necessary in the construction and fabrication process, today the person that initials the "Drawn by" box can more efficiently organize the information formerly drawn on those dozens of sheets by using AutoCAD's layers.

For those of us who have been in the business for a while, the pin-registered drafting days—when layers of mylar sheets were stacked one on top of the other to create a composite drawing—are just a faint memory. In those days a single mylar sheet was dedicated to a particular item, such as the centerline and right-of-way lines for a street. Other mylars were dedicated to other specific items, such as a mylar sheet for the curbs and gutters in the street, another for the street's underground gas lines, and another for its sewer lines. Each mylar sheet contained a series of small holes along the top, and the holes on each sheet lined up, or registered, perfectly. When someone needed a composite drawing of, say, the street's centerline, curb and gutter, and sewer line, those particular mylar sheets were stacked on top of each other on a special table with "register" pins along its top. Then a blueline print was made of the stacked mylar sheets. By layering the desired mylars on top of each other, a composite sheet was created that contained the information needed for a specific purpose.

In AutoCAD, layers mimic individual pin-registered mylar sheets. By placing specific information on a layer, the former process of placing item-specific information on a mylar sheet is emulated. Because AutoCAD can contain an unlimited number of layers, you can expand upon the idea of composite sheets and include layers for, for example, the object geometry, dimensions, and notes. By using layers to organize your drawings, you can create a single model that serves many purposes and satisfies many needs.

This chapter discusses using layers to organize your drawings and shows you how to use AutoCAD's Layer Properties Manager. This chapter explores the following topics:

- Implementing layering standards
- Controlling the drawing's layer features
- Creating and assigning a color to new layers
- Locking layers
- Setting a layer filter
- Saving and restoring layer settings
- Translating layers between drawings
- Layer Previous command

LAYER PROPERTIES MANAGER

As its name implies, the Layer Properties Manager dialog box manages layers. It also controls several other features, including lineweights, plot styles and plot/no plot status, as shown in Figure 4–1.

The Layer Properties Manager dialog box's features are explored in this chapter. Before you learn how to use the Layer Properties Manager to its fullest potential, in the next section you will first learn about the importance of standardizing layer names.

LAYER STANDARDS

Whether you are the only person working with AutoCAD in your company or one of several dozen, establishing standard layer names can increase your efficiency in layer management. By using layers to organize various elements in your drawing, you make the process of displaying the desired elements very easy, which makes editing and plotting easier. For example, if your drawing has gas lines and sewer lines, but you only want to edit the gas lines, you can make your drawing visually easier to work with by turning off the layer on which the sewer lines were drawn.

Not only does organizing your drawing's elements by layers make editing and plotting easier; it also reflects your level of skill as an AutoCAD technician. By following a preset layering standard and using it to organize objects in your drawing, you

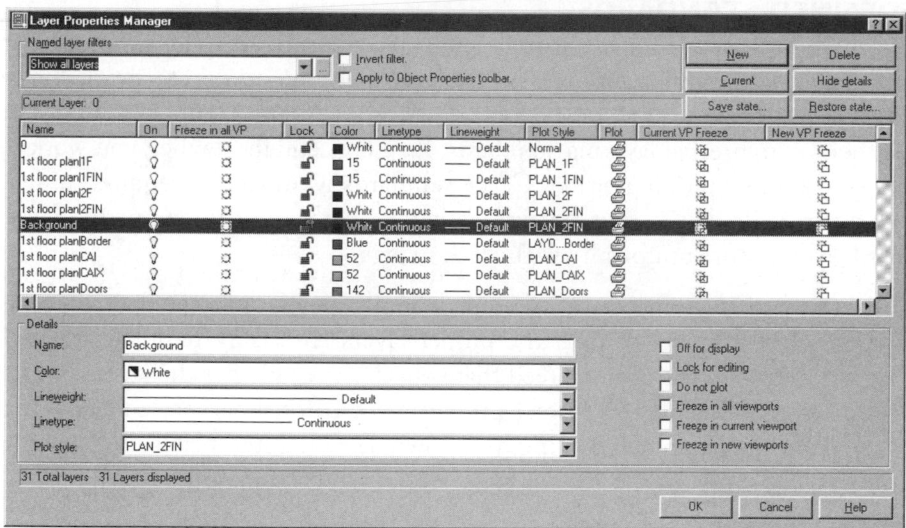

Figure 4–1 *The Layer Properties Manager controls several features.*

demonstrate your understanding of the importance of proper layer and object management. For example, the ability to control quickly which groups of objects are displayed by turning certain layers off or on can dramatically increase productivity, which shows you know what you are doing with AutoCAD.

LAYER NAME CONSIDERATIONS

The process of defining standard layer names can be daunting. Because the drawing you are creating represents only one portion of the project's life cycle, you must consider not only the number of layers, and their names, necessary to fulfill your needs, but also how many ways your drawing will be used by others. For example, will half-scale versions of your drawing be inserted into reports? If so, it may be necessary to place duplicate text on two different layers: one layer that holds text that is easy to read at full scale, and another layer that holds duplicate text that is easily read at half scale. Will other departments or companies incorporate your drawing into theirs, creating a composite drawing? If so, you must establish layering standards that you both find acceptable. What different objects must you display on the set of hardcopy plots provided to the contractor for construction? While a single model space drawing may contain both gas lines and sewer lines, your hardcopy plots must separate these two components, creating one set of plots showing only gas lines, and another showing only sewer lines. When you understand the many ways your drawing will be used, your proper development and use of layering standards can make fulfilling all needs during a project's life cycle productively easy or disastrously difficult.

PREDEFINED STANDARDS

While there are many things you must consider when developing layer standards, the good news is that in reality you probably do not have to do too much. Because AutoCAD has been around for nearly two decades, many companies have established their own internal layering standards. As a consequence, when you work for a company, you are typically issued a CAD Standards Manual and required to read it and use it in your day-to-day drafting activities. By using the company's layering standards, you ensure uniformity with all drawings.

In addition to the layering standards, many companies have created drawing template files, which already contain the proper layers. By using template files when starting a drawing, you are guaranteed that your drawing file has the layers you need, with each layer preset to the correct properties of color, linetype, and lineweight. You can learn more about creating your own drawing templates in Chapter 2, Beginning a Drawing, in the section titled Creating a Template File.

EXTENDED LAYER NAMES

In prior releases of AutoCAD, when considering layering standards, the limitations imposed by AutoCAD were factored into the process. Previously, AutoCAD imposed a limit on layer name lengths of thirty-one characters. Additionally, uppercase and lowercase characters were ignored, and spaces were not allowed. This limitation not only imposed limitations on creating descriptive layer names, it often created problems when attaching xrefs, because AutoCAD added the xref's drawing name to the beginning of xref-dependent layers, which made it easy to exceed the thirty-one-character limit. This in turn caused AutoCAD to abort opening the drawing.

AutoCAD 2000 added support for layer names with as many as 255 characters. The names can include spaces, and uppercase and lowercase characters are preserved. With these enhancements you can develop meaningful, descriptive layer names.

Note: Although xref names are still appended to the beginning of xref-dependent layers, the xref's drawing name does not count toward the 255-character limit.

While the extended features of layer names are good, you should use discretion when creating layering standards. For example, Figure 4–2 shows two different methods of using the new extended layer name features. The layer name on the bottom of the list reads, "This is the layer on which I placed all the existing sewer line text." and actually includes the period at the end. While this layer name may be very descriptive, it is not very practical in terms of layer name standards.

For example, if your drawing contains dozens, or even hundreds of layers, it is not easy to scroll down the list of layer names and locate the one for existing sewer line

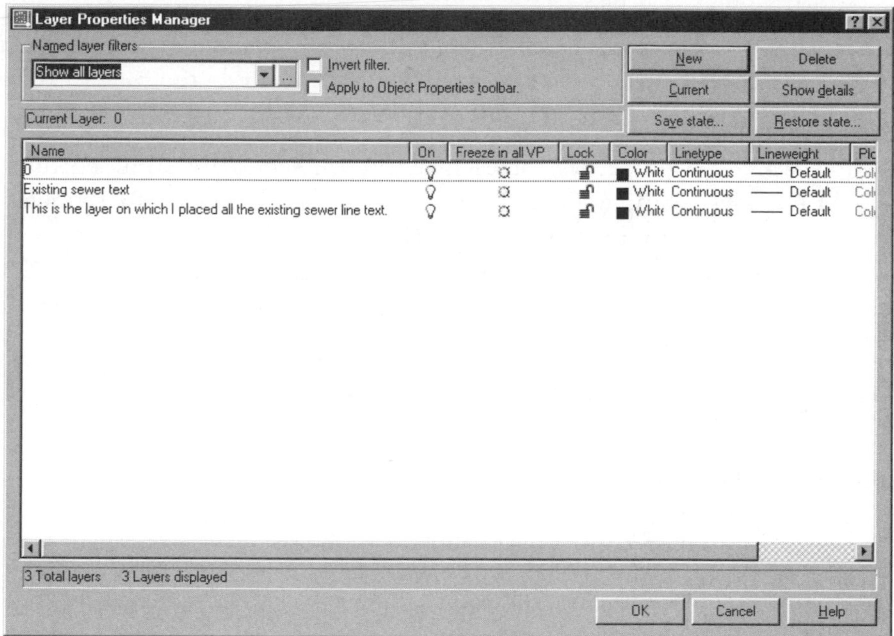

Figure 4–2 *The Layer Properties Manager supports long layer names.*

text because the layer name is so long. In contrast, the layer name in the middle of the list reads, "Existing Sewer Text." This much shorter name relates the same information as its much longer counterpart and allows users to easily peruse the layer list and find the correct layer. Because AutoCAD allows you to display names alphabetically, you can find the layer that holds existing sewer text by looking for it alphabetically. Additionally, this shorter version makes it easier to isolate the display of layer names using the Layer Properties Manager's layer filters, a feature described later in this chapter.

CONTROLLING OBJECT PROPERTIES

AutoCAD provides four object properties you can control through the Layer Properties Manager: color, linetype, lineweight, and plot style. The color, linetype, and lineweight properties affect object appearance when on-screen, and all four properties control the appearance of objects when they are plotted on paper.

Through the Layer Properties Manager you can control the values of object properties. More importantly, the Layer Properties Manager provides a method of globally controlling properties. For example, by using the Layer Properties Manager to change a layer's color to red, you automatically change the color of all objects on that layer to red. This global method of changing the color of all objects by only changing one value in the Layer Properties Manager is a tremendous timesaver, especially if there are dozens or even hundreds of objects on a layer. By using the Layer Properties Manager, you can simultaneously edit the color, linetype, lineweight, and plot style properties of numerous objects very quickly.

While this global ability to edit objects is very powerful, it does have one catch—it only works when the properties for each object are set to ByLayer. Fortunately, AutoCAD provides a simple method for setting object properties to ByLayer. By understanding and controlling how AutoCAD assigns properties to objects, you can create objects whose property values are globally controlled through AutoCAD's Layer Properties Manager.

This section discusses how AutoCAD assigns properties to an object when it is created and how to use the Layer Properties Manager to control its property values when they are set to ByLayer.

ASSIGNING PROPERTIES TO NEW OBJECTS: THE OBJECT PROPERTIES TOOLBAR

When you create an object, AutoCAD automatically assigns the object to the current layer. Additionally, AutoCAD also assigns values of color, linetype, lineweight, and plot style. These four properties that AutoCAD assigns to an object are determined by the values displayed in the Object Properties toolbar. For example, to assign new objects a layer value of zero and the color red, you select the desired values from the appropriate drop-down list in the Object Properties toolbar, as shown in Figure 4–3. Once these values are set, when a new object is created, it is assigned to layer 0 and assigned the color red. To assign new objects to a different layer, simply choose the desired layer from the layer pull-down list.

Note: AutoCAD's default layer is zero. Geometry created on layer 0 has unique properties with respect to blocks. Consequently, when creating new geometry for your project, you should typically create a new layer for the geometry, reserving layer 0 for creating blocks with special properties. For more information on the relationship of layers and blocks, refer to Chapter 12, Making the Most of Blocks.

The ByLayer Property Value

Notice in Figure 4–3 that the last three drop-down lists in the Object Properties toolbar all display the current property value as ByLayer. These last three lists are the object property values for linetype, lineweight, and plot style, respectively.

Figure 4–3 *The Object Properties toolbar sets the layer, color, linetype, lineweight, and plot style properties for new objects.*

When an object's property is set to ByLayer, it means those particular property values are controlled by the settings in the Layer Properties Manager. Therefore, with the property assignments shown in Figure 4–3, if you created a new object, its linetype, lineweight, and plot style values would be controlled by the linetype, lineweight, and plot style values in the Layer Properties Manager. Consequently, to view the values for linetype, lineweight, and plot Style, you must view the values for layer 0 in the Layer Properties Manager, as shown in Figure 4–4.

Remember, because the Object Properties toolbar is set to layer 0, all new objects are created on layer 0. It also means that any properties set to ByLayer when an object is created are controlled by the property values for layer 0. As shown in Figure 4–4, the property values for layer 0 are as follows:

- Color: White
- Linetype: Continuous
- Lineweight: 0.016 inches
- Plot Style: Normal

Note: If the lineweights in your drawing are displayed in millimeters (mm), you can switch to inches format by choosing Lineweight from the Format menu. Then, in the Lineweight Settings dialog box, select the Inches option in the Units for Listing area. For more information, refer to Chapter 5, Applying Linetypes and Lineweights.

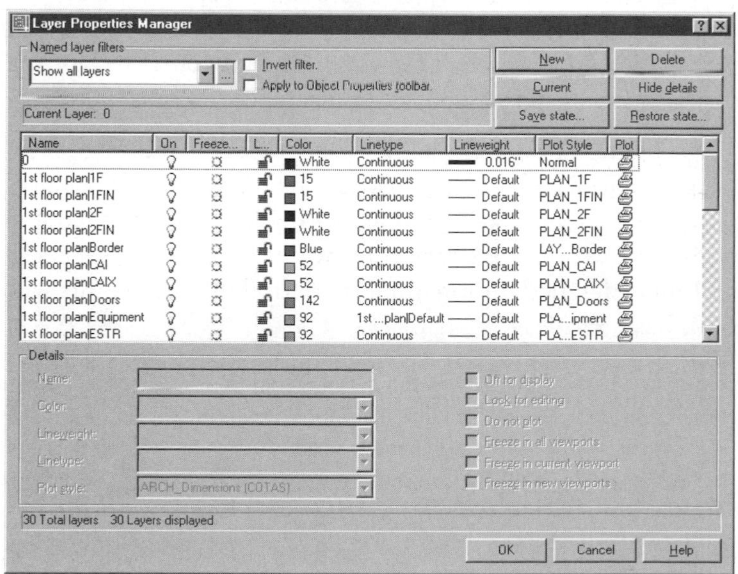

Figure 4–4 *The Layer Properties Manager controls the color, linetype, lineweight, and plot style values for all objects whose different property values are set to ByLayer.*

AutoCAD uses these first three property values when drawing an object on the screen, and AutoCAD uses all four property values when plotting an object on paper. Once again, AutoCAD only looks to the property values in the Layer Properties Manager when an object's properties are set to ByLayer.

It is important to note that even though Figure 4–4 shows that the ByLayer color value for layer 0 is white, newly created objects will be drawn in the color red. This is because the Object Property toolbar shown in Figure 4–3 is set to the color red and therefore explicitly assigns the color red to newly created objects. So while an object may reside on layer 0, and the ByLayer color value for layer 0 may be white, AutoCAD will display the object's color as red if the object's color is set explicitly in the Object Properties toolbar.

Tip: Always set an object's property values to ByLayer. This provides you with the ability to change color, linetype, lineweight, and plot style globally.

THE COLOR PROPERTY

One of the simplest properties to understand is the color property. AutoCAD provides 256 colors to choose from, and your color choice is influenced by two factors:

- How objects appear on the screen
- How objects appear on paper

For screen appearance, you should typically use various colors to help you differentiate between objects. By using an assortment of colors, you make viewing easier for objects that are drawn close together. However, when it comes to plotting objects on paper, there is more involved to an object's appearance than just color.

When plotting objects, AutoCAD allows you to assign lineweights by color. For example, when an object that is red is plotted, its lineweight may be 0.002 inches, while an object that is green may be plotted with a lineweight of 0.008 inches. The user determines these values at plotting time by setting lineweight values in the Plot dialog box. For more detailed information on controlling lineweights when plotting, see Chapter 20, Basic Plotting.

Note: Quite often, the colors you assign to objects are determined by layering standards. If your company or client has defined color assignments in their layering standards, you should use their color assignments instead of assigning your own.

THE LINETYPE AND LINEWEIGHT PROPERTIES

The linetype property allows you to set the style of a linetype. A linetype style defines whether AutoCAD draws an object with a continuous, unbroken line, or with dashed or dotted lines. You can choose from a wide variety of noncontinuous linetypes, and you can also select linetypes that have text in them. AutoCAD includes an assortment of linetypes, or you can create your own custom linetype styles.

The lineweight property, on the other hand, controls how heavy a line AutoCAD draws. In fact, the lineweight property performs the same function as the Plot dialog box described previously. However, instead of assigning lineweights by color, it assigns the lineweight as a property. Consequently, you can display the lineweight on the screen, as opposed to having to observe how wide a line is by plotting the objects. Therefore, you can set one object's lineweight very thin to make it appear subtle, set another object's lineweight much wider to make it appear bold, and then view the results on the screen, without plotting.

Note: Linetypes and lineweights are described in detail in Chapter 5, Applying Linetypes and Lineweights.

In the following exercise you will use the Object Properties toolbar and the Layer Properties Manager to control the color, linetype, and lineweight of objects.

EXERCISE: CONTROLLING AN OBJECT'S COLOR, LINETYPE, AND LINEWEIGHT

1. Open the drawing 04DWG01. The drawing displays a single horizontal red line.
2. From the Objects Properties toolbar, choose the Existing Gas Line layer from the drop-down list. (It is the fifth layer from the top.)

Note: If the entire layer name is too long to display in the layer drop-down list, hold your cursor over a layer name. After a few moments, AutoCAD displays the entire layer name in a Text Tip, as shown in Figure 4–5.

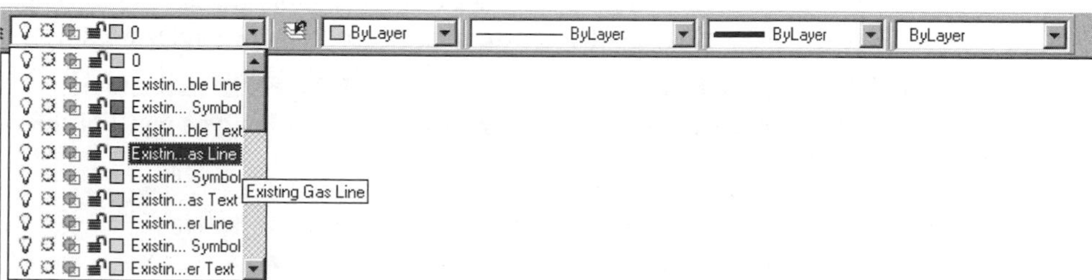

Figure 4–5 *When viewing layer names from the Object Properties toolbar's layer drop-down list, hold the cursor over a shortened layer name to display its name in the Text Tip.*

3. Draw a circle with its center on the left end of the red line, and with a radius of 1 inch. Notice that the circle is drawn with a thin, dashed, green line.
4. From the Object Properties toolbar, from the Color control drop-down list (the second list from the left), choose the color Magenta.
5. From the Object Properties toolbar, from the Linetype control drop-down list (the third list from the left), choose the Continuous linetype.
6. From the Object Properties toolbar, from the Lineweight control drop-down list (the fourth list from the left), choose the 0.016" line weight.
7. Draw a circle with its center on the right end of the red line, and with a radius of 1 inch. Notice that the circle is drawn with a heavy, continuous, magenta line, as shown in Figure 4–6.

   Notice how dramatically different the two circles in Figure 4–6 are. Even though both circles are drawn on the same layer, the circle on the right is assigned its color, linetype, and lineweight explicitly by the values in the Object Properties toolbar. However, the circle on the left has its color, linetype, and lineweight values set to ByLayer. Therefore, AutoCAD draws the circle on the left based on the values of color, linetype, and lineweight set for the Existing Gas Line layer in the Layer Properties Manager, as shown in Figure 4–7.

   Next you will use the Layer Properties Manager to modify the color, linetype, and lineweight values.

8. From the Objects Properties toolbar, choose the Layers button (the second button from the left). The Layer Properties Manager dialog box is displayed. For the Existing Gas Line layer, under the Color column, choose the green box labeled "Green." The Select Color dialog box is displayed as shown in Figure 4–8.
9. From the Select Color dialog box, choose the "blue" color, and then choose OK. AutoCAD sets the Existing Gas Line layer's color to blue.
10. For the Existing Gas Line layer, under the Linetype column, choose the DASHED linetype. The Select Linetype dialog box is displayed as shown in Figure 4–9.
11. From the Select Linetype dialog box, choose the CENTER linetype, then choose OK. AutoCAD sets the Existing Gas Line layer's linetype to CENTER.

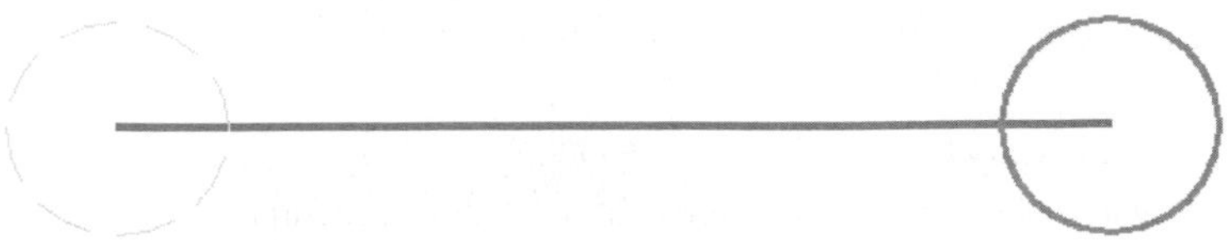

Figure 4–6 *The color, linetype, and lineweight properties can be set to ByLayer or they can be set explicitly.*

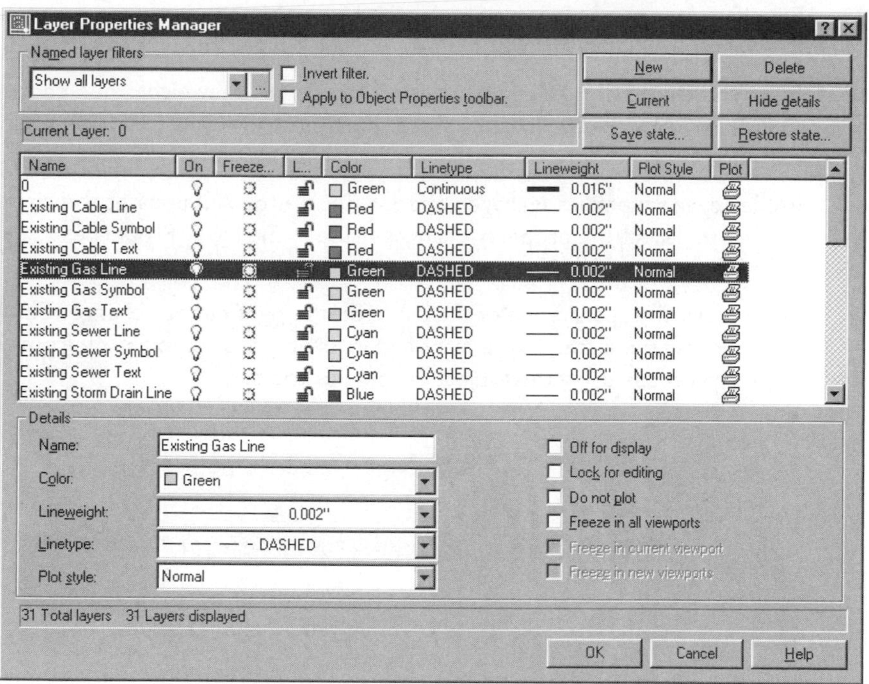

Figure 4–7 *The color, linetype, and lineweight values are highlighted for the Existing Gas Line layer in the Layer Properties Manager.*

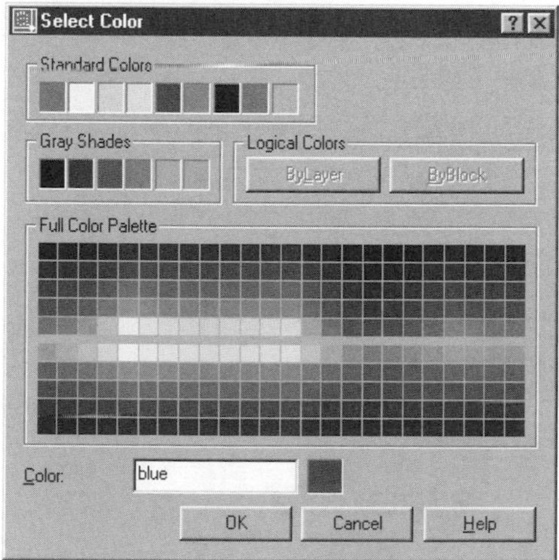

Figure 4–8 *The Select Color dialog box allows you to choose one of 256 colors.*

12. For the Existing Gas Line layer, under the Lineweight column, choose the 0.002" lineweight. The Lineweight dialog box is displayed as shown in Figure 4–10.
13. From the Lineweight dialog box, choose the 0.016" lineweight, then choose OK. AutoCAD sets the Existing Gas Line layer's lineweight to 0.016", as shown in Figure 4–11.
14. In the Layer Properties Manager, choose OK. AutoCAD updates the circle on the left to reflect the property changes made in the Layer Properties Manager, as shown in Figure 4–12.

    Notice that the appearance of the circle on the right did not change. This is true because the circle on the right had its properties set explicitly from the values you chose using the Object Properties toolbar.

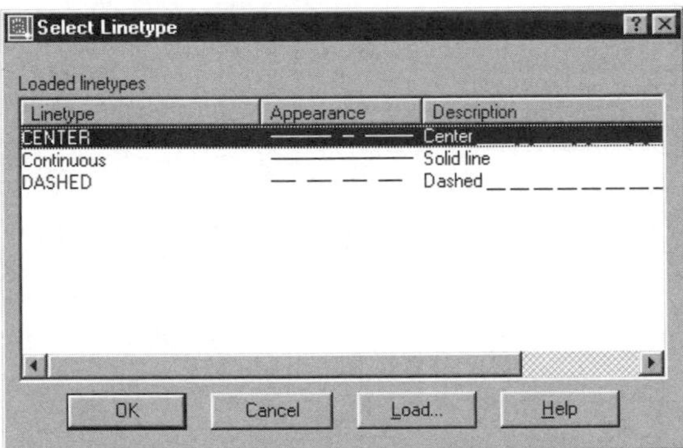

Figure 4–9 *The Select Linetype dialog box allows you to choose a layer's linetype.*

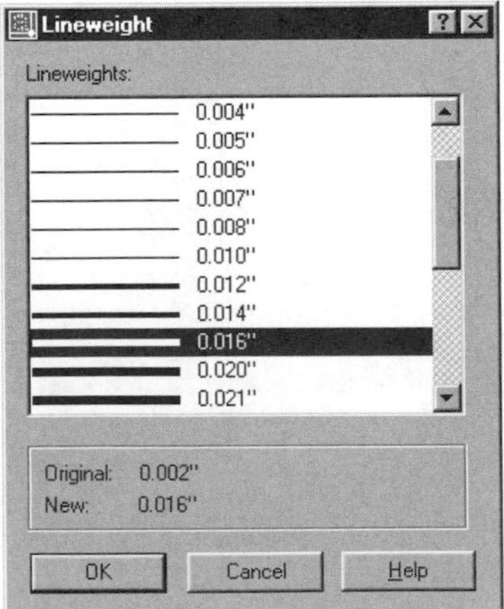

Figure 4–10 *The Lineweight dialog box allows you to choose a layer's lineweight.*

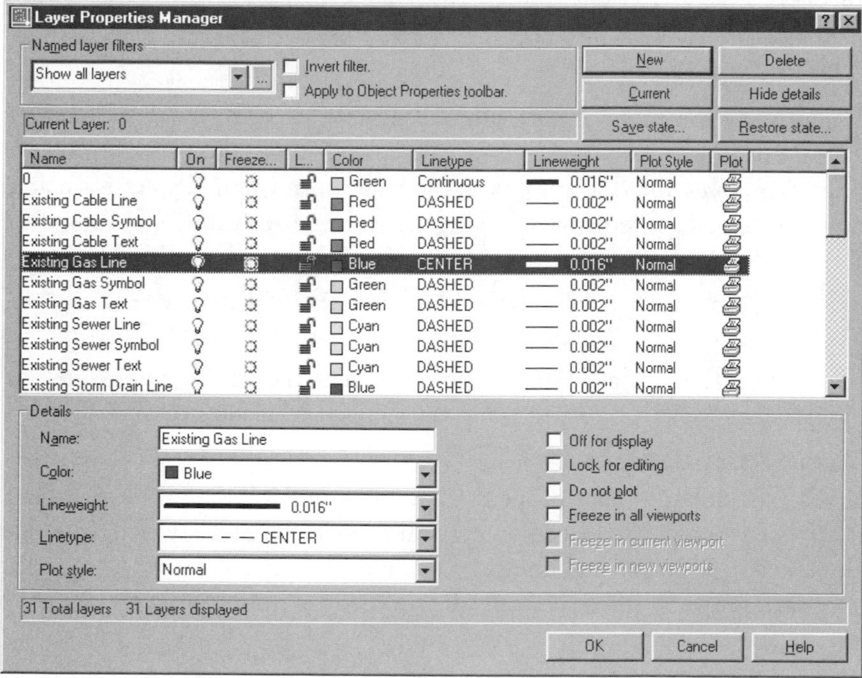

Figure 4–11 *The Existing Gas Line layer's new color, linetype, and lineweight are modified in the Layer Properties Manager.*

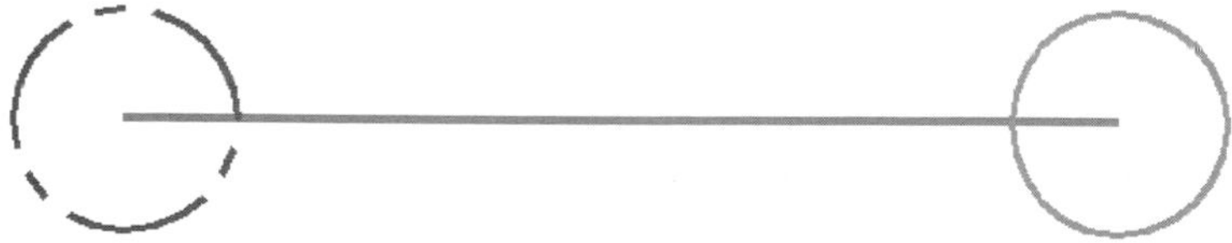

Figure 4–12 *The circle on the left displays the new property values set in the Layer Properties Manager.*

Next you will complete this exercise by reassigning the properties of the circle on the right to ByLayer.

15. Choose the magenta circle on the right. AutoCAD highlights the object, indicating that it is selected.
16. From the Object Properties toolbar, from the Color control drop-down list, choose ByLayer.
17. From the Object Properties toolbar, from the Linetype control drop-down list, choose ByLayer.

18. From the Object Properties toolbar, from the Lineweight control drop-down list, choose ByLayer. AutoCAD redraws the circle on the right, as shown in Figure 4–13.

Notice that while the circle on the right remains selected, the Color, Linetype, and Lineweight controls display ByLayer as the current type. However, if you deselect the circle by pressing the Esc key twice, the values you originally set for color, linetype, and lineweight are redisplayed. This occurs because AutoCAD shows the currently selected object's property values in the Object Properties toolbar. These include the selected object's layer and plot style assignments.

You can close the drawing without saving your changes. However, before exiting AutoCAD, be sure to deselect all objects and then change the color, linetype, and lineweight values back to ByLayer using the Object Properties toolbar.

THE PLOT STYLE PROPERTY

AutoCAD provides an object property called plot style. This property affects how objects appear when plotted by allowing you to assign a plot style to an object and override its color, linetype, and lineweight values. Additionally, plot styles allow you to specify end, join, and fill styles, as well as to control output effects such as dithering, gray scale, pen assignment, and screening.

Plot styles are intended to allow you to plot the same drawing in many different ways without making elaborate changes to the original color, linetype, and lineweight properties. By creating multiple plot style tables, you can create one plot of a drawing that displays objects with bold, heavy lines, and then plot the same drawing as a grayscale, all without making any changes to object color, linetype, or lineweight properties.

Note: For more detailed information on plot styles, refer to Chapter 20, Basic Plotting.

Figure 4–13 *Choosing ByLayer from the Color, Linetype, and Lineweight controls' drop-down lists resets the selected circle's properties.*

CONTROLLING OBJECT BEHAVIOR

In previous sections you learned how to control object properties that determine how an object looks when AutoCAD draws it on the screen or plots it on paper. In this section you will learn how to control object behavior using the Layer Properties Manager. Specifically, you will learn how to control object visibility and protect objects from accidentally being edited.

LAYER VISIBILITY

One handy feature of using layers to organize objects into logical groups is that you can use the Layer Properties Manager to manipulate groups of objects by modifying layer settings. In addition to controlling object colors and linetypes, as previously discussed, you can also control object visibility. By simply clicking your pointing device, entire groups of objects become invisible on the screen, and non-plottable.

AutoCAD allows you to control object display on the screen with two features in the Layer Properties Manager. You can turn layers off and on, and/or you can freeze and thaw layers. In either case, objects that reside on the layer that is turned off or frozen become invisible and non-plottable. Additionally, the Plot/Don't Plot feature allows you to control whether objects on a layer are plotted. This is true even though the objects are visible on the screen.

In the next few sections you will learn about the important differences between turning layers off and on, and freezing and thawing, and you will be introduced to the new Plot/Don't Plot feature.

Turning Layers Off versus Freezing Layers

When you place objects on a layer, you can control their visibility by turning the layer on and off, or by freezing and thawing the layer. When you turn a layer off, or freeze the layer, objects that are on the layer become invisible. They are not displayed on the screen, and they are not plotted.

While the end result of making objects invisible by either turning layers off or by freezing layers may seem the same, there is a very important reason why AutoCAD has the two different methods available. When a layer is turned off, even though the objects on the layer become invisible, AutoCAD still performs certain zoom and regeneration calculations on the invisible objects. In contrast, when a layer is frozen, AutoCAD does not include the objects on the frozen layer in zoom or regeneration calculations.

By freezing objects, you can dramatically reduce zoom and regeneration times, which improves your productivity. For example, suppose you have a drawing with thousands of objects. If you only need to edit objects that reside on one layer, then you can increase your productivity by freezing the layers on which all other unneeded objects reside. By freezing the layers that have unneeded objects, you eliminate those objects from AutoCAD's calculations, reducing the time AutoCAD takes to perform certain zoom and regeneration functions.

So, if freezing layers improves productivity, why not just always freeze layers instead of turning them off? Because every time you thaw a layer that has been frozen causes AutoCAD to perform a regeneration (also called a regen). (For those who are new to AutoCAD, a regeneration is often the most frustrating thing a user can experience because the user must sit idly by, unable to do anything, waiting for AutoCAD to complete its regeneration calculations.) In contrast, turning layers off (or on) does not cause a regeneration. Therefore, turning layers off makes sense when you typically need to view the objects in your drawing during an editing session, and when you want to make them invisible only temporarily. In contrast, freezing layers is the proper choice when there are objects you do not need to view during a lengthy editing session. By freezing the layers, AutoCAD visually removes the objects from the screen but no longer includes the objects on the frozen layers in future regenerations, which can dramatically reduce overall regeneration times.

Freezing Globally versus Freezing in the Active Viewport

In the previous section you learned about the differences between turning off layers and freezing layers. Though the exercise did not discuss it, the methods used to turn off and freeze layers were global. In other words, they affect all objects in all viewports. While turning off or freezing layers in all viewports is generally acceptable, there are circumstances when it is not. Specifically, when you are working with paper space layouts that have multiple viewports, you may want objects on a layer to be visible in one viewport but invisible in another. This effect is accomplished by freezing layers in the active viewport. For detailed information on freezing layers in the active viewport, refer to Chapter 19, Layouts and Paper Spaces.

The Plot/Don't Plot Feature

The Layer Properties Manager includes the Plot/Don't Plot toggle. You control whether objects on a layer are plotted by toggling this feature on or off. What makes this feature useful is that objects that are visible on the screen are prevented from being plotting if the Plot/Don't Plot feature is toggled off. Therefore, you can display and use objects such as construction lines on the screen, but prevent them from being plotted on paper.

Tip: You can use AutoCAD's Plot Preview feature to view the results of layers set to not plot. For more information, refer to Chapter 20, Basic Plotting.

The Plot/Don't Plot feature affects blocks and external references (xrefs) in unique ways. For example, you can control which objects in a block will not plot by setting their layers not to be plotted. Also, you can prevent an entire block from plotting by

Note: In previous releases you could simulate the Plot/Don't Plot feature by creating a layer called DEFPOINTS and placing on it objects that you wanted to view and edit on the screen but did not want to plot. The problem with this technique is that you could only have one DEFPOINTS layer, and therefore the layer could become cluttered with an array of objects that you needed to use on the screen but did not want to plot.

setting the layer on which it is inserted not to be plotted. This is true for xrefs as well. Therefore, you can control whether entire blocks and xrefs are plotted, or you can control whether specific objects in blocks and xrefs are plotted. For more information, refer to Chapter 20, Basic Plotting.

LOCKING LAYERS

When editing a drawing with multiple layers, you probably will mistakenly select objects that you do not want to edit. While this is frustrating, the good news is you can control which objects are editable by locking and unlocking the layers on which they reside. The ability to lock layers enables you to display the objects on a layer without selecting any objects on that layer. More importantly, while objects on a locked layer are not selectable for edit commands, they can be snapped to using object snaps.

Therefore, the layer locking feature is most useful when you are working on drawings in which you simply need to see the objects on a layer for reference and object snapping purposes, and when you want to ensure that you do not accidentally alter the objects. By locking a layer, you ensure that the objects on that layer are safe from unintended edits.

Note: You can place blocks on a layer that is locked; even though they may contain data from other layers, you will not be able to erase the blocks. You can create data on a locked layer and then be unable to modify the object just created. This can often be a trap during lisp routines that create data on locked layers.

FILTERING OBJECTS WITH LAYERS

As discussed previously, organizing objects into groups using layer names is very useful. But while it is good to use layer names to organize your drawing, if the layers become too numerous, their usefulness degrades. The usefulness of layers degrades because too many layers makes it cumbersome to use the Layer Properties Manager to locate the specific layer whose properties you need to edit or whose behavior you need to change. The more layers a drawing has, the more their usefulness degrades.

You will probably work on drawings that contain dozens, hundreds, or even thousands of layers. This is especially true when you attach xrefs to the current drawing, adding to its list of layers the layer lists of each xref. Therefore, the likelihood that you will be presented with the challenge of working with too many layers is very high.

Fortunately, AutoCAD's Layer Properties Manager provides a feature called Layer Filters, which allow you to control which layer names are displayed by defining certain parameters. By using layer filters, you can realize the benefit of organizing your drawing with layers and not be overwhelmed by viewing too many layer names in the Layer Properties Manager.

In this section you will explore how to create and apply layer filters to make working with layers easier.

APPLYING LAYER FILTERS

Applying a layer filter is pretty simple. When you open the Layer Properties Manager dialog box and then select a saved filter from the Named Layer Filters drop-down list, AutoCAD instantly adjusts the display of layers in the Layer Properties Manager's layer list.

Each AutoCAD drawing automatically includes three standard layer filters:

- Show all layers
- Show all used layers
- Show all xref-dependent layers

With these three layer filters, you can make viewing layer lists much easier. Additionally, the Layer Properties Manager includes two options, which allow you to

- Invert the current layer filter
- Apply the current layer filter to the Layer control in the Object Properties toolbar

The Show All Used Layers filter displays only names of layers on which objects reside. If a layer does not have any objects drawn on it, then it is not displayed in the Layer Properties Manager's layer list. This filter is useful in determining which layers are no longer used by objects and can therefore be deleted.

The Invert Filter check box inverts the current layer filter. When it is used in conjunction with the Show All Used Layers filter, AutoCAD displays only those layers that have no objects on them. Consequently, you could delete these layers to reduce the number of layer names cluttering the layer list.

The Apply to Object Properties Toolbar check box applies the current layer filter to the Layer control list in the Object Properties toolbar. This feature is very useful when you need to frequently switch between a few layers in a drawing that contains many layers.

The other layer filer that AutoCAD provides is the Show All Xref Dependent Layers filter. This filter only displays the names of layers that reside in xrefs. By selecting this filter, you can easily view only the layers of attached xrefs. By inverting this filter, you can display all layers that do not reside in attached xrefs. This provides you with the ability to remove unwanted xref layer names from the layer list and the Layer control list on the Object Properties toolbar.

Next you will learn how to create your own layer filters.

CREATING NAMED LAYER FILTERS

In the previous section you learned how to use the standard layer filters provided with AutoCAD. While these filters are useful, they probably will not fulfill all of your filtering needs. To satisfy all of your needs, the Layer Properties Manager allows you to create your own layer filters.

To create a custom layer filter, start by choosing the Layers button from the Object Properties toolbar. AutoCAD displays the Layer Properties Manager dialog box. In the Named Layer Filters area, choose the "..." button. (This button is next to the Named Layer Filters list's down arrow.) AutoCAD displays the Named Layer Filters dialog box. Next, enter a name, such as **Symbols Only**, for the custom filter in the Filter Name list box. Then, in the Layer Name text box, enter ***symbol**. The asterisk is a wildcard character that tells AutoCAD to accept any characters in front of the word "symbol". The Named Layer Filters dialog box appears as shown in Figure 4–14.

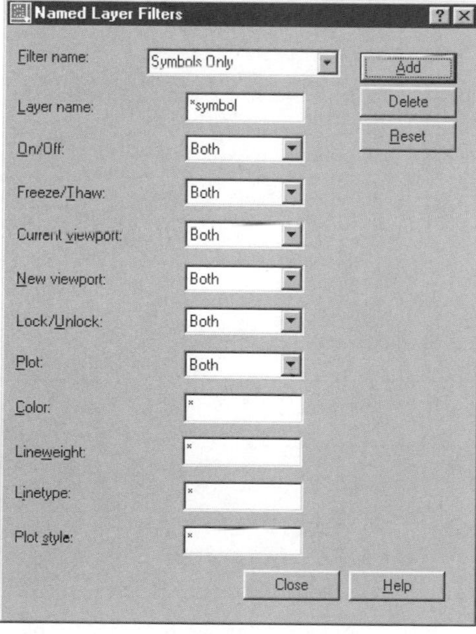

Figure 4–14 *The Named Layer Filters dialog box allows you to create custom layer filters.*

Finally, choose the Add button. AutoCAD adds the custom filter to the list of available filters and then sets the filter fields back to their default values. Choose the Close button. AutoCAD displays the Named Layer Filters dialog box.

To see how this custom filter works, choose the Symbols Only filter from the Named Layer Filters drop-down list in the Layer Properties Manager. AutoCAD invokes the selected layer filter. In the Named Layer Filters area, select the Invert Filter check box to turn this option off. AutoCAD now displays only those layer names that end with the word "Symbol," as shown in Figure 4–15.

By using the numerous fields available in the Named Layer Filters dialog box, you can precisely identify the layers you wish to display. Additionally, by inserting asterisks in the layer name, you can filter for layers whose differing names have certain words in common. You can enter multiple layer names by separating each name with a comma, as in "*symbol,*text." (Be sure to include the asterisks, and do not insert a space after the comma.)

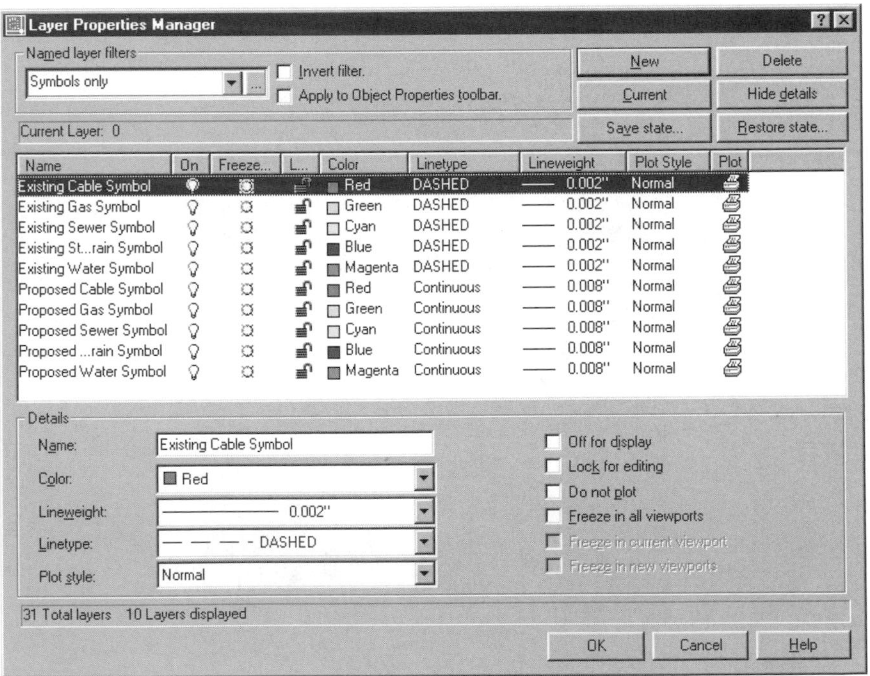

Figure 4–15 *The custom layer filter Symbols Only displays layer names that end with the word "Symbol."*

The tilde symbol (~) instructs AutoCAD to exclude layers that match the criteria that follows the symbol. For instance, "~red" means exclude all layers with the color red (see Figure 4–16).

USING WILDCARD CHARACTERS IN LAYER FILTERS

As you just learned, you can use wildcard characters such as an asterisk (*), comma (,), and tilde (~) to control which layers are displayed in the Layer Properties Manager's layer list. There are, in fact, ten different wildcard characters you can use with layer filters, and these wildcard characters can be used in combination with each other. Table 4-1 lists the available wildcard characters and the purpose of each. These wildcard characters may be used for several of the filters in the Named Layer Filter dialog box, including the following:

- Layer name
- Color
- Lineweight
- Linetype
- Plot style

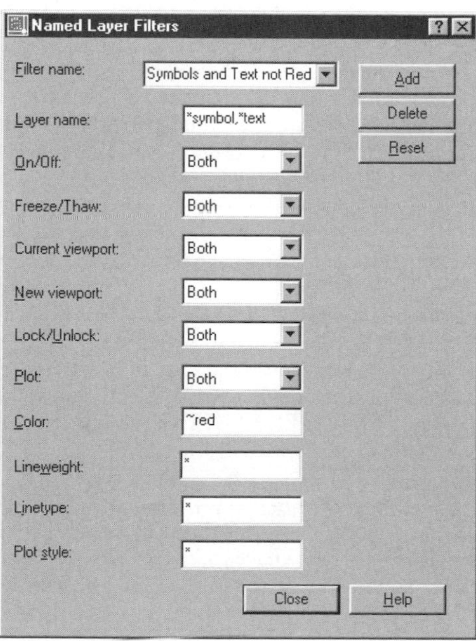

Figure 4–16 *Custom layer filters can include multiple layer names, and they can exclude certain properties when you place a tilde (~) in front of the property value.*

Table 4–1: *Acceptable Wildcard Characters for Layer Filters*

Character	Description
# (Pound)	The # symbol matches any single numeric character. For example, suppose you have layer names that are labeled with numbers 1 through 400. You can filter for layer names 200 through 299 by entering 2## as the layer name filter.
@ (At)	The @ symbol matches any single alphabetic character. For example, suppose you have two layers named NORTH and SOUTH. You can filter for both of these layer names by entering @O@TH as the layer name filter.
. (Period)	The . symbol matches any single non-alphanumeric character. For example, suppose you have layers named GAS-TXT, GAS TXT, and GAS_TXT. Notice in each case that the alphanumeric characters are separated by a hyphen, a space, and an underscore character, respectively. You can filter for these three layer names by entering GAS.TXT as the layer name filter.
* (Asterisk)	The * symbol matches any character sequence and can be used at the beginning, middle, or end of the filter. For example, suppose you have layers whose names include the word LINE. You can filter for these layer names by entering *LINE* as the layer name filter.
? (Question mark)	The ? symbol matches any single character. For example, suppose you have layers named GAS-TXT, GAS2TXT, and GASeTXT. You can filter for these three layer names by entering GAS?TXT as the layer name filter.
~ (Tilde)	If the ~ symbol is the first character in the filter, then it excludes the filter value. For example, suppose you have layers that include the name LINE. You can filter for layers that do not include the name LINE by entering ~*LINE* as the layer name filter.
[ ] (Brackets)	The [ ] symbol matches any one of the characters enclosed in the brackets. For example, suppose you have four layers whose names are 1LINE, 2LINE, 3LINE, and 4LINE. You can filter for the layers whose names begin with 1, 2, or 4 by entering [124]* as the layer name filter.
[~] (Tilde inside brackets)	The [~] symbol excludes each of the characters that are enclosed in the brackets after the tilde. For example, suppose you have four layers whose names are 1LINE, 2LINE, 3LINE, and 4LINE. You can exclude the layers whose names begin with 1, 2, or 4 by entering [~124]* as the layer name filter.
[-] (Hyphen inside brackets)	The [-] symbol allows you to specify a single-character range of values. For example, suppose you have four layers whose names are 1LINE, 2LINE, 3LINE, and 4LINE. You can filter for the layers whose names begin with 1, 2, or 3 by entering [1-3]* as the layer name filter.
, (Comma)	The , symbol separates multiple filters, allowing you to enter more than one filter in a text box. For example, suppose you have four layers whose names are 1LINE, 2LINE, 3LINE, and 4LINE. You can filter for the layers whose names begin with 1 or 3 by entering 1*,3* as the layer name filter.

By using the wildcard characters, you can develop powerful layer filters that display only the precise layers you wish to view in the Layer Properties Manager or in the Object Properties toolbar.

SAVING AND RESTORING LAYER SETTINGS

Managing the various states of AutoCAD's layers is always challenging, especially when you work with large numbers of layers. Setting layer states such as on/off or freeze/thaw, or changing a layer's color or linetype, is time-consuming when you are working with drawings containing dozens or even hundreds of layers. In addition to the time spent managing large numbers of layers, ensuring that the proper layer states are restored when plotting your final plans is not only challenging, it is essential. When editing drawings, you need the ability to quickly manipulate the layer states of large groups of layers.

Fortunately, AutoCAD 2000i added to the Properties Manager a tool for managing AutoCAD's layers. This provides convenient access to the much-needed ability to quickly save and restore layer settings. By using the Save State and Restore State buttons shown in Figure 4–17, you can quickly save and restore AutoCAD's layer settings and ensure that the proper layer states are set when you edit or plot drawings.

SAVED LAYER STATES OVERVIEW

The goal of AutoCAD's Save State and Restore State tools is simple: to easily save and then instantly restore the layer settings of large groups of layers. So, once you set the on/off and freeze/thaw states of layers, and then assign colors, linetypes, and lineweights, as well as plot styles and plot status, you can save those settings as layer states and then later restore those same layer states instantly. Using the Save State button, once you have set your layer states as desired, you can save those states as a named layer state in the Save Layer States dialog box and then later restore the layer states using the Restore State button.

Named layer states are stored in the current drawing. This means that when anyone opens the drawing for editing or plotting, they can access all the named layer states in the drawing from the Layer States Manager, which is displayed when you select the Restore State button. Additionally, you can save named layer states in

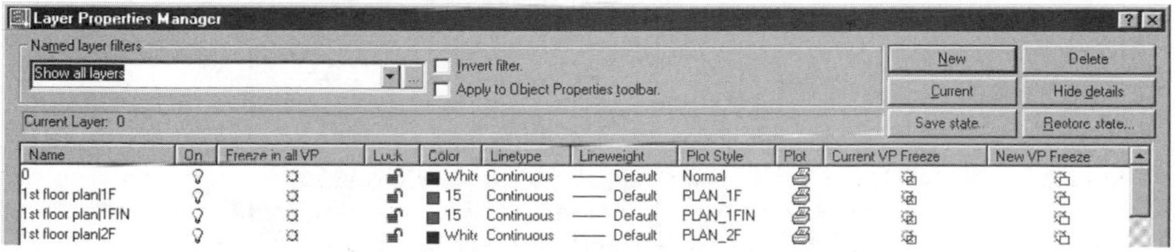

Figure 4–17 *The Save State and Restore State buttons reside in the Layer Properties Manager.*

AutoCAD's drawing template files, thereby ensuring that standard named layer states are included in each new drawing created from the template file.

SAVING AND RESTORING NAMED LAYER STATES

Saving a named layer state is very simple. Once you set the various states for your layers in the Layer Properties Manager, you then choose the Save State button. When AutoCAD displays the Save Layer States dialog box, enter a new name for the current layer state settings, select the layer states and layer properties you wish to save (see Figure 4–18), and then choose OK. AutoCAD saves the current layer settings for each layer, including the layer states for on/off and freeze/thaw states, as well as the layer properties for color, linetype, and plot style. Once they are saved, you can restore the saved layer states by opening the Layer Properties Manager and choosing the Restore State button to open the Layer States Manager, as shown in Figure 4–19. Once the Layer States Manager is open, choose the desired layer state

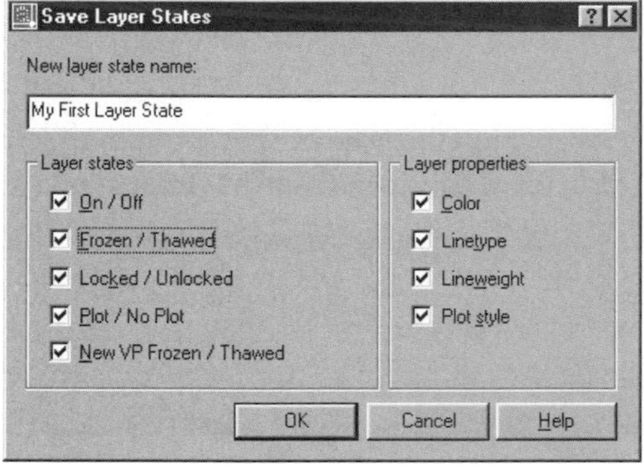

Figure 4–18 *The Save Layer States dialog box lets you select the layer states and layer properties you wish to save, and lets you assign the settings a layer state name.*

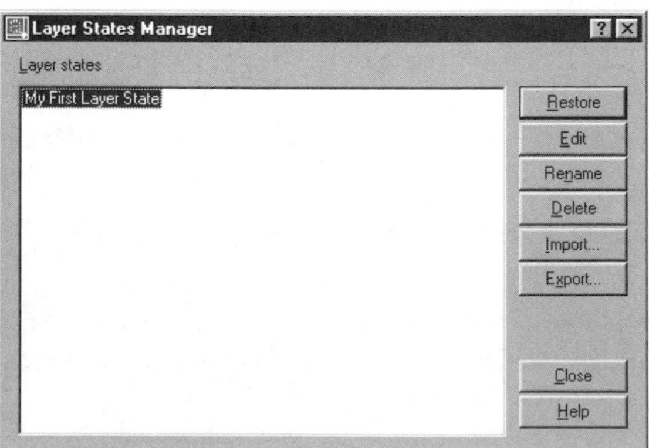

Figure 4–19 *The Layer States Manager lets you restore saved layer states.*

and then choose Restore. This causes AutoCAD to reset the various saved layer states for each layer and dismisses the Layer States Manager dialog box.

Modifying Named Layer States

Through the Layer States Manager you can edit, rename, or delete named layer states. To edit an existing named layer state, with the Layer States Manager open, select the named layer state you wish to modify, and then choose Edit. Once you do, AutoCAD displays the Edit Layer States dialog box, which is the same as the Save Layer States dialog box, and which lets you modify the layer states and layer properties settings. After you make your changes in the Edit Layer States dialog box and choose OK, AutoCAD returns to the Layer States Manager.

To rename or delete an existing named layer state, simply select the desired named layer state and choose the appropriate action. If you choose the Rename button, AutoCAD lets you edit the existing layer state name. If you choose the Delete button, AutoCAD asks you to confirm that you wish to delete the selected named layer state.

Importing and Exporting Named Layer States

The Layer States Manager lets you import and export named layer state settings. Called LAS files, these ASCII-based text files can be used to copy layer names and their current states to another drawing. This feature is not only useful for transferring the current state of layers from one drawing to another; it will also automatically create layers that do not exist. For example, if you have an existing drawing that contains standard project layer names, when you export the layer names as an LAS file and then import the LAS file into a new drawing, all the layers and their states will be automatically copied into the new drawing. This feature is very useful for ensuring that new drawings contain the project's standard set of layers.

Adding New Layers

When you add new layers to a drawing, any existing named layer states will not include the new layers. While you can still restore layer states using a named layer state, only those layers originally saved in the named layer state will be reset. If you wish to add the new layers to an existing named layer state, open the Layer Properties Manager, set the layer states as desired, and then create a new layer state using the same name as the existing saved layer state, thereby overwriting the original settings.

Note: When you save layer states, the settings of xrefs are included with the parent drawing's settings. However, if the VISRETAIN system variable is set to Off (0), then the xref's layer settings are removed from the saved layer state the next time the drawing is opened. To persistently retain both the parent's and xref's layer settings in a saved layer state, set the VISRETAIN system variable to On (1).

THE LAYER TRANSLATOR

The Layer Translator is a dialog box that provides an interface for mapping and translating layers in the current drawing to match layer properties in another drawing. When the Layer Translator dialog box is displayed, it lists the current drawing's layers in the Translate From area. Then, when you select the drawing whose layers you wish to map, the selected drawing's layers are displayed in the Translate To area. To map layers from one list to another, you simply select the layers in each list, as shown in Figure 4–20, and then choose Map. As you map the layers, AutoCAD lists them in the Layer Translation Mappings area and shows the changes AutoCAD will make to the names and properties of selected layers in the current drawing. When you choose Translate, AutoCAD makes the desired property changes to the current drawing's layers.

The Layer Translator allows you to select multiple files in the Translate To area to use for mapping the current drawing's layers. For example, by choosing the Load button, you can select multiple drawing (DWG) or drawing template (DWT) files. As you select the files, all of their layers are displayed in the Translate To area. In addition to loading layers from DWG and DWT files, you can also load saved layer translation files (DWS) or enter new layer names and properties manually. You create DWS files by first mapping layers and then by choosing Save, which saves the mapped settings.

The Layer Translator also lets you change the property settings of certain objects during the translation process. Through the Settings dialog box, shown in Figure 4–21,

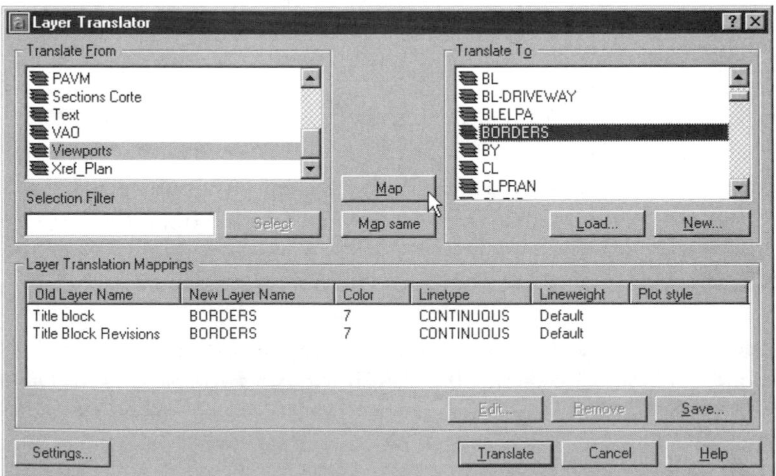

Figure 4–20 *The Layer Translator lets you convert the properties of selected layers in the current drawing to layer properties in another drawing.*

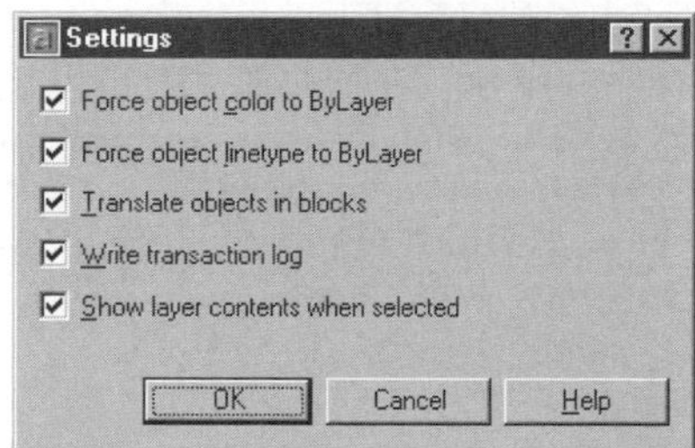

Figure 4–21 *The Layer Translator's Settings dialog box provides additional control over certain object properties during the conversion process.*

you can instruct AutoCAD to change the color and linetype properties of objects on the mapped layers to ByLayer automatically. You can also control whether the properties of objects within blocks are converted to ByLayer. The Settings dialog box also includes two other useful features. The first lets you control whether AutoCAD creates a log file that details the translation process's results. The second lets you temporarily control the display of objects in the current drawing so that only objects that reside on layers selected in the Translate From list are displayed, making it visually easier to identify which objects will be converted. So, through the Settings dialog box, you can automatically change the color and linetype properties of objects—including objects nested within blocks—to ByLayer, track translation changes with a log file, and visually confirm the layer on which objects reside.

MAPPING AND TRANSLATING LAYER PROPERTIES

The purpose of the Layer Translator is to make translating one drawing's layer property values so they match those of another drawing, template file, or layer translation file easier. This feature is very convenient for ensuring that all drawings have exactly the same settings, and it is especially useful for translating drawings provided by others outside your department or organization, thereby ensuring that their drawing's layer properties match yours.

The following examples demonstrate the Layer Translator's various features.

MAPPING AND TRANSLATING LAYER PROPERTIES

1. Open the drawing whose layer properties you wish to change.
2. From the Tools menu, choose CAD Standards>Layer Translator. AutoCAD opens the Layer Translator dialog box.

   The Layer Translator automatically loads the current drawing's layer names into the Translate From list. The next step is to create a list of layers in the Translate To list. The current drawing's layer properties will be converted to match those of the layers in the Translate To list.

In the Translate To area, AutoCAD provides two buttons for adding the desired layers to the Layer Translator dialog box. When you choose the Load button, AutoCAD displays the standard file selection dialog box, which lets you select DWG, DWT, or DWS file types that contain the desired layers. When you choose the New button, AutoCAD displays the New Layer dialog box, as shown in Figure 4–22, which lets you define your own layer properties, which are then added to the Translate To list.

Note: The Layer Translator only displays the active drawing's layers in the Translate From dialog box.

Note: The New Layer dialog box lets you define the new layer's name, color, linetype, lineweight, and plot style properties.

Tip: You can add as many layers as needed to the Translate To list, and you can mix layers from DWG, DWT, and DWS file types, as well as add custom layers using the New Layer dialog box.

3. Using the Load and New buttons, add the desired layers to the Translate To list.

   With the layers displayed in both the Translate From and Translate To lists, the next step is to map the layers from one list to the other. To map layers, select from the Translate From list a layer whose properties you wish to change, and then select from the Translate To list the layer that possesses the desired property values. Once the layers are selected, choose the Map button to add the layer mapping to the Layer Translation Mappings list.

   AutoCAD lets you simultaneously select multiple layers in the Translate From list and map those layers to the selected layer in the Translate To list. By holding down

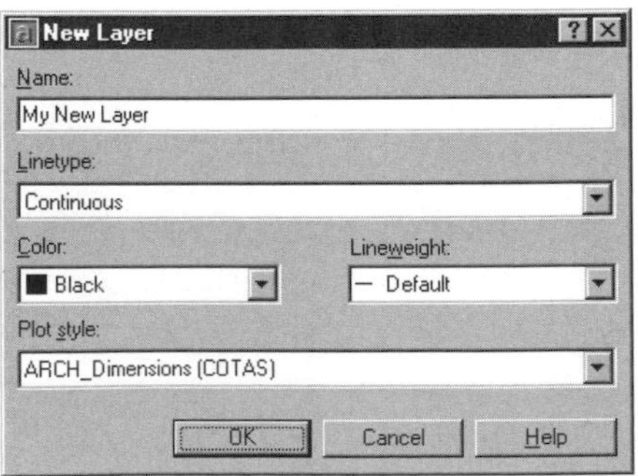

Figure 4–22 *The New Layer dialog box lets you define and add custom layer properties to the Translate To list.*

either the Shift or Ctrl keys, you can use Windows' standard selection methods to select multiple layers. In addition to using Windows' standard selection methods, you can also use the Select Filter feature in the Translate From area.

By entering a layer name in the Select Filter text box, you can quickly identify a layer in the Translate From list. Additionally, the Select Filter text box lets you use wildcard characters to indicate a layer name pattern. For example, when you enter TI* in the Select Filter text box and then choose the Select button, AutoCAD selects all layers whose name begins with TI, as shown in Figure 4–23. Table 4-2 lists acceptable wildcard characters.

Tip: If both lists contain layers with matching names, you can quickly map the layers by choosing the Map Same button, thereby updating layer properties such as color and line-type in the current drawing.

4. Choose the layers that you wish to change in the Translate From list, then choose the layer in the Translate To list with the desired properties, and then choose the Map button. AutoCAD adds the layer mappings to the Layer Translation Mappings list, as shown in Figure 4–24.

   With the layer mappings created, you can now choose the Translate button to assign the properties of the layers selected from the Translate To list to those selected from the Translate From list. However, prior to translating the mappings, AutoCAD lets you further fine-tune the layer properties. For example, when you select the Title Block Revisions mapped layer in the Layer Translation Mappings list, and then choose the Edit button, AutoCAD displays the Edit Layer dialog box, as shown in Figure 4–25, which lets you change any or all of

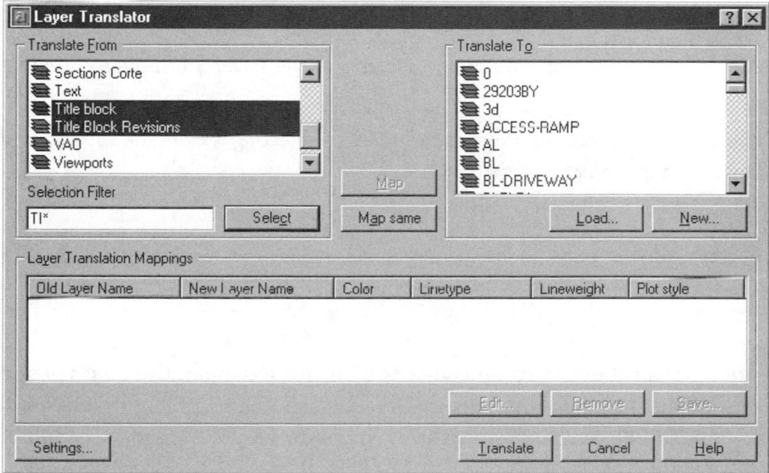

Figure 4–23 *You can quickly identify layers in the Translate From list by using the Select Filter feature.*

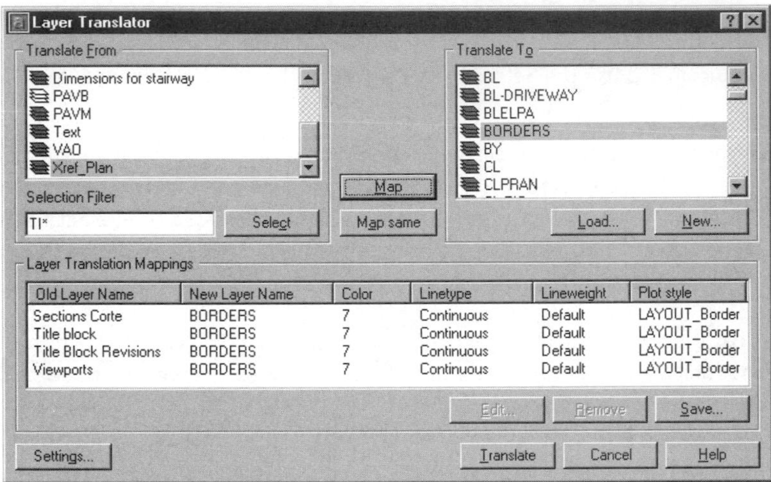

Figure 4–24 *AutoCAD displays mapped layers in the Layer Translation Mappings list.*

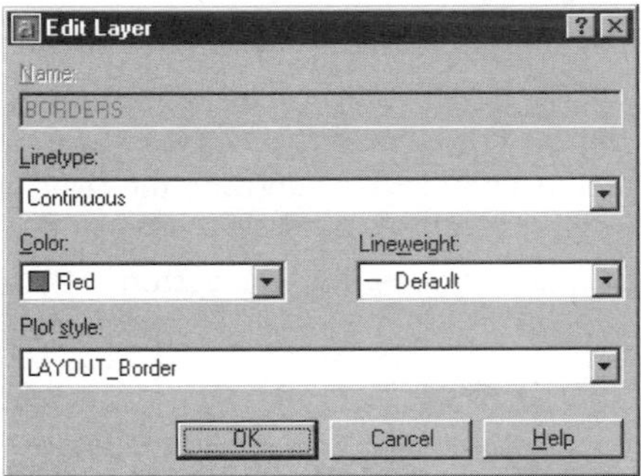

Figure 4–25 *AutoCAD displays mapped layers in the Layer Translation Mappings list.*

Note: You can remove the mapped layer by selecting it from the Layer Translation Mappings list and then choosing the Remove button.

Note: You can only modify the plot style property if all drawings displayed in the Translate From and Translate To lists use named plot styles.

the layer properties to be translated. In this example, the color property value is changed to Red. Once the properties are edited, choosing OK updates the mapped layer properties, as shown in Figure 4–26.

5. With the mapped layer properties set, choose the Translate button. AutoCAD applies the layer mappings, updating the layer properties in the current drawing, and then dismisses the Layer Translator dialog box.

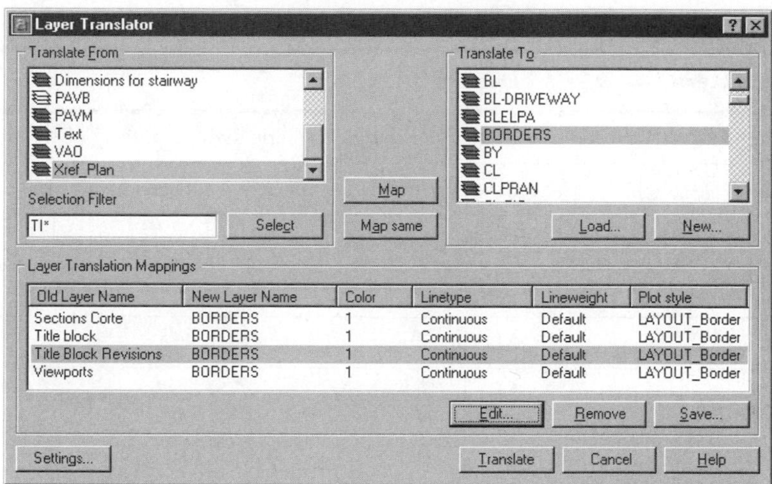

Figure 4–26 *AutoCAD lets you change mapped layer properties, such as changing the color property from 7 to 1.*

Note: When you change a mapped layer's properties, the change affects all layers in the New Layer Name column that have the same name. For example, when you edit only the Title Block Revisions mapped layer, the color properties of all BORDERS layers are changed from 7 to 1.

Tip: You can save the current mapping settings and later load them for reuse. When you choose the Save button, AutoCAD lets you save the settings to a DWS file, which you can reload at a later time. This feature is especially useful for quickly updating a series of drawings.

With the layers translated, you are now free to continue working on the drawing, knowing that your layers and their properties meet your standard layer configuration. As was demonstrated in the preceding example, through the Layer Translator, you have access to a Quality Control (QC) tool that helps ensure that your drawings are consistent and that your standards are followed.

LAYER PREVIOUS

The Layer Previous command makes it very easy to undo changes to layer states quickly. Consequently, changes made to layer states such as on/off, freeze/thaw, and color and linetype, can be undone, and the layer states can be restored to their original states. By choosing the Layer Previous button located on the Object Properties toolbar, as shown in Figure 4–27, you can restore changes made to layers during the current editing session.

The Layer Previous command is analogous to AutoCAD's UNDO command, except that the Layer Previous command only undoes changes made to layer states.

Table 4–2: *Acceptable Wildcard Characters for the Layer Translator*

Wildcard Pattern	Description
# (Pound sign)	Matches any single numeric digit
@ (At symbol)	Matches any single alphabetic character
. (Period)	Matches any single non-alphanumeric character
* (Asterisk)	Matches any character(s)
? (Question mark)	Matches any single character
~ (Tilde)	Matches every layer name except those containing the specified character(s)
[ ] (Brackets)	Matches any one of the characters enclosed in the brackets
[~] (Tilde within brackets)	Matches any character except those enclosed in the brackets after the tilde
[-] (Hyphen within brackets)	Matches any one of the characters specified in the sequential range of characters, such as A-G
` (Reverse quote)	Reads the next character literally, and is used to turn off wild-card matching (for example, `- looks for layer names containing a hyphen)

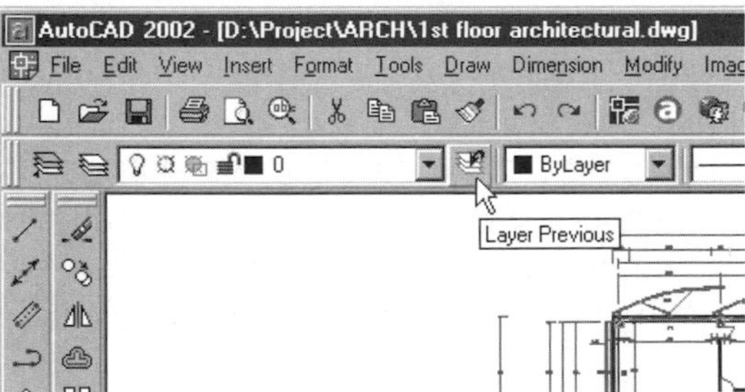

Figure 4–27 *The Layer Previous button restores changes made to layers during the current editing session.*

In other words, Layer Previous does not undo certain changes made to layers. Specifically, executing Layer Previous will not restore the original names of layers that were renamed, will not undelete layers that were deleted, and will not remove new layers that were added. Using the Layer Previous command, you can only restore layer states such as freeze/thaw, lineweights, plot styles, and plot/no plot settings.

Layer Previous will even restore the Current VP Freeze and New VP Freeze settings, both in paper space and model space viewports.

Note: While Layer Previous will restore the Current VP Freeze and New VP Freeze settings in viewports, it does not let you specify the viewport. As with the UNDO command, executing the Layer Previous command simply takes you back one set of edits, undoing the last set of edits made to layer states.

Each time you execute Layer Previous, AutoCAD restores the previous layer states. You can therefore work on a drawing, making continuous changes to layer states, and then cycle back through all changes to layer states until the original layer state settings are restored. As previous settings are restored, AutoCAD displays the message, "Restored previous layer status." When you have cycled back to the original layer state settings, AutoCAD displays the message, "*No previous layer status*."

Suspending the Layer Previous Command

The Layer Previous command restores previous layer settings because it tracks changes made to layer states. However, there may be occasions when you want to avoid undoing edits to layer states, such as when you change the Current VP Freeze settings for a particular paper space viewport. Using the LAYERPMODE command, you can suspend layer state tracking when it is not needed.

To set Layer Previous tracking on or off, at the command prompt, enter **LAYERPMODE**, then enter **ON** to turn on Layer Previous tracking, or enter **OFF** to turn off tracking.

SUMMARY

Managing objects through the new Layer Properties Manager improves your ability to organize your drawing. By using the Layer Properties Manager's features, you can control object properties such as color, linetype, and lineweight, and you can control object behavior such as object visibility.

In this chapter you learned about the importance of layer standards. You also learned about the difference between turning off layers and freezing them. You also worked with layer filters, saving and restoring layer settings, and translating layer settings between drawings.

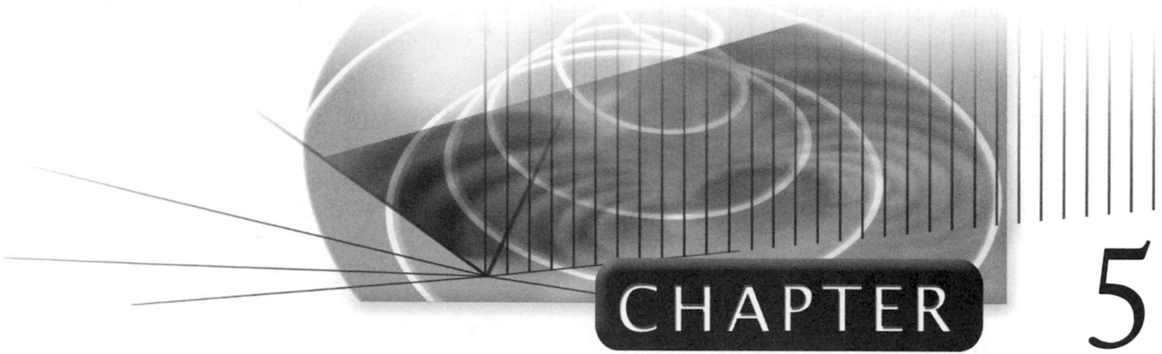

Applying Linetypes and Lineweights

Linetypes and lineweights provide a method for you to create objects that differentiate themselves from other objects. By applying different linetypes and lineweights to different objects in your drawing, you make objects distinguishable among themselves. By using linetypes and lineweights properly, you create a drawing that visually conveys its meaning to the viewer.

However, using linetypes and lineweights effectively entails more than just making a drawing look good. It requires an understanding of the features that AutoCAD provides for controlling the appearance of linetypes and lineweights. These features include setting defaults, controlling scale globally and individually, and customization. By learning about the range of features offered by AutoCAD, you can exploit the usefulness of linetypes and lineweights to their fullest. When you understand their features, you have the foundation needed to use linetypes and lineweights effectively.

This chapter covers the following subjects:

- Using default linetypes
- Scaling linetypes
- Creating and using custom linetypes
- Applying lineweights
- Controlling lineweight display

USING THE DEFAULT LINETYPES

For many years AutoCAD has provided linetypes, which allow you to create drawings whose objects are more easily discernible. A linetype is a series of dashes and/or dots that have a specified spacing that is then applied to the object. When you assign different linetypes, you make identical objects such as lines and polylines stand out as unique entities in your drawing. Linetypes are one of the most useful tools in AutoCAD for getting your drawing's meaning across to viewers.

In this section you will learn about several features of linetypes that range from the simple, such as assigning linetypes to objects, to the advanced, such as creating your own custom complex linetypes.

ASSIGNING LINETYPES

Assigning a linetype to an object is a straightforward process that can be accomplished by one of two methods. You can assign linetypes globally, through AutoCAD's Layer Properties Manager, or you can assign linetypes individually, from the Object Properties toolbar. Both methods are easy to use and accomplish the same thing: assigning a new linetype to an object.

However, while both methods accomplish the same thing, one method definitely has an advantage over the other. When you apply a linetype to a layer in the Layer Properties Manager dialog box, you control the appearance of all objects on that layer. This means you can instantly reapply a new linetype to hundreds of objects with a few simple clicks of your pointing device and change the linetype setting for a given layer.

In contrast, while assigning linetypes to objects individually does allow you to control their individual appearance, the process of changing the linetypes for dozens, or even hundreds, of objects individually can require tremendous amounts of editing time. Therefore, it is often best to avoid setting linetypes individually and instead use the Layer Properties Manager to control a layer's linetype assignment.

Tip: Always set object linetype creation mode to ByLayer using the Objects Properties toolbar. This ensures that all new object linetypes are controlled through the Layer Properties Manager, which provides a single point of control for modifying object linetypes.

For more information on controlling linetypes globally through the Layer Properties Manager, and individually through the Object Properties toolbar, refer to Chapter 4, Managing Layers, in the section titled The Linetype and Lineweight Properties.

LOADING LINETYPES

In Chapter 4, Managing Layers, in the section titled The Linetype and Lineweight Properties, you work though an exercise that shows how easy it is to assign linetypes to objects. In one of the exercise's steps, you assign a linetype globally using the

Layer Properties Manager. You select the desired linetype from the Select Linetype dialog box shown in Figure 5–1.

As you may notice, there are not a lot of linetypes to choose from in the Select Linetype dialog box. However, AutoCAD is installed with dozens of predefined linetypes that you can use. You simply need to load them into your drawing.

To load additional linetypes into the current drawing, go to the Objects Properties toolbar and choose the Layers button (the second button from the left). The Layer Properties Manager dialog box is displayed. From the Layer Properties Manager select any linetype under the Linetype column. Doing so displays the Select Linetype dialog box shown in Figure 5–1.

From the Select Linetype dialog box, choose the Load button. The Load or Reload Linetypes dialog box is displayed. Scroll down the list of available linetypes and choose one (or more) of the linetypes to load, as shown in Figure 5–2.

Choose OK when you are done selecting linetypes. The Select Linetype dialog box is display, and the newly loaded linetype(s) appear in the Loaded Linetypes list as shown in Figure 5–3.

By using the Load or Reload Linetypes dialog box, you can assign a wide range of linetypes to layers or to objects. However, assigning a linetype is only half the solution to displaying objects with linetypes. The next section discusses the other half of the solution, scaling linetypes.

SCALING LINETYPES

The ability to assign a linetype to an object is very useful for representing different things in your drawing. By assigning a linetype, you help viewers more clearly understand the meaning of different objects. However, as stated earlier, assigning a

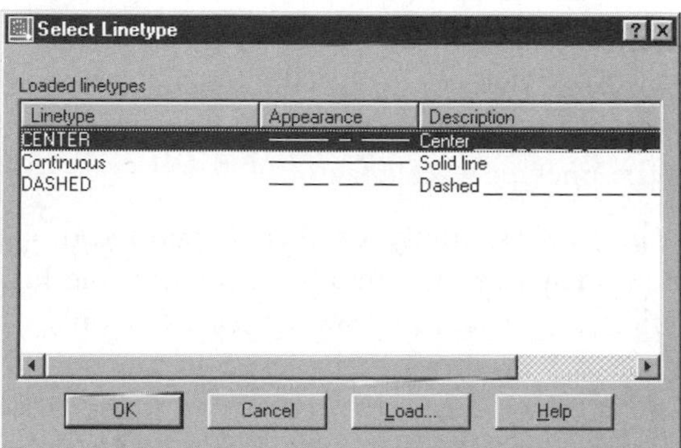

Figure 5–1 *The Select Linetype dialog box allows you to choose a layer's linetype.*

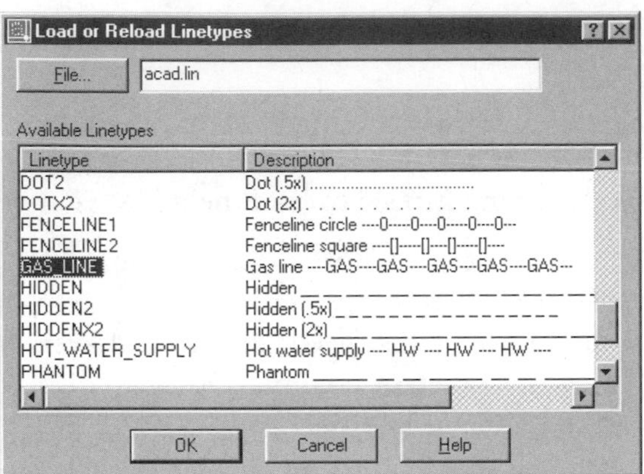

Figure 5–2 *You can choose from a variety of predefined linetypes in the Load or Reload Linetypes dialog box.*

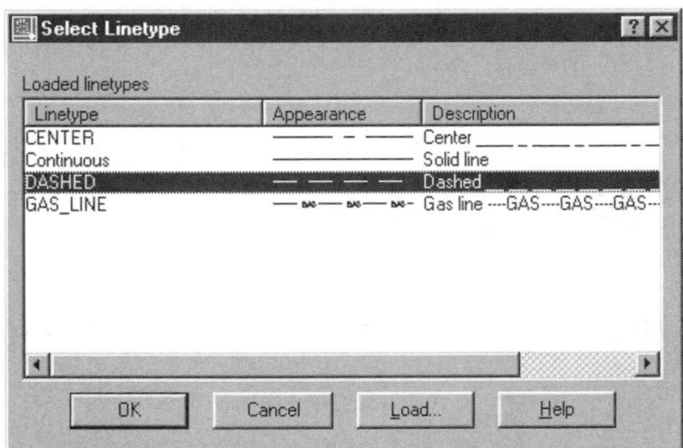

Figure 5–3 *The newly loaded GAS_LINE linetype appears in the Loaded Linetypes list and can now be assigned to a layer.*

linetype to an object, either globally or individually, is only half of the solution to making your drawings easier to understand with linetypes. The other half of the solution is assigning the linetype a scale factor.

For the most part, a linetype is simply a series of repeated dashes and spaces. The lengths of the dashes and spaces are initially defined by the linetype's description. However, the definition only specifies how many units long a dash or a space is. Depending on the scale of your drawing, the linetype may not be visible because the dashes and spaces appear too close together or too far apart. To compensate for this, AutoCAD allows you to assign a linetype scale factor to your drawing.

The linetype scale factor multiplies the current linetype style by the desired factor. Like linetypes, linetype scale factors can be assigned globally or individually. Unlike linetypes, however, you cannot assign a global linetype scale factor ByLayer using the Layer Properties Manager. When you set the linetype scale factor, it immediately affects all linetypes, no matter which layer they are on.

You change the linetype scale factor globally using the Linetype Manager dialog box, which is accessed by choosing Linetype from the Format menu. Choosing the Show Details button displays the linetype details shown in Figure 5–4.

To change the global linetype scale, enter a new value in the Global Scale Factor text box. Once you enter this value and choose OK, AutoCAD immediately assigns the global scale factor to the display of all linetypes.

To assign a linetype individually to all new objects, enter the desired linetype scale in the Current Object Scale text box. Do not be confused by the title of this particular text box. When you use this text box to define a linetype scale, you actually apply it to all new objects, not to the currently selected object.

While it may be useful to assign an individual linetype scale to new objects, it is also possible to edit the linetype scale of existing objects. To do so, choose the Properties button from the Standard toolbar, as shown in Figure 5–5. Next, select the object(s) whose linetype scale factor you wish to change. Finally, enter the new linetype scale factor to apply to the objects in the Object Properties Manager's Properties dialog box, as shown in Figure 5–6, and then press Enter. AutoCAD assigns the new line-

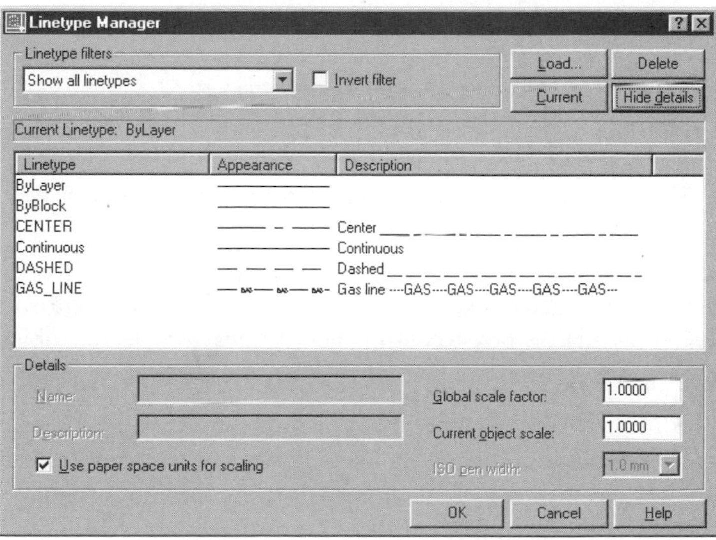

Figure 5–4 *The Linetype Manager dialog box allows you to change the linetype scale factor.*

Figure 5–5 *The Properties dialog box is accessed via the Properties button on the Standard toolbar.*

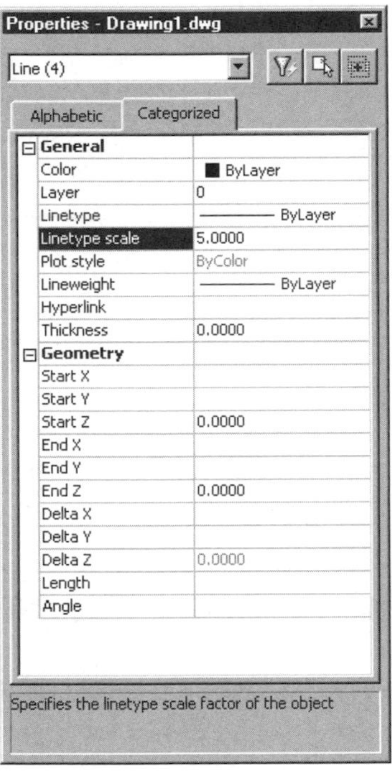

Figure 5–6 *The Properties dialog box allows you to change many values for selected objects, including their individual linetype scale factor.*

type scale factor individually to the selected objects. For more information about using the Properties dialog box to assign new values to selected objects, refer to Chapter 10, Advanced Editing.

The final tool for controlling the linetype scale factor controls how linetypes are displayed in layout viewports. This tool is necessary because AutoCAD allows you to create as many viewports in your Layout folder as you need to display your drawing. More importantly, you can apply a different zoom factor to each viewport. As a result, if one viewport is zoomed in close to an object, and another is zoomed out, the linetype will be displayed differently in each viewport. Figure 5–7 illustrates two viewports that show the same object, with each viewport zoomed at different scales.

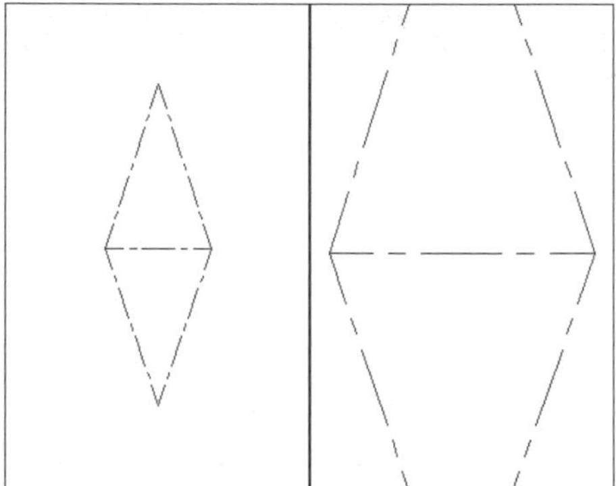

Figure 5–7 *The dashed lines of the same object appear different in each paper space viewport when the Use Paper Space Units for Scaling Feature is cleared.*

Notice that the object's dashed lines in the left viewport appear much smaller than the dashed lines in the viewport on the right. This is the effect when the Use Paper Space Units for Scaling feature is cleared (i.e, when there is no check in the box). With this feature cleared, the dashed lines are displayed proportionally to their zoom factor.

In contrast, review the same drawing with the Use Paper Space Units for Scaling feature selected, as shown in Figure 5–8. Notice that the lengths of the dashes and the spaces are the same in both viewports. This is the effect that the Use Paper Space Units for Scaling feature has on linetypes. Either method for scaling linetypes is acceptable. Whether the feature is selected or cleared depends on your needs.

There is one other property that you should be aware of when dealing with linetypes and polylines: linetype generation. Polylines are made up of a series of lines and arcs connected at their endpoints by vertices. When a linetype is assigned to a polyline, AutoCAD will either generate the linetype as one continuous line or generate the linetype anew at each vertex. In Figure 5–9 the polyline at the top has its linetype generated at each vertex, while the polyline at the bottom has its linetype generated as one continuous line. Once again, either method is acceptable, and the method you use depends on your needs.

CREATING AND USING CUSTOM LINETYPES

In a previous section in this chapter you learned how to load linetypes into the current drawing. When you load linetypes, you present yourself with a fairly large and diverse selection of linetypes to use in your drawing. By loading the desired linetypes into the current drawing, you can create a drawing that more easily conveys its meaning to the viewer.

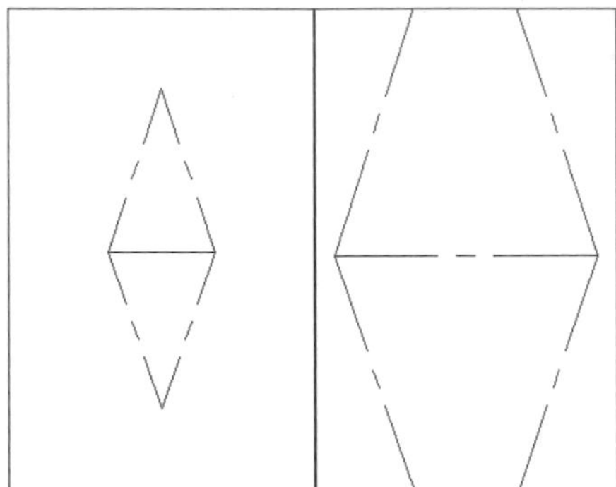

Figure 5–8 *The dashed lines of the same object appear identical in each paper space viewport when the Use Paper Space Units for Scaling feature is selected.*

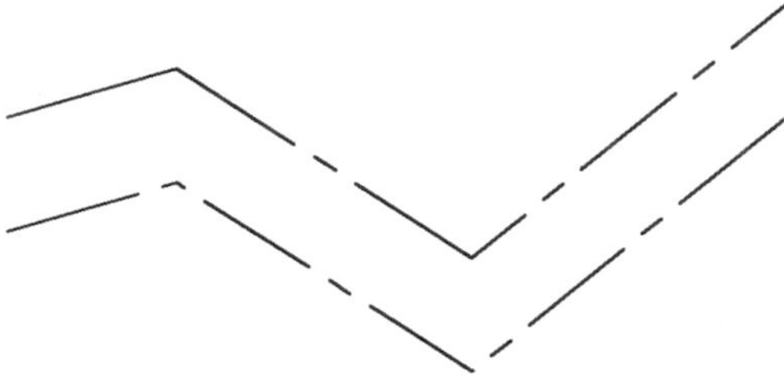

Figure 5–9 *The linetype generation property affects how linetypes are displayed when they are assigned to polylines. The property is disabled on the polyline at the top.*

Tip: The linetype generation property can be enabled or disabled using the Properties dialog box.

Unfortunately, even though AutoCAD is shipped with a fairly large number of ready-to-use linetypes, you will probably come across situations where the ready-made linetypes do not fit your needs. Whether your discipline is architecture, mechanical, electrical, or civil engineering, or surveying, your diverse needs will almost certainly exceed the variety of linetypes shipped with AutoCAD.

To compensate for the insufficient number and variety of linetypes provided, AutoCAD allows you to create your own linetypes. There is no limit to the number of linetypes you can create to meet your needs. No matter what your discipline, you can use AutoCAD to create the linetypes you need to use in your drawings.

This section discusses the two types of linetypes you can create in AutoCAD. Most of your needs can be met by creating the first type, which are called simple linetypes.

For situations where an annotated linetype is required, you can create the second type, complex linetypes. By creating your own simple and complex linetypes, you can produce a drawing that easily conveys its meaning to the viewer.

THE SIMPLE LINETYPE DEFINITION

As its name implies, a simple linetype is relatively simple to create. In fact, you can create a simple linetype during your current session of AutoCAD by starting the command line version for the Linetype command and then typing in just a few values. More importantly, you can instantly use the new linetype and apply it to objects in your current drawing.

Note: Both simple and complex linetypes are stored in ASCII text files that are appended with ".LIN". It is possible to create and edit linetypes directly in this file, as is discussed later in this chapter, in the section titled The Complex Linetype Definition.

In the following exercise you will learn how to create simple linetypes during the current AutoCAD session.

EXERCISE: CREATING SIMPLE LINETYPES

1. Open the drawing 05DWG01.DWG. The drawing displays a single horizontal line.
2. At the command line, type **-LINETYPE**, making sure you include the hyphen at the beginning. AutoCAD starts the command line version of the Linetype command and prompts you to create, load, or set a linetype.

Note: Only the command line version of the Linetype command provides the ability to create simple linetypes within AutoCAD. The Linetype Manager dialog box does not offer this capability.

3. At the command line, enter **C** to create a new linetype. AutoCAD prompts you to enter a name for the new linetype.
4. At the command line, enter **Short Dash** as the new linetype name. AutoCAD displays the Create or Append Linetype File dialog box.

   As mentioned previously, AutoCAD saves linetype definitions in ASCII text files appended with ".LIN". By default, AutoCAD lists its own standard linetype definition file, ACAD.LIN, as the file in which to save your new linetype definition. For this exercise you will create a new linetype definition file.

Tip: In practice it is best to leave AutoCAD's standard files, such as ACAD.LIN, in their original condition. Any AutoCAD customization you do should be stored in your own custom files.

5. In the Create or Append Linetype File dialog box, enter **MYLINES** in the File Name text box, as shown in Figure 5–10.

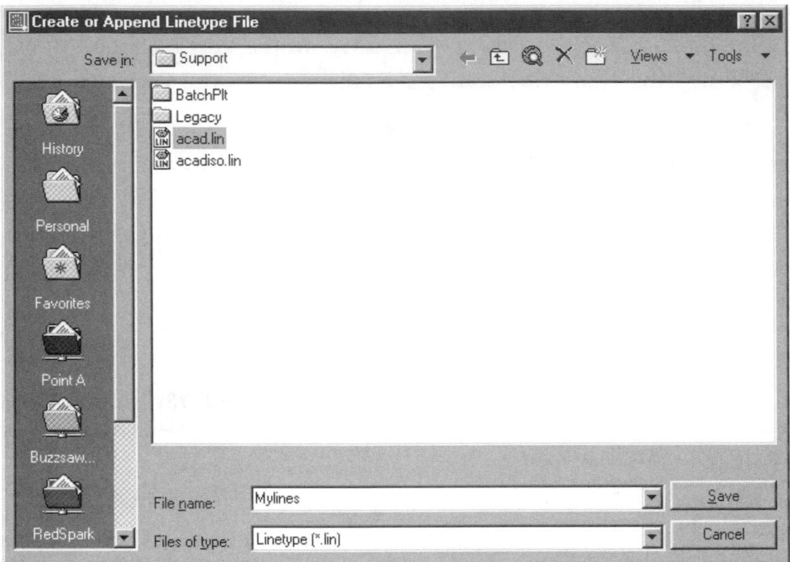

Figure 5–10 *The new linetype file is named MYLINES.*

6. Choose Save. AutoCAD creates the new linetype definition file, dismisses the dialog box, and prompts you to enter a descriptive name for the new linetype.
7. At the command prompt, type **This is a short dash** and then press Enter. AutoCAD prompts you for the linetype definition code and then begins the line of code for you.

   The linetype definition code always starts with the letter "A" followed by a comma. Because this is how all lines of code begin when you define linetypes, AutoCAD automatically specifies it for you. What AutoCAD expects you to enter now are the series of dashes and spaces you want to have to represent the linetype. All values are entered as real numbers, with positive values defining dash lengths, and negative numbers defining the length of spaces. A zero value represents a dot (a dash of zero length). A comma separates each number.
8. At the command prompt, type **0.25,-0.125** and then press Enter. AutoCAD creates the new linetype definition, adding it to the new MYLINES.LIN linetype definition file. Then it repeats the prompts to create, load, or set additional linetypes.
9. Press Enter to exit the Linetype command.

Tip: Upon completing a new linetype creation sequence, if you save the linetype to the ACAD.LIN file, AutoCAD will automatically load the linetype for immediate use.

The series of command line prompts and the appropriate responses are shown in Figure 5–11. Now that the new linetype is defined, the next step is to assign it to the line in the drawing and see what it looks like.

10. Choose the Layers button in the Object Properties toolbar. AutoCAD displays the Layer Properties Manager.
11. Choose Continuous in the Linetype column. AutoCAD displays the Select Linetype dialog box.
12. Choose the Load button. AutoCAD displays the Load or Reload Linetypes dialog box.
13. Choose the File button. AutoCAD displays the Select Linetype File dialog box.
14. Choose the MYLINES.LIN file, as shown in Figure 5–12, and then choose Open. AutoCAD displays the Short Dash linetype in the Load or Reload Linetypes dialog box, as shown in Figure 5–13.
15. Choose the Short Dash linetype and then choose OK. AutoCAD loads the new linetype into the current drawing and displays it in the Select Linetype dialog box, as shown in Figure 5–14.
16. Choose the Short Dash linetype and then choose OK. AutoCAD assigns the new linetype to layer 0.
17. Choose OK to dismiss the Layer Properties Manager dialog box. AutoCAD redraws the line with the new linetype, as shown in Figure 5–15. This drawing is used in the next exercise.

As you just learned, creating a simple linetype during an AutoCAD session truly is simple. Keep in mind that you can create as many simple linetypes as you need and, if you desire, you can save new linetype definitions in the MYLINES.LIN file. You

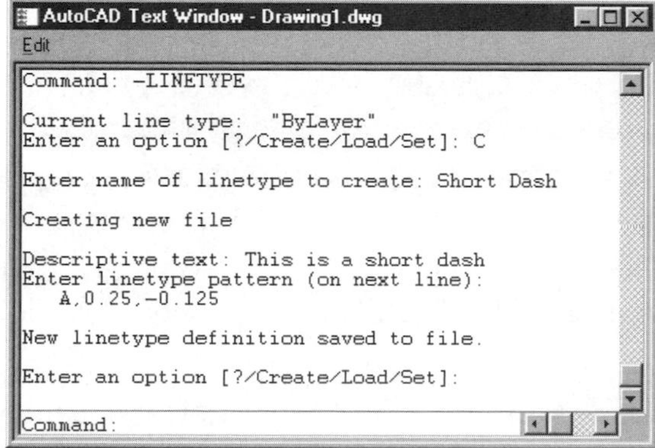

Figure 5–11 *The series of command line prompts and the appropriate responses for creating a new linetype.*

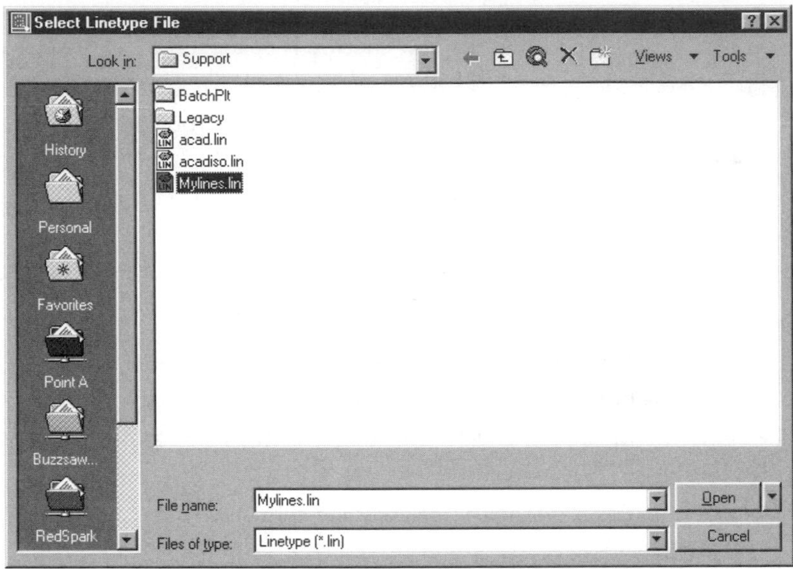

Figure 5–12 *Choose the MYLINES.LIN file to display the newly created linetype.*

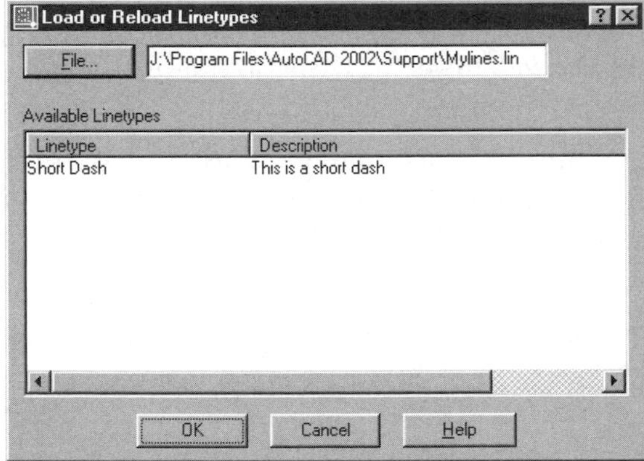

Figure 5–13 *The newly created Short Dash linetype appears.*

can also create new linetype definition files to help you organize your linetypes. For example, you can create a UTILITY.LIN linetype definition file to store all utility-related linetypes.

Now that you know how to create simple linetypes, it is time to move on to the next level. Next you will learn how to create complex linetypes.

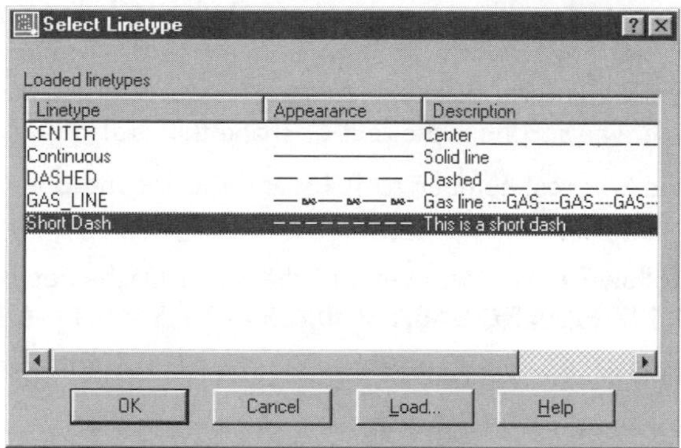

Figure 5–14 *The newly created Short Dash linetype is loaded into the current drawing.*

Figure 5–15 *The newly created Short Dash linetype is assigned to the horizontal line.*

THE COMPLEX LINETYPE DEFINITION

As its name implies, creating a complex linetype is a little bit more difficult than creating simple linetypes, but do not let its name intimidate you. The fact is that creating complex linetypes is almost as easy as creating simple linetypes. The main difference is that you cannot create complex linetypes within AutoCAD using the –Linetype command. Instead, you create them by entering linetype parameter values directly into a linetype definition file.

In the next exercise you will learn how to create complex linetypes using an ASCII text editor.

EXERCISE: CREATING COMPLEX LINETYPES

1. Continue using the drawing from the previous exercise, or open the drawing 05DWG02.DWG. The first step in creating a complex linetype is to open the MYLINES.LIN file in an ASCII text editor.
2. At the command line, type **Notepad** to open this text editor.
3. Type **MYLINES.LIN** at the request for the file to edit, and then choose Enter. Notepad opens the linetype definition file.

Note: If Notepad cannot find the MYLINES.LIN file, use Notepad to browse to the ACAD2000/SUPPORT folder in which the MYLINES.LIN file is located, then open the file.

In this exercise you will modify the Short Dash simple linetype you already created. Complex linetypes contain two main elements. The first element is the same definition as for the simple linetype; it describes the dashes and spaces in the linetype. The second is the text or shape that is also displayed in the linetype.

4. Change the first real number to **1.25**, and the second one to **–0.5**. This increases the length of the dashes and the length (or gap) of the spaces.
5. Add the following text to the end of the Short Dash linetype values: **,["OIL",STANDARD,S=0.2,R=0.0,X=-0.1,Y=-0.1],-0.5.**

   The Short Dash linetype values should now read as follows:

   A,1.25,-0.5,["OIL",STANDARD,S=0.2,R=0.0,X=-0.1,Y=-0.1],-0.5

   Recall that the first two real numbers define the length of the dash and the length of the space, respectively. Next is the code within the brackets, which defines the various parameters of the text to display within the complex linetype. After the closing bracket, another space is added; in this case, its value is set to –0.5. Notice that commas separate all values. For a detailed description of the values within the brackets, see Table 5-1.
6. In Notepad, choose File>Save. Notepad saves the modified linetype definition file.
7. In AutoCAD, choose the Layers button in the Object Properties toolbar. AutoCAD displays the Layer Properties Manager.
8. Choose the Short Dash linetype in the Linetype column. AutoCAD displays the Select Linetype dialog box.
9. Choose the Load button. AutoCAD displays the Load or Reload Linetypes dialog box.
10. Choose the File button. AutoCAD displays the Select Linetype File dialog box.
11. Find and choose the MYLINES.LIN file, then choose Open. AutoCAD displays the Load or Reload Linetype dialog box.
12. Choose Short Dash and then choose OK. AutoCAD displays the Reload Linetype dialog box, asking if you want to reload the linetype.

    Because the previous exercise used this same linetype, you will have to reload the linetype in order to use it.
13. Choose Yes. AutoCAD reloads the linetype into the current drawing.
14. Choose the Short Dash linetype, then choose OK. AutoCAD assigns the modified linetype to layer 0.
15. Choose OK to dismiss the Layer Properties Manager dialog box, type **REGEN**, and then press Enter. AutoCAD redraws the line with the new complex linetype, as shown in Figure 5–16.
16. You may close the drawing without saving your changes.

Figure 5–16 *The newly modified Short Dash linetype is assigned to the horizontal line.*

Table 5–1: *Complex Linetype Text Values*

Value	Description
OIL	The first value is the text string that is displayed in the linetype (in this case, OIL). Notice that the text value is enclosed in quotes.
STANDARD	The next value is the text style. You may enter any text style. Just be sure that the one you use is loaded in the current drawing. For this exercise, AutoCAD's default STANDARD text style is used.
S=0.2	The next value is the text's scale. This value is multiplied by the selected text's height value. Because the height value of the STANDARD text style is set to zero, AutoCAD interprets the S value literally as the height (in this case, 0.2).
R=0.0	Next is the rotation value, the amount that the text is rotated relative to the line (in this case, zero). When you set the value to zero, as the line changes direction, including through curves, the text aligns itself parallel to the line. If the value is set to any other angle, the text is rotated at each point along the line where it occurs by the specified angle.While R= indicates rotation relative to the line, you can define a rotation value with A=, which signifies absolute rotation of the text with respect to the origin. In other words, all text occurrences along the line point in the same direction, regardless of their position relative to the line. Additionally, the value can be appended with a "d" for degrees (if omitted, degree is the default), "r" for radians, or "g" for gradients. If rotation is omitted, no relative rotation is used.
X=-0.1,Y=-0.1	The X and Y values represent the offset of the text relative to the line. Typically you will set both these values as negative real numbers, and at half the complex text's scale.

In this exercise you actually accomplished two things. First, you edited an existing simple linetype. Second, you created a complex linetype. When you create both simple and complex linetypes, a little trial and error is usually necessary to achieve the desired results. Notice that once you saved the edited ASCII text file, you left Notepad running with the linetype definition file still open. Even so, you were able to reload the Short Dash linetype into AutoCAD to view it. The ability to leave a file open in Notepad while loading it into AutoCAD to view the results is very useful.

Using Shape Files in Complex Linetype Definitions

In the previous exercise you learned how to create a complex linetype with text. You can also create complex linetypes with shape files. A shape file contains code that defines shapes that you can use over and over. For example, many of AutoCAD's text fonts are actually shape files. (They are the files that end with ".SHX".) In addition to the text fonts, AutoCAD is also shipped with a shape file called LTYPESHP.SHP, which you can review and edit in an ASCII text editor. You can also create your own shape files and use them to store your custom shape definitions.

Note: Shapes files are saved as ASCII text files with a ".SHP" ending, and they can be viewed and edited in an ASCII text editor. However, so that AutoCAD uses the shape definitions in the shape file, you must compile the shape file. Compiled shape files end with ".SHX" and cannot be read by an ASCII text editor.

You define a complex linetype that uses shape files in much the same way as you did using text in the last exercise. You begin the definition with any dash and space code you may desire, and then you add the complex linetype code. The code for the shape is almost exactly the same as the code for text.

The following text string defines a complex linetype using a shape named CIRC1 from the ltypeshp.shx file:

```
A,1.25,-0.5,[CIRC1,LTYPESHP.SHX,S=0.2,R=0.0,X=-0.2,Y=0.0],-0.5
```

Notice that this complex linetype definition is almost identical to the one in the last exercise. The major differences are that the text name and the text style have been replaced. Table 5-2 defines the two major differences between the exercises' linetype definitions.

The complex linetype definition creates a linetype as shown in Figure 5–17.

In the first portion of this chapter you learned how to work with linetypes. Next you learn about lineweights.

Table 5–2: *Complex Linetype Shape Values*

Value	Description
CIRC1	The first value that is displayed is the shape definition's name (in this case, CIRC1). The shape definition resides in a shape definition file. A single shape definition file can contain many shape definitions. Notice that in the case of shapes, their names are not enclosed in quotes.
LTYPESHP.SHX	The next value is the compiled shape definition file name. The file must be in AutoCAD's search path. A good location to save this file is in AutoCAD's Support folder.

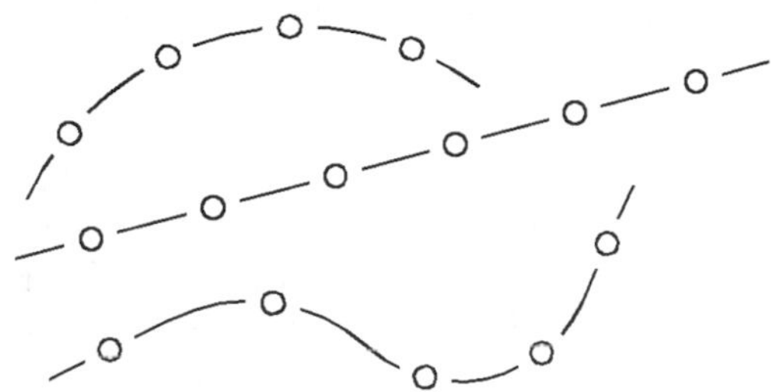

Figure 5–17 *The CIRC1 shape definition is used to create a new complex linetype.*

Tip: Shape files involve the creation of complex, cryptic code, which is beyond the scope of this book. You can read about creating shape files in Chapter 3, Shapes, Fonts, and PostScript Support, of AutoCAD's Customization Manual, in the section titled Creating Shape Definition Files.

APPLYING LINEWEIGHTS

Lineweights allow you to affect the appearance of objects. Just as linetypes make similar objects, such as lines and polylines, stand apart from each other, so, too, lineweights can make individual objects more easily identifiable.

In releases prior to AutoCAD 2000 you could assign widths to polylines, which performed a function similar to that of lineweights. The problem was that you could only assign widths to polylines—not to lines, not to circles, and certainly not to text. Another problem with assigning width to polylines was its unfriendliness. If a polyline was already created with the wrong width, you had to edit it to change the width. More importantly, unless you had access to specialized AutoLISP routines, you could only edit the width of polylines one at a time. In contrast, lineweights allow you to assign widths to a wide range of objects, including text, and assign the width individually, globally, and even ByLayer using AutoCAD's Layer Properties Manager.

In this section you will learn about assigning lineweights to objects. More importantly, you will learn how the appearance of lineweights is affected under different circumstances.

ASSIGNING LINEWEIGHTS

Assigning lineweights is much the same as assigning linetypes: you can do it either globally through the Layer Properties Manager, or individually through the Object Properties toolbar. Just as it was suggested in the argument presented in the Assigning Linetypes section, it is strongly suggested here that you avoid assigning lineweights individually, because doing so makes editing the lineweights of many objects a daunting, unproductive task.

If you are new to AutoCAD, you will probably find yourself tempted to pick a single object and reassign its lineweight individually. After all, it is a very easy thing to do. You just pick the object, choose the desired lineweight from the pull-down menu on the Object Properties toolbar, and you are done. However, chances are you will regret setting the lineweight individually as your drawing grows more and more complex, containing hundreds, or thousands, of objects. It is much better in the long run always to assign a lineweight globally by using the Layer Properties Manager, even if it means creating a new layer just for that one object and assigning that layer the desired lineweight. By doing so, you create an understanding not only with yourself but with anyone else who may work on your drawing that lineweights are always edited using the Layer Properties Manager. This understanding provides a consistent pattern for everyone to use when they edit objects.

For more information on controlling lineweights globally through the Layer Properties Manager, and individually through the Object Properties toolbar, refer to Chapter 4, Managing Layers, in the section titled The Linetype and Lineweight Properties.

CONTROLLING LINEWEIGHT DISPLAY

Lineweights are displayed differently under different circumstances. For example, when you work in model space, lineweights are displayed by a certain number of pixels. Consequently, as you zoom in closer to a line, the number of pixels displaying the lineweight does not change. If a lineweight in model space is displayed as four pixels, it is always displayed as four pixels, no matter how far out you zoom or how close in you zoom. Therefore, the lineweight always appears to be the same width. In contrast, when you work in paper space (now referred to as a layout), lineweights are displayed at their true width. If a lineweight of 0.25 mm is assigned, then AutoCAD displays the line as 0.25 mm wide in the layout. Therefore, as you zoom in closer, the line appears wider. In other words, the lineweights of objects drawn in a layout are displayed in real-world units. There is another feature of lineweights that you can control. When you are in model space, if you assign lineweights to objects, you can alter their apparent scale so they appear thinner or wider. This apparent scale does not affect lineweight widths when objects are viewed in a layout or when they are plotted. Therefore, you can dynamically alter lineweights in model space to make viewing objects easier without adversely affecting how they are plotted.

In the next exercise you will learn how lineweights act in model space and in a layout, and how to alter their apparent scale in model space.

EXERCISE: UNDERSTANDING LINEWEIGHT BEHAVIOR

1. Open the drawing 05DWG03.DWG. The drawing opens a layout, which displays objects drawn in both model space and paper space, as shown in Figure 5–18. The circle, triangle, line, and the text "Model Space" are drawn in model space. The text "Paper Space" is drawn in paper space in the Layout folder. The solid rectangle is the edge of the floating viewport, and the dashed rectangle represents the plotting limits.
2. From the View menu, choose Zoom>Window, and then pick a zoom window that surrounds the two text objects, as shown in Figure 5–19. Notice that the lineweights of both text strings appear equal in width. More importantly, they also appear wider than they did before. This is because you are viewing the layout in paper space mode. Therefore, AutoCAD displays the lineweights at their real-world size. As you zoom in closer, the lines appear wider.
3. From the View menu, choose Zoom>Previous. The view appears again as shown in Figure 5–18.

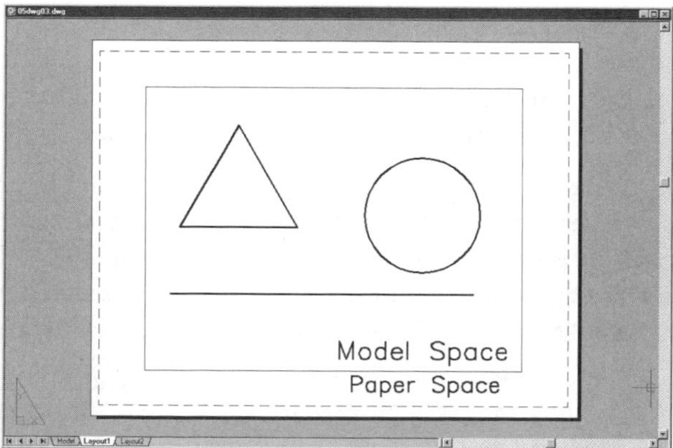

Figure 5–18 *The drawing in a layout view has several objects drawn in model space, including the text "Model Space." The text "Paper Space" is drawn in paper space.*

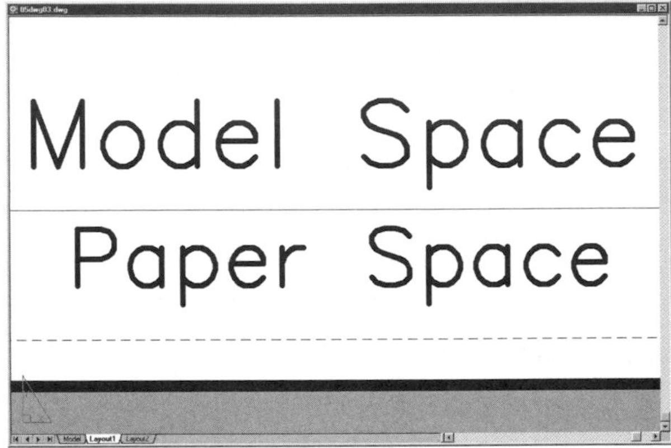

Figure 5–19 *In paper space, the lineweights of the two text objects appear wider as you zoom in closer.*

4. Choose the Model tab (just to the left of the Layout1 tab toward the bottom-left of the screen). AutoCAD switches to model space, and the model space objects are displayed as shown in Figure 5–20. (While the objects in the figure are displayed with a heavy lineweight, your drawing may look different.)
5. From the View menu, choose Zoom>Extents. AutoCAD zooms in closer to the objects, a shown in Figure 5–21. Notice that the lineweights of the objects did not get wider as you zoomed in. This is because in model space, the number of pixels that are used to display a lineweight cannot change as you zoom in closer.
6. From the Format menu, choose Lineweight. The Lineweight Settings dialog box appears, as shown in Figure 5–22.
7. Pick and drag the Adjust Display Scale button along its slide bar all the way to the right, and then choose OK. The lineweights will be updated so that they will appear wider on the screen, as shown in Figure 5–23.
8. Choose the Layout1 tab. AutoCAD switches to the Layout1 folder and displays objects in paper space, as shown in Figure 5–24. Notice that the lineweights are not as wide as they were in model space. Once again, this occurs because objects in the layout are displayed at their real-world scale. AutoCAD therefore ignores the Adjust Display Scale setting while in paper space and when plotting.

One last feature you should be aware of is how to control whether or not lineweights are displayed on the screen. You can turn off lineweight display by choosing the LWT button at the bottom of the screen in the AutoCAD status bar. By choosing this button, you toggle lineweights off and on. However, this button does not affect how lineweights are plotted.

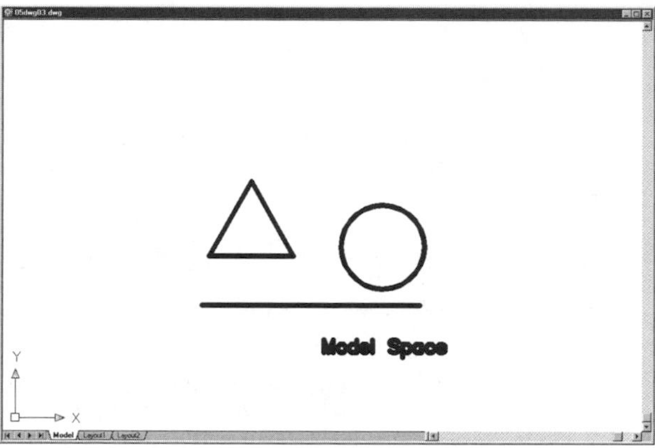

Figure 5–20 *The model space view of the objects.*

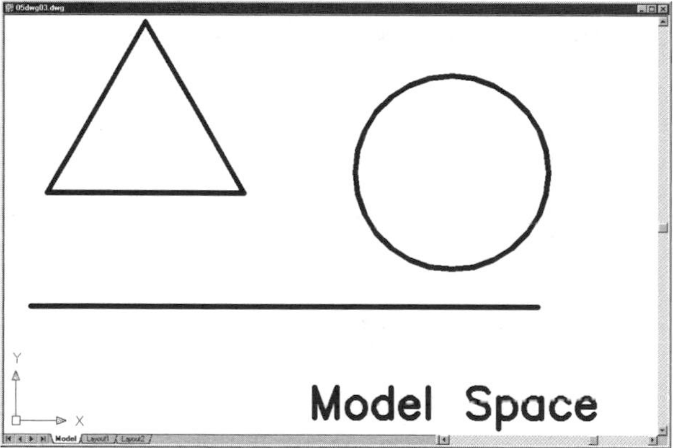

Figure 5–21 *The lineweights of the model space objects do not appear wider as you zoom in closer.*

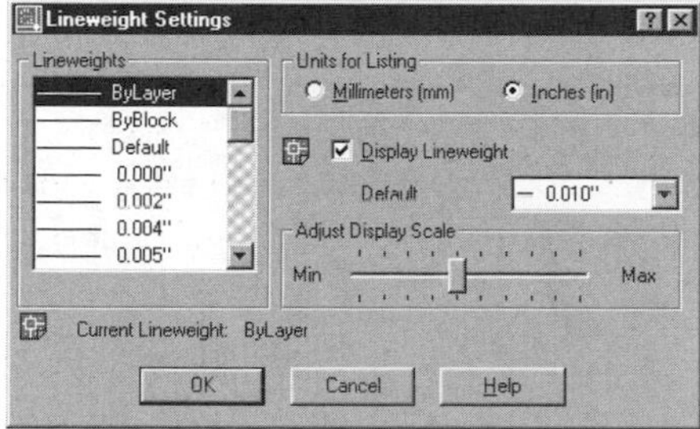

Figure 5–22 *The Lineweight Settings dialog box allows you to control various features of lineweights.*

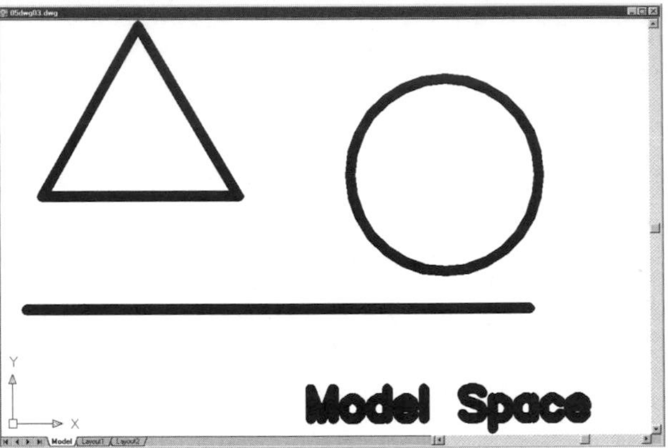

Figure 5–23 *The Adjust Display Scale feature affects how wide lineweights appear on the screen in model space.*

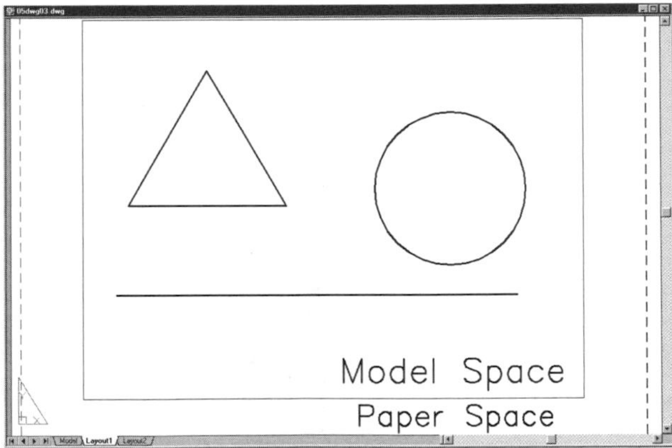

Figure 5–24 *The Adjust Display Scale feature does not affect how wide lineweights appear on the screen in paper space layouts.*

SUMMARY

In this chapter you learned how to work with linetypes and lineweights effectively. You reviewed assigning linetypes to objects, and you worked through an exercise in how to load linetypes. You reviewed how to control the linetype scale factor, both globally and individually. You also worked with custom linetypes, creating both simple and complex linetypes.

In the section discussing lineweights, you learned how to control the width of lines AutoCAD uses to draw objects. You learned that you can assign lineweights ByLayer or individually. You also learned that lineweights behave differently in model space and paper space. You worked on controlling how lineweights appear on the screen in model space, by adjusting their apparent scale without affecting their width in paper space or when they are plotted.

Linetypes and lineweights provide tools that allow you to differentiate one object from another easily, even when the objects are of similar types, such as lines and polylines. By using linetypes and lineweights effectively, you make your drawing easier to understand and convey the meaning of your drawing to the viewer.

Creating, Editing, and Manipulating Objects

Precision Drawing

To create accurate drawings with AutoCAD, you must understand how to specify and enter coordinates, and understand the points of which the drawings are composed. This, in turn, requires knowledge of AutoCAD's basic coordinate display systems, the World Coordinate System (WCS) and the User Coordinate System (UCS). AutoCAD is an extremely accurate design and drafting package with the capability of sixteen decimal places of precision stored in its database. To achieve this level of accuracy, AutoCAD supports several drawing aids that enable you to draw, place, and edit objects in your drawings. This chapter discusses AutoCAD's coordinate systems and the methods that you can use to make drawing with accuracy and precision easier.

This chapter covers the following topics:

- Understanding coordinate systems
- Coordinate point entry methods
- Manipulating coordinate systems
- Setting up drawing aids
- Object snaps
- Xlines and rays

UNDERSTANDING COORDINATE SYSTEMS

No matter what kind of drawing you create in AutoCAD, you need a systematic method of specifying points. Points define the beginnings and endpoints of lines, the center of circles and arcs, the axis points of an ellipse, and so on. The capability to place points accurately is important. When an AutoCAD command prompts you for a point, you can either specify a point on the screen with the mouse or pointing device or enter coordinates at the command line. AutoCAD uses a 3D Cartesian, or rectangular, coordinate system for entering points. Using this standard system you can locate a point in 3D space by specifying its distance and direction from an established origin measured along three mutually perpendicular axes: the X-, Y-, and Z-axes. The origin is considered to be at 0,0,0. Figure 6–1 illustrates such a coordinate system. Only two dimensions are depicted, with the Z-axis projecting up, perpendicular to the page. If you are concerned with 2D drawings only, this is the presentation of AutoCAD's coordinate system that you will see.

In Figure 6–1 the 4,6 coordinates indicate a point 4 units in the positive X direction and 6 units in the positive Y direction. Points to the left or below the origin have negative X and Y coordinate components, respectively. Figure 6–2 illustrates the same coordinate system, only now the third dimension and the Z-axis are shown. To specify 3D points, you add a third element to the coordinate designation. The point 4,6,6 in Figure 6–2 is located 4 units in the positive X direction, 6 units in the positive Y direction, and 6 units in the positive Z direction. The system of reckoning coordinates is independent of the units used so that distances can be in any measure-

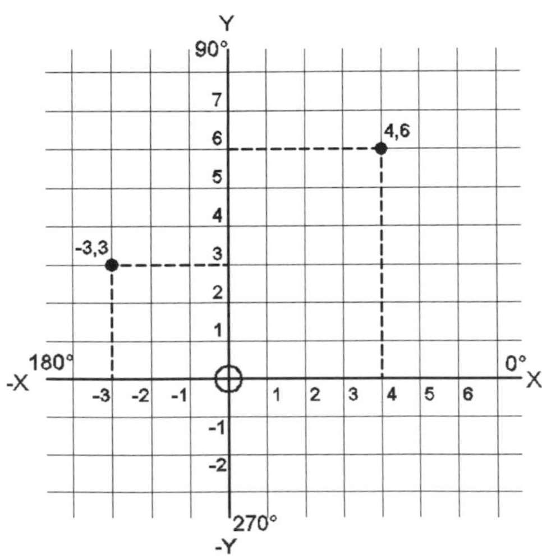

Figure 6–1 *The X-axis and Y-axis in a 2D coordinate system.*

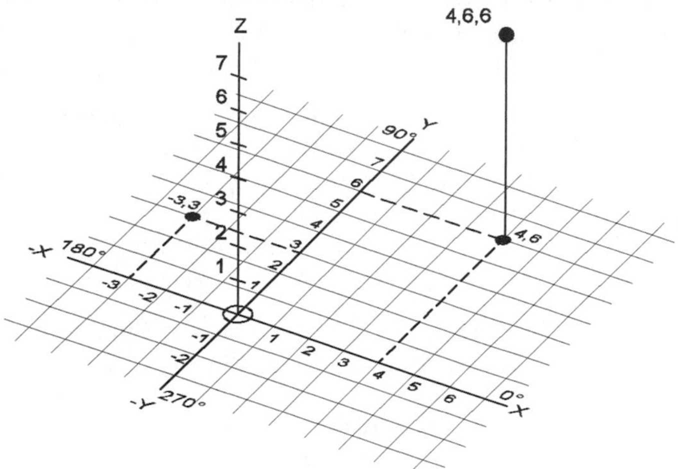

Figure 6–2 *The 3D rectangular coordinate system.*

ment. For example, the X direction could be in English feet or inches or in metric centimeters or kilometers.

Later in this chapter you will learn about the various ways that you can change the origin as well as the orientation of the three axes of AutoCAD's rectangular coordinate system. No matter how the coordinate system is oriented, you must know how to enter points.

COORDINATE POINT ENTRY METHODS

Many of the drawings you make in AutoCAD—regardless of their eventual complexity—consist of a few relatively basic AutoCAD objects such as lines, circles, or text elements. These objects require that you enter points that specify their location, size, and direction. Additionally, many editing operations also require that you specify points. There are four ways to enter points or coordinates in AutoCAD:

- Using absolute coordinates
- Using relative coordinates
- Using direct distance entry
- Using coordinate display

USING ABSOLUTE COORDINATES

Absolute rectangular coordinates are always measured from the origin point, 0,0,0. In AutoCAD you specify an absolute coordinate from the keyboard by typing in the X-, Y-, and Z-axis values separated by a comma—X,Y for 2D points, or X,Y,Z for 3D points.

You do not need to use a plus sign (+) if the displacement from the origin is positive. You do, however, need to place a minus sign (-) in front of displacements in the negative direction: -2,3 or 4,-6,3.

Absolute polar coordinates used for 2D coordinate entry are also treated as a displacement from the origin, or 0,0, but you specify the displacement as a distance and an angle. The distance and angle are separated by a left-angle bracket (<) with no spaces, as follows:

distance<angle (for example, 25<135)

Positive angles are measured counterclockwise from an assumed 0-degree angle that lies, by default, along the positive X-axis, as shown in Figure 6–3.

Tip: When entering the angle portion of polar coordinates, you can specify the angle as either positive (counterclockwise) or negative (clockwise). Thus 37<90 is equivalent to 37<-270. This applies to both absolute and relative coordinate entry.

Note: You can reverse the positive and negative directions of angles so that positive values are measured clockwise and negative values are measured counterclockwise. This is accomplished by selecting the Clockwise option from the Drawing Units dialog box. Additionally, you can control the direction of the 0-degree base line from which all angles are measured. For more information, refer to Chapter 2, Beginning a Drawing, in the section titled Controlling Drawing Units.

Absolute coordinates are adequate for creating simple objects. Although it is easy to designate the beginning of a line, measuring subsequent points in relation to a single origin point is cumbersome and often inaccurate. When lines are not orthogonal, absolute coordinates are particularly inadequate if any degree of accuracy and efficiency is desired. The use of relative coordinates solves this problem.

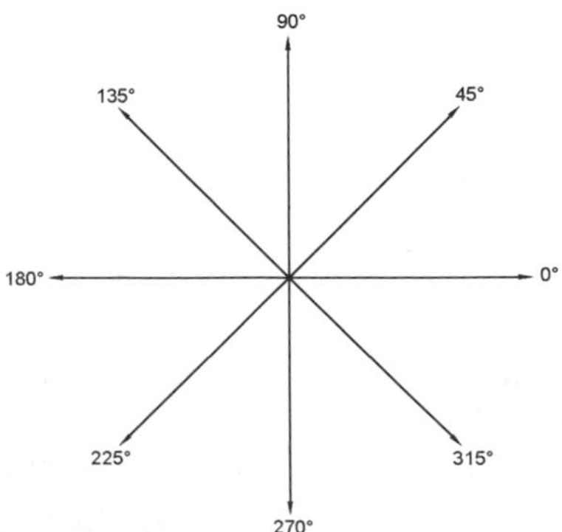

Figure 6–3 *Default angle directions.*

USING RELATIVE COORDINATES

In almost any kind of drawing, once you have established the beginning of a line, you usually know the X and Y displacement or the distance and angle to the next point. Relative coordinates do not reference the origin point but are determined relative to the last point. You can use this more straightforward method with either relative rectangular or relative polar coordinates. To distinguish relative coordinate entry from absolute entry, you precede relative coordinates with the "@" symbol. For example, you could enter @1.5,3 for relative rectangular entry or @2.6<45 for relative polar entry.

In the relative rectangular entry example, the point specified lies at a displacement of 1.5 units in the X-axis direction and 3 units in the Y-axis direction from the previous point. In the relative polar entry, the point lies at a distance of 2.6 units and at an angle of 45 degrees from the previous point.

Note: The angles involved and the availability of distance information will usually determine whether it is easier to enter the next point using relative polar or relative rectangular coordinates.

As you can see, relative coordinate point entry is much easier to use and permits more accuracy. Even when your drawing involves purely orthogonal displacements, relative coordinate entry is the superior, and usually the only accurate, method.

DIRECT DISTANCE ENTRY

A variation of relative coordinate entry, called *direct distance entry*, is supported in AutoCAD. In direct distance entry, instead of entering coordinate values, you can specify a point by moving the cursor to indicate a direction and then entering the distance from the first point. This is a good way to quickly specify a line length. This method is used primarily when the displacements involved are orthogonal and you therefore can have the ORTHO drawing aid turned on.

Direct distance entry provides a more direct and easier method of entering relative coordinates when the point lies in an orthogonal relationship to the previous point—a common situation in most drawings. Of course, if the point you want to designate lies on a snap point, whether orthogonal to the previous point or not, you can bypass keyboard entry by simply snapping the cursor to and clicking on the point. (The concept of "snapping" is covered later in this chapter.)

Note: AutoCAD 2000 introduced a new feature called AutoTracking, which allows you to simulate the effect of ORTHO mode—constricting the cursor's angular movement—but allows you to apply it in increments other than ORTHO mode's 90-degree increments. AutoTracking is discussed later in this chapter.

COORDINATE DISPLAY

The coordinate display window located at the bottom-left end of the status bar is useful when you enter coordinates, whether you type them at the command prompt or pick points on the screen with the screen cursor. Figure 6–4 shows this coordinate display in two formats.

Figure 6–4 *The coordinate display window shows coordinates in the current drawing units.*

The upper display shows decimal units, while the lower display is in architectural units. There are three types of coordinate displays, all of which can display either absolute or relative coordinates, depending on the type selected and the command in progress. You can cycle through the various display types in four ways: by pressing either F6 or Ctrl+D, or by clicking or right-clicking in the display area itself. The three coordinate display types are as follows:

- *Static display* displays the absolute coordinates of the last picked point. The display is updated whenever a new point is picked.
- *Dynamic display* displays the absolute coordinates of the screen cursor and is updated continuously as the cursor is moved. This is the default mode.
- *Distance and angle* displays the distance and angle relative to the last point whenever a command prompt requesting either a distance or an angle is active.

When static display is selected, the coordinate display appears grayed out, although the coordinates of the last selected point are still visible. At an "empty" command prompt (no command in progress) or at an active prompt that does not accept either a distance or an angle as input, you can toggle only between static display and dynamic display. At a prompt that does accept or require either a distance or angle rmat as you move the cursor.

Tip: If you have the static display active and click a grip to make it hot, the system will display the coordinate of the grip location. This is an alternative to using the ID command.

MANIPULATING COORDINATE SYSTEMS

The beginning of this chapter looked at AutoCAD's rectangular or Cartesian coordinates system from the standpoint of entering coordinates representing points in your drawing (see Figure 6–1). You learned about absolute and relative coordinate entry in both rectangular and polar formats. By using the Cartesian coordinates system and entering either absolute or relative coordinates, you can create highly accurate, detailed drawings of anything.

When you begin a new AutoCAD drawing, you are by default using a rectangular coordinate system called the World Coordinate System, or WCS. In addition to the WCS, you can create other coordinate systems called User Coordinate Systems, or UCSs. They are called UCSs because you, the user, define them to aid you in creating your drawings.

The reason AutoCAD provides the ability to create your own UCS is because a UCS often makes defining points for your model easier. For example, suppose you are creating a model of a 3D pyramid and you need to define points along the surface of one of the pyramid's faces. This is easily accomplished by aligning a UCS with the face. When you do this, you define an XY plane coincident with the face. Once the UCS is properly aligned, it is just a matter of using the Cartesian coordinate system to create the points along the pyramid's face, as shown in Figure 6–5. Chapter 27, 3D Fundamentals, presents additional UCS concepts for use when creating 3D geometry.

In this section you will learn about various UCS commands and features that you can use to make creating detailed drawings of complex models easier.

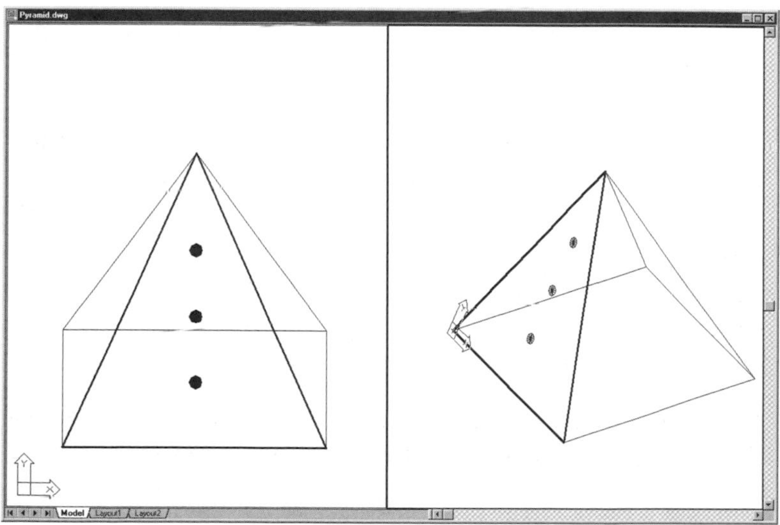

Figure 6–5 *When you properly align a User Coordinate System (UCS), developing complex 2D and 3D models becomes easier.*

WORLD COORDINATE SYSTEM

The World Coordinate System (WCS) is nothing more than a standard rectangular coordinate system with the origin in the lower-left corner of the screen, a horizontal X-axis running left to right, and a Y-axis extending vertically from the bottom to the top of the screen. The Z-axis is perpendicular to both the X- and Y-axes and is considered to extend out toward you in a direction perpendicular to the screen. To identify the WCS and establish its orientation, AutoCAD, by default, places the WCS icon at or near the origin. The WCS icon is shown in Figure 6–6. Its defining characteristic is the "W" appearing on the icon; this tells you that you are in the World Coordinate System.

USER COORDINATE SYSTEMS

You can create your own coordinate systems called *User Coordinate Systems*, or UCSs. In a UCS, the origin as well as the direction of the X-, Y-, and Z-axes can be made to move, rotate, and even align with drawing objects. Even though the three axes in a UCS remain mutually perpendicular, as they are in the WCS, a great deal of flexibility can be achieved in placing and orienting your UCS. The UCS command enables you to place a UCS origin anywhere in 3D space so that you can work relative to any point you want. You can also rotate the X-, Y-, and Z-axes in 2D or 3D space. UCSs are indispensable for working in 3D space.

The following exercise demonstrates how to create a UCS by aligning the UCS with two 2D points.

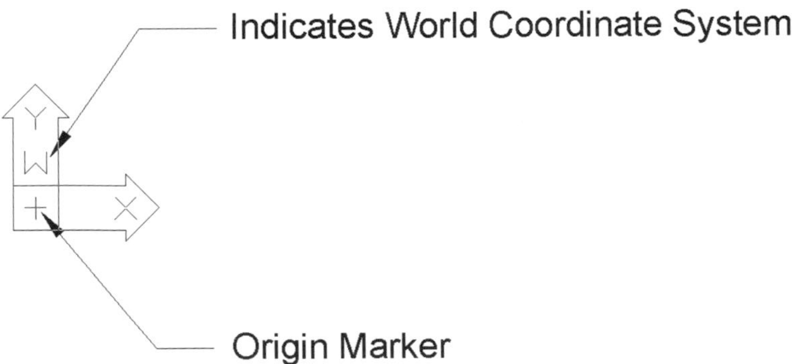

Figure 6–6 *The World Coordinate System (WCS) icon.*

EXERCISE: ALIGNING A UCS WITH A 2D OBJECT

1. Start a new drawing, using 06DWG01.DWT from the accompanying CD as a template. (See Chapter 2, Beginning a Drawing, for information on using template drawings.)
2. Your drawing should resemble Figure 6–7. Note that the "W" in the UCS icon indicates that the World Coordinate System is current.
3. Choose Tools>New UCS>3 Point. The following prompt appears:

   Specify new origin point <0,0,0>:

4. Shift+right-click to display the cursor menu, and then choose Endpoint. Then pick 1 shown in Figure 6–7. The following prompt appears. (The points given as a default may differ in your drawing.)

   Specify point on positive portion of the X-axis <6.58,2.04,0.00>:

5. Shift+right-click to display the cursor menu, choose Endpoint, and then pick 2. The following prompt appears:

   Specify point on positive-Y portion of the UCS XY plane <5.19,2.96,0.00>:

6. Pick anywhere near 3. Note that the UCS icon changes its orientation to align with the new UCS, and the "W" disappears, indicating that you are no longer in the WCS (see Figure 6–8).

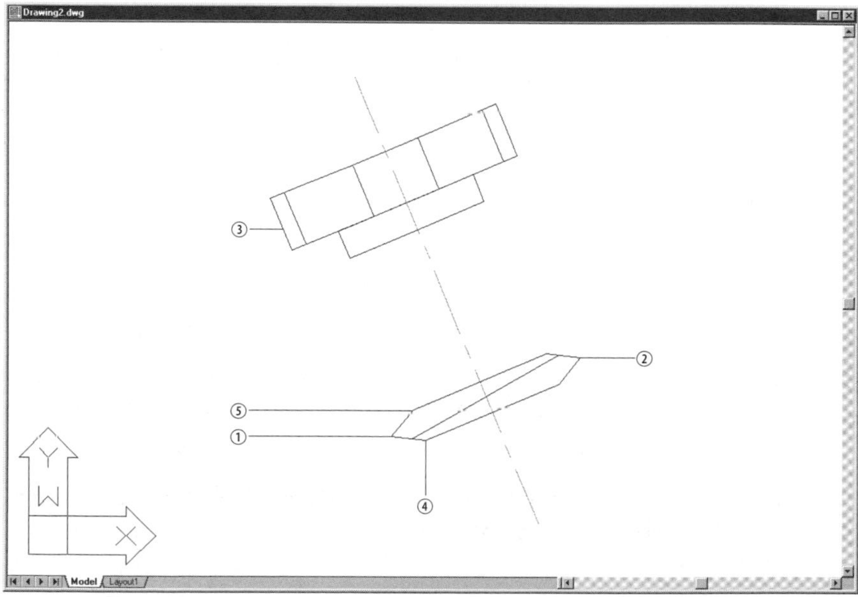

Figure 6–7 *Pick a point to change the UCS.*

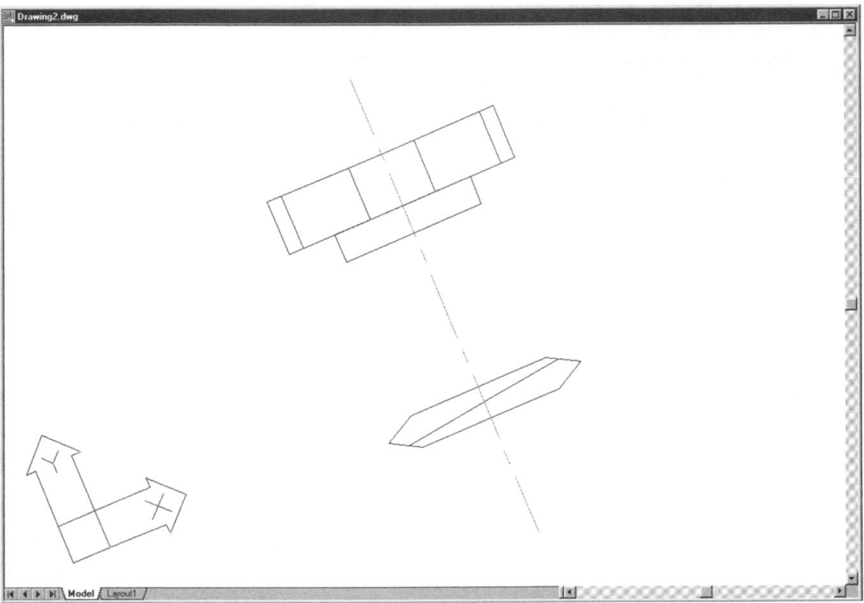

Figure 6–8 *The new UCS is displayed.*

7. To run the command line version of the Array command, type **-Array**. The following prompt appears:

   Select objects:

8. Pick anywhere on the object near 1 shown in Figure 6–7, and then end the selection process by pressing Enter.

9. Answer the following prompts as shown:

   Enter the type of array [Rectangular/Polar] <R>: **R**
   Enter the number of rows (---) <1>: **6**
   Enter the number of columns (|||) <1>:¿

   The following prompt appears:

   Enter the distance between rows or specify unit cell (---):

10. Activate the cursor menu (Shift+right-click) and choose Intersection. Click at 4. Again, activate the cursor menu, choose Intersection, and click at 5.

    The array is carried out in a direction perpendicular to the X-axis of the new UCS.

11. Return the UCS to the WCS with the UCS command. Enter **UCS** and press Enter at the command prompt. When the following prompt appears, accept the default <World>:

    Enter an option [New/Move/orthoGraphic/Prev/Restore/Save/Del/Apply/?/World] <World>:

Your drawing should resemble Figure 6–9.

12. Save this drawing and name it 06DWG01.DWG.

Tip: It is possible to change the UCS to the position and orientation you want in two or more ways. In the preceding exercise, for example, you could have rotated the UCS about its Z-axis instead of using the 3 Point option. I generally prefer to use the 3 Point option because it is more intuitive and easier to use.

Although defining new UCSs is most frequently used in 3D drafting, the preceding exercise demonstrates that the capability to change the UCS is helpful in 2D work as well. By aligning the UCS with the horizontal axis of the thread object in the drawing, a simple six-row array could be quickly carried out with the "axis" of the array perpendicular to the horizontal axis of the object.

THE UCS COMMAND

The UCS command is the key to placing, moving, rotating, and displaying UCSs. This command allows you to position UCSs appropriately to draw the elements necessary to define your 2D or 3D model properly. By understanding the various options available through the UCS command, you ease the task of defining your

Figure 6–9 *The completed array.*

model. Please note that the UCS command does not list all of the following options at the command prompt.

Most of your 2D work can be accomplished with the following subset of UCS options:

- **Origin**—Specifies a new X, Y, or Z origin point relative to the current origin.
- **3point**—Enables you to set the X- and Y-axes by specifying the origin and a point on both the X- and Y-axes.
- **OBject**—Defines a new coordinate system based on a selected object.
- **Z**—Rotates the X- and Y-axes about the Z-axis.
- **Prev**—Reverts back to the previous UCS. You can recall as many as the last ten UCSs.
- **Restore**—Sets the UCS to a previously named UCS.
- **Save**—Enables you to store the current UCS with a name you specify.
- **Del**—Removes a stored UCS.
- **?/Named UCSs**—Lists saved UCSs by name.
- **World**—Displays the WCS.

The Restore, Save, Del, and ?/Named UCSs options are tools that allow you to manage UCS configurations. By using these tools, you can save defined UCS configurations and then restore them for use later. One feature that makes these options very useful is that named UCSs are saved with the current drawing. Therefore, as you develop a series of different UCS configurations, you can save the configurations with the drawing, knowing that you can recall them later during another editing session.

Tip: You can use the AutoCAD DesignCenter (ADC) to import saved UCS configurations from other drawings into the current drawing. For more information, refer to Chapter 11, Applying AutoCAD Design Center.

AutoCAD 2000 added several new features to the UCS command. These features enhance the usefulness of UCSs by making them easier to manipulate. These features make defining and controlling UCSs simpler.

The enhanced UCS features are as follows:

- **Face**—This option allows you to configure a new UCS quickly by aligning it to the selected face object. Face objects are discussed in Chapter 29, Surfacing in 3D.
- **Apply**—This option allows you to apply the UCS configuration in one viewport to another by simply selecting the viewport.

- **Move**—This option allows you to move the origin of a UCS without redefining or renaming the UCS. For example, if you move the origin point of a named UCS, AutoCAD simply applies the new origin point to the named UCS.
- **UCSMan(ager)**—This option displays the new UCS dialog box.
- **Multiple UCSs**—This feature allows you to set a different UCS for each viewport you have open.

The following exercise demonstrates the new features and options.

EXERCISE: AUTOCAD'S ENHANCED UCS FEATURES AND OPTIONS

1. Open the drawing 06DWG02. The drawing shows four different viewports, all viewing the same cube object, and all having identical UCSs, as shown in Figure 6–10.
2. Pick anywhere inside the lower-left viewport. The lower-left viewport becomes active.
3. From the Tools menu, choose New UCS, and then choose Face. AutoCAD prompts you to select the face of a solid object.
4. In the lower-left viewport, pick the line AB, near A. AutoCAD highlights the front face of the cube. (If AutoCAD highlights the top face of the cube, choose N to switch to the adjoining face, and then press Enter to select the front face.)

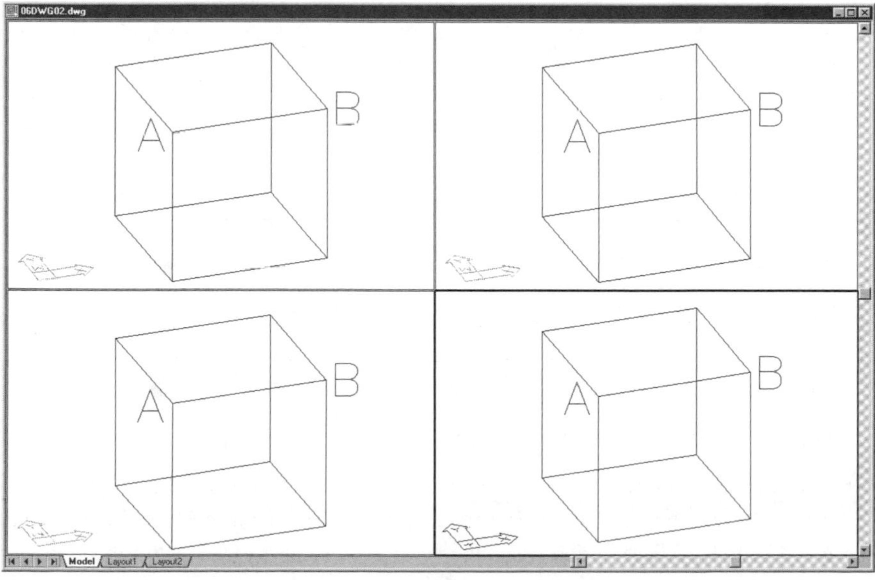

Figure 6–10 *The UCS is identical in each viewport.*

5. Press Enter to accept the front face selection. Notice that AutoCAD rotates the lower-left viewport's UCS, as shown in Figure 6–11. Also notice that no UCS is rotated in any of the other viewports.
6. From the Tools menu, choose New UCS, and then choose Apply. AutoCAD prompts you to pick a viewport to which to apply the current UCS.
7. Choose the upper-right viewport, then press Enter. AutoCAD updates the upper-right viewport's UCS so it matches the lower-left viewport's UCS.
8. From the Tools menu, choose Move UCS. AutoCAD prompts you to specify the new origin point for the current UCS.
9. In the upper-right viewport, pick a point in the approximate center of the front face. AutoCAD redefines the UCS's origin point and moves the UCS icon to the new origin point.
10. From the Tools menu, choose Named UCS. AutoCAD displays the UCS dialog box, with the Named UCSs folder displayed. Notice that the current UCS is Unnamed.
11. Right-click on the highlighted Unnamed UCS. AutoCAD displays the shortcut menu.

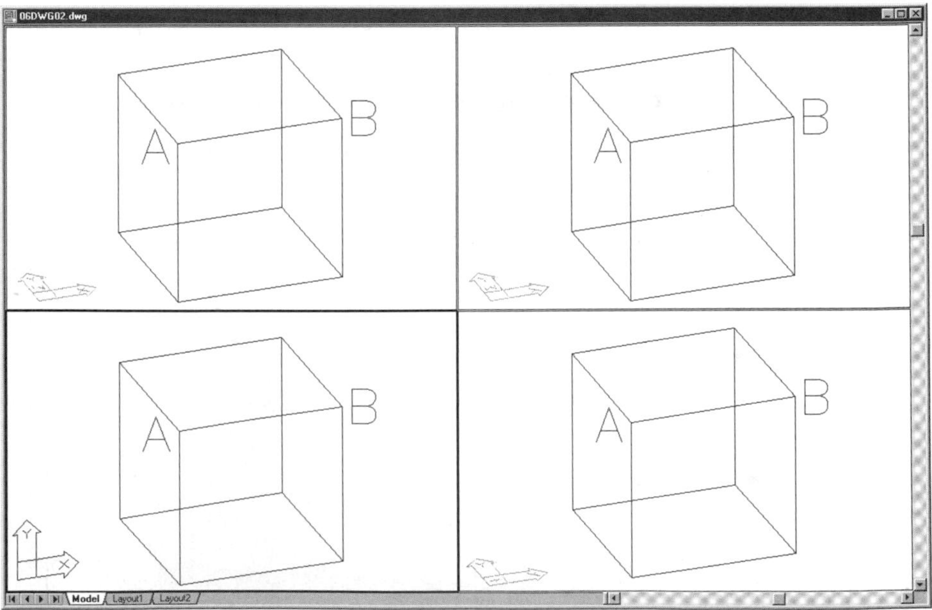

Figure 6–11 *The UCS Face option aligns the lower-left viewport's UCS with the selected front face of the cube. Notice that all other UCSs remained unchanged.*

12. Choose Rename, type **Front UCS**, and then press Enter. AutoCAD renames the UCS, as shown in Figure 6–12.
13. Choose OK. AutoCAD dismisses the UCS dialog box.
14. Choose the lower-right viewport, then choose Tools>Named UCS. AutoCAD displays the UCS dialog box.
15. Select **Front UCS**, choose Set Current, and then choose OK. AutoCAD redefines the UCS's origin point and moves the UCS icon to the new origin point in the current viewport.
16. Choose Tools>Named UCS. AutoCAD displays the UCS dialog box.
17. Choose the Orthographic UCSs tab. AutoCAD displays six predefined UCSs, as shown in Figure 6–13.
18. Choose the orthographic UCS named Left, choose Set Current, and then choose OK. AutoCAD aligns the lower-right viewport's UCS with the left face of the cube, as shown in Figure 6–14.
19. You can close the drawing without saving changes.

AutoCAD's enhanced UCS features and options make using UCSs easier than before. By applying these new tools, you can simplify object editing and increase your productivity.

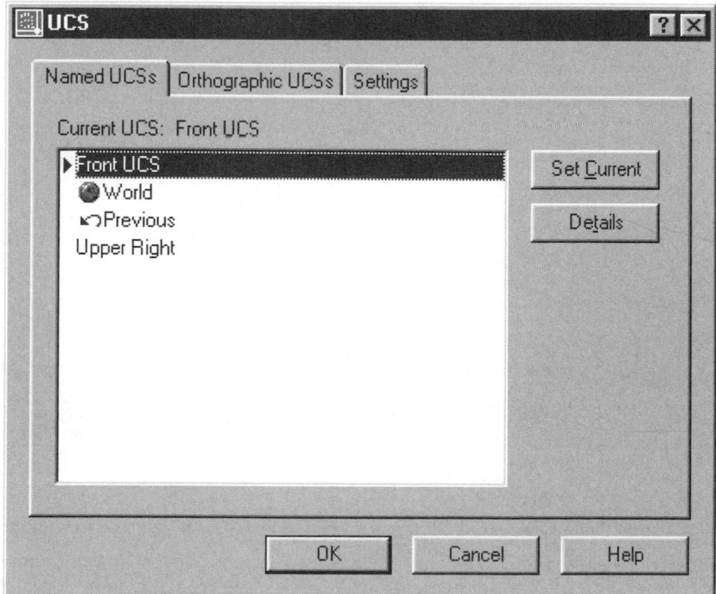

Figure 6–12 *The current UCS is renamed Front UCS.*

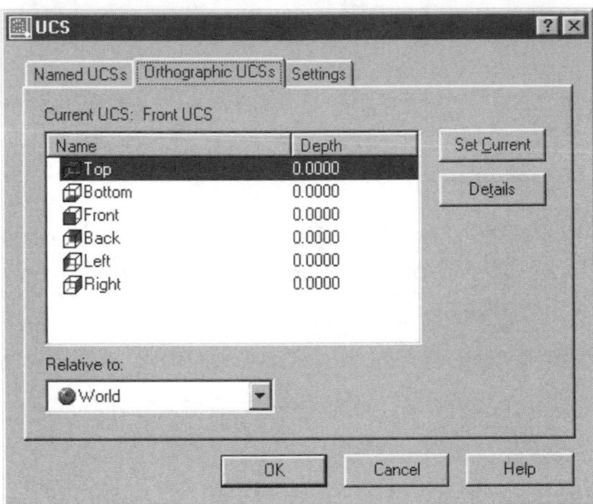

Figure 6–13 *AutoCAD provides six predefined orthographic UCSs.*

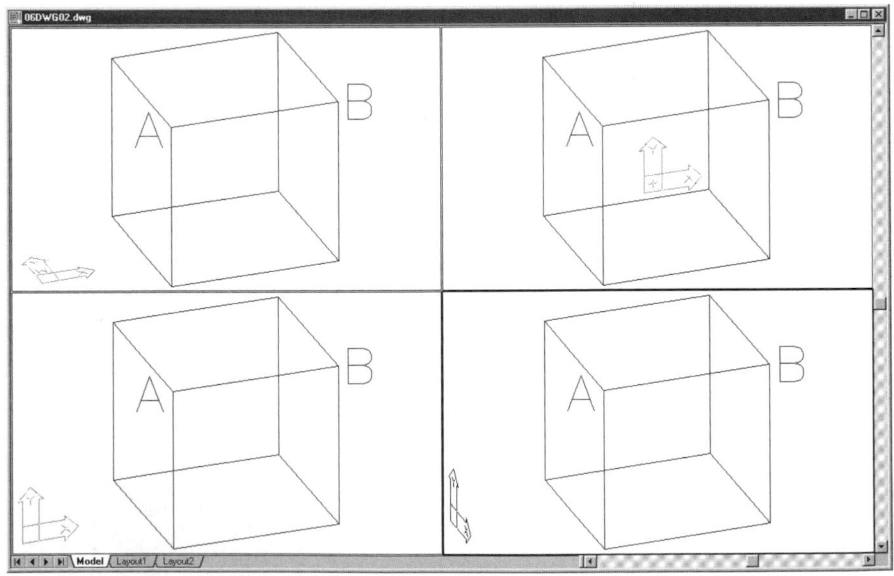

Figure 6–14 *AutoCAD applies the orthographic UCS named Left to the lower-right viewport.*

THE UCSICON COMMAND

In an earlier exercise you saw how the UCSICON command can be used to control the placement and visibility of the UCS icon. To round out the discussion of UCSs, this section explains the options for the UCSICON command. The UCSICON command displays the following prompt:

Enter an option [ON/OFF/All/Noorigin/ORigin] <ON>

The options are described as follows:

- **ON**—Turns the UCS icon on.
- **OFF**—Turns the UCS icon off.
- **All**—Applies changes to the UCS icon in all displayed viewports. Otherwise, changes affect only the current viewport.
- **Noorigin**—Displays the UCS icon in the lower-left corner of viewports.
- **ORigin**—Displays the UCS icon at the 0,0,0 origin of the current UCS, if possible. Otherwise, it displays the UCS icon in the lower-left corner of viewports.

Tip: Another capability of the UCS icon that you should know about is the system variable UCSFOLLOW. This variable controls whether or not a plan view will be automatically generated whenever you change the UCS. Setting this variable to zero will not affect the view; setting it to 1 will cause the plan view to be generated. For 2D drafting, I find the automatic plan view setting to be helpful.

The UCSICON options just described can be set from the command prompt or from the UCS dialog box. When you choose Tools>Named UCS and then select the Settings tab, AutoCAD displays the UCSICON options shown in Figure 6–15. By selecting or clearing the options, you can toggle on or off the following options:

- **ON**—When selected, this option turns the UCS icon on. When cleared, it turns the UCS icon off.
- **Display at UCS Origin Point**—When selected, this option displays the UCS icon at the 0,0,0 origin of the current UCS, if possible. When cleared, it displays the UCS icon in the lower-left corner of the viewport.
- **Apply to All Active Viewports**—When selected, this option applies changes to the UCS icon in all displayed viewports. When it is cleared, changes to the UCS icon affect only the current viewport.

In addition to the UCS icon settings, you can also control two other UCS settings, which are described as follows:

- **Save UCS with Viewport**—When selected, this option allows you to set and retain a different UCS for each viewport. When it is cleared, the viewport reflects the current UCS each time the UCS is changed, even if the viewport is not the current viewport.
- **Update View to Plan when UCS is Changed**—When this option is selected, the plan view is automatically applied when the UCS is changed. When it is cleared, the plan view is not invoked when the UCS changes. (This is the same as setting the UCSFOLLOW system variable, which was described in the last tip.)

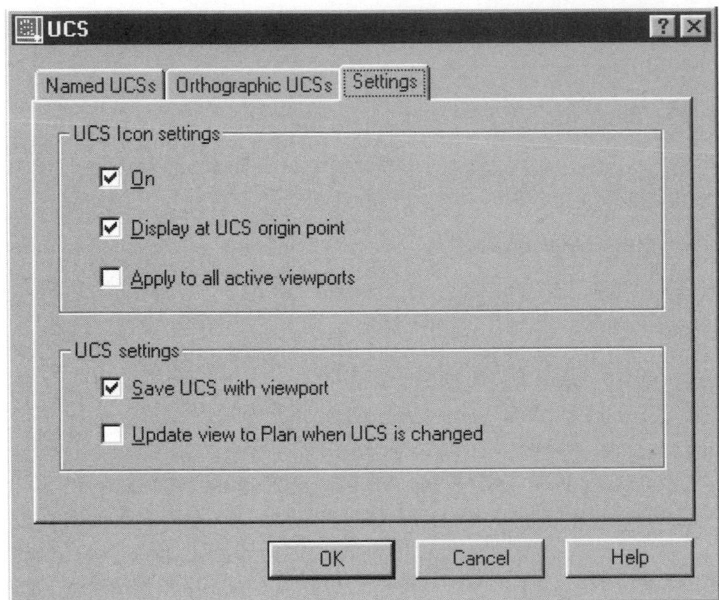

Figure 6–15 *The UCS Manager allows you to set options for the UCS icon.*

The plan view described here refers to rotating the view of the model in the current viewport so that the X-axis of the UCS appears normal (horizontal) to the screen. For example, in Figure 6–16, the image on the left has its view rotated normal to the screen, while the image on the right does not. You can invoke a plan view by using the PLAN command.

SETTING UP DRAWING AIDS

AutoCAD allows you to control a number of drafting settings that help you work efficiently and make your drawings more accurate. These settings consist of a series of system variables that you can set at the command prompt. Additionally, you can easily access them through the Drafting Settings dialog box (formerly the Drawing Aids dialog box.) The Drafting Settings dialog box, which is shown in Figure 6–17, is accessed through the Tools menu by choosing Drafting Settings. This dialog box is also accessed from the command prompt by typing **DDRMODES**.

The Drafting Settings dialog box consists of three folders. The Snap and Grid folder controls snap and grid settings, including the X and Y increment spacing, the X-axis angle, and the X and Y base point. The Polar Tracking folder controls polar tracking settings. It controls features that function similar to AutoCAD's ORTHO mode but allows you to apply the features to angles in any increment. The Object Snap folder (formerly the Osnap Settings dialog box) controls running object snaps, which are object snaps that are always turned on.

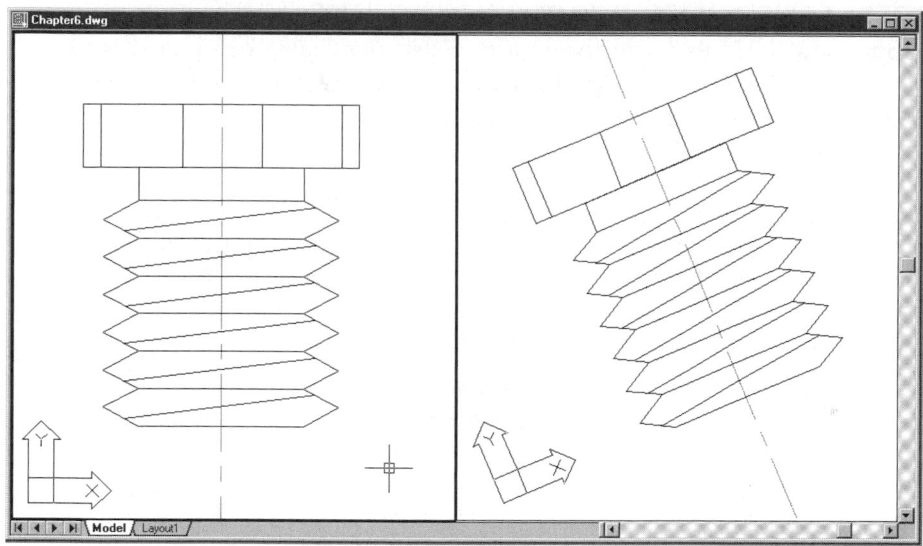

Figure 6–16 *The view of the image on the left is automatically rotated using the PLAN command.*

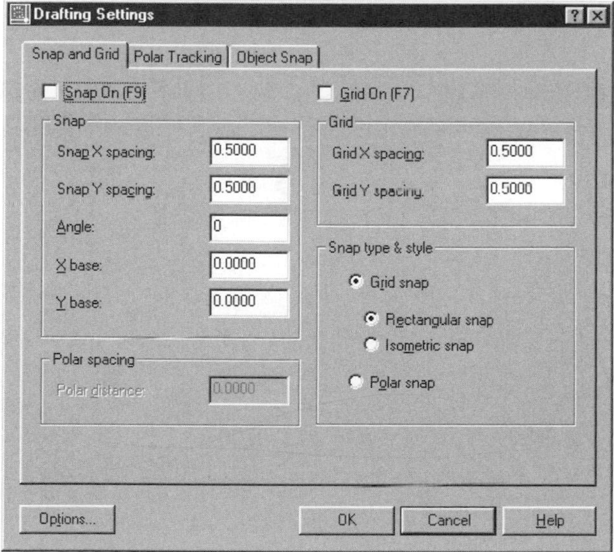

Figure 6–17 *Drafting Settings dialog box.*

This section reviews the system variables accessed through the Drafting Settings dialog box, and their related commands.

UNDERSTANDING SNAP AND GRID OPTIONS

The Snap and Grid folder of the Drafting Settings dialog box is made up of four major sections. These sections control how AutoCAD's Snap feature functions and how its grid is displayed. By understanding how these features function, you will make drawing objects with AutoCAD easier and intuitive.

Snap

Options in this section of the Drafting Settings dialog box control the snap grid (see Figure 6–18). When the snap grid is enabled (when a check mark, or "X," appears in the Snap On check box), the movement of the cross-hairs cursor is restricted to incremental displacements across a grid of invisible "snap" points. This enables you to snap to and select points on this grid with a high degree of precision. You can enter both snap X and snap Y spacing for your snap grid by typing in the input boxes in this section. The other options—Angle, X Base, and Y Base—control the angle at which the grid is oriented with respect to the current UCS and the origin's coordinates of the snap's grid.

Tip: The isometric view shown in Figure 6–18 is displayed in isoplane left mode. You can cycle through three modes—isoplane left, isoplane top, and isoplane right—by pressing F5 or Ctrl+E. As you cycle through each mode, AutoCAD realigns the cursor to match the selected mode.

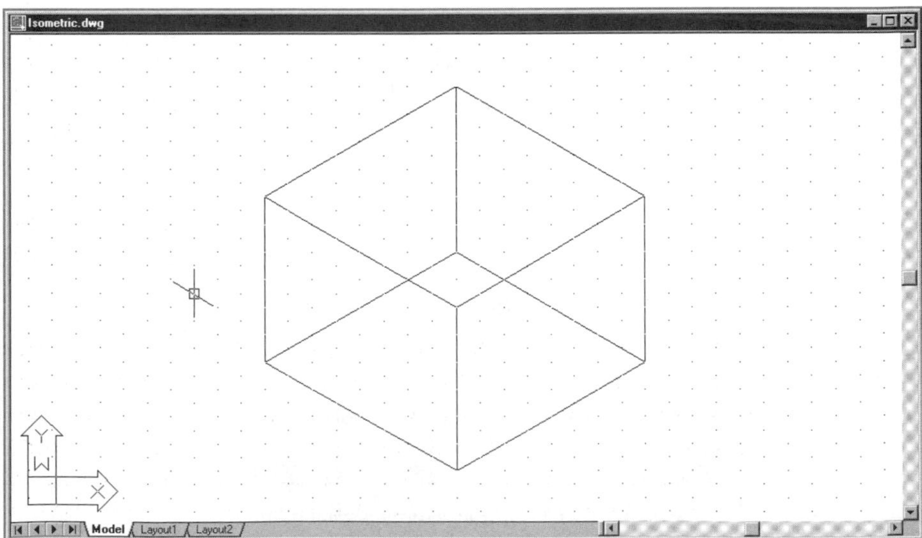

Figure 6–18 *The isometric snap mode is used to draw isometric views of objects.*

The snap grid settings can also be controlled at the command prompt with the SNAP command. You can toggle the Snap feature on and off with the F9 key, by pressing Ctrl+B, or by choosing the SNAP button on the status bar.

By carefully selecting the spacing of the snap grid, you can usually make the picking of points much easier, because you bypass the need to enter points at the keyboard.

Tip: When you enter values in the X Spacing and Y Spacing text edit boxes in both the Grid and Snap sections of the Drafting Settings dialog box, entering a value in the X Spacing box and then pressing the Enter key will automatically transfer the value to the Y Spacing box. Because I usually want both the X and Y values of both of these settings to be equal, this shortcut eliminates the need to type anything in the Y Spacing boxes.

Grid

In addition to applying a grid of invisible snap points, you can apply a grid of visible points to the drawing area. This visible drawing aid is simply called Grid. The controls found in the Grid area determine the location of the grid's visible points.

When the Grid On option is selected, grid points are made visible with the spacing specified by the values (in drawing units) that you type into the Grid X Spacing and Grid Y Spacing text edit boxes (see Figure 6–17).

It is common to link the spacing of the grid of visible points (the Grid) to the grid of invisible snap points. To establish this link, the X and Y spacing of the grid points is set to zero. AutoCAD will then use the X and Y spacing of the snap points and automatically apply these to the visible grid. You can, of course, override this 1:1 relationship by explicitly entering values other than zero for the grid spacing. Keep in mind that regardless of the setting(s) of the grid points, the origin and angle of the grid is always kept the same as the origin and angle of the snap points.

You can also control the visible grid with the GRID command, and then toggle the grid on and off by pressing the F7 key, or by pressing Ctrl+G, or by choosing the GRID button on the mode status bar.

Tip: The Grid control section of the Drafting Settings dialog box largely duplicates functions available through the basic GRID command. Using the GRID command at the command prompt, however, offers an option: grid spacing (X). Specifying a value followed by an "X" sets the grid spacing to the specified value times the snap interval. I often like to have my snap spacing be a fraction—say, ¼—of my grid value. When you use the "X" feature of the spacing setting available with the GRID command, this relationship between grid and snap remains in effect no matter how often you change the grid setting.

Snap Type & Style

The Snap Type & Style area controls whether rectangular, isometric, or polar snaps and grids are used. By selecting the Grid Snap radio button, you can choose between the rectangular or isometric snap and grid modes. The rectangular mode is AutoCAD's standard snap and grid mode. The isometric mode allows you to draw isometric views of objects, such as the cube shown in Figure 6–18, easily. Notice the alignment and skew of the cursor when isometric mode is selected.

The Polar Snap option causes the cursor to snap along polar alignment angles (which are set in the Polar Tracking folder) relative to the starting polar tracking point. The snap increment is determined by the Polar Distance setting in the Polar Spacing area, which is discussed in the next section.

THE POLAR TRACKING FEATURE

The Polar Tracking feature (introduced in AutoCAD 2000) provides the ability to constrain cursor movement within predefined polar alignment angles. This feature functions like ORTHO mode, but instead of constraining cursor movement horizontally and vertically in 90-degree increments, it allows you to set the polar alignment increments to any angle.

In the Polar Angle Settings area, you can select the desired polar alignment angle from a predefined list containing commonly used angles. The angles are selected from the Increment Angle drop-down list, which contains angles with the following degree-measures: 90, 45, 30, 22 1/2, 18, 15, 10, and 5. Additionally, you can set up to ten user-defined angles. When you select the Additional Angles option and then choose the New button, AutoCAD allows you to add your own polar alignment angles to the Additional Angles list, as shown in Figure 6–19. To remove user-defined angles, select the angle from the list and then choose the Delete button.

The Polar Angle Measurement area determines whether polar tracking alignment angles are measured absolutely or relatively. Angles measured absolutely are determined relative to the current UCS. Angles measured relatively are determined relative to the last point selected. For example, in Figure 6–20, with the Polar Angle Measurement option set to Absolute, the image on the left displays the cursor's polar position at 45 degrees, as measured from the current UCS. In contrast, with the Polar Angle Measurement option set to Relative to Last Segment, the image on the right displays the cursor's relative polar position at 90 degrees, as measured from the last line segment.

The Object Snap Tracking Settings area allows you to set tracking as orthogonal or to all the polar settings displayed in the Polar Angle Settings area. The Track Orthogonally Only option is the original option first introduced with tracking in Release 14. The Track Using All Polar Angle Settings option is new to AutoCAD 2000 and constrains cursor movement to the predefined polar alignment angles. These options are discussed in detail later in this chapter.

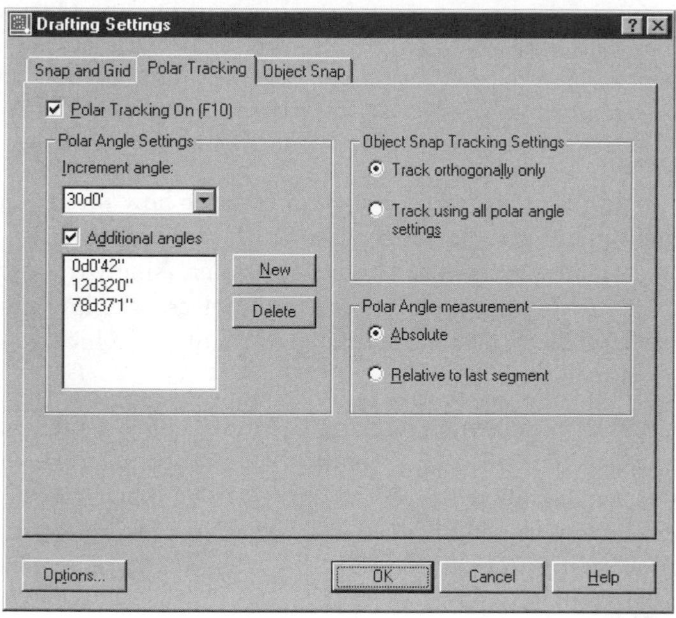

Figure 6–19 *The Polar Tracking folder allows you to define your own polar alignment angles, as shown in the list box.*

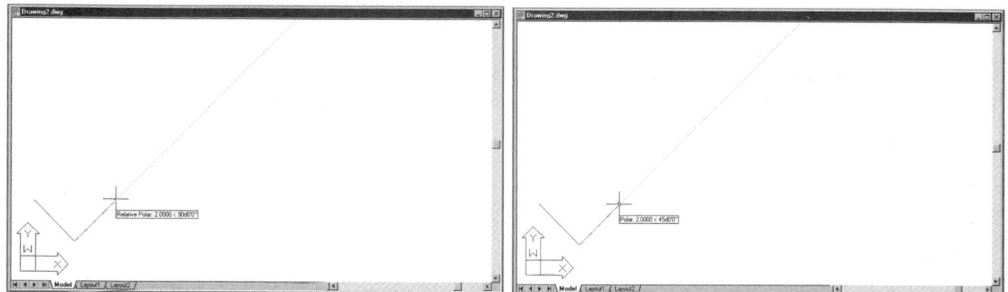

Figure 6–20 *You can track polar alignments absolutely (based on the current UCS) or relatively (based on the last point selected).*

Note: When you select an angle from the Increment Angle drop-down list, AutoCAD constrains cursor movement every time the cursor crosses an angle increment. (The cursor "snaps" to the angle increment.) In contrast, a user-defined angle entered in the Additional Angles list constrains cursor movement only when the cursor crosses the listed angle(s). Therefore, using the increment angle and additional angles shown in Figure 6–19, the cursor constrains its movement every 30 degrees (30, 60, 90, 120, etc.), and also at 0 degrees, 00 minutes, 42 seconds; at 12 degrees, 32 minutes, 00 seconds; and at 78 degrees, 37 minutes, 01 seconds.

Tip: You can temporarily constrain cursor movement to any angle not listed in the Polar Tracking folder using angle overrides. For example, to constrain cursor movement to 33 degrees during a command, when prompted to specify the next point, enter <33; AutoCAD will temporarily constrain the cursor for the next point to 33 degrees.

OBJECT SNAPS

No matter how carefully you set your snap interval or how often you change that interval, it is highly unlikely that all the points in your drawings will conveniently fall on these predefined snap points. This becomes increasingly true as your drawing becomes populated with various objects that themselves have important geometric features, such as endpoints, centers, and tangent points, to which you will want to relate other drawing objects.

Most modern CAD applications, including AutoCAD, therefore provide some means of identifying these geometric points. These tools make the construction of new geometry easier, and allow the objects created to be drawn more accurately and the results to be consistently maintained with far more precision than is possible in traditional manual drafting. In AutoCAD, this capability is called object snapping, which is also referred to as object snap or Osnap. In AutoCAD, these modes consist of a set of tools that permit accurate geometric construction.

Osnaps are used to identify directly and easily key points either on or in relation to your drawing objects. Figure 6–21 shows the Object Snap toolbar and the pop-up cursor menu.

The Osnap toolbar and the cursor menu contain the same Osnap modes and present them in essentially the same order. You can display or activate the pop-up cursor menu by simultaneously holding down the Shift button and pressing the right mouse button (commonly called Shift+right-click, or simply Shift+Enter, because the right mouse button serves as an Enter button). If you use a three-button mouse, the middle button can usually be configured to "pop up" the cursor menu. The Osnap modes are also represented on the Standard toolbar by a "fly-out" toolbar.

Note: To resize any toolbar, move your cursor to an edge of the toolbar. When the cursor is in the proper position to resize the toolbar, it turns into a double-headed arrow. Then you just click and drag to resize the toolbar as desired.

Tip: Depending on personal preference and the type of drafting you are involved with, you can use the Osnap toolbar (in either a horizontal or vertical format, or a more compact rectangular arrangement), the Osnap fly-out from the Standard toolbar, or the pop-up cursor menu. The fly-out has the advantage of being present only during Osnap selections; the Osnap toolbar can be moved around and resized; and the cursor pop-up menu requires very little cursor movement. I almost always use the cursor menu because it is quickest for me.

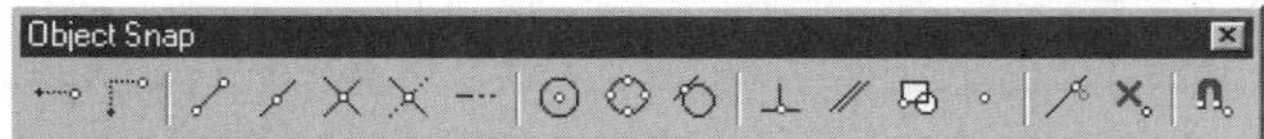

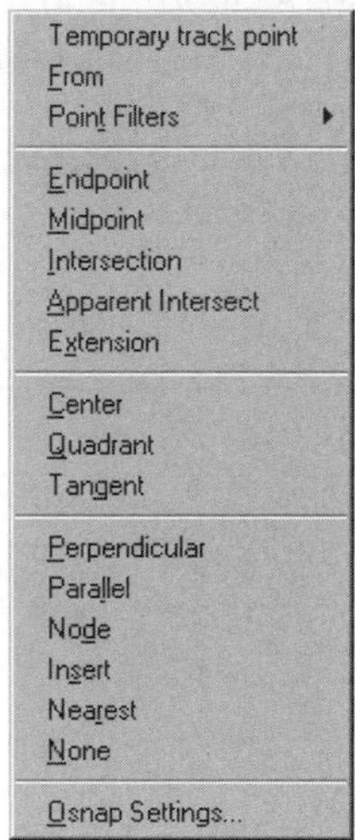

Figure 6–21 *The Object Snap toolbar and the pop-up cursor menu.*

Note: The position and size of a toolbar can be saved to a profile, which can be recalled later to restore preferred toolbar layouts. For detailed information, see Chapter 3, Fine-tuning the Drawing Environment, in the section titled Saving the Options to a Profile.

Osnap Modes

AutoCAD has sixteen object snap modes, including the two new modes. Table 6-1 gives a description of each mode.

Running Osnap Toggle and Osnap Overrides

Running Osnap Toggle and Osnap Override are two useful AutoCAD features that were first introduced in Release 14. They are important adjuncts to the overall operation of object snaps and are therefore discussed here.

Running Object Snap Toggle. Running object toggle is an Osnap enhancement that enables you to toggle any running (continuing) Osnap off prior to selecting a point without losing the running Osnap settings. This feature is accessed by choosing the OSNAP button on the mode status bar at the bottom of AutoCAD's screen. If this button is selected when no running Osnaps are set, the Object Snap folder in the Drafting Settings dialog box is displayed, giving you the opportunity to set running Osnaps.

Object Snap Override. AutoCAD provides an option that enables you to enter explicitly coordinate data that has priority over any running Osnaps that may be in effect. This enhances direct coordinate entry, and you can be certain that such entries have precedence over any other settings. To override a running Osnap temporarily, simply choose an object snap using any method described previously in the

Table 6–1: *AutoCAD 2002 Object Snap Modes*

Mode	Description
Apparent Intersect	Finds a point that represents the apparent intersection of two objects, such as two nonparallel lines that do not actually cross in 3D space
Center	Finds the center of a circle or an arc
Endpoint	Finds the endpoint of all objects except a circle
Extension	Locates a point by extending a temporary, dashed construction line from an existing arc or line
From	Establishes a temporary reference point as a basis for specifying subsequent points
Insert	Finds the insertion point of text objects and block or external references
Intersection	Locates the intersection of objects in the same plane
Midpoint	Finds the midpoint of a line, polyline, or arc
Nearest	Locates a point on an object that is nearest to the point you pick
Node	Finds the location of a point object
None	Instructs AutoCAD not to use any running Osnap modes
Parallel	Allows you to select an angle by creating a temporary, dashed construction line that is parallel to an existing object whose vector is acquired
Perpendicular	Returns a point at the intersection of the object selected and an angle perpendicular to that object from either the last or the next point picked
Quadrant	Finds the closest 0-, 90-, 180-, or 270-degree point relative to the current UCS on a circle or an arc
Tangent	Locates a point that is tangent to the selected circle or arc from either the last or the next point picked
Tracking	Specifies a point that is relative to other points, using orthogonal or polar displacements

Object Snapping section. AutoCAD uses the selected snap override for the next pick only and then reverts back to running Osnaps.

AutoSnap

AutoSnap is an important feature in AutoCAD that was first introduced with Release 14. With AutoSnap you can visually preview snap point candidates before picking a point. Depending on how you have AutoSnap features set, AutoSnap will display a Snap Tip placard similar to the toolbar's ToolTip feature. Also, a marker unique to each Osnap mode can be displayed in the color of your choice, making for easy identification of the snap location and snap type. You can also enable a "magnet" feature that snaps the marker into place much like the action of AutoCAD's Grips feature.

In the following exercise you will use some of AutoCAD's Osnap modes and the three methods of invoking them. In addition, you will see how the AutoSnap feature makes looking for and confirming Osnap points an unambiguous, efficient means of picking Osnap points.

EXERCISE: BISECTING AN ANGLE USING OBJECT SNAPS WITH AUTOSNAP

1. Open the drawing 06DWG03. The drawing shows two line segments whose end points touch, creating an angle.

   If the Object Snap toolbar is not visible, proceed with steps 2 and 3.

2. Right-click over any toolbar button. The available toolbars are displayed on the pop-up cursor menu, as shown in Figure 6–22.
3. Choose Object Snap from the cursor menu. Note that the Object Snap toolbar appears.
4. From the Tools menu, choose Options to display the Options dialog box, then choose the Drafting tab. Ensure that the Marker, Magnet, and Display AutoSnap Tooltip features are all selected, and that the Display AutoSnap Aperture Box option is cleared, as shown in Figure 6–23. Click on OK to close the dialog box.
5. From the Draw menu, choose Arc>Center, Start, End. You will see the following prompt:

   Specify start point of arc or [CEnter]: _c Specify center point of arc:

6. Click on the Snap to Endpoint tool in the displayed Osnap toolbar. Then move and rest the screen cursor at a point near ①, as shown in Figure 6–24. Notice that the AutoSnap marker is displayed at ② and the Snap Tip is also displayed, identifying the snap point as the line's endpoint.
7. With the endpoint snap marker still displayed, pick a point near ①. AutoCAD then prompts you for the arc's start point.
8. At the "Specify start point of arc:" prompt, click and hold the Osnap fly-out on the Standard toolbar (see Figure 6–21), and then choose the Snap to Nearest tool and release the pick button. Move the cursor near the lower line at ③. Note the appearance of the Nearest AutoSnap marker as you approach ③.

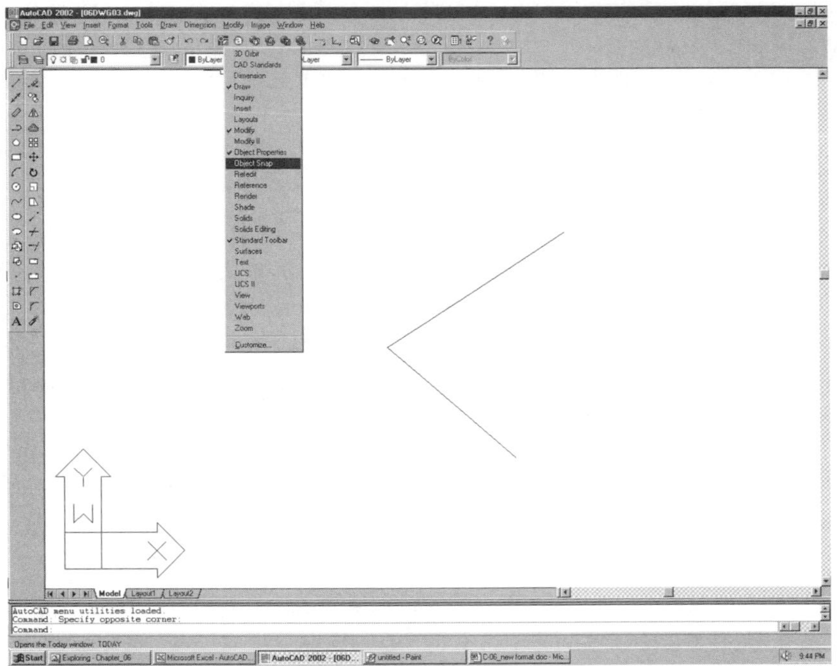

Figure 6–22 *The available toolbars appear on the cursor menu.*

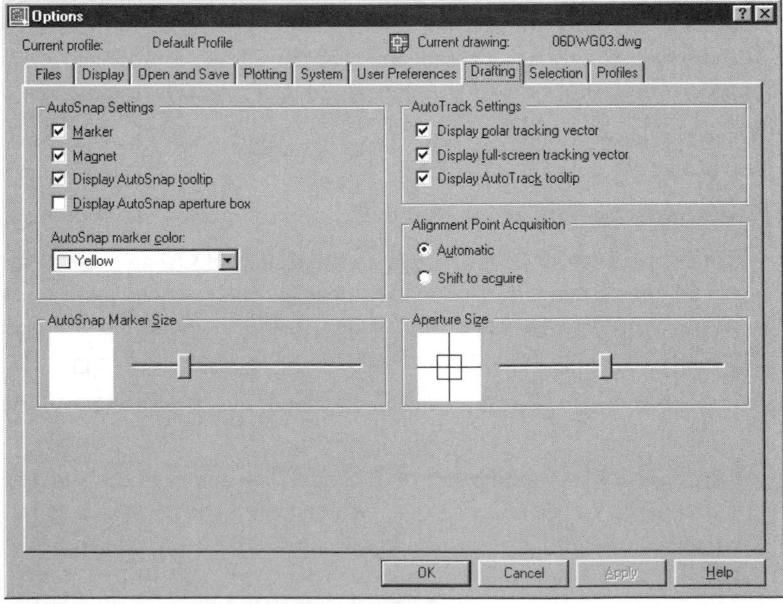

Figure 6–23 *The Drafting folder in the Options dialog box controls AutoSnap settings.*

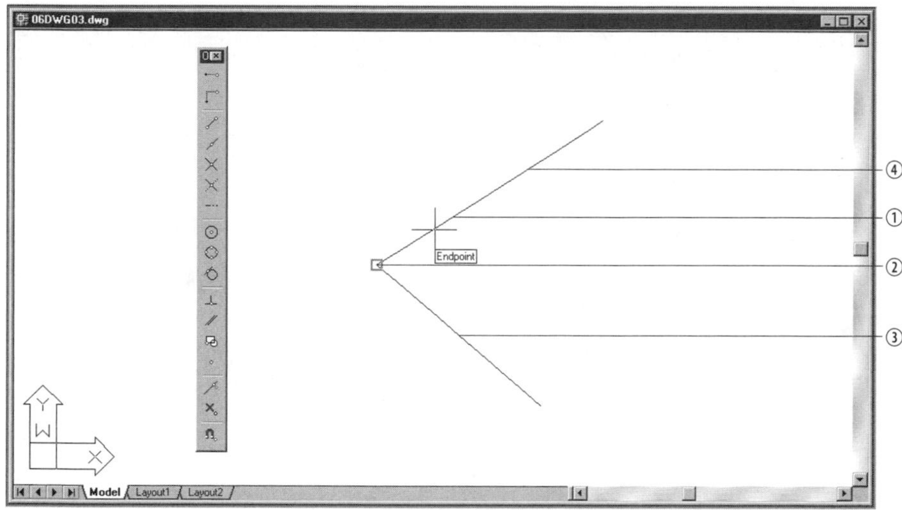

Figure 6–24 *The AutoSnap feature identifies the line's endpoint.*

9. Pick near ③. At the "Specify end point of arc or [Angle/chord Length]:" prompt, Shift+right-click to display the screen pop-up menu. Then choose Endpoint and move the cursor toward ④. Note the appearance of the Endpoint marker.
10. Pick the upper line near ④. AutoCAD draws the arc as shown in Figure 6–25.
11. Type **L** and press Enter at the command prompt. At the "Specify first point:" prompt, type **endp** and press Enter. Note the appearance of the Endpoint AutoSnap marker as you approach ⑤.
12. Click at ⑤. Then, at the "Specify next point or [Undo]:" prompt, click on the Midpoint tool on the Osnap toolbar and move the cursor to any point on the arc. Note the appearance of the Midpoint marker on the arc. With the marker showing, pick any point on the arc.
13. At the next "Specify next point or [Undo]:" prompt, press Enter to end the LINE command. The bisector line is drawn.

As seen in this exercise, the use of Osnaps and the AutoSnap feature gives a definite, unambiguous indication of the geometry to which you are snapping. Even in crowded areas of a drawing, positive identification of which point is the current snap target is possible.

AutoSnap also supports a feature that enables you to step through the object snap points of objects lying within the target aperture when it is enabled. The Tab key is used to cycle from the closest to the furthest Osnap point from the center of the aperture box. The target geometry is highlighted to aid further in identification. Repeated pressing of the Tab key cycles among the objects, highlighting the target geometry so that you can snap to the correct object's midpoint.

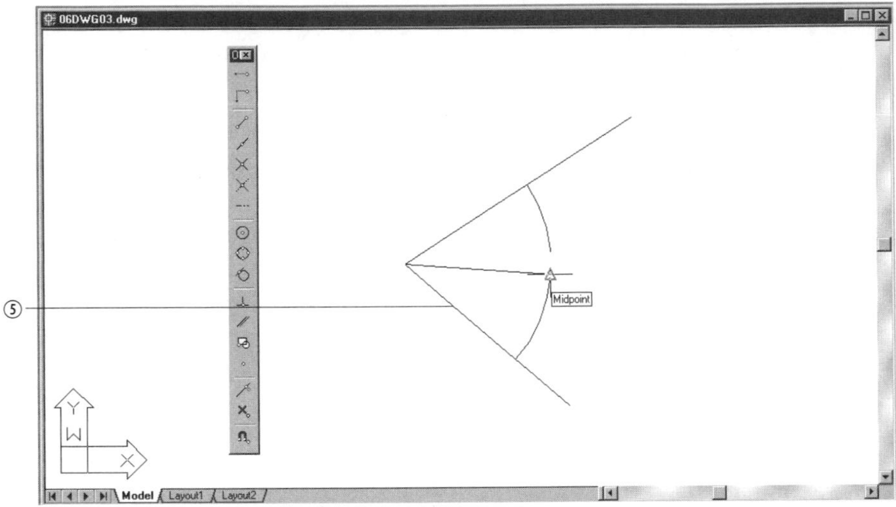

Figure 6–25 *Bisecting an angle using Osnaps.*

Using AutoTrack

AutoTrack is a combination of two features. The first, Object Snap Tracking, was originally introduced with Release 14 and was enhanced in AutoCAD 2000. The other, Polar Tracking, was added with AutoCAD 2000. Together, these two tracking features are collectively called AutoTrack.

AutoTrack creates temporary alignment paths—dashed construction lines that appear on the screen and help you identify pick points. When your cursor properly aligns with the preset polar tracking or object snap tracking angle, the dashed line appears, and your cursor is constrained to it. This makes identifying potential pick points very easy.

The difference between polar tracking and object snap tracking is this: Polar tracking creates alignment paths for points you already set, while object snap tracking creates alignment paths based on acquired Osnaps. In other words, polar tracking displays alignment paths for the last point selected and functions independently of Osnaps. In contrast, object snap tracking displays alignment paths only if you use Osnaps and move the cursor over an Osnap marker to acquire it.

This section describes both object snap tracking and polar tracking and provides a simple exercise that shows you how to use these two powerful features.

Object Snap Tracking. Although it is not an object snap in the strict sense, object snap tracking is used with standard Osnaps to enhance your ability to find points relative to another object's geometry.

You can use object snap tracking whenever AutoCAD prompts you for a point. If you try to use object snap tracking when no command is currently running, AutoCAD displays an error message.

When you start object snap tracking by acquiring an Osnap point, AutoCAD displays either orthogonal alignments or preset polar angle alignments. You control whether object snap tracking uses orthogonal or polar alignments from the Polar Tracking folder in the Drafting Settings dialog box, as shown in Figure 6–19. When you choose the Track Orthogonally Only option found in the Object Snap Tracking Settings area, object snap tracking displays only orthogonal alignment paths. If you select the Track Using All Polar Angle Settings option, alignment paths are displayed for all angles listed in the Polar Angle Settings area, which is also found in the Polar Tracking folder.

You acquire an Osnap by moving the cursor over an Osnap marker. When prompted to select a point during a command, if object snap tracking is enabled and you move your cursor over an Osnap marker, a small plus sign (+) appears in the marker, indicating that the point is acquired. Then, as you move the cursor away from the point, the orthogonal or polar alignment paths appear, and your cursor is constrained to them as you move along the paths. To clear an acquired point, move the cursor back over the Osnap marker.

Note: It is important to understand that you acquire an Autotrack point by simply moving your cursor over the Osnap marker. DO NOT pick the point. Just move the cursor over the marker, and AutoCAD will acquire the point automatically and display a small plus sign (+) in the center of the Osnap marker.

To remove the small symbol from an acquired point, move the cursor back over the small +. AutoCAD removes the symbol, and the point is unacquired.

Tip: You can set a temporary tracking point without using Osnap markers. When you type **TT** at any point prompt, AutoCAD prompts you for a temporary OTRACK point. Once a point is selected, AutoCAD places a small plus sign (+) at the point, indicating that the tracking point is set. To remove the temporary tracking point, move the cursor over the small +.

Polar Tracking. Polar tracking is the perfect compliment to object snap tracking. When it is enabled, polar tracking alignment paths appear on the screen as you move away from the point you already set. However, unlike with object snap tracking, you do not need to acquire a point to display the polar angle alignment paths. If polar tracking is enabled, the alignment paths automatically appear as you move away from the last point and prepare to pick the next point.

Polar tracking is discussed in detail in this chapter in the section titled The Polar Tracking Feature.

In the following exercise you will use the AutoTrack features.

EXERCISE: USING AUTOTRACK'S OBJECT SNAP TRACKING AND POLAR TRACKING FEATURES

1. Open the drawing 06DWG04. The drawing shows two parallel line segments, three units apart.
2. From the Tools menu, choose Drafting Settings. The Drafting Settings dialog box is displayed.
3. Choose the Object Snap tab. Make sure that the Endpoint and Midpoint object snap modes are selected, and that all other options are cleared, including Object Snap On and Object Snap Tracking On, as shown in Figure 6–26.
4. From the Drafting Settings dialog box, choose the Polar Tracking tab. Make sure the Polar Tracking On option is cleared, and that the Track Using All Polar Angle Settings and Absolute options are selected.
5. In the Polar Angle Settings area, from the Increment Angle drop-down list, choose 30.0, as shown in Figure 6–27.
6. From the Drafting Settings dialog box, choose the Snap and Grid tab. Make sure the Snap On and Grid On options are cleared.
7. In the Snap Type & Style area, choose Polar Snap. The Polar Spacing area is activates, and the Snap area is grayed.

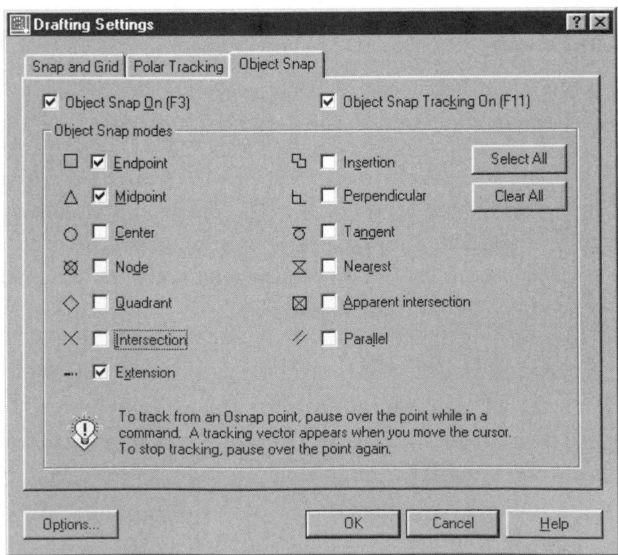

Figure 6–26 *The proper object snap settings for this exercise.*

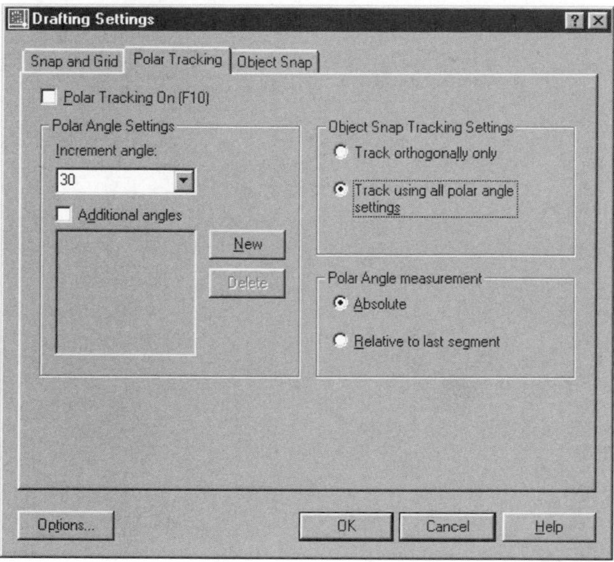

Figure 6–27 *The proper polar tracking settings for this exercise.*

8. In the Polar Spacing area, set the Polar Distance to 0.5, as shown in Figure 6–28, and then choose OK.
9. From the status bar at the bottom of the screen, choose the SNAP, POLAR, and OSNAP buttons to activate the snap grid, polar tracking, and Osnap.
10. From the Draw menu, choose Line, and then pick the endpoint of the line at location D.
11. Move your cursor away from the endpoint. Notice as you move your cursor around that dashed alignment paths appear at 30-degree increments. This occurs because you set the increment angle to 30.0 and chose the Track Using All Polar Angle Settings option, as shown in Figure 6–27.
12. Move your cursor up and to the right of the endpoint, until the 60-degree alignment path is displayed. Drag your cursor along the 60-degree alignment path. Notice that the cursor is constrained to the path and it snaps at 0.5-unit increments. This occurs because you set the Polar Distance option to 0.5 and chose the Polar Snap option, as shown in Figure 6–28.
13. Pick the line's endpoint when the AutoTrack tooltip reads "Polar: 1.0000<60," as shown in Figure 6–29, and then press Enter to end the Line command.
14. From the status bar at the bottom of the screen, choose the OTRACK button to activate the Object Snap Tracking feature. Then choose the SNAP and POLAR buttons to turn the Snap Grid and Polar Tracking features off.

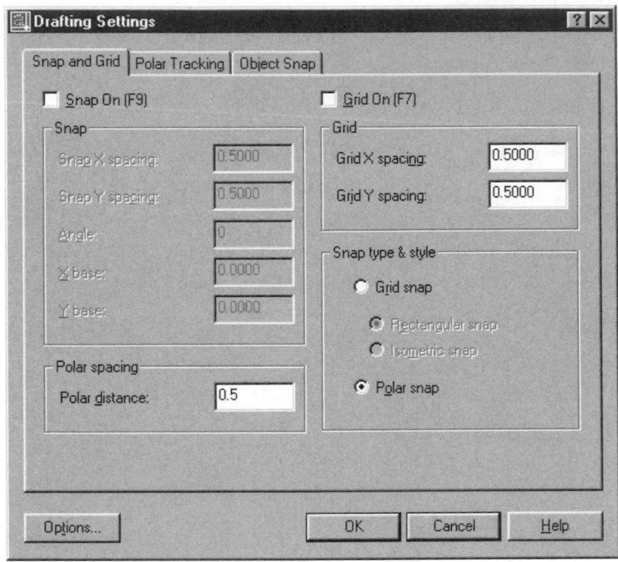

Figure 6–28 *The proper snap and grid settings for this exercise.*

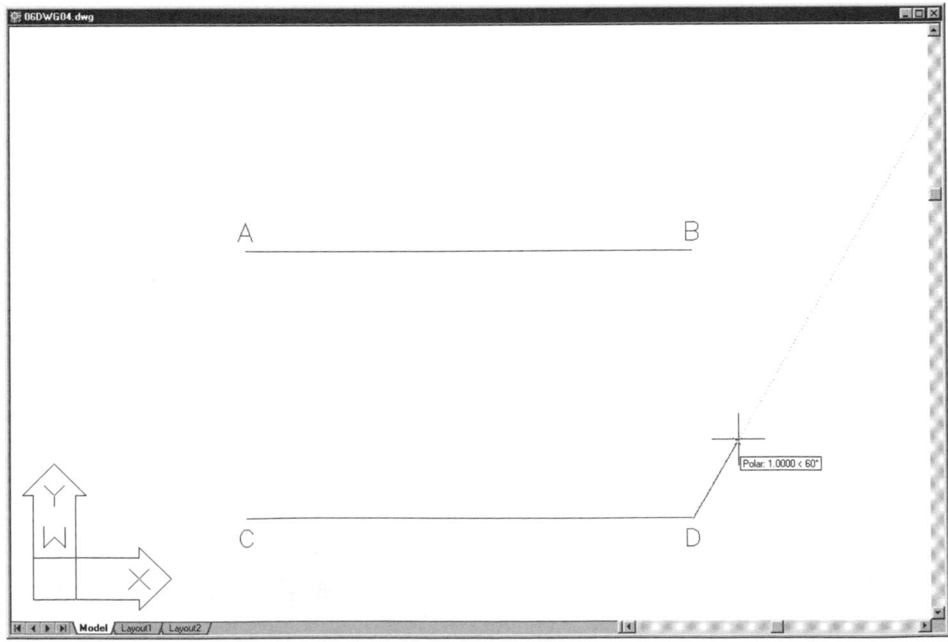

Figure 6–29 *The Polar Tracking feature displays alignment paths in the designated angle increments.*

15. From the Draw menu, choose Line and then pick the endpoint of the line at C. Move you cursor over the endpoint at C, and when the Endpoint Osnap marker appears, move the cursor into the marker until a small + appears. When the small + appears, object snap tracking is activated for this marker. DO NOT pick the endpoint.
16. Next, move your cursor to the endpoint of the first line you created. When the Endpoint Osnap marker appears, move the cursor into the marker until a small + appears. When the small + appears, object snap tracking is activated for this marker. DO NOT pick the endpoint.
17. Move your cursor to the left along the object snap tracking alignment path, toward the endpoint at C. As you near the endpoint at C, object snap tracking alignment paths appear for its marker.
18. Continue moving your cursor toward the left until the AutoTrack ToolTip reads "Endpoint:<120, Endpoint:<180," pick the point as shown in Figure 6–30, and then press Enter to end the Line command. You will continue in this drawing in the next exercise.

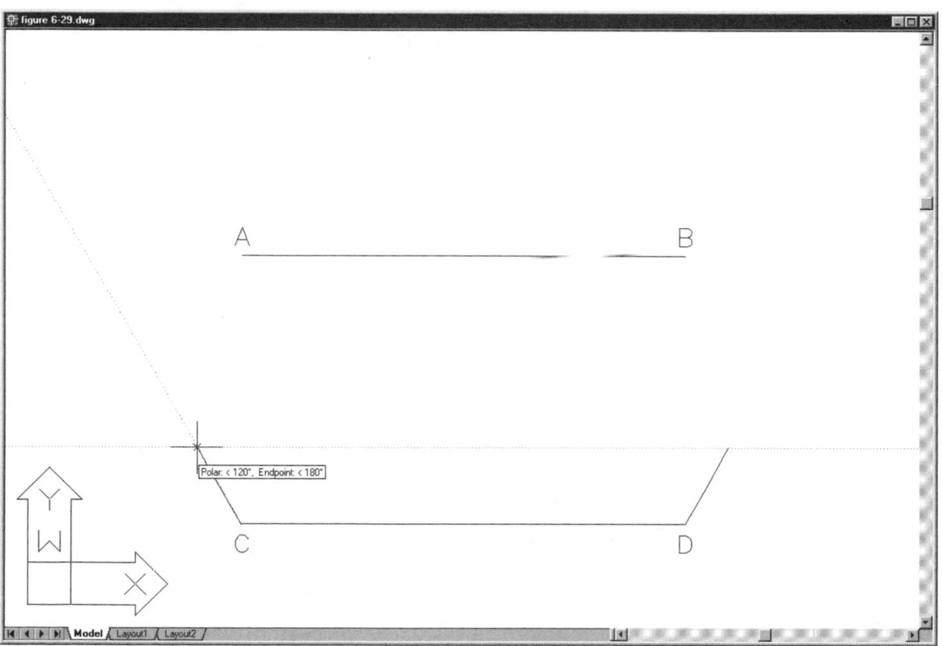

Figure 6–30 *The Object Snap Tracking feature displays alignment paths for acquired Osnap markers.*

As you just experienced, AutoTrack's Object Snap Tracking and Polar Tracking features can help you pick points by displaying temporary alignment paths. Remember that object snap tracking relies on Osnaps and sets alignment paths only when an Osnap marker is acquired. In contrast, polar tracking functions independently of Osnaps and displays alignment paths during a command as you move your cursor away from the last point set.

From and Apparent Intersection Osnaps

Much like the Object Snap Tracking feature, the "auxiliary" Osnaps, From and Apparent Intersection, supply data points that stand in some relationship to points on drawing objects. The From object snap establishes a temporary reference point as a basis for specifying subsequent points. The From object snap is normally used in combination with other object snaps and relative coordinates. For example, at a prompt for the center point of an arc, you could enter **From Endp**, select a line, and then enter @4,5 to locate a point four units to the right and five units up from the endpoint of the selected line. The center of the arc would then be located at this point.

Apparent Intersection snaps to the apparent intersection of two objects that might or might not actually intersect in 3D space. In 2D drafting, Apparent Intersection is usually involved with the projected intersection of two line elements.

The From and Apparent Intersection snap features provide tools that allow you to snap to points where no objects exist. With these tools, you set the first point of a line by offsetting it from the midpoint of an existing line, and then snap the other end of the line to the apparent intersection of two objects.

Extension and Parallel Osnaps

The Extension Osnap extends an existing object's vector, simulating the extension with a dashed construction line. Similarly, the Parallel Osnap displays a dashed line that is parallel and offset from an existing object.

EXERCISE: DRAWING A LINE USING THE EXTENSION AND PARALLEL OSNAPS

1. Open the drawing 06DWG05.
2. From the Draw menu, choose Line. AutoCAD starts the Line command.
3. At the "Specify first point:" prompt, Shift+right-click to display the cursor menu, and then choose Extension.
4. Move your cursor to the angled line extending up from point C. Move your cursor up along the line to its endpoint. AutoCAD displays a small + at the end of the line, indicating that it is acquired. DO NOT pick the endpoint.

5. Continue moving the cursor up along the imaginary extension of the line. AutoCAD extends the existing line, as shown in Figure 6–31.
6. At the "of:" prompt, type **1.5** and then press Enter. This establishes the starting point of the line.
7. At the "Specify next point or [Undo]:" prompt, Shift+right-click to display the cursor menu, then choose Parallel.
8. At the "to:" prompt, move your cursor to the line A-B. When your cursor moves over the line, the Parallel Snap marker appears on the line. Then AutoCAD places a small + in the center of the Parallel Snap marker, indicating that it is acquired. DO NOT pick the line.
9. With the parallel snap acquired, move your cursor downward, below line A-B, until a dashed construction line appears, as shown in Figure 6–32. Notice that the dashed line is parallel to line A-B.
10. At the "to:" prompt, type **4.5** and then press Enter. AutoCAD sets the endpoint of the line.
11. Press Enter to end the Line command.

Like the Osnaps discussed in the two previous exercises, the new Extension and Parallel snap features provide tools that allow you to snap to points where no objects exist. With these tools, and the tools described in the two previous exercises, you can snap to just about any point in your drawing, real or imaginary.

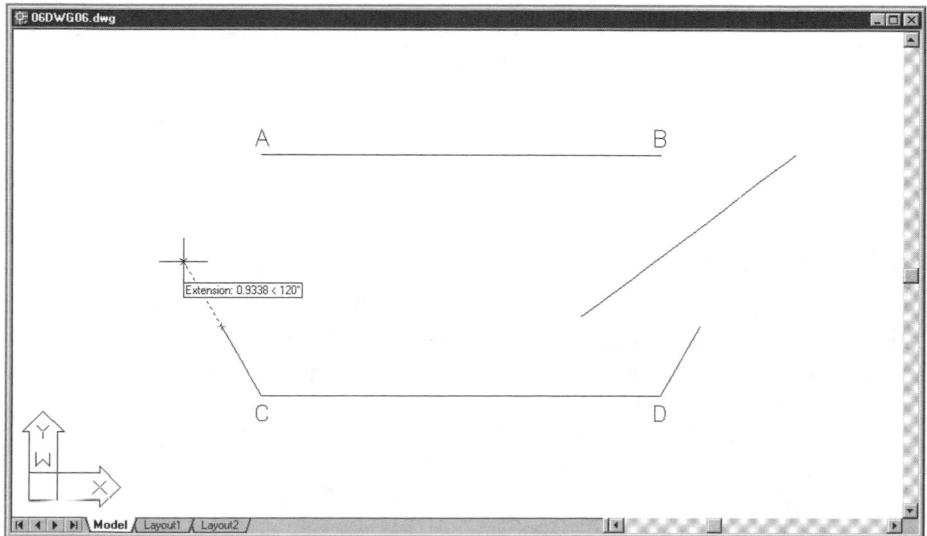

Figure 6–31 *AutoCAD displays the extension of the acquired line.*

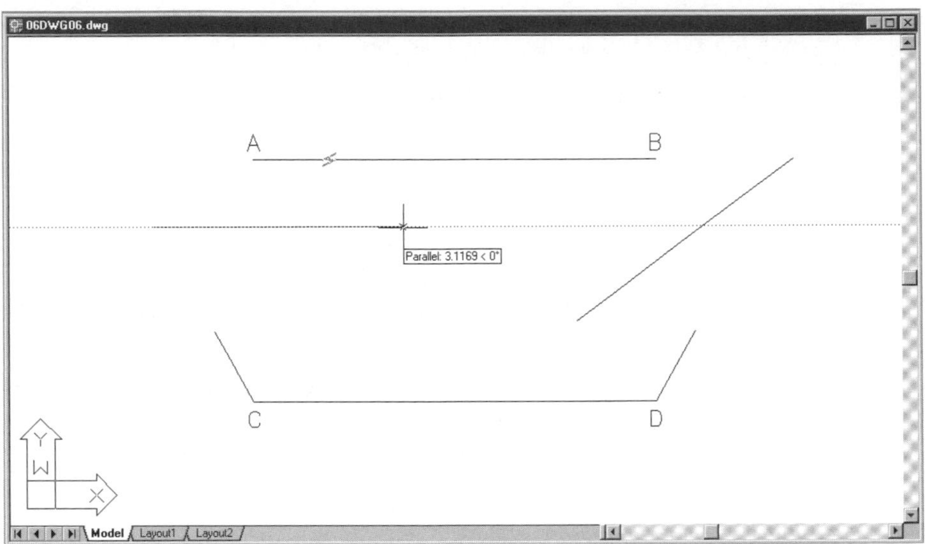

Figure 6–32 *The Parallel snap creates an imaginary alignment line parallel to the acquired object.*

XLINES AND RAYS

With the existence of Apparent Intersection, From Osnaps, and the new AutoTrack, Extension, and Parallel snap features, little need exists for the "construction lines" used in traditional "pencil" drafting. Once you become competent with these drawing aids, the time and effort required to draw and subsequently erase traditional construction lines will seem inefficient.

There may be occasions, however, when the inclusion of construction lines may be indicated to assist in visually presenting the relationship among elements of a drawing. AutoCAD has two special line objects, xlines and rays, that function as traditional construction lines.

The XLINE command creates infinite lines, which are commonly used as construction lines. Xlines can be placed vertically, horizontally, at a specified angle, offset at a specified distance, or as an angle bisector. Although xlines extend infinitely in both directions, they are ignored for the purpose of calculating the drawing's extents.

The RAY command creates "point to infinite" lines, which are commonly used as construction lines. A ray has a finite starting point and extends to infinity. As with the xline, the infinite length of a ray is ignored for the purpose of determining a drawing's extents.

In conjunction with their use as a largely visual element, both xlines and rays are often placed on separate layers with a distinctive linetype and color assigned.

SUMMARY

In this chapter you learned about AutoCAD's coordinate system and the methods you can use to enter coordinate points in your drawings. Absolute coordinate entry enables you to specify points relative to the drawing's current coordinate system's 0,0 point, or origin. Relative coordinate entry, on the other hand, enables you to specify points relative to the previous point you entered. Relative coordinates are expressed either as an X and Y distance or as a distance and angle from the last point. You also learned how to change the orientation of AutoCAD's coordinate system by creating User Coordinate Systems (UCSs). In addition, you learned how to configure drafting settings for snap and grid, polar tracking, and object snap.

This chapter also covered the important concept of snapping to specific geometry in your drawings using various Osnaps. You also learned that AutoCAD has several powerful Osnap tools, including the new AutoTrack feature, and the new Extension and Parallel snaps, which make snapping to "apparent" points both easier and less ambiguous than in previous releases.

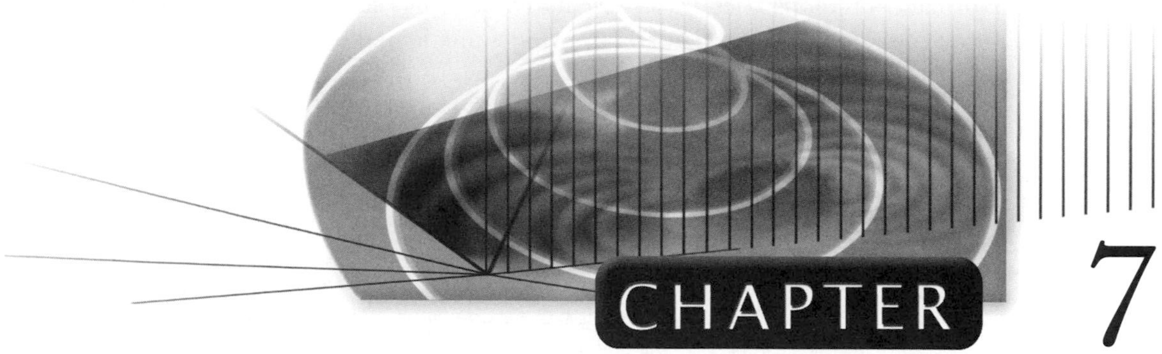

CHAPTER 7

Creating Elementary Objects

No matter how complicated a drawing is, no matter how many layers and linetypes it contains, almost every AutoCAD drawing is comprised of a few relatively basic shapes and forms. Circles, arcs, lines, rectangles, polygons, and ellipses are the basic elements from which both simple and complicated drawings are made. This chapter shows you the tools you will need to construct and control AutoCAD's basic drawing objects.

This chapter covers the following topics:

- LINE command
- ARC command
- CIRCLE command
- POLYGON command
- Creating ellipses

THE LINE COMMAND

Perhaps the most common object in a typical AutoCAD drawing is the line. In addition to representing the shortest distance between two points, lines serve a myriad of other useful purposes: Centerlines locate other geometry, border lines indicate an area's constraints, and dashed lines represent objects or boundaries that are not visible from a given point of view. All these lines are usually further identified functionally by their linetype—the periodic pattern of interruptions in the line's continuity. Normal continuous lines are representative of things such as walls or the sides of objects. Lines are very versatile, and drawing a line is one of the most basic operations in AutoCAD.

In the following exercise you will learn the basics of using the LINE command as you begin drawing a fixture base.

Note: The exercises in this chapter use the template file 07DWG01.dwt found on the accompanying CD-ROM. This drawing has most of its settings, linetypes, and layers already set or defined for you. In the first part of the chapter you learn about lines, circles, arcs, and polygons. When you finish this first section, your drawing will resemble Figure 7–1. Later in the chapter you will edit a drawing in which you will practice constructing and accurately placing ellipses. The use of template drawings is discussed in Chapter 2, Beginning a Drawing.

EXERCISE: USING THE LINE COMMAND TO DRAW A FIXTURE BASE

1. Create a new drawing named 07DWG01.DWG using the 07DWG01.DWT template drawing from the accompanying CD. Using template drawings is discussed in Chapter 2, Beginning a Drawing. Make sure that the current layer is Center.
2. Begin the LINE command by choosing Draw>Line. At the "Specify first point:" prompt, type **38,88** and press Enter.

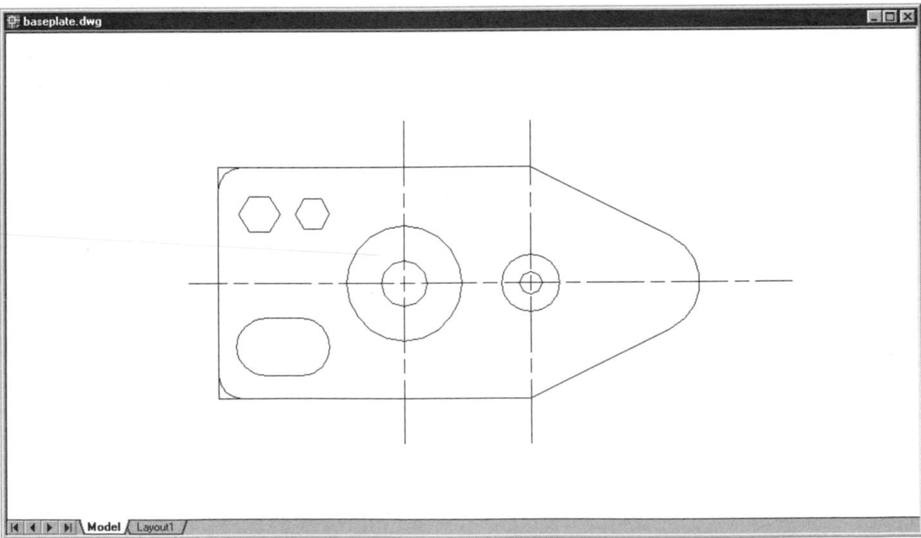

Figure 7–1 *This chapter's completed fixture base.*

3. At the 'Specify next point or [Undo]:" prompt, use relative polar coordinate entry by typing **@208<0** and pressing Enter. Note that the line segment is drawn with a linetype of Hidden. Close the LINE command by pressing Enter.
4. Restart the LINE command by pressing the spacebar. Answer the "Specify first point:" prompt by typing **112,32** and pressing Enter. Make sure that the ortho mode is on by clicking the ORTHO button on the status bar or by pressing F8.
5. Use direct distance entry by moving the cursor above the last point and answering the "Specify next point or [Undo]:" prompt by typing **112** and pressing Enter. End the LINE command by pressing Enter. Note that the line segment is drawn 112 units at 90 degrees from the first point.
6. Make sure that snap mode is on (press F9) and restart the LINE command by typing **L** and pressing Enter. Respond to the "Specify first point:" prompt by pressing F6 until the coordinate display on the status bar displays absolute coordinates. Then find and pick point 156,144.

Note: If you miss the point in the preceding step, type **U** to undo the last point placement. Restart the LINE command by pressing either Enter or the spacebar and then pick the point again.

7. At the "Specify next point or [Undo]:" prompt, type the relative coordinate **@0,-112** and press Enter. End the LINE command by pressing Enter again. Your drawing should now resemble Figure 7–2.
8. Leave this drawing open for the next exercise.

In this exercise you used several different methods of specifying coordinate point entry while using the LINE command. You also saw how the coordinate display, which is located on the status bar at the bottom of AutoCAD's display, can be helpful in locating both absolute and relative coordinate points. Keep in mind that depending on how your increment snap is set up and on the value of the actual points, the snap feature can make finding many points much easier. For points that

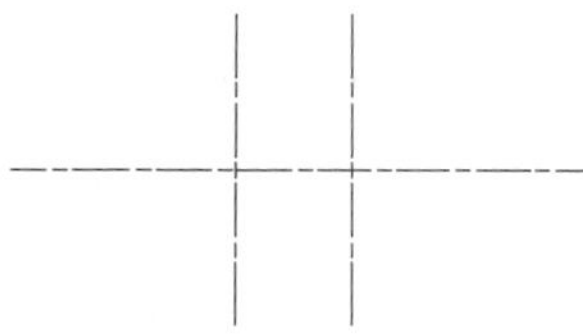

Figure 7–2 *Setting up the center lines for the fixture base.*

do not lie on your current snap pattern, entering coordinates directly through the keyboard is the only practical method of having line segments begin and end exactly where you want.

Tip: In many situations, you will find yourself frequently changing the coordinate display mode, as well as turning ortho and snap on and off. It can be more convenient to use the function key shortcuts to control these functions. F6 controls the coordinate display, F8 toggles the ortho mode on and off, and F9 toggles the snap function.

In the following exercise you will continue to use the LINE command as you outline the fixture base. After you establish the first corner of the base, you will use the efficient direct distance method for entering points that are orthogonal to the previous point.

EXERCISE: USING THE LINE COMMAND TO DRAW THE FIXTURE BASE OUTLINE

1. Continue from the previous exercise. Change the current layer to OBJECT. Refer to Figure 7–3 to help you identify the points in the steps that follow.
2. First you will deliberately enter the wrong point to see how easy it is to recover from such a mistake. Begin the LINE command, type the coordinates **150,48**, and then press Enter. Note that the point falls short of the right centerline. The X coordinate is incorrect.
3. To recover from your mistake, use the LINE command's Undo option by typing **U** and pressing Enter, and then reenter the point at ① by typing **156,48** and pressing Enter.
4. Now with ortho mode on and the cursor placed to the left of the previous point, use direct distance entry to specify the point at ② by typing **108** and pressing Enter.

Note: From now on in this chapter's exercises, instructions such as, for example, "enter **80**" and "enter **Line**" will be the same as "type **80** and press Enter" and "type **Line** and press Enter," respectively.

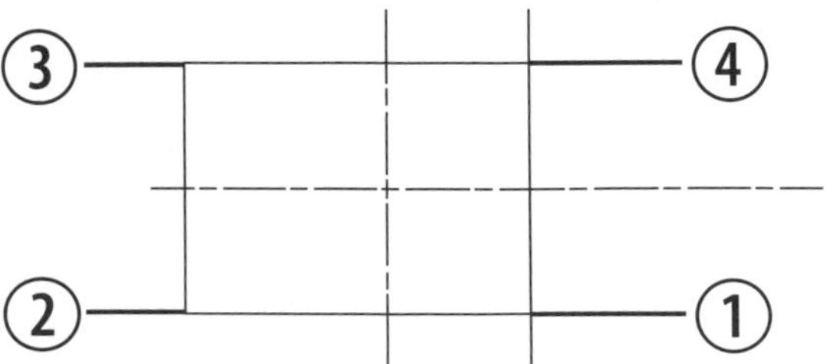

Figure 7–3 *Points used to draw the fixture base.*

5. Move the cursor above the preceding point, and enter **80** to draw a line segment to ③.

6. Move the cursor to the right of ③ and enter **108**. The line segment from ③ to ④ is drawn.

7. Now close the outline by using the Close option of the LINE command (type **C** and press Enter). This closes to the start point and ends the LINE command.

8. The line segment from ④ to ① was drawn by mistake; it is not wanted. At the command prompt, issue the U command by typing **U** and pressing Enter. Note that because the LINE command is no longer in progress, all four line segments completed during the LINE command are erased. At the command prompt, issue the REDO command by entering **REDO**. All four segments are redrawn.

Tip: You can use the REDO command to recreate the sequences undone by a UNDO.

9. To erase the last line segment, issue the ERASE command by typing **E**. At the "Select objects:" prompt, type **L** and press Enter. Note that the last completed line segment is highlighted. Now, with the "Select objects:" prompt still current, press Enter. The line is erased. Your drawing should now resemble Figure 7–4.

10. Save your drawing at this point using the name 07DWG01.DWG. If you are continuing with the next section, leave this drawing open.

THE LINE COMMAND OPTIONS

The LINE command is straightforward and easy to use. It offers the following features:

- **"Specify first point:"**—At this prompt, your input places the first point of the first line segment. If you press Enter at this prompt, the line segment will start from the last specified endpoint of the last-drawn line segment or arc.
- **"Specify next point:"**—At this prompt, your input places the point to which the current line segment is drawn.

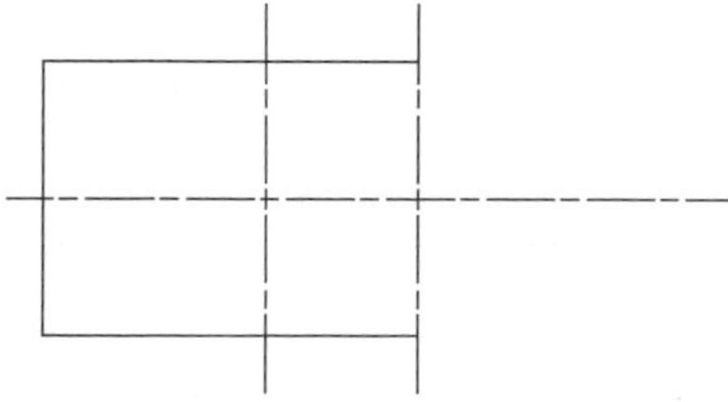

Figure 7–4 *Setting up the center lines for the fixture base.*

- **[Close/]**—After two successive line segments have been drawn, you can enter **C** (Close) to close the series. A line is drawn from the last endpoint to the first point of the series.
- **[Undo]**—At any "Specify next point:" prompt, you can enter **U** (undo) to undo the last line segment drawn. Repeating the U option will step back through multiple line segments.

Tip: If you are constructing a long series of line segments using the LINE command, break the continuity of the chain and occasionally restart the command by pressing Enter three times. This has the effect of ending the series, restarting the LINE command, and beginning a new series from the end of the last. Then if you perform a U option at a command prompt, the option will undo only the last LINE command's series of line segments, instead of all the segments since the very first start point.

You can end or exit the LINE command at any time by pressing Esc or by pressing Enter at any "Specify next point:" prompt.

THE ARC COMMAND

The ARC command is used to draw portions of circles known as arcs. Its several options make constructing an arc with a variety of known parameters—such as center, start point, chord length, radius, and so on—much easier than with traditional manual drafting methods.

In the following exercise you will draw an arc by specifying three points on its circumference. After you complete the arc, you will erase it and then draw two small arcs to round the corners of the fixture base. Use the coordinate display on the status bar and the Snap feature to identify points.

EXERCISE: USING ARC TO DRAW CIRCULAR ARCS

1. Continue with 07DWG01.DWG from the preceding exercise. If necessary, turn the snap mode on by pressing the F9 key. Begin the ARC command by selecting the Arc tool from the Draw toolbar. Respond to the "Specify start point of arc or [Center]:" prompt by specifying the point at ① in Figure 7–5 by entering its coordinates, **156,48**.
2. At the "Specify second point of arc or [CEnter/End]:" prompt, specify the point at ② by entering **196,88**. Note that the arc passes through ② and is dragged as you move the cursor.
3. At the "Specify end point of arc:" prompt, specify the point at ③ by entering **156,128**. The ARC command draws the arc and is terminated. Your drawing should resemble Figure 7–5.
4. At the command prompt, issue the U command by typing **U** and pressing Enter. The arc is deleted.

Before starting the next step, you may want to zoom in to enlarge the view. To do this, type **Z** and press Enter. Then type **W** and press Enter. Click and drag a window around the left side of the base plate, then click to zoom.

5. Start the ARC command again by typing **A**. At the "Specify start point of arc or [Center]:" prompt, enter the coordinates **56,128 (**see ④ in Figure 7–6).
6. Respond to the next prompt by typing **CE** and pressing Enter to select the Center option. Then, at the "Specify center point of arc": prompt, specify the point at ⑤ by entering **56,120**.
7. Make sure that ortho mode is on (press F8 if necessary) and note that as you move the cursor, the arc snaps in 90-degree increments. Move the cursor to the left of ④ and pick. The ARC command draws the arc; ortho mode forces it to 90 degrees.
8. Restart the ARC command by pressing Enter. At the "Specify start point of arc or [Center]:" prompt, enter the coordinates **48,56** to start the arc at ⑥.

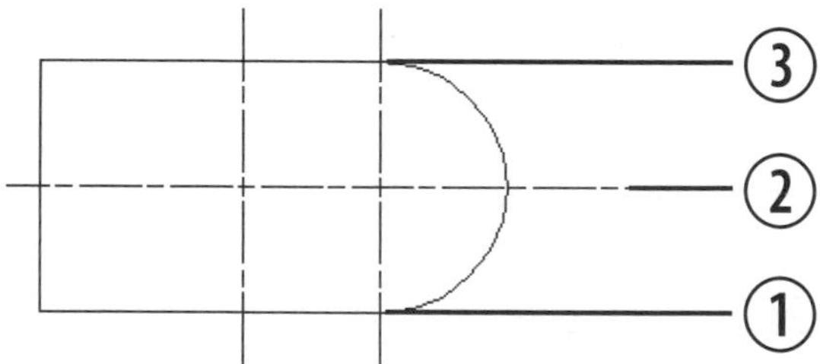

Figure 7–5 *Drawing an arc using the three-point method.*

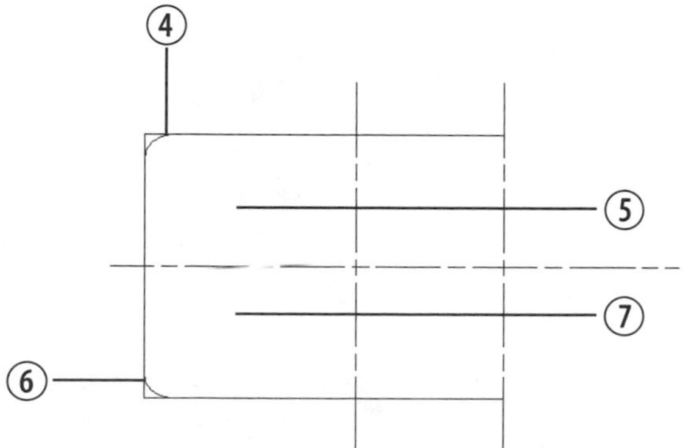

Figure 7–6 *Creating rounded corners with the ARC command.*

9. At the "Specify second point of arc or [CEnter/End]:" prompt, type **CE** and press Enter. This indicates that the next point you supply will be the center of the arc.
10. At the "Specify center point of arc:" prompt, enter the relative rectangular coordinates **@8,0**. This places the center of the arc at ⑦. With ortho mode on, the arc jumps to 90 degree increments as you move the cursor around.
11. Disable ortho mode by pressing F8; notice that the arc is now dragged with the cursor. Answer the "Specify end point of arc or [Angle/chord Length]:" prompt by entering **A**, indicating that next you will supply the included angle for the arc.
12. Turn ortho mode back on and move the cursor anywhere above the center point. Note that the arc now snaps to 90-degree points again. Pick any point directly above the center point near ⑤ to complete the arc and end the ARC command.
13. Save the changes you have made by pressing Ctrl+S. Leave the drawing open for the next exercise.

Whenever you draw an arc, you know either its center or its start point and can supply the other necessary information from existing geometry in the drawing. Figure 7–7 shows the Arc submenu in AutoCAD's Draw menu. This submenu conveniently lists the various choices that can be made depending on the information supplied. Starting the ARC command by using this submenu provides a shortcut through the longer prompt choices provided at the command prompt.

THE ARC COMMAND OPTIONS

The ARC command requires three pieces of information to complete an arc, one of which must be either the center of the arc or its start point. The other required

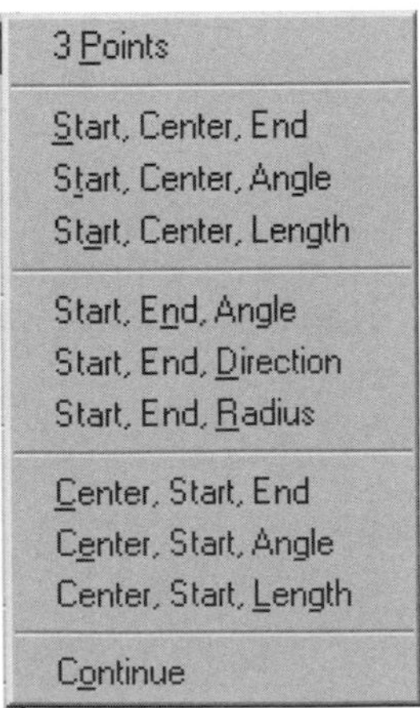

Figure 7–7 *The Arc submenu offers eleven methods of drawing an arc.*

parameters can be supplied in various combinations. The following list explains the ARC command's three-piece combination options:

- **3 Points**—This method creates an arc that passes through three points that you supply. The first point is considered the start point, the second point is the endpoint, and the third point is any other point between these two. This is the default method of constructing arcs.
- **Start, Center**—This method requires the arc's starting and center points. The third piece of data can be the endpoint, an included angle, or the length of the chord. Counterclockwise arcs are drawn if the included angle is supplied as a positive angle; clockwise arcs are drawn if the angle is supplied as a negative angle. A positive chord length draws a minor arc (less than 180 degrees), and a negative chord length creates a major (greater than 180 degrees) arc.
- **Start, End**—This method enables you to supply the start point and endpoint of the arc and specify how to draw the arc. You define the arc with an angle, direction, radius, or center point. When supplied with a positive angle, AutoCAD draws a counterclockwise arc; when supplied with a negative angle, it draws a clockwise arc. If you choose the radius option, AutoCAD always draws the arc counterclockwise. A negative radius forces a major arc, and a positive radius forces a minor arc.
- **Center, Start**—This method enables you to identify first the center of the arc, and then the start point. Supplying either the angle, chord length, or endpoint completes the arc. When you supply a chord length, a negative length creates a major arc, and a positive length creates a minor arc. If you supply an angle, a negative angle draws the arc clockwise; a positive angle draws the arc counterclockwise.
- **Continue**—This method is the built-in default. You invoke this option by pressing Enter at the first arc prompt. It begins a new arc tangent to the last line or arc drawn.

Tip: There probably is no other AutoCAD command that seems at times to be as uncontrollable as the ARC command. The trick in drawing arcs with other than the basic and simple 3-point or Start, Center, End options is knowing how to force the arc in the direction you desire. Understanding that AutoCAD, by default, thinks of arcs as developing counterclockwise from the start point is the key to controlling arcs. To force an arc to proceed clockwise requires you to input a negative angle, for example, or to supply a negative distance for a chord length parameter. The same type of entries control whether an arc is drawn as a minor or major arc. Armed with this knowledge and a little practice, your arcs can come out correctly—usually on the first try.

The ARC command's Continue option is often convenient to use when arcs are associated with line segments. The following exercise demonstrates this feature of the ARC command.

EXERCISE: PUTTING LINES AND ARCS TOGETHER

1. Continue in the drawing in the preceding exercise. Make sure ortho mode is on. Refer to Figure 7–8 for this exercise. It will be helpful to zoom in to provide a larger view of the base plate features you have drawn so far.
2. Start the LINE by entering **L**. At the "Specify first point:" prompt, specify the point at ① by typing **64,56** and pressing Enter. Then at the "Specify next point or [Undo]:" prompt, enter the relative coordinates **@12,0** (see ② in Figure 7–8 for reference).
3. Choose the Arc tool from the Draw toolbar to cancel the LINE command and start the ARC command.
4. At the "Specify start point of arc:" prompt, press Enter. This activates the Continue option of the ARC command and starts an arc tangent to the endpoint of the line you just drew. Now specify the point at ③ by entering **76,76**.
5. Now restart the LINE command by selecting the Line tool from the Draw toolbar. At the "Specify first point:" prompt, press Enter. This activates the Continue option of the LINE command and starts the line tangent to the arc.
6. At the "Length of line:" prompt, enter **12**. This draws a line to ④.
7. Again, start the ARC command and press Enter. This again cancels the LINE command and starts an arc at the endpoint of the line.
8. At the "Specify endpoint of arc:" prompt, enter the relative polar coordinates **@20<270**. This completes the arc at ①.
9. Your drawing should now resemble Figure 7–8. Press Ctrl+S to save the drawing in its present form. If you are not continuing immediately with the next section, you can close this drawing and reopen it later.

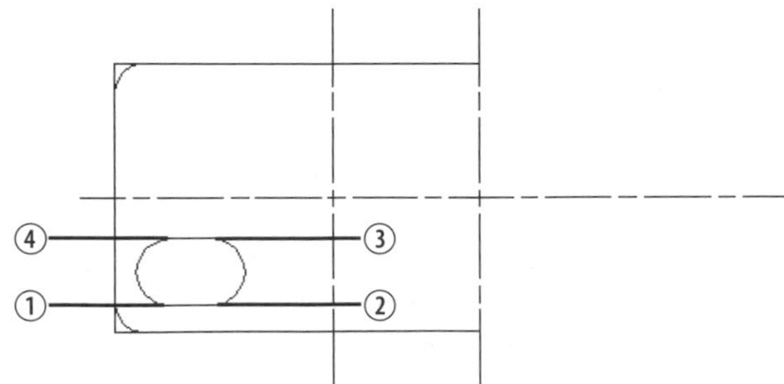

Figure 7–8 *Use the Continue feature to draw slots.*

The drawing shows the fixture base with rounded corners on the left side. After this exercise, you should be familiar with the continuing features of both the LINE and ARC commands. The "trick" is to disregard the first "Specify first (or start) point:" prompt of these two commands and press Enter instead. When you draw an alternating series of line segments and arcs, this method can save significant amounts of time spent selecting start points.

Note: Chapter 10, Advanced Editing, explains the use of the FILLET command, which provides an easy alternate method of creating arcs tangent to lines.

THE CIRCLE COMMAND

Another basic AutoCAD shape is the circle. Circles are used to represent holes, wheels, shafts, columns, trees, and so on. Several methods of drawing circles exist. Unlike manual drafting, constructing circles in AutoCAD is quick and accurate. Circles have centers, diameters, radii, and tangent and quadrant points. Providing a combination of these parameters will enable you to draw any circle and place it anywhere you want.

In the following exercise you will construct a few basic circles as you add holes and sleeves to the fixture base.

EXERCISE: ADDING HOLES AND SLEEVES BY DRAWING CIRCLES

1. Continue in the drawing from the previous exercise. Refer to Figure 7–9 during this exercise.
2. Start the CIRCLE command by typing **C** and pressing Enter or by selecting the Circle tool from the Draw toolbar.
3. At the "Specify center point for circle or [3P/2P/Ttr (tan tan radius)]:" prompt, specify the point at ① by entering **156,88**. Move the cursor and watch the radius and circle be dragged. The coordinate display at the lower-left of the AutoCAD window should show changing X,Y coordinates. If it does not, click in the display pane until it does. Now click twice in the coordinate display to change to a polar coordinate display. While you watch the polar coordinate display, slowly move the cursor and pick when the display reads 4.00@<0.00. AutoCAD draws a circle passing through ②.
4. Press Enter or the spacebar to restart the CIRCLE command. At the circle prompt, type @ and press Enter. The @ symbol, when entered at any AutoCAD "point prompt," automatically enters the last point entered—in this case, the center point of the circle you just drew.
5. At the "Specify radius of circle or [Diameter]:" prompt, enter **D** to specify a diameter, and then enter **20**. This specifies the diameter and draws the circle.

6. Now, from the Draw menu, choose Circle, then 3 Points. At the "Specify first point on circle:" prompt, enter the point **104,88**.
7. At the "Specify second point on circle:" prompt, enter the point **112,96**. At the "Specify third point on circle:" prompt, enter the point **120,88**. AutoCAD draws the circle using the three points you picked on the circumference.
8. Now press Enter to restart the CIRCLE command with the "Specify center point for circle or [3P/2P/Ttr (tan tan radius)]" default prompt. Enter **112,88** for the center point.
9. Now experiment with the effects of ortho mode. Turn ortho mode off by pressing F8. Move the cursor around while watching the polar coordinate display. Pick when a radius of 20 is displayed. AutoCAD draws the circle with a radius of 20 units.
10. The base plate should now resemble Figure 7–9. You will continue with this drawing in the next section.

CIRCLE COMMAND OPTIONS

The CIRCLE command provides you with several options to control the sequence in which you create circles. In addition to the default center-point, radius mode, you can create a circle by specifying three points on the circumference, or by selecting two objects (lines, circles, or arcs) to which the circle is to be tangent and then specifying a radius. The CIRCLE command offers the following options:

- **Center Point**—Type or pick the center point and you will be prompted for a radius or diameter. Radius is the default option. You override the radius default by entering **D**. When using either the radius or diameter option, you can enter a value or pick two points to show the distance.
- **3P (3 Points)**—Use this option to specify the circumference by entering or picking three points that will lie on the circumference.

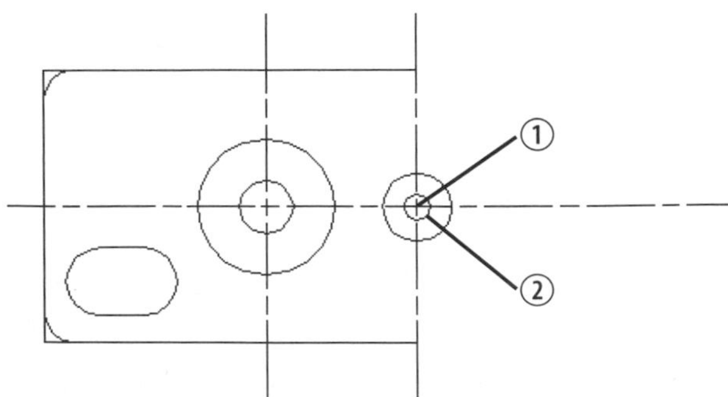

Figure 7–9 *Drawing and placing circles.*

- **2P (2 Points)**—Use this option to specify two diameter points on the circumference.
- **TTR (Tangent-Tangent-Radius)**—Use this option to select two lines, arcs, or circles (any combination) that form tangents to the circle. Then specify the circle's radius.

Note: Note the difference between Center Point/Diameter and the 2P option. Both options enable you to specify a diameter, but if you pick the second point with Center Point/Diameter, it merely indicates the diameter's distance, and the circle does not pass through the second point. If you pick two points with 2P, a circle appears between those two points, and the distance is the diameter. The 2P option enables you to draw a diameter of a circle the way most of us intuitively think of a diameter.

In the following exercise you will practice using the TTR option of the CIRCLE command.

EXERCISE: PRACTICING CIRCLES WITH THE TTR OPTION

1. Continue in the drawing from the preceding exercise. Refer to Figures 7-10 and 7-11 during this exercise. Make sure object snap is on (press F3 if necessary, and check the command line to see the Osnap status).
2. Start the CIRCLE command by selecting the Circle tool from the Draw toolbar. Respond to the "Specify center point for circle or [3P/2P/Ttr (tan tan radius)]:" prompt by entering **T**, indicating the tangent:, tangent, radius option.
3. Respond to the "Specify point on object for first tangent of circle:" prompt by resting the cursor anywhere on the circumference of the circle at ②. Note the appearance of the tangent osnap symbol and the label "Deferred Tangent" on the circle's circumference. Pick to establish the tangent object.

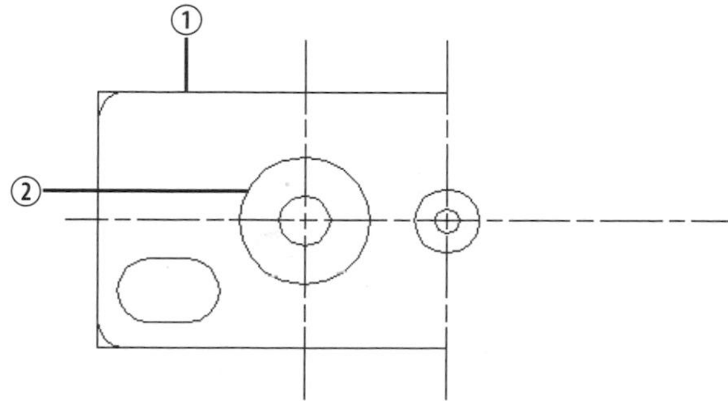

Figure 7–10 *Placing circles using the TTR method.*

4. At the "Specify point on object for second tangent of circle:" prompt, rest the cursor anywhere on the line at ①. Note the appearance of the tangent osnap symbol and the label "Deferred Tangent" on the line. Pick to establish the second tangent object.
5. At the "Specify radius of circle:" prompt, enter **20**.
6. AutoCAD draws the only circle that meets the requirements of being tangent to both the circle at ② and the line at ① and having a radius of 20 units (see Figure 7–11). Type **U** and press Enter to delete the circle.
7. Now repeat steps 2 through 5, but at step 5 type **50** and press Enter. AutoCAD draws the only possible circle ⑥ that is tangent to both the circle at ③ and the line at ④ and has a radius of 50 units (see Figure 7–11).
8. Issue the Undo command again to delete the last circle drawn. It is not needed to complete the fixture.
9. Save the changes you have made by pressing Ctrl+S.

Note: When specifying the TTR option, you may encounter the message Circle does not exist. This indicates that a circle with the radius you specified or that was tangent to the two points you chose, or both, does not exists. Most often, the radius specified is too small.

THE POLYGON COMMAND

In AutoCAD, the POLYGON command is used to create regular polygons with sides of equal length. You can draw a polygon composed of from three to 1,024 sides. After you specify the number of sides, several options are available to complete the polygon:

- **"Enter number of sides:"**—At this prompt, enter the number of sides (3 to 1,024).
- **"Specify center of Polygon or [Edge]:"**—At this prompt, choose whether you want to define the polygon by specifying its center or the endpoints of an edge.

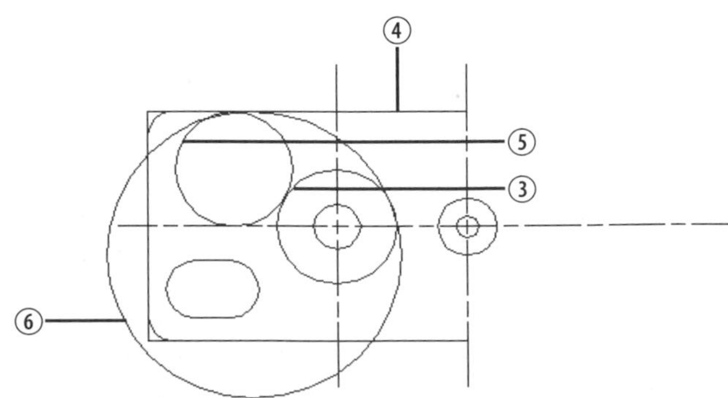

Figure 7–11 *Drawing circles with the TTR option.*

- **"Enter an option [Inscribed in circle/Circumscribed about circle]:"**—This prompt appears if you chose the center option. If you choose Inscribed in Circle, all vertices of the polygon fall on the circle; if you choose Circumscribed about Circle, the radius equals the distance from the center of the polygon to the midpoints of the edges. If you use the pointing device to specify the radius, you dynamically determine the rotation and size of the polygon. If you specify the radius of the circle by typing a specific entry, the angle at the bottom edge of the polygon equals the current snap rotation angle (usually zero degrees).
- **"Specify first endpoint of edge:" and "Specify second endpoint of edge:"**—If you place the polygon by specifying its edge, these prompts enable you to specify the endpoint of one edge.

In the following exercise you will practice drawing a polygon representing a mounting hole on the fixture base.

EXERCISE: ADDING A MOUNTING HOLE WITH THE POLYGON COMMAND

1. Continue in the drawing from the preceding exercise. Start the POLYGON command by choosing Polygon from the Draw menu.
2. At the "Enter number of sides:" prompt, enter **6**.
3. At the "Specify center of polygon or [Edge]:" prompt, enter **62,112** to specify the center of the polygon.
4. At the "Enter an option [Inscribed in circle/Circumscribed about circle]" prompt, enter **C**.
5. At the "Specify radius of circle:" prompt, enter **6**. AutoCAD draws the hexagon.
6. Press Enter to restart the POLYGON command. Repeat steps 3 through 6 but specify a center coordinate of **80,112** and enter **I** to indicate the inscribed option. Specify a radius of **6**.
7. AutoCAD draws the second hexagon. Note the difference in size of the hexagons drawn with the inscribed and circumscribed options. Your drawing should now resemble Figure 7–12.
8. Save the changes you made in this exercise.

AutoCAD's POLYGON command provides a convenient way to draw regular (equilateral) multi-sided polygons, including triangles. Several options make sizing and placing the final polygon relatively easy.

The fixture base drawing that has gradually grown during this chapter is almost complete. In the following exercise you will use the line/arc Continue method of "rounding corners" that you learned earlier in this chapter to complete the drawing.

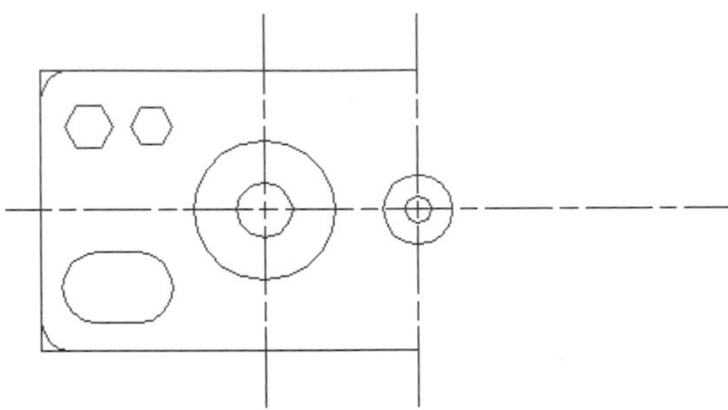

Figure 7–12 *Adding mounting holes with the POLYGON command.*

Tip: AutoCAD's POLYGON command produces polygons that are composed of polylines. Polylines are beneficial because they can be exploded into individual line segments for editing. Their width can also be changed using the PEDIT command. In addition, you can fillet and chamfer all edges with one command sequence. You will learn about both of these techniques later in this book.

EXERCISE: COMPLETING THE FIXTURE BASE WITH A ROUNDED CORNER

1. Continue with the drawing from the preceding exercise. Make sure that the coordinate readout is active. If necessary, press F6 to make it active. Also, turn ortho mode off and snap mode on.
2. Start the LINE command by typing **L** and pressing Enter. At the "Specify start point:" prompt, use the coordinate readout to find and pick the point **156,128** ①.
3. At the "Specify next point:" prompt, enter the relative coordinate: **@48,Ð24**. A line is drawn to ②.
4. End the LINE command by pressing Enter. Start the ARC command by typing **A** and pressing Enter. Press Enter again to take advantage of the Continue feature. At the Specify endpoint of arc: prompt, type the relative polar coordinate: @32<270 and press Enter. AutoCAD draws the arc to ③.
5. Now start the LINE command again and press Enter to activate the Continue feature. At the "Length of line:" prompt, use the coordinate readout to find and pick point **156,48** ④. (If necessary, press F6 until the coordinate display shows absolute coordinates.) Press Enter to end the LINE command.

This completes the fixture base. Your drawing should look like Figure 7–13 and Figure 7–1 at the beginning of this chapter. You are now finished with this drawing.

So far in this chapter you have learned about the basic AutoCAD drawing elements of lines, circles, arcs, and polygons. You will use these elements over and over again in most of your drawings. You have also learned how AutoCAD gives you a large amount of flexibility in constructing and placing these basic elements. Next you will learn how to draw ellipses.

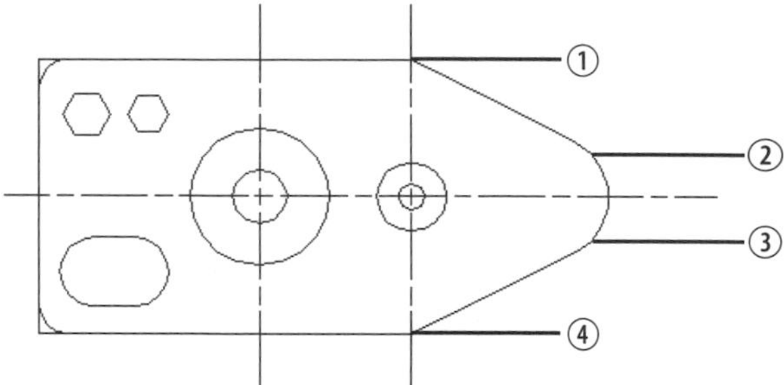

Figure 7–13 *Finishing the fixture base.*

CREATING ELLIPSES

AutoCAD 2000 introduced true ellipses and elliptical arcs, both of which are exact mathematical representations of ellipses. The default method of drawing an ellipse is to specify the endpoints of the first axis and the distance, which is half the length of the second axis. The longer axis of an ellipse is called the major axis, and the shorter one is the minor axis. The order in which you define the axes does not matter.

Note: AutoCAD is also capable of constructing elliptical representations of ellipses using polylines. The system variable PELLIPSE determines the type of ellipse drawn. A value of 1 creates a polyline representation; a value of zero creates a true ellipse.

Tip: The default method of drawing ellipses is to draw true ellipses. Unless you have a specific reason to use the less-accurate polygon approximation, leave the system variable PELLIPSE set to zero and draw true ellipses. Polygon ellipses offer little, if any, advantage.

Ellipses are somewhat complicated geometric figures, but if you have a basic understanding of the geometry of an ellipse, AutoCAD enables you to draw them easily. Figure 7–14 illustrates the major axis and minor axis of an ellipse. Although from a mathematical point of view an ellipse has two "centers," or foci, AutoCAD considers the geometric center to be the intersection of the two axes. The quadrants of an ellipse are the points of intersection between the axes and the ellipse. In AutoCAD, you can use both the quadrants and center of an ellipse as object snap points.

AutoCAD offers you several ways to specify the various parameters of an ellipse.

ELLIPSE COMMAND OPTIONS

When you issue the ELLIPSE command, the following prompt appears:

Specify axis endpoint of ellipse or [Arc/Center]:

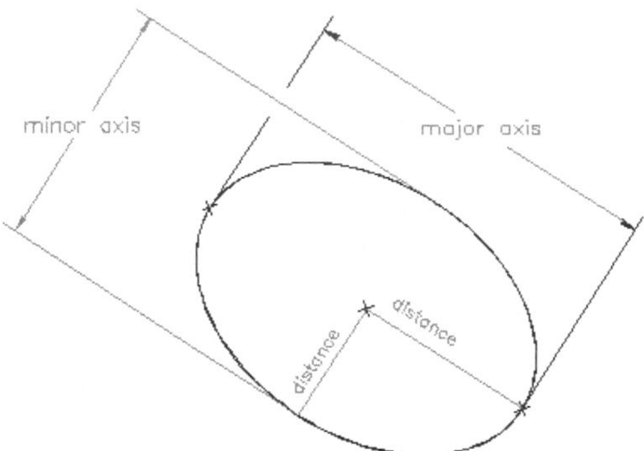

Figure 7–14 *The geometry of an ellipse.*

At this point your options include the following:

- **Axis endpoint**—Specify an axis (major or minor) endpoint. The "Specify other endpoint of axis:" prompt will appear, at which you specify the second axis endpoint. The "Specify distance to other axis or [Rotation]:" prompt then appears. If you specify the other axis distance, AutoCAD draws the ellipse. If you specify Rotation (by typing **R** and pressing Enter), AutoCAD prompts you for an angle and then completes the ellipse.
- **Center**—If you choose the Center option, the "Specify center of ellipse:" prompt appears, at which you specify the center point. The "Specify endpoint of axis:" prompt then appears. Specify an endpoint. The "Specify distance to other axis or [Rotation]:" prompt appears. If you specify the other axis distance, AutoCAD draws the ellipse. If you specify Rotation (by typing **R** and pressing Enter), AutoCAD prompts you for an angle. The angle specifies the ratio of the major axis to the minor axis. An angle of zero degrees defines a circle. The maximum angle acceptable is 89.4 degrees, which yields a "flat" ellipse.
- **Arc**—If you choose the arc option (by typing **A** and pressing Enter), the "Specify axis endpoint of elliptical arc or [Center]:" prompt appears, which requests the same information as the prompt for a full ellipse. After you answer the prompt sequence for the full ellipse, the "Specify start angle or [Parameter]:" prompt appears. Specifying a point defines the start angle of the arc and causes the "Specify end angle or [Parameter/Included angle]:" prompt to appear. Specifying an end angle draws the arc. Specifying **I**, for Included Angle, enables you to specify an included angle for the arc, beginning with the start angle.

In the following exercises you will practice drawing ellipses in a three-view mechanical drawing. During the exercises you will utilize the object snap techniques learned in Chapter 6, Accuracy in Creating Drawings. In the next exercise, first you will use a full set of construction lines to draw an ellipse in one view. Then a reduced set of construction lines will provide the information you need to draw an ellipse in another view.

Note: In reckoning angles for elliptical arcs, the direction of the first point of the major axis is considered zero degrees. If the system variable ANGDIR is set to zero (the default), angles for the elliptical arc are measured counterclockwise; if ANGDIR is set to 1, angles are measured clockwise. If the minor axis is defined first, the major axis zero point is 90 degrees in a counterclockwise direction. If you choose the Included Angle option, the angle is measured from the start point, not from the zero-degree point.

Tip: When specifying elliptical arc angles, it is helpful to set the coordinate display on the status bar to indicate polar coordinates. You cycle through the coordinate display modes by pressing the F6 key.

EXERCISE: DRAWING ELLIPSES IN A THREE-VIEW DRAWING

1. Open the drawing 07DWG02.DWG found on the accompanying CD-ROM. This drawing, shown in Figure 7–15, shows an unfinished three-view drawing of a mechanical mounting bracket. A shaded isometric view is included. The hole in the angled portion of the bracket has already been projected onto the Front view. You will use the Object Snap Tracking feature of AutoCAD (discussed in Chapter 6) to draw the ellipse and the elliptical arc representing the hole in the Top view.

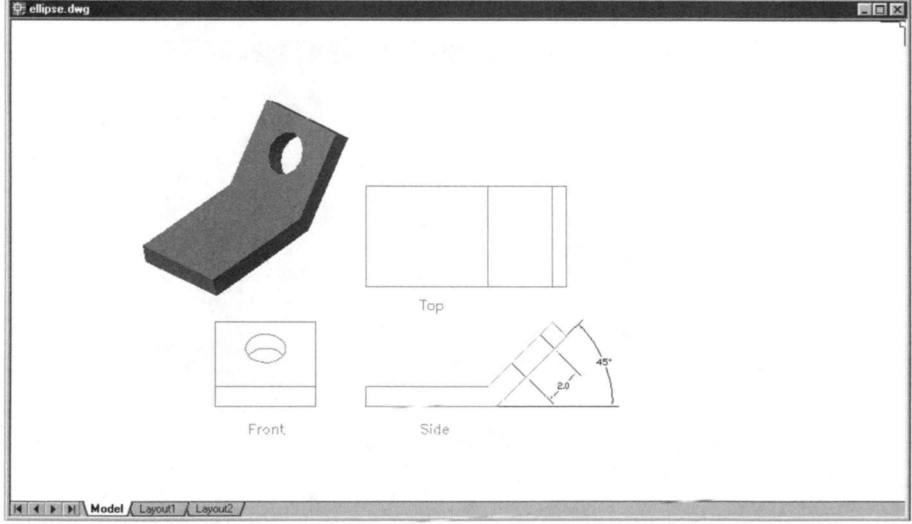

Figure 7–15 *An incomplete three-view drawing of a mounting bracket.*

2. First, establish the settings you will need for object snap tracking. Right-click on the OTRACK button on the mode status line at the bottom of the display and select Settings from the shortcut menu.
3. In the Drafting Settings dialog box, in the Object Snap folder, choose Object Snap On and Object Snap Tracking On. Under Object Snap Modes, select Endpoint, Midpoint, Quadrant, and Intersection (see Figure 7–16). Choose OK to close the dialog box.
4. To zoom to a closer view, from the View menu, choose Named Views. In the View dialog box, select view AAA, select Set Current, and then choose OK to close the dialog box and establish the new view shown in Figure 7–17.
5. Referring to Figure 7–17, you will use the intersections of the hidden lines with the surface of the bracket at ① and ②, and the midpoint of the top edge at ③, to establish AutoTracking points to position the ellipse accurately in the Top view. From the Draw menu, choose Ellipse, then select Axis, End.
6. At the "Specify axis endpoint of ellipse or [Arc/Center]:" prompt, move the cursor over ① (see Figure 7–17) and hold it stationary until the AutoTrack symbol appears, as shown in Figure 7–18a.
7. Move the cursor up into the Top view and hold it over ③ until the Midpoint AutoTrack symbol appears. Then move the cursor back to the left until the tracking from ① appears. AutoTrack will indicate the acquired point as Intersection <90, Midpoint <180, as shown in Figure 7–18b. Pick this point as the axis endpoint of the ellipse.

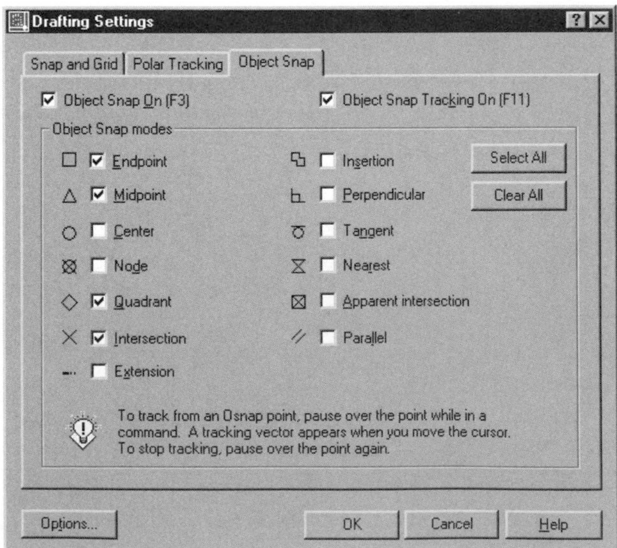

Figure 7–16 *The settings needed for Object Snap Tracking.*

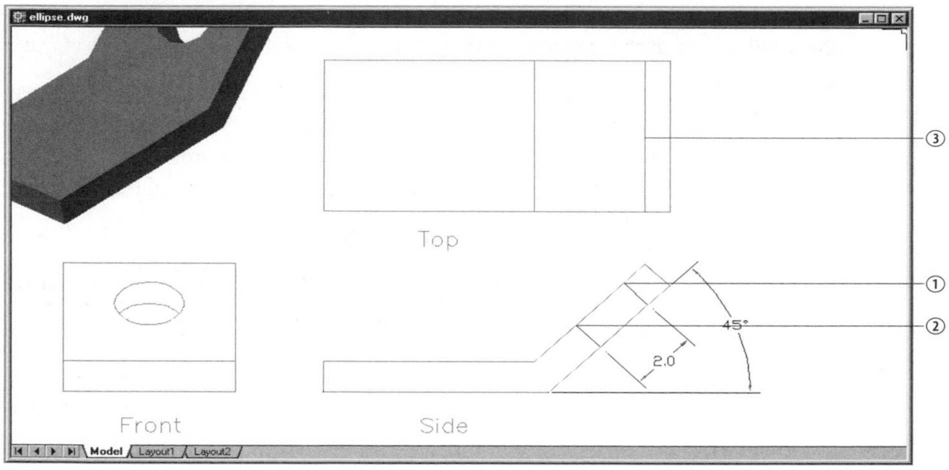

Figure 7–17 *A close-up of the Top, Side, and Front views.*

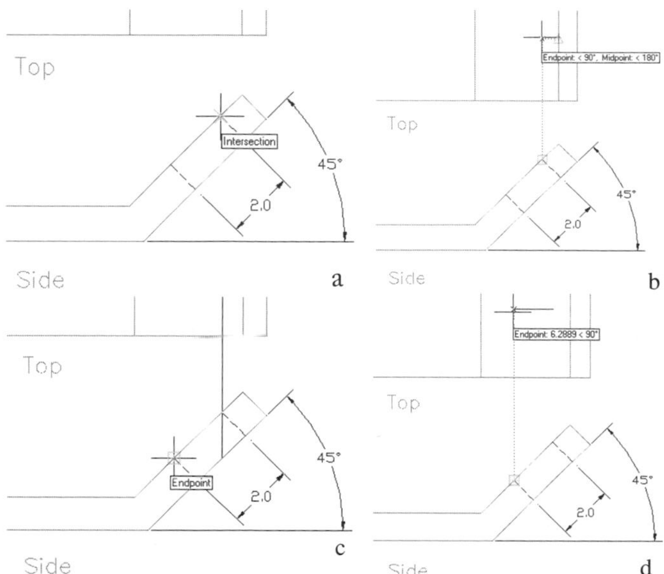

Figure 7–18 *Auto Tracking points used to place the ellipse.*

8. At the "Specify other endpoint of axis:" prompt, move the cursor over ② (see Figure 7–17) and hold it stationary until the Endpoint AutoTracking symbol appears, as shown in Figure 7–18c.

9. Now drag the cursor up into the Top view again until the AutoTracking point is acquired. AutoTracking will report "Endpoint: 6.2889 < 90°," as shown in Figure 7–18d. Click on this point to establish the second axis endpoint. AutoCAD

draws a tentative ellipse using the axis endpoints supplied thus far. Note that the ellipse center point is now established.

10. At the "Specify distance to other axis or [Rotation]:" prompt, type **1** and press Enter. AutoCAD complete the ellipse. This distance is referred to as the ellipse center point, and the dimension in the Side view gives the diameter of the hole as 2 units, and the radius as 1 unit. Your drawing should now resemble Figure 7–19.
11. Keep this drawing open for use in the next exercise.

In the following exercise you will use the Arc option of the ELLIPSE command to draw a partial ellipse, or an elliptical arc, to complete the ellipses in the Top view of the mounting bracket drawing.

EXERCISE: DRAWING ELLIPTICAL ARCS

1. Continue in the drawing of the mounting bracket from the preceding exercise. You will now place an elliptical arc "inside" the arc you drew in the last exercise. You will again use AutoTracking to place the arc. You will require the same Auto Tracking settings that you used in the last exercise.
2. From the Draw menu, choose Ellipse>Arc. At the "Specify axis endpoint of elliptical arc or [Center]:" prompt, rest the cursor over ① (see Figure 7–20a) until the Endpoint AutoTracking symbol appears.
3. Move the cursor straight up into the Top view, then to the right to acquire the midpoint of the bracket edge, as shown in Figure 7–20b.
4. Slowly move the cursor to the left until the AutoTracking symbol appears as shown in Figure 7–20c. Select this point.

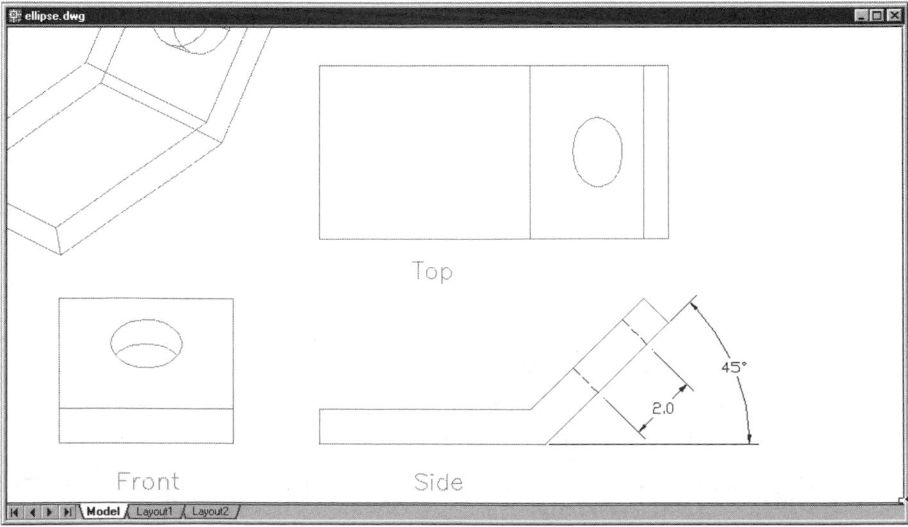

Figure 7–19 *The completed Top view ellipse.*

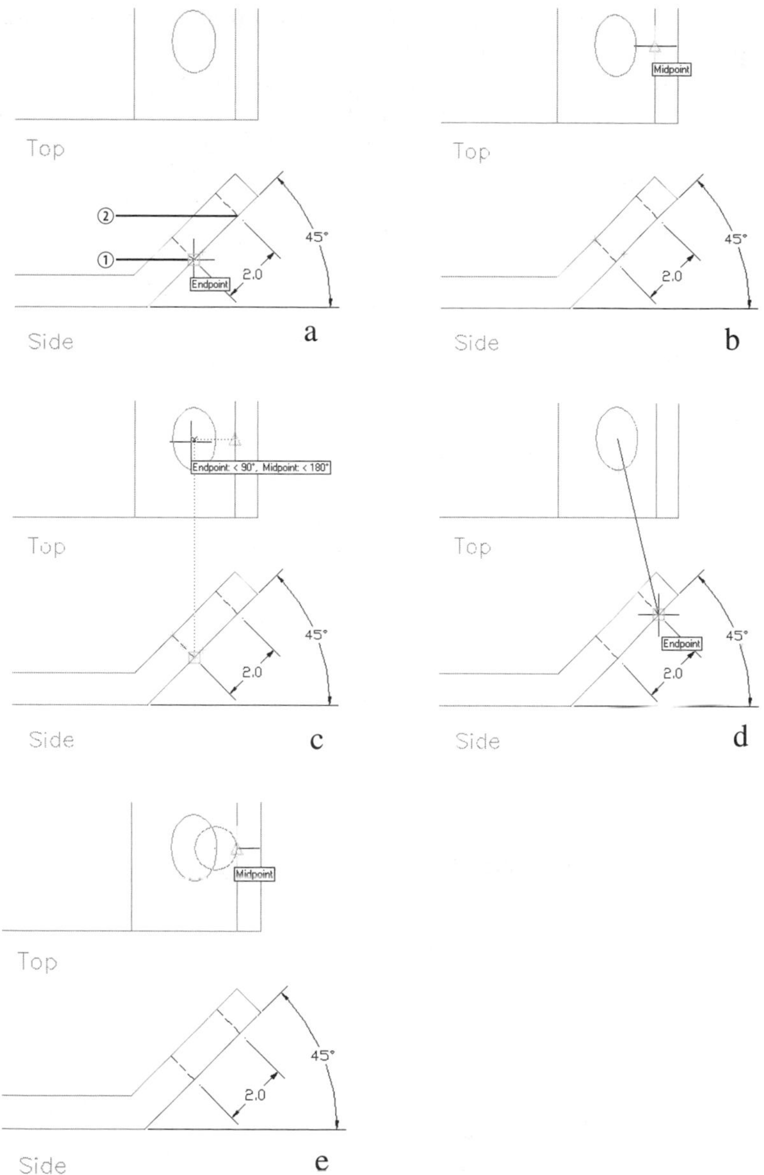

Figure 7–20 *AutoTracking points used to draw the elliptical arc.*

5. At the "Specify other endpoint of axis:" prompt, move the cursor over ② (see Figure 7–20a) until you acquire the tracking point, as shown in Figure 7–20d.
6. Move the cursor back into the Top view to acquire the midpoint again. Select this point as shown in Figure 7–20e. AutoCAD draws a tentative ellipse based on the endpoints supplied thus far.

7. At the "Specify distance to other axis or [Rotation]:" prompt, type **1** and press Enter. AutoCAD draws the complete ellipse. At the "Specify start angle or [Parameter]:" prompt, carefully place the cross hairs at ① in Figure 7–21 and pick. Then drag the cross hairs around the ellipse in a clockwise direction to ② and pick. This completes the elliptical arc. The full ellipse and the elliptical arc should resemble Figure 7–22.
8. This completes your work in this drawing.

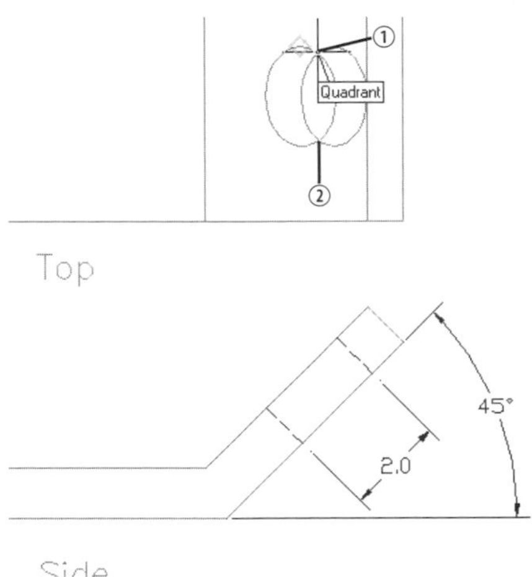

Figure 7–21 *Specifying the included angle for the elliptical arc.*

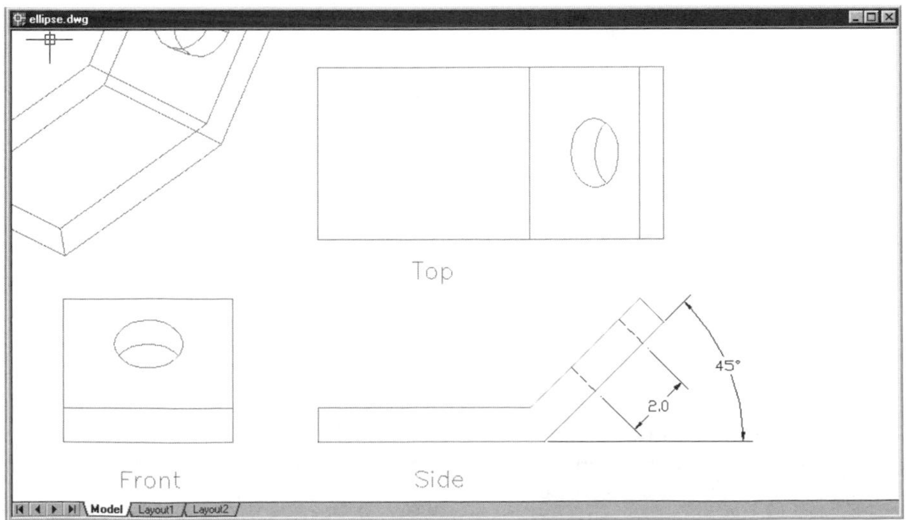

Figure 7–22 *The completed base plate fixture with an ellipse and elliptical arc.*

SUMMARY

In this chapter you learned to use the commands AutoCAD provides for drawing basic objects, from simple lines to the more complicated arc segments. Mastering their use is the basis of acquiring the range of skills required to use AutoCAD effectively.

Although the LINE command involves few options, when used in combination with the other basic objects and drawing aids, such as ortho mode and object snaps, it is a fundamental component of most AutoCAD drawings.

Several methods are available for drawing circles and arcs. The method you select depends on the information that you have available. You can often use more than one of the methods to accomplish the task. Understanding how and when to use each method enables you to construct complex designs quickly.

Polygon objects are common figures found in many AutoCAD drawings. Knowing how to construct them quickly and accurately increases your efficiency.

True mathematical ellipses and elliptical arcs are possible in AutoCAD. Ellipses have true center and quadrant points to which you can snap. These objects are also frequently used to represent circular objects viewed in axonometric views.

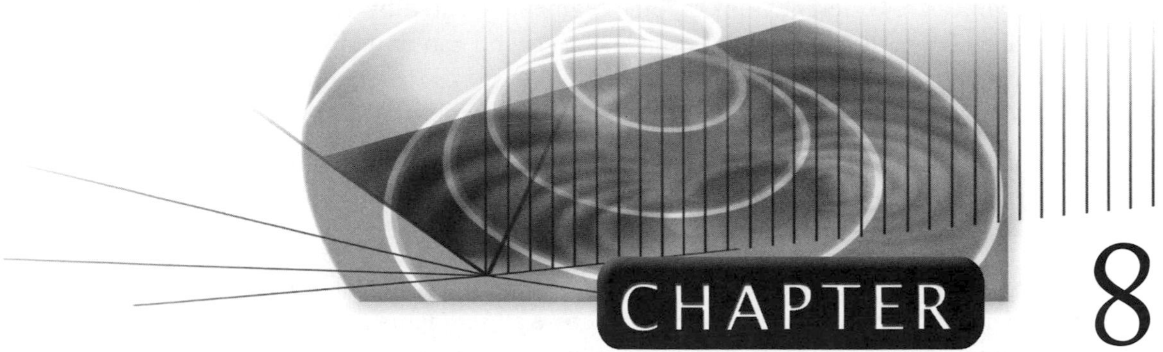

8

Creating Polylines and Splines

In the previous chapter you learned about lines and arcs and how to use them to draw straight-line segments and circular arcs. Lines and arcs are separate entities, even if they are connected end to end. A box drawn with the LINE command, for example, is really four separate line segments whose endpoints share an endpoint with another line segment. Polylines, however, are multi-segmented objects; they can be composed of multiple straight or curved segments. A polyline, no matter how many segments it is composed of, acts as a single, multi-segmented line.

In most AutoCAD drawings, you use traditional elements such as lines, arcs, polylines, and circles. Sometimes, however, you may want to draw smooth, irregular curves. AutoCAD provides the spline object for drawing free-form irregular curves. Splines are useful when you need to draw map contours, roads, walkways, or other smooth, flowing objects.

This chapter covers the following topics:

- How polylines are different from lines
- Creating polylines
- Two types of polylines
- Editing polylines
- Spline curves
- The SPLINEDIT command

HOW POLYLINES ARE DIFFERENT FROM LINES

Polylines are different from the line segments created by AutoCAD's LINE command. Polylines are treated as single objects and can include both line and arc segments connected at their vertices. AutoCAD stores information about the vertices, and you can access this information and edit the appearance of the polyline.

Polylines offer two advantages over lines. First, polylines are versatile; they can be straight or curved, thin or wide, and even tapered. Figure 8–1 shows some of the various forms that a polyline can take.

Second, editing polylines is easier; you can select any segment of a polyline and all other segments will be also selected because all segments are connected. This makes editing operations, such as moving or copying, faster and more accurate. Objects drawn with lines and arcs may appear to be connected but, depending on how they were drawn, they may actually have gaps or discontinuities that make using them as boundaries for cross hatching difficult.

CREATING POLYLINES

Polylines are created using the PLINE command. The PLINE command enables you to draw two basic kinds of polyline segments: straight lines and arcs. Some PLINE command prompts, therefore, are the same as those you find for the LINE and ARC commands prompts. If, for example, you draw straight polyline segments, you will find options such as Specify Next Point, Close, and Undo. You see these options in the standard polyline prompt for line segments:

Specify next point or [Arc/Close/Halfwidth/Length/Undo/Width]:

In addition, certain prompt options, such as Width, are specific to polylines. The prompt options are described as follows:

- **Arc**—Switches from drawing polylines to drawing polyarcs and issues the polyarc options prompt.
- **Close**—Closes the polyline by drawing a segment from the last endpoint to the initial start point and exits the PLINE command.

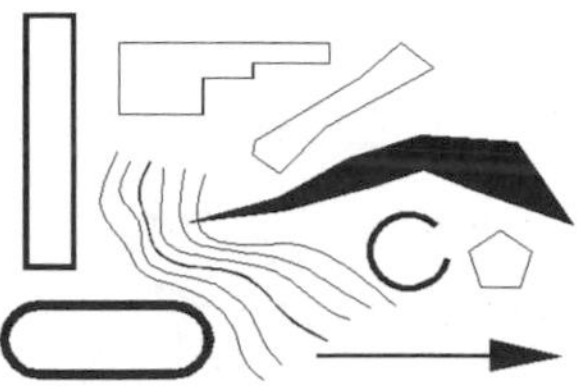

Figure 8–1 *Polylines can assume various useful forms.*

- **Halfwidth**—Prompts you for the distance from the center to the polyline's edge (half the actual width). See the Width option.
- **Length**—Prompts you for the length of a new polyline segment. AutoCAD draws the segment at the same angle as the last line segment or tangent to the last arc segment. The last line or arc segment can be that of a previous polyline, line, or arc object.
- **Undo**—Undoes the last drawn segment. It also undoes any arc or line option that immediately preceded drawing the segment. It does not undo width options.
- **Width**—Prompts you to enter a width (by default, zero) for the next segment and enables you to taper a segment by defining different starting and ending widths. AutoCAD draws the next segment with the ending width of the previous segment. Unless you cancel the PLINE command prior to drawing a segment, the ending width is stored as the new default width.
- **"Specify start point"/"Specify next point"**—Prompts you for the starting point or the next point of the current line segment. Once you begin a new polyline, "Specify next point" is the default option.

If you select the Arc option, the arc mode options prompt appears as follows:

[Angle/CEnter/CLose/Direction/Halfwidth/Line/Radius/Secondpt/Undo/Width]

This prompt contains some of the same options as the ARC command, and others specific to the drawing of polylines. The arc options include the following:

- **Angle**—Prompts you for an included angle. A negative angle draws the arc clockwise.
- **CEnter**—Prompts you to specify the arc's center.
- **CLose**—Closes the polyline by using an arc segment to connect the initial start point to the last endpoint, and then exits the PLINE command.
- **Direction**—Prompts you to specify a tangent direction for the segment.
- **Halfwidth**—Prompts for a halfwidth, the same as for the Line option.
- **Line**—Switches back to the line mode.
- **Radius**—Prompts you for the polyarc's radius.
- **Secondpt**—Selects the second point of a three-point polyarc.
- **Undo**—Undoes the last drawn segment.
- **Width**—Prompts you to enter a width, the same as in line mode.
- **"Specify endpoint of arc"**—Prompts you for the endpoint of the current arc segment. This is the default option.

Tip: You can also close a polyline of two or more segments after the fact. Use the PEDIT command's Close option. Using the PEDIT Close option draws a line between the last point of the last segment drawn and the start of the first segment. The PEDIT command is covered later in this chapter.

Although using the PLINE command to draw lines and arcs is similar to using the LINE and ARC commands to draw similar objects, there are several important differences:

- You get all the line or arc mode prompts every time you enter a new polyline vertex.
- Prompt options for Halfwidth and Width control the width of the segment.
- You can switch back and forth from line segments to arc segments as you add additional segments to the polyline.
- You can apply linetypes continuously across vertices.
- You can apply the MEASURE and DIVIDE commands to polyline polylines.
- You can use the AREA command to report the total length of a polyline and to calculate the area enclosed by certain polylines.

In the following exercise you will practice drawing a single polyline using many of the options for both the line and arc mode options of the POLYLINE command.

EXERCISE: PRACTICING WITH THE POLYLINE COMMAND OPTIONS

1. Create a new drawing named CHAP08.DWG using the 08DWG01.DWT template drawing from the accompanying CD. (See Chapter 2, Beginning a Drawing, for a discussion of template drawings.)
2. Start the POLYLINE command by typing **PL** and pressing Enter. The initial polyline prompt "Specify start point:" appears. Respond by entering the point **2.5,1.0**. The line mode options prompt appears.
3. Use the default endpoint option by entering the relative polar coordinate **@5<90**. AutoCAD draws the first segment of the polyline from ① to ② in Figure 8–2. The line mode options prompt reappears.
4. Switch to the polyarc mode by typing **A** and pressing Enter. The arc mode options prompt appears.
5. Use the default "Specify endpoint of arc:" option by typing **@.25<45** and pressing Enter. AutoCAD draws the polyarc to ③. The arc mode options prompt reappears.
6. Switch back to the line mode by choosing the Line option. Type **L** and press Enter. The Line mode options appear. Choose the Length option by entering **L**. At the "Specify length of line:" prompt, type **@3<0** and press Enter. AutoCAD

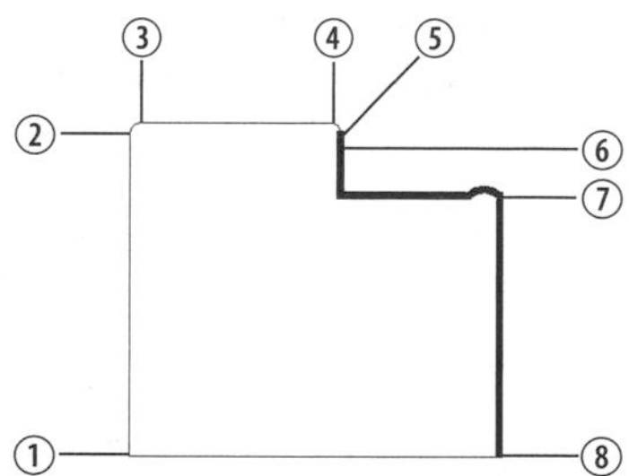

Figure 8–2 *Drawing a multi-segment polyline.*

draws the next segment ③ units long, to ④. Now switch back to arc mode by entering **A**. The arc mode options prompt appears.

7. Choose the angle option by typing **A** and pressing Enter. At the "Specify included angle:" prompt, type **-90** and **E**. This specifies a 90-degree clockwise included angle to the point at ⑤. The "Specify endpoint of arc or [Center/Radius]:" prompt appears.
8. Choose the Radius option by typing **R** and pressing Enter. When prompted with "Specify radius of arc:," enter the value **0.125**. Respond to the "Specify direction of chord for arc:" prompt by typing **315** and pressing Enter.
9. Switch to the line mode by entering **L**. Choose the Width option by entering **W**.
10. Respond to the "Specify starting width:" prompt by typing **0.1** and pressing Enter. Accept the "Ending width <0.1>:" value by pressing Enter. Respond to the line mode's "Specify next point:" prompt by entering **@1<-90**. AutoCAD draws the next segment with a uniform width of 0.1 units (see ⑥).
11. Now, at the "Specify next point:" prompt, enter the relative coordinate **@2,0**. AutoCAD draws the next segment at the current width of 0.1 units.
12. Again, switch to the Arc mode by entering **A**. Respond to the arc mode prompt by typing **CE** and pressing Enter. At the "Specify center point of arc:" prompt, enter **@0.25,-0.25** to specify the center of the next arc section.
13. Respond to the "Specify endpoint of arc or [Angle/Length]:" prompt by entering **A**. Specify an included angle of **-90** (see ⑦).
14. Switch to the line mode (type **L** and press Enter) and enter the relative polar coordinate **@4.052<270**. AutoCAD draws the next line segment at the current 0.1 width (see ⑧).
15. Now change the width for the next segment by typing **W** and pressing Enter. Respond to the "Specify starting width:" prompt by entering **0**. Accept the default ending width by pressing Enter.

16. The standard "Specify next point or [Arc/Close/Halfwidth/Length/Undo/Width]:" prompt appears. Choose the Close option by entering **C**. AutoCAD draws the next segment to the start point of the first segment of the series with zero width 1.
17. Your drawing should now resemble Figure 8–2. You will continue using this drawing in the next exercise. For now, press Ctrl+S to save the changes you have made.

Note: Polylines can consist of a number of segments, with the possibility of significant changes in the polyline's characteristics taking place from segment to segment. The endpoint of one segment is the start point of the next. Later, when we discuss editing polylines, we will refer to these points collectively as vertices, or singly as a vertex.

As you saw in this exercise, you can easily switch back and forth between the line mode and the arc (polyarc) mode within the PLINE command. Each mode has its own set of prompts that are repeated after you draw each segment. You also saw that both line and arc segments can have widths other than zero, and that when you specify a new width, it becomes the default width for the next segment. In other words, both the mode (line versus polyarc) and the width remain in effect unless and until you explicitly change them. This facilitates drawing a long series of either straight line or curved arc segments.

In the following exercise you will construct an arc leader using both the line and arc modes. You will also take advantage of the ability to vary the width of a single polyline segment.

EXERCISE: DRAWING AN ARC LEADER WITH A POLYLINE

1. Continue in the drawing from the preceding exercise. Refer to Figure 8–3 for this exercise.
2. Begin the PLINE command by choosing the Polyline tool from the Draw menu. The "Specify start point:" prompt appears.
3. Pick a point near ① in Figure 8–3. At the "Specify next point:" prompt, choose the Width option by typing **W** and pressing Enter.
4. Specify a starting width of **0.0** and an ending width of **0.1**.
5. Respond to the "Specify next point:" prompt by entering **@.2<115**.
6. AutoCAD draws the first segment with a tapered width. You may not see the taper until the next step is completed. At the "Specify next point:" prompt, enter **W** again.
7. Specify a starting width of **0** and an ending width of **0**.
8. Respond to the line mode prompt by choosing the Arc option.

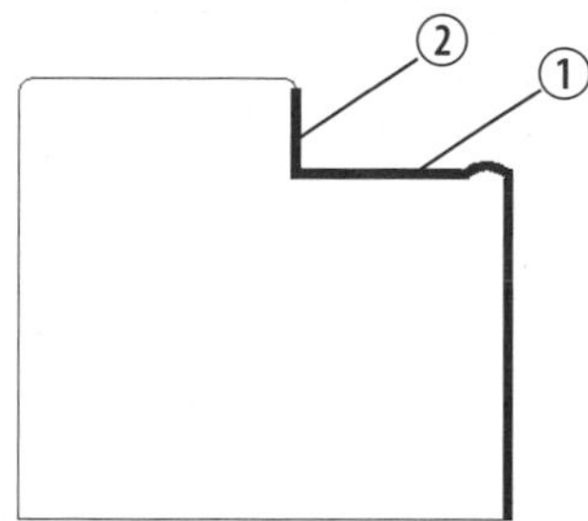

Figure 8–3 *Picking the points for a polyline arc leader.*

9. The arc mode prompts appear. Note that the arc is dragged as you move the cursor near ②. Pick a point near ②. AutoCAD draws the arc. Press Enter to exit the PLINE command.

Your drawing should resemble Figure 8–4.

This exercise demonstrated the versatility of polylines. Not only can polylines have varying width, but each segment can have different starting and ending widths. The current polyline width is stored in the system variable PLINEWID.

CONTROLLING POLYLINE APPEARANCE WITH FILL

A drawing that consists of a large number of polylines with widths other than zero can severely increase the time AutoCAD takes to redraw the screen or to plot the drawing. AutoCAD provides the FILL command to enable you to control the visibility of the filled portion of wide polylines. When you turn FILL off, AutoCAD displays or plots only the outline of filled polylines. You must regenerate the drawing before you can see the effect of the FILL command. Figure 8–5 shows the effect of having FILL on and off. In addition to polylines, the FILL command controls the appearance of all hatches and of objects created using the SOLID, TRACE, and MLINE commands.

TWO TYPES OF POLYLINES

In AutoCAD Release 13 and earlier, polylines were generated and stored in AutoCAD's database differently. The current PLINE command produces a "leaner" optimized polyline—sometimes referred to as a "lightweight polyline." Some of the functionality and control available in the older 2D polylines objects are not available in the current 2D optimized polyline, including arc-fit curve data, spline-fit data, and curve-fit tangent direction data.

To provide compatibility with the older 2D polyline objects of pre-AutoCAD Release 14, the system variable PLINETYPE is provided. PLINETYPE specifies whether AutoCAD uses the newer optimized 2D polylines. It controls both the creation of new polylines with the PLINE command and the conversion of existing

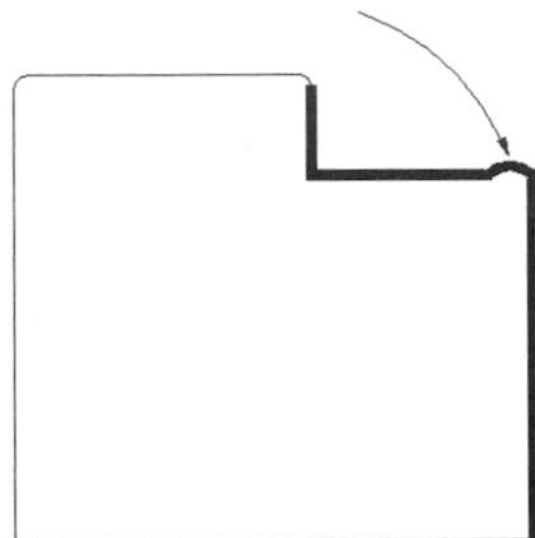

Figure 8–4 *The completed polyline arc leader.*

Figure 8–5 *Fill can be turned on (left) and off (right) using the FILL command.*

Note: The FILL command controls the setting of the system variable FILLMODE. A setting of 1 is the same as having FILL on; a setting of 0 is the same as having FILL off. You can also control the setting of FILLMODE in the Options dialog box's Display folder, under the Apply Solid Fill option.

polylines in drawings from previous releases. PLINETYPE can have one of three settings, with the following meanings:

- **0**—Polylines in older drawings are not converted when opened; PLINE creates old-format polylines.
- **1**—Polylines in older drawings are not converted when opened; PLINE creates optimized polylines.
- **2**—Polylines in older drawings are converted when opened; PLINE creates optimized polylines. This is the default setting.

An optimized polyline can have the following characteristics:

- Straight line segments
- Arc segments
- Constant and variable width
- Thickness.

Note: PLINETYPE also controls the polyline type created by using the following commands: BOUNDARY (when object type is set to Polyline); DONUT; ELLIPSE (when PELLIPSE is set to 1); PEDIT (when selecting a line or arc); POLYGON; and SKETCH (when SKPOLY is set to 1).

You also can manually convert the older 2D polylines by using the CONVERT command, which is described next.

CONVERT COMMAND

The CONVERT command allows you to optimize manually 2D polylines created in AutoCAD Release 13 and earlier. The prompt for the CONVERT command is as follows:

Enter type of objects to convert [Hatch/Polyline/All] <All>:

Enter **H** for hatches, **P** for polylines, or **A** for both. The next prompt appears:

Enter object selection preference [Select/All] <All>:

Enter **S** to select objects or **A** to convert all candidate objects in the drawing. Depending on your response to these two prompts, AutoCAD displays one or both of the following messages (where "number" is the number of objects converted):

number hatch objects converted.

number 2d polyline objects converted.

In most cases you do not need to convert polylines with the CONVERT command, because the PLINETYPE system variable (described in the previous section) specifies that polylines are updated automatically whenever you open an older drawing. However, older, non-optimized polylines may be created in your AutoCAD 2000 drawing by third-party applications, or they may be contained in an older drawing that has been inserted into your current drawing and then exploded.

Note: Polylines that contain curve-fit or splined segments always retain the old-style format, as do polylines that contain extended entity data.

THE 3DPOLY COMMAND

The 3DPOLY command produces three-dimensional polylines. 3D polylines are not as versatile as the standard (2D) polylines produced with the PLINE command because they can only contain straight line segments with continuous linetype and no width information. They are more versatile than 2D polylines, however, because their vertices can be placed in 3D space. That is, you can draw them with a nonzero Z coordinate, in addition to the required X and Y coordinates. 3D polylines are discussed in more detail in Chapter 23, Drawing in 3D.

EDITING POLYLINES

As you have seen so far in this chapter, polylines are complex objects that may consist of a collection of arc and line segments, which may contain width information. AutoCAD, therefore, provides the PEDIT command, a command devoted to editing these complex entities. PEDIT does not differentiate between the newer, optimized polylines and the older, pre-Release 14 polylines.

EDITING ENTIRE POLYLINES WITH PEDIT

PEDIT contains a large number of subcommands or options for the various polyline properties. To manage this large number of options, AutoCAD divides them into two groups of editing functions. The primary group operates on the polyline as a whole, while a secondary group is devoted to the vertices that mark the beginnings and ends of polyline segments. Several editing option are available for both 2D and 3D polylines. The primary group of PEDIT options includes the following:

- **Multiple**—(This option must be chosen before you select polylines to edit.) Enables you to select multiple polylines for editing. AutoCAD 2002 is the first version of AutoCAD that lets you edit multiple polylines simultaneously.
- **Close/Open**—Adds a segment (if required) and joins the first and last vertices to create a continuous polyline. If the polyline is open, the prompt shows Close; if the polyline is closed, the prompt shows Open. A polyline can be open, even if the first and last points share the same coordinates. A polyline is open unless the polyline Close option is used when you draw it or you later use the Close option to close it.
- **Join**—(2D only) Enables you to add selected arcs, lines, and other polylines to an existing polyline. Endpoints must be exactly coincident before you can join them. You can join lines, arcs, 2D polylines, and lightweight polylines to lightweight polylines or 2D polylines. Specifically, the following will occur:
  - If you are editing a lightweight polyline, AutoCAD converts all joined segments to a single lightweight polyline at the end of the command.
  - If you are editing an old-style 2D polyline, AutoCAD converts all joined segments, as in previous releases. This would be the case for 2D polylines that were not converted when the drawing was opened.
- Width—(2D only) Prompts you to specify a single width for all segments of a polyline. The new width overrides any individual segment widths already stored. You can edit widths of individual segments by using a suboption of the Edit vertex option.
- **Edit vertex**—Presents a prompt for a set of options that enable you to edit vertices and their adjoining segments.
- **Fit**—(2D only) Creates a smooth curve through the polyline vertices. AutoCAD converts lightweight polylines to 2D polylines before computing the curve.

- **Spline curve**—Creates a curve controlled by, but not necessarily passing through, a framework of polyline vertices. AutoCAD converts lightweight polylines to 2D polylines before computing the curve.
- **Decurve**—Undoes a fit or spline curve, restoring it back to its original definition. Selecting the Decurve option has no effect on lightweight polylines because they do not support curve or spline fitting.
- **Ltype gen**—(2D only) Controls whether linetypes are generated between vertices (Ltype gen OFF) or between the polyline's endpoints (Ltype gen ON), spanning vertices (see Figure 8–6). AutoCAD ignores this option for polylines that have tapered segments.
- **Undo**—Undoes the most recent PEDIT function.
- **eXit**—(the default, <X>) Exits the PEDIT command.

When PEDIT prompts you with "Select polyline:," you can pick a polyline line, polyarc, or other polyline object. The PEDIT command operates on only one object at a time. Before you select a wide polyline, you must select either an edge or a vertex. You can use a window or crossing selection, but you must first enter **W** or **C** because PEDIT does not support the implied windowing selection feature. For convenience, you also can use the Last, Box, Fence, All, Wpolygon, and Cpolygon (but not Previous) selection methods. Selection ends as soon as PEDIT finds a line, arc, or polyline. If your selection method includes more than one valid object, PEDIT selects only one. If the first object you select is not a polyline but is capable of being converted into one (a line or arc, for example), AutoCAD asks if you want to turn it into one.

In the following exercise you will practice using some of the options of the primary PEDIT command options.

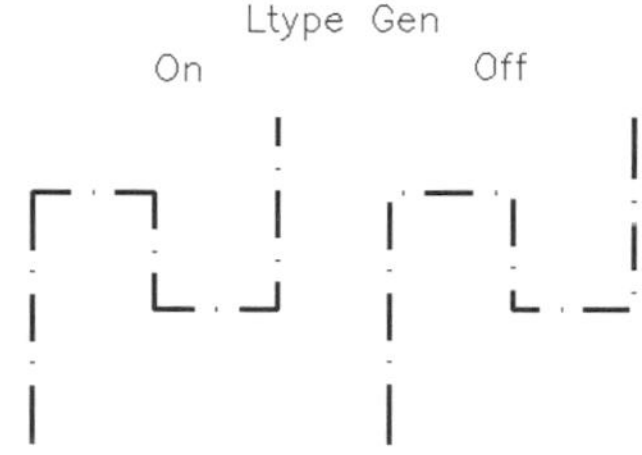

Figure 8–6 *The effect of turning Ltype gen on and off.*

EXERCISE: USING THE PRIMARY OPTIONS OF THE PEDIT COMMAND

1. Open 08DWG02.DWG from the accompanying CD and save it as Parking.dwg for use in this exercise (see Figure 8–7).
2. You will first join two separate polylines. Start the PEDIT command by typing **PE** and pressing Enter. At the "Select polyline or [Multiple]:" prompt, pick the line at ①. The primary PEDIT prompt appears as follows:

   Enter an option [Close/Join/Width/Edit vertex/Fit/Spline/Decurve/Ltype gen/Undo]:
3. Right-click to display the PEDIT shortcut menu and then choose the Join option. At the "Select objects:" prompt, pick the polyline at ② and close the selection process by pressing Enter. AutoCAD joins the two lines into a single polyline, and the primary prompt returns.
4. Right-click again and choose the Width option from the shortcut menu. Answer the "Specify new width for all segments:" prompt by typing **9** and pressing Enter.
5. Note that the polyline changes to a new width of 9 inches and the prompt returns. The line is too wide. Choose Undo from the shortcut menu. This undoes the previous editing operation. Repeat step 4, specifying a width of 3. Then right-click and choose Cancel to cancel the PEDIT command.
6. Zoom into the area around ③. Next you will convert an arc object into a polyarc and join it to two other polylines. Restart the PEDIT command. At the "Select polyline:" prompt, pick the arc object at ③. The following prompts appear:

   Object selected is not a polyline

   Do you want to turn it into one? <Y>

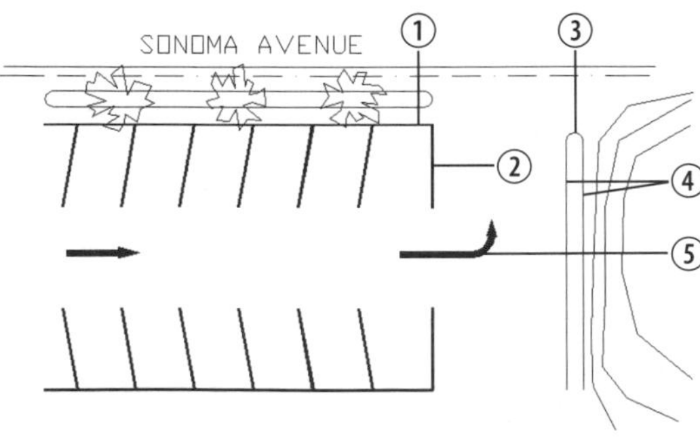

Figure 8–7 *Editing entire polyline segments.*

7. Accept the default option, <Y> (for yes), by pressing Enter. The primary PEDIT prompt appears. Right-click and choose the Join option. At the "Select objects:" prompt, pick the two polylines at ④. AutoCAD joins the three polylines. Press Enter to end the PEDIT command.
8. Next you will explode a polyline using AutoCAD's EXPLODE command. To start the EXPLODE command, type **EXPLODE** and press Enter. The "Select objects:" prompt appears.
9. Pick the arrow at ⑤ and press Enter. Note that the width information for this polyline disappears after it is exploded. Actually, AutoCAD destroys the polyline, demoting it to a line object.
10. Restore the polyline arrow with the U command by choosing Undo from the right-click menu.
11. You will use this drawing in the next exercise. For now, press Ctrl+S to save your changes.

In this exercise you saw that you can work with one polyline at a time. The primary PEDIT prompt, with its several options, returns after each edit operation on the assumption that you may want to edit another polyline parameter. You must specifically dismiss the prompt by accepting the default eXit option, or pressing Esc to cancel the command, or choosing the Cancel option from the right-click menu. Also note that several of the options in the primary prompt undo other options. For example, after performing a fit (curve) edit, you can undo the operation with the Decurve option. The Undo option undoes the last edit operation and returns the primary prompt. The EXPLODE command destroys a polyline and reduces it to a lower-order object, either a line or arc.

Tip: You can produce a wide polyline with an apparent mitered end by using a trick. After specifying the last vertex, set the width so it tapers from the current full width to zero, and then draw another very short segment using a typed relative polar coordinate. The "miter" will appear perpendicular to the angle you type.

For example, enter **@0.00001<45** to miter the top of a vertical wide polyline to an angle of 135 degrees (45 + 90). The end of the last segment is actually pointed, but the extremely short length has the effect of ending the previous segment with a miter.

USING THE PEDIT FIT AND SPLINE OPTIONS

PEDIT provides two options for making a polyline that passes through or is influenced by control points (see Figure 8–8). A fit curve actually passes through vertex points and consists of two arc segments between each pair of vertices. A spline-fit curve interpolates between control points, but the curve does not necessarily pass through the points.

To help you visualize a spline-fit curve, AutoCAD provides the system variable SPLFRAME. If you set SPLFRAME to a value of 1, the reference frame with control points appears. Figure 8–8 shows only the control points. In the case of the original polyline, the control points are coincident with the vertices of the polyline. In the instance of the fit curve example, the straight line segments between control points (vertices) have been replaced with arc segments but still pass through the control points. The spline-fit curve uses the control points as guides to influence the shape of the curve.

AutoCAD can generate two types of spline-fit polylines: a quadric b-spline and a cubic b-spline. The system variable SPLINETYPE controls the type of curve generated. A SPLINETYPE value of 5 approximates a true quadric b-spline; a value of 6 approximates a true cubic b-spline. In addition, the system variable SPLINESEGS controls the fineness of the b-spline. The numeric value of SLINESEGS sets the number of line segments in the control frame.

In the following exercise you will use both the Fit and Spline options of PEDIT to generate curves from polylines.

EXERCISE: GENERATING CURVES WITH PEDIT

1. Continue in the drawing from the previous exercise.
2. First, zoom in to the contour lines area of the parking lot as shown in Figure 8–9.

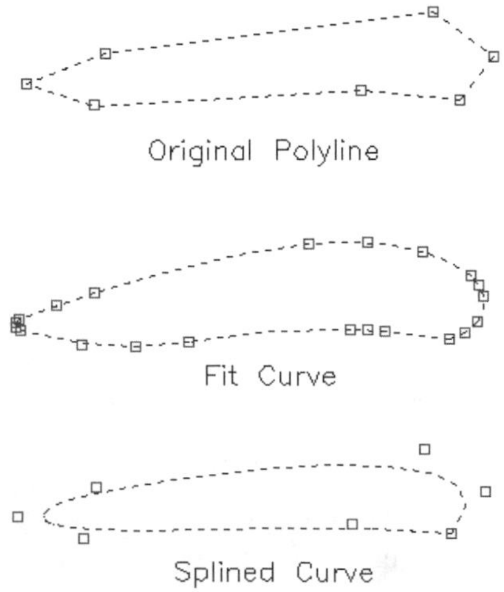

Figure 8–8 *Creating curves from straight polylines.*

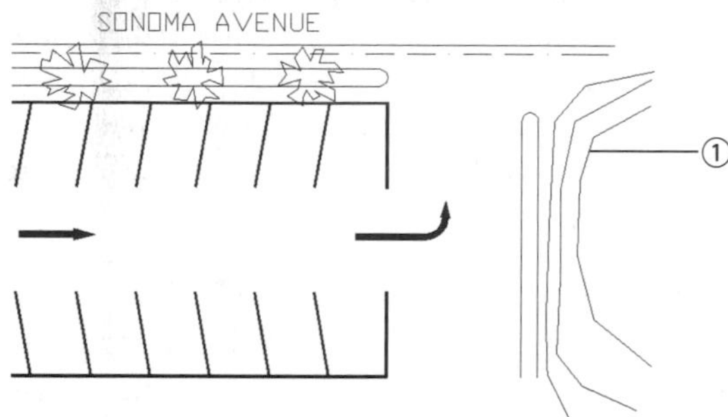

Figure 8–9 *Creating smooth curves with PEDIT.*

3. You will smooth the contour line at ① first. Start the PEDIT command by typing **PE** and pressing Enter. At the "Select polyline:" prompt, pick the polyline at ①. The primary PEDIT options appear. Type **F** (for Fit) and press Enter.
4. AutoCAD performs a fit-curve smoothing on the polyline, and the prompt returns. This does not smooth the curve as you want it to, so right-click and choose the Undo option from the shortcut menu. AutoCAD undoes the fit curve operation. The PEDIT prompt returns.
5. Now right-click again and choose the Spline option. AutoCAD spline fits the polyline. Note that this curve more closely approximates the contour. Exit the PEDIT command by pressing Enter.
6. By turning on the spline frame feature, you can examine the original data points for this new curve. At the command prompt, enter **SPLFRAME** and respond to the prompt by typing **1** and pressing Enter.
7. For the spline frame to be displayed, you must regenerate the drawing. At the command prompt, type **REGEN** and press Enter. The spline frame is displayed. Note that with an open polyline such as this contour line, the spline-fit curve passes through the spline frame at the start point and endpoint.
8. Turn the spline frame off by repeating step 6, but set the SPLFRAME variable to zero. Then perform another regeneration to clear the frame.
9. You will use this drawing in the next exercise. For now, save the drawing by pressing Ctrl+S.

In this exercise you saw how the PEDIT command takes data points in the form of polyline vertices and transforms them into a close approximation of a true b-spline curve. In many instances, these polyline spline-fit curves are adequate for representing data such as contour lines. At other times, the smoothing procedure of the fit-curve procedure will suffice to remove the angles present at polyline vertices.

Note: Optimized polylines do not support either the fit curve or spline fit options of the PEDIT command. Whenever you choose either of these options, AutoCAD, if necessary, converts optimized polylines into the older-style 2D polyline and then carries out the transformation to a curve. If the curve is subsequently changed back to its original shape using the Decurve option of the PEDIT Command, AutoCAD converts the 2D polyline back into an optimized polyline.

EDITING POLYLINE VERTICES WITH PEDIT

Each polyline segment belongs to and is controlled by the preceding vertex. The Edit vertex option of the primary PEDIT set of options displays another prompt with a separate set of options:

Enter a vertex editing option

[Next/Previous/Break/Insert/Move/Regen/Straighten/Tangent/Width/eXit]
<N>:

When you use these options, AutoCAD marks the current vertex of the polyline with an X to show the vertex you are editing. Move the X (by pressing Enter to accept the <N>, or next, prompt) until the X marks the vertex you want to edit.

The Edit vertex option of the PEDIT command includes the following options:

- **Next/Previous**—Moves the X marker to a new current vertex. Next is the initial default.
- **Break**—Splits the polyline in two or removes segments of a polyline at existing vertices. The first break point is the vertex on which you invoke the Break option. Use Next/Previous to access another vertex for the second break point. The Go option performs the actual break.
- **Insert**—Adds a vertex at a point you specify, following the vertex currently marked with an X. You can combine this option with the Break option to break between existing vertices.
- **Move**—Changes the location of the current (X-marked) vertex to a point you specify.
- **Straighten**—Removes all intervening vertices from between the two you select and replaces them with one straight segment. This option also uses the Next/Previous and Go options.
- **Tangent**—Sets a tangent to the direction you specify at the currently marked vertex to control curve fitting. You can see the angle of the tangent at the vertex with an arrow and you can drag it with the screen cursor or enter the angle at the keyboard.
- **Width**—Sets the starting and ending width of an individual polyline segment to the values you specify.
- **eXit**—Exits vertex editing and returns to the primary PEDIT prompt.

Tip: It is usually easier to edit the position of a polyline vertex by using AutoCAD's Grips feature. This is especially true for spline-fit polylines. Chapter 10, Advanced Editing, covers grips editing.

In the following exercise you will perform vertex editing on polylines in the current exercise drawing.

EXERCISE: EDITING POLYLINE VERTICES WITH PEDIT

1. Continue in the drawing from the previous exercise (see Figure 8–10). You will first move the first vertex of the polyline contour at 1. Begin by starting the PEDIT command, and then pick the contour line at ①.
2. At the primary PEDIT prompt, choose the Edit vertex option by typing **E** and pressing Enter. Note that an X appears at the endpoint and first vertex of the line, and the Edit vertex prompt appears.
3. If the vertex you want to move is already selected (marked with an X), respond to this prompt by typing **M** and pressing Enter. Respond to the "Specify new location for marked vertex:" prompt by typing **@0,_12!** and pressing Enter. (Note: Watch the vertex as you press Enter.)
4. The vertex moves 12 units in the -Y direction. The Edit vertex prompt returns.

Tip: In the next step, if you go too far, type **P** and press Enter to return to the previous vertex. Keep in mind, too, that all the prompt options are available from the PEDIT right-click shortcut menu.

5. For the purposes of this exercise, assume that you want to insert a new vertex between the current fourth and fifth vertices, as shown at ② in Figure 8–10. Step the current vertex down the line by pressing Enter to accept the default Next (vertex) option. You will need to press Enter three times. The X is now at vertex 4 at ③ in Figure 8–11.

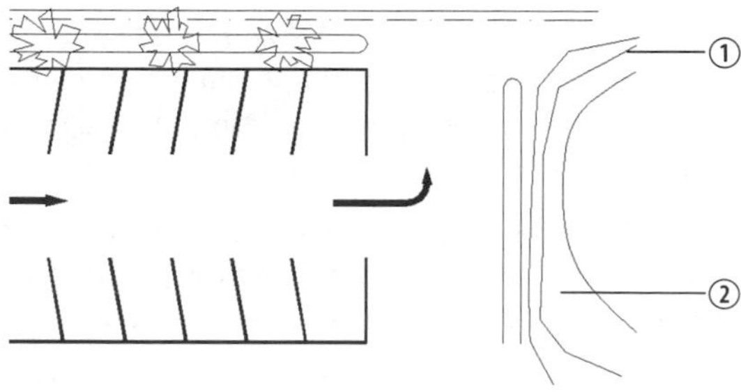

Figure 8–10 *Moving a polyline vertex.*

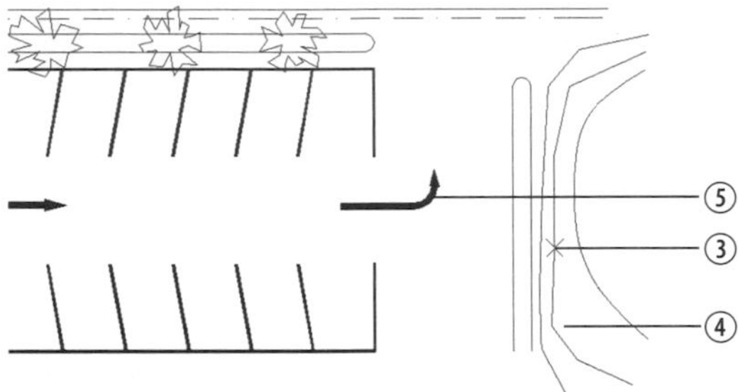

Figure 8–11 *Inserting a new polyline vertex.*

6. Now choose the Insert option by typing **I** and pressing Enter. Answer the "Enter location of new vertex:" prompt by picking a point near ④. AutoCAD inserts a new vertex and the Edit vertex prompt returns.
7. Exit the Edit vertex prompts by typing **X** and pressing Enter. The primary PEDIT prompt returns. Select the Spline option by typing **S** and pressing Enter. AutoCAD spline fits the curve.
8. Close the PEDIT command by pressing Enter.
9. Restart the PEDIT command by pressing the spacebar. At the primary PEDIT prompt, pick the polyline at ⑤. You will change the width of the vertex at the base of the arrowhead.
10. Choose the Edit vertex option of the primary PEDIT prompt and then choose Next until the active vertex is at the base of the arrowhead. This is the third vertex.

Note: In the next step, observe the width of the arrowhead as you press Enter.

11. Choose the Width option by typing **W** and pressing Enter. At the "Specify starting width for next segment:" prompt, enter the new value by typing **20** and pressing Enter. AutoCAD changes the width of the arrowhead base. At the "Specify ending width for next segment:" prompt, type **0** and press Enter.
12. Exit the Edit vertex prompt by typing **X** and pressing Enter. Then exit the PEDIT command by pressing Enter again. Your drawing should now resemble Figure 8–12.

In this exercise you saw how to make changes to an existing polyline by editing parameters at its vertices. When combined with the primary editing option, the Edit vertex option provides a great degree of editing capability for polylines.

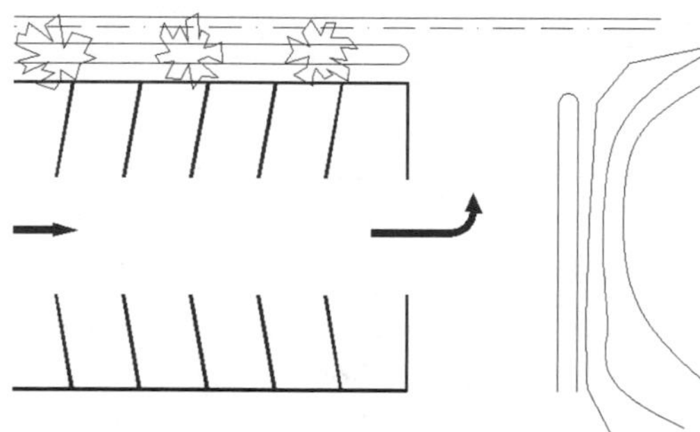

Figure 8–12 *The completed parking lot drawing.*

SPLINE CURVES

To create true spline curves in AutoCAD, you use the SPLINE command. Splines can be either 2D or 3D objects. You draw splines by specifying a series of fit-data points (vertices) through which the curve passes. The fit-data points determine the location of the spline's control points. Control points contain the curve information for the spline. AutoCAD's spline objects are true splines, unlike the spline approximations formed by spline-fit polylines. The SPLINE command draws nonuniform rational B-splines (NURBS). AutoCAD NURBS are mathematically more accurate than spline-fit polylines. Even though spline objects are more accurate than spline-fit polylines, they actually require less memory for storage and result in smaller drawings, with all other factors being equal.

You can use the SPLINE command to convert 2D and 3D spline-fit polylines into true splines. Unlike spline-fit polylines, which may be either quadric or cubic spline approximations, depending on how the SPLINETYPE system variable is set, true spline objects produced by the SPLINE command are not affected by the SPLINETYPE system variable.

The SPLINE command offers the following options:

- **Specify first point or [Object]**—The default option of the main prompt specifies the starting point of the spline. After you specify a starting point, AutoCAD prompts you to specify a second point. Spline objects must consist of a minimum of three points.
- **Object**—This option enables you to convert existing spline-fit polylines to true spline objects. After you specify this option, AutoCAD asks you to select objects.
- **Specify next point or [Close/Fit tolerance] <start tangent>**—AutoCAD displays this prompt after you specify a second point. The default option is to continue to specify additional data points for the spline that

you are drawing. If you press Enter, AutoCAD prompts you to specify the tangent information for the start point and endpoint, and then it ends the command.

- **Close**—This option causes the start and end of the spline to be coincident and to share vertex and tangent information. When you close a spline, AutoCAD prompts you only once for tangent information.
- **Fit Tolerance**—This option controls how closely the spline curve follows the data points. The distance of the Fit Tolerance is expressed in current drawing units. The smaller the tolerance, the closer the spline fits the data points. A value of zero causes the spline to follow the data points exactly.
- **(Undo)—**Although this option does not appear in the prompt, you can enter U after any point to undo the last segment.

In the following exercise you will use the SPLINE command to draw the outline of a mechanical part.

EXERCISE: CONSTRUCTING A SPLINE

1. Open 08DWG03.DWG from the accompanying CD and save it as Spline.dwg for use in this exercise (see Figure 8–13).

In the next step you will use the pre-positioned data points shown in the drawing with X's. You need to set the node and endpoint to be running Osnaps so you can easily snap to the data points.

2. From the Draw menu, choose Spline to start the SPLINE command. The "Object/<Specify first point>:" prompt appears.
3. Use the default option of specifying a first point. Snap to the endpoint at ① and enter the first point by clicking. The "Specify next point:" prompt appears.

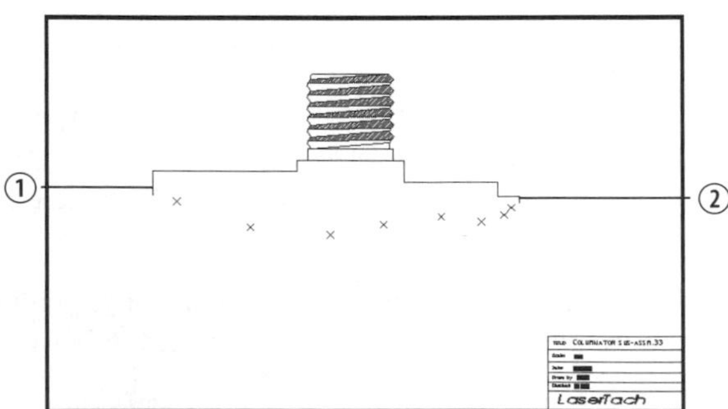

Figure 8–13 *Fitting a spline to data points.*

4. Continue to specify points, snapping to each successive X node at each "Specify next point:" prompt. Finally, specify the last point by snapping to the endpoint at ②.
5. Press Enter three times to end the SPLINE command. Your drawing should now resemble Figure 8–14.
6. You will use this drawing in the next exercise. For now, save your work by pressing Ctrl+S.

In this exercise you used pre-positioned points to help construct the spline. Depending on the type of work you want to do, you will usually base splines on some form of preliminary data points rather than just draw the curve in free-form style, although both methods are available with AutoCAD. At best, splines are "tricky" objects to construct, and once they are drawn, you frequently need to alter or refine them.

THE SPLINEDIT COMMAND

The SPLINEDIT command enables you to edit a spline's control points and, if they are present, fit-data points. When you draw a spline, you pick fit-data points. AutoCAD uses these points to calculate the location of the spline's control points. You can add additional control or fit-data points or move points already present. You can change the weight, or influence, of control points, as well as the tolerance of the spline. You can also close or open a spline and adjust the tangent information of the start point and endpoint.

Control points usually are not located on the spline curve (except at the start point and endpoint), but they control the shape of the spline. AutoCAD uses the fit-data points to calculate the position of the control points. After AutoCAD determines the control points, it no longer needs the fit-data points. If you remove the fit data from a spline, though, you cannot use any of the fit-data editing options of the SPLINEDIT command to edit the shape of the spline further.

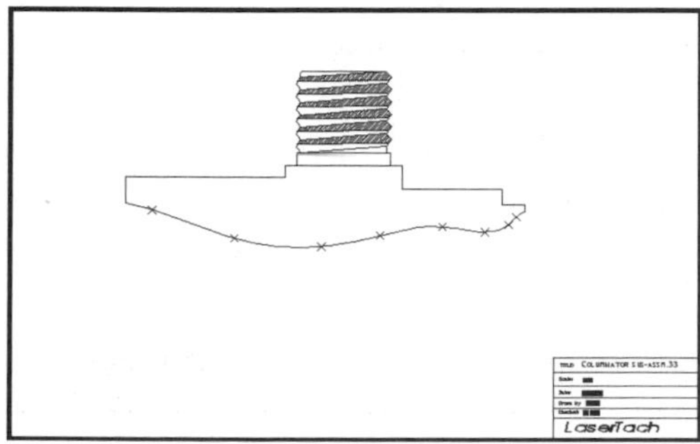

Figure 8–14 *The completed spline curve.*

When you select a spline for editing, AutoCAD displays its control points just as it does if you use grips editing, although grips editing is not available through the SPLINEDIT command. The SPLINEDIT command works on only one spline object at a time.

The SPLINEDIT command has the following options:

- **Fit Data**—Enables you to edit the fit-data points for spline objects that have them, in which case AutoCAD displays another prompt of fit-data options, which are described following this list.
- **Close**—Closes an open spline. Adds a curve that is tangent between the start and end vertices for splines that do not have the same start point and endpoint. If the spline does have the same start point and endpoint, the Close option makes the tangent information for each point continuous. If the spline is already closed, the Close option is replaced by the Open option.
- **Open**—Opens a closed spline. If the spline did not have the same start point and endpoint prior to being closed, this option removes the tangent curve and removes tangent information from the start point and endpoint. If the spline shares the same start point and endpoint before you close it, the Open option removes the tangent information from the points.
- **Move Vertex**—Enables you to move the control vertices of a spline. You specify a vertex to edit by moving the current vertex to the next or previous vertex.
- **Refine**—Displays suboptions that enable you to add control points and adjust the weight of control points. You can add individual control points in areas where you want finer control of the curve. Performing any Refine operation on a spline removes fit data from the spline. You can reverse the effect of the Refine option and restore the fit data with the Undo option before ending the SPLINEDIT command. The following list describes the Refine options.
- **Add control point**—Enables you to add a single control point to a spline. AutoCAD locates the new control point as close as possible to the point you pick on the spline. Adding a control point does not change the shape of the spline.
- **Elevate Order**—Elevates the order of the spline's polynomial, which adds control points evenly over the spline. This option does not change the shape of the spline. The order of the polynomial cannot be reduced once it has been increased.
- **Weight**—Controls the amount of tension that pulls a spline toward a control point.
- **eXit**—Returns you to the main SPLINEDIT prompt.
- **rEverse**—Changes the direction of the spline.

- **Undo**—Undoes the most recently performed SPLINEDIT option.
- **eXit**—Ends the SPLINEDIT command.

The following are the suboptions of the Fit Data option for editing spline object fit data. When you choose the Fit Data option, the grips-like boxes change to highlight the fit-data points.

- **Add**—Adds additional fit-data points to the curve. Adding fit-data points changes the shape of the spline curve. Added fit-data points obey the current tolerance of the spline.
- **Close**—Performs the same function as the control point Close option using fit-data points.
- **Open**—Performs the same function as the control point Open option using fit-data points. The Open option replaces the Close option if the spline is already closed.
- **Delete**—Removes fit-data points and redraws the spline to fit the fit-data points that remain.
- **Move**—Enables you to move fit-data point vertices of a spline. You specify a vertex to edit by moving the current vertex to the next or previous vertex. You cannot edit fit-data information with AutoCAD's grips editing feature.
- **Purge**—Removes all fit data from the spline.
- **Tangents**—Enables you to change the tangent information of the start point and endpoint of a spline.
- **toLerance**—Changes the tolerance of the spline's fit-data points and redraws the spline. A spline loses its fit data if you change the tolerance and move a control point or open or close the spline.
- **eXit**—Returns to the control point editing prompt.

In the following exercise you will edit the spline you drew in the preceding exercise using some of the options of the SPLINEDIT command. First you will reduce the number of fit-data points, and then you will add additional control points to the spline by changing its order.

EXERCISE: EDITING A SPLINE

1. Continue from the previous exercise. Remove the running node Osnap by right-clicking on the OSNAP window on the status line at the bottom of the display. Choose Settings and, in the Drafting Settings dialog box, remove the check next to the Node option. Press OK to exit the dialog box.
2. Press F3 to disable the OSNAP function. From the Modify menu, choose Spline to start the SPLINEDIT command. Select the spline you drew in the preceding

exercise. The control points appear (see Figure 8–15). Note that the control points do not necessarily fall on the spline curve itself. The following prompt appears:

Enter an option [Fit Data/Close/Move Vertex/Refine/rEverse/Undo]:

3. Right-click to display the SPLINEDIT shortcut menu. Choose the Fit Data option. AutoCAD changes the control points to show the fit-data points. Notice that the fit-data points are not the same as the control points. The Fit Data prompt appears:

   Enter a fit data option

   [Add/Close/Delete/Move/Purge/Tangents/toLerance/eXit] <eXit>:

4. Right-click and choose the Delete option and then pick a couple of points. Note that the points disappear as you pick them. Press Enter after you remove the points. The prompt returns. Choose the Tangents option. The "Specify start tangent or [System default]:" prompt appears.
5. Press Enter to keep the current tangent. Then at the "Specify end tangent or [System default]:" prompt, pick a point near ①, as shown in Figure 8–15.
6. When the prompt returns, press Enter to return to the main control point prompt. The control points are displayed and the control point prompt appears.
7. Right-click and choose the Refine option. Then, at the new prompt, right-click again and choose the Elevate order option.
8. At the "Enter new order <4>:" prompt, type **6** and press Enter. This elevates the order of the polynomial and adds control points. The prompt returns. Respond by pressing Enter to exit to the main prompt.
9. Press Enter again to exit the PLINEDIT command.
10. From the layer control box on the Object Properties toolbar, turn off the layer Frame. Your drawing should now resemble Figure 8–16, depending on which points you removed in step 4. You are finished with this drawing.

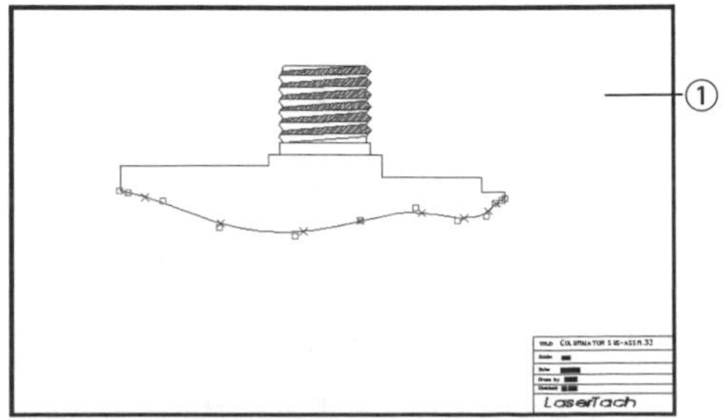

Figure 8–15 *SPLINEDIT displays a spline's control points.*

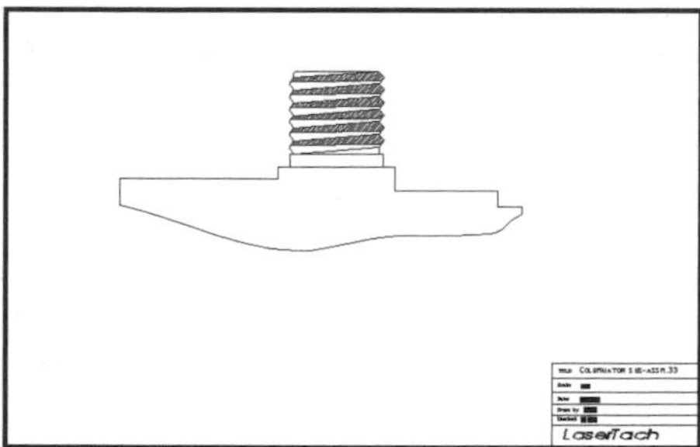

Figure 8–16 *A completed spline-fit curve after modification.*

The options and suboptions of the SPLINEDIT command offer a great deal of control. In this exercise you had the opportunity to investigate several ways you can modify and "tweak" a spline object. Splines are quite complex, but they offer the advantage of also being quite accurate and flexible.

Tip: As with polylines, spline objects can be modified using AutoCAD's grips editing feature. Chapter 10, Advanced Editing, explains more about grips editing.

SUMMARY

In this chapter you explored the versatility of polylines. You saw how you can edit polylines in a variety of ways. You also saw how you can convert polylines into two different types of curves. You learned about the versatility and accuracy of true NURB splines, and the many ways of editing and shaping them into complex smooth curves. Many of the shapes you may be required to draw in AutoCAD are not composed of straight lines, and being able to fit polylines and splines to complex curves is an important skill to have.

Basic Editing

Most of your time using AutoCAD will be spent editing existing objects. With this in mind, it is critical that you understand and efficiently implement the different methods for editing. Editing commands are grouped into two classes: editing commands that can be used on a variety of objects, and editing commands that are designed for a specific type of object. This chapter concentrates on the selection methods, standard editing commands, and processes. Chapter 10, Advanced Editing, concentrates on enhanced editing methods and object-specific editing commands.

This chapter discusses the following skills:

- Selecting objects
- Removing objects from a selection set
- Undoing changes made in the drawing
- Resizing objects
- Relocating and duplicating objects
- Adding chamfers and fillets to objects

ASSEMBLING A SELECTION SET

Many of the editing commands start by displaying the "Select Objects:" prompt, which signals the beginning of the process in which you can assemble a selection set by selecting the desired objects using a variety of methods. Most of the selection options covered in the following list are invoked by typing a letter or two at a "Select Objects:" prompt. The typical selection process is open-ended, which means that you can invoke any of the options listed, as many times as you want and in any order that you want. When an object is selected, it is highlighted on the screen as a visual confirmation. When you finish selecting objects for any given command, press the Enter key or spacebar, or right-click on the input device. Then you will pass the selection on to the command that is currently active.

Note: The highlighting of objects is enabled by default. You can control this feature with the HIGHLIGHT system variable. By typing HIGHLIGHT at the command line, you can set the variable's value to 1 (on) or 0 (off).

The typical selection process provides you with the following options:

- Picking the objects directly
- Using an implied window, a boundary window, and a crossing window
- Selecting the last object
- Selecting all objects
- Using a fence
- Using a window and crossing polygon
- Selecting the previously selected objects
- Using the grouping option
- Undoing the last selection option
- Removing and adding objects to the selection
- Using object cycling

These options are discussed in detail in the sections that follow.

PICKING OBJECTS DIRECTLY

When you are prompted to select objects, the normal pointer is replaced with a box cursor, which is referred to as the "pickbox." To select an object directly, position the pickbox over the object and select it. The size of the pickbox is controlled through the Selection folder in the Options dialog box (see Figure 9–1).

Figure 9–1 *The Selection folder in the Options dialog box lets you control the size of the pickbox.*

USING IMPLIED WINDOWING, BOUNDARY WINDOWS, AND CROSSING WINDOWS

The basic method for selecting large groups of objects is to create a rectangular window around the objects. There are two types of windows: boundary and crossing. The boundary window selects only objects that are totally enclosed within the window. The crossing window selects objects that are enclosed within the boundary or that cross the window's boundary. You can easily initialize the type of selection window you wish to use through AutoCAD's Implied Windowing feature.

Implied Windowing

When you position the pickbox over an empty portion of the drawing and pick a point, the system assumes that you want to place the first corner of a rectangular selection window at that point. You determine the selection window's size by moving the cursor to the opposite corner and then picking a second point (the diagonally opposite corner's point, relative to the initial point). When the window is defined from left to right, a boundary window is initialized, and all objects completely enclosed in the window are selected. The drawing on the left in Figure 9–2 uses a boundary window to select just the circle and vertical line. (The rectangle is the boundary window.)

When the window is defined from right to left, a crossing window is initialized, and all objects that are completely enclosed in the window or that merely cross the boundaries of the crossing window are selected. A crossing window is displayed with a dashed line, whereas a boundary window is drawn with a continuous line. As you can see from the drawing on the right in Figure 9–2, using a crossing window selects the circle and the vertical line, which are enclosed within the window. Additionally, the horizontal line is also selected, because it crosses the window's boundary line.

Explicitly Using the Boundary Window or Crossing Window Method

Selection window types include the typical boundary window and crossing window. They are accessed by typing **W** or **C**, respectively, at the "Select Objects:" prompt. When you use the W option, you explicitly define a window with which to select objects. Unlike an implied window, the first point selected does not have to be located in an empty portion of the drawing. Furthermore, it does not matter whether the window is defined from left to right or from right to left. Using an

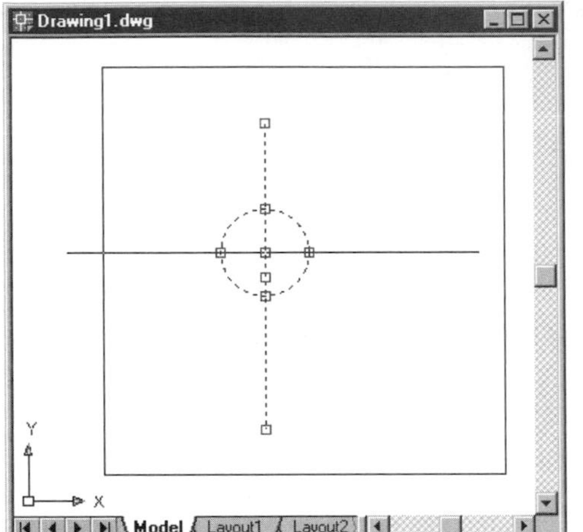

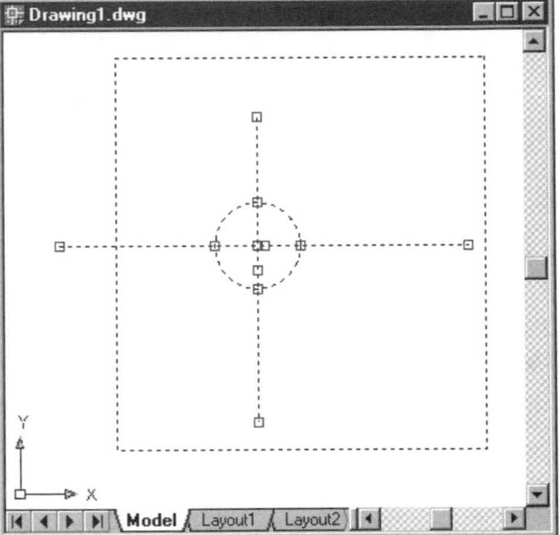

Figure 9–2 *In the drawing on the left, a boundary window is used to select objects; in the drawing on the right, a crossing window is used to select objects.*

Note: By default, the Implied Windowing selection option in the Selection folder of the Options dialog box is enabled. If it is disabled, the only way you can define a boundary window or a crossing window is to use the Window or Crossing Window options explicitly.

Note: All window type selections are relative to the display, not to the current UCS. This means that you cannot select objects in isometric views using a skewed isometric selection window.

explicit boundary or crossing window is superior to using an implied window when you are dealing with a crowded drawing and encounter difficulties finding an empty area of the drawing in which to anchor the first point of the implied window.

SELECTION MODES AVAILABLE FOR SELECTING OBJECTS

To display the Selection folder in the Options dialog box (see Figure 9–3), choose Options from the Tools menu and choose the Selection tab. In this dialog box you will find the settings that control various aspects of the selection process. The following sections discuss these controls in more detail.

Noun/Verb Selection

The first option in the Object Selection Settings dialog box is Noun/Verb Selection, which is enabled by default. With this setting enabled, you have the option to select the objects to be manipulated prior to invoking a command. When the Noun/Verb option is *disabled*, you must invoke a command first and then select the objects to be modified. This option affects only those commands that begin with the "Select Objects:" prompt (such as MOVE, COPY, ERASE, and LIST).

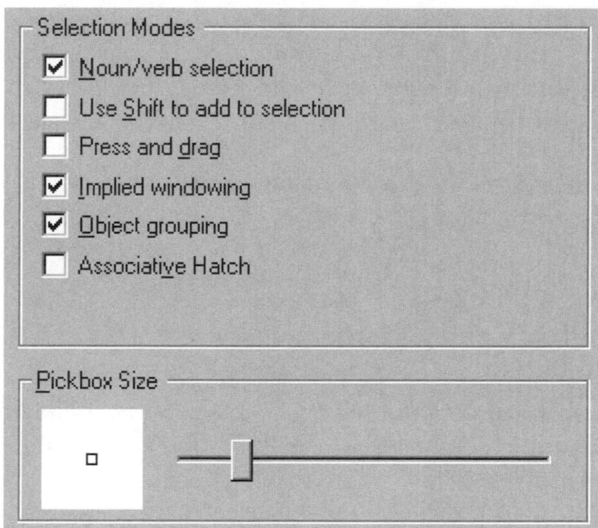

Figure 9–3 *Selection modes and pickbox size options in the Selection folder of the Options dialog box.*

Tip: Working with the enabled Noun/Verb Selection option can be confusing. You should practice using the option until you are comfortable with the processes of selecting objects and using the various editing commands. If you leave Noun/Verb Selection enabled and then accidentally select objects before choosing a command, press the Esc key to unselect the selected objects.

Use Shift to Add to Selection

The Use Shift to Add to Selection option is disabled by default. Therefore, when you select additional objects, the objects are automatically added to the current selection set. If you enable the Use Shift to Add to Selection option, to add more objects to the current selection set, you must hold down the Shift key as you select the additional objects.

Tip: This option is provided for compatibility with the way other Windows applications deal with selecting objects.

Press and Drag

The Press and Drag option is disabled by default. Consequently, when you define a selection window, you do so by pressing and releasing your pick button to pick the first corner of the selection window, and then picking the selection window's second point by again pressing and releasing your pick button. However, when Press and Drag is enabled, you first pick the initial corner point and then hold your pick button as you drag to the opposite corner of the rectangular selection window. You establish the second point by releasing the pick button. Enabling this option makes the process of selecting objects with a rectangular window in AutoCAD similar to the process of selecting objects using standard Microsoft Windows selection techniques.

Tip: Using the Press and Drag option may be faster than using implied windowing. With implied windowing, if you miss the intentional selection of an object and pick nothing, an implied window will be started. With Press and Drag enabled in the same situation, you can simply continue to select objects with the pickbox.

OTHER USEFUL SELECTION OPTIONS

In some cases a typical selection method will not suffice. Fortunately AutoCAD provides several more methods of selecting objects. The following list describes these secondary selection methods. For each one, the character(s) in parentheses can be entered to invoke the respective option.

- **Last (L)**—The Last option automatically selects the last object drawn that is visible in the display. This option can select an unexpected object, so you should use it only when you are working in a contained area.
- **ALL (ALL)**—The All option selects all objects not residing on a locked or frozen layer. This option selects objects even if they are not visible in the current view.
- **Fence (F)**—This option enables you to define a series of continuous line segments, which is referred to as a "fence." All objects that the fence crosses are selected.
- **Wpolygon (WP)**—The Wpolygon (Window Polygon) option is similar to the Window option in that it selects only objects that lie within the window's boundary. The only difference is that with Wpolygon you can define an irregular polygon-shaped window as opposed to a rectangular window. You can define the window polygon with as many points as necessary. The Wpolygon option automatically draws the closing segment back to the beginning point.
- **Cpolygon (CP)**—The Cpolygon (Crossing Polygon) option is similar to the Crossing option in that it selects objects that lie within or cross the window's boundary. The only difference is that with Cpolygon you can define an irregular polygon-shaped crossing window instead of a rectangular crossing window.
- **Previous (P)**—Use this option to reselect the objects in the previous selection set, which was created during the most recent editing command.
- **Group (G)**—The Group option enables you to specify the name of a predefined group.
- **Undo (U)**—When you enter **U** at the "Select Objects:" prompt, AutoCAD undoes the last selection option performed. Choosing the Undo tool cancels the current command when it is chosen at the "Select Objects:" prompt.

In addition to the options listed above, you can use several other options, including BOX, AUto, and Single. Programmers commonly use these options for AutoLISP and ARX-based programs, or for menu and toolbar macros.

Note: When you use the Multiple option (which is invoked by typing **M** at a "Select Objects:" prompt), the selected objects are not highlighted until you press the Enter key or the spacebar. In addition, other selection options such as Previous, Last, All Window Types, Remove, and Undo do not function when the first object has been picked in multiple-selection mode. You must press Enter to finish the multiple-selection mode and highlight the objects.

REFINING THE SELECTION SET

It is often impossible to get your initial selection set to include exactly the objects needed. This is especially true when your drawing is crowded with overlapping objects. To control the object selection set better, you can use a few tools that AutoCAD offers for refining your selection set.

Note: The selection options previously discussed are only available if you choose the command first and respond to the command's prompt to select objects. The exceptions to this are implied windowing, using Shift to deselect objects, object cycling, and the normal pickbox selection.

By default, when you are selecting objects, you are working in the add mode so that any objects selected are automatically added to the current selection set. We will now discuss how to remove objects from a selection set, the use of the Shift key, and object cycling.

Removing Objects from the Selection Set

The Remove option switches you from the default add mode to the remove mode. In this mode all objects selected are removed from the selection set. Undo is often used to remove the objects just selected from the selection set, but the Remove option enables you to select the specific objects that you want to remove from the selection set. You remain in the remove mode until you either end the selection process (by pressing the Enter key or the spacebar) or invoke the Add option by pressing the A key.

Holding Down the Shift Key

Instead of using the Remove option to deselect objects, you can hold down the Shift key and click on objects that are already in the selection set (i.e., already highlighted). This removes those objects from the current selection set. You can hold down the Shift key in combination with any of the aforementioned methods of selecting objects. When you release the Shift key, you are immediately placed back into the current selection mode.

Note: The Shift-key method of deselecting objects works only in the add mode; it will not add objects while you are in the remove mode. To add objects that you have removed from the selection set, simply release the Shift key. In addition, the Shift-key method does not support the Window Polygon or Crossing Polygon (WP or CP) selection options.

Using Object Cycling

In a dense drawing, it can be difficult to select an object directly without inadvertently picking another nearby object. This is especially true when objects lie on top of each other. While you can zoom into the area, making it easier to select the desired object, alternatively, you can use AutoCAD's Object Cycling feature.

To use object cycling, start an edit command such as MOVE and position the pickbox over a dense area in the drawing so your pickbox is on or touching more than one object. While holding down the Ctrl key, pick that location and then release the Ctrl key. This starts object cycling so that only a single object is highlighted. If more than one object was found in the area of the pickbox, "<Cycle on>" appears at the prompt line. The next time you left-click (without holding down the Ctrl key), another object occupying the area of the pickbox is highlighted, as shown in Figure 9–4. If the highlighted object is not the required object, you can pick again, and the next object found in the area of the original pick location is highlighted. Every time you left-click, the next object found in the area of the original pickbox location is highlighted, cycling among all objects in that original pickbox area. When the desired object is highlighted, you can end the cycling by pressing the Enter key or the spacebar, or by right-clicking.

Note: As you continue to left-click to cycle and highlight each object within the pickbox, the physical location of the current pickbox is immaterial. The location of the pickbox at the time you initiate object cycling is what defines the area that is searched. Also, any transparent command can be used during this mode; for example, you can use ZOOM to magnify the display.

RECOVERING FROM UNINTENTIONAL EDITS

Often when you are using AutoCAD you will need to go back to an earlier state of the drawing, either by erasing unneeded objects or by restoring removed objects. The following sections discuss methods of recovering edits made to the current drawing.

RECOVERING ERASED OBJECTS

You can use the ERASE command to remove selected objects from a drawing. The ERASE command is found on both the Modify toolbar and on the Modify menu. Once you have selected the desired objects, press the Enter key, or press the spacebar, or right-click to end the general selection process. Because the ERASE command does not require any further information, the objects are then erased.

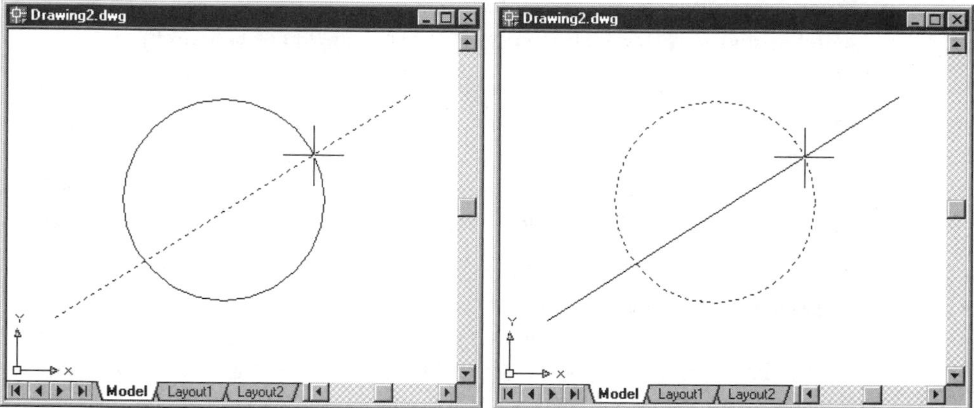

Figure 9–4 *When the cursor's pickbox encompasses multiple objects with object cycling on, AutoCAD selects one object at a time, cycling through the objects as you continue to pick. The image on the left shows the line selected first, and the image on the right shows the circle selected next.*

Tip: If Noun/Verb Selection is enabled, you can select the objects first and then choose the ERASE command, which immediately erases the highlighted objects. In addition, in noun/verb mode, pressing the Delete key on the standard keyboard will erase the selected objects.

Tip: Using the OOPS command to "unerase" objects is preferable to using the UNDO command. OOPS simply restores the last erased objects without undoing non-erase commands.

If you erase any objects by accident, you can use the UNDO command to reinsert the erased objects. Additionally, you can use the OOPS command to unerase objects removed with the last ERASE command, without affecting any other edits you have made since the last ERASE command.

UNDOING CHANGES

In previous sections you learned to use the Undo tool to undo the effects of the last command. UNDO can be invoked by simply entering **U** at the command line. However, a more powerful version of the U command also exists, and is executed by typing UNDO at the command prompt. The following list outlines the UNDO command's options and their functions:

- **Number**—With Undo, you can specify the number of commands to be undone by typing at the prompt the number of commands you want to undo. The REDO command undoes the effects of the UNDO command regardless of the number of options you specify within a single UNDO.

- **Auto**—Auto enables you to undo a complete menu selection as a single command sequence. (This is also reversible with a single U command.)
- **Control**—Using Control, you can limit AutoCAD's ability to undo commands to undoing just one command. Typically, the only time you must deal with the options available through Control is when you are critically short of drive space (because keeping track of commands takes up hard drive space). In such situations, it is better to allocate more free drive space than to limit AutoCAD's ability to undo commands.
- **Begin/End**—The Begin option begins the process of grouping a series of commands. All commands following that option become part of the group until you use the End option to close the group. Undo and U treat grouped commands as a single operation.
- **Mark/Back**—Use the Mark option of the UNDO command to mark (much like a bookmark) a place in the Undo information file that is maintained by AutoCAD 2000. Later, during the same drawing session, you can use the Back option to undo all commands issued since the last Mark option was specified. In addition to undoing the commands, the Back option also removes the last mark. You can issue the Mark option as many times as you want during a drawing session to place multiple "bookmarks" throughout your drawing process.

Tip: Issue the Mark option when you are unsure if your edits will be correct. Then, if you decide the edits were not correct, you can issue the UNDO command and use the Back option.

The following section covers various methods of resizing drawing objects.

RESIZING OBJECTS

A number of commands can be used to modify existing objects. The commands that can be used to resize objects are SCALE, STRETCH, LENGTHEN, TRIM, EXTEND, and BREAK, all of which are covered in this section. These commands are located on both the Modify toolbar and the Modify pull-down menu.

SCALING OBJECTS

The SCALE command is used to scale objects up or down in size. This command employs the general selection process, which gives you many options for selecting the objects. After selecting the objects to be scaled, you are prompted to pick a base point. You then can enter a scaling factor or graphically pick a distance, or you can specify the Reference option. When you graphically scale the selection set, the length of the rubber-banding line is used as the scaling factor.

The Reference option of the SCALE command provides a very useful control. After you request the option, you are prompted to specify a reference length. You usually specify it graphically along an existing object or distance. You are then prompted for a new length value.

STRETCHING OBJECTS

The STRETCH command is used to stretch an object's length by relocating a portion of the object. Although the command does issue the general "Select Objects:" prompt, you must select the desired objects using one of the crossing selection methods (an implied crossing window, a crossing window, or Cpolygon). You can use only one crossing window per occurrence of the STRETCH command. If you define more than one crossing window in a single STRETCH command sequence, only the objects selected with the last crossing window are stretched.

After you select the objects, you are prompted to select a base point or to enter a displacement. If you choose to pick a base point, you also must specify a second point for the new location of the base point. As an alternative to picking a base point, you can specify a displacement.

The displacement is defined as the delta-X, delta-Y, and delta-Z (or the distance and angle from the start point to the end point, which is essentially the distance the object(s) are to be stretched and the direction you want to apply to the objects). If you know the exact displacement you want to use, you can enter the displacement at the prompt for the base point and then press Enter twice. The displacement you enter can be an absolute Cartesian coordinate, a relative distance from the base point, or a polar coordinate. The STRETCH command will move all objects that are completely enclosed in the crossing window. Any objects not completely enclosed within the crossing window are stretched by repositioning those endpoints inside the crossing window while maintaining the position of those endpoints outside the crossing window, as shown in Figure 9–5.

Note: The STRETCH command cannot be used to stretch circles or ellipses; those objects are resized using grips.

TRIMMING OBJECTS

With the TRIM command, you begin the trimming process by selecting the object(s) that will define the "cutting edge," then selecting the object(s) to be trimmed against the selected cutting edge(s), selecting those objects which are to be trimmed away or removed. If you want to select more than one object at a time, you can use the Fence option to select multiple cutting edges or multiple objects to be trimmed. The TRIM command is effective only if an object is left over; otherwise, the ERASE command should be used.

Any edge object, such as a line, circle, or objects in a block, can be used as a cutting edge or can be the object to be trimmed. However, an mline object can be selected

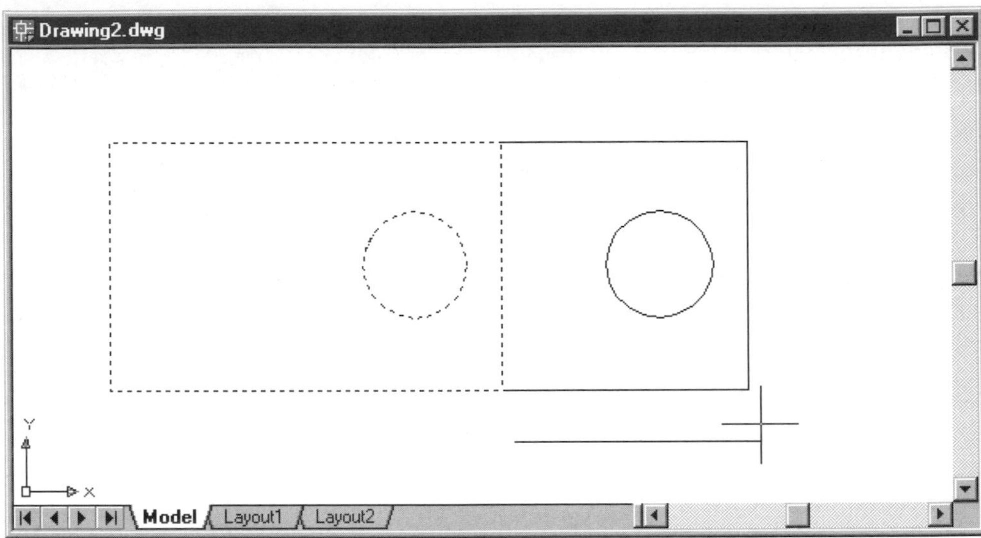

Figure 9–5 *The STRETCH command stretches objects that cross the selection window, and moves objects that lie within the selection window. In this drawing, the selection window crossed the rectangle, and the circle was within the selection window.*

Tip: Using the Fence option can return incomplete results. This may be due to the display scale, the object's linetype, any running Osnap modes, or the selection of the object(s) to be trimmed more than once. If the AutoCAD 2000 fence line does not cross each object at a point where the object is displayed (not the gaps in a dashed linetype), then command cannot find the objects to trim. In addition, running Osnaps sometimes interferes when they do not apply to the type of objects selected. In all cases, performing several passes of the process or increasing or decreasing the display scale around the objects usually resolves the problem.

as a cutting edge but cannot be trimmed. It is worth mentioning that when you are trimming out a segment of a circle, after the first trim, the object is converted into an arc object.

Starting from the point that is used to select the object, use the TRIM command to proceed in one direction along the object until you encounter either an endpoint or a cutting edge. The TRIM command then proceeds from the original pick point in the opposite direction until it encounters either an endpoint or another cutting edge. The resulting defined portion of the object is then removed. Under no circumstances can an object be trimmed in such a way that nothing is left of the object. This condition exists when you trim a circle with a cutting edge that does not cross the circle completely.

Tip: If you press Enter without selecting any cutting edges, all edge objects on the screen are automatically selected as valid cutting edges but are not highlighted. In addition, any object that is selected as a cutting edge can be trimmed (if permissible) with the same TRIM command.

Note: When using the TRIM command, you can temporarily switch to the EXTEND command by holding down the Shift key.

EXTENDING OBJECTS

The EXTEND command is used to elongate an edge object to an existing boundary. EXTEND is the complement to the TRIM command, so it has the same options. Instead of prompting you to select the cutting edge(s), the EXTEND command prompts you to select the boundary edge(s). The Project and Edge options are settings that are saved and shared by the TRIM and EXTEND commands.

Tip: When working with an arc or partial ellipse object, you cannot extend any endpoint of the object to a point that would cause it to close on itself. If you attempt such a move, AutoCAD returns the prompt "Object Does Not Intersect An Edge" and then asks for another object selection.

Using the Edge Option to Extend the Cutting Edges

By default, the object to be trimmed must physically intersect the cutting-edge object. You can bypass this requirement with the Edge option. Choosing the extend mode of the Edge option extends the cutting edges so that the object(s) being trimmed do not need to intersect the cutting edge.

Choosing the Edge option to Extend sets the EDGEMODE system variable to 1. This option controls just how the TRIM and EXTEND commands find cutting and boundary edges. With a setting of zero, it uses the selected edge without any implied extension. With a setting of 1, it extends or trims the selected object to an implied extension of the selected cutting or boundary edge. This system variable can be used with the TRIM and EXTEND commands. When the setting has been made in either command, it affects the operation of both.

Note: When using the EXTEND command, you can temporarily switch to the TRIM command by holding down the Shift key.

BREAKING OBJECTS

To remove a portion of an object, consider using the BREAK command. BREAK offers two advantages over TRIM for removing a portion of an object. First, you do not need any cutting objects to use BREAK. Second, the BREAK command can be used to break a non-closed object at a single point, thereby breaking the object into two objects.

After you select the object to be broken, the default option requests a second point on the object. Then the portion of the object that lies between the point used to select the object and the second point is removed. Notice that when you first select the object to break, it is not necessary to press Enter to go to the next prompt.

Sometimes, however, the point used to select the object is not where you want to begin the break. In such a case, you use the First option to specifically define the beginning point of the break and the second point of the break.

Note: When you are breaking curved edge objects, such as a circle, the removal process always proceeds from the first point to the second point in a counterclockwise direction.

Note: When you use BREAK, if the second break point does not actually intersect the object to be broken, AutoCAD projects the picked point to a location perpendicular to the object, and the projected point becomes the second break point.

If you simply want to break an object into two objects without actually removing a portion of the object, define the second point at the same location as the first point of the break. The easiest way to do this is to enter @ as the second point. When you enter @, AutoCAD passes the last point recorded directly into the current request for a point location.

Sometimes you need to edit objects by repositioning them at new locations. The following section covers commands for repositioning objects.

RELOCATING OBJECTS USING MOVE AND ROTATE

The MOVE and ROTATE commands allow you to relocate and/or reorient selected objects. MOVE and ROTATE are found on both the Modify pull-down menu and the Modify toolbar.

MOVING OBJECTS

When you use the MOVE command to relocate selected objects to a new location, after you select the objects, you must determine the base point for the selection set. For the base point, you typically choose a point on one of the objects being moved. You then pick a second point, which is the new location of the base point.

In reality, the MOVE command simply calculates the distance and direction from the first point to the second point and then uses that information to move the objects. Picking the base point on one of the objects being moved can make it easier for you to visualize the end result of the MOVE command. As an alternative to picking a base point, you can specify a displacement.

The displacement is defined as the delta-X, delta-Y, and delta-Z (or the distance and angle from the base point to the second point). Essentially, this is the amount of movement you want to apply to the objects. If you know the exact displacement desired, you can enter the displacement at the prompt for the base point.

You can enter the displacement as an absolute Cartesian coordinate or as a polar coordinate. You can force the MOVE command to interpret the numbers as a displacement by pressing Enter at the prompt for the second point. (In other words, do not define a second point.)

ROTATING OBJECTS

You use the ROTATE command to rotate selected objects about a particular point—the base point of the rotation. After you pick the base point, the default option enables you to specify a rotation angle by typing the rotation angle, or picking a point to rotate the objects graphically, or selecting the Reference option. If you pick a point, the angle of the rubber-banding line is used as the rotation angle.

Using the Reference Option for the ROTATE Command

A valuable feature of the ROTATE command is the Reference option. Occasionally you might draw an object at an incorrect or unknown angle. You can use the Reference option to correct the angle without having to investigate it.

To use the Reference option, first you select the objects and specify the base point, then you type **R** for reference. AutoCAD asks for a reference angle. At this point you can enter the angle if you know it, or you can pick two points on the screen to graphically specify the unknown angle. You are then prompted for a new angle, at which point you can either graphically place a point or enter the desired angle for the selected objects.

DUPLICATING OBJECTS

You can use the COPY, OFFSET, MIRROR, and ARRAY commands—and even the Clipboard—to make copies or exact duplicates of selected objects.

COPYING OBJECTS

With the COPY command, you can make exact duplicates of selected objects. After you select the objects to copy, you are prompted to specify the base point of the displacement and a second point. The distance and direction from the first point to the second point are calculated and used to locate the object copies. To help you visualize the results, start the COPY command, select an object (such as a circle), and then pick a base point on the object (such as the circle's center.) The second point then becomes the copy's base point, which is analogous to the original object's base point.

Just as you can use the displacement function with the MOVE command, if you know the exact displacement, you can enter it at the prompt for the base point when you use the COPY command. You force the COPY command to interpret your entry as a displacement by pressing Enter again at the prompt for the second point. (In other words, you do not define a second point.)

By default the COPY command makes one copy of the selected objects. You can, however, use the Multiple option to make multiple copies of the selected objects. After you select the Multiple option and choose an initial base point, the COPY command repeatedly prompts you to select a second displacement point at which to locate the copies. Press Enter to end the multiple-copy process.

DUPLICATING OBJECTS WITH OFFSET

With the OFFSET command you can create a copy of the selected object and have AutoCAD offset it a specified distance from the original object. At the initial prompt you have the choice of entering the offset distance or using the Through option. To enter a specific offset distance, type the distance (or pick two points on the screen) at the "Specify offset distance:" prompt. Thereafter you can select one object at a time to create an offset from the duplicate and choose the side of the original on which you want the duplicate made.

If you choose the Through option, you pick a point the offset object is to go through after you select the object you want to copy. The distance along a perpendicular from the point that you pick to the original object serves as the offset distance. The copy made by the OFFSET command might or might not be an exact duplicate of the original. Table 9-1 lists the various types of objects you can choose with the OFFSET command, and the shape of the resulting copy. The differences between the sizes and lengths of the resulting copies can be attributed to the side of the original object specified for the offset copy.

CREATING A MIRROR IMAGE

With the MIRROR command you can create a mirror-image copy of the selected objects. After you select the objects to be mirrored, you are prompted to pick two points to define the mirror line. The mirror line is the line, or axis, about which the mirror image is created, as shown in Figure 9–6. The mirror line itself does not have to be a physical line in the drawing.

The only option you have with the MIRROR command is whether or not to delete the original objects. The default is not to delete the original objects.

By default, the copy of text and mtext objects will appear backward, as if you were holding a page of text up to a mirror. To prevent text from being reversed in the mirrored objects, set the system variable MIRRTEXT to zero before you begin the MIRROR command.

Note: When a block is mirrored, it is assigned a negative X or Y scale factor, depending upon the mirror-line position.

Table 9–1: *Objects and Resulting Duplicates Created with OFFSET*

Original Object	Resulting Duplicate
Arc	The new arc is created so that it has the same included angle and center point as the original arc, but the arc length will change.
Circle, ellipse	The new circle or ellipse is created so that it has the same center point as the original circle or ellipse. The radius of the new circle on the axis lengths of the new ellipse will be different from the original object's radius on the axis lengths.
Line, ray, xline	The new line, ray, or xline is an exact duplicate of the original.
Lwpolyline	The lengths of the line and arc segments of the new lwpolyline are adjusted so that the endpoints of the new polyline are located along a direction perpendicular to the corresponding endpoints on the original open lwpolyline. For an intermediate vertex point, the new vertex points are located along a direction that bisects the angle between the segments on either side of the vertex point.
Spline	The length and shape of the new spline are adjusted so that the endpoints of the new spline are located along a direction perpendicular to the corresponding endpoints on the original open spline.

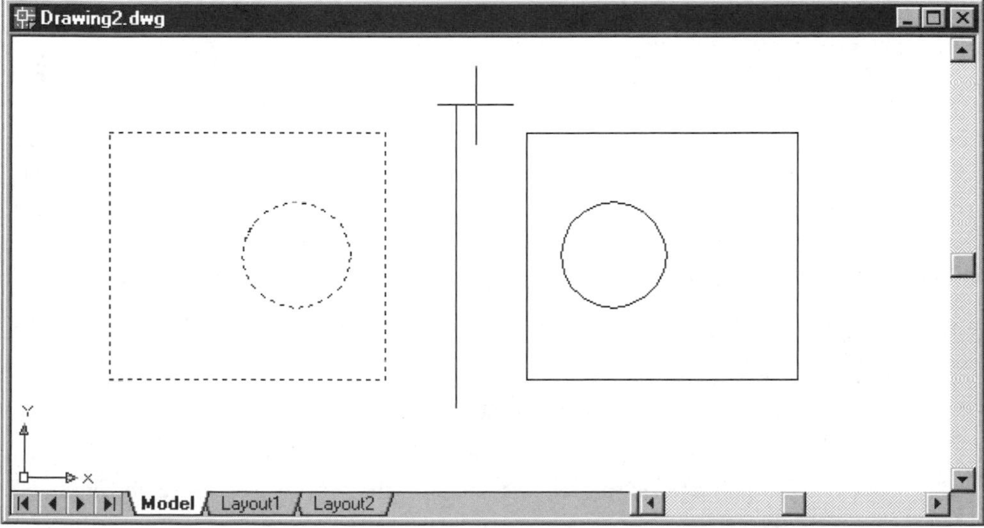

Figure 9–6 *The MIRROR command creates a mirror image of the selected objects.*

CREATING ARRAYS OF OBJECTS

The ARRAY command is designed to make multiple copies of selected objects simultaneously. When the copies are simultaneously created, the ARRAY command arranges the copies either in rows and columns (called a rectangular array) or in a circular pattern (called a polar array). By using the ARRAY command, you can instantly create dozens, or even hundreds, of copies of selected objects.

The ARRAY command includes a dialog box interface designed to simplify the construction of both rectangular and polar arrays. The main dialog boxes for the rectangular and polar array options are shown in Figure 9–7 and Figure 9–8.

Array Dialog Box—Rectangular Option

If you choose to create a rectangular array, you are prompted to enter the number of rows and columns, the distance between the rows (Row Offset) and columns (Column Offset), and the array's angle.

Generally you enter the row and column distances in the dialog box as shown in Figure 9–7. You can, however, specify the distances with a window (referred to as a "unit cell") by choosing the Pick Both Offsets button, and then picking the two corner points of the unit cell window on the screen, thereby simultaneously setting the row and column distances. The height of the window is used as the distance between rows, and the width of the window is used as the distance between columns. Alternatively, you can indicate only the row offset by choosing the Pick Row Offset button, or only the column offset by choosing the Pick Column Offset button, and then pick the distance between rows or columns on the screen.

If you enter a negative value for the distance between columns, the columns of the array are created in the negative direction along the X-axis; otherwise, they are created in the positive X direction. Likewise, if you enter a negative value for the distance between rows, the rows are created in the negative direction along the Y-axis; otherwise, they are created in the positive Y direction. If you choose to use a unit cell to specify the distances, the direction in which the rows and columns are created is determined by the direction from the first pick point to the second pick point.

Array dialog box features for the Rectangular Array option include the following:

- As you specify the number of rows and columns, an example window is dynamically updated to show you the array configuration.
- You specify both row and column offsets by entering values in edit boxes or by exiting the dialog box to show distances on the screen. The example window is updated to show you the relative offsets.
- Using the Angle of Array parameter, you can specify the angle by either entering a value or by choosing the Pick Angle of Array button, and then picking the

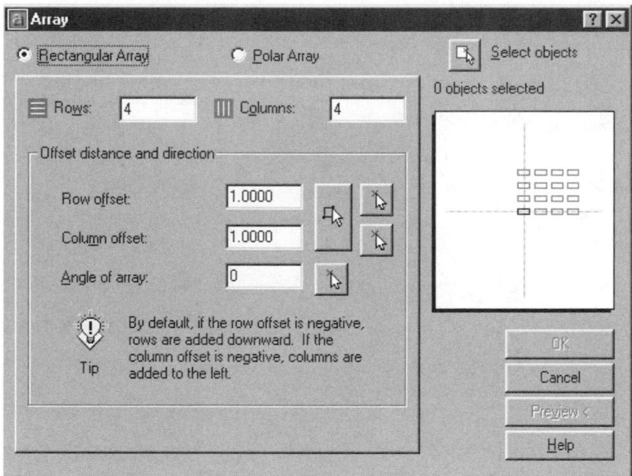

Figure 9–7 *Main dialog box for the ARRAY command—Rectangular Array option.*

angle on the screen. The example window is updated to show you the angle configuration.

- Once you have supplied all required parameters, a Preview button dismisses the dialog box to show you the actual array on the screen. A dialog box allows you to accept the array as shown or modify the current values by returning to the Array dialog box.

The built-in dialog box Tip reminds you that if you enter negative row offsets, the array is downward; if your column offset is negative, columns are added to the left.

Array Dialog Box—Polar Option

If you choose to create a polar array, you are prompted to specify the center point about which the copies are made, the number of items (or copies) to create (which includes the original object), and the angle to fill, as shown in Figure 9–8.

The angle to fill is the angle you want to occupy with your copies. AutoCAD uses the value you enter for the angle fill to determine the angular separation between adjacent items. For example, if you specify six items and enter 180 degrees as the fill angle, the angular separation between adjacent items is 180 divided by 6, or 30 degrees. If you specify a positive fill angle, the copies are made in a counterclockwise direction; otherwise, the copies are made in a clockwise direction by default.

Additionally, the polar array gives you the choice of rotating or not rotating the copies. If you select the Rotate Items as Copied option, the copies are rotated about the specified center point of the array. If you clear the Rotate Items as Copied option, the copies are not rotated about the reference point of the selection set. The reference point of the selection set is determined from the last object selected. If a win-

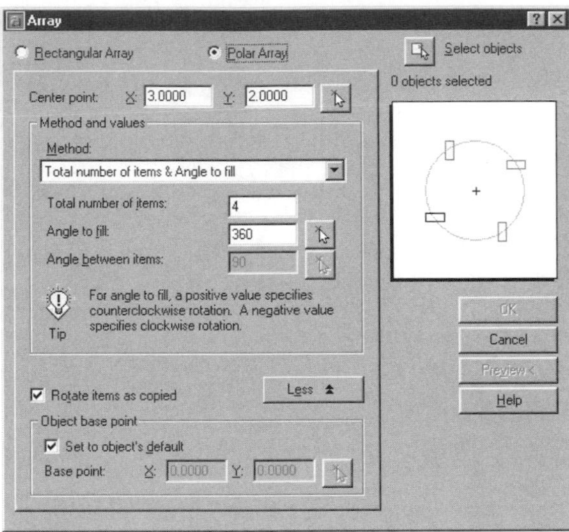

Figure 9–8 *Main dialog box for the ARRAY command—Polar Array option.*

dow of some type is used to select the objects, the last object in the selection set is picked arbitrarily. Therefore, the reference point that is selected is based on the type of object (see Table 9-2).

Array dialog box features for the Polar Array option include the following:

- You specify the center point of the array by entering values or by picking a point on the screen.
- The Method drop-down list lets you choose from three methods of creating the array:

  1. Total Number of Items & Angle to Fill
  2. Total Number of Items & Angle between Items
  3. Angle to Fill & Angle between Items

  Method 3 represents added functionality to the ARRAY command. Depending upon which method you choose, the other parameters' edit boxes in this section will become active. Where appropriate, you may either enter values, or dismiss the dialog box to pick values on the screen.
- As array parameters are specified, the example window is updated dynamically to show the proposed array orientation.
- Once you have supplied all required parameters, a Preview button dismisses the dialog box to show the actual array on the screen. A dialog box allows you to accept the array as shown or modify the array's current values by returning to the Array dialog box.

Table 9–2: *Point on Object Used As the Reference Point for a Polar Array*

Object	Reference Point Used
Block insertion, text, mtext	Insertion point
Dimension objects	One of the definition points of the dimension object
Lines, rays, traces, mlines	One of the endpoints
Arcs, circles, ellipses	The center point
Lwpolylines, splines	The first vertex point
Xlines	The point connecting an imaginary line perpendicular to the xline with the center point of the polar array

CHAMFERING CORNERS

If your design requires you to draw a beveled corner, use the CHAMFER command. The CHAMFER command is used to bevel corners formed by two nonparallel lines, rays, xlines, or the joined line segments of a single polyline. To issue this command, you choose Chamfer from the Modify pull-down menu or from the Modify toolbar. To use the CHAMFER command, you first set the parameters defining the bevel to be generated, and then you select the two line segments that form the corner.

DEFINING THE BEVEL

To obtain the desired bevel, you first define one of two sets of parameters. One set of parameters, accessed with the Distances option, enables you to define beveling operation with two distances—one along the first selected line, and the other along the second selected line. Both distances are measured from the corner, or intersection, of the two lines.

The other set of parameters, accessed with the Angle option, consists of a distance measured from the corner point along the first selected line and the angle of the new line relative to the first selected line and the angle of the new line relative to the first selected line.

You can use either one or both of the Distances and Angle options, depending on what design information is available to you. The CHAMFER command uses the most recently defined set of parameters. If both sets of parameters are defined, you can switch between them by using the Method option.

The two lines you bevel do not have to intersect at a corner point. CHAMFER automatically trims or extends the two selected lines to an intersecting point before generating the bevel line. A quick way to trim or extend two lines to a corner point is to use the CHAMFER command with the distances set to zero.

If the two lines selected are on the same layer and have identical color and linetype properties, the new bevel line is drawn with the same properties. If a particular property of the two selected objects differs, the bevel line takes on the drawing's current object property value. For example, if the two selected lines are drawn on different layers, the new bevel line is drawn on the current layer. If the two selected lines are drawn with a different color, the bevel line is drawn with the current color property. If the two selected lines are drawn with a different linetype, the bevel line is drawn with the current linetype property.

Note: If two objects are drawn on two different layers and displayed in two different colors, it does not mean that the color property of the two objects is different. If the two layers have different colors assigned to them, and the color property of the two objects is BYLAYER, the two objects are drawn with the color assigned to the layer on which the objects reside. Obviously, if the two layers have different assigned colors, the BYLAYER color setting causes the objects to be drawn in two different colors; however, both objects have the identical BYLAYER color property. The BYLAYER setting also affects the linetype used to display objects in the same manner.

DEALING WITH POLYLINES IN THE CHAMFER COMMAND

To bevel all the corners of a polyline simultaneously, specify the Polyline option and select the target polyline. Be aware, however, that when you generate a bevel line at any angle other than 45 degrees relative to the selected lines, the result will not be symmetrical.

This asymmetric result is produced because the polyline segments are processed in the order in which they are drawn. To produce a symmetrically beveled shape, you must bevel one corner at a time. By explicitly selecting the first and second line segments, you control how much each segment is trimmed.

TO TRIM OR NOT TO TRIM

As previously stated, by default, CHAMFER extends or trims the lines to a corner point before applying the chamfer distances and/or angle. If, however, you want to draw the bevel line without any modifications to the original lines, choose the Trim option. At this point you can choose between the Trim and No Trim settings. If you do not want the original lines modified, choose the No Trim setting.

FILLETING OBJECTS

If your design includes rounded corners, use the FILLET command, which you choose from the Modify pull-down menu or from the Modify toolbar. With FILLET, you not only create rounded corners between two lines, rays, xlines, or line segments of a polyline, you also draw an arc segment between any combination of two lines, rays, xlines, circles, ellipses, arcs, elliptical arcs, or splines. The generated arc is always drawn so that it starts and ends tangent to the two selected objects.

CREATING AN ARC USING FILLET

To draw the arc, you first use the Radius option to set the radius, and then you select the two objects. As with CHAMFER, if the two objects are nonparallel lines, the lines are trimmed or extended to a corner point, and the arc is drawn so that the tangent lengths are equal.

With the FILLET command (unlike the CHAMFER command), the two lines do not have to be nonparallel lines. If the lines are parallel, FILLET automatically draws a semicircle between the ends of the two lines, using the endpoint of the first selected line to determine how far to trim or extend the second selected line. The radius of the generated semicircle is set automatically to half the distance between the two parallel lines.

Using FILLET with any objects other than line objects (such as arcs) can produce surprising results. The governing rule is that the generated arc must be drawn in such a way as to start and end tangent to the selected objects.

If the two selected objects reside on the same layer and have identical color and linetype properties, then the new arc is drawn with those properties. If a particular property of the two selected objects differs, then the arc takes on the drawing's current object property value. For example, if the two selected objects are drawn on different layers, the new arc is drawn on the current layer. If the two selected objects are drawn with different colors, the arc is drawn with the current color property. If the two selected objects are drawn with different linetypes, the arc is drawn with the current linetype property.

Tip: A quick and easy way to extend or trim two lines to a corner point is to use FILLET with a zero radius. This method also requires that the Trim mode option be current (as opposed to No Trim).

Note: If you want to round all the corners of a polyline simultaneously, choose the Polyline option and then select the polyline. If an arc segment separates two line segments, the arc segment is automatically removed and replaced by a new arc based on the current FILLET radius setting.

Generally, if you use FILLET on two objects that are not closed (such as on any object other than a circle or ellipse), the two objects are trimmed or extended as necessary so that the arc can be drawn correctly. If you do not want the original objects to be trimmed, choose the Trim option, and then choose No Trim. This Trim option is the same as the Trim option in the CHAMFER command. This setting is common to both commands, so setting Trim in FILLET affects CHAMFER, and vice versa.

SUMMARY

In this chapter you learned about the general commands and tools used to select objects and edit them. For the most flexibility in selecting objects, you should use the command line version of the various editing commands. In addition to the editing operations you can use, AutoCAD offers a number of other editing commands, such as TRIM and EXTEND. The next chapter covers the remaining editing commands not covered in this chapter, and some of the more advanced editing methods.

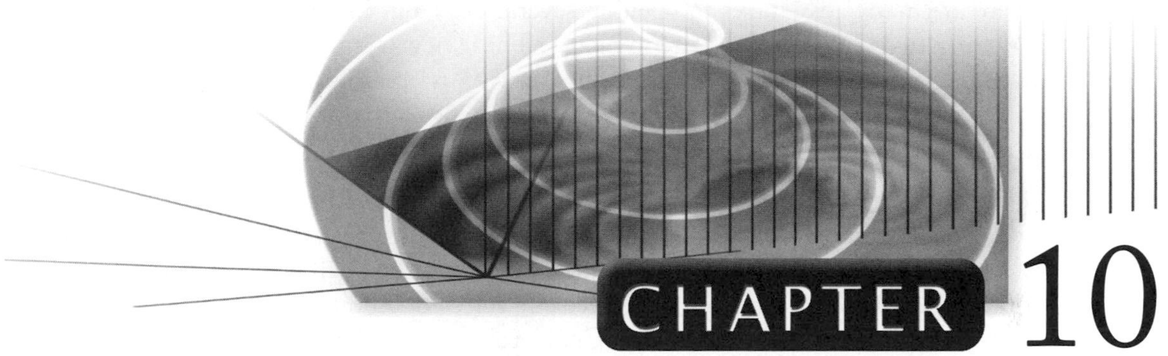

Advanced Editing

In Chapter 9, Basic Editing, you learned the basic commands and tools needed to make changes to existing objects. In this chapter you will build on that foundation and learn about the following topics:

- Using cut and paste
- Using drag and drop
- Grip editing commands
- Understanding the Object Properties Manager
- Using the Matchprop tool
- Using Quick Select
- Using object-selection filters
- Creating and editing groups
- How to explode objects
- Specialized object editing
- Lengthening and shortening objects
- Aligning objects
- Renaming and purging named objects

WINDOWS FUNCTIONALITY IN AUTOCAD

There are many advantages to opening several drawings in a session of AutoCAD, including two features that relate to blocks. These new features allow you to cut and paste or drag and drop objects from one drawing to another. Additionally, you are given the option to insert the objects as a block.

COPYING WITH THE CLIPBOARD

To copy objects to the Windows Clipboard, choose Copy from the Edit pull-down menu. This will copy *selected* objects to the Clipboard. Alternatively, choose Copy Link to copy the current *view* to the Clipboard. Copy Link copies all objects in the current drawing, regardless of whether the objects are selected. Additionally, you can use the CUT command to copy objects to the Clipboard and remove them from the drawing.

Note: When you paste objects copied to the Clipboard into a drawing using the Copy Link command, all the drawing's objects are pasted. However, when you paste the objects into another file, such as a Word document, only the drawing's view is pasted. Therefore, only the objects visible in the view are displayed in the Word document.

The Copy with Base Point option allows you to define a coordinate and save it with the copy stored in the Clipboard. This feature is useful when you paste objects using either the Paste as Block or Paste to Original Coordinates commands. The Paste as Block command uses the base point as the block's insertion point. The Paste to Original Coordinates command uses the base point to insert the objects into exactly the same coordinate positions they had in the original drawing. This is very useful in transferring data from one drawing to another while maintaining a specific location.

Note: The Paste to Original Coordinates command only activates when pasting objects into a drawing other then the one from which they were copied.

CUT-AND-PASTE BLOCK INSERTION

The phrase "cut and paste" refers to the process of selecting an object or group of objects and cutting (or copying) them to the Windows Clipboard. When selected objects are cut, they are copied to the Clipboard and erased from the original document. In contrast, when you copy objects, they are copied to the Clipboard, and the selected objects are left untouched in the original document.

AutoCAD's Cut-and-Paste Block Insertion feature allows you to identify the base point of an object when you copy it to the Clipboard. This feature provides you with the control to identify specifically the new insertion point of the selected object, and it is very powerful. Once the object(s) are pasted as a block, you can later explode the block to remove the arbitrary block name assigned by the Paste as Block command.

When you choose Paste from the Edit menu in an AutoCAD drawing, the contents of the Clipboard are inserted as individual objects. Alternatively, you can paste an entire drawing into the current drawing by dragging the drawing file's icon from Windows Explorer into the current drawing.

When you insert a block from the Clipboard, you can either choose to paste it using a selected base point, or you can paste it using the block's original coordinates. The Paste to Original Coordinates option found on the shortcut menu copies the coordinate location of the objects in the original drawing, and then pastes the objects into the new drawing using those same coordinates. This option is useful for copying objects from one drawing to another when both drawings use the same coordinate system. You must have previously saved a base point with the copied objects, and be in a drawing other than the one from which the objects were copied, to have access to the Paste to Original Coordinates option.

Note: To use the Copy with Base Point feature, you must be in MDE mode. If you use only Single Drawing Interface, the Paste to Original Coordinates command will be unavailable.

DRAG-AND-DROP BLOCK INSERTION

When you drag and drop objects, you move them from one place to another. When you drag and drop objects within the same AutoCAD drawing, the result is the same as using the MOVE command to move the object. When you drag and drop objects from one drawing to another, the result is similar to copying the object.

The Drag and Drop feature is useful for quickly moving or copying objects. However, this feature does not provide an option for accurately selecting the base point for pasting the objects as a block. Consequently, to select objects in one drawing and paste them as a block in another while controlling the new block's base point, use the Copy and Paste feature described in the previous section.

You invoke AutoCAD's Drag and Drop feature by right-clicking on a selection set and dragging and dropping it into either the same drawing or a different drawing. Once you drop the selection set, AutoCAD displays a shortcut menu that presents several options for indicating the selection set's new position. The following options are available on the shortcut menu:

- **Move Here**—This option appears when the selected objects are dragged and dropped within the same drawing.
- **Copy Here**—This option appears when objects are dragged and dropped within the same drawing or between two drawings.
- **Paste as Block**—This option appears when objects are dragged and dropped within the same drawing or between two drawings.

- **Paste to Orig Coords**—This option appears only when objects are dragged and dropped between two drawings.
- **Cancel**—This option appears when objects are dragged and dropped within the same drawing or between two drawings.

Note: You can also drag and drop files from Windows Explorer into a drawing.

While the Drag and Drop feature does not provide an option for accurately selecting the base point for pasting the objects, if you select objects and enable their grips, you can use one of the grips to define the selections set's base point, as discussed later in this chapter in the section titled Grip Editing.

DRAG-AND-DROP HATCH

In AutoCAD 2002 you can use AutoCAD DesignCenter (discussed in Chapter 11) to drag and drop hatch patterns directly into an open drawing. Once you master the method, it is perhaps the quickest way to apply a hatch pattern. The steps involved are as follows:

1. Open AutoCAD DesignCenter by clicking on the AutoCAD DesignCenter icon on the Standard toolbar (or press Ctrl+2).
2. Using the tree view of DesignCenter, navigate to the \AutoCAD 2002\Support folder and open (click) the acad.pat file. The hatch patterns in this file appear in the right pane of DesignCenter, as shown in Figure 10–1.
3. Find the pattern you want, drag it into the drawing, and drop it onto the object you want to hatch (see Figure 10–2).

GRIP EDITING

Grip Editing is a feature that integrates object snap points with the most commonly used editing commands and then places the combined capabilities literally at your fingertips. With grips, it is possible to edit objects and select specific object snap points without ever having to pick a tool, use a menu command, or choose a running Osnap. In the following sections you will learn how to enable the Grips feature, activate grips, and make use of the various options available with grips.

ENABLING GRIPS

Grips are an optional facility that you can choose to use. By default, grips are enabled. You can disable grips with the Enable Grips toggle in the Selection folder of the Options dialog box (see Figure 10–3), which you display by choosing Options from the Tools pull-down menu.

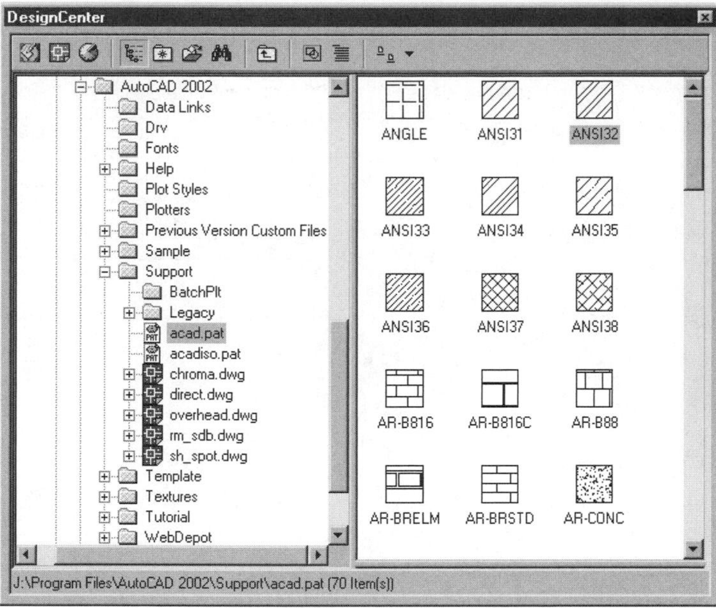

Figure 10–1 *Displaying hatch patterns in DesignCenter.*

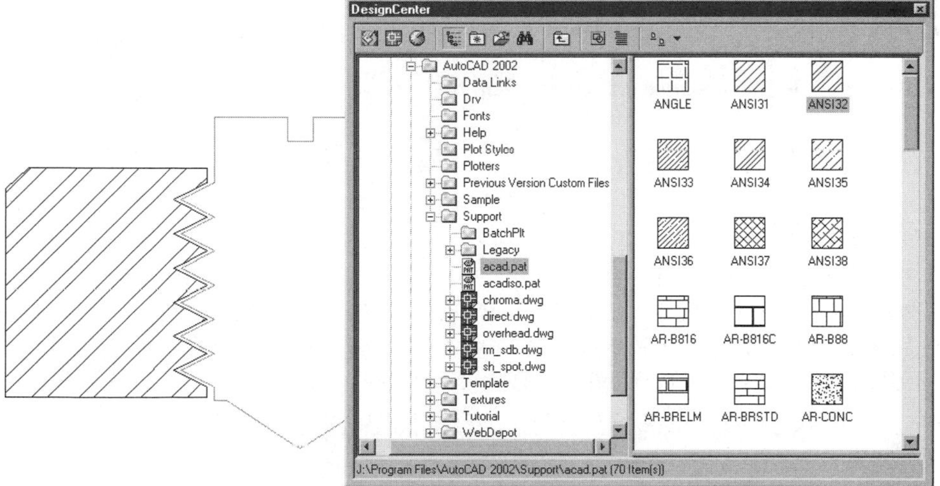

Figure 10–2 *Dragging and dropping a hatch pattern onto an object.*

Tip: Once an object has a hatch pattern applied, you can double-click on the pattern to display the standard Hatch Edit dialog box to alter the hatch parameters. You can also right-click on the hatch pattern in DesignCenter to display the standard Boundary Hatch dialog box.

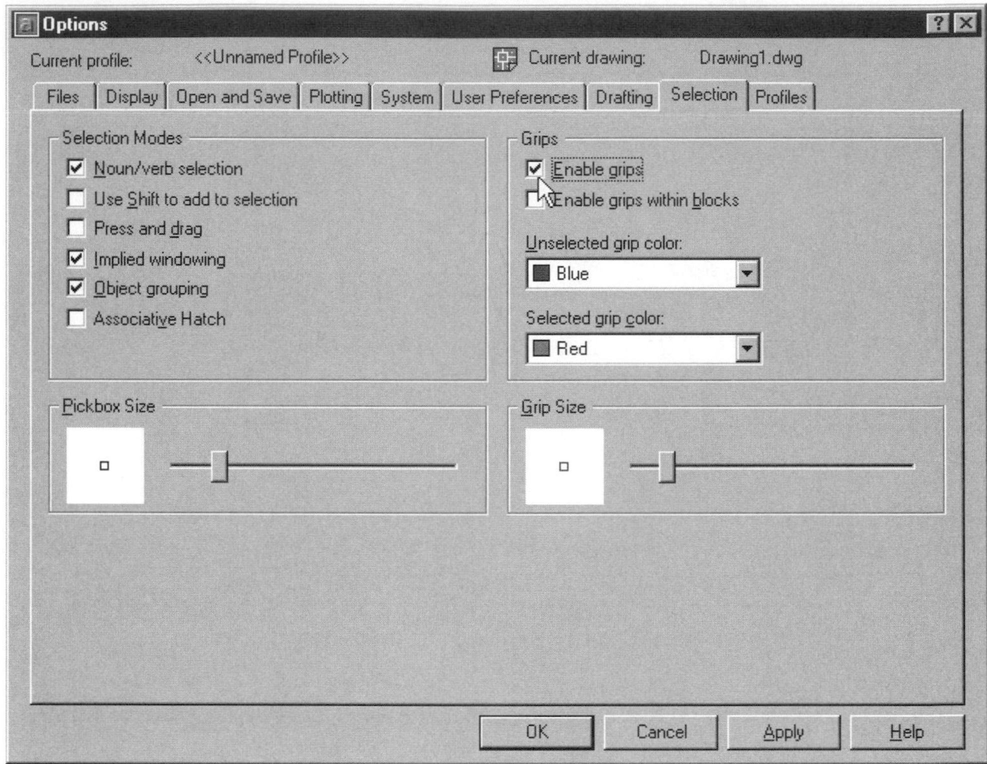

Figure 10–3 *You enable grip editing from the Selection folder in the Options dialog box.*

With grips enabled, you start the process of using the grip editing modes by selecting the objects you want to edit at the command prompt. In other words, you do not initiate any commands. Instead, you simply select the objects by picking them or by using implied windowing. After you select the objects, their grips are displayed as blue squares. The displayed grips correspond to the control points of the objects, and, for the most part, the grip locations are the same as the object snap points for the various types of objects. The major exceptions to this rule are presented in Table 10-1.

Note: The color and size of grips is controlled in the Grips area located in the Options dialog box's Selection folder, which is shown in Figure 10–3.

Like object snaps, grips enable you to choose a very specific point on an object easily. After the grips are displayed, you can choose one grip to activate the grip editing modes.

Table 10–1: *Specific Grip and Object Snap Discrepancies*

Object	Description
Arc	Only three grips exist for an arc: its two endpoints and its midpoint. In contrast, object snap points include the center point and any valid quadrant points. Grip editing on an arc can be used to shorten to an intersection.
Block insertion	By default, only one grip is displayed at the insertion point of each block insertion. However, when you enable the Enable Grips Within Blocks setting from the Selection folder in the Options dialog box, then the grips of all the component objects are also displayed. Grip editing a block will relocate the insertion point.
Elliptical arc	Grips correspond to the arc's endpoints, midpoint, and center points, but not to its visible quadrant points. Grip editing an ellipse will relocate it or resize it about its center.
Mline	Grips exist at the points used to locate the mline object. In contrast, endpoint and midpoint object snap points can exist on each visible segment. Grip editing an mline will change the vertex points.
Mtext	Four grips exist on an mtext object: one at each corner of the imaginary box that surrounds an mtext object. In contrast, only one insertion object snap point can be shown on an mtext object. Grip editing an mtext object can be used to relocate or resize the bounding box affecting paragraph appearance.
Spline	A grip exists at every point used to define the spline, known as the spline's control points. Object snap points include only the endpoints. Grip editing can be used to redefine the curves as well as the start point and endpoint of the spline.

ACTIVATING THE GRIP EDITING MODES

When you select one or more objects, grips are displayed. You may then pick any grip and initiate the grip editing mode for the selection set. Selecting a grip affects the cursor much like Osnaps do: the grip acts as a magnet and pulls the cursor to the grip. By default, the unselected grip box color is blue.

After a grip is selected, by default it is displayed as a red box, and it is referred to as a hot or selected grip. The selected grip subsequently is used as the base point for the various grip editing modes: Stretch, Move, Rotate, Scale, and Mirror. Initially the Stretch grip mode is activated, but you can press Enter or the spacebar to cycle through the other grip commands. Alternatively, right-click and pick the desired grip mode from the shortcut menu that appears (see Figure 10–4). The various grip editing mode options are discussed in the sections that follow.

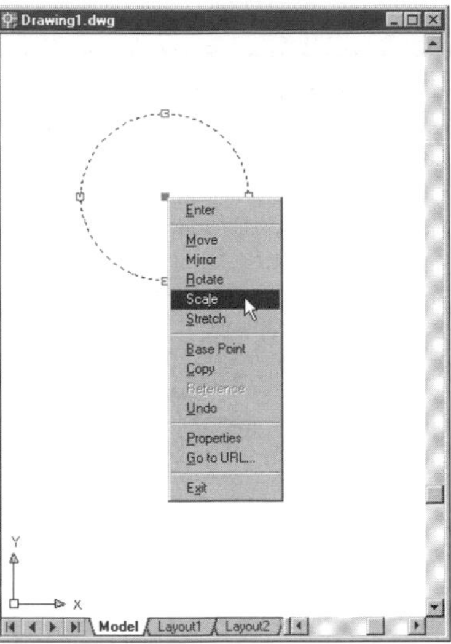

Figure 10–4 *You can pick the desired grip mode from the shortcut menu.*

Note: The color used to fill a selected (or hot) grip is set from the Grips area of the Options dialog box's Selection folder, which is shown in Figure 10–3.

Note: To clear the grips from selected objects, press the Esc key. This also clears the objects from the selection set.

Tip: You can clear objects from the selection set, yet still display their grips, by holding down the Shift key and choosing objects in the selection set. This feature is useful for using the grips as snap points on unselected objects.

USING THE STRETCH MODE

The default editing mode for working with grips is Stretch, which enables you to relocate the selected grip. This in turn affects only the object or objects defined by the selected grip. For example, if the selected grip is the endpoint of a line, then just the endpoint is moved (or stretched) to a new position. If the selected grip is the endpoint at which two lines meet, then the endpoints of both lines are stretched to the new location.

Stretching Multiple Points at the Same Time

If you want to stretch more than one point on the selected object(s) at a time, you must initiate a modified procedure to activate grips. First you must hold down the Shift key while selecting all the grips you want to edit during the stretching procedure. Then, after you release the Shift key, pick the grip that you want to use as the base point of the stretch.

Note: When you use the grip Stretch and Move modes and/or their Copy option, if you know the exact delta-X, delta-Y, delta-Z, or distance and angle you want to apply to the selected grip, you can specify relative or polar coordinates rather than picking the new location for the edit. You also can define the distance and direction of the stretch with direct distance entry.

USING THE MOVE MODE

Using the Move grip editing mode simply moves the selected objects to the new location. Unlike the Stretch grip editing mode, in which only the selected grip(s) are moved, the Move grip editing mode moves all selected objects.

USING THE ROTATE MODE

The Rotate grip editing mode enables you to rotate the selected objects about the selected grip.

You can specify the amount of rotation to apply to the selected objects graphically by either moving your cursor or by typing the specific value. The angle entered is relative to the drawing's 0-degree angle set in the Units dialog box. Alternatively, you can specify the Reference option.

To use the Reference option you first specify a reference angle by picking two points that define that angle or by typing an angle value. Then you indicate the desired rotation angle by dragging the reference line or by typing an angle, which is measured from the reference line, to rotate the object.

USING THE SCALE MODE

The Scale grip editing mode enables you to scale the selected objects about the selected grip. You can either type the scale factor or pick a point on the screen. Picking a point is subsequently used as the scale factor. The grip point is the static point about which the objects are rescaled. Similar to the Rotate grip editing mode, the Scale grip mode has a Reference option.

To use the Reference option you specify an initial reference length by picking two points or by typing a known length value. Then you specify the scale factor by indicating a second length, which can be defined by typing the second length or by picking a third point to define the second length. Once the second length is indicated, AutoCAD determines the scale factor by dividing the second reference length by the first.

Tip: One advantage that Scale grip editing mode has over the SCALE command is that Scale grip editing mode enables you to use the Copy option to scale and make copies of the selected objects simultaneously.

USING THE MIRROR MODE

The Mirror grip editing mode enables you to mirror the selected objects about the mirror line that is anchored at the selected grip.

The mirror line is the imaginary line about which all the selected objects are flipped. Text and mtext objects also are flipped so that they appear backwards. By default, the MIRRTEXT system variable is set to On, which produces backward text objects during mirroring. If you want the text and mtext objects to remain readable, type **MIRRTEXT** at the command prompt, and enter **0** to turn off the mirroring of text objects.

INVOKING THE GRIP EDITING BASE POINT OPTION

The Base Point option of the grip editing modes enables you to specify a new anchor point for the selected grip editing mode command. In the Stretch grip editing mode, relocating the base point does not affect the grip that is stretched. A Base Point option exists in each grip editing mode. This option is very useful in the Mirror grip editing mode when either of the mirror line points does not coincide with a grip location.

INVOKING THE GRIP EDITING COPY OPTION

A Copy option also exists in each of the five grip editing modes. When this option is invoked, the original objects are left unchanged, and any changes are made to copies of the original. An alternative when using the Copy option is to hold down the Shift key after you have placed the first copy. Be aware, however, that if you continue to press the Shift key, AutoCAD will use the distance and direction between the original object and the first copy point to create a set of invisible snap points. For the Stretch, Move, and Scale grip editing modes, the snap points are arranged into a grid, with one of the grid axes running from the base point to the first selected point. For the Rotate and Mirror grip editing modes, the snap points are placed in a circular arrangement such that the angular displacement between adjacent snap points is equal. When used with the Shift key, this process presets the available angles in increments equal to the first angle selected.

So far, this chapter has discussed how to use grips to modify objects. The following section covers AutoCAD's object modification tools.

CHANGING AN OBJECT'S PROPERTIES

The properties of an object consist of values such as its layer, color, and linetype. You can use the Object Properties toolbar, the Properties window, or the Match Properties tool to view or change an object's properties.

USING THE OBJECT PROPERTIES TOOLBAR

The Object Properties toolbar (see Figure 10–5) is one of AutoCAD's standard toolbars. What is handy about this toolbar are two things. First, it displays the current

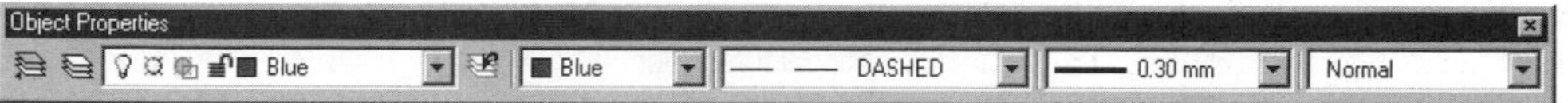

Figure 10–5 *You can view or change certain object properties through the Object Properties toolbar.*

values of certain properties of the selected object, which is useful for quickly identifying the property settings. Second, it lets you quickly change the object's properties.

When you select an object, the Object Properties toolbar updates its controls (the drop-down lists) to reflect the following property values:

- Layer property setting, including the on/off, freeze/thaw, current viewport freeze/thaw, and lock/unlock states
- Color property setting
- Linetype property setting
- Lineweight property setting
- Plot style property setting

All the properties listed here can be changed from the Object Properties toolbar by selecting the desired value from the control's drop-down list. When you select multiple objects, the Object Properties toolbar's control only displays those properties whose values are identical. For example, when you select two circles whose color, linetype, lineweight, and plot style properties are the same, the values of those properties are displayed in the appropriate control. If, however, the circles resided on different layers, then the Layer control would appear blank, displaying no value.

When no objects are selected, the Object Properties toolbar displays AutoCAD's current object creation values. These values are used when you create a new object in the drawing. Therefore, the current settings for layer, color, linetype, lineweight, and plot style are applied to a newly created object. Changing these values in the Object Properties toolbar when no objects are selected updates the property applied to new objects as they are created.

USING THE PROPERTIES WINDOW

The Properties window (see Figure 10–6) is the next generation of a properties editor. With it comes everything from the power to change a simple layer setting to the ability to manipulate individual dimension variables of a selected set of dimensions. The Properties window can be accessed either from within the Tools pull-down menu under Properties, by selecting the Properties toolbar button (located on the Standard toolbar); or by selecting an object and choosing Properties from the shortcut menu.

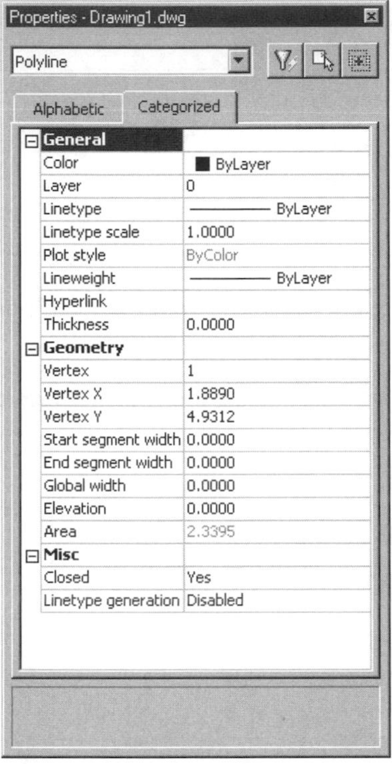

Figure 10–6 *You can view or change an object's properties through the Properties window.*

The Properties window is composed of two primary folders: Alphabetic and Categorized. In the Alphabetic folder, all properties are listed in alphabetical order, which is useful if you know the name of the property you want to change. The Categorized folder organizes the same information into groups, which lets you find a particular property under a group name. For example, coordinate values for the endpoint of lines, the center point of circles, and the insertion point of blocks and text are all found under the Geometry group heading.

When you select multiple objects, the Properties window filters the available options down to those that are common for the selected objects. For instance, selecting two circles would allow editing of the radius. But selecting a line and a circle would limit editing to only those property types that these objects have in common, such as layer and color.

At the top of the Properties window is a drop-down list that provides a listing of the type and quantity of object(s) selected. Additionally, it lets you modify similar objects from within a multiple selection. For example, if you selected a circle and two lines, this list would let you change all general properties such as layer and color, and it would also let you choose a specific object type whose object-specific properties you can then modify. So, if you selected a circle and two lines, to modify properties specific to the circle, such as its center point or radius, you would select Circle from the list, as shown in Figure 10–7.

Picking Objects from the Properties Window

The purpose of the Properties window is to change the property values of objects, especially multiple objects, quickly. Through the Properties window you can instantly change the property values of all selected objects. For example, if you needed to change the radius of hundreds of circles, once the circles were selected, you would simply type in the new radius value in the Properties window and then press Enter. AutoCAD would instantly update the radius property of all selected circles to the new value. This is a very powerful feature.

Note: In the Properties window, if the text in an edit box is blank, you can simply type in new values. Some edit boxes will expand to a list box of their value options, and some edit boxes are unavailable. Lastly, edit boxes that are blank generally indicate that the objects selected do not share identical values for that property.

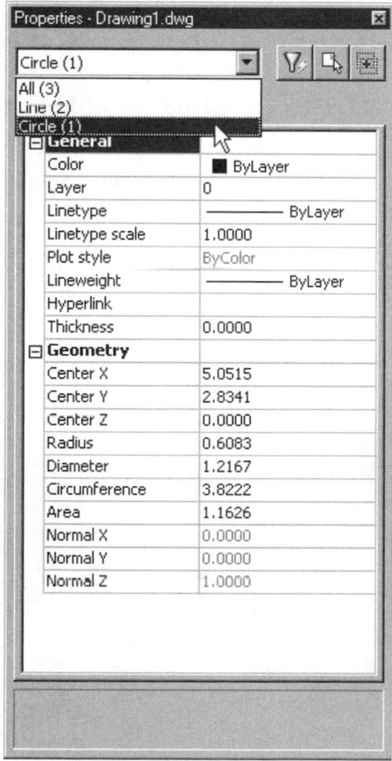

Figure 10–7 *You can select specific object types from within a selection set and modify their object-type properties.*

So, because the Properties window can instantly change the values of all selected objects, the trick, then, is to select all the desired objects quickly. Using the previous example of changing the radius of multiple circles, chances are the circles whose radii you need to change represent a subset of all the objects in your drawing, and are probably interspersed throughout the drawing, intertwined with all the other objects. This can make selecting the particular circles you need to edit very difficult. Fortunately, the Properties window provides access to three tools that you can use to filter a selection set of objects quickly to limit the selection set to only the objects you need to modify.

The first tool is the Quick Select button (the button with the filter and lightning icon), which is discussed in the section titled Using Quick Select later in this chapter. The next tool is the Select Objects button (shown in Figure 10–8), which lets you create an object selection set. Once you choose the Select Objects button, AutoCAD prompts you to select objects and then lets you continue selecting objects until you end the object selection mode.

The third tool is the Toggle Value of PICKADD Sysvar button (shown in Figure 10–9), which turns the PICKADD system variable on and off, and which complements the Select Objects button. When you choose the Select Objects button to start a selection set, if the PICKADD system variable is toggled on, each object selected, either individually or by windowing, is added to the current selection set. (You can remove objects from the selection set by holding down the Shift key while selecting.) In contrast, if the PICKADD system variable is toggled off, as you select new objects, either individually or by windowing, the newly selected objects replace the current selection set, becoming the new selection set. (You can continuously add objects to the selection set by holding down the Shift key while selecting.)

By using the Properties window, you can very quickly edit multiple objects, and thereby instantly update common properties.

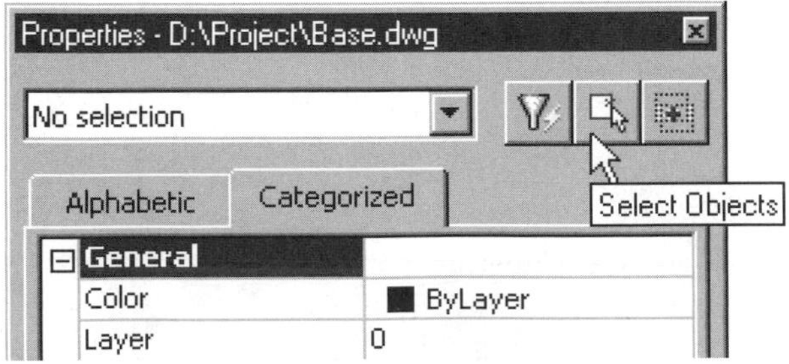

Figure 10–8 *The Select Objects button lets you create an object selection set.*

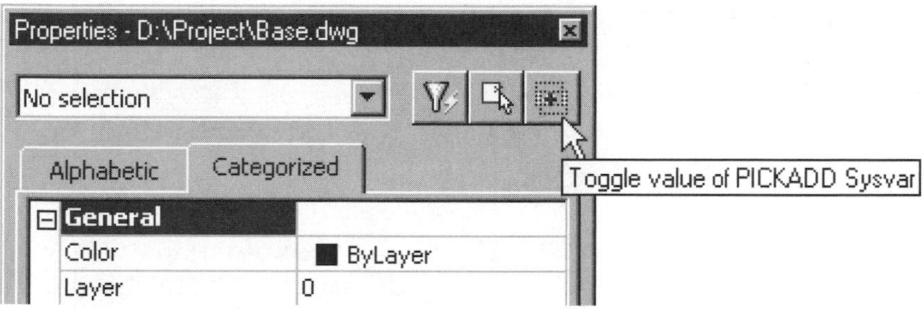

Figure 10–9 *The Toggle Value of PICKADD Sysvar button toggles the PICKADD system variable on and off.*

THE MATCH PROPERTIES TOOL

The Match Properties tool, accessed from either the Standard toolbar or the Modify pull-down menu, lets you copy the properties of one object (the source) and then apply the properties to selected objects (the destination objects). When selecting the destination objects, you can use any of the object selection methods available.

The Match Properties tool lets you control which source object properties are applied to destination objects. When you set the desired properties to apply in the Property Settings dialog box (see Figure 10–10), you select the property values that AutoCAD applies to destination objects. To set the properties to apply to destination objects, launch the Match Properties command, then select the source object, and then right-click to display the shortcut menu. From the shortcut menu, choose Settings to display the Property Settings dialog box.

USING QUICK SELECT

The Quick Select tool allows you to filter quickly for a selection set based on objects within the current drawing. Additionally, from the Apply To drop-down list, you can apply filter criteria either to currently selected objects or to all objects in the drawing. For example, if you have no objects selected, the Quick Select window applies the filter criteria to all objects in the drawing. However, if you already have objects selected, the Quick Select window gives you the option of applying the filter criteria to the currently selected objects or to all objects in the drawing.

The Quick Select tool allows you to refine your filter criteria further by applying the criteria to either multiple objects or specific object types. For example, if you have both line and circle objects in a drawing, you can define the filter to include only lines, or to include only circles, or to include both.

Once you identify the type of object(s) you wish to select, you can refine your filter by including specific property values. For example, while you can initially create a filter that includes all objects in the drawing, you can refine that filter to include only circle objects. Once you identify the circle as the type of object, you can further

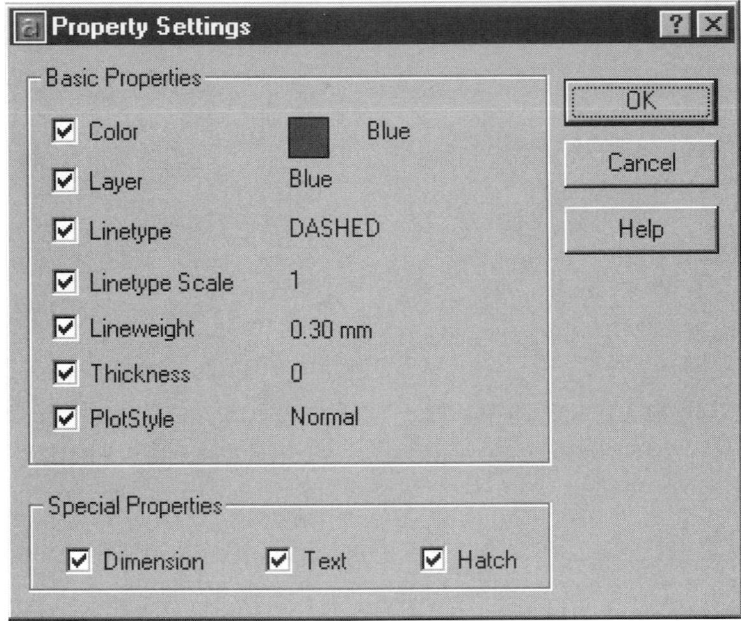

Figure 10–10 *The Property Settings dialog box lets you select the properties to apply to destination objects.*

refine the filter to include only circles whose radius property is less than 1.0, as shown in Figure 10–11.

While it is possible to refine filter criteria down to a specific property, this level of filtering is optional. For example, it is possible to instruct the Quick Select tool simply to ignore the circle object's properties and simply select all circles.

Another very powerful feature of the Quick Select tool is that it allows you to apply filter criteria several times in succession to refine an existing selection set further. This means that while you may initially create a selection set of circle objects whose radii are less than 1.0, you can then define additional filter criteria that only select circle objects on a specific layer, and apply this new filter to the current selection set. Additionally, once you further define filter criteria, you can either include or exclude the objects from the current selection set.

USING OBJECT-SELECTION FILTERS

Object filtering enables you to search for objects based on certain attributes. For example, you could use object filtering to select all circles in a drawing with a specific radius. While AutoCAD provides the Quick Select tool so you can quickly filter for a subset of objects, there is another tool you can use to invoke object filtering.

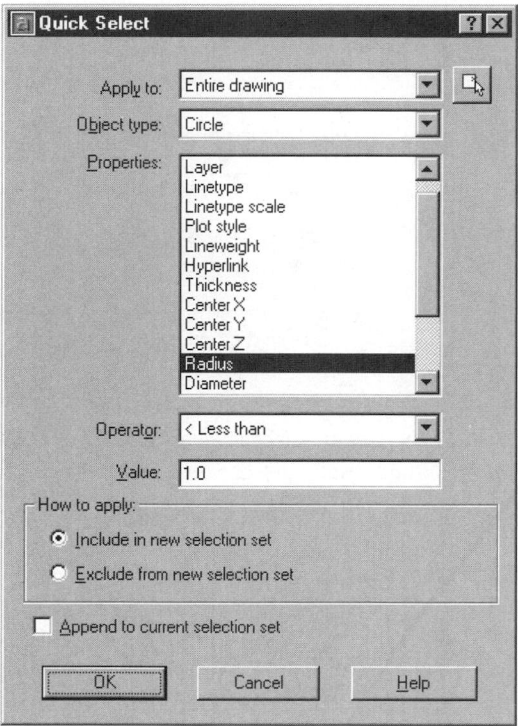

Figure 10–11 *The Quick Select tool lets you filter objects based on specific properties.*

Note: You access the Quick Select tool either by choosing Quick Select from the Tools menu, or by right-clicking in the drawing to display the shortcut menu and then choosing Quick Select. You can also access the Quick Select tool from the Properties window.

When you type **FILTER** at the command line, you display the Object Selection Filters dialog box shown in Figure 10–12.

You can assemble a list of the properties, also known as filter criteria, with which you want to conduct the search. Then, by clicking the Apply button, you can select a group of objects within which you want to find those objects that meet your list of criteria. The following sections discuss how to define a filter list.

DEFINING SIMPLE SELECTION CRITERIA FOR FILTERS

A filter can be a type of object, a characteristic of that type of object, or defined from an object property. For example, you can search for arcs in general or arcs that have a specific radius. The list of available filters is extensive and is displayed in the Filters drop-down list. If you choose a characteristic of an object, you must also supply the specific value of the characteristic that you seek in the X edit box, which is located below the filter list. For some filter selections, you can click the Select button to

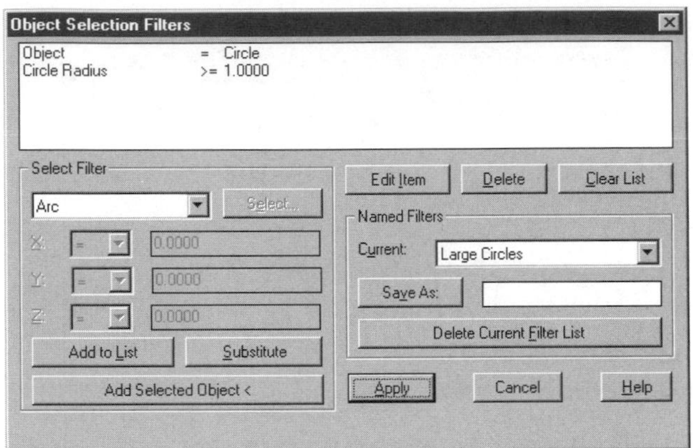

Figure 10–12 *The Object Selection Filters dialog box lets you save and restore object filter criteria.*

choose the specific value from a list of existing or valid values. For other properties, you must type that value in the edit box.

After you select the filter and its associated value (if any), click the Add to List button to add the selected filter to the list at the top of the dialog box. To select the objects to which your filter criteria will be applied, click the Apply button.

To remove a specific filter criterion from the list, choose the filter criterion and click Delete. To edit the value of a specific filter criterion in the list, select the filter criterion and click Edit Item. After you change the value of the property, click Substitute to replace the old property with the revised property.

DEFINING A COMPLEX SELECTION CRITERIA FOR FILTERS

The search criteria employed can be a complex set consisting of multiple filters. By default, when you assemble a list of filters, only objects that meet all the individual filters in the list are selected. For example, you could choose to select only arcs that reside on the layer CURVES by choosing the Arc and Layer filters. In doing so you assemble a list of properties that must be met; this is referred to as an AND conditional. When you assemble a list of properties, the system assumes that you are assembling an AND conditional filter list. Other options do exist, however.

The most common option is to create an OR conditional filter list. In an OR conditional, the objects must meet only one of the conditions, not all of them. For example, you could assemble a list of properties such that any object that is on the CURVES layer OR that is an arc is selected. You begin an OR conditional by choosing the **Begin OR filter. Then you assemble the various properties in which you are interested. You end the list of properties with the **End OR filter.

The list of filters can consist of AND and OR conditionals nested within each other, but for most users simple search criteria consisting of a single conditional filter is enough.

SAVING AND RESTORING THE CRITERIA FOR FILTERS

The chief advantage that the Object Selection Filters dialog box has over the Quick Select tool is its ability to save and restore filter lists. To save a list of filters you have assembled so that the list can be reused at a later date or in another drawing, type a name in the Save As edit box and click Save As. The next time you want to use that filter, simply select its name from the Current drop-down list. To delete a named filter, select the name from the Current drop-down list and click Delete Current Filter List.

Note: Named filter lists are saved in the file FILTER.NFL, which is located in the AutoCAD 2002 directory.

In addition to searching for objects that share certain properties, you can also gather up objects and place them in named groups for later retrieval. The next section explains this process in detail.

CREATING AND EDITING GROUPS

AutoCAD's Groups feature lets you combine different objects into a single object that you can then manipulate. For example, once you create a group of objects, you can move the group just as you would a single entity. Once a group is created, you can select all the member objects of the group by choosing just one member of the group or by naming the group. Assembling the objects into a group, however, does not prevent you from editing the member objects individually. To create and edit a group, you use the GROUP command, which you initiate by typing **Group** at the command line (see Figure 10–13).

The following sections discuss how to use the GROUP command for specific tasks.

CREATING A GROUP

Every group must have a name. To create a group, first type a name in the Group Name edit box. Then click New and select the objects you want to combine into a group. To complete the creation of the group within the drawing, you must click OK in the Object Grouping dialog box. If you do not want to name the group, enable the Unnamed option, and AutoCAD will give the group an arbitrary name that begins with an asterisk. Unnamed groups also are created when you duplicate a group using commands such as copy or ARRAY. To include unnamed groups in the list of groups displayed in the dialog box, enable the Include Unnamed option.

By default, any group you create is selectable. This means that a group of objects can be selected by name or by member. If you turn off the Selectable option before you

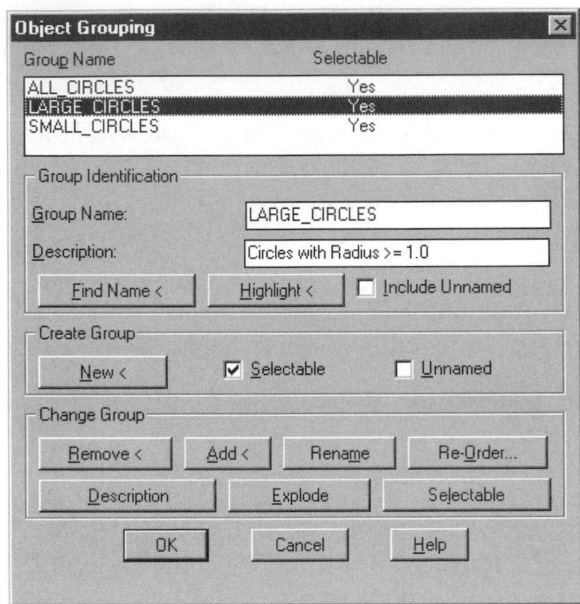

Figure 10–13 *The Object Grouping dialog box lets you group objects into a single entity.*

create a group, the group will not be selectable. The individual group members will still be listed as members of the group, but they will be selectable only as individual objects. You might want to create a non-selectable group, for example, if you want to associate various objects with each other for use with custom programs (created by you or a third-party developer) that interact with the drawing database but are not for use with AutoCAD editing commands.

A group can have as many members as you desire, and an individual object can be a member of more than one group. The group description is an optional piece of information that you use to describe the contents of the group or the relationship between the member objects.

SELECTING A GROUP TO EDIT

After you create a selectable group, you can select all members of the group simply by selecting one group member or by naming the group. The Object Grouping setting controls whether all members of a group are selected when one member is picked. This setting is found in the Selection folder of the Options dialog box. Even with the Object Grouping setting disabled, you can select the members of a selectable group by typing G at the select objects prompt and then entering the group name.

INQUIRING ABOUT A GROUP'S MEMBERSHIP

If you ever forget whether an object is a member of a group or which objects are members of a particular group, you can use the following two buttons to find this information through the Object Grouping dialog box:

- **Find Name**—Use the Find Name button to determine the group, if any, to which a selected object belongs. This allows you to select an object within a group and display the group name(s) of which it is a member.
- **Highlight**—Select a group name from the list of group names and click the Highlight button to highlight all the members of the selected group on the screen.

Note: You also can use the LIST command to see the contents of a group. When you do so, all the group objects are highlighted, and descriptions of the objects contained within are listed.

MODIFYING AN EXISTING GROUP

To modify the makeup of a particular group, first select the group name from the list of groups. The following list describes the buttons you use to modify the group you select:

- **Remove and Add**—You can use these buttons to remove an object from or add an object to an existing group.
- **Rename and Description**—You can rename a group or change the group's description by selecting the group name, typing in the new name or description, and then selecting the corresponding button.
- **Selectable**—Use this button to change the selectable status of the selected group.
- **Re-Order**—Selecting this button displays the Order Group dialog box, which enables you to change the order in which the member objects are arranged in a group. You can use this option to visually control how the objects in the group are ordered internally.

DELETING A GROUP

To remove or undefine a group, click Explode in the Object Grouping dialog box. Exploding a group dissolves the associations between the member objects but does not erase the member objects.

EXPLODING COMPOUND OBJECTS

Several objects are considered compound objects—meaning that the objects themselves are composed of other AutoCAD objects. Compound objects can be exploded, or broken down, into their constituent parts with the EXPLODE command. You usually explode a compound object to modify one or more of its constituent objects in a way that cannot be done with the compound object itself.

You issue EXPLODE by choosing Explode from the Modify pull-down menu or toolbar. Table 10-2 lists the types of 2D compound objects covered in this book and describes briefly how EXPLODE affects the objects, and some reasons why you would consider exploding the object.

Tip: An exploded object can only be returned to its original unexploded form by the U or UNDO command.

An additional option to the standard EXPLODE command is XPLODE. This command allows the user to control what happens to the objects after they are exploded. You can opt to apply the changes individually or globally. Properties that can be controlled are color, layer, lineweight, and linetype. These properties can also be gathered from the object being exploded.

Table 10–2: *2D Compound Objects and EXPLODE*

Object Type	Result of EXPLODE
Block insertions	An insertion of a block is replaced with duplicates of the block's component objects. Component objects originally drawn on layer 0 are redrawn onto layer 0. A block insertion is usually exploded because you want to modify the component objects themselves. This is usually, but not always, done in the context of redefining the block definition.
Dimensions	A dimension is replaced by a combination of lines, mtext, points, solids, and block insertions. Dimensions usually are exploded so that you can further manipulate their component objects. Generally, because exploded dimensions are no longer associative, you should avoid exploding dimensions.
Hatch	Hatch is replaced by its component lines. An exploded hatch is no longer associative. Again, because of the loss of associativity, exploding a hatch is not normally a good idea.
Mline	A multiline (mline) is replaced by its component lines. In this way you can work around editing commands, such as EXTEND and TRIM, that do not work with mlines. By replacing the Mline object with its component lines, you can then trim or extend those lines.
Polylines	A polyline is replaced by a series of lines and arcs. If the polyline has a width, the replacement lines and arcs will have no width. Polylines are drawn with the PLINE, POLYGON, RECTANG, and DONUT commands.
Region	A region is replaced by the edge objects (such as lines and circles) that define the loops (closed shapes) in the region.

DOUBLE-CLICK OBJECT EDITING

Double-click editing provides an efficient means of both selecting an object for editing and simultaneously displaying either the Properties window or another object-specific editing dialog box. Double-clicking on a circle object, for example, will display a Properties window similar to that shown in Figure 10–14.

Double-clicking some objects, such as a hatch or text object, displays their specialized editing dialog boxes, letting you carry out your editing operation. A new system variable, DBLCLKEDIT, determines whether double-clicking is enabled. With DBLCLKEDIT set to On, double-clicking on a hatch object, for example, displays the Hatch Edit dialog box, as shown in Figure 10–15.

The dialog box that is displayed depends on the object type on which you double-click, as described in the following:

- **Attribute**—Displays the Edit Attribute Definition dialog box
- **Attribute within a block**—Displays the Edit Attributes dialog box
- **Block**—Displays the Reference Edit dialog box

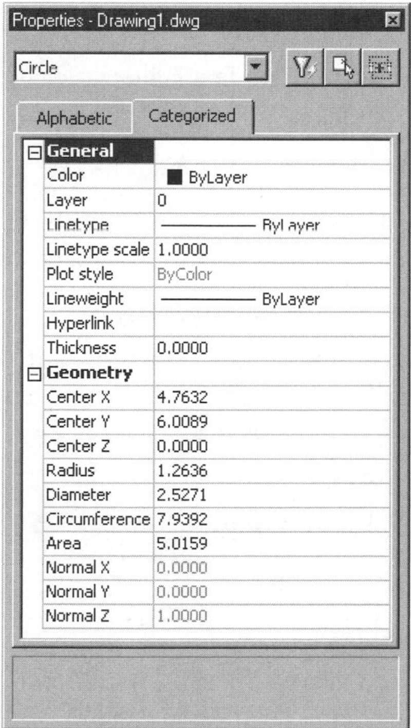

Figure 10–14 *Double-clicking on a circle object automatically displays the Properties window.*

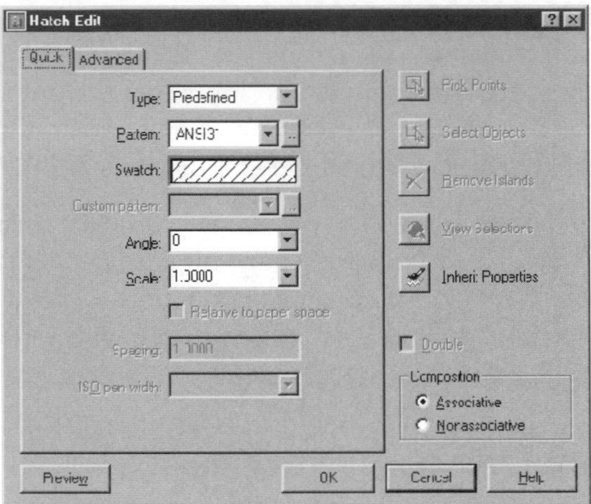

Figure 10–15 *Double-clicking on a hatch object displays the Hatch Edit dialog box.*

- **Hatch**—Displays the Hatch Edit dialog box
- **Leader text**—Displays the Multiline Text Editor dialog box
- **Mline**—Displays the Multiline Edit Tools dialog box
- **Mtext**—Displays the Multiline Text Editor dialog box
- **Text**—Displays the Edit Text dialog box
- **Xref**—Displays the Reference Edit dialog box

Regardless of whether the Properties window or an object-based dialog box is displayed, double-click editing provides a means of quickly editing a drawing object.

SPECIALIZED EDITING COMMANDS

Within AutoCAD 2002 exist numerous commands that provide the ability to edit very specific object types. These unique objects have characteristics that can be exposed through specialized commands made just for them.

EDITING ATTRIBUTE VALUES

There are two editing commands available that are used specifically on attributes contained within an attributed block. The commands are DDATTE and -ATTEDIT, both of which are discussed in the following sections.

Using DDATTE on Attributes

If you want to change the text values of a variable attribute that is part of an inserted block, use DDATTE. You issue this command by entering **DDATTE** at the command line. After you select the block, the Edit Attributes dialog box is displayed. It shows the attribute prompts and the current text values of the attributes (see Figure 10–16).

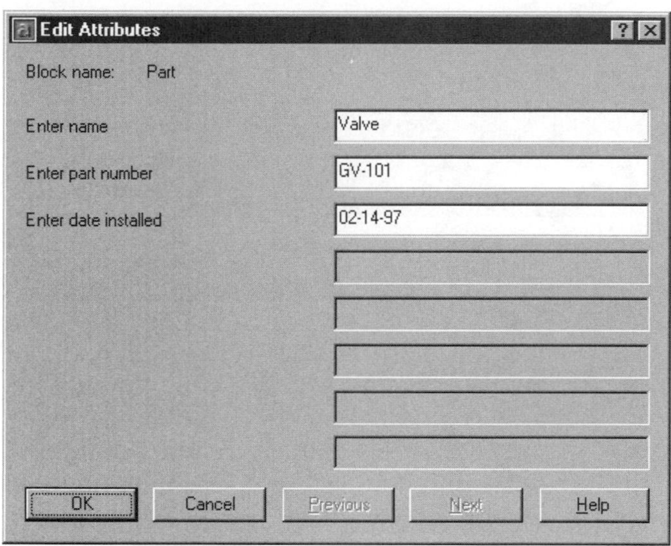

Figure 10–16 *The Edit Attributes dialog box lets you change the text value of an attribute.*

If more attributes exist than can be displayed in the dialog box, use the Next and Previous buttons to display the additional sets of attributes.

Using -ATTEDIT on Attributes

Whereas DDATTE enables you to change the text values of attributes, the -ATTEDIT command enables you to change other properties of inserted attributes. You issue this command by entering **-ATTEDIT** at the command line. You are prompted to indicate whether you want to edit attributes one at a time. If you answer No, you can perform a test search-and-replace on the text string of the selected attributes. If you answer Yes to editing attributes one at a time, you can change the value, position, height rotational angle, style, and color of the selected attributes.

Whether you answer Yes or No, you also have the option of filtering the selected attributes by block name, attribute tag name, or attribute value. The default value for all three filters is an asterisk (*), which indicates that no filters should be used and that the attributes the user selects are to be accepted.

The Enhanced Attribute Editor

The Enhanced Attribute Editor allows you to edit attributes in a single selected block instance. It largely replaces the DDATTE/ATTEDIT commands, but unlike these commands, it permits you to alter virtually any attribute property, including the attribute value, layer, color, text style, etc. Changes are automatically displayed as they are made. Figure 10–17 shows the Attribute folder of the Enhanced Attribute Editor dialog box.

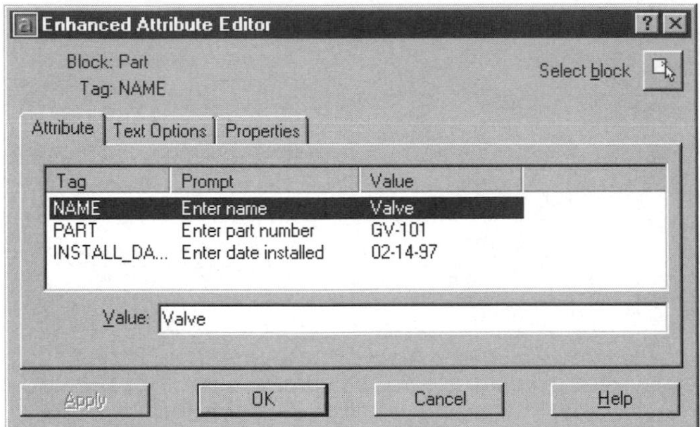

Figure 10–17 *Enhanced Attribute Editor—Attribute folder.*

Tip: To launch the Enhanced Attribute Editor dialog box, simply double-click on the block object you wish to edit.

Note: If you double-click on a block that does not have attributes, the Reference Edit dialog box is displayed.

USING WITH IN-PLACE REFERENCE EDITING

AutoCAD 2002 provides a feature called In-Place Reference Editing, which allows you to redefine block references without requiring you first to insert the block and explode it into its individual AutoCAD objects. In-place reference editing also allows you to redefine external references (xrefs) from within the current drawing. Additionally, you can edit block and xrefs that are nested, no matter how deep the nesting is. In fact, you can edit blocks that are nested within other blocks, which are in turn nested within xrefs. By using in-place reference editing, you can easily edit and update blocks and xrefs, including nested blocks and xrefs, no matter how deep the nesting is.

In-place reference editing works the same for blocks and xrefs. The tools and procedures are identical because the principal behind block and xref objects is the same. In fact, the only real difference between a block and an xref is that a block reference resides within the current drawing, whereas an xref (external reference) resides outside the current drawing.

To demonstrate how the new In-Place Reference Editing feature works, the following discussion focuses on editing a block reference. However, remember that the tools and features used for editing blocks are the same for editing xrefs.

Updating a Block Reference

A very common use of blocks is for the title blocks of plans. Title blocks typically include information that is duplicated on each sheet, and therefore they represent an ideal situation for inserting a block reference. To demonstrate the power of in-place reference editing, we will update the legend within a drawing's title block.

The drawing's legend is a block nested within another (parent) block. The parent block acts as the title block for the drawing, as shown in Figure 10–18. The title block consists of several items, including the client's logo, a sheet index, the current sheet's information (such as its sheet number), and the legend. Each item is actually an individual block that is nested within the parent title block.

Typically the process of editing the legend would involve inserting a temporary copy the legend block, identifying its insertion point, exploding the block into its individual AutoCAD objects, and then finally making the desired edits to the legend. Once the edits have been completed, the updated block's various components are selected, its insertion point is correctly identified, the legend block reference is updated, and the temporary copy is deleted. When the block reference is updated, AutoCAD automatically updates all insertions of the legend block.

Virtually the entire process just described is avoided by using in-place reference editing. Inserting and exploding a copy of the block, then reselecting components and identifying the block's insertion point, are unnecessary steps when blocks are edited using in-place reference editing.

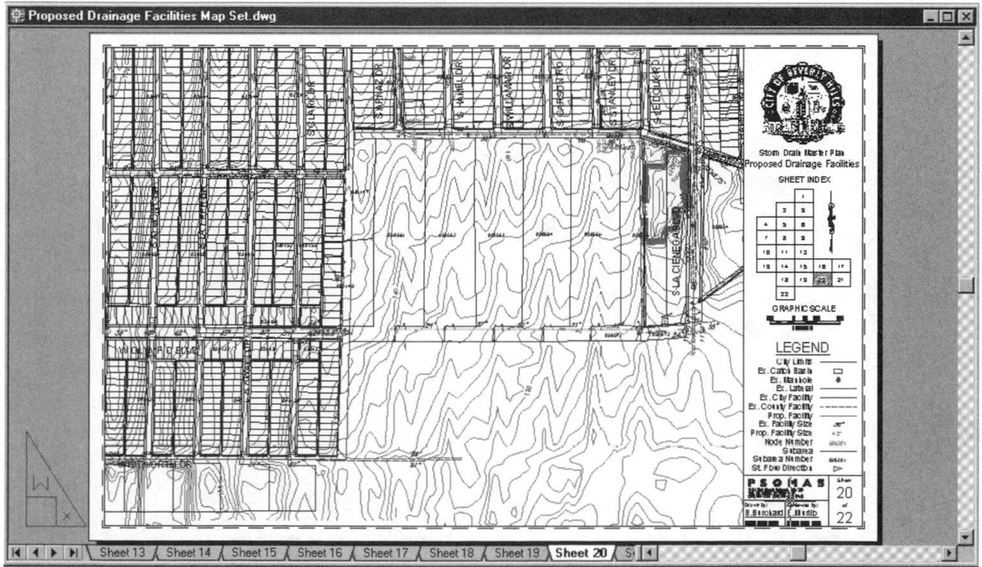

Figure 10–18 *The title block consists of several nested blocks, including the legend block.*

Using In-Place Reference Editing

For this demonstration, assume that the assignment is to add an additional symbol to the legend. As noted previously, the legend is a block that is nested within the title block. Using in-place reference editing, you can directly edit the legend block without exploding it into its individual components.

You start in-place reference editing from the Modify menu by choosing In-Place Xref and Block Edit>Edit Reference. Once it is started, AutoCAD prompts you to select the reference you wish to edit. Then the legend block is selected.

Once the legend block has been selected, AutoCAD displays the Reference Edit dialog box, as shown in Figure 10–19. Notice that the dialog box displays two block references. The first block is named Title Block and it represents the parent block. The second block is named Legend-Proposed Drainage Map Set and is the legend block that is selected for editing. Notice that the Reference Edit dialog box displays nested blocks in a hierarchical format similar to how Windows Explorer displays nested folders. The hierarchy indicates that the legend block is nested within the title block.

The Reference Edit dialog box allows you to indicate if unique layer and symbol names may be created, and if the selected block's attribute definitions should also be displayed for editing. Because the legend block does not contain any attributes, and because there is no need in this particular case to assign copied block objects a unique layer name, these two options are cleared.

After you indicate that the legend block is the block reference to edit, and then select the OK button, AutoCAD prompts you to select the individual AutoCAD objects within the legend block that you wish to edit. In this example, because a new

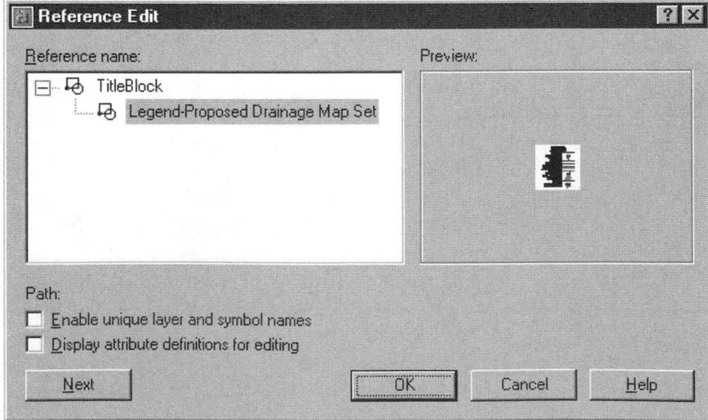

Figure 10–19 *The Reference Edit dialog box allows you to select the nested block you wish to edit.*

Note: When working with xrefs, it is generally good practice to select the Enable Unique Layer and Symbol Names check box, especially if you intend to create new objects to add to the xref. When you select the check box, AutoCAD creates temporary layer names, which are duplicates of the layer names in the xref, in the current drawing. The temporary layer names are prefixed with ?0?, similar to how AutoCAD prefixes layer names when you bind an xref to a drawing.

When you need to add new objects to the xref you are editing, this feature lets you place the new objects on the temporary layers. Because AutoCAD does not allow you to assign objects to an xref's layers, if you assign the objects to the xref's temporary layers, when you save back your edits, the new objects will be placed on the layers in the xref that correspond to the temporary layers.

symbol will be added to the legend, several symbols must be moved to make room for the new symbol. The new symbol will simply be a copy of an existing symbol, which is then edited as desired. Consequently, the upper six symbol descriptions and the legend text are selected with a crossing window, as shown in Figure 10–20.

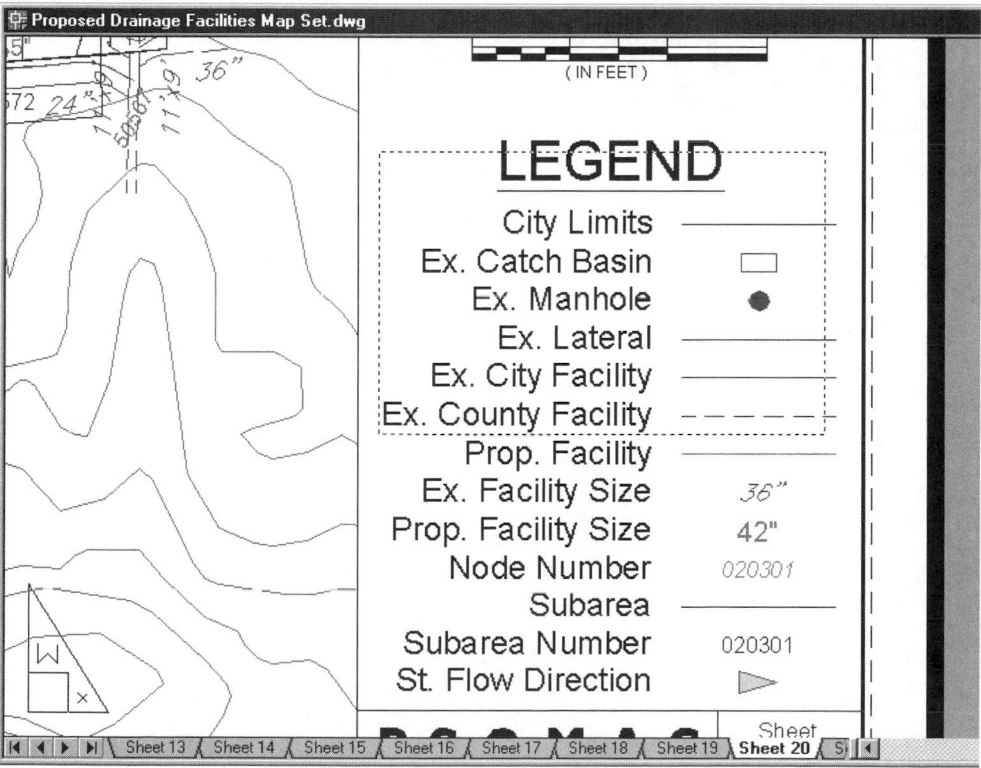

Figure 10–20 *The legend block's individual objects to be edited are chosen.*

Once the objects within the legend block are selected, AutoCAD invokes in-place reference editing mode and adds the objects to the working set. Once this mode is invoked, only the objects in the working set may be edited. Therefore, all other objects are grayed, indicating that they are not part of the working set and may not be edited, as shown in Figure 10–21. Also, AutoCAD displays the Refedit toolbar, which allows you to save or discard edits to the selected block reference.

Note: You can adjust the fading intensity from the Display folder found in the Options dialog box, which you display by choosing Options from the Tools menu.

Once in-place reference editing mode is invoked, you may use AutoCAD's commands as you would normally, thereby editing, adding, or deleting objects as desired. Once the modifications to the block are complete, choose the Save Back Changes to Reference button, as shown in Figure 10–22, to save your edits and redefine the legend block reference.

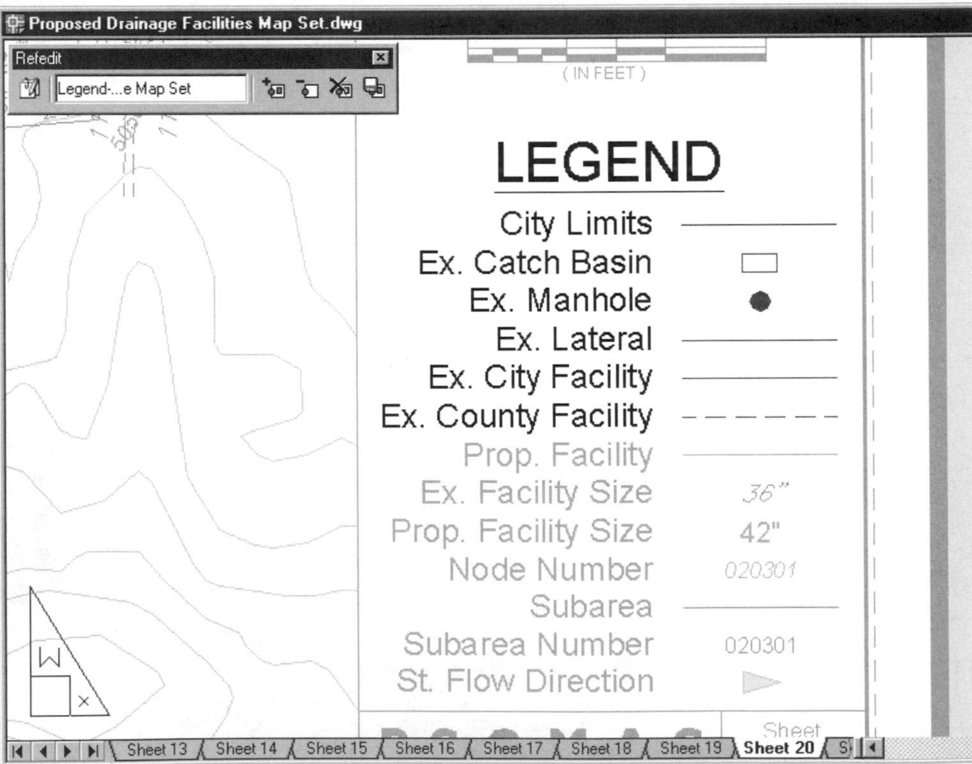

Figure 10–21 *Objects that cannot be edited are grayed.*

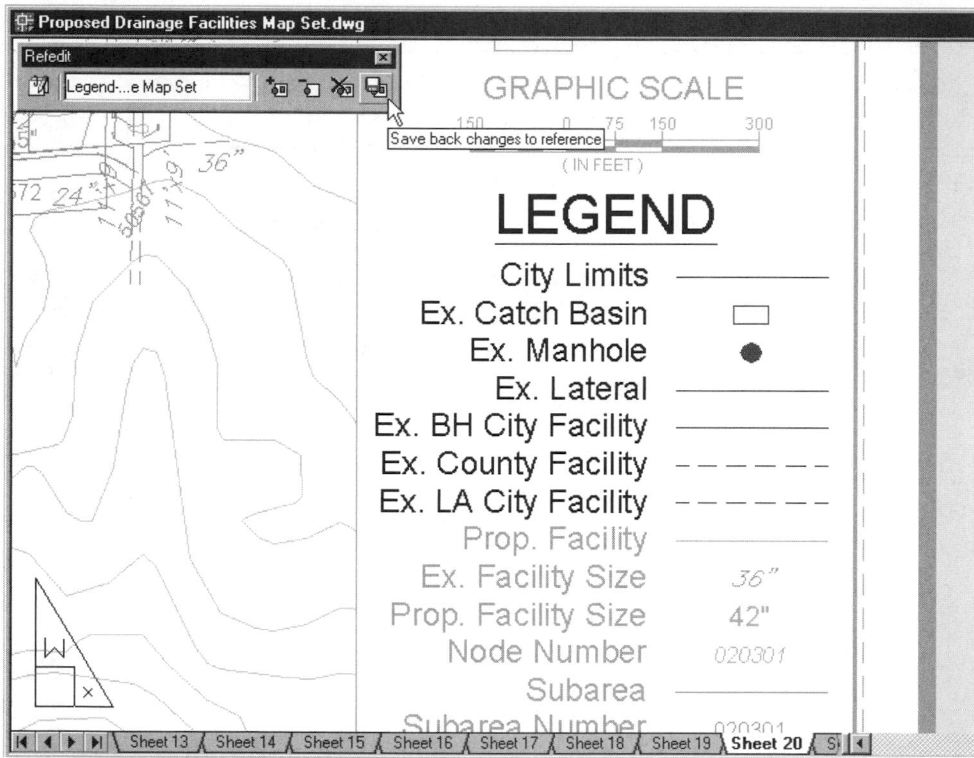

Figure 10–22 *The edits to the legend block are saved back to the block's reference.*

Once the edits are saved back to the block's reference, AutoCAD regenerates the drawing and updates the display, reflecting the changes made to the legend block, as shown in Figure 10–23. Once the block reference is updated, all block insertions are automatically updated, reflecting the new edits.

Using the new in-place reference editing feature, the block reference was easily updated, while the more cumbersome aspects of block editing were avoided (such as exploding blocks and duplicating the edited block's original rotation and insertion points when the block is redefined). By using in-place reference editing, you can quickly select a block or xref for editing, make the desired changes, and then save your changes back and automatically update the original block or external reference.

The Refedit Toolbar

The Refedit toolbar includes features that allow you to add objects to or remove objects from the working set. You add additional objects by choosing the Add Objects to Working Set button, and you remove objects by choosing the Remove Objects from Working Set button.

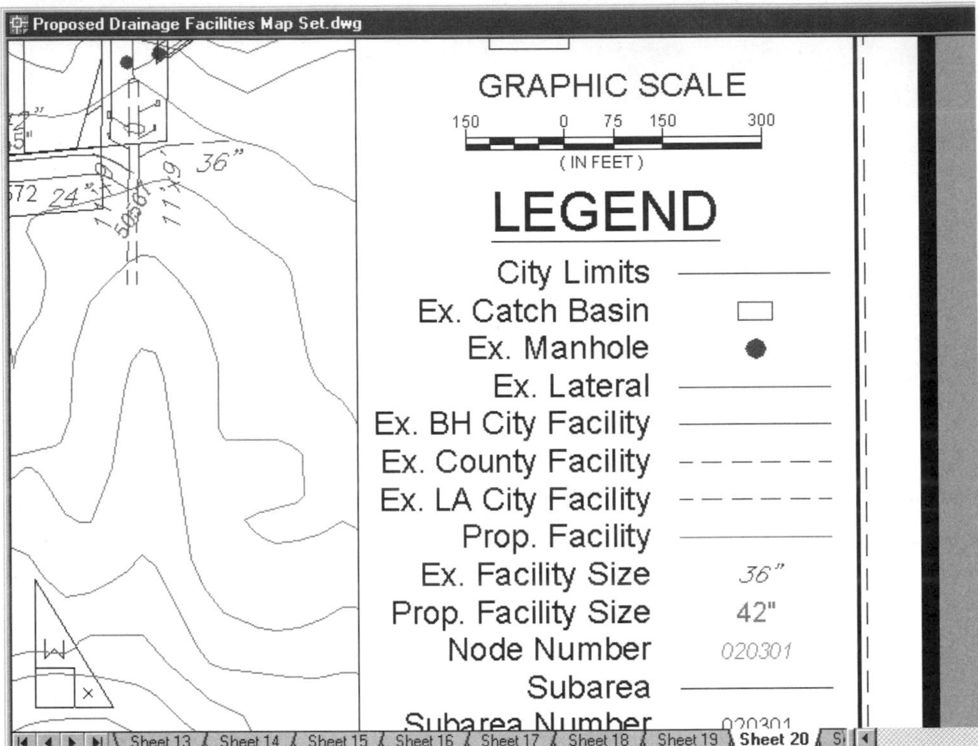

Figure 10–23 *The legend block reference is updated and the new edits are displayed.*

While the reference editing mode is active, if you create new objects, they are almost always added to the working set. There are some situations in which this is not true, such as when AutoCAD generates an arc object during execution of the FILLET command. Consequently, it is a good idea to select all new objects and add them to the working set.

While AutoCAD fades objects that are not part of the working set, objects in the current drawing may still be edited while in-place reference editing mode is invoked. Consequently, you should exercise caution when selecting and editing objects.

Another feature of the Refedit toolbar is the Discard Changes to Reference button. After you edit a block or xref, if you decide you do not want to save edits back to the original reference, choose the button to discard your changes and end in-place reference editing mode.

EDITING RASTER IMAGES

There are three editing commands available that are used specifically on images, which are typically inserted as raster images: IMAGECLIP, IMAGEADJUST, and TRANSPARENCY. In addition to these commands, the system variables

IMAGEFRAME and IMAGEQUALITY affect the display of images. The commands and system variables are discussed in the following sections.

Clipping Images

You can clip portions of an image just as you can clip the display of an external reference. The equivalent of XCLIP (used for xrefs) is IMAGECLIP, which is designed for use on images. To issue the IMAGECLIP command, choose Image from the Modify>Clip submenu. With IMAGECLIP you can define a new rectangular or irregular polyline clipping boundary. You can also use the command to turn on or off the clipping boundary. To display the clipping frame of all images, turn on the system variable IMAGEFRAME by choosing Frame from the Modify>Object>Image submenu. The frame must be displayed in order for you to select and edit the boundary object.

Adjusting the Image

Several additional editing commands are available in the Modify>Object>Image submenu. Choosing Adjust issues the IMAGEADJUST command and displays the Image Adjust dialog box, where you can adjust the brightness, contrast, and fade settings for the selected image.

You can accelerate the display of images by setting the system variable IMAGEQUALITY to the draft setting. IMAGEQUALITY, accessed by choosing Quality from the Modify>Object>Image submenu, affects only the on-screen display of raster images, not the plotting of images; raster images are always plotted at the high-quality setting.

Controlling Transparency

Some raster image file formats support a transparency setting for pixels. When transparency is enabled, the graphics on the display show through the transparent pixels of the overlaid raster image. By default, images are inserted with transparency turned off. You can turn this setting on or off for the selected images by using the TRANSPARENCY command, which you issue by choosing Transparency from the Modify>Object>Image submenu.

EDITING MULTILINES

The MLEDIT command is designed specifically to enable you to perform specialized editing operations on mline objects. To issue the MLEDIT command, choose Multiline from the Modify>Object submenu. Figure 10–24 shows the Multiline Edit Tools dialog box.

With MLEDIT you can clean up various types of intersections of two mlines; remove or add a vertex point in an mline, and insert or heal breaks in an mline.)

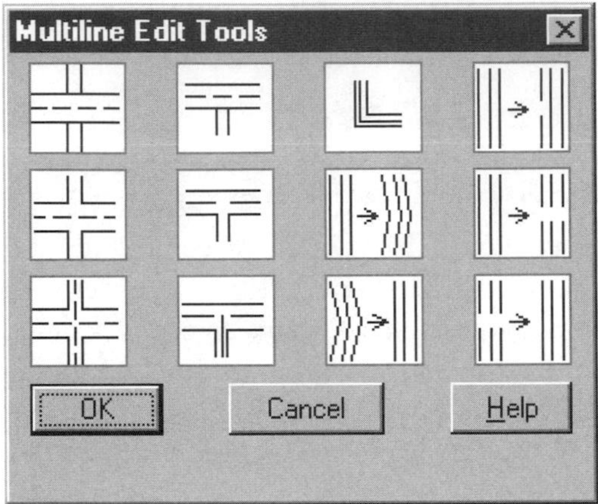

Figure 10–24 *Multiline Edit Tools dialog box.*

EDITING POLYLINES

The PEDIT command is designed for the editing of polylines, and you issue it by choosing Polyline from the Modify>Object submenu. With PEDIT you can accomplish the following tasks:

- Create a polyline from a selected line or arc
- Close an open polyline (Close option) or open a closed polyline (Open option)
- Join additional segments to the selected polyline (Join option)
- Change the polyline's width (Width option)
- Set the polyline's Ltype generation setting (Ltype gen option)
- Fit a curve to the polyline (Fit option)
- Fit a spline to the polyline (Spline option)
- Delete the curve or spline fitted to the polyline (Decurve option)
- Move, delete, or add vertex points in the polyline (Edit vertex option

Enhanced Polyline Editing

Those familiar with AutoCAD's optional Express Tools (formerly, Bonus Tools) will appreciate the new capability of the PEDIT command to edit more than one polyline object at a time. The first prompt of the PEDIT command now includes a Multiple option, as shown here:

Command: PEDIT (Enter)

Select polyline or [Multiple]:

If you choose the M (Multiple) option, you can select any number of 2D or 3D polylines, or 3D polygon meshes. You can, for example, change the width or spline multiple polylines with one PEDIT operation, as shown in Figure 10–25.

When you use the Multiple option, you can now also join two polylines whose adjacent endpoints are not coincident, including polyline segments whose ends do not meet, as well as segments whose ends overlap. In each instance you supply a "fuzz factor" and a "Jointype":

- **Fuzz factor**—Specifies the maximum distance (gap or overlap) that the endpoints of the polylines can be from each other. In order for a join to occur, the fuzz factor distance must be greater than the distance separating the endpoints.
- **Jointype**—The method of joining polylines. The Extend method extends (or trims) the segments to the nearest common endpoint. The Add method adds a straight segment between the two endpoints. The Both option attempts to extend or trim, but otherwise adds a segment between the nearest endpoints.

The new joining capabilities of PEDIT are shown in Figure 10–26.

EDITING SPLINES

The SPLINEDIT command is designed for the editing of splines, and you issue it by choosing Spline from the Modify>Object submenu. With SPLINEDIT you can accomplish the following tasks:

- Edit the fit points of the spline (Fit Data option)
- Open or close a spline (Open and Close options)
- Move the vertex points of the spline (Move option)
- Control the number or weighting of the control points (Refine option)
- Reverse the direction of the spline (Reverse option)

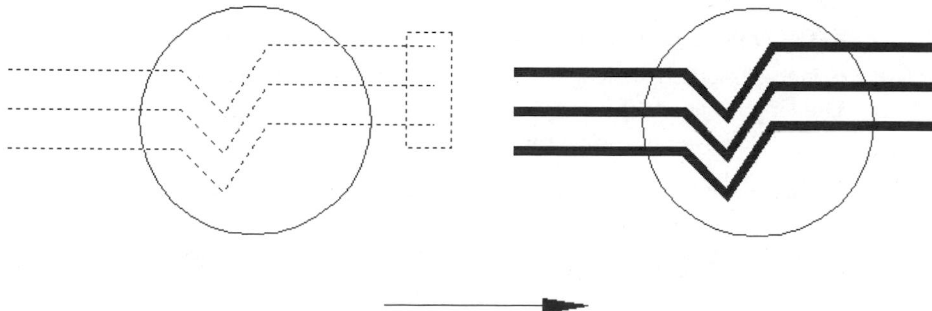

Figure 10–25 *The PEDEIT command lets you edit multiple polylines.*

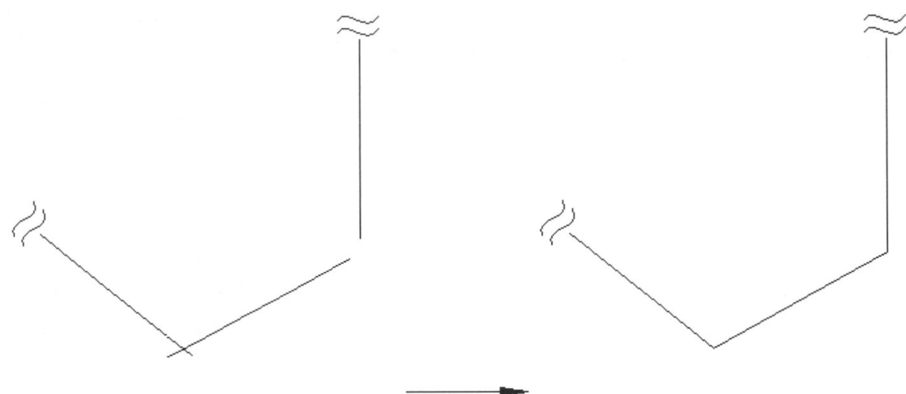

Figure 10–26 *Joining overlapping or gapped polyline segments.*

EDITING TEXT AND MTEXT

The DDEDIT command enables you to edit text and mtext objects (as well as the value of associative dimensions). You issue the DDEDIT command by choosing Edit from the Modify>Object>Text submenu. If a text object is selected, the Edit Text dialog box is displayed. If an mtext object is selected, the Multiline Text Editor dialog box is displayed.

Note: By double-clicking on the text or mtext object, you can display the Edit Text or Multiline Text Editor dialog boxes.

LENGTHENING AND SHORTENING OBJECTS

Any open object, such as a line or an arc, can be lengthened or shortened with the LENGTHEN command. The following listing details the options available at the initial prompt:

- **Select Object**—The default option involves selecting an object. When an object is selected, its length is displayed and the initial prompt is redisplayed. The length value is shown using the current units setting for both style and precision. However, this value can contain round-off error.
- **DElta**—Use this option to specify the length by which the object is to be lengthened or shortened. Enter a positive value to lengthen the object, or enter a negative value to shorten the object. If the object to be affected is an arc, you have the option of entering a change in the arc length (the default) or a change in the included angle.

- **Percent**—Use this option to define the change as a percentage, where 100 percent is the original length. Enter a percentage greater than 100 percent to lengthen the object, or a percentage less than 100 percent to shorten the object.
- **Total**—Use this option when you know the specific length you want the object to have.
- **DYnamic**—Use this option to drag the endpoint dynamically to the desired location. Dragging the endpoint of the object ensures that the alignment of the object cannot change.

After defining the amount of change to be applied, pick the object you wish to modify. The endpoint nearest the point used to select the object is the endpoint that is moved, so pick closer to the endpoint that you want to affect when you select the object.

ALIGNING OBJECTS

The ALIGN command initially was conceived as a 3D editing command, which explains why it is found in the 3D Operation submenu of the Modify pull-down menu. In 2D work, however, ALIGN can also be very useful. In effect, it is a combination of the MOVE, ROTATE, and SCALE commands. ALIGN typically is used to align one object with another object

After you select the objects to align, you are prompted to specify up to three pairs of points. Each pair consists of a source point and a destination point. The source point is a point on the object to be aligned, and the destination point is the corresponding point on the object to which you want to align.

You need to specify only two pairs of points in 2D work. When prompted for the third pair, simply press Enter. The selected objects are moved from the first source point to the first destination point. Then the objects are rotated such that the edge defined by the first and second source points is aligned with the edge defined by the first and second destination points.

Finally, you have the option to scale the objects such that the length defined by the first and second source points is adjusted to be equal to the length defined by the first and second destination points. In effect, this scaling option serves the same function as the Reference option of the SCALE command.

RENAMING NAMED OBJECTS

AutoCAD objects fall into two categories: named and unnamed objects. Named objects are items that you name when you create them, and they are referred to by their assigned names. Examples of named objects include layers, block definitions, and text styles. Unnamed objects are objects such as lines, circles, and arcs that cannot be assigned individual names.

Sometimes you need to rename a layer or a block because of changing conditions or simple typographic errors committed when you initially created the objects. To rename a named object, you can use the Rename dialog box (see Figure 10–27), which you invoke by choosing Rename from the Format pull-down menu.

After you choose the named object that you wish to rename, a list of the existing objects of this type is displayed. To change a name in the Items list, first select the specific object to be renamed, which displays the object's name in the Old Name edit box. Type the new name in the edit box to the right of the Rename To button, then click the Rename To button to rename the object.

Any changes entered into the Rename dialog box will not actually be processed until you click OK and the dialog box closes. Every valid modification you make will visually change in the dialog box, but it will not actually occur if you exit the dialog box by clicking the Cancel button.

PURGING NAMED OBJECTS

Sometimes you will find unneeded layers or linetypes. You should delete these items from the drawing to remove unnecessary information from the drawing database.

Note: Layer 0 is the only layer that cannot be renamed, which accounts for why this layer is never displayed as part of the list of layers that can be renamed. However, you may rename any default objects named Standard.

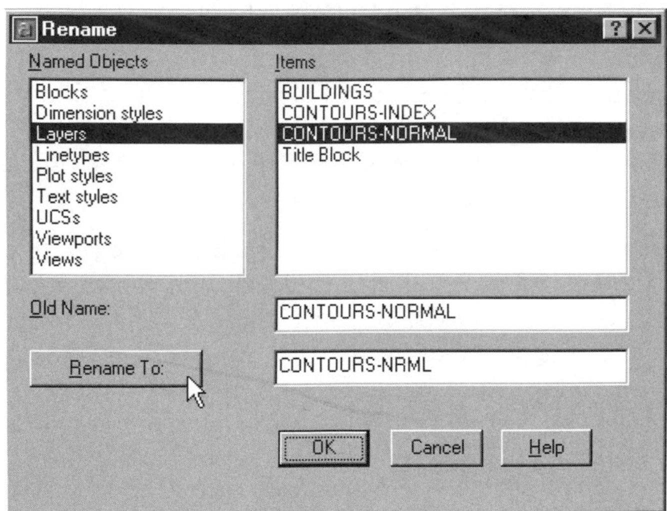

Figure 10–27 *You can rename objects such as blocks and layers using the Rename dialog box.*

The act of removing a named object from the database is referred to as "purging the object." This action is performed with the PURGE command, which you invoke by choosing Purge from the Drawing Utilities submenu of the File menu. Once you select Purge, AutoCAD displays the Purge dialog box shown in Figure 10–28.

The Purge dialog box functions in two display modes. Both modes use a standard Windows tree control to display purgeable item categories. The interface features are summarized as follows:

1. **View Items You Can Purge**—Sets the tree view to display a summary of named objects in the current drawing that you can purge.
   - **Items Not Used in Drawing**—Shows a tree view of all named object categories (blocks, layers, etc.) in the current drawing. A plus sign (+) appears next to the object categories containing items that you can purge. Clicking on the plus sign or double-clicking on an object category expands the tree view, displaying all unused named objects that exist for the category. To purge all unused named objects, select All Items in the tree view and then choose Purge All. To purge a specific named object category, select the category in the tree view and then choose Purge.

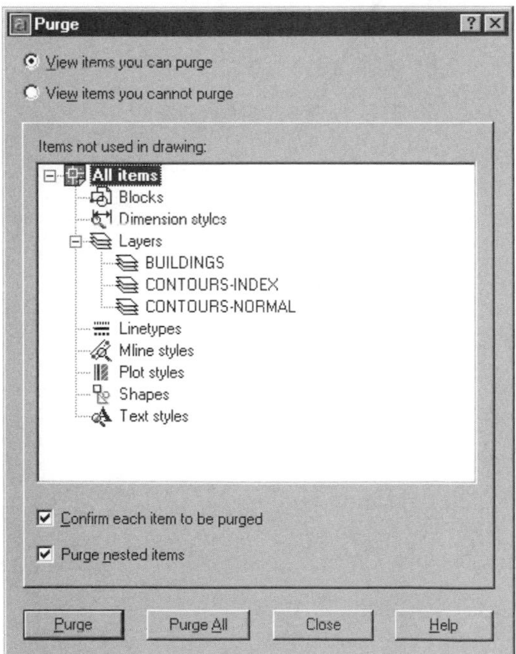

Figure 10–28 *The Purge dialog box lets you delete unneeded named objects.*

- **Confirm Each Item to Be Purged**—Displays the Verify Purge dialog box whenever you purge an item.
- **Purge Nested Items**—Removes all unused named objects from the drawing even if they are contained within or referenced by other unused named objects. The Verify Purge dialog box is displayed, and you can cancel or confirm the items to be purged.

2. **View Items You Cannot Purge**—Changes the tree view to display a summary of named objects in the current drawing that you cannot purge. The Purge and Purge All buttons are disabled in this mode.

   - **Items Currently Used in Drawing**—Displays a tree view of all named object categories (blocks, layers, etc.) in the current drawing. A plus sign (+) appears next to the object category names that you cannot purge, as shown in Figure 10–29. Clicking on the plus sign or double-clicking on a named object category expands the tree view, displaying all named objects that cannot be purged in the category. When you select individual named objects, information about why you cannot purge the item is displayed below the tree view. This information may be helpful in altering the drawing so that an item can be made purgeable.

Figure 10–29 *The new PURGE command's dialog box, with categories expanded.*

SUMMARY

This completes the discussion of the advanced editing commands available in AutoCAD 2002. This chapter introduced you to grip editing and the available object-specific editing commands. To learn more about how and when to create and edit the objects discussed with the object-specific editing commands, refer to the appropriate chapters.

Applying AutoCAD DesignCenter

The power of AutoCAD lies in its ability to reuse existing data. Objects such as blocks and xrefs, as well as layers, text styles, and linetypes, can be used over and over once they are defined. By using existing data and placing it into your drawing, you avoid duplicating tasks, which saves time and increases productivity.

AutoCAD 2002 includes a feature called AutoCAD DesignCenter, which allows you to quickly locate, view, and import a variety of existing AutoCAD objects into the current drawing. In essence, you can look inside a drawing to see the blocks it contains, and you can even identify its xrefs. You can view its defined text styles, dimension styles, and linetypes. You can also identify its layers and its layouts. Once the desired objects are located, AutoCAD DesignCenter allows you to place duplicates of the objects into the current drawing, thereby instantly populating your drawing with valuable data from other drawings. When you use AutoCAD DesignCenter, you take advantage of AutoCAD's real power, the power to reuse existing, valuable data.

In this chapter you will learn about the following subjects:

- Exploring the DesignCenter interface
- Loading DesignCenter with content
- Inserting content into drawings

EXPLORING THE DESIGNCENTER INTERFACE

DesignCenter is composed primarily of two windowpanes, as shown in Figure 11–1. The pane on the left is the navigation pane, or *tree view* interface, and the pane on the right is the contents pane, or *palette* interface. The tree view allows you to locate source objects, and the palette allows you to view the contents of the source objects. For example, in Figure 11–1 the tree view is used to navigate to the My Computer folder, and the folder's contents are displayed in the palette. By using the tree view and the palette, you can locate and view source objects.

EXPLORING DESIGNCENTER'S TREE VIEW

DesignCenter's tree view allows you to navigate through a directory structure easily. It is similar to Windows Explorer, allowing you to expand or collapse folders to control the display of subfolders. By using the tree view, you can easily navigate to the desired location.

While the tree view allows you to view and navigate a directory structure, you are not required to use it. For example, you can turn the tree view display off by choosing the Tree View Toggle button, as shown in Figure 11–2. Toggling off the tree view display is useful once you have located the desired folder and no longer need the tree view. When you toggle off the tree view, the palette automatically expands, making viewing source objects easier.

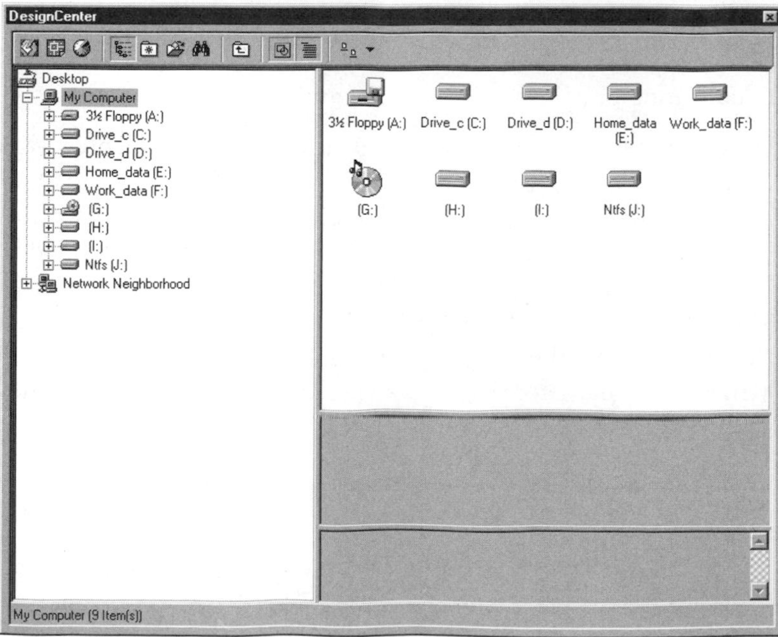

Figure 11–1 *DesignCenter is composed of tree view and palette windowpanes.*

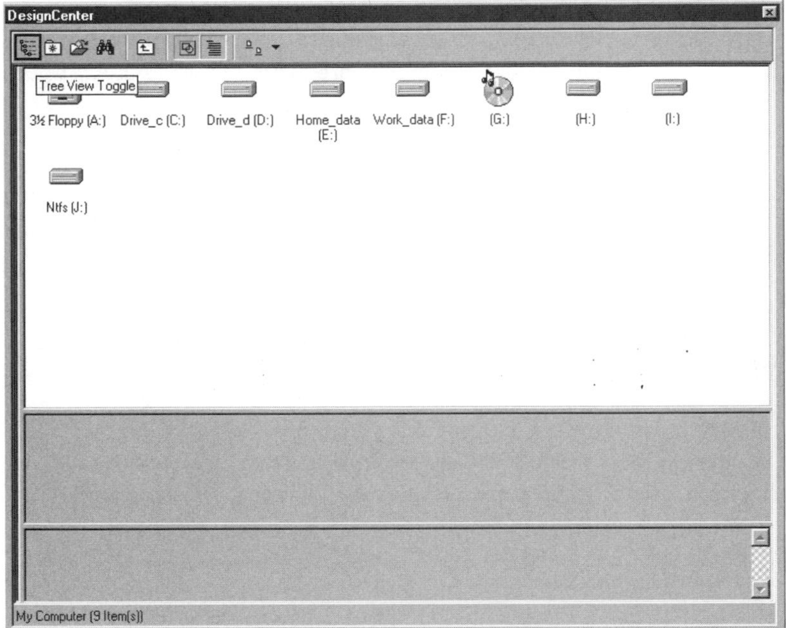

Figure 11–2 *You can toggle off the tree view's display.*

Note: You can navigate through directories using the palette by double-clicking on a folder to display its contents, or by choosing the Up button to move up one level in the directory.

Tip: While you can use the palette to navigate through a directory, it is better to use the tree view, because the tree view makes it easier to identify your location in a directory structure. When you press the [+] or [-] buttons, directories open and close their contents, respectively.

The tree view can display views in four different modes. Using a particular mode can assist you in locating the desired content source more quickly. The four modes are as follows:

- **Desktop view**—Allows you to locate source data on local or network drives
- **Open drawings view**—Lists all opened drawings in the current AutoCAD session
- **History view**—Displays the last twenty locations of source objects accessed through DesignCenter
- **Custom content view**—Lists the currently registered applications used to create custom objects, if custom objects are present

By selecting the proper mode, you can quickly find the locations that contain the desired source objects.

Note: The Custom Content button is only displayed when there are applications currently registered with your AutoCAD application. When applications that contain custom content, such as Object ARX applications, are registered with AutoCAD, DesignCenter displays the Custom Content button, allowing you to locate and then view the registered application's contents.

You can switch to the desired mode by choosing the appropriate view button. The buttons are displayed when the tree view is toggled on, and they are located in the upper-left corner of DesignCenter, as shown in Figure 11–3. The current mode in Figure 11–3 is History, which lists the most recent drawings from which source objects were queried. By choosing the proper button, you can more quickly display the locations of desired source objects.

Note: In history mode, DesignCenter automatically turns off the palette. This mode is intended to allow you to locate quickly the most recent locations from which you copied source objects. To redisplay the palette, double-click on a drawing displayed in the tree view.

Tip: You can refresh the tree view and palette display by right-clicking in the palette and then choosing Refresh from the shortcut menu.

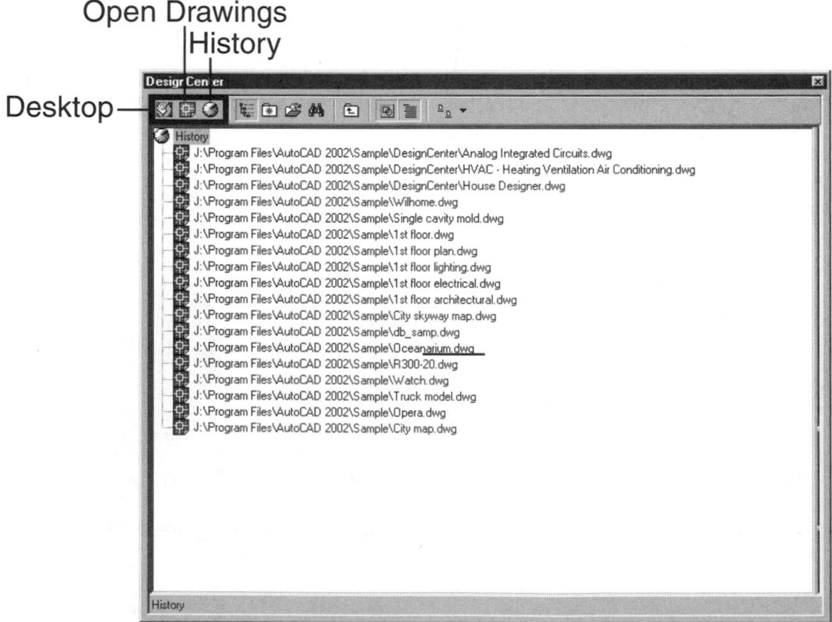

Figure 11–3 *The tree view's different display modes are accessed using the buttons in the upper-left corner.*

EXPLORING DESIGNCENTER'S PALETTE

DesignCenter's palette displays the source objects found in a particular location. For example, when a location is selected using tree view, the location's source objects are displayed in the palette, as shown in Figure 11–4. By using the palette, you can easily view the available source objects.

The palette can display source objects in one of four views. The view you prefer largely depends on the source objects you are viewing. For example, when you view blocks in a drawing, it is useful to use the large icons view to see each block's thumbnail image better, as shown in Figure 11–4. However, when you view drawing files in a folder, choosing the details view may be more desirable, as shown in Figure 11–5. The four available views are as follows:

- **Large icons**—Displays source objects using a large object icon, and uses thumbnail images if they are available
- **Small icons**—Displays source objects using a small object icon, and does not use thumbnail images even if they are available
- **List**—Displays source objects as a simple list, without file detail information
- **Details**—Displays source objects as a list that includes file information such as its size and type, if it is available

By selecting a particular view, you can preview the source objects in the desired format.

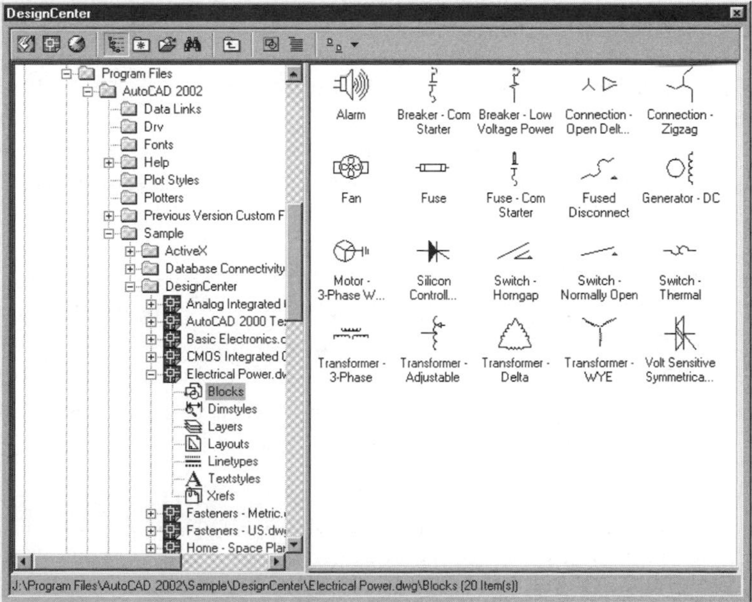

Figure 11–4 *The palette displays the selected drawing's source objects.*

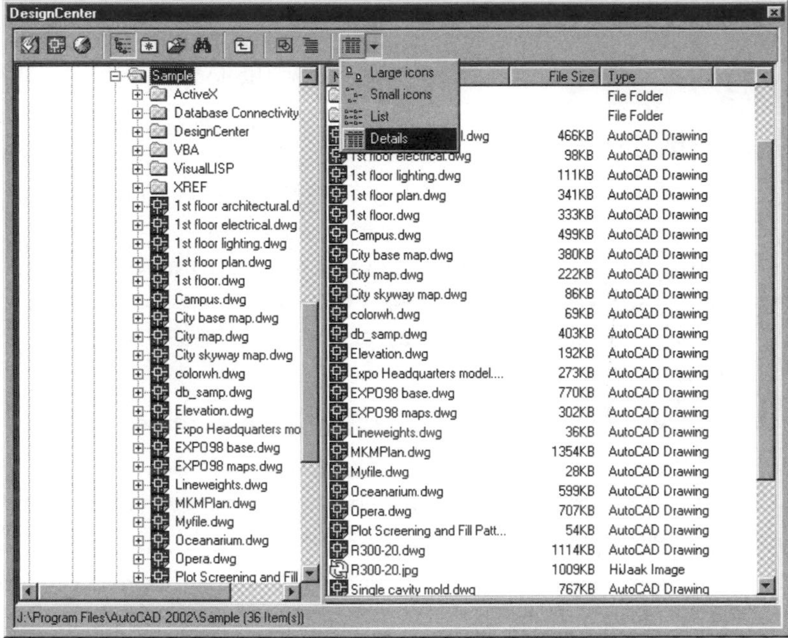

Figure 11–5 *The palette displays source objects using the details view.*

Tip: By right-clicking in the palette, you can select the desired view display by choosing View from the shortcut menu. Alternatively, you can select the desired view display by choosing the Views list button.

Tip: You can control the sort order of objects displayed in the details view by choosing the button at the top of each column. For example, in Figure 11–5, if you choose the File Size button at the top of the palette, you can sort the list of objects by their file size in ascending order. If you choose the button a second time, you re-sort the objects by size in descending order.

VIEWING SOURCE OBJECTS

There are two additional windowpanes you can open in DesignCenter: the preview pane and the description pane. The preview pane displays an image of the selected source object, and the description pane displays its description, if such information was saved with the source object. By using these panes, you can better identify the contents of a source object before its contents are copied to the current drawing.

The preview pane allows you display an image of the selected object. You activate the pane by choosing the Preview button, as shown in Figure 11–6. The preview pane is resizable and may be expanded so you can better view the source object's image. By using DesignCenter's preview pane, you can visually see the source object prior to inserting it into the current drawing.

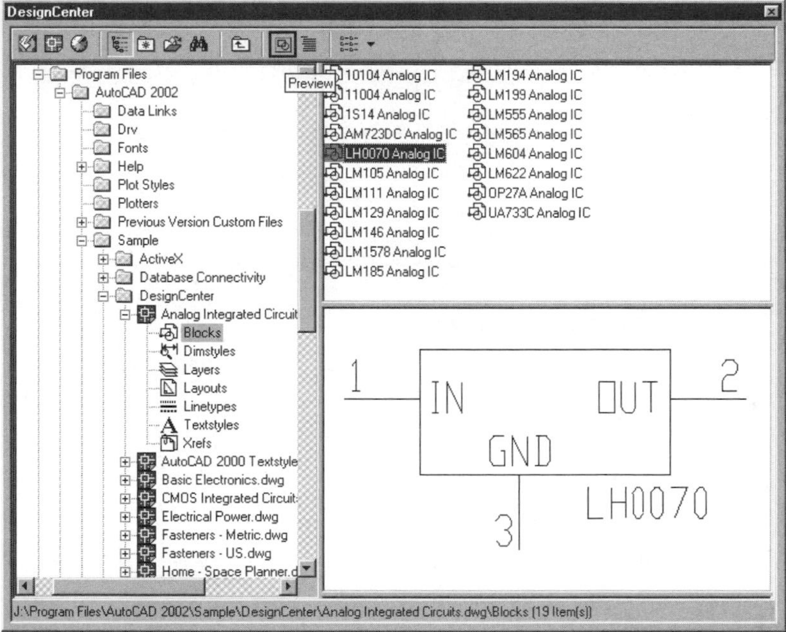

Figure 11–6 *The preview pane (lower right) displays an image of the selected source object.*

Note: You can automatically generate preview images for blocks that do not have preview images by using the BLOCKICON command, which will generate preview images for block references defined in the current drawing.

The description pane displays text that describes the selected source object. When you choose the Description button, if a description was provided, the description pane displays the description below the palette, as shown in Figure 11–7. By using the DesignCenter's description feature, you can better determine whether or not the selected source object is the one you need to copy into the current drawing.

By using the preview and description features provided in DesignCenter, you can better identify the contents of a source object before its contents are copied to the current drawing.

LOADING DESIGNCENTER WITH CONTENT

When you use the tree view to locate source objects and then display those objects in the palette, you are actually *loading content* into the palette. Using the tree view, you can load content into the palette using the desktop, open drawings, history, and custom content modes. By using the tree view's simple interface, you can quickly locate and load content into the palette.

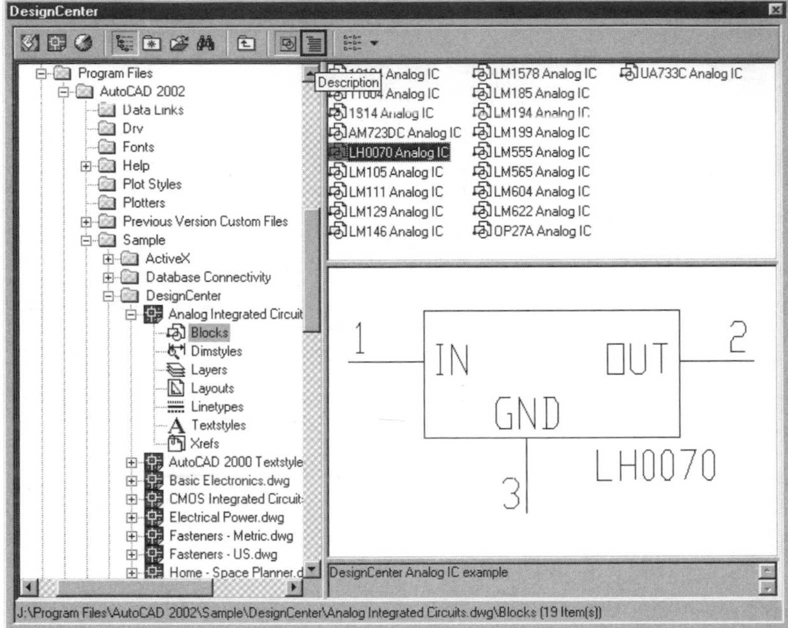

Figure 11–7 *The description pane (directly below the preview pane) displays notes describing source objects.*

While the tree view's simple interface can load content into the palette, its capabilities are limited. For example, the tree view does not allow you to define the file type you wish to locate when you browse for source objects. Additionally, the tree view does not provide an automated find feature that locates content based on a keyword search. The tree view's features for locating source objects are useful but limited.

To compensate for the tree view's limited abilities in locating source objects, DesignCenter has additional methods for loading content that are very powerful. For example, you can locate drawings using the Load DesignCenter Palette feature, which allows you to search for files by indicating a file type. You can also use the Find feature, which is a powerful search feature that can search local and network drives for files, as well as source objects within files, using search criteria. By using these features, you can more easily locate the desired source objects, especially when you are not sure where to look.

EXPLORING THE LOAD DESIGNCENTER PALETTE FEATURE

DesignCenter provides the Load DesignCenter Palette feature so you can more easily locate and load content into the palette. By using the Load DesignCenter Palette dialog box, you can browse for files on local and network drives, in your Favorites folder, or over the Internet. You can also use this feature to automate searching for a particular file based on its name or its file type. By using the Load DesignCenter Palette dialog box, you can use specialized tools to locate, preview, and load content into the palette.

You open the Load DesignCenter Palette dialog box by clicking on the Load button located between the Favorites and Find buttons in DesignCenter, as shown in Figure 11–8. After you click the Load button, the Load DesignCenter Palette dialog box appears, as shown in Figure 11–9. Once it is opened, you can use the various features of the Load DesignCenter Palette dialog box to locate files.

SEARCHING FOR CONTENT USING THE FIND FEATURE

DesignCenter's Find feature is very powerful, allowing you to locate the specific source objects you need. While you can search for drawing files by name, keyword, or description, you can also search inside drawings for a variety of source objects, including blocks, layers, linetypes and text styles. By using DesignCenter's Find feature, you can locate the source objects you need, even if they reside within a drawing.

To illustrate the ability to search for content inside of drawings, let us suppose that you know that a block named ALARM existed within a drawing, but you do not know in which drawing the block resides. Using the Find feature, you can search

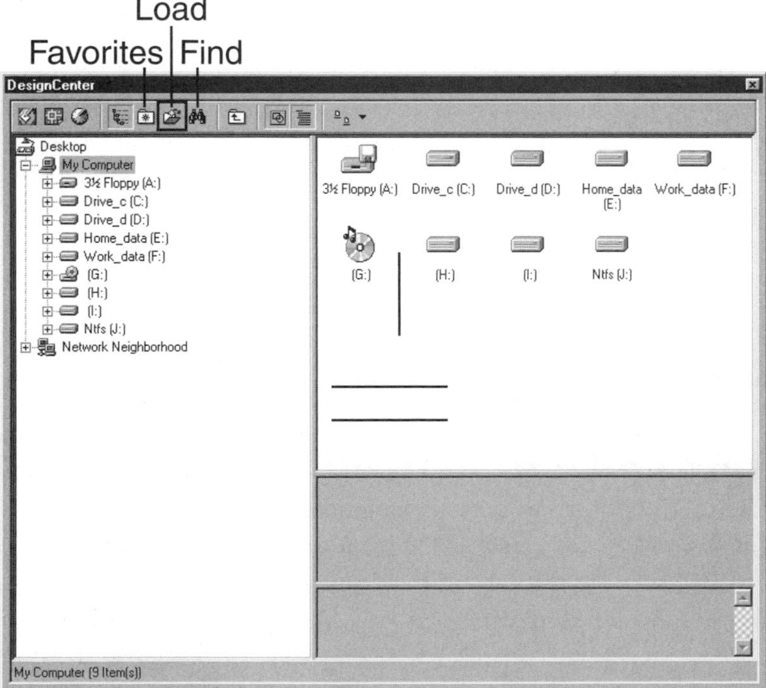

Figure 11–8 *DesignCenter's Favorites, Load, and Find buttons.*

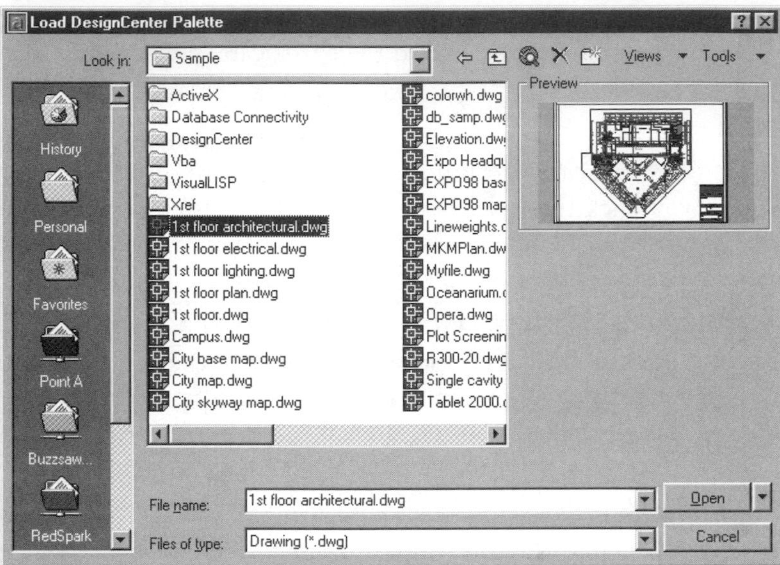

Figure 11–9 *The Load DesignCenter Palette dialog box displays source files.*

Note: You can also load content into the palette using Windows Explorer by dragging a file from Explorer to DesignCenter's palette.

through all drawings in a specified set of folders, searching specifically for a block named ALARM. Once the search is completed, the Find dialog box displays the drawings in which the block is located, as shown in Figure 11–10.

The Find dialog box allows you to locate files and source objects within files by entering search criteria. You can use its three folders to locate files by type, by date modified, or by specific text. You can search for a variety of object types, such as blocks, layers, layouts, and xrefs, and you can search for objects based on text values for block names, drawing and block descriptions, and attribute tags and values. When you use DesignCenter's Find feature, you have access to a powerful set of tools that help you pinpoint the source objects you need.

INSERTING CONTENT INTO DRAWINGS

The purpose of DesignCenter is to make locating existing AutoCAD objects easy. Consequently, DesignCenter provides several methods for locating and identifying pre-made source objects. As discussed in previous sections, by using the features available in DesignCenter, you can quickly locate existing sources of AutoCAD objects to copy into the current drawing.

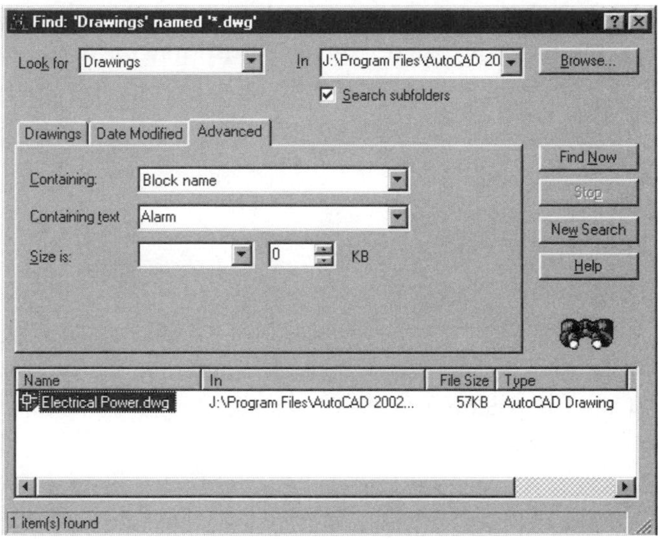

Figure 11–10 *The Find dialog box locates source objects, such as blocks, within drawing files.*

Tip: You can make drawings and blocks easier to find using two features introduced with AutoCAD 2000. To make drawings easier to find, add keywords to the current drawing by using the Drawing Properties dialog box, which you display by choosing Drawing Properties from AutoCAD's File menu. To make blocks easier to find, AutoCAD allows you to add a description to the block in the Block Definition dialog box. By adding keywords and descriptions to drawings and blocks, you can use DesignCenter's Find feature to quickly locate the desired source object.

Tip: You can avoid searching for frequently accessed source objects by right-clicking on the source object and then choosing Add to Favorites from the shortcut menu. AutoCAD adds a shortcut to the selected source object in the Autodesk Favorites folder. You can then load the source object into the palette by choosing the Favorites button in DesignCenter and then selecting the source object's shortcut icon.

Once you have located a desired source object, you must copy it into the current drawing. Once it is copied into the current drawing, it becomes a part of the current drawing and can be used just as if it were originally created in the current drawing. Therefore, objects like blocks and xrefs, layers and layouts, and text styles and dimensions styles can be copied into the current drawing from DesignCenter's palette.

Typically, once you have identified a source object and displayed it in DesignCenter's palette, you can copy the object to the current drawing by simply dragging it from the palette into the current drawing's window. Objects that you can copy by simply dragging them from the palette to the current drawing are as follows:

- Dimstyles
- Layers
- Layouts
- Linetypes
- Textstyles

You can select multiple source objects of the types shown in the list. Once they are copied, the object(s) become part of the current drawing. By dragging the desired source object(s) from the palette into the current drawing, you can quickly add content to the current drawing.

Not all individual objects can be simply dragged into the current drawing. Source objects such as drawings, blocks, images, and xrefs can be copied only one at a time. Additionally, you must define their scale, rotation, and insertion point. Therefore, when inserting drawings, blocks, images, and xrefs, you must tell AutoCAD their scale, rotation, and location.

Note: DesignCenter will not copy an object if an object with an identical name already exists in the current drawing. To copy such an object using DesignCenter, you must change the name of the object that already resides in the current drawing.

When you copy a block from DesignCenter into a drawing, the Insert dialog box is displayed, prompting you to specify the block's insertion values. However, if you only want to copy the block's definition information, you simply choose Cancel to close the Insert dialog box.

Note: When you choose Cancel to close the Insert dialog box, you insert a block definition into the current drawing without actually inserting a block reference into the current drawing's model or paper space. The block definition may then be used at a later time to insert a block reference using the INSERT command.

By using DesignCenter, you can easily locate source objects and copy their contents into the current drawing. When copying objects such as blocks and xrefs, you can copy the objects only one at a time. When copying objects such as layers, linetypes and text styles, you can select multiple objects and copy them simultaneously.

SUMMARY

In this chapter you learned how to use DesignCenter to locate source objects and copy their contents into the current drawing. You reviewed Design-Center's Load and Find features, which are used to search for source objects using search criteria. You also learned how to use the tree view to locate source objects, and how to display the contents of source objects in the palette. You reviewed how to drag different types of content from the palette, and copy it into the current drawing. You learned how to use DesignCenter to find and use existing data, which allows you to work more productively by reusing existing AutoCAD objects located in other drawings.

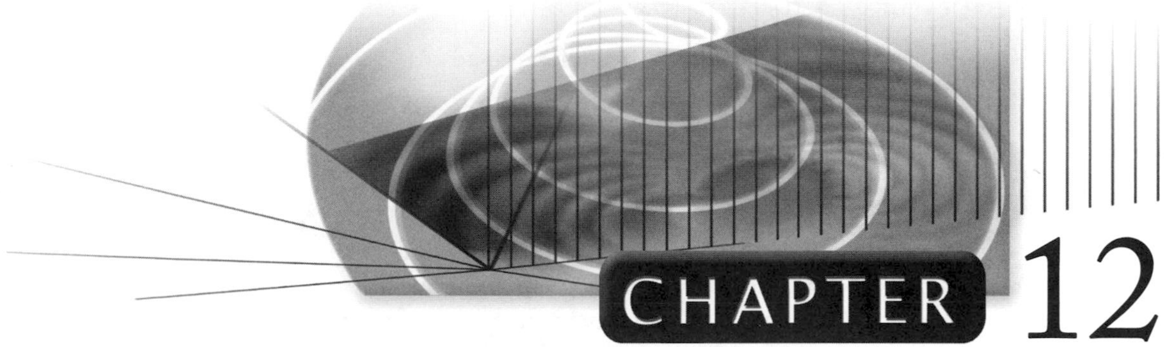

Making the Most of Blocks

Blocks are a very powerful feature of AutoCAD. They enable you to define an object or collection of objects that can be inserted into a drawing over and over without having to create the objects again from scratch. They also provide the capability to reduce a drawing's file size significantly. More important, although a drawing may contain hundreds of insertions of a particular block, if it becomes necessary to edit the blocks, AutoCAD requires that you edit only a single block definition. Once it is redefined, the hundreds of instances of the inserted block will automatically be updated; i.e., the new changes will appear instantly.

Additionally, attributes containing user-defined textual information can be attached to a block, providing a means to create, locate, and then extract useful data unique to a particular block insertion. Block attributes are discussed in detail in Chapter 14, Extracting Information from a Drawing.

To use the power of blocks to its fullest, it is necessary to first understand the nature of blocks. By understanding how blocks work, and how to properly manage blocks, you will learn how to make AutoCAD do tedious, repetitious tasks automatically, thereby increasing your everyday productivity.

This chapter discusses the following subjects:

- Defining and inserting blocks
- Modifying blocks with in-place reference editing
- Nesting blocks within blocks
- Managing all of these blocks

UNDERSTANDING BLOCKS

A *block* is a collection of individual objects combined into a larger single object. Think of the block as the parent of a family, and the individual objects as the parent's children. Although the children have identities of their own (color, layer, lineweight, and linetype), they are also under the control of their parent, which also has its own color, layer, lineweight, and linetype properties.

The fact that both the block (parent) and its individual objects (children) have their own properties makes it important to understand how these properties are affected by certain conditions. For example, assume that a block has been created from several child objects, and that each child object was originally created on its own layer. The layers on which the child objects were created can be frozen individually. If one of these layers is frozen, the child object that resides on that layer is also frozen and becomes invisible. However, the other child objects in the block remain visible because the layers they are on are still thawed. In contrast, if the parent block is inserted on a layer, and that layer is then frozen, all of its child objects will become frozen. This is true even though the layers on which the child objects reside are thawed.

For example, Figure 12–1 shows an inserted block made up of three objects: a rectangle, a triangle, and a circle. The block (parent) is inserted on the layer Parent. When the block was defined, the rectangle, triangle, and circle objects (children) were on the layers Rectangle, Triangle, and Circle, respectively.

When the Triangle layer is frozen, as shown in Figure 12–2, only the triangle object disappears, and all other objects remain visible. This is true even though the triangle is part of another object—the inserted block object. In other words, because the triangle child's layer was frozen, only it is affected. None of the other children or the parent are affected.

In contrast, all the objects disappear when the Parent layer is frozen, as shown in Figure 12–3. This demonstrates the difference between freezing a child object's layer and freezing its parent's layer.

This example is just one of several different conditions that can influence the behavior and appearance of an inserted block. Understanding the rules that govern these conditions is essential to implementing the power of blocks and increasing your productivity.

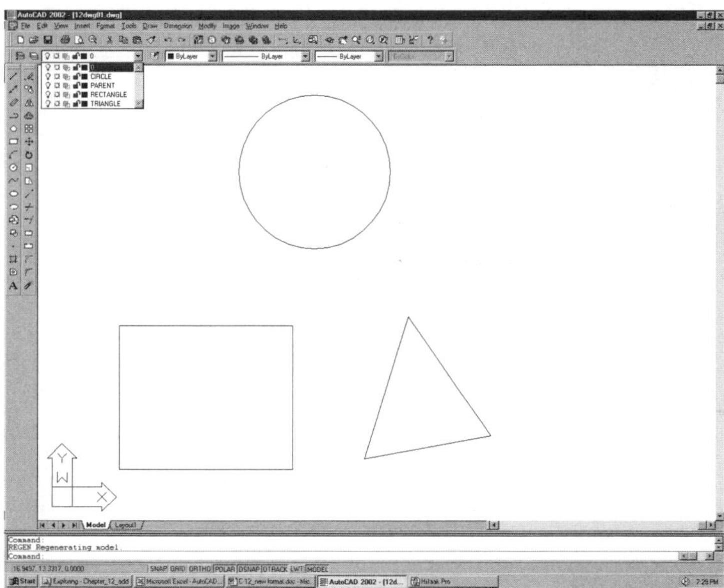

Figure 12–1 *The parent block is made up of three child objects.*

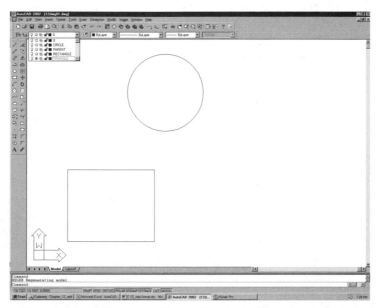

Figure 12–2 *The parent block with only the Triangle layer frozen.*

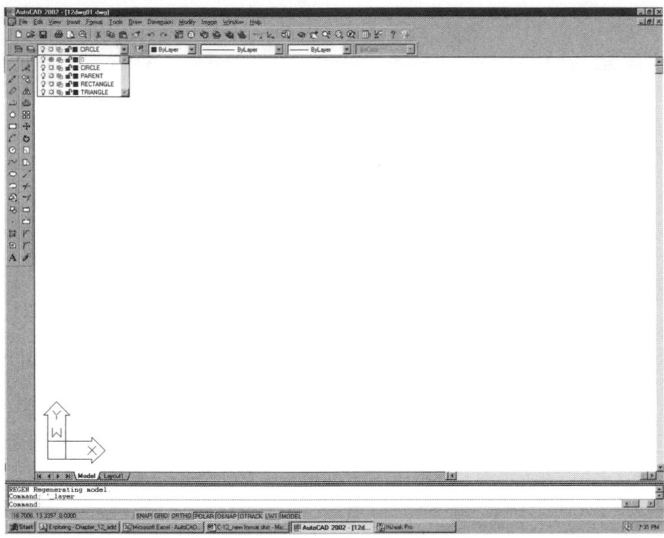

Figure 12–3 *The parent block, and its child objects, disappear when the Parent layer is frozen.*

THE BLOCK DEFINITION DIALOG BOX

Using the Block Definition dialog box is the most common method of creating new blocks. AutoCAD 2000 updated the dialog box, adding new features and enhancing its functionality. You access the Block Definition dialog box (shown in Figure 12–4) from the Draw pull-down menu by choosing Block>Make. Its features are described in the following sections.

Name

This is where you specify the name of the block. To assign a new name, type the name in the edit box.

This feature is improved over its predecessor by the addition of a drop-down list. The list displays the names of all currently defined blocks. In addition, by selecting a name, you can display all of its current settings and then redefine any or all of its values.

Base Point

This area allows you to define the X,Y,Z insertion coordinates for the block. You can enter the coordinate values in the appropriate edit boxes, or you can choose the Pick Point button and specify the insertion base point by selecting a point on the screen.

The base point value is saved with the block and represents the point in the block that AutoCAD 2000 uses to define the block's position when it is inserted into the drawing.

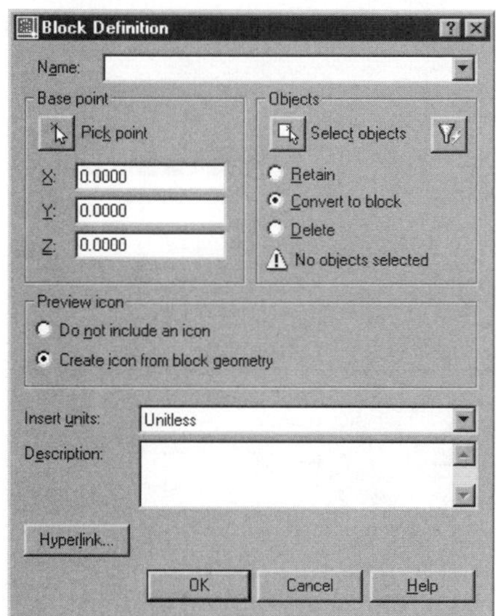

Figure 12–4 *The Block Definition dialog box.*

Objects

When you select objects, this area controls various options that define a new block, and it displays the number of objects selected. After you choose the Select Objects button, you then choose the objects that make up the block definition on the screen.

The Quick Select button displays the Quick Select dialog box (shown in Figure 12–5), which allows you to select objects based on filter criteria. For example, you can select all circle objects whose color is blue.

Refer to Chapter 10, Advanced Editing, for more information on using the Quick Select option.

The Retain, Convert to Block, and Delete radio buttons tell AutoCAD what to do with the selected objects once the block is defined. When you retain the objects, they are left in the drawing and are not converted to a block after AutoCAD uses them to define the new block. When you convert the objects to a block, the original objects are erased and then reinserted as a single block in their original positions in the drawing. When you delete the objects, AutoCAD erases them from the drawing once the new block is defined.

Preview Icon

This feature allows you to control whether AutoCAD creates an image of the new block, and it saves the image with the block definition. The image is displayed when you view blocks by selecting a block name, and when you use AutoCAD Design-Center, as discussed later in this chapter.

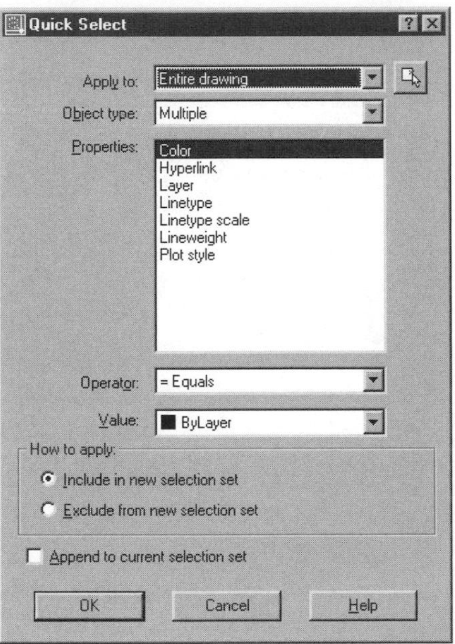

Figure 12–5 *The new Quick Select dialog box.*

Insert Units

This feature specifies the units to which a block is scaled when it is inserted from DesignCenter. You can choose from numerous unit types, including feet, inches, or millimeters. This feature is discussed in the section titled AutoCAD DesignCenter later in this chapter.

Description

This feature allows you to provide a detailed description of the block definition. The description is displayed when you select it in the block command, and when blocks are viewed using the new DesignCenter, as discussed later in this chapter.

DEFINING BLOCKS

What happens inside a drawing when a new block is defined? If you have created blocks before, you know that once you select the child objects that make up the block, they all disappear from the screen. This happens because AutoCAD automatically erases them once they have been used to define a block.

Tip: Use the OOPS command to unerase the objects. The OOPS command can always be used to unerase the most recently erased selection set. This is true even if several other commands have been executed after an object was erased.

Note: I always thought it silly that AutoCAD erased the block's child objects. After all, I typically created the objects where I needed them in the first place. I was only trying to make a copy of the objects to put them some other place in my drawing. Why, then, would AutoCAD erase them? The following paragraph explains why.

There is a logical reason why AutoCAD erases the objects. The block command is more than another COPY command. Instead, it is a way of making copies of a collection of child objects that use less file space. It keeps the file size of an AutoCAD drawing smaller by storing each child object's property data in a place AutoCAD calls the block table. It stores this information under the name of the parent block. When a block is inserted into AutoCAD, instead of duplicating the property data of each child object (as the COPY command does), AutoCAD simply refers back to the property data stored in the block table. It then draws the child objects based upon this data. This enables AutoCAD to store each child object's property data in just one place, the block table. You can therefore insert multiple copies of a block, duplicating the child objects where needed. In each case, AutoCAD refers back to the block table for the data it needs to draw the child objects. Consequently, AutoCAD erases the original objects once they are used to define a block because it assumes you will want to reinsert those objects as a block, reducing the file size of your drawing.

The Effect of the Current UCS on Block Definitions

When you create a block, you must define its insertion base point. This point's coordinates are relative to the block object and are set to its 0,0,0. Consequently, when you define a block's insertion base point, even though the current UCS (User-defined Coordinate System) coordinates may be 100,100,100 when you pick them, AutoCAD ignores these values and stores the block's insertion base point as 0,0,0. This is true in both paper space and model space. If this block is then exported out as its own drawing using the WBLOCK command, the insertion base point will be at 0,0,0 of the WCS (World Coordinate System) in the new drawing. This feature enables predictable insertion of blocks.

Tip: AutoCAD remembers the series of commands you enter at the command prompt. You can recall the previously entered commands by pressing the up arrow key on your keyboard. Each time you press the up arrow key, you move back through the previous commands. You can also press the down arrow key to move forward through the previous commands. Once the desired command is displayed at the command prompt, press Enter to execute the command.

Note: AutoCAD ignores the current UCS coordinate values in paper space and model space when you define a block's insertion base point.

Tip: When you define the insertion base point of a new block, simply imagine that AutoCAD is temporarily redefining the UCS origin as the point you pick.

In addition to understanding how AutoCAD deals with the current UCS's coordinates when you define a block, you must also understand the effect that the current UCS's X-axis orientation has on the angle that a block assumes when it is inserted into a drawing.

When creating a block, AutoCAD uses the current UCS to determine its insertion angle. This angle is oriented relative to the current UCS's X-axis at the time the block is defined. When inserting a block, AutoCAD sets the insertion angle relative to the current UCS's orientation at the time of insertion.

INSERTING BLOCKS

Several commands can be used to insert blocks. Understanding the unique features of these commands is important in selecting the right tool for a particular task.

INSERT and -INSERT

The INSERT and -INSERT commands are used to insert blocks. The INSERT command prompts you for insertion information using the Insert dialog box, whereas -INSERT prompts you for information at AutoCAD's command line. The Insert dialog box interface makes it easy to select blocks already stored in the current drawing's block table. The Insert dialog box also makes it easy for you to search for blocks stored outside the current drawing by choosing the Browse button.

Tip: When using script files to perform repetitious commands, AutoCAD automatically uses the command line version of the commands. Therefore, you can use the -INSERT command to understand the series of options the script file will need to address to ensure that the script functions correctly.

Note: To use the standard command alias of the Insert command, at the command prompt, simply type **I** to open the Insert dialog box. Type **–I** at the command prompt to start the command line version of the Insert command.

Tip: When you use the command line version of the Insert command, you can preset the scale and rotation of a block before it is inserted. This feature is useful when you want to see the effect of a scale or rotation angle prior to inserting the block. To take advantage of this feature, type **-INSERT** at the command prompt, then select the block to be inserted. When AutoCAD prompts you for the insertion point, type **S** to preset the scale or **R** to preset the rotation.

MINSERT versus ARRAY

Sometimes it may be necessary to insert a block as a rectangular array. Two options exist to accomplish this.

The MINSERT Command. First, the MINSERT command combines the INSERT and ARRAY commands. When it is executed, the first command line prompts are the typical ones for inserting a block. Next appear the typical command line prompts for creating an array. The exception is that MINSERT can create only rectangular arrays. Therefore, no option is available to select a polar array.

One drawback to this command is that the MINSERT object cannot be exploded, nor can its individual objects be moved or edited. The advantage to using this command, however, is that the MINSERT object requires less file space to define it, thereby reducing the file size of your drawing. It is important to note that the reduction in file size can be dramatic. For example, a simple block inserted as an array of 100×100 using the MINSERT command will have little impact, if any, on the file's size. In contrast, the same number of blocks inserted using the ARRAY command can increase the file's size by ½ MB or more.

If you have a situation in which you need to insert an array of blocks, and you will not need to explode or edit the objects, I suggest using the MINSERT command. Otherwise, use the ARRAY command discussed in the next section.

The ARRAY Command. The ARRAY command accomplishes the same thing as the MINSERT command but affords you more control over the inserted objects. With the ARRAY command you have the option of rectangular or polar arrays. Also, once the array is created, you can explode the inserted objects individually or move them independent of the other inserted objects. To use the ARRAY command with a block, however, you must first use the INSERT command to create the first object, and then use the ARRAY command to create the desired array.

The disadvantage of using the ARRAY command is that multiple insertions of the block object are made, which therefore increases your drawing's file size.

MEASURE and DIVIDE

The MEASURE and DIVIDE commands provide a method of inserting a block along a path.

MEASURE Command. The MEASURE command enables multiple insertions of a block along a line, arc, or polyline at a given distance. To demonstrate this, the next exercise will create a series of rectangles along the centerline of a street design. By creating a block of a rectangle and inserting it at the appropriate distances along a centerline path, a series of viewport guides can quickly be created. For this particular drawing, the guides can be used to define the various plan view sections for street improvement plans.

EXERCISE: USING MEASURE TO SET A SERIES OF BLOCKS ALONG A PATH

1. Open the 12DWG01A.DWG drawing file from the accompanying CD. This file contains a typical street centerline with right-of-way lines. A block called Viewport exists in the block table. This block consists of a rectangle with an insertion base point located in the center of the rectangle. The rectangle is 400 feet wide and 1,000 feet long.

   You will use the MEASURE command to insert a series of this block every 800 feet along the centerline.

2. From the Draw pull-down menu, select Point>Measure. AutoCAD prompts you for the object to measure.
3. Select the red centerline.
4. At the "Specify length of segment or [Block]:" prompt, type B to select Block, then press Enter.
5. At the "Enter name of block to insert:" prompt, type the block name **Viewport**, then press Enter.
6. At the "Align block with object? [Yes/No] <Y>:" prompt, type **Y**, and then press Enter.
7. At the "Specify length of segment:" prompt, type a segment length of **800**, then press Enter. AutoCAD creates many Viewport blocks along the centerline path, placing one every 800 feet, as shown in Figure 12–6.

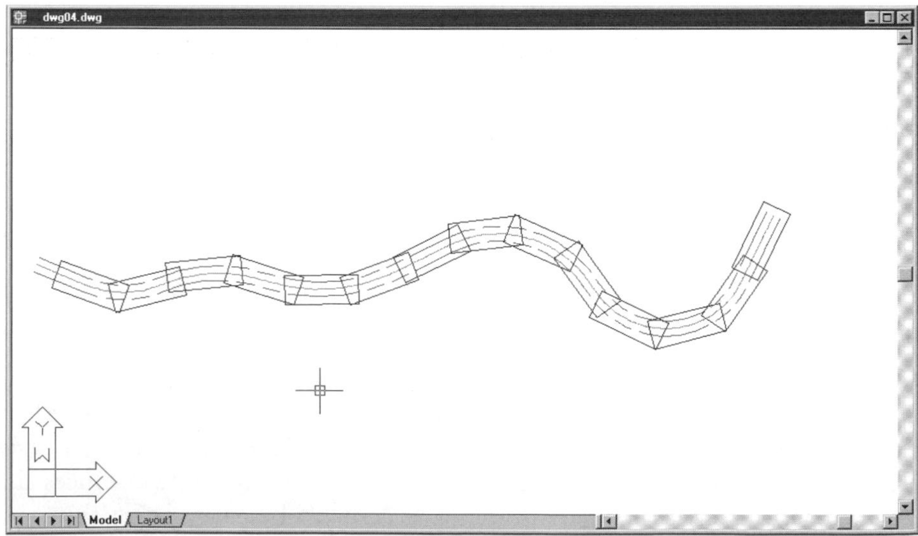

Figure 12–6 *The MEASURE command places the Viewport blocks along a path.*

DIVIDE Command. The DIVIDE command allows multiple insertions of a block along a line, arc, or polyline. Suppose you must draw a series of manholes along the street centerline in the previous example. With the Manhole block already created, you can use the DIVIDE command to insert thirty copies of it along the centerline path.

EXERCISE: USING DIVIDE TO INSERT THIRTY MANHOLE BLOCKS ALONG A PATH

1. Continue with the previous drawing or open the 12DWG01B.DWG drawing file from the accompanying CD.
2. From the Draw pull-down menu, select Point>Divide. AutoCAD prompts you for the object to divide.
3. Select the red centerline.
4. At the "Enter the number of segments or [Block]:" prompt, type **B** to select Block, then press Enter.
5. At the "Enter name of block to insert:" prompt, type the block name **Manhole**, then press Enter.
6. At the "Align block with object? [Yes/No] <Y>:" prompt, type **Y** to align the block with the selected object, then press Enter.
7. At the "Enter the number of segments:" prompt, type a segment number of **30**, then press Enter. AutoCAD draws thirty evenly spaced Manhole blocks along the centerline path, as shown in Figure 12–7.
8. You may close the drawing without saving it.

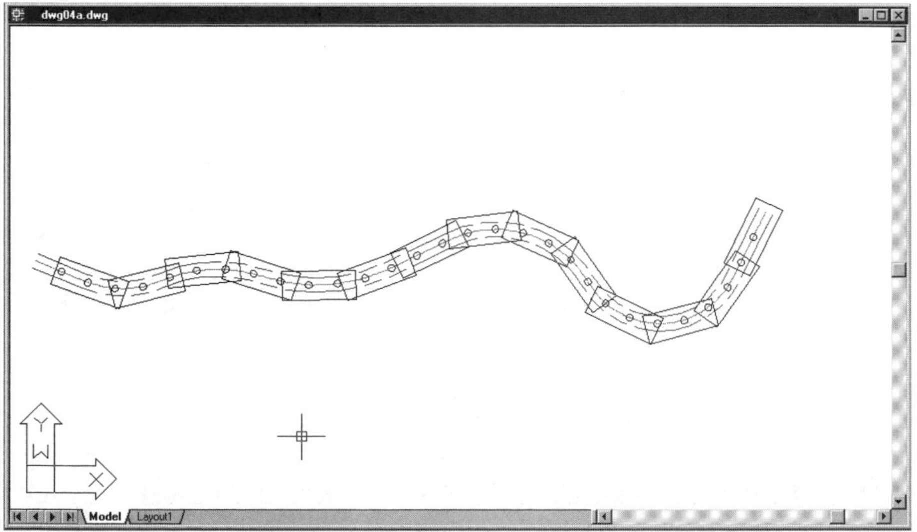

Figure 12–7 *The DIVIDE command evenly spaces the Manhole blocks along a path.*

Note: When you insert blocks into a drawing, it is important to remember that AutoCAD aligns the block's WCS parallel to the current UCS. This feature affects not only the insertion angle of the block; it also affects the rotation angle. If the rotation angle is assigned when a block is inserted, the rotation angle is relative to the current UCS. This is true in both paper space and model space.

Windows-Based Insertion Features

Introduced with AutoCAD 2000, the Multiple Document Environment (MDE) allows you to open multiple drawings in a single AutoCAD session. This feature works in the same fashion as other Windows applications, such as Microsoft Word and Excel.

There are many advantages to opening several drawings in a session of AutoCAD, and two that relate to blocks. Along with MDE, AutoCAD 2000 also introduced two features that take advantage of multiple open drawings. These features allow you to either cut and paste, or drag and drop objects from one drawing to another. Additionally, you are given the option to insert the objects as a block.

Cut-and-Paste Block Insertion. The phrase "cut and paste" refers to the process of selecting an object or group of objects and cutting or copying them to the Windows Clipboard. When selected objects are cut, they are copied to the Clipboard and erased from the original document. In contrast, when you copy objects, they are copied to the Clipboard and left untouched in the original document.

When you insert the block from the Clipboard, you can choose either to paste it using a selected base point, or to paste it using the block's original coordinates. The Paste to Original Coordinates option, found on the shortcut menu, copies the coordinate location of the objects in the original drawing, and then pastes the objects into the new drawing using those same coordinates. This option is useful for copying objects from one drawing to another when both drawings use the same coordinate system. You must have previously saved a base point to have access to the Paste to Original Coordinates option. For more information, refer to Chapter 9, Basic Editing.

Drag-and-Drop Block Insertion. When you drag and drop objects, you are moving them from one place to another. When you drag and drop objects within the same AutoCAD drawing, the result is the same as using the MOVE command to move the objects. When you drag and drop objects from one drawing to another, the result is similar to copying the object from the original drawing to the target drawing.

The Drag and Drop feature is useful for quickly moving or copying objects. However, this method does not provide a method for accurately selecting the base point for pasting the objects as a block. Consequently, to select objects in one drawing and paste them as a block in another while controlling the new block's base point, use the Copy and Paste feature described in the previous section. For more information, refer to Chapter 9, Basic Editing.

Note: You can also drag and drop files from Windows Explorer into a drawing.

BLOCK REFERENCE

When you insert a block, you are actually creating an Insert object. The Insert object refers to a particular set of block data in the block table. This is called a block reference. AutoCAD uses the block reference to find the data stored in the block table. It uses this data to draw the child objects that make up the Insert object.

Although only one set of data in the block table is used to define a block, there can be multiple block references referring to that data. In fact, there is no limit to the number of Insert objects that can be created. In each case, AutoCAD uses the block reference to find the data it needs to draw the Insert object.

Behavior of Block Properties

There are three properties of blocks that behave in different ways depending on their settings when the block is defined. The color, linetype, and lineweight properties can behave in different but predictable ways when they are defined on layer 0, as opposed to other layers. Also, you can define these properties explicitly, by selecting particular values, or implicitly, by defining them as BYLAYER or BYBLOCK.

The Effect of Creating Blocks on Layers Other than Layer 0. The simplest way to control the appearance of a block is to define it on a particular layer and explicitly define its color, linetype, and lineweight. For example, suppose you have created a circle object on a layer called Circle. To explicitly define its color, from the Modify menu, choose Properties. When AutoCAD displays the Properties window, select the circle. When the Properties window displays the circle's properties, select a color from the Color property drop-down list, or choose the Other option from the list to display the Select Color dialog box. The Properties window now lists the color you chose as the property of the circle, and it changes the circle object's color to reflect the modified property. You have just defined the color value explicitly. As a consequence, if the circle object is used to define a block, and the block is inserted into the drawing, its color will be constant. It will always be the color you explicitly defined.

In contrast, if an object's color, linetype, and lineweight are defined implicitly by choosing BYLAYER, altering the properties of the layer that the original object was on when it was defined as a block will change the object's appearance in the block. For example, suppose the circle object in the previous example is used to define a block. Also suppose that the circle's color, linetype, and lineweight are defined as BYLAYER, and the circle object is on a layer called Circle. When the block containing the circle object is inserted into the drawing, altering the color, linetype, or lineweight of the Circle layer will change the circle object's color, linetype, and lineweight. This is true no matter the layer on which the block is inserted. A child object with its color, linetype, and lineweight properties set to BYLAYER has those properties determined by the values of the child object's original layer.

The Effect of Creating Blocks on Layer 0. Layer 0 has a unique feature. When a block is defined from child objects created on layer 0, AutoCAD assigns special properties to that block if its color, linetype, and lineweight properties are set to BYLAYER or BYBLOCK. This feature can be very powerful.

Note: Another property that may be set to BYLAYER or BYBLOCK is the plot style property. The Plot Style property is discussed in Chapter 20, Basic Plotting.

If BYLAYER is used to define a child object's color, linetype, and lineweight, the layer on which the block is inserted controls the child object's color, linetype, and lineweight properties.

The following exercise demonstrates the effects of inserting a block whose color, linetype, and lineweight properties have been set to BYLAYER.

EXERCISE: INSERTING A BLOCK WITH BYLAYER PROPERTIES

1. Open the 12DWG02.DWG drawing file from the accompanying CD. The screen is blank, containing no objects. In this drawing file, two blocks are already defined. The block C1 is a circle created on layer 0 with its color, linetype, and lineweight properties set to BYLAYER. The block C2 is a circle created on layer 0 with its color, linetype, and lineweight properties set to BYBLOCK.

   Note that on the Object Properties toolbar, the current layer is BLUE, and the layer's color, linetype, and lineweight properties are set to BYLAYER.
2. From the Insert pull-down menu, choose Block. The Insert dialog box is displayed.
3. From the Name drop-down list, choose the C1 block.
4. In the Insertion Point area, make sure the Specify On-screen option is selected. In the Scale and Rotation areas, make sure the Specify On-screen option is cleared.
5. Choose OK to close the Insert dialog box.
6. Choose a location on the left side of the screen to insert the block.The C1 block is inserted and assumes the color, linetype, and lineweight of the BLUE layer's property values. Remember, this occurs because the block's child objects were created on layer 0 and their property values for color, linetype, and lineweight were set to BYLAYER.

In contrast, if BYBLOCK is used to define a child object's color, linetype, and lineweight, the current object creation values control the child object's color, linetype, and lineweight values. This is true no matter the layer on which the block is inserted. These values are controlled from the Object Properties toolbar.

The following exercise demonstrates the effect of inserting the C2 block, whose color, linetype, and lineweight properties have been set to BYBLOCK.

EXERCISE: INSERTING A BLOCK WITH BYBLOCK PROPERTIES

1. Continue with the 12DWG02.DWG drawing file. From the Object Properties toolbar, change the color property to Magenta, change the linetype property to Hidden2, and change the lineweight property to 0.020 inches.
2. From the Insert menu, choose Block. The Insert dialog box is displayed.
3. From the Name drop-down list, choose the C1 block.
4. Choose OK to close the Insert dialog box.
5. Choose a location on the right side of the screen to insert the block.

Your screen should look similar to Figure 12–8. Notice that the C1 block acquired the color, linetype, and lineweight property values based on the layer's values, while the C2 block acquired the color, linetype, and lineweight property values set by the Object Properties toolbar.

Note: DEFPOINTS is another layer that AutoCAD deals with uniquely. AutoCAD automatically creates this layer any time you draw associative dimensions. The unique property of this layer is that objects residing on this layer will not be plotted.

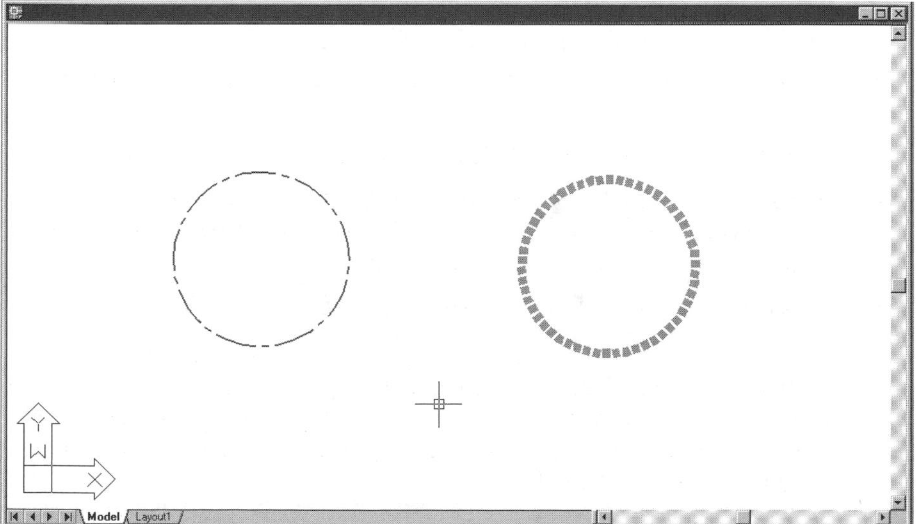

Figure 12–8 *The different effects BYLAYER and BYBLOCK have on inserted blocks.*

Note: Objects that would not be plotted have occasionally frustrated me, even though I could see them on the screen. The problem was that the objects, including blocks, were accidentally placed on the DEFPOINTS layer.

IN-PLACE REFERENCE EDITING OF BLOCKS

The ability to redefine blocks with in-place reference editing was introduced with AutoCAD 2000. This feature allows you to edit an inserted block, altering its child objects and automatically redefining all block insertions. While this feature allows you to edit inserted blocks, its real power is found in its ability to edit attached external references in the current drawing and save the changes back to the original drawing. (For more information, see Chapter 13, Working with Large Drawings and External References.) Nonetheless, you can use it to edit an inserted block quickly without having to explode the block.

EXERCISE: EDITING A BLOCK WITH IN-PLACE REFERENCE EDITING

1. Open the 12DWG03.DWG drawing file from the accompanying CD. The drawing displays four insertions of the same block.
2. From the Modify menu, choose In-Place Xref and Block Edit>Edit Reference. AutoCAD prompts you to select the reference to edit.
3. Choose the block named Terri in the upper-left corner. AutoCAD displays the Reference Edit dialog box. Notice that the block's name is listed as Circle in a Square.
4. Select the Display Attribute Definitions for Editing option, as shown in Figure 12–9, because the name in the center of the block insertion is an attribute. Selecting this option allows you to edit the block's attribute.
5. Choose OK. AutoCAD prompts you to select nested objects.
6. Select the circle in the block named Terri, then press Enter. The circle is added to the list of objects to edit. AutoCAD then displays the Refedit toolbar, as shown in Figure 12–10. There is no need to try to select the attribute.
7. Choose the circle. AutoCAD highlights the circle and displays its grips.
8. Choose the grip in the bottom quadrant of the circle, then drag the grip down to the bottom line in the square. Use the midpoint snap to snap the circle to the midpoint of the line. The circle's radius increases.
9. Right-click anywhere in the drawing area and, from the shortcut menu, choose Deselect All. The highlighted circle is unselected.
10. Choose the attribute, then drag and drop it toward the lower half of the circle. Now click on the fourth (far-right) icon on the Refedit toolbar to save the changes back to the reference block. Notice that the enlarged circle object is displayed in all the blocks. However, the attribute you moved toward the lower half of the circle does not appear to have changed, as shown in Figure 12–11. This is because changes made to edited attributes do not take effect in previously inserted blocks; they only take effect in new block insertions.

11. From the Insert menu, choose Block. The Insert dialog box is displayed.
12. Make sure the Circle in a Square block is selected in the Name drop-down list, choose OK, and then insert the block on the right side of the drawing. Enter your name when prompted. AutoCAD inserts the revised block. Notice that the attribute appears in its modified position in the lower half of the circle, as shown in Figure 12–12.

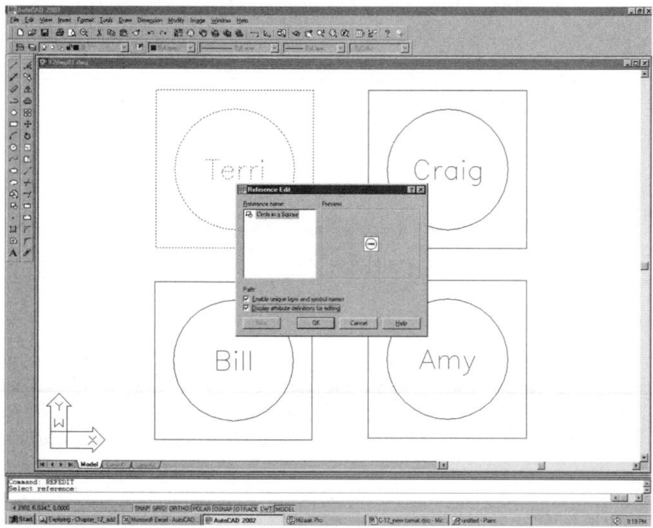

Figure 12–9 *The Reference Edit dialog box allows you to edit inserted blocks.*

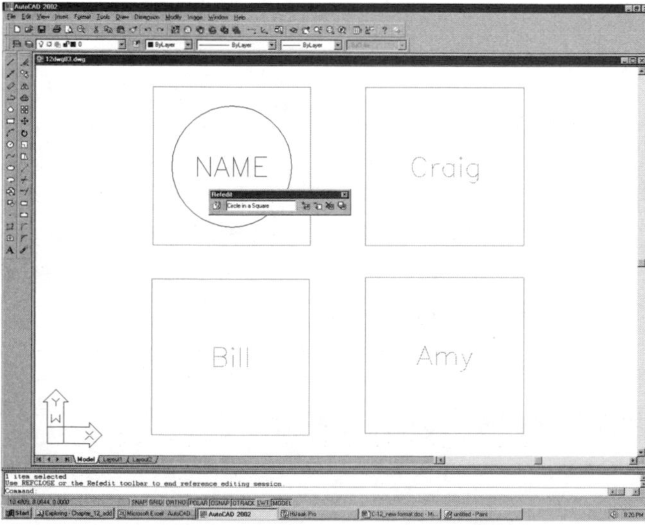

Figure 12–10 *The Refedit toolbar.*

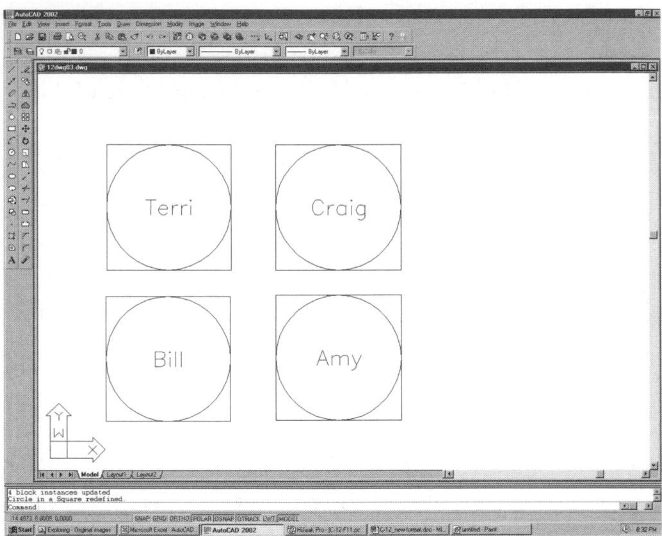

Figure 12–11 *The circle object is updated in all inserted blocks, but the block's attribute is not.*

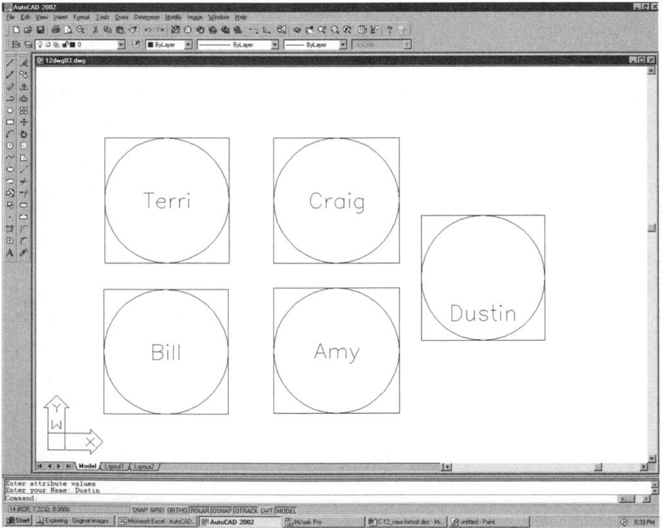

Figure 12–12 *The redefined block attribute appears in its new position only with new block insertions.*

NESTING BLOCKS WITHIN BLOCKS

As indicated earlier in this chapter, two significant reasons exist for using blocks. The first is to reduce a drawing's file size. The second is to quickly update all the insertions of a particular block. For example, suppose you have a block that is made up of a circle with a text object in its center, and you have inserted this particular block hundreds of times. If the text value is currently the letter M, but it needs to be changed to S, you can simply redefine the block with the correct letter. Once the block is redefined, the hundreds of block insertions are instantly updated with the S text value. This is a very valuable feature that can save you hours of editing time.

The previous example demonstrated a powerful feature of blocks. This power can be expanded through the use of nested blocks. A *nested block* is simply a block that contains other blocks and objects.

Nested blocks increase the power of blocks by making it easier to redefine blocks. For example, suppose you have a large block made up of numerous objects. Also suppose that one object in the large block occasionally changes its color. Instead of redefining the large block and all of its objects every time the object changes its color, you can create a small block of that one object and insert it as a nested block in the large block. Then, when it becomes necessary to change the object's color, you can redefine it in the small block. Once it is redefined, the small block is automatically updated in the large block. The important point in this example is that the large block reflects the new color of the nested block object, without being redefined itself.

Note: There is a problem you must be aware of when you redefine nested blocks. To redefine a nested block, you must redefine it explicitly in the current drawing. If you redefine a nested block in its parent block outside of the current drawing, and then use the -INSERT command to redefine the parent block in the current drawing, the nested blocks will not be updated. AutoCAD redefines the parent block only when you use the -INSERT command. The nested block definitions in the current drawing always take precedence over nested block definitions inserted from another drawing. The nomenclature for this is block name=file name.

Tip: Using the -INSERT command with the name=filename nomenclature is a technique that enables you to redefine a nested block. Simply WBLOCK the updated nested block to its own drawing, and use the -INSERT command to redefine the nested block in the necessary drawing.

MANAGING ALL OF THOSE BLOCKS

As you learn how to take advantage of the power of blocks, you will eventually develop hundreds of blocks, possibly more. You can further enhance the power of blocks by managing those blocks in a fashion that enables you and other users to find the desired block definition quickly. If this is not done, productivity will be lost in one of two ways. First, significant time will be spent simply trying to find the appropriate block. Second, if the block cannot be found, time will be spent recreating the block from scratch. It is therefore necessary to establish criteria that everyone follows to create and store blocks properly for future use.

WBLOCK COMMAND

When you create a block library, the most important component is the block itself. Chances are that you probably already have a wealth of predefined blocks residing in existing drawing files. These existing files are the first place you should go to develop your block library because they probably contain the blocks that your organization frequently uses.

Once you find useful predefined blocks in an existing drawing, you must export them as their own, individual drawings. The WBLOCK command is a convenient way to extract blocks quickly from the current drawing.

Typing WBLOCK at the command prompt displays the Write Block dialog box, as shown in Figure 12–13. This dialog box makes the process of exporting a block very intuitive. You can export an existing block by selecting the Block option, export the entire drawing, or select a group of objects and export them as a new block. Once the desired block or objects are identified, you can assign a new file name, determine the location to save the new block, and select its "insert units." Once these desired options are selected and you choose the OK button, AutoCAD exports the data and creates the new drawing. Once it is created, the new drawing file may be inserted as a block into any drawing.

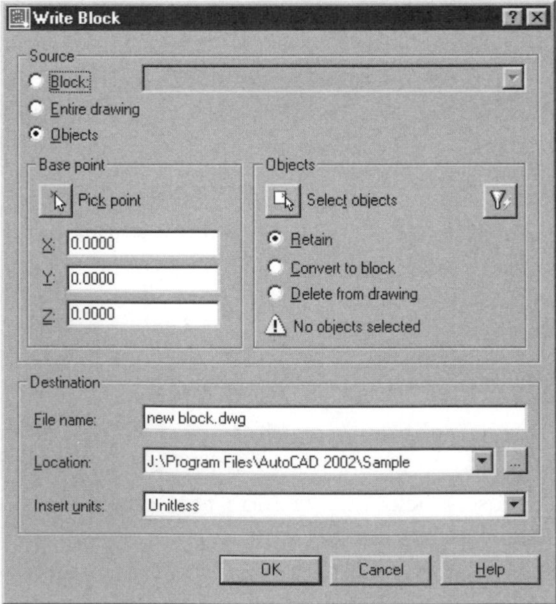

Figure 12–13 *Write Block dialog box.*

ORGANIZING BLOCKS

The key to managing your block library is to organize the block locations using a well-thought-out path structure. Store blocks in a standard location on each computer or on a network server, such as on the C drive under a subdirectory called WORK\BLOCKS. You can further organize blocks into classes and subclasses. The organizational structure should reflect a class structure used in your industry. For example, in civil engineering, it may be useful to organize standard storm drain junction structures using the following path structure:

C:\WORK\BLOCKS\STANDARDS\ORANGE_COUNTY\STORM_DRAIN\

JUNCTION_STRUCTURE-201A\STD-OC-SD-JS-201A.DWG

Following this type of structure, a CAD technician could easily follow the path to find a particular block. If the block is not found with this path structure, then the block probably was not created yet. Therefore, it can be created in the current drawing, and then WBLOCK-ed out to the appropriate path location.

AUTOCAD DESIGNCENTER

AutoCAD 2000 introduced AutoCAD DesignCenter, a powerful drawing content management tool. DesignCenter helps you manage drawing content by providing you with tools for searching, locating, and viewing blocks created in other drawings. By using these tools, you can quickly locate a desired block among hundreds of drawings and then insert it into the current drawing.

It is important to understand that what makes this feature so powerful is that DesignCenter actually looks "inside" AutoCAD drawings for the desired block. For example, suppose you need to use a block of a sewer manhole, and you know that the block already exists in another drawing. Also suppose you do not know which drawing the block is in, and there are hundreds of drawings you must search. By using DesignCenter's Find feature, you can quickly search through all drawings for a block whose name or description contains the word "manhole."

DesignCenter allows you to view blocks inside the selected drawing, and it displays both the block's graphic image and its description if they exist, as shown in Figure 12–14. Once the desired block is found, you may drag and drop it into the current drawing. Additionally, if the block definition includes an "insert units" value, and if the current drawing's units are also defined in the Drawing Units dialog box, AutoCAD will automatically scale the block to its proper size in the current drawing.

As briefly discussed in this section, DesignCenter provides powerful tools for managing blocks. However, its capabilities go beyond simply managing blocks; it allows you to manage other drawing content, such as external references, text styles, dimension styles, and more. For detailed information on using DesignCenter's features, refer to Chapter 11, Applying AutoCAD Design Center.

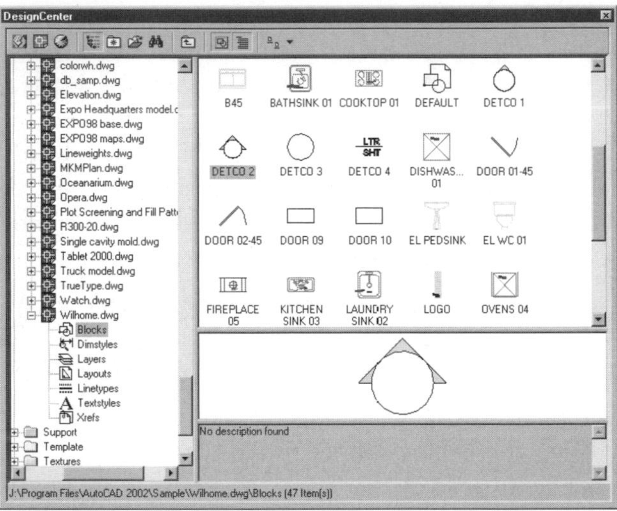

Figure 12–14 *The new DesignCenter makes finding blocks embedded in a drawing very easy.*

SUMMARY

Blocks are a very powerful feature of AutoCAD. This chapter has shown you how to extract the power of blocks by explaining the nature of blocks. You learned what happens to AutoCAD's database when a block is defined, and how the current UCS affects a block when it is being defined or inserted. You learned how AutoCAD stores a block definition in the block table, and how it refers to the block table to create the insert object. You saw the effects of defining blocks on a normal layer and on layer 0, and the difference between explicitly and implicitly defining the color and linetype properties of a block. The advantages of creating complex blocks from simpler blocks were discussed, and the steps necessary to redefine nested blocks with the -INSERT command were explained. Several techniques for managing block libraries, which enable you to quickly find the block you need, were presented.

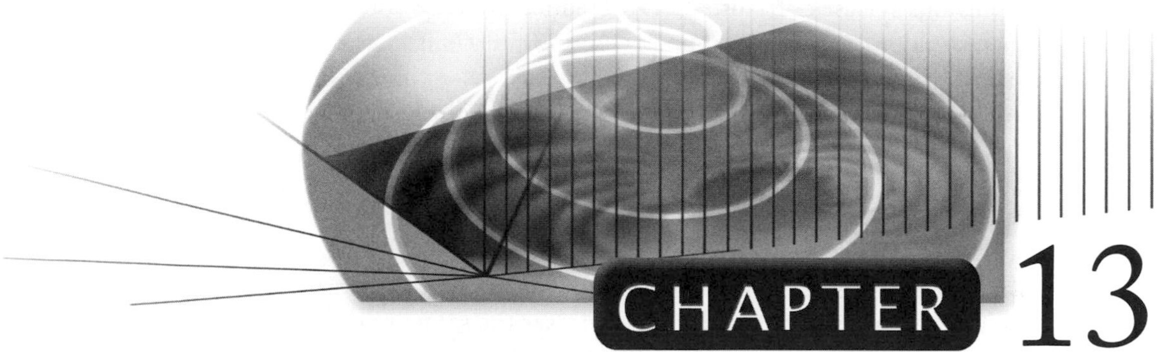

Working with Large Drawings and External References

AutoCAD 2000 introduced three features that enhance how you work with drawings and external references. The first two features, Partial Open and Partial Load, let you load only the portions of a drawing you need to edit. The third feature, In-Place Reference Editing, allows you to edit external references from the current drawing. Understanding how to use these features can increase your productivity.

This chapter shows you how to use these features and also provides an in-depth review of AutoCAD's external references. This chapter discusses the following subjects:

- AutoCAD's Partial Open and Partial Load features
- When to use xrefs instead of blocks
- Xrefs: Attach versus Overlay
- Xrefs: Bind versus Xbind
- Clipping boundaries
- Demand loading, and layer and spatial indexes
- Tracking xrefs
- In-Place Reference Editing feature

WORKING WITH LARGE DRAWINGS

In previous releases of AutoCAD, to edit a drawing you simply opened it. While this technique worked fine in many editing situations, it was a slow, tedious process if you were working with large drawing files. As a consequence, many technicians developed procedures that helped limit a drawing file's size, thereby increasing AutoCAD's response time when opening drawings and executing commands. One of these techniques included separating a large drawing into a group of smaller drawings. In some situations, a drawing was divided into a tiled grid, and each grid was exported as its own drawing file. In other situations, a drawing was isolated by different layers, with each layer and all of its objects saved as a single drawing file. Whichever method was used, the intent was the same: to decrease the drawing's file size and thereby increase productivity through faster response by AutoCAD.

The Partial Open and Partial Load features provide the ability to simulate the advantages achieved by separating a drawing into smaller files without having to divide the original drawing into groups of smaller drawings. These features allow you to open only a portion of a drawing, thereby loading only the elements you need to edit. Objects may be loaded based on selected layers, by choosing a predefined view, or by windowing in on an area. Once you have loaded objects, simply save the drawing as normal. This capability allows you to maintain a high level of productivity even while you work on large drawings, and it avoids the management problems encountered by separating a single drawing into many smaller drawings.

THE PARTIAL OPEN FEATURE

The Partial Open feature allows you to open a portion of a drawing by selecting the drawing's extents, the last view when the drawing was saved, a predefined view, if one exists. Once you select the desired view option, you identify the object geometry you wish to load by selecting the layer(s) on which they reside.

To access the Partial Open feature, choose Open from the File menu to display the Select File dialog box, as shown in Figure 13–1.

Choosing the Partial Open button displays the Partial Open dialog box. From here you select the view geometry and the layer geometry to load, as shown in Figure 13–2.

AutoCAD opens the drawing, loading only the objects that lie within or pass through the view, and that reside on the selected layers. At this point you can edit the loaded geometry, create new objects, and zoom into or pan the drawing. You can also create and modify data outside the area you opened. You are not limited by the initial view when you work with partially opened drawings.

When you use the Partial Open feature, you must select a view, but it is not necessary to select layers. You can leave all the check boxes next to the layer names cleared. If you do, when you choose Open, AutoCAD issues a warning that no

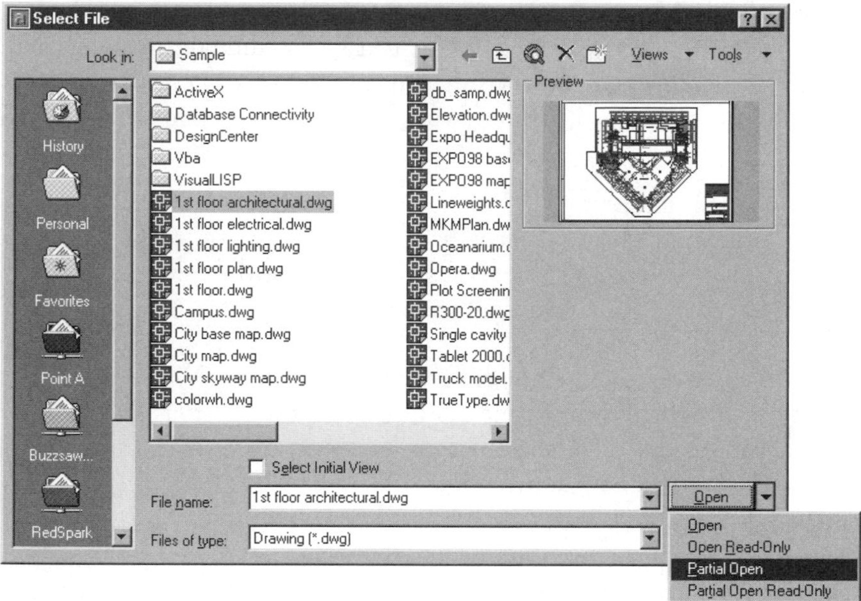

Figure 13–1 *You access the Partial Open feature by clicking on the arrow next to the Open button in the Select File dialog box.*

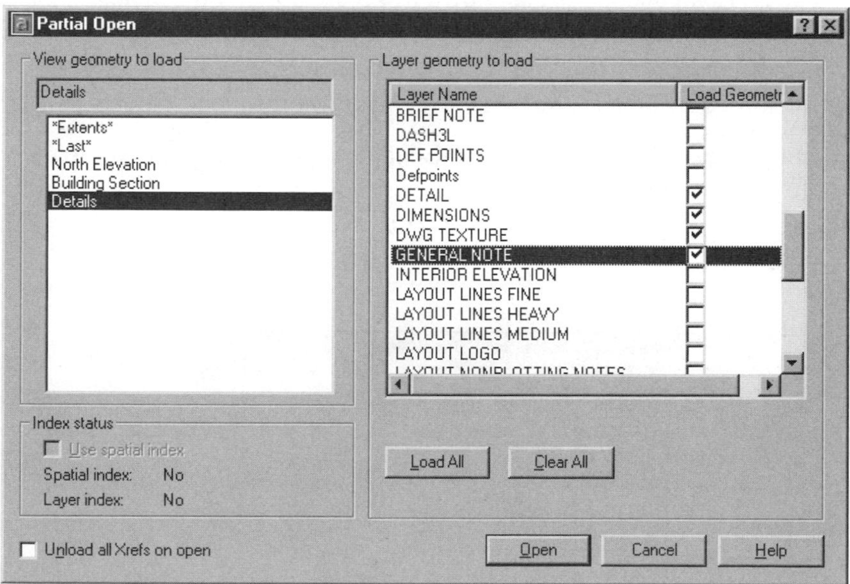

Figure 13–2 *The Partial Open dialog box allows you to identify the object geometry to load by selecting the view and the layers on which the objects reside.*

objects will be loaded. However, AutoCAD does load all the layers, and it also loads all the other named objects, such as blocks, text styles, dimension styles, and linetypes. Once they are partially opened, it is possible to use these named objects to create new object geometry in the drawing without actually loading any existing geometry.

One other behavior unique to the Partial Open feature occurs when you save and then reopen a drawing that was partially opened. When you reopen the drawing, AutoCAD displays a warning that notes that the drawing was partially opened when it was last saved. This warning presents three options, as shown in Figure 13–3. When you choose the Fully Open button, you load all of the drawing's object geometry. This is how AutoCAD normally opens a drawing. When you choose the Restore button, you partially open the drawing using the previous Partial Open settings. When you choose the Cancel button, you stop AutoCAD from loading the drawing and you create a new drawing instead.

THE PARTIAL LOAD FEATURE

The Partial Load feature allows you to load additional geometry into a drawing that is already partially opened. By using this feature, you can further refine the selection set of objects you load into the current editing session.

After you partially open a drawing, choose Partial Load from the File menu to display the Partial Load dialog box, as shown in Figure 13–4. This dialog box is almost identical to the Partial Open dialog box. The main difference is that the Partial Load dialog box includes the Pick a Window button, which allows you to drag a selection window around the area from which you want to load objects.

AutoCAD's Partial Open and Partial Load features allow you to load specific objects into the current editing session. By using these features, you can reduce the time it takes to open a drawing and to execute AutoCAD's commands, and thereby increase productivity.

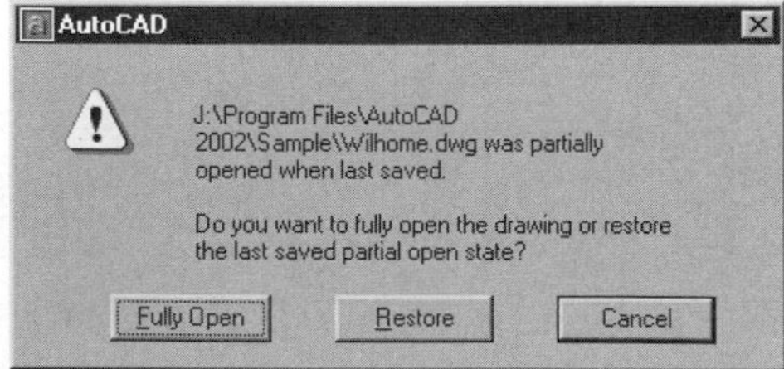

Figure 13–3 *AutoCAD displays a warning when you select a file that was partially opened when it was last saved.*

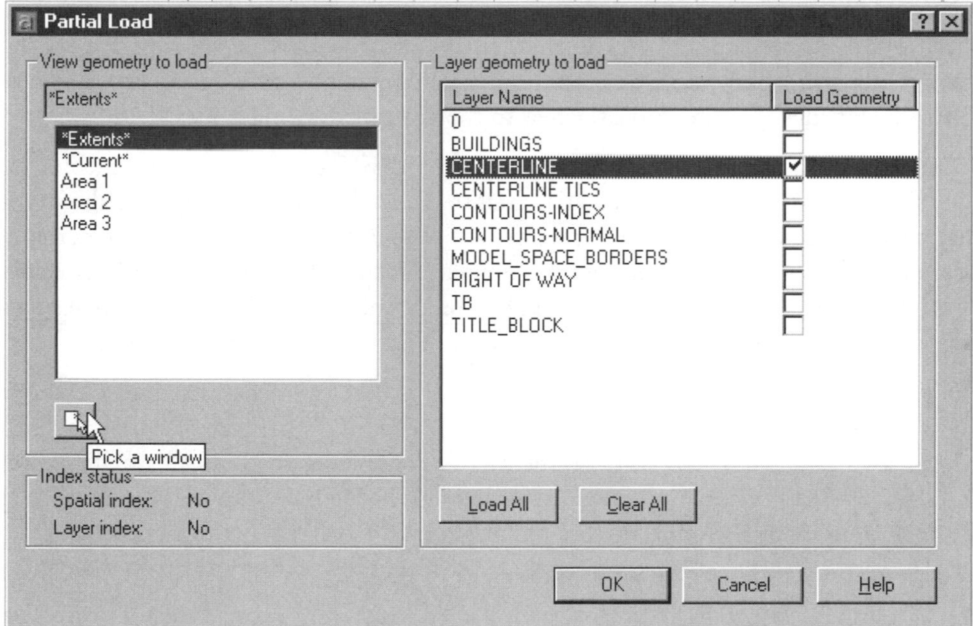

Figure 13–4 *The Partial Load dialog box allows you to pick an area from which to load named objects.*

USING EXTERNAL REFERENCES

External references, or xrefs, represent a powerful feature of AutoCAD. They provide the capability to create a composite drawing using other saved drawings, even while those other drawings are being edited. In a multidisciplinary work environment, you can attach another discipline's drawings to see the impact their design will have on your design. The drawings can be attached temporarily, or inserted permanently as a block. You can permanently insert the entire xref or just its named objects, such as text styles or dimension styles. You can attach an entire xref or just the portions you need to review. You can even define an irregularly shaped polygon as the clipping boundary for the portion of the xref you want to attach. By attaching small clipped portions of an xref, you can dramatically reduce regeneration times.

Additionally, AutoCAD 2000 introduced a feature called In-Place Reference Editing, which allows you to edit an attached xref from the current drawing and save the changes back to the original drawing file. These versatile features make AutoCAD's xref capabilities a very powerful tool.

XREFS VERSUS BLOCKS

External references have behavior that is similar to that of blocks. They can be inserted—or "xref-ed"—into a drawing and used to display a group of objects as a single object. They can be copied multiple times, and you can change all the insertions of a block or an xref by editing the original reference file. The only real difference between xrefs and blocks is that blocks are inserted permanently into the current drawing, becoming part of the drawing, whereas xrefs exist externally as independent files that are only attached to the current drawing.

So, if xrefs behave like blocks, then when should you use xrefs instead of blocks? One situation in which you should use xrefs is when the objects in external drawings that you need to view are undergoing changes. When edits are made to an externally referenced file, you can reload the xref to update it to reflect the most recent condition of the xref. Additionally, AutoCAD automatically loads the latest version of an xref-ed drawing when you open your drawing. This is not true with blocks.

Another situation in which to use xrefs instead of blocks is when the xref-ed drawing is large. Not only can you keep the current drawing's file size low by attaching large drawings as xrefs, you can also instruct AutoCAD to load only a small portion of an xref instead of the entire xref-ed drawing. This reduces the number of objects in the current drawing and therefore reduces file size and regeneration times.

INSERTING AN XREF: ATTACH VERSUS OVERLAY

You can link an xref to the current drawing in two different ways: attach it or overlay it. Both methods enable you to turn layers on and off, or to freeze and thaw layers. Both enable you to change the color, linetypes, and lineweights of layers of xref-ed drawings. Both methods are virtually identical, with one exception: Attached xrefs appear when they are nested in other xref-ed drawings, whereas overlays do not.

The following exercise demonstrates the difference between attaching and overlaying an xref.

EXERCISE: ATTACHING VERSUS OVERLAYING AN XREF

1. Open the 13DWG01C.DWG drawing file from the accompanying CD to display a Tentative Tract Map consisting of right-of-way lines, property lines, street centerlines, and proposed building pads.

   Next you will insert two xrefs. One you attach, and the other you overlay.

2. From the Insert menu, choose Xref Manager, and then choose the Attach button. The Select Reference File dialog box opens.
3. From the Select Reference File dialog box, open the 13DWG01A.DWG drawing file. AutoCAD displays the External Reference dialog box.
4. In the External Reference dialog box, under Reference Type, choose Overlay.
5. In the Insertion point area, the Scale area, and the Rotation area, clear any checked Specify On-screen check boxes, as shown in Figure 13–5.
6. Choose OK. AutoCAD overlays the drawing and displays its existing contours.

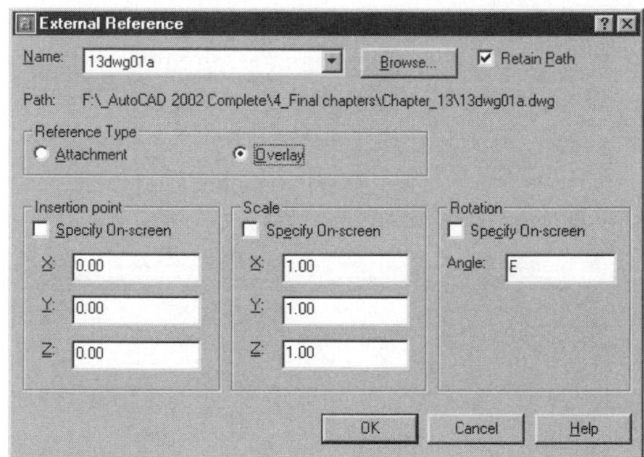

Figure 13–5 *The Overlay option is selected in the External Reference dialog box.*

7. From the Insert menu, choose External Reference. The Select Reference File dialog box opens.
8. From the Select Reference File dialog box, open the 13DWG01B.DWG drawing file. AutoCAD displays the External Reference dialog box.
9. In the External Reference dialog box, under Reference Type, choose Attachment.
10. In the Insertion point area, the Scale area, and the Rotation area, clear any checked Specify On-screen check boxes.
11. Choose OK. AutoCAD attaches the drawing and displays its existing trees, as shown in Figure 13–6.
12. Save the file in the ACAD2002\SAMPLE directory.
13. Close the 13DWG01C.DWG drawing file.
14. Open the 13DWG01D.DWG drawing file on the accompanying CD. AutoCAD displays the drawing, which contains building footprints.

    Next you will insert the Tentative Tract Map drawing, with its two xrefs.
15. From the Insert menu, choose External Reference. The Select Reference File dialog box opens.
16. From the Select Reference File dialog box, open the 13DWG03C.DWG drawing file from the ACAD2002\SAMPLE directory. AutoCAD displays the External Reference dialog box.
17. In the External Reference dialog box, under Reference Type, choose Attachment.
18. In the Insertion point area, the Scale area, and the Rotation area, clear any checked Specify On-screen check boxes.
19. Choose OK.
20. You may now exit the drawing without saving your changes.

AutoCAD attaches the drawing, as shown in Figure 13–7. It is important to realize that the existing Contours and Trees drawings are already referenced in the Tentative Tract Map drawing, and are therefore nested xrefs. Note that the Trees xref is displayed but the Contours xref is not. This occurs because the Contours drawing was overlayed, whereas the Trees drawing was attached to the Tentative Tract Map drawing.

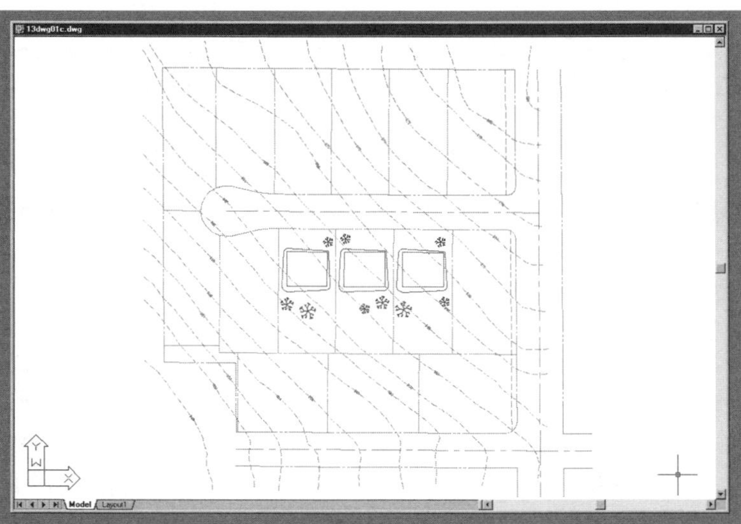

Figure 13–6 *One xref is attached, and the other is overlayed in the current drawing.*

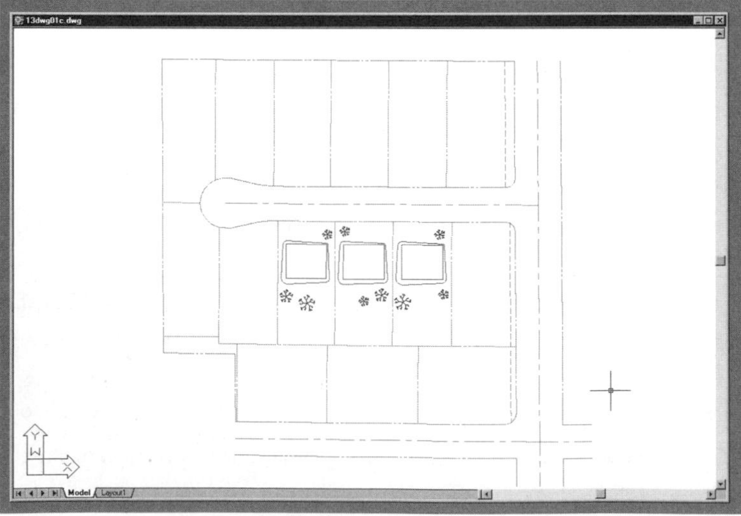

Figure 13–7 *The nested Trees xref is displayed, but the nested Contours xref is not.*

By using the attach and overlay features as shown in the preceding exercise, you can more easily manage the visibility of xref-ed drawings.

Note: If the xref file you attach or overlay has objects organized with the DRAWORDER command, the objects will not be displayed in their proper drawing order. This is because the DRAWORDER command works only on objects in the current drawing, not on objects in xrefs.

Tip: You can force xref objects to be displayed in their desired drawing order by opening the xref and using the DRAWORDER command to arrange objects as desired, and then using the WBLOCK command to write the entire xref out as a new drawing. AutoCAD retains the desired drawing order when the xref is recreated using the WBLOCK command.

UNLOADING AND RELOADING XREFS

Once you have attached or overlaid xrefs, you may then unload and then reload the xrefs as desired. This ability allows you to remove xrefs temporarily from a drawing so that AutoCAD does not spend time calculating the positions of an xref's objects during regenerations. When you unload an xref, AutoCAD removes the xref from the current drawing but leaves its path. To display the xref in the drawing again, you can reload the xref.

In addition to reloading xrefs that have been unloaded, you can also reload xrefs that are already loaded. This is useful when you know an xref-ed drawing is being edited, and you want to refresh the drawing's display to see any new changes that have taken place since you originally loaded the xref.

Tip: Loaded xrefs can significantly increase regeneration times. If you are editing a drawing and do not need to see a loaded xref, use the UNLOAD command to remove it temporarily from the drawing. This will increase your productivity.

MAKING XREFS PERMANENT: BIND VERSUS XBIND

Occasionally you may need to make an xref drawing a permanent part of the current drawing. When archiving files for permanent storage, or when submitting a file to another technician, you may find it useful to make xrefs a part of the drawing to which they are attached, thereby combining all the drawings into a single DWG file. By combining the xrefs into a single drawing, you ensure that all of the drawing data exists in a single file. You can make an xref a permanent part of another drawing by binding it to the drawing, which inserts the entire xref into the drawing as a block.

When you choose to bind an xref to the current drawing, AutoCAD prompts you to select the type of bind to use, Bind or Insert. When the Bind option is selected, AutoCAD inserts the xref into the current drawing and prefixes all named objects with the xref's drawing name. Therefore, named objects such as layers, blocks, and text styles are prefixed by the xref's drawing name and then inserted into the current drawing. However, if you choose the Insert option, AutoCAD inserts the drawing as a normal block, and does not prefix named objects with the xref's drawing name. Consequently, any duplicate named objects in the xref are ignored, and the named objects in the current drawing hold precedence.

Note: When you attach an xref to a drawing, the xref's drawing name prefixes its layer names, which are separated by the pipe (|) symbol. When you bind the xref to the drawing, the pipe symbol is replaced by 0, where the zero represents the first instance of a bound xref. If another xref with the same name as the first is also bound to the drawing, the zero increases to one, and 1 appears. This feature avoids potential problems when you bind xrefs from different departments or companies that may coincidentally use the same xref name.

This means, for example, that if a block in the xref has the same name as a block in the current drawing, the block in the current drawing will take precedence and be substituted in place of all sam-name block insertions in the xref. While this feature eliminates the redundancy of duplicate layer names, it may give unexpected results if you are not aware of duplicate named objects.

While binding an xref into the current drawing is useful under certain conditions, there are occasions when it is more useful to bind only an xref's named objects. You can accomplish this with the XBIND command.

To display the Xbind dialog box, choose Object>External>Reference>Bind from the Modify menu. If you double-click on an xref name, AutoCAD displays the named object headings in the selected xref drawing. There is a small box containing a plus sign (+) next to any named object heading that includes dependent objects. Click on a box with the plus sign; AutoCAD displays the dependent objects. Choose one of the dependent objects and click the Add button. The selected object appears in the Definitions to Bind text box, as shown in Figure 13–8. Click on OK; AutoCAD binds the selected object to the current drawing.

By using AutoCAD's Bind and Xbind features, you can permanently insert the entire xref, or just specific named objects, such as text styles and linetypes.

CLIPPING BOUNDARIES

The XCLIP command allows you to use rectangles and irregularly shaped polygons to define clipping boundaries for xrefs. The polygons can be created on the fly or by selecting an existing 2D polyline. After the clipping boundary has been chosen, AutoCAD removes from display any portion of the xref that lies outside the clipping boundary.

The following exercise demonstrates how to use the Select Polyline feature of the XCLIP command to define the xref clipping boundaries with an irregular polygon.

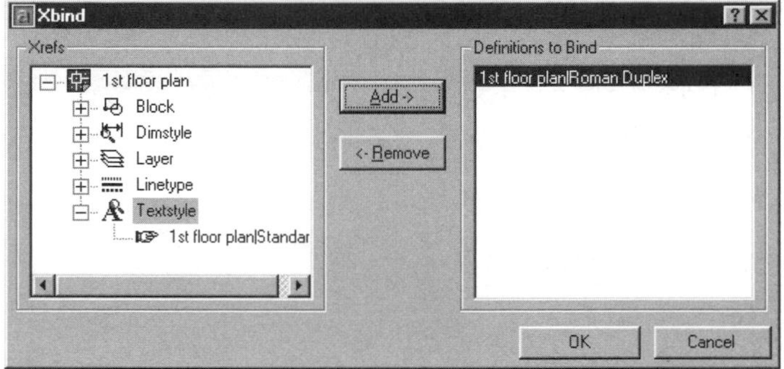

Figure 13–8 *The XBIND command allows you to insert named objects such as text styles from xref-ed drawings into the current drawing.*

Note: AutoCAD always prefixes bound named objects with their original drawing's filename.

Note: The Xbind feature does not permit you to load view objects from the attached drawing.

EXERCISE: USING THE SELECT POLYLINE FEATURE OF THE XCLIP COMMAND

1. Open the 13DWG02B.DWG drawing file on the accompanying CD. The 13DWG02A.DWG drawing file is already attached as an xref.
2. From the Modify menu, choose Clip>Xref.
3. Select the xref and then press Enter.
4. Press Enter to accept the default for a new boundary.
5. Type **S** to choose Select Polyline.
6. Select the large, green polyline.

   AutoCAD determines the limits of the clipping boundary and then redisplays only the portion of the xref that is inside the clipping boundary, as shown in Figure 13–9.
7. Keep the drawing open for the next exercise.

This exercise demonstrated how to use the Select Polyline feature of the XCLIP command. In some cases, however, defining only one clipping boundary for the xref may not be enough. The following section leads you through the steps necessary to create multiple boundaries.

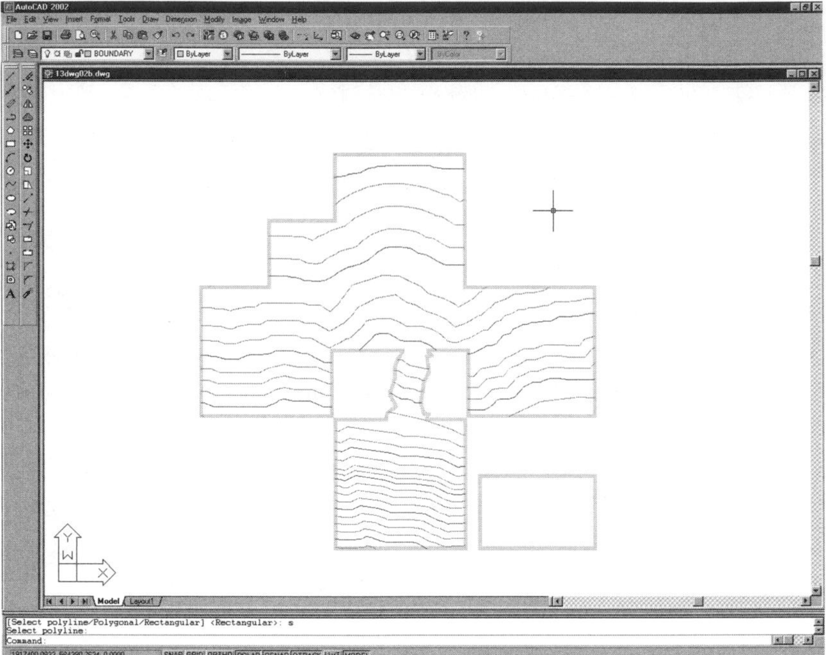

Figure 13–9 *The xref is clipped using the polyline.*

Creating Multiple Clipping Boundaries

One limitation of the XCLIP command is that an xref can have only one clipping boundary. But what if you want to clip the same xref with more than one polygon? How do you create multiple clipping boundaries? One answer is to insert the same xref more than once.

The following exercise demonstrates how to use two separate polygons to create two clipping boundaries for the same xref.

EXERCISE: CREATING MULTIPLE CLIPPING BOUNDARIES FOR THE SAME XREF

1. Continue with the 13DWG02B.DWG drawing from the previous exercise.
2. Choose Copy from the Modify pull-down menu and select the xref 13DWG02A, which has been clipped.
3. At the "Specify base point or displacement, or [Multiple]:" prompt type @ and press Enter twice.
4. AutoCAD "attaches" the xref again, creating a duplicate xref on top of the existing clipped xref.

Next you will use the Select Polyline feature of the XCLIP command to clip the xref by selecting the small rectangular polygon.

5. From the Modify menu, choose Clip>Xref.
6. Select one of the xrefs in the large area and then press Enter.

   Because the first xref has been clipped, it can be selected only from inside the large polygon.
7. Press Enter to accept the default, New boundary.
8. At the "Delete old boundary(s)? [Yes/No] <Yes>:" prompt press Enter to accept the default, Yes.
9. Enter **S** to choose Select Polyline.
10. Select the small, green rectangle.
11. You may now close the drawing without saving it.

AutoCAD determines the limits of the clipping boundary and then redisplays only the portion of the copied xref that is inside the small rectangular clipping boundary, as shown in Figure 13–10.

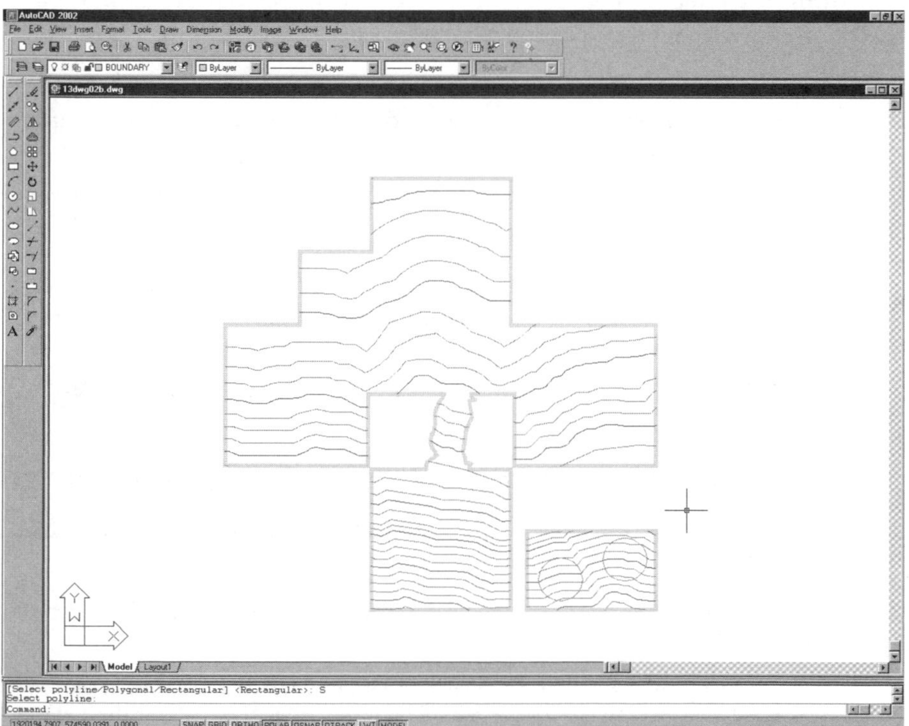

Figure 13–10 *The same xref is inserted twice, and each xref is clipped separately.*

Tip: As an alternate method of inserting a copy of an xref, you could use the RENAME command to rename the original xref insertion, then insert the xref again and then use the XCLIP command on the new insertion. This method would then create two uniquely named xrefs, which would be useful for modifying different xref layer properties.

Tip: The xref can be renamed from the Xref Manager dialog box. Simply highlight the xref to be renamed and then press F2. After you change the name, press Enter.

DEMAND LOADING

Demand loading works in conjunction with layer and spatial indexes, and it reduces regeneration times. When you enable demand loading when you use xrefs, AutoCAD loads only specific objects from the xref into the current drawing. By loading only a portion of the xref's objects, the number of objects in the current editing session is minimized, thereby increasing AutoCAD's performance.

Demand loading is controlled by a system variable named XLOADCTL, which controls how AutoCAD uses layer and spatial indexes that may exist in xrefs. When you enable demand loading (by setting XLOADCTL to either 1 or 2), AutoCAD loads only objects on layers that are thawed when the xref has layer indexes, and only objects within the clipping boundary when the xref has spatial indexes. Table 13-1 shows the variable's three settings and their effects.

In most networked environments, a setting of zero is preferred. With today's newer systems, the timesavings provided by a setting of 1 is negated by the problems with xref-ing drawings that others have open or have xref-ed. A setting of 2 can be useful due to its timesavings, but it can result in increased hard-disk activity through the duplication of the DWG file.

Note: The XLOADCTL system variable's settings may be controlled through the Options dialog box, from the Open and Save folder. For more information, refer to Chapter 3, Fine-tuning the Drawing Environment, in the section titled External References.

Table 13–1: *The Demand Loading Settings*

Setting	Effect
0 – Disabled	Turns off demand loading.
1 – Enabled	Turns on demand loading and prevents other users from editing the drawing file while it is xref-ed.
2 – Enabled with copy	Turns on demand loading and creates a copy of the drawing that it xrefs. This allows other users to edit the xref's original drawing.

LAYER AND SPATIAL INDEXES

In the previous section you learned about the system variable that controls demand loading. AutoCAD has another system variable, called INDEXCTL, that controls layer and spatial indexing and works in conjunction with the demand loading system variable, XLOADCTL.

When layer indexing is enabled, AutoCAD does not load xref objects residing on layers that are frozen in the external reference drawing. When spatial indexing is enabled, AutoCAD will not load xref objects that reside outside of the clip boundary. In both cases, fewer objects are brought into the current drawing, and regeneration times are reduced.

When you enable the INDEXCTL system variable, AutoCAD can enhance its performance by reducing the regeneration times of drawings with xrefs. Table 13-2 shows the variable's four settings and their effects.

Spatial indexes work three-dimensionally by defining a front and back clipping plane. The front and back clipping planes are defined via the XCLIP command's Clipdepth option. By creating a clipping boundary and specifying the clip depth, you can greatly limit the number of xref objects that AutoCAD loads into the current drawing session.

Layer and spatial indexes are created in a drawing when the INDEXCTL system variable is set to the desired value and the drawing is then saved. If INDEXCTL is set to 3, for example, both layer and spatial indexes are created when the current drawing is saved. The indexes are saved with the drawing file. Consequently, if you attach the drawing as an xref to a new drawing that has demand loading enabled, AutoCAD uses the xref's layer and spatial indexes to load only those objects that are on thawed layers and lie within the clipping boundary.

Note: Setting the INDEXCTL system variable to a value other than 0 enables layer or spatial indexing (or both). Consequently, when the drawing is saved, AutoCAD adds to it the additional layer and spatial index data, thereby increasing the drawing's file size.

Table 13–2: *The Layer and Spatial Indexes' Settings*

Setting	Effect
0 - None	Both layer and spatial indexing disabled
1 - Layer	Only layer indexing enabled
2 - Spatial	Only spatial indexing enabled
3 - Layer & spatial	Both layer and spatial indexing enabled

Tip: Leave the INDEXCTL system variable set to its default value of zero to help keep the drawing's file size minimal. Set the variable to a value other than zero only when the file you are saving is to be used as an xref, and the layer and spatial indexes will be used by the Demand Loading feature.

TRACKING XREFS

The advantage of using xrefs is that they provide the capability to create composite drawings that have relatively small file sizes and are easily updated. Unfortunately, on large projects involving multiple disciplines, keeping track of xref drawings can be difficult. Proper xref management is critical to ensure that composite drawings can find the latest versions of xrefs on standalone stations or over networks. Features available in AutoCAD can make managing xrefs easier.

THE XREF MANAGER

The Xref Manager makes the task of managing xrefs easier. The dialog box's diagrams and intuitive button commands are great visual aids, as is its display of such pertinent data as the xref's name and path, current load status, whether the xref is attached or overlayed, and the xref's file size and last modification date.

Displaying Xrefs with List View versus Tree View

When the Tree View feature is selected, it displays any nested xrefs that may exist and a diagram of the hierarchy of xrefs. This display makes it easy to see which xrefs have been attached and how they relate to one another.

Note: The Tree View feature instantly displays a visual diagram of xrefs and any nested xrefs. More importantly, the nested xrefs are actually shown attached to their parent xref.

The following exercise demonstrates the Tree View feature.

EXERCISE: ACCESSING THE TREE VIEW DISPLAY

1. Open the 13DWG03A.DWG drawing file on the accompanying CD.

   The drawing contains two xrefs, each of which also contains two xrefs. When the drawing opens, the hierarchy of the xrefs and nested xrefs is displayed on the screen, as shown in Figure 13–11.

2. From the Insert menu, choose Xref Manager. The Xref Manager dialog box opens. Initially, it opens in list view mode. Two buttons appear in the upper-left corner of the dialog box. The one on the left, the List View button, is grayed. The one on the right is the Tree View button.

3. Click on the Tree View button.

   The text box below the buttons changes and now displays the hierarchy of the xrefs, as shown in Figure 13–12. From this display you can easily manage the xrefs. For example, you can unload a nested xref that is no longer needed.

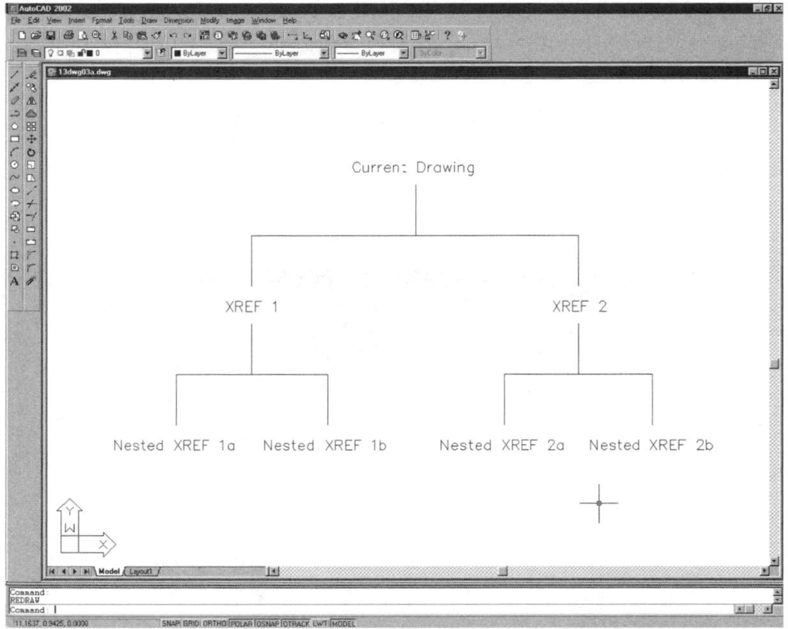

Figure 13–11 *The hierarchy of the xrefs and nested xrefs.*

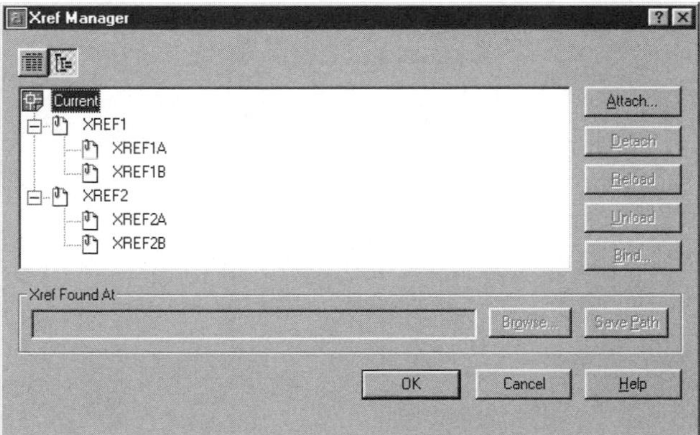

Figure 13–12 *The Xref Manager in tree view mode.*

Tip: In list view mode you can sort the xrefs in the display box in ascending or descending order. This is true for any of the displayed data. To sort the data, choose a column's title bar. AutoCAD sorts the data in ascending order based on the selected column. Select the column's title bar again to sort the data in descending order.

4. In the Xref Manager, choose XREF1A. Several buttons in the Xref Manager become active, and the xref's path and drawing file name appear, as shown in Figure 13–13.
5. Choose the Unload button, and then choose OK.

AutoCAD unloads the nested xref XREF1A and redisplays the drawing, as shown in Figure 13–14.

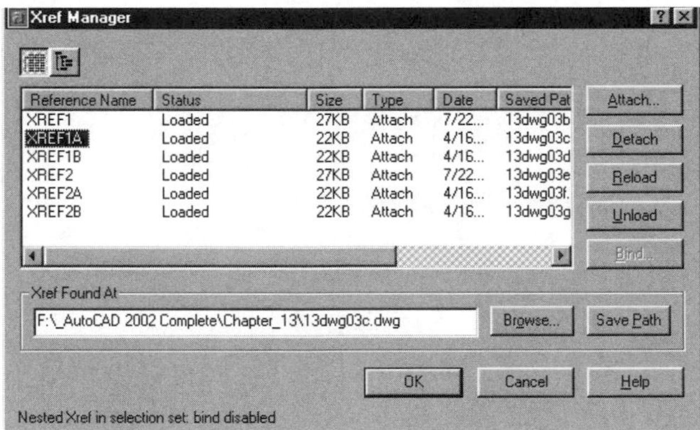

Figure 13–13 *The Xref Manager's buttons are activated when XREF1A is selected.*

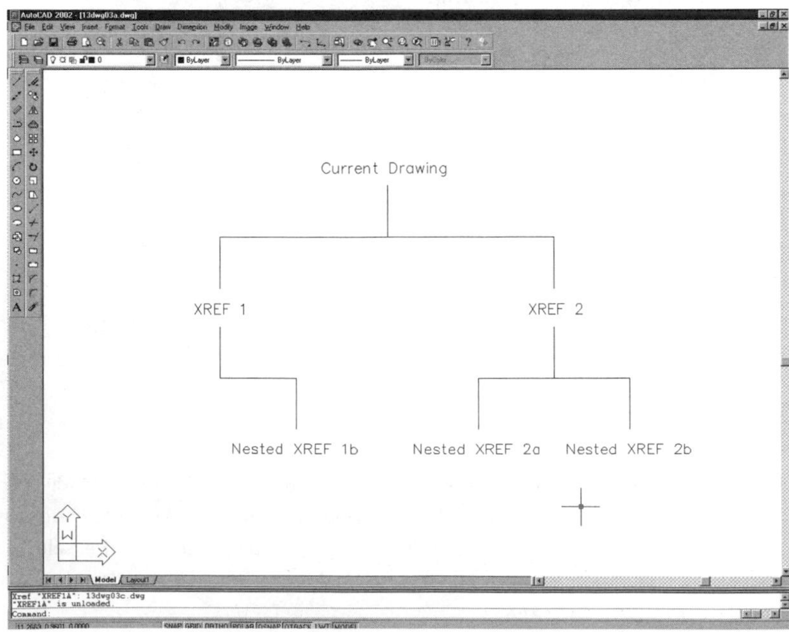

Figure 13–14 *The tree view illustrated in the current drawing shows the drawing after XREF1A is unloaded.*

Tip: The best way to eliminate the display of unwanted nested xrefs is to overlay an xref, but you can achieve the same effect—reducing regeneration time—by unloading an unwanted nested xref.

Note: In releases previous to AutoCAD 2000, xref-dependent layer names in the current drawing could be only thirty-one characters long. Because AutoCAD would add the xref's drawing name to its layer names, if the layers names were long, AutoCAD would abort loading the xref. Since the release of AutoCAD 2000, layer names can be 255 characters long. More importantly, the xref's drawing name does not count toward the 255-character limit.

Note: It is not recommended that you use long layer names if the data is to be used by Release 14 or earlier, because the long names cannot be read properly by the older versions.

EXPLICIT AND IMPLICIT XREF PATHS

An xref's path can be defined explicitly or implicitly. Explicit paths include the entire directory path and end with the xref's file name. Implicit paths contain only a partial subdirectory path and the xref's file name. AutoCAD saves both explicit and implicit path data with the drawing.

There is an advantage to saving the xref paths implicitly. If the drawing is opened on another workstation, AutoCAD will successfully resolve the xref as long as the implicit path hierarchy exists at the new workstation. In contrast, if the path is explicitly defined, AutoCAD will probably not find the xref file.

For example, suppose that a drawing lies in the following directory:

G:\WORK\JOB-ONE\13DWG07A.DWG

Also suppose that this drawing has an xref attached that lies in the following directory:

G:\WORK\JOB-ONE\XREFS\13DWG07B.DWG

Both of these xref paths are explicitly defined. Notice that the entire drawing's path is shown, including its root directory. This means that while you edit the 13DWG07A.DWG file at the workstation on which it resides, AutoCAD can successfully resolve the xref because it will find it in the explicit path. In other words, the xref file is still located on the G drive, and in the subdirectory shown.

But what happens if the drawing and xref are moved to another workstation? Suppose that the files are moved to the following hard drive and directory:

C:\ACAD2002\SAMPLE\13DWG04A.DWG

C:\ACAD2002\SAMPLE\XREFS\13DWG04B.DWG

The 13DWG04A.DWG file can still be opened in AutoCAD on the new workstation once it is located, but if the xref's path is not in the explicitly defined path, AutoCAD issues the following error message:

Resolve Xref XREF1: D:\WORK\JOB-ONE\XREFS\13DWG04B.DWG

Can't find D:\WORK\JOB-ONE\XREFS\13DWG04B.DWG

AutoCAD indicates that it cannot find the xref indicated by the explicitly defined path. Consequently, AutoCAD opens the 13DWG04A.DWG drawing without resolving the attachment of the xref (which it could not find). To avoid the problem of unresolved xrefs, you can redefine the xref's path implicitly, as described in the following exercise.

EXERCISE: IMPLICITLY DEFINING AN XREF'S PATH

1. Create a new directory folder called XREFS in the ACAD2002\SAMPLE subdirectory.
2. Copy the 13DWG04A.DWG drawing file from the accompanying CD into the ACAD2002\SAMPLE subdirectory.
3. Copy the 13DWG04B.DWG drawing file from the accompanying CD into the ACAD2002\SAMPLE\XREFS subdirectory.
4. Open the 13DWG04A.DWG drawing file from the ACAD2002\SAMPLE directory.

   The drawing opens and then issues the warning that it cannot find the xref. (By pressing F2, you can toggle the AutoCAD text window on and view the information.)
5. From the Insert pull-down menu, choose Xref Manager. The Xref Manager is displayed, as shown in Figure 13–15. Notice that the XREF1 drawing file's status is Not Found. Also notice that AutoCAD looked for the xref in the explicitly defined path location listed in the Saved Path column.
6. Select the reference name XREF1. The Xref Found At text box becomes active.
7. Choose the Browse button. The Select New Path dialog box opens.
8. Open the 13DWG04B.DWG drawing file from the ACAD2002\SAMPLE\XREFS directory.

   The Xref Manager is displayed again. The xref's path is now displayed in the Xref Found At text box, as shown in Figure 13–16. Notice that the XREF1 status is still Not Found.
9. Choose the Reload button. The XREF1 status changes to Reload.

   Next you will redefine the xref's path implicitly.
10. In the Xref Found At text box, highlight the beginning portion of the path, from the root directory listing up to the Xrefs folder, as shown in Figure 13–17.

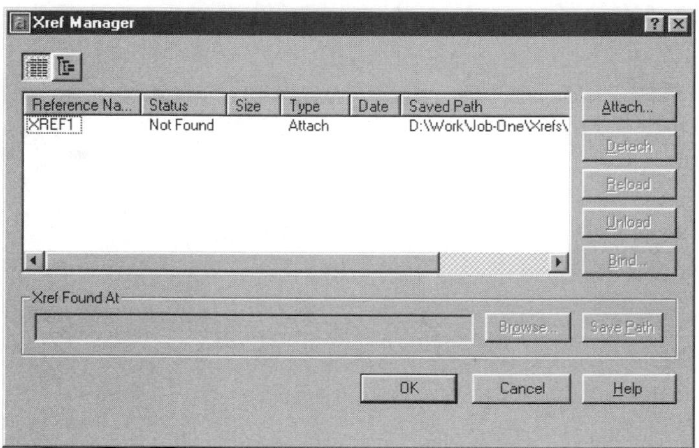

Figure 13–15 *AutoCAD cannot find the xref using its explicitly defined path.*

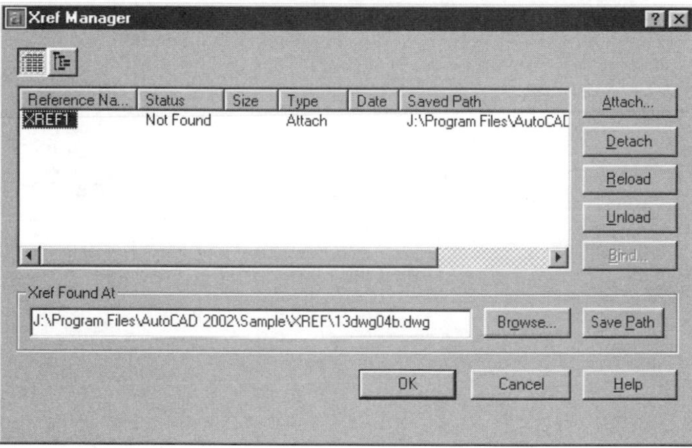

Figure 13–16 *The XREF1 file's location is displayed in the Xref Found At text box.*

11. Press the Delete key. AutoCAD deletes the highlighted portion of the path.
12. Press Enter to save the modified path as the implicit path. AutoCAD redefines the path as the implicit path and displays it in the Saved Path column, as shown in Figure 13–18.
13. Choose OK.
14. You may close the drawing without saving your changes.

AutoCAD resolved the xref and loaded it into the current drawing. The reason AutoCAD was able to resolve XREF1 using the implicit path is because the 13DWG04A.DWG and the XREFS folder were in the same folder. In this exercise, both were in the SAMPLE folder.

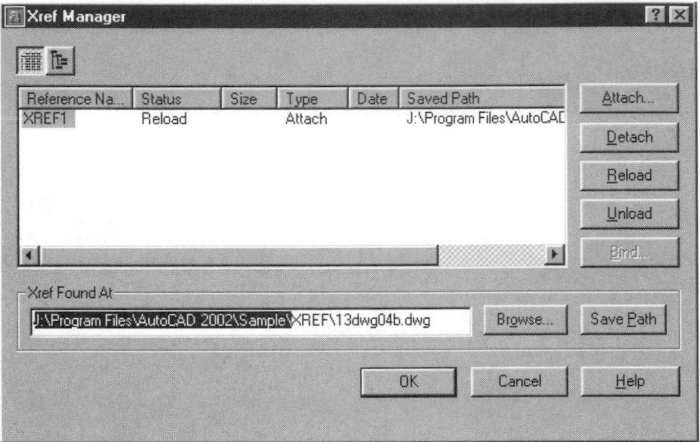

Figure 13–17 *The beginning portion of the XREF1 file's path is selected and then deleted.*

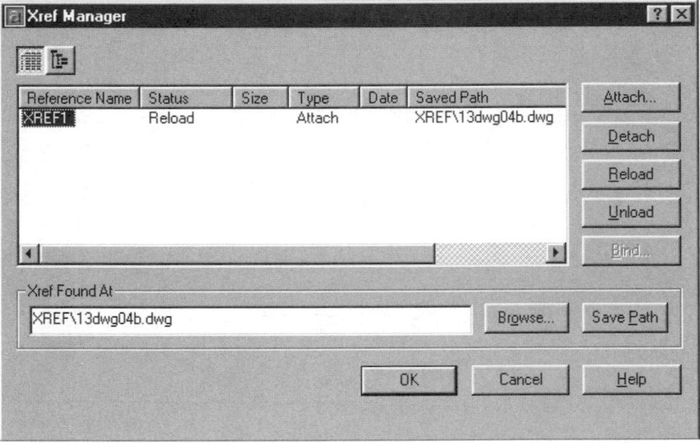

Figure 13–18 *The XREF1 file's path is redefined as implicit.*

With the xref's path now implicitly defined and saved with the drawing, if the 13DWG04A.DWG file is moved to a different folder, and the 13DWG04B.DWG is moved and placed in a subfolder named XREFS, AutoCAD will resolve the xref when the 13DWG04A.DWG file is opened. This is true as long as 13DWG04A.DWG and the XREFS subfolder are placed in the same folder.

By using the preceding technique to define xref paths implicitly, you can avoid the problem of unresolved xrefs when you transfer drawing files from one workstation to another.

Tip: You can use the PROJECTNAME system variable to assign xref search paths to your workstation. By assigning the search paths to a project name, you can resolve xrefs even though their paths are defined explicitly or implicitly. For more information, refer to the section titled Using PROJECTNAME to Specify Xref Search Paths in this chapter.

CIRCULAR XREFS

Circular xrefs occur when two drawings are inserted as xrefs into each other. For example, suppose you have drawing A and drawing B. A circular xref occurs when drawing A is attached as an xref into drawing B, and then drawing B is attached as an xref into drawing A.

In releases prior to Release 14, AutoCAD would issue a warning about the circular reference and abort the XREF command. Now, AutoCAD loads the xref up to the point where the circularity exists. It stops at the point of circularity because a drawing cannot load itself as an xref.

USING PROJECTNAME TO SPECIFY XREF SEARCH PATHS

The PROJECTNAME system variable allows you to assign a project name to a drawing. The project name can also be declared in AutoCAD on each workstation and be assigned search paths for xrefs. By assigning xref search paths to a project name on a workstation, you can load a drawing that has the same project name assigned to its PROJECTNAME variable, and thereby use the workstation's search paths to resolve xrefs.

AutoCAD saves the project name in the workstation's system registry and includes the search paths assigned to the project name. AutoCAD also saves the project name to the current drawing, but it does not save the search paths in the drawing. What this means is that a particular project name can be assigned to a drawing that has xrefs attached, and each CAD technician can assign different xref paths to the same project name on their workstation. When the drawing is opened, no matter what xref paths are defined in the drawing, each workstation will resolve the drawing's xrefs. This feature avoids the problem of managing xref paths, either explicitly or implicitly, in order to resolve xrefs.

This feature also allows you to create on your workstation multiple project names, each of which can contain a specific set of xref search paths. You can add as many paths as necessary to each project name, and you can create as many project names as you need.

Note: When AutoCAD searches for xrefs, it first searches for xrefs in the current drawing's folder, then it searches using explicitly and implicitly defined paths. Next it searches using the current PROJECTNAME search paths; finally, it uses AutoCAD's default search paths.

When a project name is highlighted as shown in Figure 13–19, when you choose the Set Current button, you set the current drawing's PROJECTNAME variable to the highlighted project name. The highlighted project name's search paths then become the current paths AutoCAD uses to find xrefs.

XREF LAYERS, COLORS, LINETYPES, AND LINEWEIGHTS

When an xref is attached to the current drawing, AutoCAD duplicates the xref's layer names in the current drawing. AutoCAD prefixes the layer names with the xref's name, followed by the pipe symbol (|). Then AutoCAD assigns these new layers the same colors, linetypes, and lineweights as those in the xref drawing.

The only time AutoCAD does not assign the same colors, linetypes, and lineweights as those in the xref is when the objects are created on layer 0 in the xref. Just like blocks, these xref objects have special properties. If their color, linetype, and lineweight properties are set to BYLAYER, they assume the color, linetype, and lineweight of the layer on which the xref is inserted. If their color, linetype, and lineweight properties are set to BYBLOCK, they assume the color, linetype, and lineweight properties that are currently defined for the creation of new objects in the current drawing, as displayed in the Object Properties toolbar. Finally, if their color, linetype, and lineweight properties are explicitly defined in the xref, those properties remain fixed.

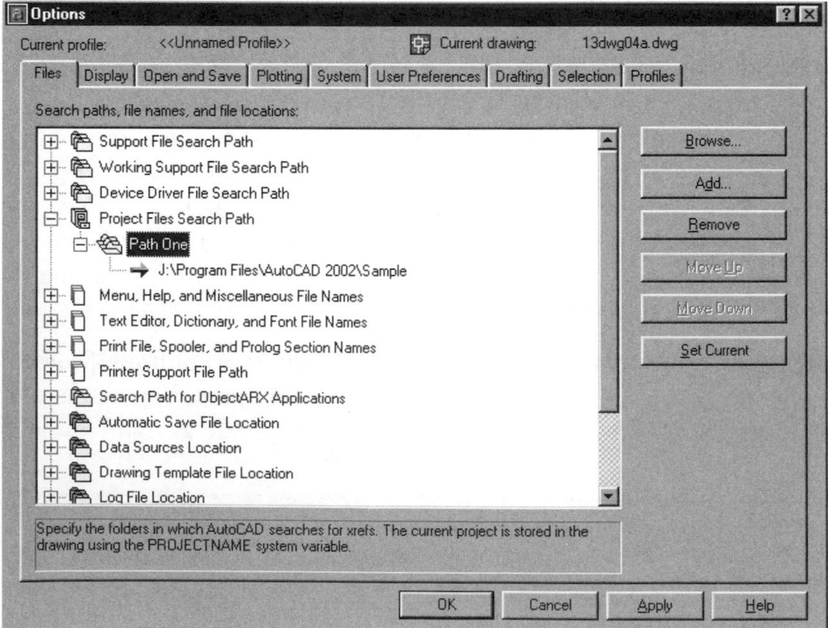

Figure 13–19 *Highlighting the project name and choosing the Set Current button sets the drawing's PROJECTNAME system variable.*

Except for colors, linetypes, and lineweights assigned explicitly in the xref, you can change the color, linetype, and lineweight of an xref's layers from the Layer Manager. These changes appear in the current drawing and do not affect the color, linetype, and lineweight in the original xref file. After you exit the drawing, however, any changes to the color, linetype, and lineweight properties are lost. When the drawing is opened again, the colors, linetypes, and lineweights assume the settings in the original xref.

Tip: To save changes you make to an xref layer's color, linetype, and lineweight properties with the current drawing, set the system variable VISRETAIN to 1. When you open the drawing, the changes you made to the xref layer's color, linetype, and lineweight properties in a previous editing session are restored.

IN-PLACE REFERENCE EDITING

AutoCAD 2000 introduced a powerful feature called In-Place Reference Editing. This feature allows you to edit objects in an inserted xref file from the current drawing and then save the changes back to the xref file. You can use in-place reference editing to quickly edit an inserted xref without having to open the xref file.

In-place reference editing is intended to allow you to make modest edits to an xref file from the current drawing. While you can make significant changes to the xref file from the current drawing, it is more efficient to open the xref file to directly perform significant edits. Using in-place reference editing to make significant changes to xref files can temporarily increase the current drawing's file size, thereby increasing regeneration times and slowing productivity. Therefore, use this new feature to make modest edits to xrefs quickly.

Once you select the xref, you specify the objects you want to edit. AutoCAD temporarily brings the selected objects into the current drawing for modification and makes them part of a working set. Once the objects are modified, the working set of objects is saved back to the xref file.

The following exercise demonstrates how to use in-place reference editing.

EXERCISE: EDITING AN XREF WITH IN-PLACE REFERENCE EDITING

1. Copy the files 13DWG05A.DWG and 13DWG05B.DWG from the accompanying CD to the ACAD2002\SAMPLE subdirectory.

   Before you continue with this exercise, you must clear the read-only attribute so you can edit and save changes back to the files.

2. From Windows Explorer, select the 13DWG05A.DWG and 13DWG05B.DWG drawing files in the ACAD2002\SAMPLE directory, and then right-click to display the shortcut menu.
3. From the menu, choose Properties. The Properties dialog box is displayed.

4. From the Properties dialog box, clear the Read-Only check box, then choose OK. The read-only attribute is cleared from the files.

   Next you will start a new drawing and insert the 13DWG05A.DWG file as an xref.

5. From AutoCAD, start a new drawing from scratch.
6. From the File menu, choose Save. AutoCAD displays the Save Drawing As dialog box.
7. Name the file MYFILE and save it to the ACAD2002\SAMPLE subdirectory.
8. From the Insert menu, choose External Reference. The Select Reference File dialog box is displayed.
9. Open the 13DWG05A.DWG drawing file from the ACAD2002/SAMPLE directory. AutoCAD displays the External Reference dialog box.
10. In the External Reference dialog box, under Reference Type, choose Attachment.
11. In the Insertion point area, the Scale area, and the Rotation area, clear any checked Specify On-screen check boxes.
12. Choose OK. AutoCAD attaches the drawing and displays two text objects inside a rectangle object.

    The upper text object is located in the 13DWG05A.DWG xref file. The lower text object is located in the 13DWG05B.DWG file and is a nested xref.

13. Zoom in close to the objects to view them better.
14. From the Modify menu, choose In-place Xref and Block Edit>Edit Reference. AutoCAD prompts you to select a reference.
15. Select the lower text object. AutoCAD displays the Edit Reference dialog box, which shows the two xref file names.

    Because the 13DWG05B.DWG file is nested within the 13DWG05A.DWG file, and because you selected the text object that is in the nested xref, AutoCAD displays both names. At this point you will select the file whose objects you wish to edit.

16. In the Reference name area, choose the 13DWG05B.DWG file. AutoCAD highlights the selected xref name, as shown in Figure 13–20.
17. Make sure the Enable Unique Layer and Symbol Names check box is selected and that the Display Attribute Definitions for Editing check box is cleared.
18. Choose OK. AutoCAD prompts you to select the nested objects.

Note: AutoCAD assigns temporary layer names to the current drawing when the Enable Unique Layer and Symbol Names check box is selected.

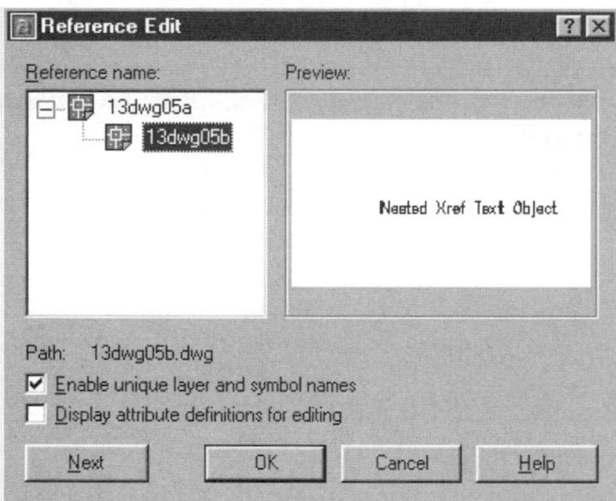

Figure 13–20 *The Edit Reference dialog box allows you to select the xref to edit.*

19. Choose the lower text object, then press Enter to end object selection. AutoCAD visually moves the selected object into the current drawing and displays the Refedit toolbar.

    Notice that AutoCAD fades objects that are not being edited, as shown in Figure 13–21. This makes it easier to identify the objects you selected for editing.

    Next you will edit the lower text object and save the changes back to its xref.
20. Select the lower text object. AutoCAD highlights the object.
21. Right-click to display the shortcut menu, then choose Text Edit. AutoCAD displays the Text Edit dialog box.
22. In the text box, replace the highlighted text with the word **MODIFIED**.
23. Choose OK, then press Enter to end object selection. AutoCAD modifies the text object.
24. Choose the Save Back Changes to Reference button, as shown in Figure 13–22. AutoCAD issues a warning noting that all reference edits will be saved.
25. Choose OK. AutoCAD saves the changes back to the xref file and then exits the reference edit mode.
26. You may close the drawing without saving your changes.

At this point, if you open the 13DWG05B.DWG drawing from the ACAD2002\SAMPLE directory, you will notice that the text object has been changed to MODIFIED.

Note: You can adjust the fading intensity from the Display folder found in the Options dialog box, which you display by choosing Options from the Tools menu.

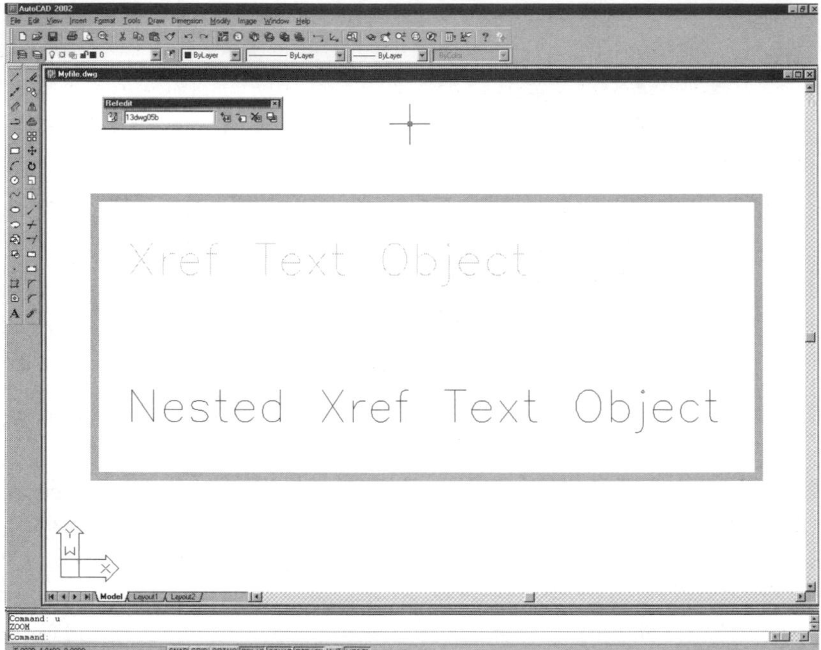

Figure 13–21 *AutoCAD fades objects that are not being edited.*

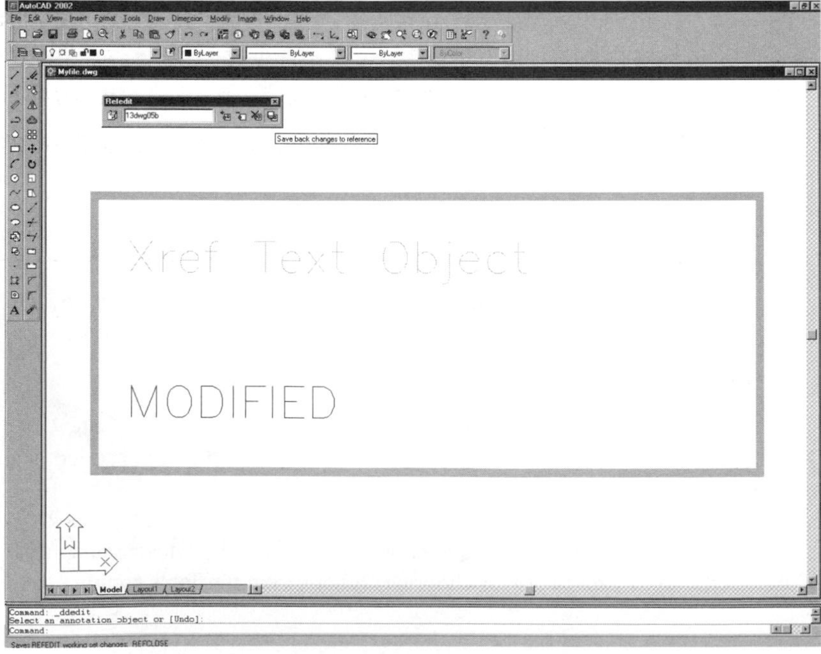

Figure 13–22 *The edits to xref objects are saved back to their original xref file.*

AutoCAD allows you to add or remove additional objects to the working set. This is done by choosing the Add Objects to Working Set and Remove Objects from Working Set buttons, respectively, that are located on the Refedit toolbar. If you decide you do not want to save edits back the original xref, you can choose the Discard Changes button.

While in reference edit mode, if you create new objects, they are almost always added to the working set. There are some situations in which this is not true, such as when AutoCAD generates an arc object during the FILLET command.

While AutoCAD fades objects that are not part of the working set, the objects may still be edited, so caution should be taken. For example, if the MyFile drawing contained another text object, you could edit its text string. The changes you made to it, however, would be saved in the current drawing only.

SUMMARY

In this chapter you learned about AutoCAD's Partial Open, Partial Load, and In-Place Reference Editing features. You also learned about the differences between attaching and overlaying xrefs, and between binding and "xbinding" xrefs. You reviewed how to create clipping boundaries, and you learned how to increase productivity with demand loading and spatial and layer indexes. The way AutoCAD deals with circular xrefs was covered, as was the PROJECTNAME system variable and the way it stores xref's paths. You also learned about the Xref Manager and how it is used to manage xrefs and nested xrefs.

AutoCAD's xref capabilities are a powerful tool. You can save regeneration time and increase your productivity by using these xref features to manage xrefs better.

Extracting Information from Drawings

AutoCAD stores information about every object in a drawing behind the scenes in the drawing database. The type and extent of data depends on the particular object. As an example, a listing of the data associated with a circle would include layer; color; linetype; linetype scale; plot style; lineweight; thickness; X,Y,Z coordinates for its center point; radius; diameter; circumference; and area. The data for other objects would include block names, insertion points, rotation angles, endpoints, and volumes of 3D objects. Block attributes are a very special kind of data—user-defined textual information attached to blocks. AutoCAD 2002 introduces new tools that let you easily extract this block attribute information from drawings. To realize the full power of AutoCAD, you need to mine the hidden data in your drawings.

This chapter covers the following topics:

- Object Properties toolbar
- Inquiry tools
- Understanding block attributes
- Block Attribute Manager
- Enhanced Attribute Editor
- Extracting block attributes

QUERYING OBJECTS

The capability to query a drawing and its objects for data is a powerful feature of AutoCAD. Many of the querying tools are grouped together for easy access in the pull-down menus. In addition, the Object Properties toolbar displays many common properties of selected objects. We will explore the Object Properties toolbar first.

OBJECT PROPERTIES TOOLBAR

AutoCAD's Object Properties toolbar (see Figure 14–1) automatically displays a selected object's layer, color, linetype, lineweight, and plot style—when appropriate. As objects are added to the selection set, only properties that are the same for all selected objects are listed. When the selected objects have different properties, such as different colors, the property value is left blank.

Note: The Properties dialog box (see Figure 14–2) gives you an instant look at much of an object's hidden data. After you select an object, AutoCAD offers four methods for starting the Properties dialog box: double-clicking on the object, choosing Properties from the right-click menu, choosing Modify>Properties from the pull-down menu, or typing DDMODIFY at the command prompt. The Properties dialog box also lets you directly edit many of the listed properties. See Chapter 10, Advanced Editing, for more details.

INQUIRY TOOLS

AutoCAD's collection of inquiry tools provide information such as area; length; mass properties; and point list of objects; as well as the X,Y,Z coordinates of points. The inquiry tools are conveniently grouped together in a single location. From the pull-down menu choose Tools>Inquiry. AutoCAD displays all of its querying tools (see Figure 14–3). To access one of the tools, simply click on it.

Distance

The DIST command reports the distance between any two selected points in a drawing. When you use the DIST command, remember that it measures distances three-dimensionally. If you select two points with different Z values, the total distance will be based on the 3D vector between the two points. To ensure that the distance is based on the current two-dimensional UCS, use X,Y,Z point filters when you select the two points.

Note: The last distance queried by using the DIST command is saved as a system variable. To view it, type **DISTANCE** at the command prompt.

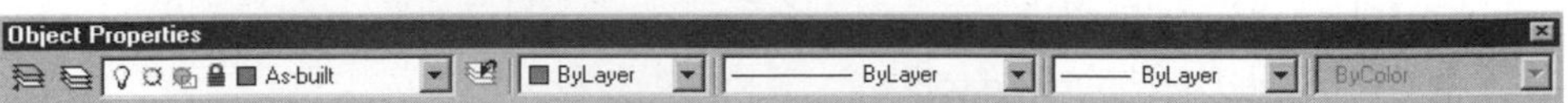

Figure 14–1 *The Object Properties toolbar displays many common properties of selected objects.*

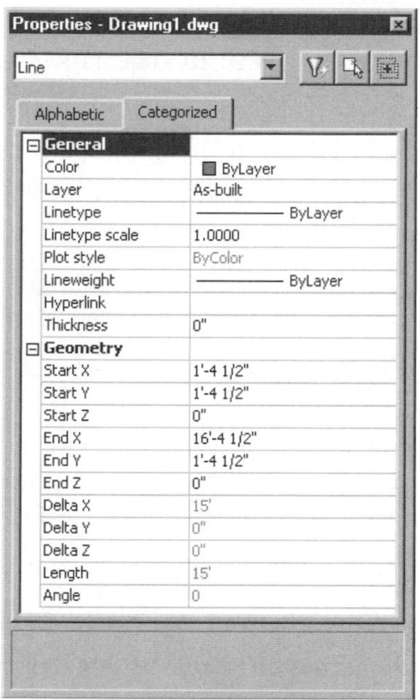

Figure 14–2 *The Properties dialog box displays even more properties of selected objects.*

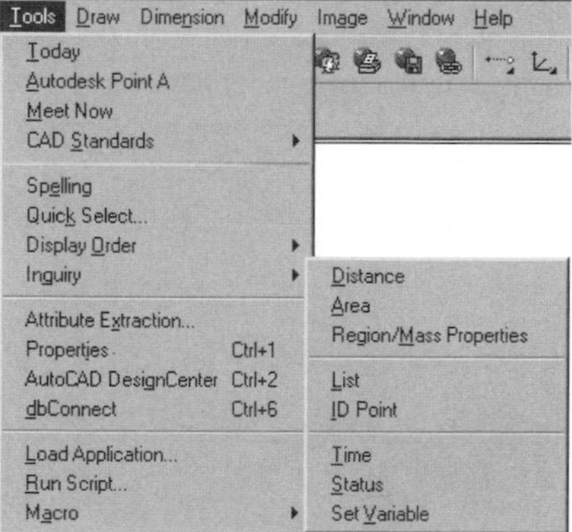

Figure 14–3 *AutoCAD's querying tools can be found on the Inquiry menu of the Tools pull-down menu.*

QUERYING AREAS OF BLOCKS AND XREFS

The AREA command works with most AutoCAD objects, including circles, ellipses, splines, regions, polylines, and polygon objects. Using the AREA command's Object option, AutoCAD calculates and lists the object's area and perimeter.

If you select an object that is inserted as a block or xref, however, AutoCAD warns you that the selected object does not have an area. How can you calculate the area of objects inserted as blocks or xrefs?

You can find the area of blocks and xrefs using the BOUNDARY command. With this command, AutoCAD creates a region object that you can then use to get area and perimeter information.

The next exercise demonstrates how to calculate the area of objects contained within blocks and xrefs.

EXERCISE: FINDING THE AREA OF XREF OBJECTS

1. Load the drawing file 14DWG01.DWG from the accompanying CD-ROM. The drawing contains four objects. The mtext object is part of the current drawing, while the other three objects are part of an xref (14DWG01A.DWG).
2. From the Draw menu, choose Boundary. The Boundary Creation dialog box appears.
3. From the Object Type drop-down list, select Region.
4. Click on the Pick Points button; the dialog box disappears to let you select points.
5. Pick inside the square. AutoCAD determines a boundary based on the square polygon and then highlights it.
6. Select the interior of the remaining xref objects. Pick inside the circle but outside the inner square, and then pick the area shared by the circle and the triangle. Finally, pick inside the triangle but outside the circle. When you finish picking inside the xref objects, all objects will be highlighted, as shown in Figure 14–4.

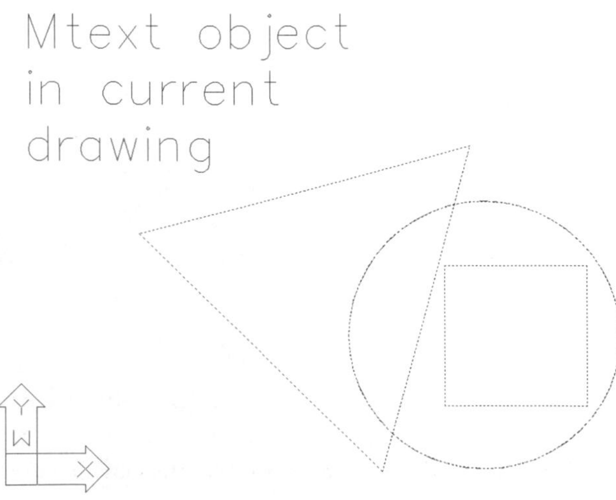

Figure 14–4 *The three objects from the xref are highlighted after object selection.*

7. Press Enter to end object selection.

   The BOUNDARY command created five regions. It created two regions for the square: one because you picked inside the square, and the second because you picked inside the circle but outside the square. The BOUNDARY command's Island Detection option, which was enabled in the Boundary Creation dialog box, automatically detected and excluded the interior square when it created the second boundary.

   You can now use the AREA command to calculate the area and perimeter of each region. You can also create a single composite region object and calculate its area. This is accomplished by using Boolean operations.

8. From the Modify menu, choose Solids Editing> Union.
9. At the "Select objects:" prompt, type **F**. This begins the Fence selection method.
10. Select all the region objects except the square by picking inside the circle as shown in Figure 14–5.
11. Drag the fence line left, pick inside the triangle, and then press Enter to complete the command.

    AutoCAD creates a new composite region by combining the three smaller regions. You can calculate the new region's area and perimeter using the AREA command.

    Next you will subtract the remaining circle region from the composite region.

12. From the Modify menu, choose Solids Editing>Subtract. AutoCAD prompts you to select the solids and regions from which to subtract a region.

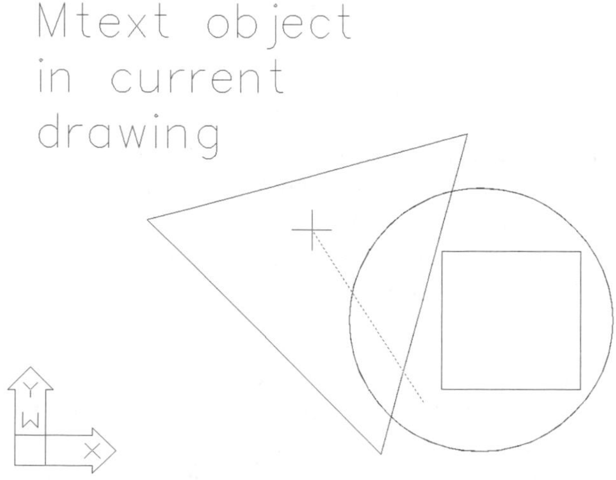

Figure 14–5 *Use the Fence selection method to choose all of the region objects except the square.*

13. Choose the composite region and press Enter. AutoCAD then prompts you to select the solids and regions to subtract.
14. Using the fence selection method, choose the square region, and then press Enter. AutoCAD subtracts the square region from the composite region, creating a new composite region.

    Next you will use the AREA command to calculate the area and perimeter of the new composite region.
15. From the Tools menu, choose Inquiry>Area.
16. When prompted, type **O** for Object, and then choose the composite region. AutoCAD calculates the region's area and perimeter and displays them as follows:

    Area = 845166.7464, Perimeter = 3979.1006

This example showed you how to quickly find the area and perimeter of inserted block and xref objects.

UNDERSTANDING BLOCK ATTRIBUTES

One of the most powerful capabilities of AutoCAD block references is their ability to carry attached text data in the form of attributes. This data can be a constant value, defined during the creation of the block, or it can be input by the user when the block is inserted into a drawing. The attribute data can be edited and extracted at any time. Each block can include an unlimited number of attributes.

Tip: A drawing title block is an ideal place to use attributes. When the title block is inserted into the current drawing, it prompts the user for information particular to the drawing, such as sheet number, drawing title, the project engineer's name, the CAD technician's name, and so on. This guarantees that the appropriate data is added to the drawing automatically.

One of the quirks of attributes has to do with the order in which they are selected during the creation of a block. When inserting a block with attributes, AutoCAD starts with the *last* attribute added to the block and then proceeds in reverse order. In most cases, the order in which the user is prompted for attribute data is important. In these cases, you must define the attributes in the reverse order of the order in which you wish them to be presented.

To demonstrate how the order of definition affects the sequence in which AutoCAD prompts you for attribute values, the following exercise creates two block definitions using a rectangle and four attributes. For the first block, the attributes are selected from top to bottom. For the second, the attributes are selected from bottom to top. Finally, the two blocks are inserted to demonstrate the order in which you are prompted to define values for the attributes.

EXERCISE: EXPLORING THE ORDER ATTRIBUTES PROMPT FOR VALUES

1. Load the drawing file 14DWG02.DWG from the accompanying CD-ROM. The drawing contains a rectangle and four attributes that you will use to define two blocks.
2. From the Draw menu, choose Block>Make. The Block Definition dialog box is displayed.
3. In the Name list box, type **Down,** as shown in Figure 14–6.
4. Choose the Pick Point button. The Block Definition dialog box is temporarily dismissed.
5. Using endpoint Osnap, select the upper-left corner of the rectangle. AutoCAD uses that point as the block's base (insertion) point. The Block Definition dialog box reappears.
6. Click on the Select Objects button. The Block Definition dialog box is again temporarily dismissed.
7. Select the rectangle object first, then select each attribute from the top down, and then press Enter. Once again the dialog box reappears.
8. In the Objects area, choose the Retain option button, and then click OK.

   This completes the first block. Next you will create the second block.

Tip: You can use upper- and lower-case characters, as well as spaces, in block names.

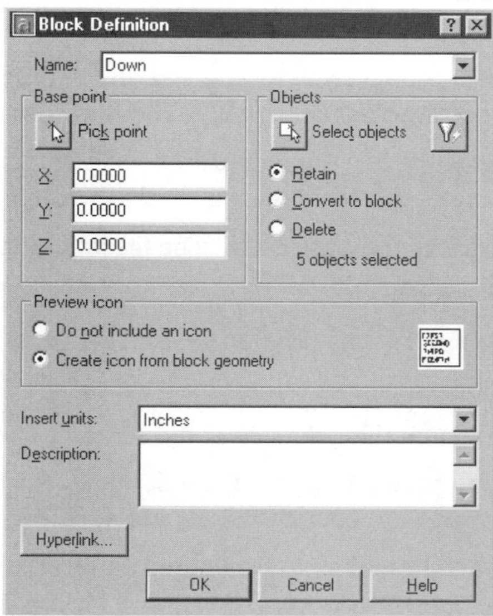

Figure 14–6 *The Block Definition dialog box is used to name and define blocks.*

9. From the Draw menu, choose Block>Make. The Block Definition dialog box is displayed.
10. In the Name list box, type **Up**.
11. Click on the Pick Point button. The Block Definition dialog box is temporarily dismissed.
12. Using endpoint Osnap, select the upper-left corner of the rectangle. AutoCAD uses this point as the block's base point. The dialog box returns.
13. Choose the Select Objects button. The Block Definition dialog box is temporarily dismissed.
14. Select the rectangle object first, then select each attribute from the bottom up, and then press Enter. The dialog box reappears.
15. In the Objects area, make sure the Retain option button is still selected, and then click OK.

    AutoCAD creates the second block, leaving the objects used to define the blocks. Next you will insert the two blocks to observe the order in which AutoCAD prompts you to fill in the attributes.
16. From the Insert menu, choose Block. The Insert dialog box opens.
17. Choose Down from the Name drop-down list, if is not already displayed.
18. In the Insertion Point area, make sure the Specify On-screen option is selected. In the Scale and Rotation areas, make sure the Specify On-screen option is cleared.
19. Click OK to close the Insert dialog box.
20. Choose a location near the center of the screen to insert the block. AutoCAD prompts you for the attribute values at the command line.
21. When AutoCAD prompts you for the first attribute value, type **1**. At each subsequent prompt, type the numbers **2**, 3, and **4**, respectively.

    AutoCAD inserts the block and its attributes. The numbers 1, 2, 3, and 4 appear in reverse order. Next you will insert the second block.
22. From the Insert menu, choose Block. The Insert dialog box opens.
23. Click on the Block button.
24. Choose Up from the Name drop-down list.
25. Click OK to close the Insert dialog box.
26. Choose a location on the right side of the screen to insert the block.
27. When AutoCAD prompts you for the first attribute value, type **1**. At each subsequent prompt, type the numbers **2**, **3**, and **4**, respectively.

    AutoCAD inserts the block. Notice this time, however, that the numbers 1, 2, 3, and 4 appear in the correct order, as shown in Figure 14–7. This occurs because you selected the attributes in reverse order when you defined the second block.

FINDING AND REPLACING ATTRIBUTES

AutoCAD 2000 introduced a tool that lets you find and replace attribute text values in a drawing. From the Edit menu, choose Find to display the Find and Replace dialog box, which is shown in Figure 14–8. To locate a text value, enter the value in the Find Text String list box. To replace the found text string with a new one, enter the new text value in the Replace With list box.

To search only block attributes for a listed text value, click on the Options button to display the Find and Replace Options dialog box, and then clear all options except Block Attribute Value, as shown in Figure 14–9. The Options dialog box offers other options, including Match Case and Find Whole Words Only, for searching for the specified text.

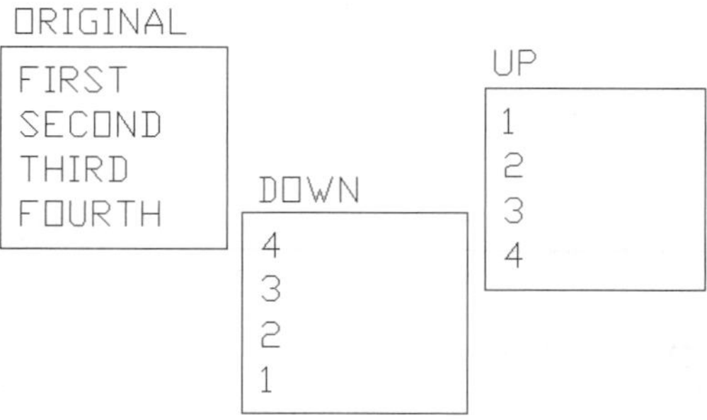

Figure 14–7 *The effect of the order in which attributes are selected.*

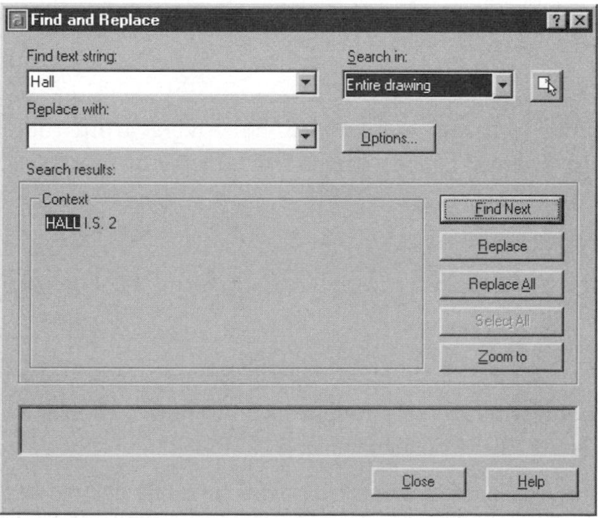

Figure 14–8 *The Find and Replace dialog box lets you locate text values in a drawing.*

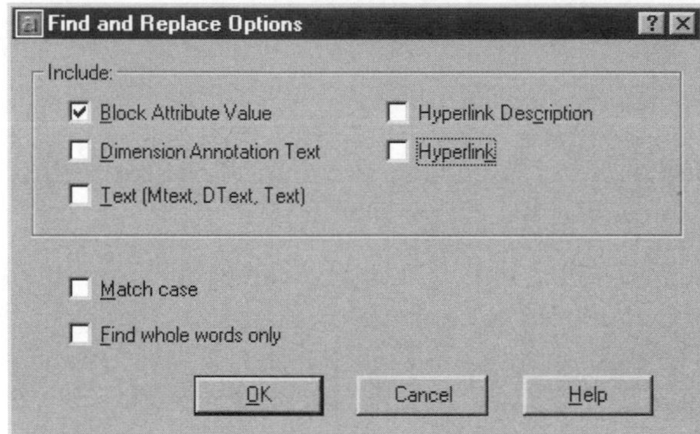

Figure 14–9 *The Find and Replace Options dialog box allows you to search for text values in block attributes only.*

After you type in the find and replace text and set the appropriate search options, click on the Find button. AutoCAD displays the found text (if any), including the entire text string it was embedded in, in the Context window. To replace the text, click on the Replace or Replace All buttons. Once a text string is found, click on the Zoom button to zoom to its location in the drawing.

BLOCK ATTRIBUTE TOOLS

As powerful as block attributes are, getting at stored data has traditionally been a difficult and complex task. AutoCAD 2002 introduces a set of powerful block attribute tools that specifically address this problem, making the presentation, management. and extraction of attribute information easier than ever before. The tools consist of the Block Attribute Manager, the Enhanced Attribute Editor, and the Attribute Extraction wizard.

THE BLOCK ATTRIBUTE MANAGER

The Block Attribute Manager, which you can invoke from the Modify menu by choosing Object>Attribute>Block Attribute Manager or by typing **BATTMAN** at the command prompt, provides a dialog box to manage virtually all aspects of a block's attributes (see Figure 14–10).

The Block Attribute Manager allows you to perform the following attribute management operations:

- Remove an attribute from a block definition
- Change the order of attribute prompts during insertion and when using the DDATTE or ATTEDIT commands
- Update all instances of a block with the currently defined attribute properties without affecting the values associated with attributes in each block

- Change an attribute's tag name, prompt string, mode, or default value
- Change an attribute's text settings—justification, style, etc.
- Change an attribute's non–text properties—layer, linetype, color, etc.

Figures 14-11, 14-12, and 14-13 show the Edit Attribute dialog box's folders, which let you edit the highlighted attribute of the current block selected in the main Block Attribute Manager dialog box. Editing a block attribute is a simple, straightforward process. Figure 14–14, for example, shows the "default" text style for an attribute changed to a text style definition using a bold font. In a similar manner, virtually any property or characteristic of an attribute can be altered from "outside" the block itself.

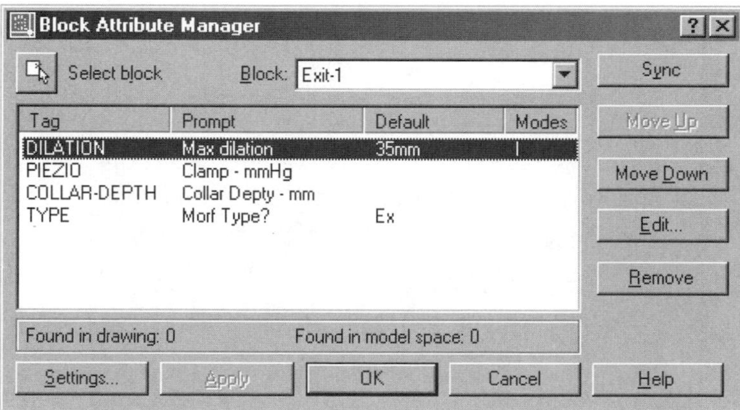

Figure 14–10 *Block Attribute Manager—main dialog box.*

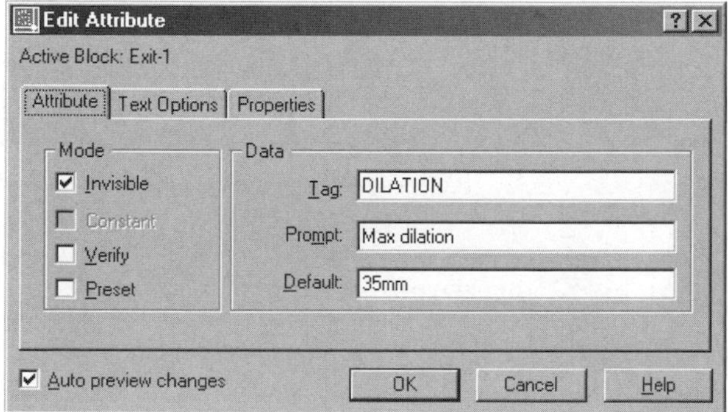

Figure 14–11 *Edit Attribute dialog box—Attribute folder.*

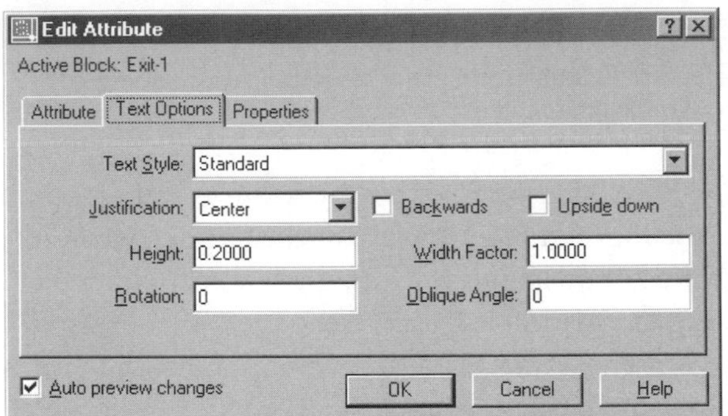

Figure 14–12 *Edit Attribute dialog box—Text Options folder.*

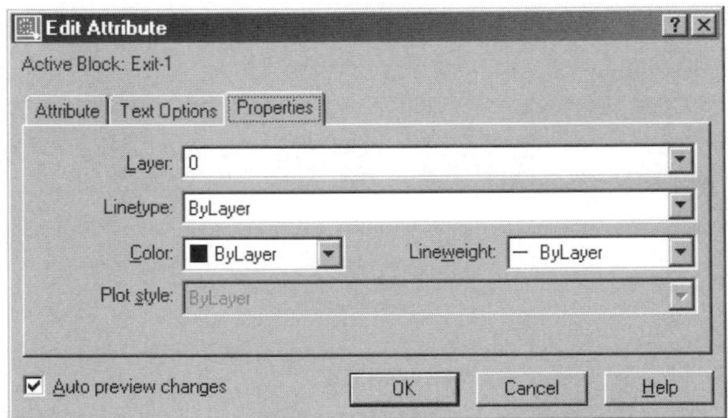

Figure 14–13 *Edit Attribute dialog box—Properties folder.*

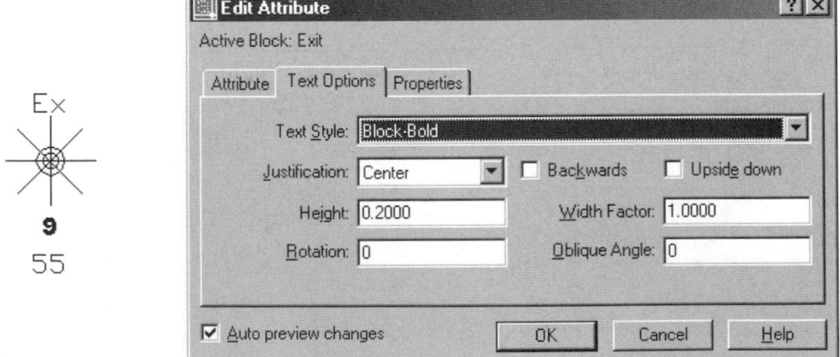

Figure 14–14 *Changing the property of a single attribute.*

THE ENHANCED ATTRIBUTE EDITOR

The Enhanced Attribute Editor dialog box allows you to edit attributes in a single selected block instance. You can invoke this dialog box from the Modify menu by choosing Object>Attribute>Single, or by typing **EATTEDIT** at the command prompt, or by simply double-clicking on a block with attributes. The Enhanced Attribute Editor dialog box shares much of the functionality and interface features of the Block Attribute Manager but acts on only the currently selected block instance. It largely replaces the DDATTE/ATTEDIT commands but, unlike these commands, it permits you to alter virtually any attribute property, including the attribute value, layer, color, text style, etc. Changes are automatically displayed as they are made. Figure 14–15 shows the Attribute folder of the Enhanced Attribute Editor dialog box. Notice the resemblance to the equivalent dialog box of the Block Attribute Manager.

EXTRACTING BLOCK ATTRIBUTES

The Attribute Extraction wizard allows you to extract block attribute data easily to several formats, including comma-separated (.cvs), Excel (.xls), and Access (.mdb) files. You can start the wizard from the Tool menu, by choosing Attribute Extraction, or by typing **EATTEXT** at the command prompt. The Attribute Extraction wizard largely automates the extraction of attribute data, making the process both fast and virtually error-free. You can attach alias names to blocks and attributes, publish data from multiple drawings and xref attachments, and save templates of selected blocks, attributes, and alias information for reuse in other drawings or sets of drawings.

The Attribute Extraction wizard makes the tedious and error-prone process of extracting the data from attributed blocks simple, efficient and virtually error-free. The first half of the following exercise explores the process of extracting attribute data from blocks representing door tags in an architectural floor plan. The second half uses Microsoft Excel to create a simple door schedule in the drawing.

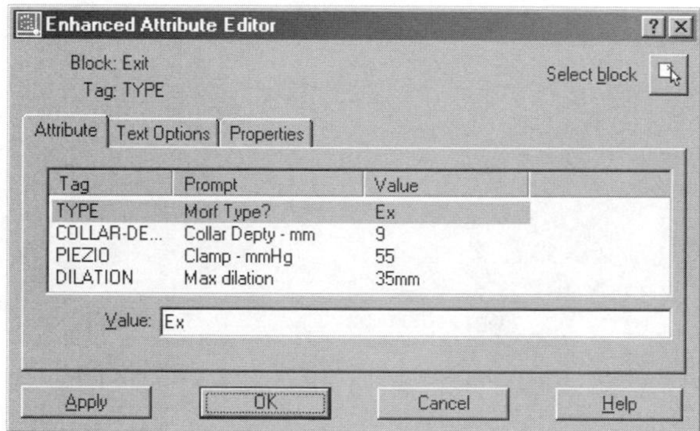

Figure 14–15 *Enhanced Attribute Editor—Attribute folder.*

EXERCISE: EXPLORING THE ATTRIBUTE EXTRACTION WIZARD

1. Load the drawing file 14DWG03.DWG from the accompanying CD-ROM.

   The drawing contains a simple architectural floor plan with fifteen pairs of doors (shown in blue) and door tags (shown in red). For this exercise we will concentrate on the door tags.

2. From the Tools menu, choose Attribute Extraction. The opening page (Select Drawing) of the Attribute Extraction wizard is displayed, as shown in Figure 14–16.

3. Accept the default setting of Current Drawing in the Drawings area and then click Next. The Settings page is displayed, offering options Include xrefs and Include Nested Blocks. These are both enabled by default and can be left on for this exercise.

4. Click Next to display the Use Template page.

   Templates are used to store settings from the Attribute Extraction wizard for later use and reuse. Near the end of this exercise you will learn how to create a template. For now, accept the default of No Template.

5. Click Next to display the Select Attributes page. In the Blocks area, click Uncheck All, and then select DoorTag at the bottom of the list.

6. Click Uncheck All in the Attributes for block area. Scroll through the list and select the following attributes: DONO, DOROOM, DOTYP, DOHEIGHT, DOCONS, and DORATE. Your dialog box should look like Figure 14–17.

7. Click Next to display the View Output page of the Attribute Extraction wizard. The table shows the results of the current query. Click Alternate View to switch to a view that lists one door tag per line, as shown in Figure 14–18.

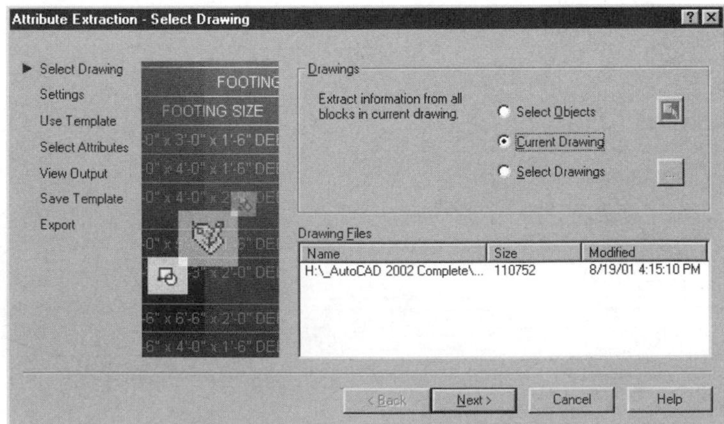

Figure 14–16 *Attribute Extraction wizard—opening page.*

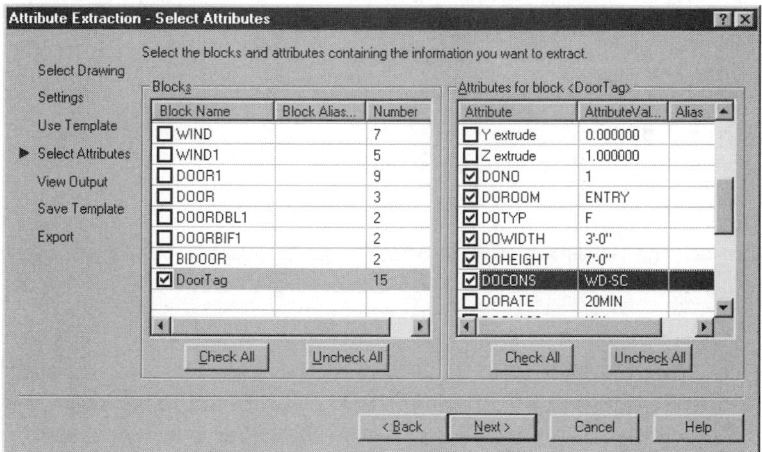

Figure 14–17 *Attribute Extraction wizard—Select Attributes page.*

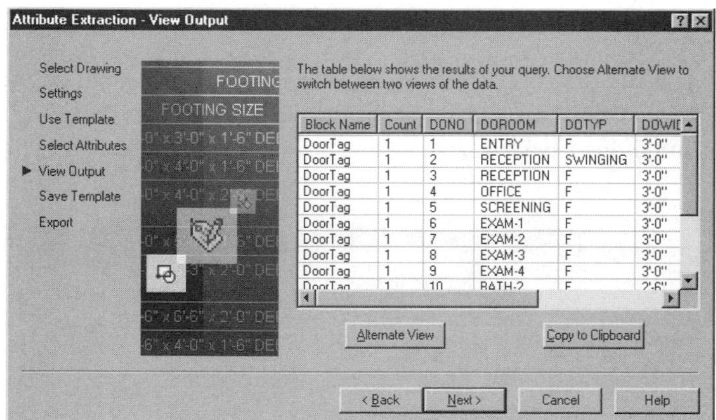

Figure 14–18 *Attribute Extraction wizard—View Output page.*

8. Click Next to move onto the Save Template page. Click Save Template, type **Door Schedule** as the File Name, and then click Save. AutoCAD saves the current query settings to a file named Door Schedule.blk.
9. Click Next to display the Export page, the final page of the Attribute Extraction wizard.
10. Select Microsoft Excel (*.xls) from the File Type drop-down list and type **14DWG03 Door Schedule** as the file name. Click Finish to extract and save the attribute data.

Note: The next time you create a door schedule, you can save time by selecting the Door Schedule template from the Use Template page of the Attribute Extraction wizard.

Note: The Attribute Extraction wizard supports four export file formats: CSV (Comma delimited, *.csv); Tab Delimited File (*.txt); Microsoft Excel (*.xls); and Microsoft Access Database (*.mdb).

Tip: For this exercise we will use the Microsoft Excel format because it is ideal for working with columnar data such as door schedules.

Note: The remainder of this exercise requires that you have Excel installed on your computer and that you have a basic understanding of its operation.

Now that you have saved the data to an Excel file, let us format it a bit and import it back into the drawing as a simple door schedule. Leave the current drawing open in AutoCAD; you will be returning to it momentarily.

11. Open the Excel file you just saved. It should look similar to Figure 14–19.
12. Perform the following operations in Excel:
    - Delete the first two columns (Block Name and Count).
    - Edit the top-row headings for the remaining six columns so they read as follows: Door Number, Door Room, Door Type, Door Width, Door Height, and Door Construction.

14DWG03 Door Schedule.xls

	A	B	C	D	E	F	G	H	I	J
1	Block Nam	Count	DONO	DOROOM	DOTYP	DOWIDTH	DOHEIGHT	DOCONS		
2	DoorTag	1	1	ENTRY	F	3'-0"	7'-0"	WD-SC		
3	DoorTag	1	2	RECEPTIC	SWINGING	3'-0"	3'6	WD-SC		
4	DoorTag	1	3	RECEPTIC	F	3'-0"	7'-0"	WD-SC		
5	DoorTag	1	4	OFFICE	F	3'-0"	7'-0"	WD-SC		
6	DoorTag	1	5	SCREENIN	F	3'-0"	7'-0"	WD-SC		
7	DoorTag	1	6	EXAM-1	F	3'-0"	7'-0"	WD-SC		
8	DoorTag	1	7	EXAM-2	F	3'-0"	7'-0"	WD-SC		
9	DoorTag	1	8	EXAM-3	F	3'-0"	7'-0"	WD-SC		
10	DoorTag	1	9	EXAM-4	F	3'-0"	7'-0"	WD-SC		
11	DoorTag	1	10	BATH-2	F	2'-6"	7'-0"	WD-SC		
12	DoorTag	1	11	PATIENT E	F	3'-0"	7'-0"	WD-SC		
13	DoorTag	1	12	LOUNGE	F	3'-0"	7'-0"	WD-SC		
14	DoorTag	1	13	LAB	F	3'-0"	7'-0"	WD-SC		
15	DoorTag	1	14	PATIENT E	BI-FOLD	2'-6"	7'-0"	WD-SC		
16	DoorTag	1	15	LOUNGE	BI-FOLD	2'-6"	7'-0"	WD-SC		
17										
18										
19										
20										
21										
22										
23										
24										
25										
26										

~cs5F

Figure 14–19 *The raw extracted DoorTag attribute data in Excel.*

- Select the six columns with data and change their width by choosing Format>Column>AutoFit Selection from the menu.
- Finally, with the six columns still highlighted, click Align Left from the Formatting toolbar.

Your spreadsheet should resemble Figure 14–20. Be sure to save the spreadsheet after you make your changes.

13. Select the data (six columns by sixteen rows) in the Excel spreadsheet and choose Edit>Copy from the menu.
14. Switch back to 14DWG03.DWG and move the drawing down to make room for a door schedule above the floor plan.
15. From the Edit menu, choose Paste. The data is pasted into the upper-left corner of the drawing, and the Ole Properties dialog box is displayed. Make sure the Lock Aspect Ratio option in the Scale area is enabled, and then swipe-and-type **85** in the Height box. Click OK to complete the paste operation. Move the door schedule to an open area and zoom in for a closer look. Your drawing should resemble Figure 14–21.

This exercise showed how the Attribute Extraction wizard walks you through the entire process of extracting attribute data from blocks. Also, it showed how easy it is to use that data to automate the task of creating a door schedule.

14DWG03 Door Schedule.xls

Door Number	Door Room	Door Type	Door Width	Door Height	Door Construction
1	ENTRY	F	3'-0"	7'-0"	WD-SC
2	RECEPTION	SWINGING	3'-0"	3'6	WD-SC
3	RECEPTION	F	3'-0"	7'-0"	WD-SC
4	OFFICE	F	3'-0"	7'-0"	WD-SC
5	SCREENING	F	3'-0"	7'-0"	WD-SC
6	EXAM-1	F	3'-0"	7'-0"	WD-SC
7	EXAM-2	F	3'-0"	7'-0"	WD-SC
8	EXAM-3	F	3'-0"	7'-0"	WD-SC
9	EXAM-4	F	3'-0"	7'-0"	WD-SC
10	BATH-2	F	2'-6"	7'-0"	WD-SC
11	PATIENT ED.	F	3'-0"	7'-0"	WD-SC
12	LOUNGE	F	3'-0"	7'-0"	WD-SC
13	LAB	F	3'-0"	7'-0"	WD-SC
14	PATIENT ED.	BI-FOLD	2'-6"	7'-0"	WD-SC
15	LOUNGE	BI-FOLD	2'-6"	7'-0"	WD-SC

Figure 14–20 *The extracted DoorTag attribute data after basic formatting.*

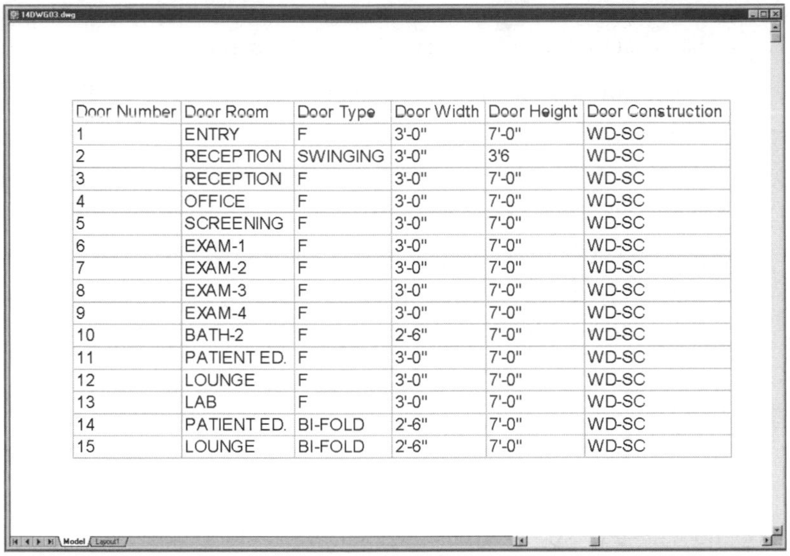

Door Number	Door Room	Door Type	Door Width	Door Height	Door Construction
1	ENTRY	F	3'-0"	7'-0"	WD-SC
2	RECEPTION	SWINGING	3'-0"	3'6	WD-SC
3	RECEPTION	F	3'-0"	7'-0"	WD-SC
4	OFFICE	F	3'-0"	7'-0"	WD-SC
5	SCREENING	F	3'-0"	7'-0"	WD-SC
6	EXAM-1	F	3'-0"	7'-0"	WD-SC
7	EXAM-2	F	3'-0"	7'-0"	WD-SC
8	EXAM-3	F	3'-0"	7'-0"	WD-SC
9	EXAM-4	F	3'-0"	7'-0"	WD-SC
10	BATH-2	F	2'-6"	7'-0"	WD-SC
11	PATIENT ED.	F	3'-0"	7'-0"	WD-SC
12	LOUNGE	F	3'-0"	7'-0"	WD-SC
13	LAB	F	3'-0"	7'-0"	WD-SC
14	PATIENT ED.	BI-FOLD	2'-6"	7'-0"	WD-SC
15	LOUNGE	BI-FOLD	2'-6"	7'-0"	WD-SC

Figure 14–21 *The finished door schedule pasted into the drawing.*

SUMMARY

In this chapter you learned how to use the Object Properties toolbar to query objects quickly for data such as layer, color, and linetype, as well as how to query properly for 2D distances and areas of blocks and xrefs. You also learned how blocks and attributes work, and that AutoCAD 2002's new improved tools for manipulating and extracting block attributes let it do tedious, repetitious work automatically, greatly increasing your productivity.

Adding the Finishing Touches

Annotation with Text

Text is a very important part of any drawing. On any given drawing, you may need to draw a single word, a single sentence, or dozens of paragraphs of text. Being able to draw efficiently and edit text directly affects your productivity. In this chapter you will learn how to do the following:

- Create and modify single lines of text
- Define and use text styles to control the appearance of your text
- Create and modify paragraphs of text
- Spell-check your text
- Explore more text options

CREATING SINGLE-LINE TEXT

A single line of text can consist of a single character, a word, or a complete sentence. The easiest way to draw such text is to use the DTEXT command. To insert a single line of text, from the Draw menu, choose Text>Single Line Text. The initial prompt displayed in the command window presents several options:

Specify start point of text or [Justify/Style]:

The default option is to specify the lower-left corner, which is otherwise known as the start point, of the new line of text. After you pick the start point, you are prompted to supply the height (unless the height is set in the style being used) and the rotation angle of the text, and the new text to be drawn. As you type the text, it is displayed in your drawing. If you make a typographical error, you can use the Backspace key to delete the error and retype the text. You signify the end of the line of text by pressing the Enter key, after which you may begin a new line of text immediately below the line of text just created. To stop adding any additional lines of text, press the Enter key without typing any new text. You may also relocate the next line of text by picking a point with the cursor.

Note: AutoCAD 2000 replaced the TEXT command with the DTEXT command. Therefore, when you type TEXT at the command prompt, you are actually starting the DTEXT command.

When you enter text, you can take advantage of the command line buffer to repeat previously entered text by using the up and down arrow keys to scroll through the buffer.

Tip: The spacing between successive lines of text is fixed at a factor of approximately 1.67 times the text height. However, because each line of text created with the DTEXT command is a separate object, you can use the MOVE command to rearrange the lines. You also can pick a new justification point at the "Text:" prompt prior to typing the new line of text, thus enabling you to override the default line spacing.

Typing the text you want to create is the easy part. It is also important to know how to format the text according to your needs. The following sections discuss how to choose the correct text height, justification, and text style. You will also learn how to continue text below the previous line, use special formatting codes and symbols, and edit text.

CHOOSING THE CORRECT TEXT HEIGHT

The hardest part of drawing text is deciding on the correct text height for the scale to which the drawing is set. Unfortunately, because AutoCAD does not have a built-in mechanism for storing and using the drawing scale to set the correct text height for full-size drawings, it is necessary to take into account the drawing scale in specifying the text height. Use Tables 15-1 and 15-2 to help specify the correct text height. To use these tables, go to the row associated with your drawing scale. Then move along the row to the column associated with the height that you want your text to have on your plot.

CHOOSING A JUSTIFICATION

The default option of DTEXT is to specify the left endpoint, or the start point, of the line of text. Specifying the Justify option at the initial DTEXT prompt displays the following prompt:

```
Enter an option [Align/Fit/Center/Middle/Right/TL/TC/TR/ML/MC/MR/BL/BC/BR]:
```

Table 15–1: Text Heights for Architectural Scales

Drawing Scale	Plotted Text Heights				
	3/32"	1/8"	3/16"	1/4"	3/8"
1/16"=1'	18"	24"	36"	48"	72"
3/32"=1'	12"	16"	24"	32"	48"
1/8"=1'	9"	12"	18"	24"	36"
3/16"=1'	6"	8"	12"	16"	24"
1/4"=1'	4.5"	6"	9"	12"	18"
1/2"=1'	2.25"	3"	4.5"	6"	9"

Table 15–2: Text Heights for Decimal Scales

Drawing Scale	Plotted Text Heights				
	3/32"	1/8"	3/16"	1/4"	3/8"
1:10	0.9375 d.u.*[a]	1.25 d.u.	1.875 d.u.	2.5 d.u.	3.75 d.u.
1:20	1.8750 d.u.	2.50 d.u.	3.750 d.u.	5.0 d.u.	7.5 d.u.
1:50	4.6875 d.u.	6.25 d.u.	9.375 d.u.	12.5 d.u.	18.75 d.u.
1:100	9.3750 d.u.	12.50 d.u.	18.750 d.u.	25.0 d.u.	37.5 d.u.

a. * d.u. stands for drafting units

Figure 15–1 shows the various justification options and their corresponding locations.

Unlike the justification options illustrated in Figure 15–1, the Align and Fit options require that you define two points.

Use the Align option when you want to specify the left and right endpoints of the text and do not care about the resulting height. The text height is automatically set to make the text fit between the specified points. Also, the angle from the first point to the second point is used as the rotation angle of the text.

Figure 15–1 *The possible justification points for a line of text.*

Use the Fit option when you want to specify the left and right endpoints and the height of the text. To make the text fit between the specified points, the height-to-width ratio of the text characters is varied. Therefore, you may end up with skinny-looking characters on one line and very fat-looking characters on the next.

Tip: You can enter the desired text justification option when the DTEXT command prompts you for the start point, which eliminates the need for first selecting the Justify option.

When the text is initially drawn with one of the alternate justification options specified, it is drawn left-justified, as if the default justification were being used. Upon completion of the DTEXT command, however, the text is redrawn with the correct justification.

Note: You can snap to the justification point of an existing line of text using the INSERT object snap mode.

This concludes the discussion of text justification options. The next section deals with the topic of text styles.

CHOOSING A TEXT STYLE

The appearance of the text drawn by DTEXT is controlled via a named group of settings referred to as a text style. The default text style supplied in the template drawings ACAD.DWT and ACADISO.DWT is STANDARD. In other template drawings, several text styles are predefined for you. Use the Style option to set the style you want to use to display text. The process of actually defining new styles and modifying existing ones is discussed later in this chapter in the section titled Defining Text Styles.

CONTINUING BELOW THE PREVIOUS LINE

If after you end the DTEXT command you want to draw an additional line of text below the last drawn line of text, you can easily do so by issuing the DTEXT command and pressing the Enter key rather than picking a new start point. DTEXT will then draw the new line of text right below the last one, using the style, height, and rotation angle of the previously drawn text.

Tip: To help you spot the last line of text drawn, that line is highlighted when you begin DTEXT. The highlighting, however, may not be apparent if the text is too small on the screen.

USING SPECIAL FORMATTING CODES AND SYMBOLS

You can do a limited amount of formatting with the DTEXT command. For instance, you can draw a line under or above the text simply by adding the codes %%u (underlining) and %%o (overlining) to the text as you enter it. The codes act as toggle switches. The first time you include the code in a line of text, it turns that effect on and is applied to the successive text characters. The second time the code is encountered in the same line of text, the effect is turned off. If the code is not encountered a second time in the line of text, then the effect is continued to the end of the text line but is not continued to the next line. For example, to underline a portion of a line of text, type in **%%uThis portion is underlined%%u and the rest is normal**. You can even combine codes. To create a line of text that is both underlined and overlined, type in **%%o%%uThis line of text is underlined and overlined**.

In addition to underlining and overlining, you also can draw symbols that are in the font file but are not on the keyboard. Table 15-3 shows several formatting codes and the resulting symbols.

The codes are not case-sensitive. In addition to using the codes in Table 15-3, you can use the code %%nnn, where nnn is a three-digit integer, to draw any character in a font file.

Table 15–3: *Additional Formatting Codes*

Formatting Code	Symbol	Meaning
%%c	Ø	Diameter
%%d	°	Degree
%%p	±	Plus/minus

A much easier way of drawing a symbol is simply to use Window's Character Map program.

To use the Character Map program in place of the %%nnn code, simply start the Character Map program (usually found in the Accessories group of programs) and select the font file you have specified in the current text style. Then select the character you want to draw and copy it to the Clipboard. You can now paste the character into the text you are typing.

Note: Not all font files contain the same characters, which is why it is important that the font file you choose to copy from in the Character Map program is the same font file specified in the text style you are drawing with in AutoCAD. What the Character Map program actually copies when you copy a character to the Clipboard is the character's position number in the font chart. When you paste that character in AutoCAD, the character corresponding to the position number recorded in the Clipboard is drawn; if you are using a different font file in AutoCAD, you may end up with a different character altogether.

After drawing and formatting the initial text, you may want to change the wording or appearance of the text. The following section covers the commands you will need to do this.

MODIFYING SINGLE-LINE TEXT

Two commands are of particular use for editing existing text: DDEDIT and Properties. DDEDIT is easier to use than Properties when all you want to do is change the text string. Properties is more powerful than DDEDIT in that it displays the new Properties dialog box, which enables you to change several properties of the selected text.

USING DDEDIT

From the Modify menu, choose Text to issue the DDEDIT command. After you select the text object to be changed, the Edit Text dialog box appears, displaying the selected text (see Figure 15–2).

Initially, the entire line of text is highlighted, and it will be replaced by whatever you type. If you want to edit a specific portion of the text, it is necessary to position the cursor at the desired point in the text and pick it. You can then use the Insert, Delete, and Backspace keys to add and delete characters.

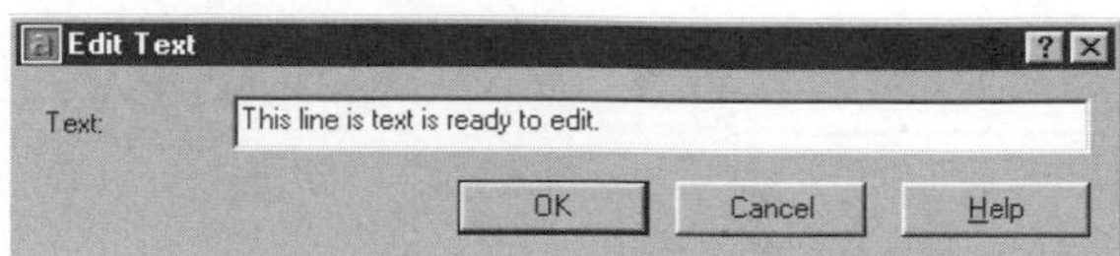

Figure 15–2 *Edit Text dialog box.*

USING PROPERTIES

Choose Properties from the Standard toolbar or Modify menu to launch the Properties dialog box, which is shown in Figure 15–3. The Properties dialog box enables you to change the text string, the style, the justification point, and the various settings that control the appearance of the text object.

See the section titled Defining Text Styles later in this chapter for a clearer explanation of the text settings that you can change.

DEFINING TEXT STYLES

A text style is a named group of settings that controls the appearance of text in a drawing. By defining different text styles, you can quickly select the text style you need for a particular text object, and thereby automatically assign such properties as font type and text height. By creating and using text styles, you can easily control the way your text looks.

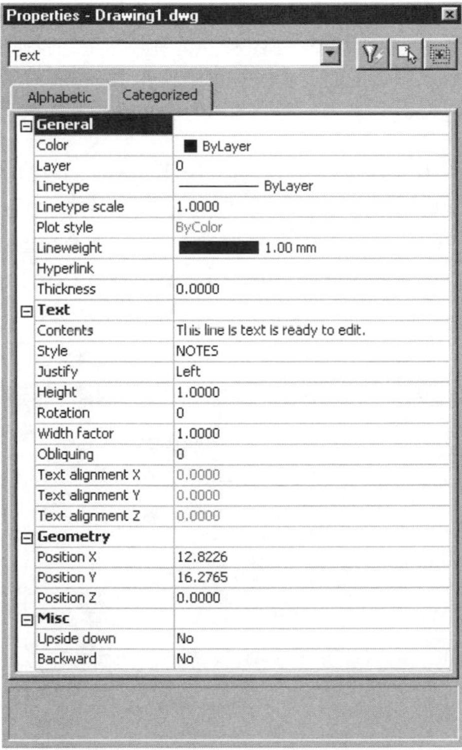

Figure 15–3 *Properties dialog box.*

The default text style, and the only defined style, in the templates ACAD.DWT and ACADISO.DWT is named Standard. You can, however, have as many text styles defined as you want in a drawing. (The other template files provided with AutoCAD each have several styles predefined.) Text styles are defined and modified with the STYLE command, which you issue by choosing Text Style from the Format menu. Figure 15–4 shows the Text Style dialog box. The various settings within the Text Style dialog box are explained in more detail in the following sections.

To create a new style, you actually begin by making a copy of the current style. If the current style is not the style you want to begin with, then select the desired style from the list of existing styles (thereby making it the current style).

Click on the New button. Specify a name for the new style. A duplicate style is created from the selected style. To rename an existing style, select the style from the list of existing styles, click on the Rename button, and then enter a new name. To delete an existing style, highlight the name from the list of existing styles and click on the Delete button. The Standard text style cannot be renamed or deleted.

Note: When a text object is created, the style it is created with is recorded with the object. A text style can only be deleted if no existing text objects reference the style.

Text styles are stored in the drawing in which they are defined. If you want to have multiple styles immediately available in a new drawing, define the styles in your template drawings. If you want to import a style from another drawing, use AutoCAD's new DesignCenter. For more information on importing object tables into the current drawing, refer to Chapter 11, Applying AutoCAD DesignCenter.

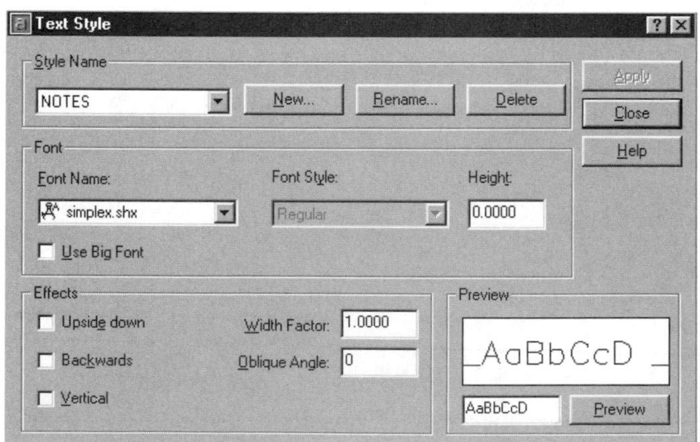

Figure 15–4 *Text Style dialog box.*

When you define a new style or modifying an existing style, you must choose a font file, the special effects you want enabled, a text height, a width factor, and an oblique angle. Choosing these settings and previewing the results of these settings are covered in the following sections.

PREVIEWING THE TEXT STYLE SETTINGS

The Preview area enables you to view a sample of the selected style and the results of changing the various settings. To view your own sample text, type your sample text in the text edit box and click on the Preview button.

CHOOSING A FONT AND STYLE

The font file is the file that contains the information that determines the shape of each character. Table 15-4 lists the different types of font files supplied with AutoCAD.

You may use PostScript files in AutoCAD. To do so, you must first use the COMPILE command to compile the PostScript (PFB) font file into a shape file.

In addition to the TrueType font files supplied with AutoCAD, you can also use the TrueType fonts supplied with Windows and other Windows applications.

AutoCAD also supports TrueType font families, which means that for some TrueType fonts, you can choose a font style such as regular, italic, bold, or bold italic. Note that not all TrueType fonts have more than the regular style defined.

There are two system variables that affect the plotting of text drawn with TrueType fonts: TEXTFILL and TEXTQLTY. When TEXTFILL is disabled, the characters are plotted in outline form only. If TEXTFILL is enabled, the characters are filled in.

The value of TEXTQLTY affects the smoothness of the characters at plotting time. The value of TEXTQLTY can be set from 0 to 100, with the default value set to 50. The higher the value, the better the resolution of the characters, but the longer it will take to process the drawing for plotting. Both system variables can be typed at the command prompt and then set to the desired value.

Table 15–4: *Types of Font Files*

File Name Extension	Font Type
SHX	AutoCAD's native font file, known as a shape file
TTF	TrueType font file

Tip: Using the most simply shaped characters will minimize the drawing size and speed up opening and working with the drawing file. Simply shaped characters are those that use very few elements, or line segments, to define the character's shape. The characters in the Simplex and Romans font files are quite simple in appearance and are similar to the Simplex characters used in board drafting. Some shape files contain the alphabet of foreign languages, such as GREEKS.SHX, or even symbols, such as SYMUSIC.SHX.

After you change the font file associated with an existing style, all text drawn with the modified style is updated to reflect the change. If you want to draw text with more than one font file, you must create one style per font file and switch between the styles as you draw the text.

SETTING TEXT HEIGHT

Also found in the Font area of the Text Style dialog box is the Height setting. The default text height of zero dictates that the user sets the text height at the time the text is drawn. A height other than zero sets the text height for that particular style to that height. The style is then referred to as a fixed height style, and the text height prompt for the DTEXT command is suppressed.

Changing the text height setting of an existing style does not affect the appearance of existing text objects.

SPECIFYING SPECIAL EFFECTS

In the Effects section of the Text Style dialog box are the Upside Down, Backwards, Vertical, Width Factor, and Oblique Angle settings. These settings are covered in detail in the following sections.

Upside Down, Backwards, and Vertical Text

In the Effects area you can enable the Upside Down, Backwards, and Vertical settings. Although the Upside Down and Backwards options work with all font files, the Vertical setting works only with SHX files.

Tip: If you want to draw text upside down, you do not have to enable the Upside Down option. Instead, specify a text rotation angle of 180 degrees. The Backwards option is useful if you want to plot text on the back side of a transparent plot sheet so that the text is readable when the sheet is viewed from the front. The Vertical option is useful when you need to draw text down the side of a vertical surface such as a building.

Unlike changing the font file setting, changing the Upside Down and Backwards settings of an existing style does not result in the existing text being automatically updated to reflect the changes. Changing the Vertical setting, however, does affect existing text objects, so you may want to create a new style before you change the Vertical setting.

Setting a Width Factor

The Width Factor setting determines the width-to-height ratio of the drawn characters. A factor of 1 results in the characters being drawn with the width-to-height ratio defined in the font file used. A factor greater than 1 results in fatter characters; while a factor less than 1 results in skinnier characters.

Tip: Drawing text with a width factor that is less than 1 may make it easier to squeeze text into an already crowded drawing but still keep the text readable.

Setting an Oblique Angle

The Oblique Angle setting affects the slant of the characters. It is often used to draw italic text when the characters in the font file being used are not naturally italic. Unlike the text rotation angle, the oblique angle of 0 degrees refers to a vertical direction. A positive text value makes the letters lean to the right, and a negative value makes the letters lean to the left.

USING MTEXT TO CREATE PARAGRAPHS OF TEXT

While the DTEXT command can be used to draw multiple lines of text, each line is drawn as a separate object. Sometimes you will want to draw multiple lines of text as a single unit, such as a paragraph of text. At such times, use the MTEXT command (see Figure 15–5), which you issue by choosing Text from the Draw toolbar, or by choosing Multiline Text from the Text submenu of the Draw menu.

After you issue the MTEXT command, you are prompted to select the first corner point of a window. This window is used to determine the direction in which the mtext object will be drawn. When the window is dragged to the right, the mtext object is drawn to the right; when the window is dragged to the left, the mtext object is drawn to the left. Within the window, the mtext object is drawn with a top-left justification. If you want, you can change the justification type to one of eight others: TC (Top Center), TR (Top Right), ML (Middle Left), MC (Middle Center), MR (Middle Right), BL (Bottom Left), BC (Bottom Center), or BR (Bottom Right). These justification types are similar to those available with the

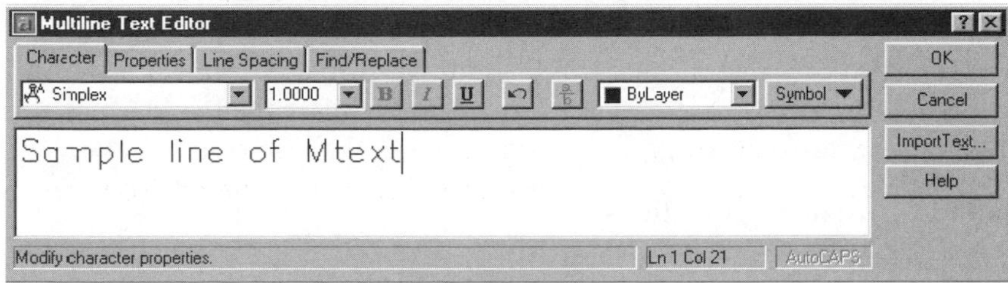

Figure 15–5 *Multiline Text Editor.*

DTEXT command (refer to Figure 15–1), except that they apply to the whole mtext object frame, not just a single line of text.

If you want, you can also choose to use the first window point as the justification point by specifying the Justify option and choosing a justification option. Then choose the Width option and numerically enter the desired width of the mtext window rather than graphically specifying its width.

A width of zero will disable the word-wrap feature of the Multiline Text Editor, and you will have to press the Enter key every time you want to begin a new line of text.

Several other command line options appear, but these are easier to set through the Multiline Text Editor dialog box. The Multiline Text Editor is divided into two parts. The bottom part is the screen editor, and the top part is divided into four folders: Character, Properties, Line Spacing, and Find/Replace, all of which are described in detail in the following sections.

If you have text in an existing ASCII or RTF file, use the Import Text button to import the file into the editor and then edit the text as you want.

USING THE CHARACTER FOLDER

The Character folder controls the text's properties. The settings can be used in either of two ways. First, you can use the settings to control the appearance of the text you type. You also can change the properties of selected text through these settings, thereby creating various special effects. To select text, position the cursor at the beginning of the text, left-click, and then drag the cursor to the end of the text you want to select. You can select a word by double-clicking on the word, or you can select the entire body of text by triple-clicking anywhere in the text.

Changing MTEXT's Font File and Text Height

After selecting the text to be affected, you can change the font file to be used and even the height of the text. The text height drop-down list is actually a combination drop-down list/text edit box. You can enter a new text height in the text edit box, or you can select a height that was previously entered from the drop-down list.

Setting the Bold, Italic, and Underline Text Properties

The Bold and Italic buttons enable you to bold or italicize the text, but only if the chosen font file is a TrueType font. You can use the Underline button to underline any selected text regardless of the font file used. All three buttons act as toggles, and you can turn their properties on or off by simply selecting the desired text and choosing the appropriate button.

Stacking and Unstacking Fractions

The Stack/Unstack button was added with AutoCAD 2000, and it is used to stack or unstack the selected text. For example, you can stack selected text by using a carat

(^), a forward slash (/), or a pound sign (#) character between the characters you want stacked. Text to the left of the character is stacked on top of the text to the right of the character. To unstack stacked text, select it and choose the Stack button.

AutoCAD provides three stacked text types, which are described as follows:

- **Carat (^):** Converts selected text to left-justified tolerance values
- **Forward Slash (/):** Converts selected text to center-justified text separated by a horizontal bar
- **Pound Sign (#):** Converts selected text to a fraction separated by a diagonal bar

You can edit stacked text and change the stack type, its alignment, and the size of the stacked text in the Stack Properties dialog box, as shown in Figure 15–6. To display the Stack Properties dialog box, from the Multiline Text Editor dialog box, select the stacked text, then right-click and choose the Properties option from the shortcut menu.

When you first create stacked text, AutoCAD displays the AutoStack Properties dialog box, which allows you to control default settings for stacked text, as shown in Figure 15–7. Ideally, you should set the values for enabling AutoStacking, removing leading blanks, and creating horizontal versus diagonal stacking as desired, and then select the check box to keep the AutoStack Properties dialog box from being displayed each time you create stacked text. If you need to change settings, you can display the AutoStack Properties dialog by box choosing the AutoStack button on the Stack Properties dialog box.

Color Settings

The Text Color list in the Multiline Text Editor's Character folder allows you to set the color for selected text. You can set the color to ByLayer, ByBlock, or any one of AutoCAD's 255 other colors.

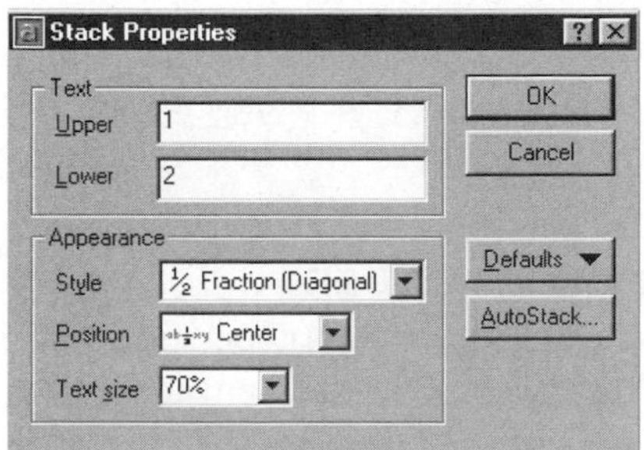

Figure 15–6 *Stack Properties dialog box.*

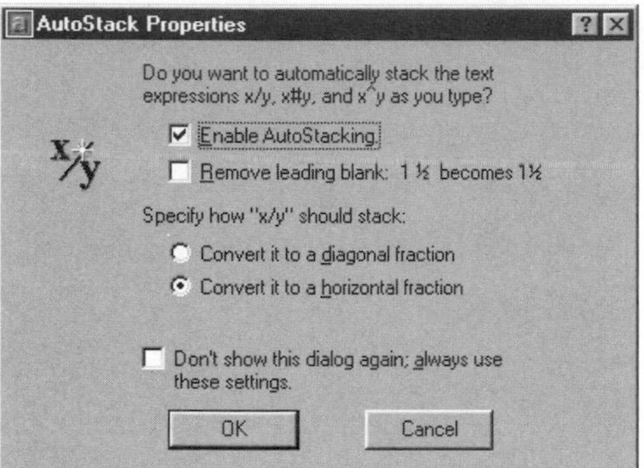

Figure 15–7 *AutoStack Properties dialog box.*

Using Special Symbols

Use the Symbol drop-down list to insert the degree, plus/minus, or diameter symbol. To insert any other symbol, choose Other from the list to invoke the Character Map program. Inserting a non-breaking space prevents the Multiline Text Editor from making a break at that point when it is deciding where to break the line of text (using the word-wrap feature) to continue to the next line.

USING THE PROPERTIES FOLDER

Choosing the Properties folder enables you to set the text style, justification option, width, and rotation angle of the overall mtext object. Remember that if you use a window to define the location of the mtext object, the justification used is TL, or Top Left, and that the width of the window is the width used for the mtext object. Using the settings on the Properties folder, you can modify these values.

USING THE LINE SPACING FOLDER

From the Line Spacing folder, you control line spacing for new or selected mtext. You control line spacing by selecting the desired line spacing properties from two lists. The first list determines if the selected line spacing is exactly as indicated or if the line spacing is at least the spacing indicated. The second list determines the line spacing value, which can be one of three settings: a single line space, a 1.5 line space, or a double line space. By using these two lists together, you control the spacing between lines of text.

USING THE FIND/REPLACE FOLDER

Use the Find/Replace folder to search for a specific combination of characters and even to replace the found text with a replacement text string. If the Match Case setting is enabled, only text that exactly matches the case of the find string is found. If the Whole Word setting is enabled, only words that exactly match the find string are found; otherwise, even words that simply contain a fragment of the find string are located. After you specify the settings, use the Find button to start the search.

EDITING MTEXT OBJECTS

You can edit mtext objects by choosing Text from the Modify menu, and then choosing the mtext text object you wish to edit. When you start the command, AutoCAD prompts you to select the text object. If you choose an mtext object, AutoCAD displays the Multiline Text Editor. You can then modify the text in the mtext object, or use the Multiline Text Editor's various options to modify the mtext object's appearance.

Additionally, you can use grips to move or change the width of the mtext object. When you select the grip point that corresponds to the justification point, the mtext object can be moved. Selecting any other grip point enables you to stretch the width of the mtext object.

SPELL-CHECKING YOUR TEXT

To check the spelling of text in your drawing, issue the SPELL command by choosing Spelling from the Tools menu. When you start the command, AutoCAD prompts you to select the text objects to check. If the SPELL command encounters an unknown word, the Check Spelling dialog box is displayed, and you must choose to replace the word, or ignore the discrepancy, or add the word to your supplemental dictionary. If no errors are found, a message box appears informing you that the spell-check has been completed.

Spell-checking in previous releases of AutoCAD was limited to text and mtext objects and block attribute values. AutoCAD 2002's SPELL command also verifies the correct spelling of text strings that are part of block definitions. Therefore, if you insert a block reference that contains a text object, the spell-checker will find and correct misspelled words in the block. While the spell-checker instantly updates the drawing's display to reflect the corrections made to block attribute values and text and mtext objects, it does not instantly display any corrections made to text objects within the block. When text objects reside in a block reference, while the spell-checker will find and correct misspelled words, the correction will not be displayed until the drawing is regenerated.

Note: When the spell-checker corrects text objects within a block reference, it actually corrects the block's definition. Therefore, when additional references to the corrected block definition are inserted into the drawing, the new insertions reflect the correct spelling.

Note: If you want to check the spelling of text in model space, you must repeat the SPELL command with the model space viewport current.

SPECIFYING THE DICTIONARIES

The SPELL command looks up words in as many as two dictionaries at any given time: a main and a supplemental dictionary. Several main dictionaries are supplied with AutoCAD, with the default being the American English Dictionary. The default supplemental dictionary is SAMPLE.CUS. (SAMPLE.CUS contains a number of AutoCAD command words and terms). To change the dictionaries used by SPELL, issue the OPTIONS command. In the Options dialog box, in the Files folder, change the Main Dictionary and Custom Dictionary File settings located under Text Editor, Dictionary, and Font File Names.

Unlike the supplemental dictionary, the main dictionary file cannot be modified or expanded. You can, however, add words to update the current supplemental dictionary, or you can select a new supplemental dictionary.

CREATING A SUPPLEMENTAL DICTIONARY

A supplemental dictionary file is a text file that contains the additional words that you want SPELL to use when checking words for correct spelling. The supplemental dictionary is as an ASCII text file that contains one word per line. You can create as many supplemental dictionaries as you want, but you can use only one at time. When you create a supplemental dictionary, be sure to use a CUS file name extension and place the file in one of the folders listed in the Support Files Search Path setting in the Options dialog box.

Tip: If you use abbreviations a lot, be sure to add them to your supplemental dictionary.

MORE TEXT OPTIONS

The following sections cover several optional text-handling features that may prove useful to you. These features will enable you to speed up the display of text, handle missing font files, and insert text files into the current drawing.

SCALETEXT: GLOBALLY SCALING TEXT

The SCALETEXT command, introduced with AutoCAD 2002, lets you quickly change the height of text and mtext objects, as well as block attributes. By using SCALETEXT, you can select multiple objects—including different objects types, such as mtext and block attributes—and then change their text heights to a value you specify, all in a single command.

You access the SCALETEXT command from the Modify menu, by choosing Object>Text>Scale, or by typing **Scaletext** at the command prompt. After you select the text objects to scale, SCALETEXT prompts you for the base point from which each text object is to be scaled. The base point is AutoCAD's standard justification point—such as left, center, or right—for text strings. The default option is Existing, which scales each text object based on its individual insertion point. Next, SCALETEXT prompts you for the new text height. While you can enter an explicit height such as 0.40, SCALETEXT offers two other text-scaling methods. The first is the

Match Object option, which prompts you to choose a text object whose height is then applied to the selected text objects. The second is the Scale Factor option, which lets you provide a value that is multiplied by each string's text height.

JUSTIFYTEXT: GLOBALLY JUSTIFYING TEXT

AutoCAD 2002 also introduced the new JUSTIFYTEXT command, which lets you instantly redefine the justification point for multiple text objects—without changing their actual location. As with the SCALETEXT command, JUSTIFY-TEXT lets you quickly change the justification of text and mtext objects, as well as block attributes. However, text objects within block references, or text within dimensions, are ignored.

You access the JUSTIFYTEXT command either from the Modify menu by choosing Object>Text>Justify, or by typing **Justifytext** at the command prompt. After you select the text objects to scale, JUSTIFYTEXT prompts you for the justification option. Its default justification option is Left.

ENABLING THE QUICK TEXT DISPLAY

When AutoCAD opens or regenerates a drawing, if there are numerous text objects, it may take quite some time to complete the regeneration process, especially if the text is drawn with complex fonts. If you want to speed up the regeneration of the drawing, and you do not actually need to see the existing text, enable the Quick Text mode. You enable the Quick Text mode from the Display folder in the Options dialog box by selecting the Show Text Boundary Frame Only check box. With Quick Text enabled, text and mtext objects are displayed as simple rectangles that contain no characters. If you enable Quick Text mode after opening a drawing, issue the REGEN command to redisplay the text as empty rectangles.

Note: Even with Quick Text enabled, new text objects are displayed as text characters while the DTEXT or MTEXT commands are active, rather than as rectangles, making adding text easier.

SPECIFYING AN ALTERNATE FONT FILE

Font files are not stored with the drawing file. If a font file that is referenced in the drawing is not available when the drawing is opened, an error message is displayed. You are then prompted to choose a replacement font file. If you want to bypass all such error messages, you can specify a font file that is automatically used whenever a needed font file cannot be found. This alternate font file is specified in the Alternate Font File setting under Text Editor, Dictionary, and Font File Names in the Files folder of the Options dialog box. The default alternate font file is SIMPLEX.SHX.

MAPPING FONTS

If you need to specify more than one alternate font file, specify a font mapping file. A font mapping file is a text file where in which each line specifies the font file to be

Note: A couple of possible problems with using an alternate font can occur. If the missing font file contains special characters that the alternate font file does not have, the text on the drawing may end up incomplete. Furthermore, because the space that a line of text occupies is dependent on the font file used to generate the text, you may find that when the alternate font is applied, the text on the drawing looks out of place or does not fit properly. The best solution is to obtain the correct font files and use them, unless you are sure you have a suitable alternate font file.

replaced and its substitute font file (separated by a semicolon). The default font map file is ACAD.FMP. You can identify a different font map file by changing the Font Mapping File setting under Text Editor, Dictionary, and Font File Names in the Files folder of the Options dialog box.

DRAWING TEXT AS ATTRIBUTES

An alternate method to drawing text objects that are to be incorporated into block definitions is to draw attributes. Attributes behave much like text objects but they have additional functions beyond displaying text. Attributes are discussed in more detail in Chapter 14, Extracting Information from a Drawing.

DRAGGING AND DROPPING TEXT FILES

In Windows you can drag a text file from the desktop or from Windows Explorer and then drop it into your drawing. AutoCAD automatically inserts the file as an mtext object, using the current text settings for the text height, rotation angle, and text style.

COPYING TEXT USING THE CLIPBOARD

You also can copy text from any application to your Clipboard and then paste the contents into your drawing. If you use the PASTE command, the contents are dropped into your drawing as an embedded object. If you use the PASTESPEC command, you can choose to paste the Clipboard contents as text, in which case the text is drawn as an mtext object.

The Clipboard operations depend on OLE (Object Linking and Embedding). For more information on AutoCAD's OLE features, refer to Chapter 27, Taking Advantage of OLE Objects in AutoCAD.

CREATING CUSTOM SHAPE FILES

You have the option of creating your own shape file containing the characters you want to use for text objects. Creating each character is a laborious procedure because you have to break each character into a series of short line segments, and then enter the codes for those line segments into the new font file. In the earlier versions of AutoCAD that did not support the use of TrueType fonts, defining your own shape file was the only way to add to the font files supplied with AutoCAD. However, since Release 14, AutoCAD supports TrueType font families, which are supplied with Windows and are much easier to use than creating your own fonts from scratch. You can also purchase additional fonts from a number of software vendors at a very low cost.

SUMMARY

AutoCAD provides a variety of tools to deal with drawing and editing text. This chapter covered the basic steps needed to deal with single lines of text in your drawings, as well as how to add multiple paragraphs of text using the MTEXT command. Editing text and defining and changing text styles to control the appearance of text were also covered.

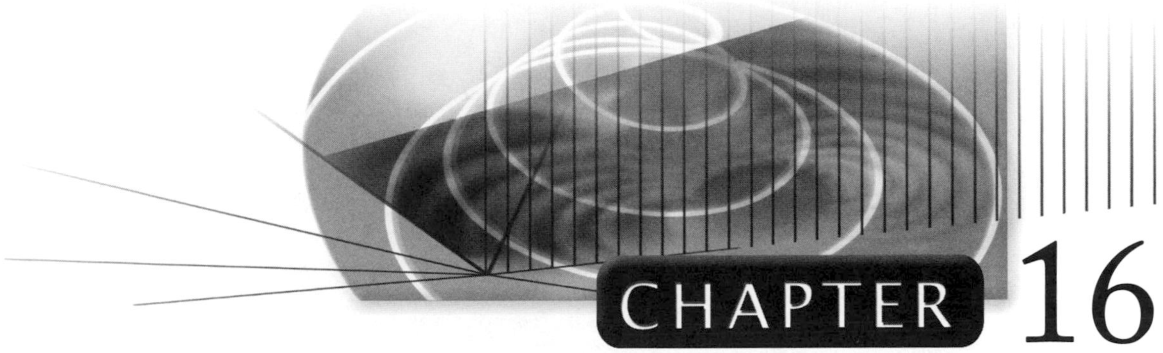

Using Hatch Patterns

When you need to fill an area with a repetitive pattern or a solid fill, you can use the BHATCH command to create a hatch object. In this chapter you will learn how to

- Use BHATCH to create hatch patterns
- Define hatch boundaries
- Edit hatch patterns
- Turn hatches on and off with FILLMOSE
- Create your own custom hatch patterns
- Use the BOUNDARY command to create outlines of complex areas

USING BHATCH TO CREATE HATCH PATTERNS

You draw hatch patterns, *including solid fills*, to highlight an area of your drawing, or to separate visually areas of your drawing that share common boundaries, or to convey information about an area of your drawing. For example, you might have a map on which a hatch pattern identifies a type of terrain, or that might use slightly different patterns to separate contiguous land plots (see Figure 16–1).

You use the BHATCH command to draw hatch patterns. This section introduces the BHATCH command and also discusses how to specify hatch patterns, define

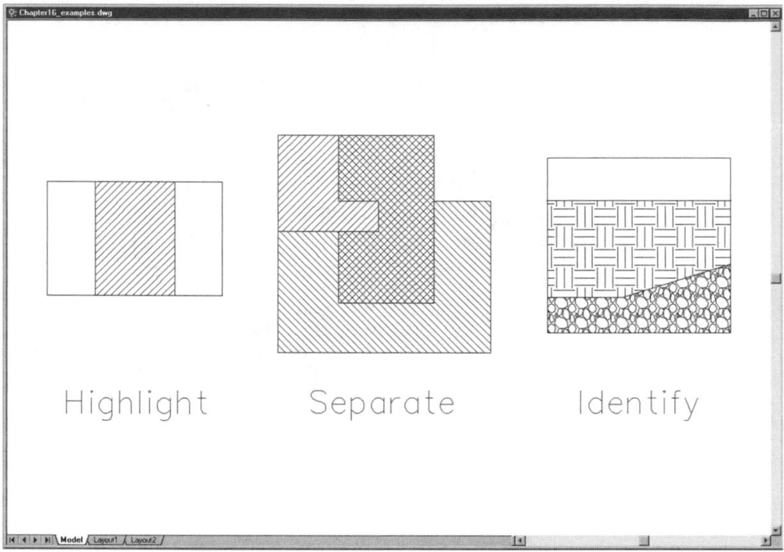

Figure 16–1 *Examples of the uses of hatch objects.*

hatch boundaries, set attributes, work with boundary islands, and use the BHATCH command's advanced settings.

To issue the BHATCH command, from the Draw toolbar, choose Hatch. The Boundary Hatch dialog box appears (see Figure 16–2).

To add a hatch to an object, you first specify a pattern and its parameters and then define the limits of the area to be hatched.

SPECIFYING A PATTERN

When you select a hatch pattern, you have several choices. First, you may choose from any of the predefined patterns that come with AutoCAD. Second, you may make a basic line pattern "on the fly" using the current linetype. Finally, you may select a pattern that is defined in any custom PAT file that you have added to the AutoCAD search path. All of these methods will be discussed in the following sections.

Predefined Patterns

AutoCAD comes with a large number of predefined patterns. A number of these are standard patterns established by the American National Standards Institute (ANSI) and are used widely in North America. Another group of predefined patterns is derived from patterns established by the International Standards Organization (ISO), the organization that sets international drafting standards in all fields except electrical and electronics. Yet a third group of predefined patterns includes many useful and traditional patterns used worldwide. Figure 16–3 shows a sampling of these patterns.

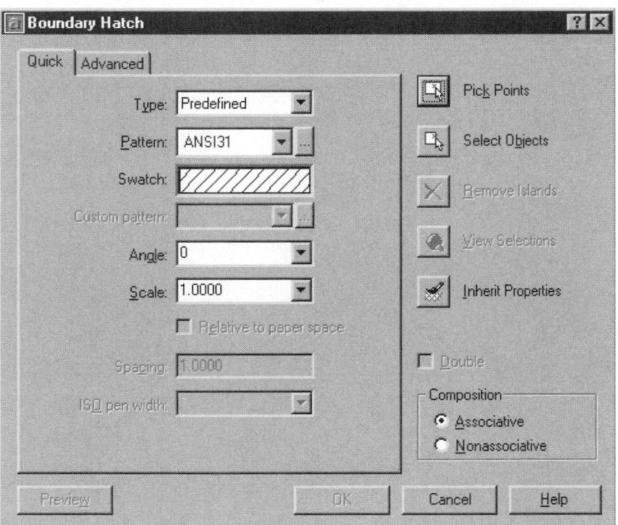

Figure 16–2 *The BHATCH command's Boundary Hatch dialog box—Quick folder.*

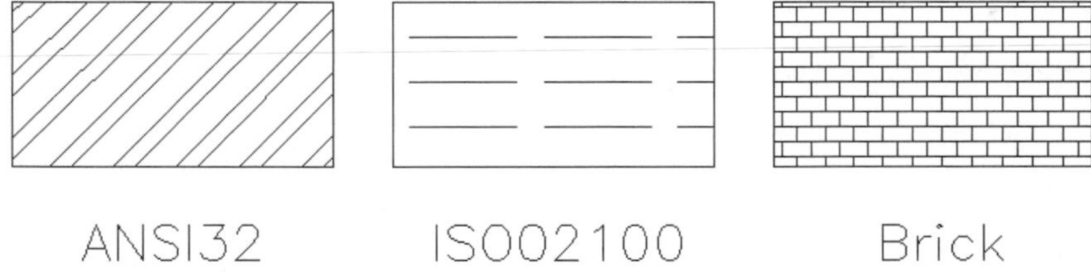

Figure 16–3 *Some sample predefined hatch patterns supplied with AutoCAD.*

To select a predefined hatch pattern, in the Boundary Hatch dialog box, in the Quick folder, make sure that Predefined is selected from the Type drop-down list. From the Pattern drop-down list, select a pattern name. Notice that a representation of the selected pattern is displayed in the Swatch display box. You can also select a pattern by type (ISO, ANSI, etc.) by choosing the "…" button next to the Pattern drop-down list. Doing so displays the Hatch Pattern Palette dialog box. You can select a pattern from any of the folders by either double-clicking on the pattern or by selecting the pattern and closing the dialog box by choosing OK (see Figure 16–4).

Choosing Scale and Angle Settings. After you choose a pattern, you may also adjust the angle and scale settings as desired. The angle setting allows you to change the orientation of the pattern by rotating it. The scale setting provides a scaling factor used to scale the pattern's size, much as LTSCALE is used to control the generation of linetypes. Figure 16–5 shows the effect of changing scale and rotation.

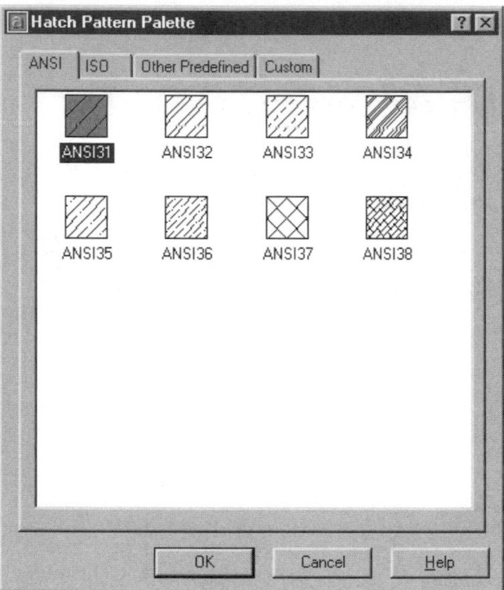

Figure 16–4 *Hatch Pattern Palette dialog box of BHATCH.*

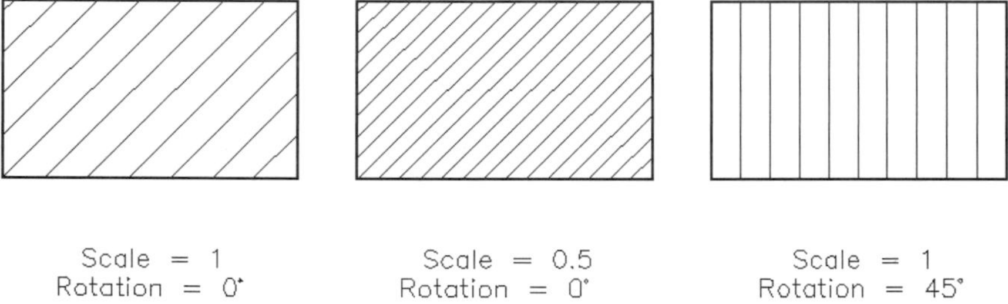

Figure 16–5 *The effect of changing hatch pattern scale and rotation.*

Some of the predefined patterns are intended to represent real-world materials and are defined with appropriate dimensions. Drawing these patterns at a scale factor at or near 1.0 will yield realistic results in full-scale drawings. The pattern AR-B88, for example, represents 8" x 8" building blocks and, when drawn at a scale of 1.0, the blocks measure 8" x 8". Other patterns, such as ANSI31, are simply standard drafting hatching symbols that can be scaled to give the best visual results consistent with the plot or dimension scale factor for which the drawing is set up.

Tip: If you set the scale of a pattern too small, it will take an inordinate amount of time to generate and may be plotted unsatisfactorily. If you set the scale to an overly large value, the pattern may be so large that it does not appear in the area being hatched. Use the Preview feature of the BHATCH dialog box to visually fine-tune the pattern's usable scale.

In the following exercise you will use the BHATCH command to carry out some basic hatch placements.

EXERCISE: HATCHING AREAS WITH BHATCH

1. Open the drawing 16DWG01.DWG. The layer BODY will be current. Perform a Zoom/All. Your screen should resemble Figure 16–6.
2. From the Draw menu, choose Hatch. In the Boundary Hatch dialog box's Quick folder, make sure that the pattern type is Predefined and that the pattern is ANSI31. Confirm the settings for angle 0 and scale 1.0000.
3. From the right side of the dialog box, choose Pick Points and, in the drawing, pick ① (see Figure 16–6). Notice that BHATCH automatically determines and highlights the boundary. Right-click to display the BHATCH shortcut menu. Notice that you can change many of the BHATCH parameters in the shortcut menu. Choose Preview. The completed hatch pattern is previewed. Right-click again to return to the dialog box. You can now either accept the hatch pattern as previewed or change the pattern parameters.
4. Display the preset values in the Scale input box by choosing the down arrow, and then choose the 0.7500 scale factor. Likewise, in the Angle input box, choose an angle of 90.
5. Choose Preview to view the effect of the new scale factor and angle. Right-click or press Enter to return to the dialog box, and then choose OK to accept the hatch object.
6. Save your work. You will continue work in this drawing in a later exercise in this chapter.

ISO (Metric) Patterns. The ISO hatch patterns are designed to be used with metric drawings. These patterns begin with the prefix "ISO." They appear alphabetically in the drop-down list of predefined patterns and can be viewed in the ISO folder of the Hatch Pattern Palette dialog box. These patterns are defined with the millimeter as the unit of measure. Choosing an ISO pattern enables the ISO Pen Width input box and drop-down list in the Quick folder of the Boundary Hatch dialog box. Choosing a pen width sets the initial value for the pattern's scale setting equal to the pen width. You can, of course override this initial scale setting.

User-Defined Patterns

You can define a simple line hatch pattern "on the fly" by choosing the User Defined option from the Type drop-down list in the Boundary Hatch dialog box. This disables the Pattern and Scale input boxes and enables the Spacing input box and the Double check box. User-defined patterns are simple and consist of either one or two sets of parallel lines. The spacing and angle of the lines are set in the Spacing and Angle input boxes, respectively. If you check the Double check box near the lower-right side

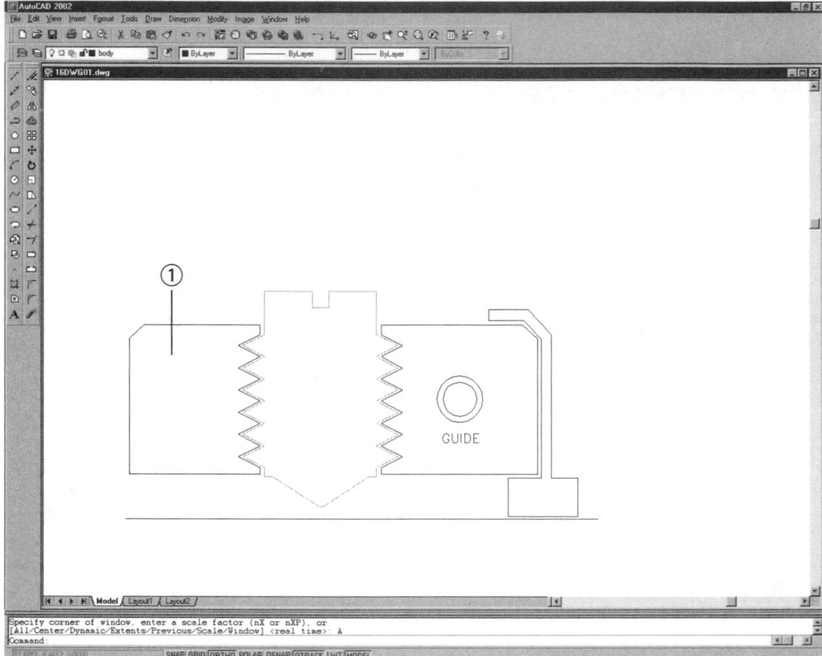

Figure 16–6 *You can pick a point and let AutoCAD determine the hatching boundary.*

of the Boundary Hatch dialog box, a second set of parallel lines perpendicular to the first will be generated. The spacing of the second set is the same as that of the first set.

Although user-defined patterns lack a great deal of variety, their simplicity and the speed with which they can be drawn make them useful for quickly hatching an area.

Custom Patterns

You can define additional patterns similar to the patterns supplied with AutoCAD. These additional pattern definitions are either added to the ACAD.PAT file (or ACADISO.PAT file for metric patterns), or you can save each definition as an individual PAT file. These individual hatch pattern files are referred to as "custom pattern" files. To access one of these custom pattern files, from the Type drop-down list in the Quick folder of the Boundary Hatch dialog box, choose Custom. This enables the Custom Pattern input box, where you enter the name of the custom pattern you want to use. The scale and angle parameters for a custom pattern function as they do for a standard predefined pattern.

Procedures for creating your own custom hatch pattern are discussed in the section titled Creating Custom Hatch Patterns, near the end of this chapter.

Inherit Properties Button

By selecting the Inherit Properties button on the right side of the Boundary Hatch dialog box, you can copy the parameters of an existing hatch object to other objects in the drawing. After you select this button, the dialog box is temporarily dismissed and a special Inherit Properties pick box and icon appear in the drawing, as shown in Figure 16–7. Select the hatch pattern whose properties you want to duplicate. Another special "painter" cross hair and icon appear. Use these cross hairs to select an internal point of the area to which you want to apply the hatch pattern (see Figure 16–7).

Tip: After selecting the hatch object whose properties you want the new hatch object to inherit, you can right-click and use the shortcut menu to toggle between the Select Objects and Pick Internal Point options to create boundaries.

ASSOCIATIVE VERSUS NON-ASSOCIATIVE HATCH OBJECTS

By default, BHATCH generates an associative hatch object. An associative hatch object is one that conforms to its boundary object(s) such that modifying the boundary objects automatically adjusts the hatch. If, for example, the boundary object(s) defining an associative hatch are edited, then the hatch object's size is automatically adjusted to fill the new boundary area.

Tip: Unless there is a compelling reason to generate a non-associative hatch object, use the default associative hatch generation method. Associative hatch objects offer the significant advantage of allowing you to edit the boundary object without having to reselect the boundary and re-create the hatch.

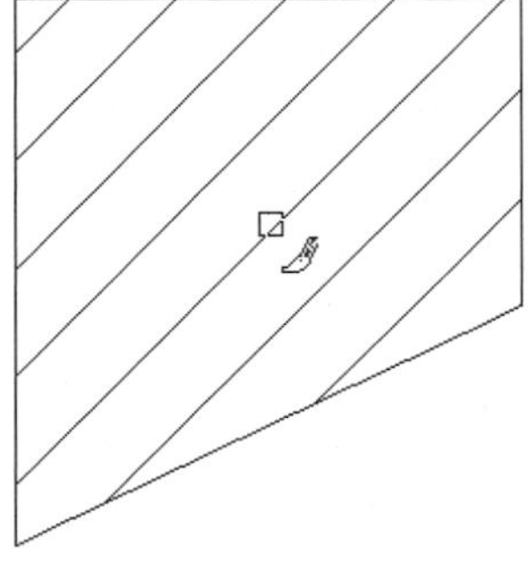

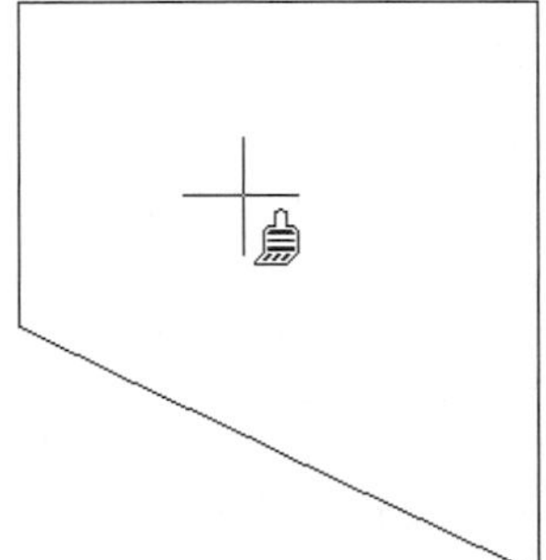

Figure 16–7 *Two distinctive icons: Inherit Properties and Transfer Properties.*

DEFINING HATCH BOUNDARIES

After you select a hatch pattern and specify its parameters, you need to define the area(s) you want to fill with the pattern. The area must be completely enclosed by one or more objects.

In Figure 16–8, a single closed object, a circle, defines the boundary of area A. A series of lines and curves that meet end to end define the boundary for area B. Two lines and two arcs define the boundary for area C. The lines and arcs of area C overlap and do not meet end to end, but together they define an enclosed area. Objects that define the area to be filled are called *boundary objects.*

With the BHATCH (boundary-hatch) command, you can define the boundaries of the area you want hatched by either choosing a point or points within the desired boundary or by selecting an object or objects that define an outer boundary.

When you choose the Pick Points button in the Boundary Hatch dialog box, the overall hatch boundary can be automatically calculated by BHATCH. You merely pick a point that lies within the area that you want filled. This point is referred to as an *internal point*. Using this method of establishing a hatch boundary offers a major advantage: If multiple objects are involved in establishing the boundary, the boundary objects do not have to meet end to end (refer to area C in Figure 16–8).

Selecting Objects

Another method for indicating the boundary of an area you want to hatch is provided by choosing the Select Objects button in the Boundary Hatch dialog box. With this method, you select the individual boundary objects that define the area to be hatched. This method is adequate for simple areas that are bounded by a single closed object, such as a circle, for example. But if more than one object will make up the boundary, this method requires that the objects meet end to end, as shown in area B of Figure 16–8. Using the Select Objects method with an intended boundary, such as in area C in Figure 16–8, will yield erroneous results because the boundary objects do not meet end to end.

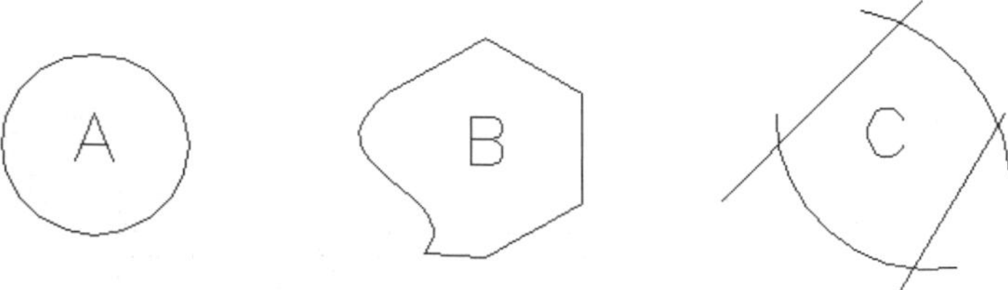

Figure 16–8 *Closed areas that define permissible areas to hatch.*

Tip: If the area you want to hatch is enclosed by a single boundary object, and no areas within the boundary need be excluded from hatching, using the Select Objects method may be faster than picking an internal point. In all but the simplest situations, however, picking an internal point and letting the BHATCH command calculate the boundary generally yields the best results.

DEALING WITH ISLANDS

It is not uncommon to have enclosed areas within the overall hatching area. These areas are referred to as *islands*. There can even be islands within islands. Text and mtext objects lying within an area to be hatched can behave like islands.

If you use the Pick Points method of defining the hatch boundary, AutoCAD automatically detects islands. This is one of the advantages of using the Pick Points method. On the other hand, if you use the Select Objects method, you must explicitly indicate those internal boundary objects that you want to be considered as islands [em]; otherwise, BHATCH will not recognize their presence. One or more objects can define islands in the same way boundary objects define the overall hatch area.

The ways islands are detected and treated by BHATCH are controlled by settings found in the Advanced folder of the Boundary Hatch dialog box (see Figure 16–9).

In the Advanced folder, in the Island Detection Style area, you control the way BHATCH treats successive levels of nested islands. The three styles are Normal,

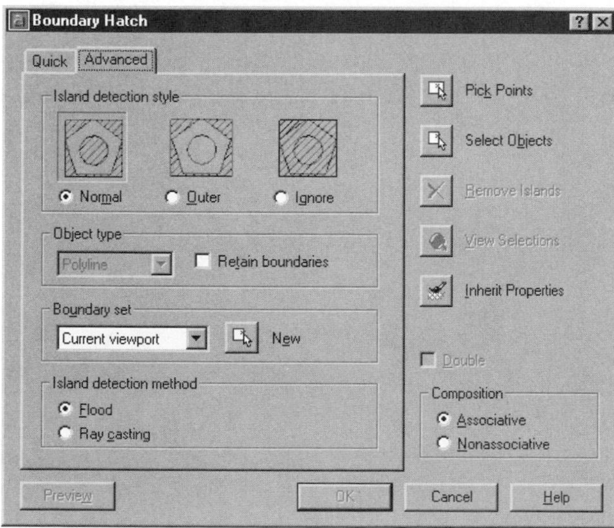

Figure 16–9 *Boundary Hatch dialog box—Advanced folder.*

Outer, and Ignore. Figure 16–10 shows how the same outer boundary and inner islands are treated using these three detection styles. The styles are described as follows:

- **Normal**—Hatches inward from the outermost boundary. Hatching is turned off at the first internal island and then turned on at the next detected island, hatching alternate areas.
- **Outer**—Hatches inward from the outermost boundary. Hatching is turned off and remains off at the first detected island.
- **Ignore**—Hatches all of the area within the outermost boundary, ignoring all internal islands.

The default style is Normal, which is applicable in most situations. The Outer detection style is useful if you want to hatch overlapping islands with different patterns. Figure 16–10 shows how this can be done. Using the Outer style, pick in area 1 and apply a pattern. Next, repeat BHATCH and, again using the Outer style, pick island 2 and apply a different pattern. Continue with this method until all islands are hatched. In Figure 16–11, islands 2 and 3 have different patterns applied; the outer area and island 4 have the same pattern with a different angle setting applied.

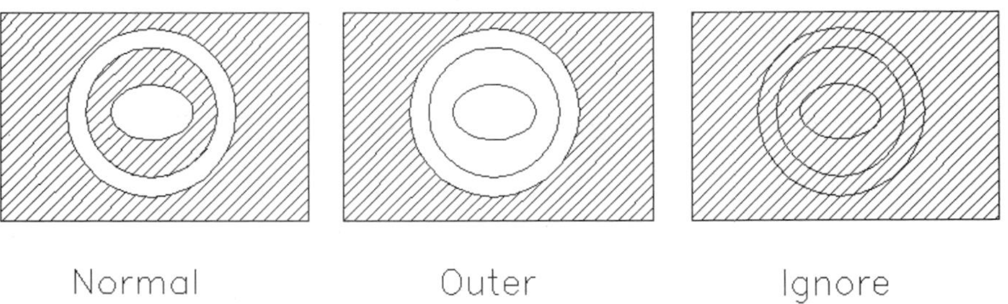

Figure 16–10 *The three styles of island detection.*

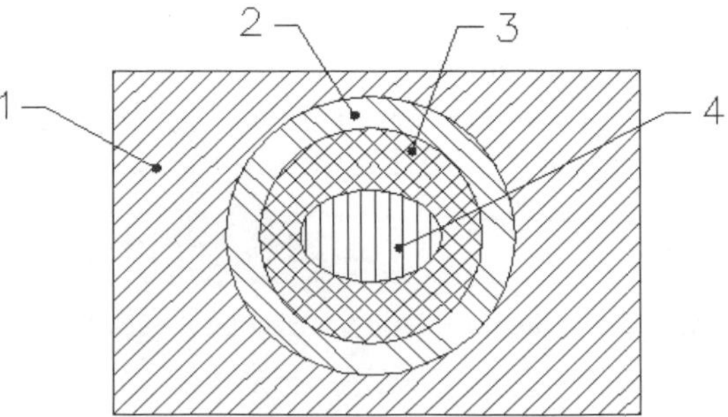

Figure 16–11 *Using the Outer style to hatch overlapping areas.*

BHATCH can also be used on regions. Islands within a region are detected by BHATCH and treated according to the current setting of the island detection style. A benefit of using a single region is that it effectively saves the boundary set into a object that can be selected quickly for later use. Regions are discussed in Chapter 25, Modeling in 3D.

Tip: You can also access the Normal, Outer, and Ignore options from a shortcut menu by right-clicking in the drawing area while you specify points or select objects to define your boundaries.

Defining the Boundary Set

Normally the Pick Points method of defining the boundary set of an area to be hatched examines all the objects in the current viewport. You can, however, choose the New button in the Boundary Set area in the Advanced folder of the Boundary Hatch dialog box and explicitly select a smaller set of objects to be examined for valid hatch boundaries. This option is useful when you have a crowded drawing and want to speed up the search mechanism by restricting the number of objects examined.

If you use the New button to create a new set of objects to be examined, the Boundary Set drop-down list will list an Existing set in addition to Viewport. There can be a maximum of two search sets at any time: the entire viewport (which is the default set and is always available), and an existing set (if you have defined one). Defining a new set replaces any pre-existing set.

Retaining Boundaries

When hatch area boundaries and internal islands are defined, AutoCAD uses temporary polylines to delineate these areas. These polylines are normally removed after the hatch object has been generated. When you check Retain Boundaries in the Object Type area of the Advanced folder of the Boundary Hatch dialog box, the temporary polylines are retained on the current layer rather than removed. You can save them as either closed polylines or as regions, depending on the option you select from the drop-down list.

Enabling the Retain Boundaries option is useful when the hatching area is delineated by multiple objects and you want polylines or regions to represent the hatch area. If you subsequently use the AREA command on the resulting polylines, or you use the MASSPROP command on the resulting region, you can easily measure the hatched area.

Detecting Islands

The controls located in the Island Detection Method area of the Advanced folder of the Boundary Hatch dialog box allow you to turn island detection on or off. There are two choices:

- **Flood**—Includes islands as boundary objects
- **Ray Casting**—Runs a line from the point you specify to the nearest object, then traces the boundary in a clockwise direction, excluding islands as potential boundary objects

If you use the Ray casting method, you must be careful where you pick points, because AutoCAD casts a ray to the *nearest* object. In Figure 16–12, for example, point A is a valid point, while point B is not. The object nearest point A is a line that qualifies as part of a potential boundary of which point A is *inside*. Point B, however, is closest to a line that qualifies as a part of a potential boundary of which B is *outside*. Point B will cause BHATCH to issue an error message as shown in Figure 16–12.

Other Boundary Hatch Controls

There are several other buttons shared by both the Quick and the Advanced folders of the Boundary Hatch dialog box:

- **Remove Islands**—Allows you to remove individual islands from the boundary set when you use Pick an Internal Point. You may also remove an island by pressing Ctrl and picking inside the selecting island.
- **View Selections**—Displays the currently defined boundaries with highlighted boundary objects. This option is unavailable when you have not yet specified points or selected objects.
- **Preview**—Displays the currently defined boundaries with the current hatch settings. This option is not available when you have not yet specified points or selected objects to define your boundaries.

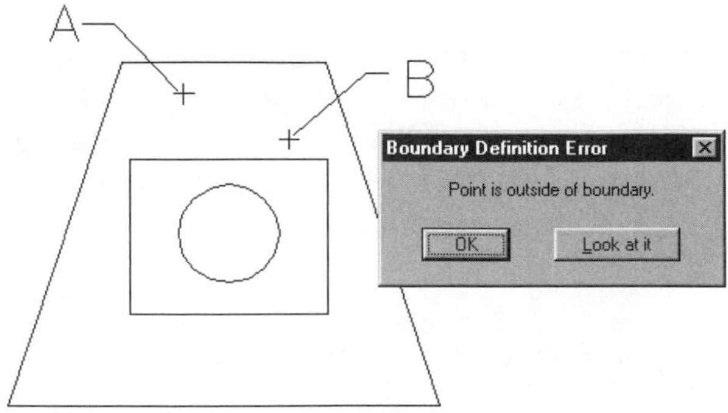

Figure 16–12 *Selecting valid points with the Ray casting option.*

In the following exercise you will continue drawing hatch objects in CHAP15.DWG.

EXERCISE: HATCHING AREAS WITH BHATCH (CONTINUED)

1. Continue with or reopen 16DWG01.DWG. Set the layer LOCK-SCREW current. Start the BHATCH command by opening the Draw menu and choosing Hatch. In the Quick folder of the Boundary Hatch dialog box, select pattern ANSI33. Check that the angle is set to zero and the scale is set to 1.000.
2. Select the Select Objects button. The lock screw is drawn on its own layer and is composed of a pair of mirrored polylines with ends meeting, forming an enclosed space. Pick the two polylines at ① and ② of Figure 16–13.
3. Right-click and choose Preview from the shortcut menu. Right-click to return to the dialog box and select a scale of 0.75. Choose Preview in the lower-left corner of the dialog box. Right-click and choose OK to apply the hatch object. Notice that the hatch is drawn on the current layer. Your drawing should resemble Figure 16–14.
4. Set layer BODY current. Restart the BHATCH command and, on the right side of the dialog box, choose Inherit Properties. In the drawing, place the Inherit Properties cursor in the area you hatched earlier and pick a point (see ① in Figure 16–14).
5. Right-click and select Pick Internal Point from the shortcut menu. Place the Transfer Properties cursor in the body to the right of the lock screw at ② and pick a point. BHATCH calculates and highlights the boundary and internal islands. Right-click and choose Preview. Notice that the text object is treated as an island and excluded from hatching.

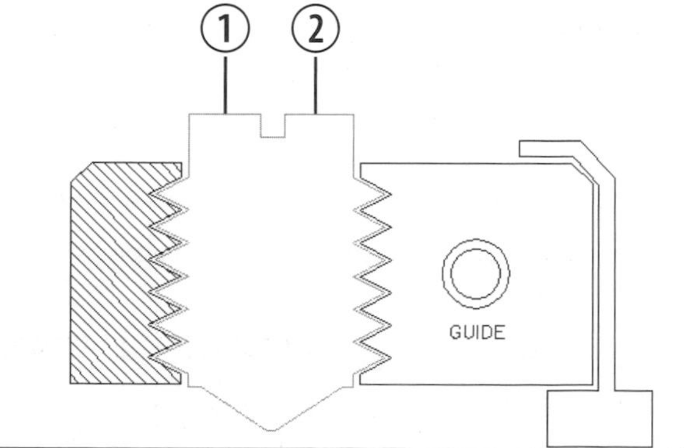

Figure 16–13 *Select the two polylines that compose the lock screw.*

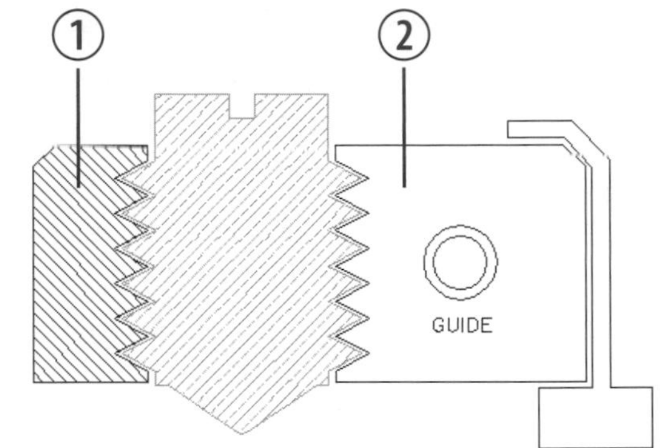

Figure 16–14 *Hatching the lock screw.*

6. Assume you do not want the inner circle that represents a guide bar to inherit the current hatch properties. Right-click to return to the dialog box and choose Remove Islands. In the drawing, pick the inner circle to remove it from the boundary set. Right-click and choose Preview. Right-click and choose OK to accept the hatch.
7. Right-click and choose Repeat BHatch from the shortcut menu. In the Boundary Hatch dialog box, select User Defined from the Type drop-down list. In the Spacing input box, enter **0.1**. Select the Double option check box.
8. Choose Pick Points and, in the drawing, pick a point inside the inner circle. Right-click and choose Preview. Right-click and choose OK to accept the new hatch. Your drawing should resemble Figure 16–15.
9. Right-click and repeat the BHATCH command. From the Type drop-down list, select Predefined. From the Pattern drop-down list, select Solid. Notice that there are no angle or scale parameters available with the Solid pattern.
10. Choose Pick Points and, in the drawing, pick inside the bracket on the right of the assembly. Right-click and preview the boundary set. Right-click and choose OK to accept the solid hatch. Your drawing should now resemble Figure 16–16.
11. Save your work. If you plan to continue immediately with the next section, leave your drawing open.

EDITING HATCH PATTERNS

The HATCHEDIT command lets you modify hatch patterns or replace an existing pattern with a different one. Hatch objects drawn as associative hatch objects (the default type) automatically adjust to modifications in the boundary that defines them.

USING HATCHEDIT

With HATCHEDIT you can change the pattern of a hatch object or the parameters that control the generation of the pattern. To issue the HATCHEDIT command,

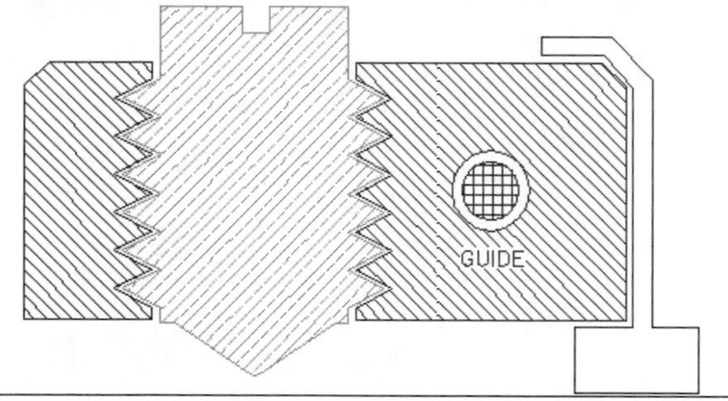

Figure 16–15 *Hatching the guide bar with a user-defined pattern.*

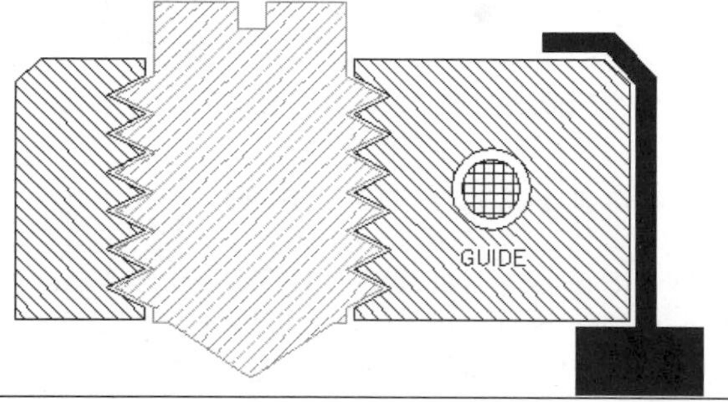

Figure 16–16 *Hatching the mounting bracket with a solid pattern.*

Note: You can also access the BHATCH command from the Draw toolbar or by typing **BHATCH** on the command line.

Tip: The Object Property Manager provides a comprehensive interface to all adjustable settings of an associative hatch pattern and can be accessed easily through the grip, right-click menu under Properties.

from the Modify menu, choose Hatch. The Hatch Edit dialog box is identical to the Boundary Hatch dialog box, but it has several settings disabled and unavailable (see Figure 16–17).

In the following exercises you will perform an edit on a previously drawn hatch object and investigate the behavior of associative hatches.

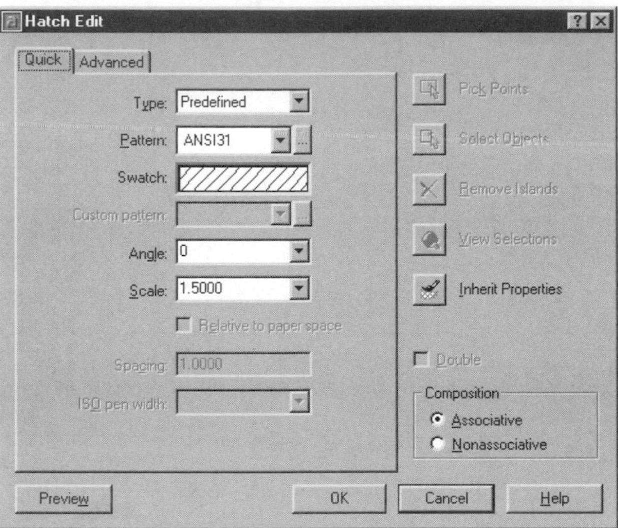

Figure 16–17 *HATCHEDIT dialog box.*

EXERCISE: EDITING A HATCH PATTERN.

1. Continue with or open 16DWG01.DWG. To start the HATCHEDIT command, from the Modify menu, choose Hatch. Select the solid hatch pattern by picking anywhere inside the hatch object.
2. In the Hatch Edit dialog box, notice that the name and parameters of the pattern you just selected are now displayed in the appropriate edit and drop-down boxes.
3. In the Pattern drop-down list, select ANSI31. Choose the Preview button.
4. To further modify the pattern, right-click to return to the Hatch Edit dialog box. Set the angle to 30 and the scale to 0.500. Choose Preview. Right-click and choose OK to accept the edits. Your drawing should resemble Figure 16–18. You will continue in this drawing in the next section.

EDITING HATCH BOUNDARIES

If you stretch or otherwise modify the scale or shape of the boundary objects defining the overall area of an associative hatch object, the hatch object automatically adjusts to fit the modified boundaries. If you move, delete, or stretch any of the islands within the overall hatch boundary, the hatch object is also adjusted.

If you delete any of the boundary objects defining the overall hatch area or islands (resulting in an open rather than a closed area), the associativity is removed from the hatch object, and the hatch loses the capability to adjust to modifications to the boundary. In the following exercise you will see how associative hatch objects automatically adjust to changes in their boundaries.

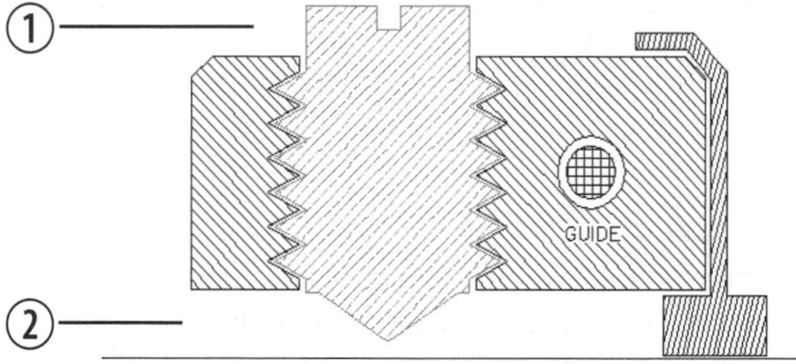

Figure 16–18 *Modifying a hatch pattern with HATCHEDIT.*

It is impossible to "repair" an associative hatch pattern after editing removes the associative property. However, you may use the Undo (U) command to correct a drafting mistake immediately afterward.

EXERCISE: UNDERSTANDING HATCH ASSOCIATIVELY

1. Continue with 16DWG01.DWG. From the Modify menu, choose Stretch and pick first at ①, then at ② (see Figure 16–18). Right-click to end the object selection process.
2. Type in the displacement **–1,0** and press Enter twice to carry out the stretch. Notice that the hatch pattern automatically adjusts to fill the new boundary geometry.
3. Save your work. You will continue in this drawing in the next section.

OTHER HATCHING CONSIDERATIONS

There are several other considerations to keep in mind when you create and work with hatch objects. These include the ability to "hide" or turn off hatching, exploding hatch objects into their constituent lines, and controlling whether associative hatch objects include their boundaries when they are selected for an editing operation.

ALIGNING HATCH OBJECTS

Areas hatched with the same pattern and with the same scale and angle parameters will have the corresponding elements in the pattern lined up in adjacent areas. All hatch patterns are referenced from the snap origin, which usually coincides with the drawing's origin, 0,0. If you wish to realign a hatch pattern, change the snap origin before you draw the hatch. The snap origin is controlled by the system variable SNAPBASE, which stores the value of a point. Set SNAPBASE to a point other than 0,0, such as the corner of a rectangular area, to cause the next hatch object drawn to align with the current snap origin.

The reason for this stems from the fact that families of lines in the pattern were defined with the same base point and angle and this is true no matter where the patterns appear in the drawing. This causes hatching lines to line up in adjacent areas. All hatch patterns are referenced from the snap origin, which by default is the same as the drawing's 0,0 origin.

EXPLODING HATCH OBJECTS

You can explode a hatch object into its constituent lines with the EXPLODE command. Exploding a hatch object removes any associativity. Additionally, the grouped set of line objects that make up the pattern replaces the single hatch object. While exploding a hatch object does enable you to edit the individual lines of the hatch, in most cases you lose more than you gain.

Note: Because a hatch object is composed of lines, you can use the same object snap modes (such as endpoint and midpoint) that you use on line objects on the individual lines in hatch objects, associative or exploded.

TURNING HATCHES ON AND OFF WITH FILLMODE

You can control the visibility of all hatch objects in a drawing by setting the FILLMODE system variable to zero. With FILLMODE off (set to zero) all hatch objects become invisible, regardless of the status of the layers on which the hatch objects reside. You must issue the REGEN or REGENALL command after you change FILLMODE in order for the change in visibility to take effect.

The disadvantage of using FILLMODE to control hatch object visibility is that FILLMODE also controls the fill of other objects such as wide polylines and multilines. If you wish to "hide" hatching more selectively, place hatching objects on separate layers so that the layers may be turned on and off without affecting other elements of the drawing.

SELECTING HATCH OBJECTS AND THEIR BOUNDARIES

Usually you will want to select both an associative hatch and its boundary for editing operations such as moving, mirroring, or copying. By default, however, AutoCAD treats the two elements separately during the object selection process. To speed hatch and boundary selection, you can change the value of the PICKSTYLE system variable from its default value of 1 to a value of 3. The PICKSTYLE system variable controls the selection of groups and hatch elements. A setting of zero or 1 disables simultaneous hatch and boundary element selection. A setting of 2 or 3 enables simultaneous hatch and boundary selection.

USING POINT ACQUISITION WITH THE HATCH COMMAND

The older version of the BHATCH command is HATCH. Although BHATCH replaced HATCH in functionality and, especially, in the ability to calculate boundaries, HATCH is still supported. The principle disadvantage of the HATCH command is that it can create only non-associative hatch objects.

Despite its drawbacks, HATCH does have an option that you may find useful: the Direct Hatch option. The Direct Hatch option enables you to define an area to be hatched "on the fly," negating the necessity to draw boundary objects before drawing the hatch object. The Direct Hatch option, or point acquisition method, is useful when you want to "suggest" large hatch areas by hatching only a few representative patches. This method is shown in the following exercise.

EXERCISE: UNDERSTANDING HATCH ASSOCIATIVELY

1. Continue with or open 16DWG01.DWG. Start the HATCH command by entering **HATCH** at the command prompt. Press Enter to accept the default ANSI31 hatch pattern.
2. At the next two command line prompts, enter **0.75** for the pattern scale and then press Enter to accept the default of zero for the pattern angle.
3. At the "Select objects:" prompt, press Enter to indicate that you will be specifying points instead of objects.
4. Enter **N** to indicate that you want to discard the polyline boundary after the hatch is completed.
5. Referring to the left portion of Figure 16–19, use an endpoint Osnap to pick point 1, and then pick points 2 through 9. Use an endpoint Osnap to pick point 10, enter **C**, and then press Enter to create and close the polyline boundary "on the fly."
6. The hatching has been completed. Your drawing should resemble the right portion of Figure 16–19.
7. Save your work and close the drawing.

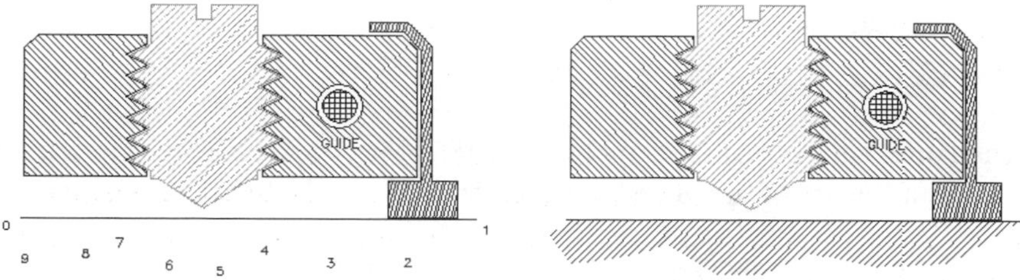

Figure 16–19 *Using the HATCH command and the Direct Hatch option.*

CUSTOMIZING HATCH PATTERNS

It is possible to add new patterns to the ones supplied with AutoCAD. You can add these new patterns (called *custom hatch patterns*) to the file ACAD.PAT (or ACADISO.PAT for metric patterns) or define each new pattern and assign it to its own PAT file. ACAD.PAT and ACADISO.PAT are found in the AutoCAD 2002\SUPPORT folder of a standard AutoCAD installation. If you choose to store each custom pattern in its own PAT file, the file must have the same name as the pattern. The new custom pattern files should be placed in one of the directories or folders defined in your installation's support file search path (see the OPTIONS command). Because hatch pattern files are ASCII files, a text editor is all you need to create these custom pattern files and add them to the ACAD.PAT file.

Note: If you choose to add custom hatch patterns to the ACAD.PAT file, you should first make a backup copy of this file. This will allow you to revert to a functional file should ACAD.PAT become corrupted. Creating a folder called SAFE, for example, under your AutoCAD 2002 installation provides a convenient place to locate such backup files.

A hatch pattern consists of one or more families of parallel pattern lines. The rules for defining a pattern line are the same as those for defining a new linetype except that no text or shapes can be included in the definition of a hatch pattern line. (See Chapter 5, Applying Linetypes and Lineweights, for information on linetypes.) A pattern definition can be broken down into two components: the header and the definition body. These will be explained in the next section.

Note: Although the rules for defining a hatch pattern are relatively straightforward, implementing the rules can take time, effort, and patience. A much easier and more cost-effective solution might be to purchase any of several third-party hatch pattern sets. If you definitely want to define your own pattern, read on.

DEFINING THE HEADER LINE

The first line in any pattern definition is called the header line. The format for the header line is as follows:

```
*Pattern-name [, description]
```

"Pattern-name" cannot contain any blank spaces. As shown, the description and the preceding comma are optional and are used only by the HATCH command.

Tip: Although the description portion of the header line is optional, you may always want to include a descriptive phrase. Doing so may be especially helpful when you work in the ACAD.PAT file, where the pattern names are frequently short and non-descriptive.

The header line is followed by one or more pattern line descriptors, one for each family of lines. A pattern line has the following syntax:

Angle, x-origin, y-origin, delta-x, delta-y [,dash-1, dash-2, ...]

The following line descriptor, for example, would result in the hatch shown on the left in Figure 16–20:

*L45, 45 degree lines @0.25 units

45,0,0,0,0.25

Each family of lines starts with one line, and the line's angle and origin are specified by the first three numbers of the line descriptor. In the L45 example, the first line is drawn at a 45-degree angle through the point 0,0. The family of lines is generated by offsetting each successive line by delta-X and delta-Y offsets, with delta-X measured along the line and delta-Y measured perpendicular to the lines. In the L45 example, each successive line is offset zero in the X direction and 0.25 in the Y direction. With no other *dash* specifications included in the definition, AutoCAD draws the lines with the current linetype.

The pattern shown on the right in Figure 16–20 is found in AutoCAD's ACAD.PAT file. It has the following definition:

*TRIANG, Equilateral triangles

60, 0,0, .1875,.324759526, .1875,-.1875

120, 0,0, .1875,.324759526, .1875,-.1875

0, -.09375,.162379763, .1875,.324759526, .1875,-.1875

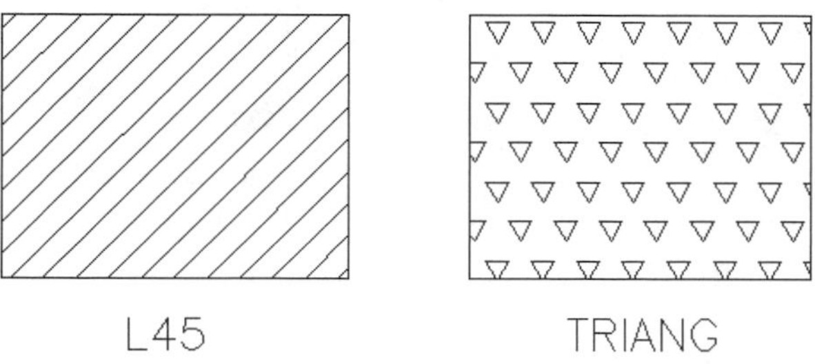

Figure 16–20 *The L45 and TRIANG patterns.*

In this example, the pattern consists of three families of lines: one family drawn at 60 degrees, another drawn at 120 degrees, and a third drawn at 0 degrees. The dash specifications (the last two numbers of each line) indicate that each line is to consist of a repetitive pattern of a 0.1875 dash and a 0.1875 space.

USING BOUNDARY TO DELINEATE AREAS AND ISLANDS

As you saw earlier in this chapter, when you calculate the boundary for a hatch object using the Pick Internal Point(s) method, the BHATCH command constructs a polyline or region to delineate the boundary set. The BOUNDARY command is a variation of the BHATCH command that creates objects delineating an overall area and the islands, if there are any, within that area. BOUNDARY also offers the option of creating region objects from the calculated polylines. The Boundary Creation dialog box is nothing more than the Advanced folder of the BHATCH Boundary Hatch dialog box (see Figure 16–21). Use the BOUNDARY command when you wish to delineate an area and its internal islands without applying hatch objects.

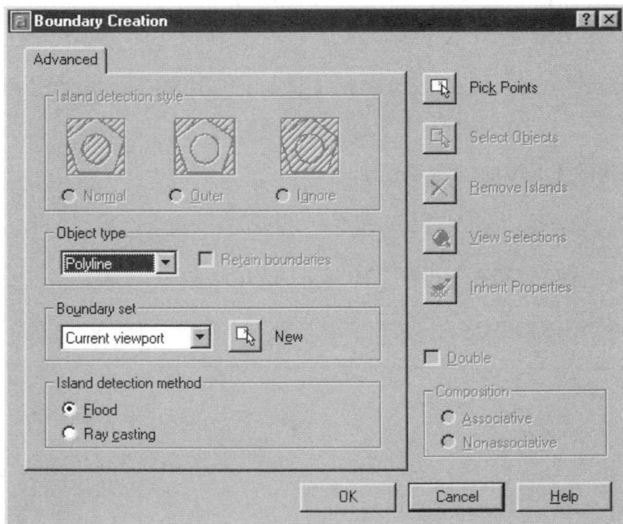

Figure 16–21 *Boundary Creation dialog box of the BOUNDARY command.*

SUMMARY

Hatching is a powerful tool for clarifying the meaning of your drawing or for conveying information to the reader. Effective use of hatching is one of the more powerful features of computer-aided design. Hatching is easy to apply using the BHATCH command and just as easy to edit with the HATCHEDIT command. Creating associative hatch objects lets you modify the hatched area easily after its creation. You can even design your own hatch patterns and add them to those that are supplied with AutoCAD.

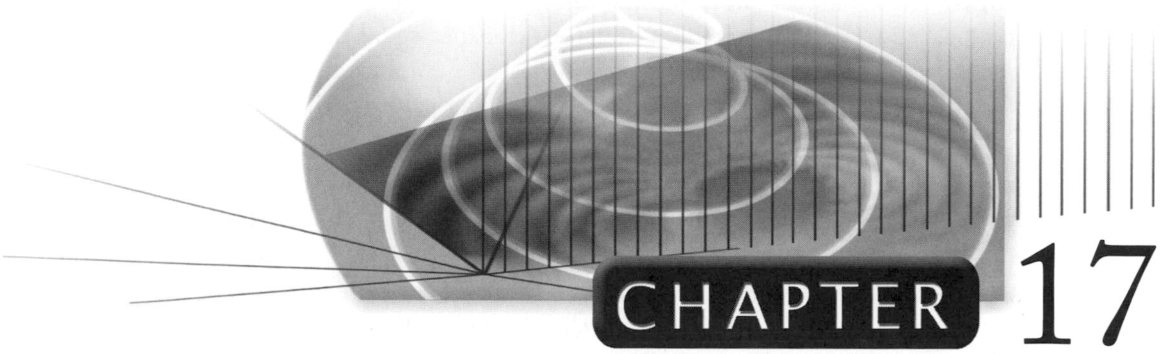

Dimensioning Basics

In a production environment, one of the most important aspects of a drawing is the dimension. AutoCAD provides you with many tools to dimension your drawings quickly and easily, especially if they are drawn accurately and to scale. In this chapter we will explore the tools that make dimensioning such a productive tool. In the following chapter we will explore how to modify dimensions and styles. This chapter covers the following topics:

- True associative dimensions
- Geometry-driven dimensions
- Trans-spatial dimensions
- The dimension subsystem
- Linear dimensioning types
- Other dimension types
- Dimensioning in layout view versus model space

TRUE ASSOCIATIVE DIMENSIONS

Probably one of the most substantial improvements to AutoCAD 2002 is the new true associative dimension object, which consists of two key features. The first feature is geometry-driven dimensioning, and the second is trans-spatial dimensioning.

Geometry-driven dimensioning automatically updates the dimension as the object is edited. Trans-spatial dimensioning lets you dimension model space objects from paper space, and it automatically updates the paper space dimension as the model space object is edited. Together, these two features enhance the value of the original associative dimensioning tool and make annotating your drawings very easy and virtually foolproof.

GEOMETRY-DRIVEN DIMENSIONING

While associative dimensions exist in previous releases, this updated version no longer uses only defpoints to indicate the point to which a dimension's extension line is tied. Instead, the dimension understands the points on an object with which its extension line defpoints are associated. Therefore, when you resize an object by moving its grips, the geometry-driven dimension automatically resizes itself, moving its extension line along with the grip's new position, and then updating the dimension's value to indicate the new measurement. The key improvement is that you no longer need to move a dimension's defpoint to maintain accurate dimensions. In AutoCAD 2002, geometry-driven dimensions update themselves dynamically as measured objects are resized.

TRANS-SPATIAL DIMENSIONING

As important an enhancement as geometry-driven dimensions are to AutoCAD 2002, a new feature of true associative dimensions that is even better is what Autodesk calls trans-spatial dimensioning, which allows you to dimension objects in model space from paper space. The key to this improvement is that the paper space dimension is automatically updated when the model space object is edited. This means that if you use a true associative dimension in paper space to measure an object in model space, AutoCAD maintains the association between the object and the dimension as the object is edited. Consequently, as you zoom and pan, or as you edit the model space object's geometry through the paper space viewport, AutoCAD instantly updates the true associative dimension's position and dimension value in paper space. Therefore, you can now dimension all model space objects from paper space and be assured that dimensions are always accurate. If you have not yet taken advantage of associative dimensions in previous releases, trans-spatial dimensioning provides a compelling reason for you to rethink how you dimension objects in your drawings.

Note: Trans-spatial dimensioning applies not just to model space objects, but to referenced drawings as well. Therefore, you can dimension xref objects through a paper space viewport, and, if the xref's objects are edited, the paper space dimensions are automatically updated to reflect the changes.

UPDATED DIM VARIABLES

To accommodate the new geometry-driven and trans-spatial dimensioning features, AutoCAD 2002 includes one new system variable and three new commands. The DIMASSOC system variable replaces DIMASO and allows you to determine the type of dimension created. The three possible settings for DIMASSOC are as follows:

- **0**—Exploded dimensions, which draw dimensions using a collection of separate objects, such as lines and text
- **1**—Non-associative dimensions, which are not automatically updated (similar to associative dimensions in previous releases)
- **2**—True associative dimensions, which are automatically updated as the geometry of objects changes (default)

Note: You can toggle between true associative and non-associative dimension creation from the Options dialog box, by selecting or clearing the Make New Dimensions Associative check box located in the Associative Dimensioning area of the User Preferences folder.

The three new dimensioning commands are the following:

- **DIMREASSOCIATE**—Allows you to convert a non-associative dimension into a true associative dimension and lets you reassociate a true associative dimension's extension line with a different point on an object
- **DIMDISASSOCIATE**—Removes true associativity from selected dimensions, changing them to non-associative dimensions
- **DIMREGEN**—Updates the locations of all associative dimensions, which is sometimes necessary when you zoom or pan in a viewport using your mouse wheel, or when you reload xrefs measured by true associative dimensions

CREATING TRUE ASSOCIATIVE DIMENSIONS

The process of creating true associative dimensions is the same as in previous releases and is fairly straightforward. You simply set current the dimension style that you wish to use, and then you choose the desired dimension command.

To demonstrate the power of true associative dimensions, the following example takes you through the process of dimensioning a model space object from paper space, where the object resides in an xref. In the example, the dimension is created in the Parent1 drawing, and it measures an object in the referenced Child1 drawing.

DIMENSIONING AN XREF OBJECT FROM PAPER SPACE

1. With the Parent1 drawing displayed, the Linear dimension command is selected from the Dimension menu, and then the xref's object is dimensioned from paper space, as shown in Figure 17–1. AutoCAD creates the true associative dimension in paper space, measuring the rectangle, which resides in an attached xref.

2. In the Child1 drawing, the rectangle is resized, as shown in Figure 17–2.
3. In the Parent1 drawing, the edited Child1 xref is reloaded, and then the true associative dimension is regenerated by issuing the DIMREGEN command, which updates the rectangle's image and its dimension, as shown in Figure 17–3.

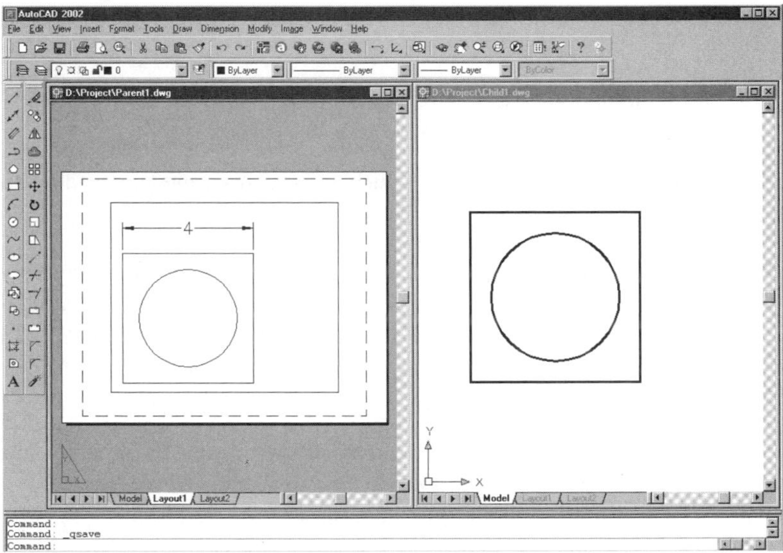

Figure 17–1 *The referenced drawing's rectangle is measured from paper space using true associative dimensioning.*

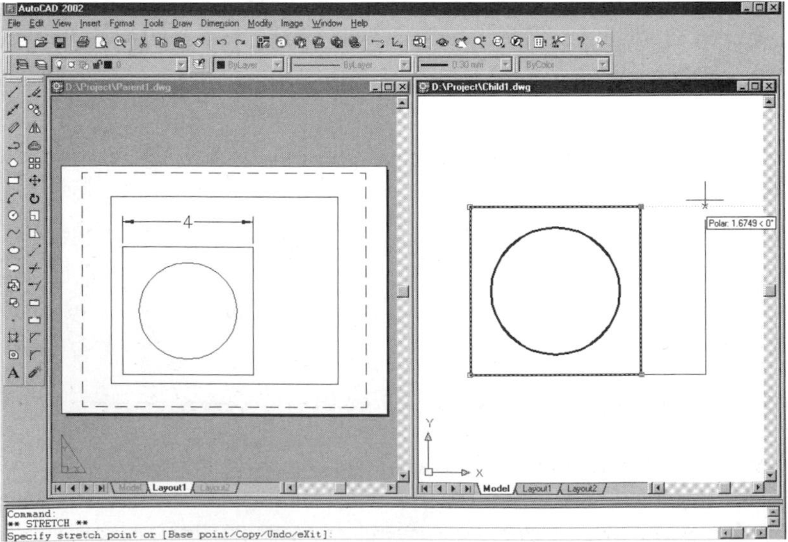

Figure 17–2 *The rectangle is stretched in the reference (Child1) drawing.*

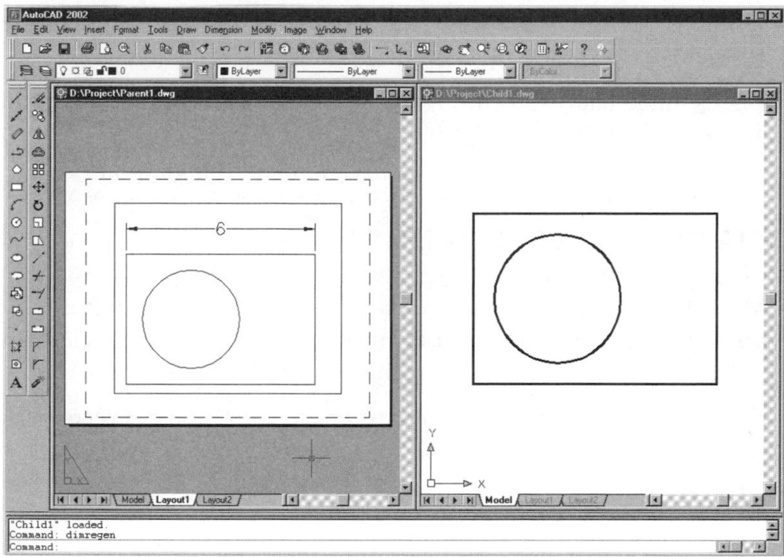

Figure 17–3 *When the referenced drawing is reloaded and the dimension is regenerated, the true associative dimension is updated to reflect the resized rectangle's new length.*

In Figure 17–1, the Parent1 drawing (on the left) references the Child1 drawing and displays the xref's objects through a floating viewport in paper space. When we dimension the reference file's rectangle using a true associative dimension, when the Child1 drawing's rectangle is edited, and then the referenced file is reloaded in the Parent1 drawing, the rectangle's new size is displayed in the Parent1 drawing. To update the true associative dimension, the DIMREGEN command is executed. This example demonstrates that true associative dimensions created in paper space retain their association with the objects that they measure, even if those objects lie in model space and belong to referenced drawings.

CONVERTING LEGACY DIMENSIONS TO TRUE ASSOCIATIVE DIMENSIONS

If you have existing projects whose legacy drawings contain non-associative dimensions, then you will likely wish to convert the legacy drawing's dimensions to true associative dimensions. The conversion process involves issuing the DIMREASSOCIATE command, selecting the non-associative dimension that you wish to convert, and then indicating the points on the object that the dimension's extension lines should measure. This means that the process of converting a legacy dimension to a true associative dimension requires that you pick the point to which a dimension's extension line measures, one extension line at a time. You can also use the DIMREASSOCIATE command to redefine the point to which an existing true associative dimension's extension line is tied. In essence, you use the DIMREASSOCIATE command to reassociate a single extension line, for either non-associative or true associative dimensions.

The following example takes you through the process of first converting a non-associative dimension to a true associative dimension, and then reassociating one extension line of the newly created true associative dimension to a different point on an object.

REASSOCIATING EXTENSION LINES

1. With the non-associative dimension and the object it measures displayed in your drawing, from the Dimension menu, choose Reassociate Dimensions. AutoCAD prompts you to select the dimension to reassociate.
2. Choose the dimension and then press Enter to end object selection. AutoCAD displays the current point on the object that the extension line measures, as shown in Figure 17–4. The "X" symbol indicates that the extension line belongs to a non-associative dimension.
3. Using Osnaps, pick the extension's new endpoint. In this case, the same point on the square is selected because the dimension is being converted from non-associative to true associative. Once you select the extension's new endpoint, AutoCAD moves to the next dimension line, prompting you to pick its new endpoint, as shown in Figure 17–5.

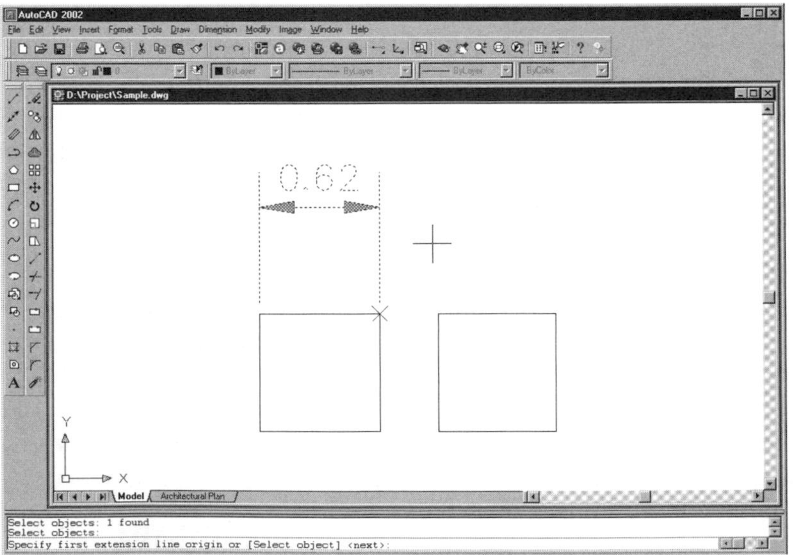

Figure 17–4 *AutoCAD prompts you to select the first extension line's new endpoint. The "X" symbol indicates the extension's current endpoint.*

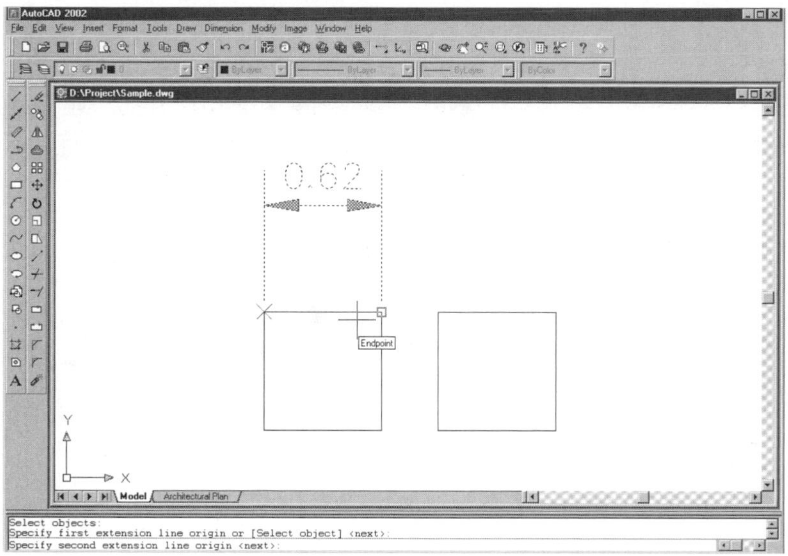

Figure 17–5 *After you select the first extension line's new endpoint, AutoCAD prompts you to select the second extension line's new endpoint, as indicated by the "X" symbol.*

4. Using Osnaps, pick the extension's new endpoint. Once you select the final extension line's endpoint, AutoCAD converts the non-associative dimension to a true associative dimension.

   Next you will reassociate one extension line to a point on another object.

5. From the Dimension menu, choose Reassociate Dimensions. AutoCAD prompts you to select the dimension to reassociate.

6. Choose the dimension and then press Enter to end object selection. AutoCAD displays the current point on the object that the extension line measures, as shown in Figure 17–6. The boxed "X" symbol indicates that the extension line belongs to a true associative dimension.

7. Using Osnaps, pick the extension's new endpoint. In this case, a new point is selected on a different object. Once you select the extension's new endpoint, AutoCAD moves to the next dimension line, prompting you to pick its new endpoint.

8. In this example there is no need to reassociate the second extension line. Consequently, you skip it by pressing Enter. Once you skip the final extension line's endpoint, AutoCAD updates the true associative dimension, as shown in Figure 17–7.

Once a non-associative dimension is converted to a true associative dimension, any time the measured object is edited, using either grips or AutoCAD editing commands, the dimension is automatically updated to reflect the edits. In this example, if either of the square's dimensioned corners moves, the true associative dimension will be updated to reflect the new measured value.

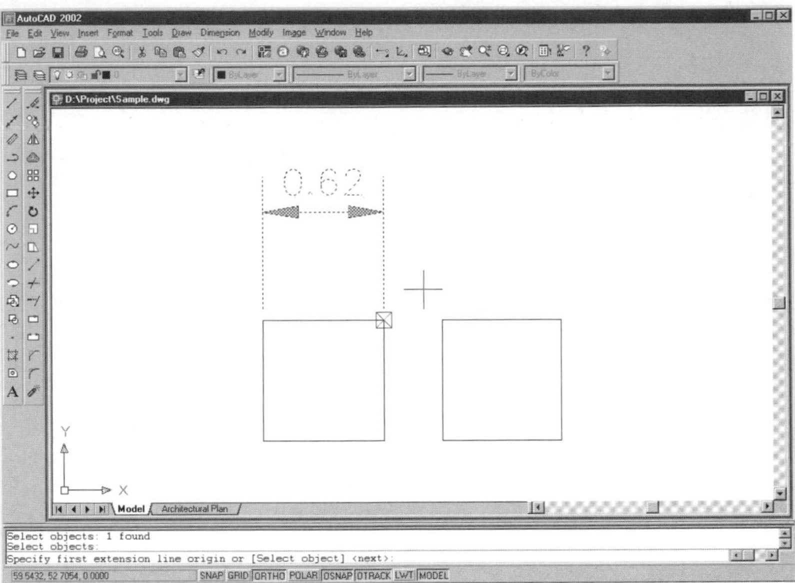

Figure 17–6 *The boxed "X" symbol indicates that the extension line belongs to a true associative dimension.*

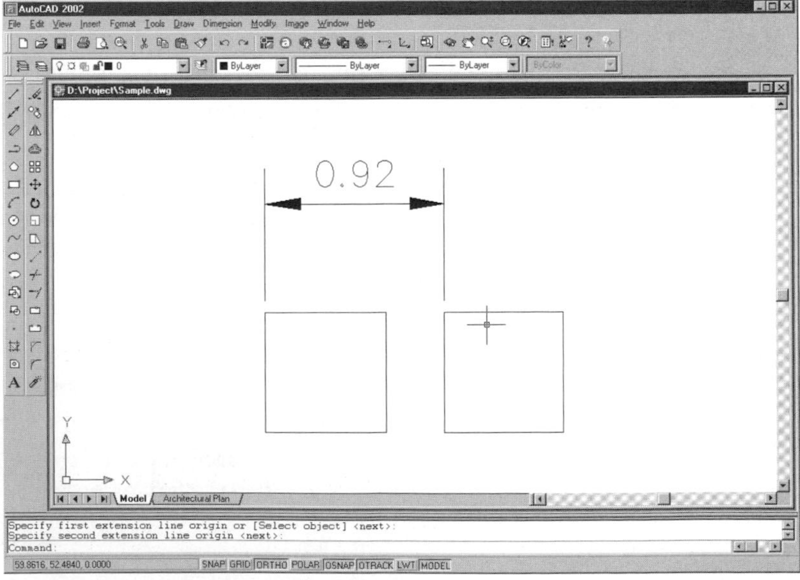

Figure 17–7 *When an extension line is reassociated with a new point, AutoCAD updates the true associative dimension to reflect the new measurement.*

THE DIMENSION SUBSYSTEM

AutoCAD provides an entire subsystem for dimensioning. Most of the time you will access these commands from the pull-down menus or toolbars. But you can also access them by typing in **DIM** at the command prompt. The prompt will change to a "DIM:" prompt, indicating that you are now in the dimension subsystem. Now, instead of having to type in the full name of a command, such as LEADER, you can just type in **L** to access the same command. As such, the keyboard shortcuts are slightly different in the dimension subsystem. When you are in regular command mode, L stands for the LINE command, not LEADER, so watch your command prompts when you work directly in the dimension subsystem. To exit the subsystem, just type in **Exit** and press Enter. The prompt will return to a standard AutoCAD prompt.

The subsystem method is an older way of accessing the dimension commands, but it is occasionally still useful to know. Now let us take a look at the basic dimensioning commands.

LINEAR DIMENSIONING TYPES

Linear dimensions, of course, define a specific length, whether it is horizontal, vertical, or aligned to the object you are dimensioning. AutoCAD provides you with five different linear dimensioning commands: DIMLINEAR, DIMBASELINE, DIMCONTINUE, DIMALIGNED, and DIMROTATED.

LINEAR DIMENSIONS

The linear dimensioning commands are accessed from the Dimension pull-down menu (see Figure 17–8), the Dimension toolbar (see Figure 17–9), or the command prompt. You should access this command using the method you are most comfortable with. For example, DIMLINEAR is the command prompt command for a linear dimension, but it appears as "Linear" on the pull-down menu and on the toolbar. Both are the same command.

The base linear command, DIMLINEAR, is fairly straightforward and easy to use. But you may not be aware of one or two options of the command, which are covered in the next section.

Linear Options

The DIMLINEAR command is based on selecting three points to create the dimension. These points are the start point and endpoint, as well as the location of the dimension line. Alternatively, under certain circumstances, you can create a linear dimension by selecting only two points on the screen.

When you choose the DIMLINEAR command, you are prompted to select the first extension line. Instead of selecting the first extension line, you can press Enter to select the line you want to dimension. Then, all you have to do is select the line and

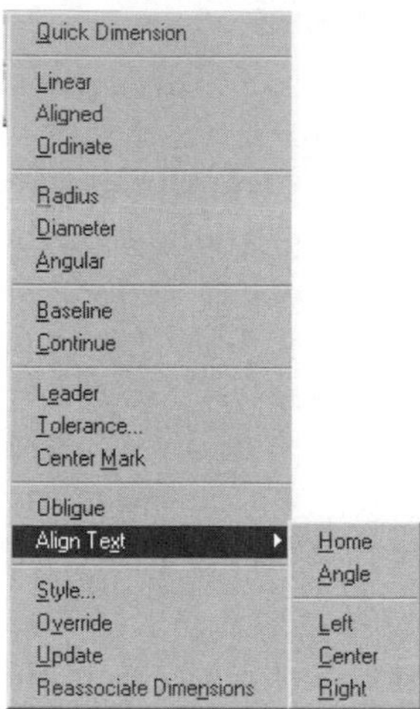

Figure 17–8 *The Dimension pull-down menu. where you can access all the dimensioning commands.*

Figure 17–9 *The Dimension toolbar, where you can select dimensioning commands instead of using the pull-down menu.*

place the dimension. The endpoints of the dimension are automatically determined from the endpoints of the line.

This option works well when you are dimensioning a single line, arc, circle, or polyline segment that is precisely the length you need it to be. This method does not work in the layout window, but it does work on internal lines of blocks and xrefs, simplifying the dimensioning process a bit.

If you use this method with a multi-segmented polyline, the segment you click on will be the segment that is dimensioned. If you use this method with a circle, you can dimension the diameter of the circle with a linear dimension. DIMLINEAR will recognize objects that it cannot dimension and then issue the following informational message: "Object selected is not a line, arc, or circle." Figure 17–10 shows you some sample dimensions created with two clicks.

The DIMLINEAR selection option does not, however, solve every situation for linear dimensions. In instances where using DIMLINEAR with the selection option does not work, you may still need to resort to using construction lines in conjunction with object snap modes.

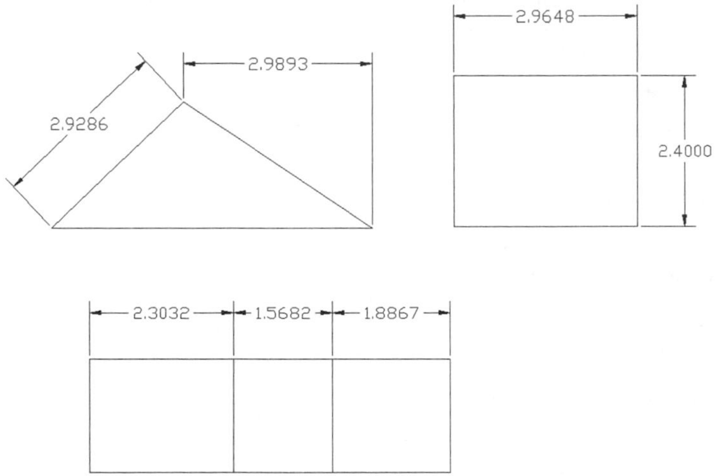

Figure 17–10 *Various types of linear dimensions.*

To get more productivity when you create linear dimensions, a couple of other commands are very useful. In particular, the DIMBASELINE and DIMCONTINUE commands are great timesavers. Both commands are used after you create an initial linear dimension to create additional dimensions quickly. These commands are discussed in the following sections.

BASELINE DIMENSIONS

Baseline dimensions are used to create a series of dimensions from a single base point quickly and easily. If you want to dimension various objects along a wall, but, for example, you want all the dimensions to be measured from one end of the wall, Baseline is the method to use.

To make use of the DIMBASELINE command, you must first create a linear, aligned, or rotated dimension. After you have the initial dimension, choose Baseline from the Dimension pull-down menu or from the Dimension toolbar. The first thing to do is to select the endpoint of the next dimension. Each dimension is then automatically placed above the previous dimension and spaced appropriately. AutoCAD remembers the last dimension placed when you use the DIMBASELINE and DIMCONTINUE commands. You can perform any non-dimensioning command by using DIMLINEAR and DIMBASELINE.

If you want to baseline a dimension that was not the most recently based dimension, you can press the Enter key at the "Specify a second extension line origin or (<select>/Undo):" prompt. This will enable you to select the dimension you want to baseline. This will work with the "continue" dimension type as well.

Figure 17–11 shows the result of using the baseline dimensioning command.

The following exercise shows you how to dimension an object quickly and efficiently using baseline dimensions.

EXERCISE: CREATING LINEAR DIMENSIONS BY USING BASELINES

1. Load the file 17DWG01.DWG from the accompanying CD.
2. Choose Linear from the Dimension pull-down menu, or select Linear from the Dimensioning toolbar.
3. Once you are in the LINEAR command, press Enter so you can select a single line segment.
4. Click on the top segment of the first step of the object. This creates the dimension.
5. Now click again at approximately 7.5,10.5 to place the dimension line. Your drawing should look similar to Figure 17–12.

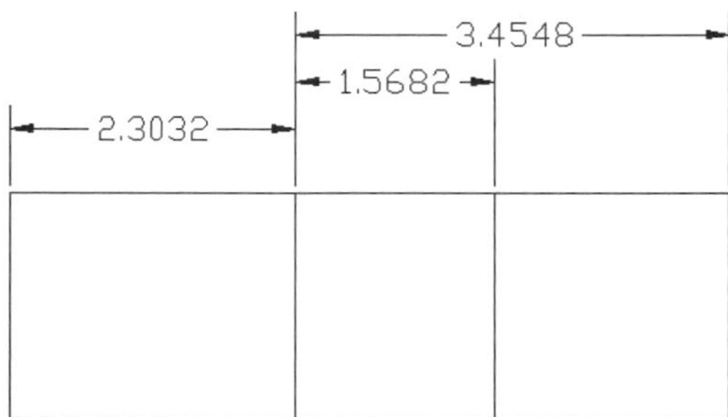

Figure 17–11 *The result of using the baseline dimensioning command, DIMBASELINE.*

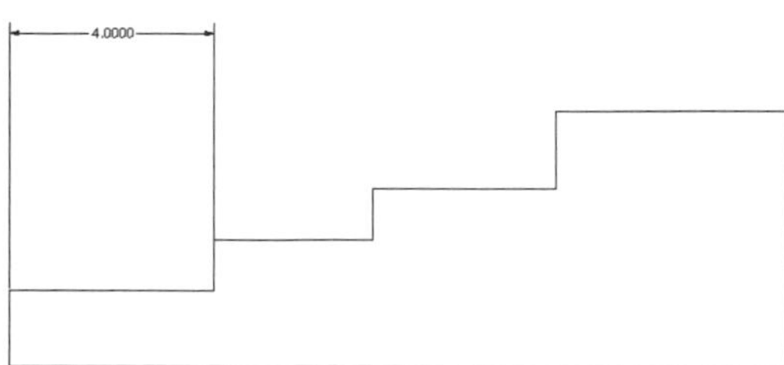

Figure 17–12 *The object after you add the first dimension.*

6. Choose Baseline from the Dimension pull-down menu.
7. Select the corner to the right of the first dimension. A new dimension is added just above the original. The result looks similar to Figure 17–13.
8. Select the next corner up and to the right.
9. Repeat step 9 to finish the dimensioning process.
10. Press Esc to exit the command when you are done. Figure 17–14 shows the final dimensioning of the object.

As you can see in this exercise, after you have created the first linear dimension—regardless of whether the dimension was created using DIMLINEAR, DIMALIGNED, or other linear commands—the block is dimensioned quickly with a minimal amount of mouse operations.

To test the DIMBASELINE command further, try rotating the plate 45 degrees before you create the first dimension. Then create a DIMALIGNED dimension for the first dimension. When you use the DIMBASELINE command again, you will see that it works perfectly.

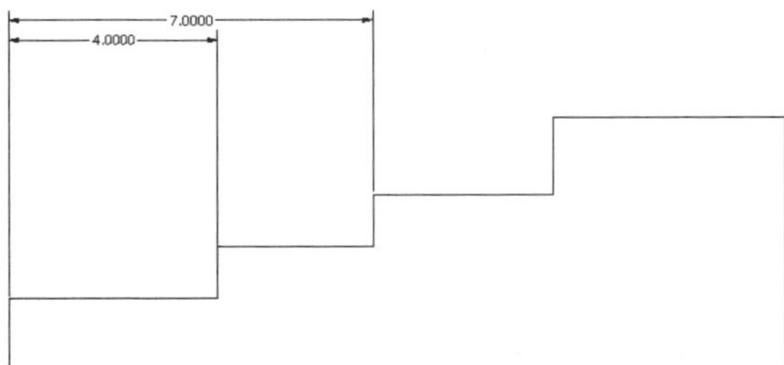

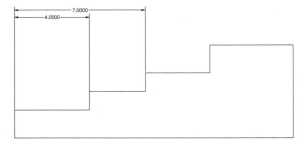

Figure 17–13 *The object after you add the second dimension.*

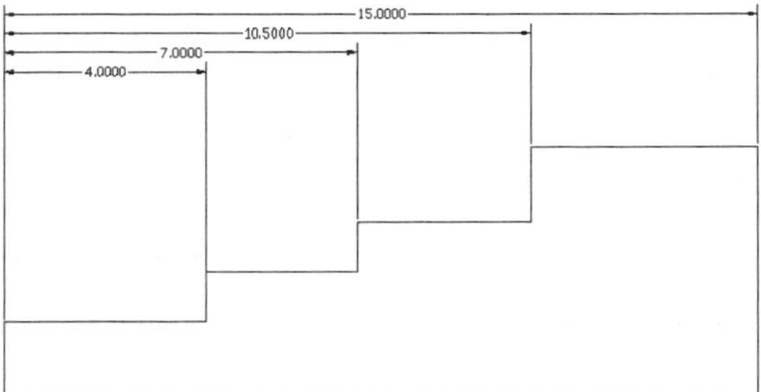

Figure 17–14 *The object with all the dimensions in place.*

QUICK DIMENSIONS

Before you look at some of the other dimensioning commands available in AutoCAD 2002, let us take a look at quick dimensions. Quick dimensions, or QDIM, is an automated system for quickly dimensioning a series of objects. It works by selecting the objects you want to dimension and then placing the dimension line. AutoCAD automatically figures out which parts to dimension and then creates the dimensions accordingly. To illustrate this, we will run through the previous exercise again, but this time we will use QDIM to create the dimensions.

EXERCISE: DIMENSIONING THE OBJECT WITH QUICK DIMENSIONS

1. Load the file 17DWG02.DWG from the accompanying CD.
2. Using a cross selection, select the entire top part of the object.
3. Choose Quick Dimension from the Dimension pull-down menu.
4. Place the dimension line somewhere above the object. You should see a result similar to Figure 17–15.
5. Select each of the vertical segments of the steps of the object.
6. Choose Quick Dimension again and place the dimension line to the left of the object. Figure 17–16 shows the final dimensions.

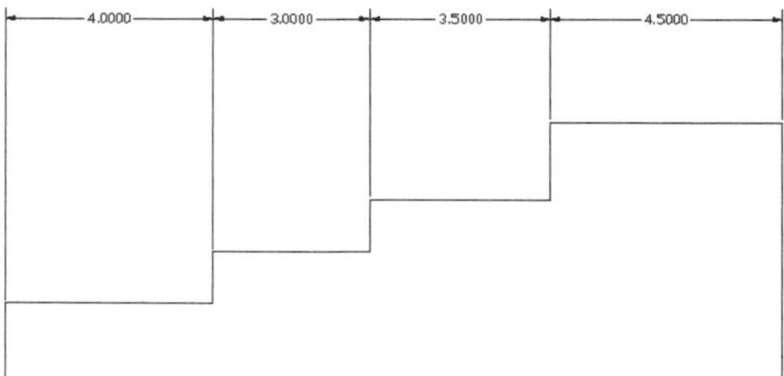

Figure 17–15 *The object after you add the first set of quick dimensions.*

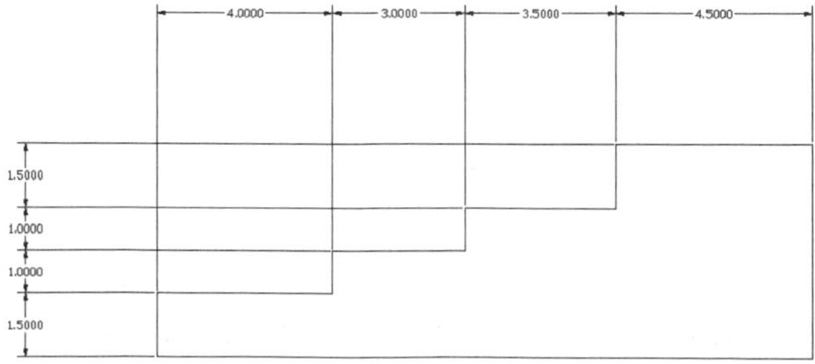

Figure 17–16 *The object with the second set of quick dimensions.*

As you can see in this exercise, QDIM is very quick and easy to use to create a set of linear dimensions. In the first set of steps in the last exercise, the location where you placed the dimension line determined whether or not you created horizontal or vertical dimensions. You could have just as easily created the vertical dimensions first. As you work with AutoCAD 2002, you will find that this command is a great timesaver.

CONTINUE DIMENSIONS

Continue dimensions are very similar to baseline dimensions, with one exception: instead of being based on a single point, continue dimensions are based on the endpoint of the last dimension drawn. Continue dimensions automatically line up the dimension lines to create crisp, clean dimensions. For example, a wall is generally dimensioned from centerline to centerline of the components, such as doors and windows, of the wall. A continue dimension type makes this very easy, whereas the earlier example of a baseline dimension would base all dimensions on a single point in the wall.

If you have to create a series of dimensions one after the other on a single dimension line, use the DIMCONTINUE command because it automates the placement of additional dimensions, much like the DIMBASELINE command does. Figure 17–17 shows an example of a continue dimension.

Like DIMBASELINE, DIMCONTINUE requires that you have one linear dimension type already created before you issue the command.

For an exercise that uses this command, repeat the baseline exercise but choose **Dimension>Continue instead of Dimension>Baseline.** Both commands work the same way; they just produce different results. As you may have guessed, in a lot of ways QDIM can replace continue dimensions, but there will still be instances in which it will be easier to use continue dimensions instead.

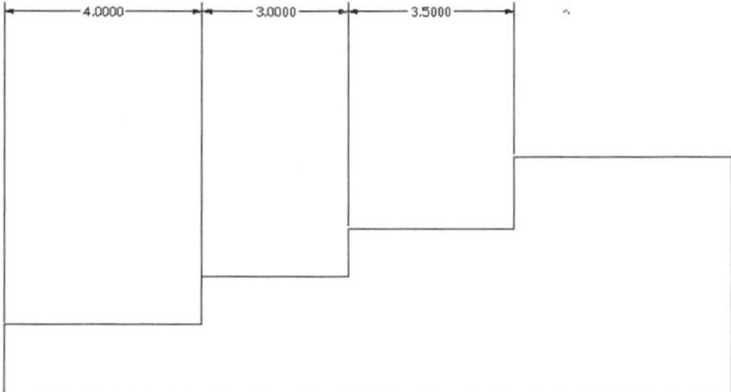

Figure 17–17 *A set of continue dimensions. They appear to be very similar to quick dimensions, as both are linear dimensions.*

ALIGNED AND ROTATED DIMENSIONS

The last two linear dimension types are aligned and rotated. Both of these types are similar to each other in that they are neither horizontal nor vertical dimensions. Aligned and rotated dimensions are the only linear dimensions in which the dimension line is not horizontal or vertical.

Aligned dimensions arrange the dimension line to match the angle defined by the start point and endpoint of the dimension. Rotated dimensions have the dimension line rotated a specific angle measure before the start point and endpoint are selected. Figure 17–18 shows examples of both types of dimensions.

As you can see in Figure 17–18, you can use the DIMROTATED command to create linear dimensions with any orientation. The DIMALIGNED command, however, is forced to align itself along the start point and endpoint of the dimension. Also note that a different dimension is measured, even though both dimensions use the same endpoints.

You can access DIMALIGNED through the Dimension pull-down menu, from the Dimension toolbar, or at the command prompt. The DIMROTATED command, however, is only available at the command prompt. Simply type in **DIMROTATED** to access the command.

The following exercise shows you how to use the DIMROTATED command to create a few dimensions. The exercise also shows you that you need to be careful when you create rotated dimensions because a rotated linear dimension may end up with a different measured length than the original.

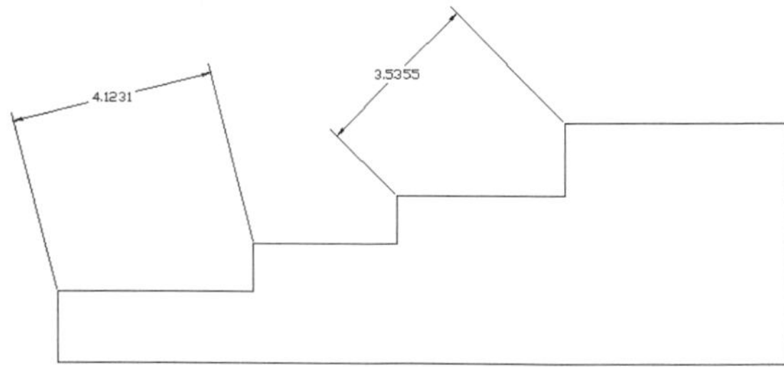

Figure 17–18 *An aligned dimension (left) and a rotated dimension (right). The rotated dimension is set to an angle of 45 degrees.*

EXERCISE: CREATING ALIGNED AND ROTATED DIMENSIONS

1. Load the file 17DWG03.DWG from the accompanying CD. This is the same drawing that you worked on earlier.
2. Choose Aligned from the Dimension pull-down menu.
3. Select the first and second points on the first step. Place the dimension line up and to the left. Figure 17–19 shows the dimension line, which is now aligned with the points you selected.
4. Now we will redo the dimension using a rotated dimension. Choose Undo.
5. Type in **DIMROTATED** at the command prompt.
6. Type in **45** as the angle for the line.
7. Select the same two points. Figure 17–20 shows you the resulting dimension, which is at the 45-degree angle specified.

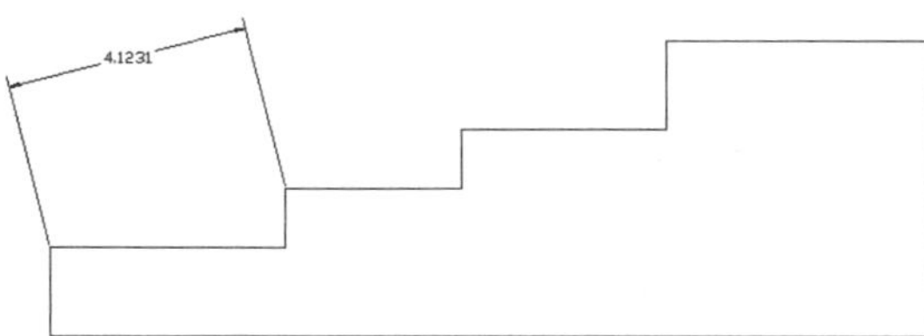

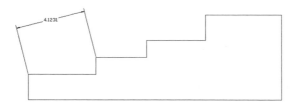

Figure 17–19 *The aligned dimension properly placed up and to the left. Notice how the dimension line is parallel to the points selected.*

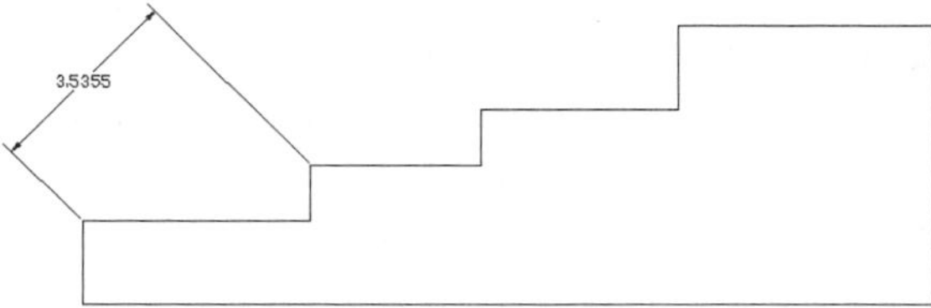

Figure 17–20 *A rotated dimension using 45 degrees as the angle of rotation.*

Notice that in both cases the distance measured is accurate for the angle that was selected.

As you can see, over 80 percent of the dimension types that you will create in AutoCAD will be one form of linear dimension or another. Knowing each of these commands can save you enormous amounts of time when it comes to creating dimensions.

Now let us take a look at a few other dimensioning types that are available in AutoCAD.

OTHER DIMENSIONING TYPES

Several other dimensioning types are worth exploring. These dimension types are not linear, and they serve specific purposes. Depending on your discipline, you may have a use for some of these types. For example, a mechanical part designer will make heavy use of radius and diameter dimension types, whereas a civil engineer will make use of datum dimension types.

RADIUS AND DIAMETER DIMENSIONS

Radius and diameter dimensions are used to dimension the size of an arc or circle, regardless of the object type. If you create a polyline with an arc in it, for example, you can use either dimension type to dimension the arc. If you select the Center Mark check box in the Dimension Styles dialog box, the center mark will automatically be used with this type of dimension when the dimension text is placed outside the circle or arc. Other than that, radius and diameter dimension types are straightforward. With these dimension types, you simply pick the arc or circle to dimension and then you pick the dimension line location.

ANGULAR DIMENSIONS

Angular dimensions are used to dimension the angle between two nonparallel lines. Of course, when you dimension angles between two lines, four angles are possible, one on each side of the intersection point of the two lines. Where you place the dimension line determines which angle is measured. Like radius and diameter dimensions, angular dimensions are straightforward.

ORDINATE DIMENSIONS

Ordinate dimensions are used to dimension a specific coordinate, such as a point from a civil survey. For example, a civil survey relies upon a set of three-dimensional data points on which to base topography. These coordinates are labeled using an ordinate dimension type, which labels the point's exact X and Y coordinates.

When you use ordinate dimensions, you may dimension the X- or Y-axis points, called datums. You also have the option to create a leader-like ordinate dimension that has text before or after the coordinate. Figure 17–21 shows an ordinate dimension.

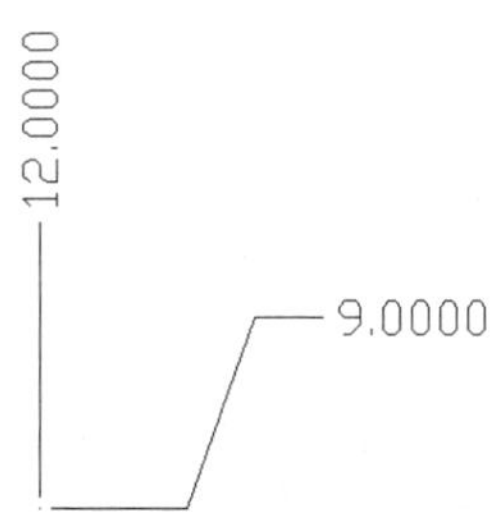

Figure 17–21 *An ordinate dimension with both X and Y datums visible.*

You can access the Ordinate Dimension command from the Dimension pull-down menu, from the Dimension toolbar, or at the command prompt. When you select this command, you are prompted to select the feature. AutoCAD is looking for you to tell it the coordinate to dimension. After you select the coordinate, you can select the type of ordinate dimension you want to use.

The four types of ordinate dimensions are X datum, Y datum, mtext, and text. X datum and Y datum produce the corresponding coordinate. Mtext pops up the Multiline Text Editor dialog box so you can add text before and after the datum dimension. The datum dimension appears as the <> symbol in the Multiline Text Editor dialog box. Figure 17–22 shows the Multiline Text Editor dialog box when it is used with the ordinate Mtext option. The Text option enables you to modify the text of the datum dimension without using the Multiline Text Editor.

You should not delete this <> marker; if you do, the actual coordinate will not appear in the dimension.

The following exercise shows you how to use ordinate dimensions to dimension several survey points. In this exercise the PDMODE system variable has been set to 2 so that points appear as crosses.

EXERCISE: CREATING ORDINATE DIMENSIONS

1. Load the file 17DWG04.DWG from the accompanying CD. It contains two points.
2. Choose Ordinate from the Dimension pull-down menu or from the Dimension toolbar.
3. Set your Osnap Type to Node.
4. Click on one of the crosses.
5. Place the dimension out to the right.
6. Select the same node again and then place the dimension above the point. Figure 17–23 shows the final result.
7. Repeat steps 4 through 6 for the other point in the scene.

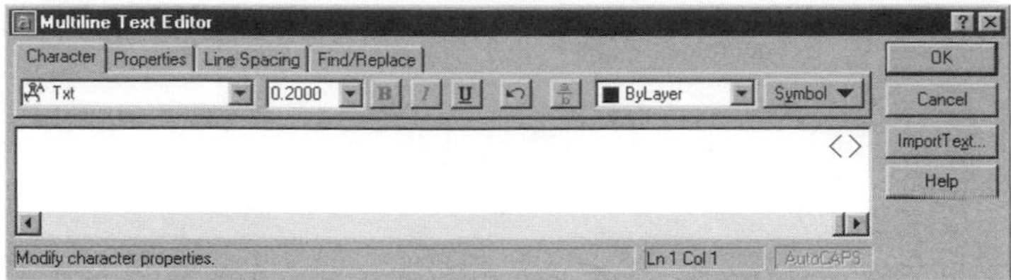

Figure 17–22 *The Multiline Text Editor with an ordinate dimension showing. The dimension appears as the <> symbol and should not be changed or deleted.*

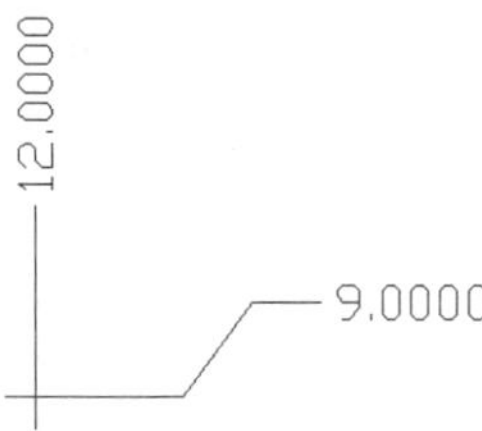

Figure 17–23 *The scene after you add the first set of ordinate dimensions.*

Ordinate dimensions can be slow to add because you have to add two for each point. Most civil engineering applications for AutoCAD provide shortcuts for creating these automatically.

Now let us take a look at tolerances.

WORKING WITH TOLERANCES

Another dimension type is the tolerance dimension. Tolerances are used to provide constraints within which you can construct the drawn object. For example, you might construct a mechanical part and specify that its length may be 2.0 cm ± 0.001 cm.

AutoCAD provides you with two methods of creating tolerance dimensions. One method is to specify the tolerances in the Dimension Styles dialog box. The tolerances are then automatically added to the dimension text as you place dimensions. The second method is to use the tolerance command and place tolerance symbols on the drawing. The second method is discussed in the following section. Figure 17–24 shows a standard tolerance symbol inside of AutoCAD.

Placing Tolerance Symbols in a Drawing

On the Dimension menu and toolbar you will find a Tolerance option. Choosing this option displays the Geometric Tolerance dialog box shown in Figure 17–25. The Geometric Tolerance dialog box is used to select the appropriate type of tolerance you want to use, through the use of industry-standard tolerance symbols.

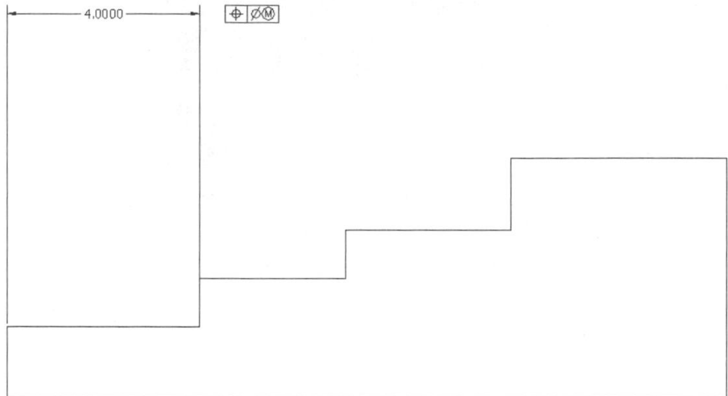

Figure 17–24 *A dimension with a tolerance created to the right of it. Tolerances define conditions for the dimension based on a material or other conditions.*

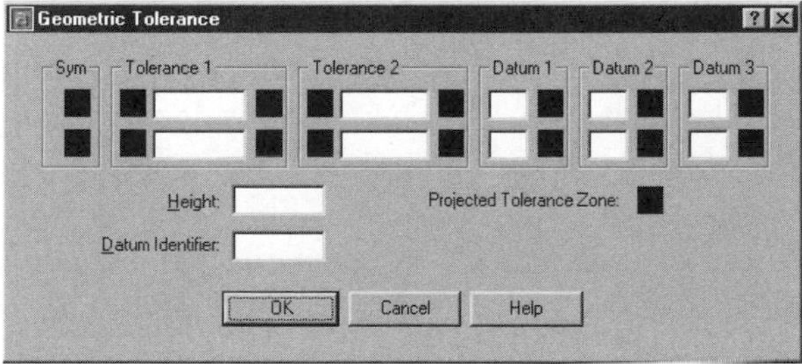

Figure 17–25 *The Geometric Tolerance dialog box enables you to specify the tolerances.*

In the Geometric Tolerance dialog box you can specify values for tolerances 1 and 2, as well as round symbols. You can also specify up to three datums, such as a material condition and a value for that condition. You can also specify height, projected tolerance zones, and datum identifiers.

At the far left of the Geometric Tolerances dialog box you will find two black boxes for symbols. Click on one to display the Symbol dialog box (see Figure 17–26). In this dialog box, notice the several different symbols, each representing a different geometric characteristic. When you choose one of the symbols, the selected tolerance method is then placed in the tolerance dimension itself. After you choose a geometric tolerance type and click on it, you are transferred back to the Geometric Tolerance dialog box, where you may then enter the values for the tolerances.

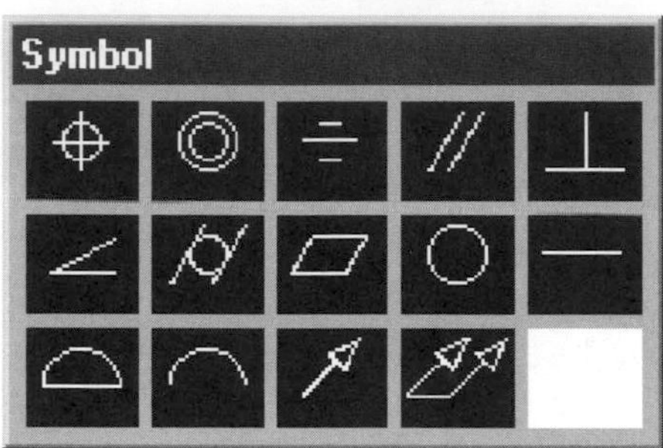

Figure 17–26 *The Symbol dialog box, where you can select the type of tolerance you want to use.*

CREATING LEADERS

Leaders are a very popular method of adding notes and pointing out specific aspects of a drawing. A leader is a line with an arrowhead pointing to a specific feature, with some sort of text or graphics at the end of the line. For example, you might create a wall of a house and use leaders to point out specific materials in the wall.

Selecting the Leader command from the Dimension pull-down menu easily creates a leader. When you are prompted for the first point, select the point where you want the arrowhead of the leader to appear. Then simply draw as many straight leader segments as you like. When you are done, simply press Enter, enter your text, and press Enter twice more to exit the command. This chapter will focus on several more advanced features of leaders, such as using the Multiline Text Editor dialog box to enter multiline text, and using splines instead of straight-line segments in your leaders.

Leader Options

In AutoCAD 2002, leaders have several more options than those found in previous versions of AutoCAD. When you select Leader from the Dimension pull-down menu or toolbar, you launch the QLEADER command. QLEADER is a quick leader command, much like the quick dimension command, and it is the commonly used version of the Leader command today. The standard Leader command is still available, and it uses the older methods for defining how the leaders look. In this book we will explore only the QLEADER command because you should use it instead, as it is more flexible.

When you choose QLEADER, before you choose your first point, a prompt appears asking you about settings. If you press Enter, the dialog box shown in Figure 17–27 appears.

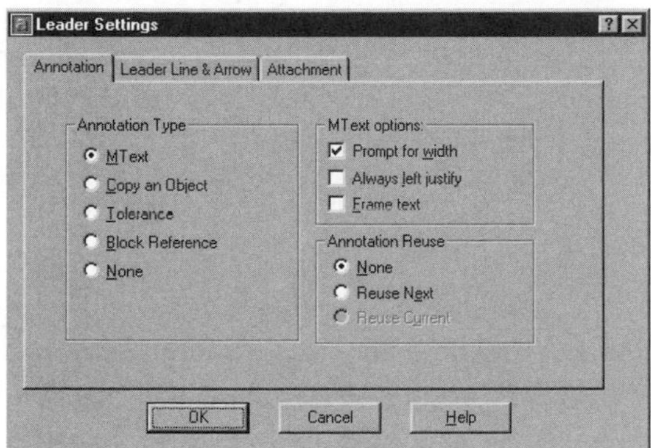

Figure 17–27 *The Leader Settings dialog box, where you can define how the leader is created.*

This dialog box consists of three different folders:

- Annotation
- Leader Line & Arrow
- Attachment

The last folder, Attachment, is only available when Mtext is selected as the annotation type. Figure 17–27 shows the Annotation folder. Here you can set what is shown at the end of the leader. Most of the time this is text, and you will use the Mtext option. But you can use the following options as well:

- **Copy an Object**—When this option is selected, you create the leader as normal. You will be prompted to select an object to copy. This object will then be copied and placed at the end of the leader.
- **Tolerance**—This option enables you to add a tolerance notation at the end of the leader.
- **Block Reference**—This option enables you to add an internal or external block to the end of the leader.
- **None**—Selecting this option enables you to create just the leader line and arrow, without additional annotation.

Depending upon which of these options you choose, various controls become available to the right of the Annotation folder. For example, if you choose Mtext, you will see three Mtext checkboxes available to you. These enable you to control the text finely. For example, you can have the system prompt you for the width of the mtext. You can set the system to left-justify the text all the time, or you can have it frame the text.

Additionally, you now have options for annotation reuse, which simply reuses the annotations you used in previous Leader commands. For example, if you need to create leaders to add a bunch of tolerances to your drawing, using leaders with the annotation reuse options makes life a lot easier.

The Leader Lines & Arrow folder (see Figure 17–28) enables you to define how the lines and arrowheads appear in the drawing.

There are four groups of controls here. The first is the Leader Line group. Here you can define whether the line of the leader is a straight line or a spline. Below this group is the Number of Points group of controls, where you can define the number of points that can be used to draw the leader line. The default is 3 points. You can set this to any number you like or just turn on the No Limit checkbox. Figure 17–29 shows a leader with straight segments, and a leader with splines.

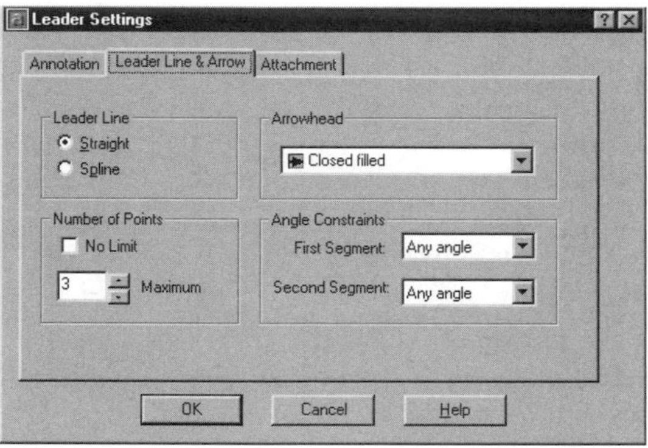

Figure 17–28 *The Leader Line & Arrow folder of the Leader Settings dialog box. Here you can define the look of the leader.*

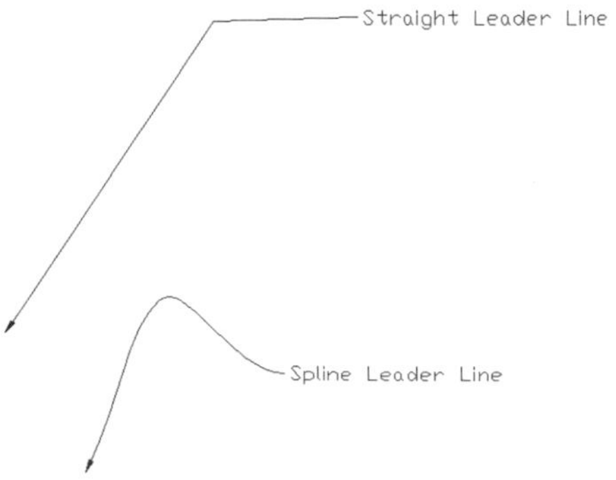

Figure 17–29 *Two leaders, one with a straight setting and the other using splines.*

The next group of controls consists of the Arrowhead option. Here you can define different types of arrows to be drawn at the start of the leader. Figure 17–30 shows the drop-down list of all arrow types that are available. In addition to those shown in Figure 17–30, you can also choose User Defined and then select any block that you have drawn and use that as an arrowhead. You should choose whichever arrowhead suits your style of drawing.

The last group of options you can set in the Leader Line & Arrow folder is the Angle Constraints group. Here you can force the leader lines always to be drawn at a particular angle. This is very useful for creating consistency in your drawings and your dimensions, making them easier to read and navigate. The default is any angle, which means you define the angle as you draw. You can also select Horizontal, 90, 45, 30, and 15. The angle constraints can be applied to the first and second segments of the leader line independently.

The last folder of the Leader Settings dialog box is the Attachment folder. This folder, shown in Figure 17–31, is only available when Mtext is selected as the attachment type, and it simply enables you to define the location of the text at the end of the leader line.

MTEXT OPTIONS

Once you have gone through the settings section of QLEADER, you place the points to define the leader. At the end, you will be prompted for the width of the text. A value of zero will create the width as needed. Then you will be prompted for the annotation options, with Mtext being the default. Simply press Enter here to bring up the standard Multiline Text Editor, where you can create any type of text you might need. Figure 17–32 shows the Multiline Text Editor dialog box.

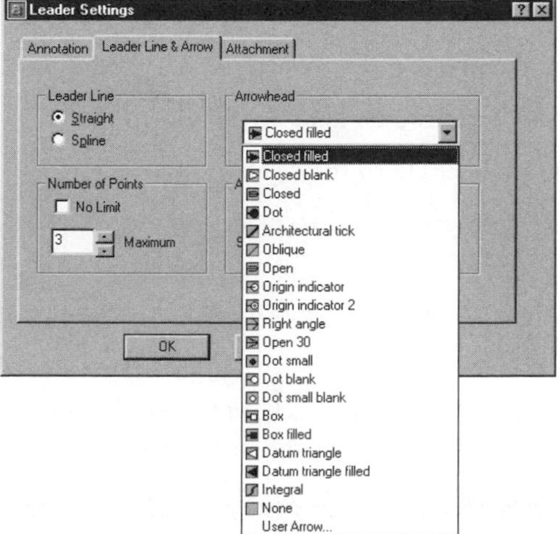

Figure 17–30 *The different types of arrowheads that you can select.*

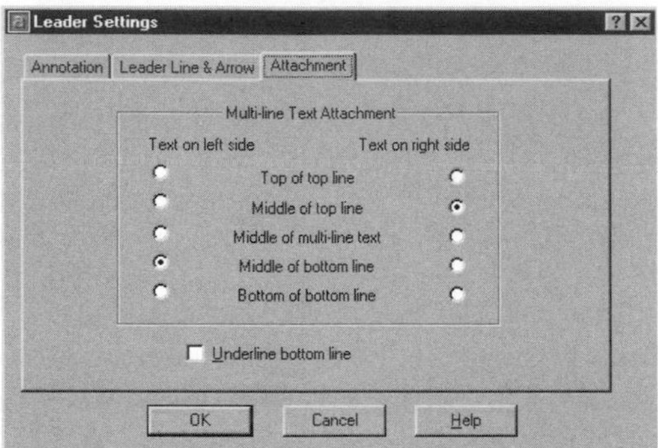

Figure 17–31 *The Attachment folder of the Leader Settings dialog box. Here you can define how mtext is placed at the end of the leader, depending upon the type of justification used in the mtext.*

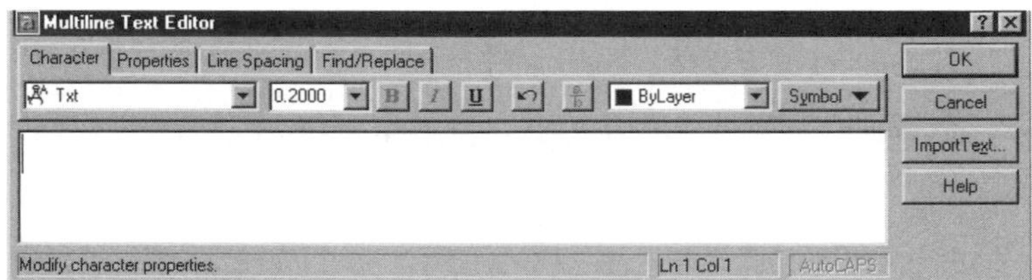

Figure 17–32 *The MultiLine Text Editor dialog box, where you can add and format text for use with your leader.*

The following exercise shows you how to create leaders on a simple architectural wall section.

EXERCISE: CREATING LEADERS IN AUTOCAD

1. Load the file 17DWG05.DWG from the accompanying CD. This file contains the same drawing from earlier exercises, but it contains two circles.
2. Choose Leader from the Dimension pull-down menu.
3. Click on the larger circle to the right.
4. Click at approximately 22,10.5 to place the second point, then click at 23,10.5 to place the third point.
5. When you are prompted for a width, press Enter.
6. When you are prompted for annotation, press Enter to start the Mtext command.
7. Type in **Radius +/- 0.125** and click on OK to close the Multiline Text Editor dialog box. Figure 17–33 shows the drawing at this point.
8. On your own, create a spline leader for the smaller hole.

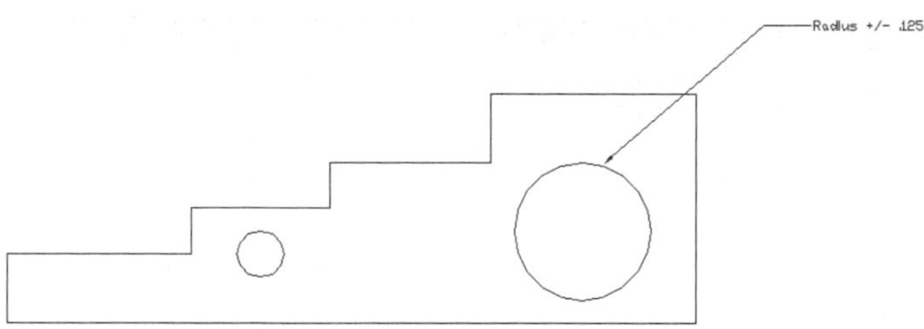

Figure 17–33 *The leader line with a note.*

LAYOUT VIEW VERSUS MODEL SPACE DIMENSIONING

When you look at productivity in terms of dimensioning a drawing, one other factor to consider is where you are placing your dimensions. You have two choices in AutoCAD: model space and the layout window. Each space has pros and cons.

PROS AND CONS OF MODEL SPACE

Most users today dimension their drawing in model space. This comes naturally because the drawing is actually created in model space. The advantages and disadvantages are listed and briefly described here.

The following are some advantages of creating dimensions in model space instead of in the layout window:

- You can use quick intuitive dimensioning directly on the drawing.
- When you use associative dimensioning, you can stretch both the geometry and dimensions at the same time, enabling both the geometry and the dimensions to be updated at the same time.
- You can use the object selection dimensioning method.

The following are some disadvantages of creating dimensions in model space instead of in the layout window:

- If you have a sheet with drawings created at different scales, you must use different scale dimensions as well.
- For dimensions to be plotted correctly, all dimensions must be scaled by a scale factor that is equivalent to the output plot scale.

Overall, the biggest reason to place your dimensions in model space is that you do not understand the layout window and how it works. If you are not comfortable working in the layout window yet, create your dimensions in model space until you do feel you are ready to work in the layout window.

If you work in an environment where you constantly create drawing sheets with varying drawing scales, you should strongly consider using the layout window and its associated dimensioning methods.

PROS AND CONS OF LAYOUT VIEW DIMENSIONING

With layout window dimensioning, you place your dimensions in the layout window, thus separating them from the drawing. Like model space, the layout window also has advantages and disadvantages in dimensioning.

The following are some advantages of creating dimensions in the layout window instead of in model space:

- Layout dimensions are separate from the drawing, which makes it easy to switch over to model and view a clean drawing.
- All layout dimensions make use of the same dimension scale factor: 1.
- Dimensions can be placed on sheets more easily with multiple scales.

The following are some disadvantages of creating dimensions in the layout window instead of in model space:

- You cannot stretch layout dimensions and model space geometry at the same time.
- You cannot use the object selection dimensioning method.

Ultimately, the decision of whether to use layout window dimensioning or not depends on your comfort with and understanding of the layout window itself. If you are not comfortable with it, continue to place dimensions in model space.

IMPROVING PRODUCTIVITY: TIPS AND TECHNIQUES

The following are a few techniques to help you increase your dimensioning speed when you create dimensions. Editing dimensions is covered in Chapter 18, Managing Dimensions.

- Create keyboard shortcuts for most of the dimension commands. For example, DIMLINEAR can be shortened to DL, which is much quicker to type.
- Create a chart of dimension scales for standard plot scales. That way, you create consistency in your drawing when you work with multiple draftsmen.
- Create a variety of dimension styles and save them to AutoCAD template files. Then all you have to do is assign the appropriate style as the current one and begin dimensioning.
- Whenever possible, use the object selection method for dimensioning because it is quickest.
- If you are going to create a series of dimensions, consider using baseline or continue dimensions to help automate and speed up the process.
- If you have a third-party program, consider using that program's dimensioning routines, if it has any. These routines will probably be quicker than the standard AutoCAD commands.

SUMMARY

As you saw in this chapter, many of the dimensioning commands that you might use are quite powerful, yet very easy to use. Linear, angular, diameter, and leader dimension types are used in almost all drawings created.

Managing Dimensions

One of the most powerful aspects of dimensioning in AutoCAD is the ability not only to edit dimensions after you have created them, but to define styles as well. Dimension styles are used to define the overall look of dimensions such as colors or arrowhead types. By defining a good set of styles, you can ensure easy readability of your drawings as well as consistency in the drawing. This chapter covers the following topics on dimensions:

- Editing dimensions
- Editing dimension text
- Updating dimensions
- DimEdit
- Dimension styles
- Dimension variables

EDITING DIMENSIONS

After you have created your dimension styles and created a variety of dimensions in your drawing, you will eventually need to be able to edit those dimensions. Some reasons why you might need to modify your existing dimensions include the following:

- The drawing plot scale changes.
- You make a change to already dimensioned geometry.
- You want to specify a different dimension text and override the AutoCAD measurement.
- You want to reposition the dimension text to help clean up the drawing.
- You want to change the style of a dimension without having to recreate the dimension.

The following sections discuss the various techniques for modifying existing dimensions.

Note: As mentioned in the last chapter, AutoCAD 2002 has a new Associativity feature for all dimensions except quick dimensions. This means that in a lot of cases, editing a dimension just means editing the underlying geometry.

GRIP EDITING

One of the most powerful methods of editing in AutoCAD is grip editing. Just like you can grip edit most objects in AutoCAD, you can grip edit dimensions as well. Of course, you can only use grip editing if PICKFIRST and GRIPS are enabled and set to a value of 1.

To grip edit a dimension, simply click on the dimension to highlight it. You will immediately see the grip boxes appear on a linear dimension, as shown in Figure 18–1.

Of course, the exact location and effect of each grip will differ from dimension type to dimension type. Figure 18–2 shows the grip layout for a radius dimension.

Figure 18–1 *A dimension that has been selected for grip editing. Notice the boxes, where you can edit portions of the dimension.*

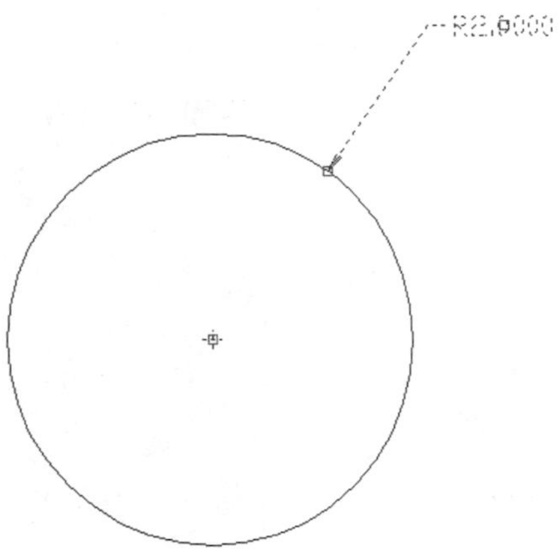

Figure 18–2 *A radius dimension that has been selected for grip editing.*

To edit a grip, simply click on one of the blue grip boxes. The box will turn red to indicate that it is selected. Then right-click on the box to view the grip editing pop-up menu shown in Figure 18–3.

With the enhanced right-click capabilities in AutoCAD 2002, you will find that not only can you select options for working with the grip; you can also find options for editing the dimension itself. This enables you to control options such as the placement of the dimension text, the precision of the measurement, and even the dimension style itself.

Most of the time you will use the Move option to reposition the dimension text, the dimension line, or the start point or endpoint of the dimension. After you select the option you want, simply grip edit the dimension just like you would any other object.

A couple of problems you might encounter are as follows:

- If you select the grip that is nearest to the dimension text and select Rotate, the dimension will rotate around the text. You will not rotate the text itself. You must use a special dimension editing command to rotate the text and not the dimension line.
- If you are working with a radius or diameter dimension, you can grip edit the center point of the dimension. If you reposition the center point, the dimension text will change. AutoCAD does not maintain a link between the dimensioned object's center and the dimension itself. Always make sure you move the point back to the center of the dimensioned object.

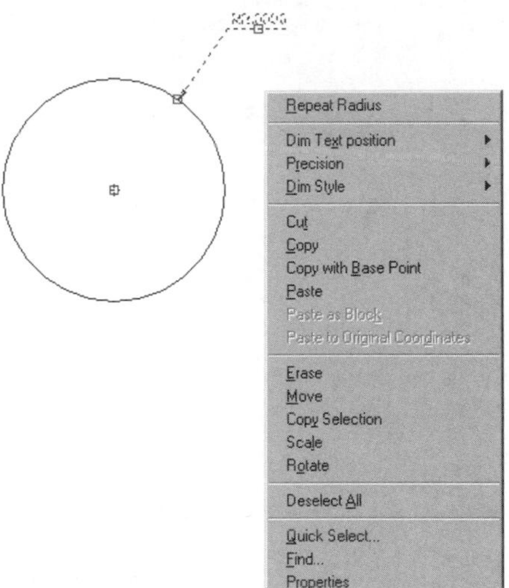

Figure 18–3 *The right-click menu that appears when you right-click on a selected dimension.*

The following exercise shows you how to make use of grip editing with dimensions.

EXERCISE: GRIP EDITING A DIMENSION

1. Load the file 18DWG01.DWG from the accompanying CD. This file is the same figure of steps that was used in the last chapter.
2. Select the dimension. You should see the grips appear.
3. Select the grip at the endpoint of the dimension. It will turn red to indicate that it is selected.
4. Move the grip to the end of the next step and select that endpoint.
5. The dimension will stretch and the value will change as shown in Figure 18–4.
6. Right-click on the same grip and choose Precision>0.00 from the pop-up menu. This changes the precision of the readout.
7. Hit Esc a couple of times to get out of the command.

EDITING DIMENSION TEXT

One of the most common editing tasks for a dimension is changing the dimension text after the dimension has been created. The easiest way to edit the text is simply to choose Text from the Modify pull-down menu. This executes the DDEDIT command. If you select the dimension object, the Multiline Text Editor will appear, as shown in Figure 18–5.

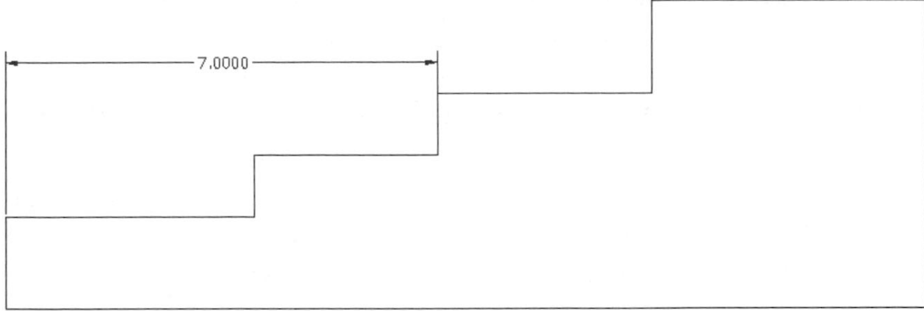

Figure 18–4 *The dimension after you stretch it with grip editing.*

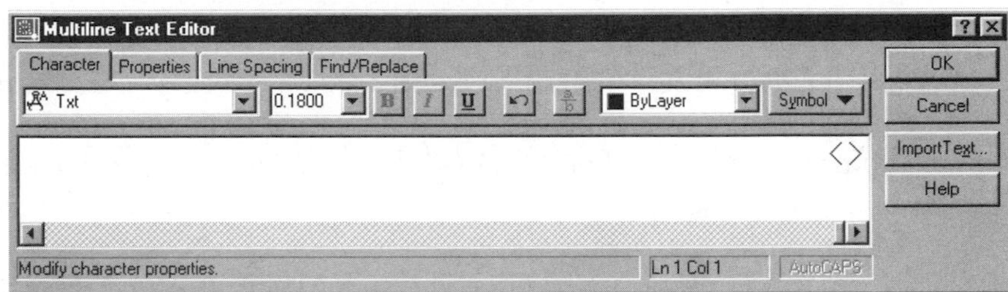

Figure 18–5 *The Multiline Text Editor dialog box when it is used with a dimension.*

The only thing that appears in the dialog box is the <> symbol. This symbol indicates the measured AutoCAD value. To replace the measured value, delete the <> and replace it with the value you want. Otherwise, add the text before and/or after the symbol as you see fit. Just make sure to pay attention to how large the text will be inside the dimension. You do not want to put more text than the amount that will fit.

When it comes to leaders, the DDEDIT command works just fine for editing the text.

In addition to changing the value of the text in a dimension, you can also rotate and reposition the text. The fastest and easiest way to reposition text is simply to grip edit the dimension. Alternatively, you can also use the DIMTEDIT command, which you access by clicking on the Dimension Text Edit button on the Dimension toolbar or by choosing Align Text from the Dimension pull-down menu. When you use the pull-down menu, each DIMTEDIT option is listed individually on the Align Text flyout.

DIMTEDIT enables you to reposition the text as well as align it to the left or right side of the dimension. If you make a mistake, DIMTEDIT also has a Home option that you can use to move the text back to the original position it was in when the dimension was created. The last DIMTEDIT option is Rotate, which enables you to rotate the text of a dimension without rotating the dimension itself.

The following exercise shows you how to edit the text of a dimension.

EXERCISE: EDITING THE TEXT OF A DIMENSION

1. Load the file 18DWG02.DWG from the accompanying CD. This file contains the same simple object with which we have been working.
2. Choose Modify>Object>Text>Edit from the pull-down menu. This brings up the DDEDIT command. Alternatively, you can also just type in **DDEDIT** at the command prompt.
3. Select the dimension. The Mtext dialog box should appear.
4. Click after the <> symbol. Put in a space and the word **FEET**.
5. Close the Mtext dialog box. Figure 18–6 shows the resulting dimension.

Dimedit

DIMEDIT is another AutoCAD dimension editing tool. You access this command by typing **DIMEDIT** at the command prompt, or by selecting the Dimension Edit tool from the Dimension toolbar. The DIMEDIT command is not available from the Dimension pull-down menu.

DIMEDIT enables you to reposition the dimension text back to the home position, rotate the text, and replace the dimension text, just like DIMTEDIT does. What is new is the capability to add an obliquing angle to a dimension. An obliquing angle forces the vertical extensions lines off from the vertical position by the angle specified. This is more of a cosmetic adjustment that makes a dimension look more interesting. Obliquing a dimension does not affect the text, dimension line, arrowheads, or origin points. It affects the extension lines only.

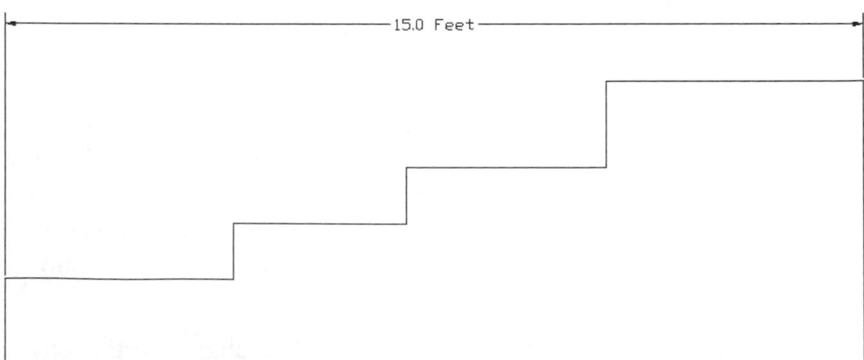

Figure 18–6 *The dimension after you modify the text to include the word "FEET."*

The following exercise shows you how to use DIMEDIT.

EXERCISE: USING DIMEDIT ON A DIMENSION

1. Load the file 18DWG03.DWG from the accompanying CD. This contains the drawing from the last exercise.
2. At the command prompt, type in **DIMEDIT**
3. Type in **R** at the command prompt to activate the Rotate command.
4. Enter **45** as a value.
5. Select the dimension. The text is then rotated 45 degrees, as shown in Figure 18–7.
6. Hit Enter to bring up DIMEDIT again.
7. Type in **O** at the command prompt to enter Oblique mode.
8. Select the dimension.
9. Again enter 45 degrees as the value.
10. The entire dimension is shifted so the vertical lines of the dimension are at a 45-degree angle, as shown in Figure 18–8.

Dimension Styles

Dimension styles are the primary method for controlling how a dimension appears. When you create a dimension style, you define exactly how that dimension is going to appear in the drawing. This includes the dimension scale, the types of arrowheads, whether or not the dimension lines appear, and, if they do, what color the dimension lines will be.

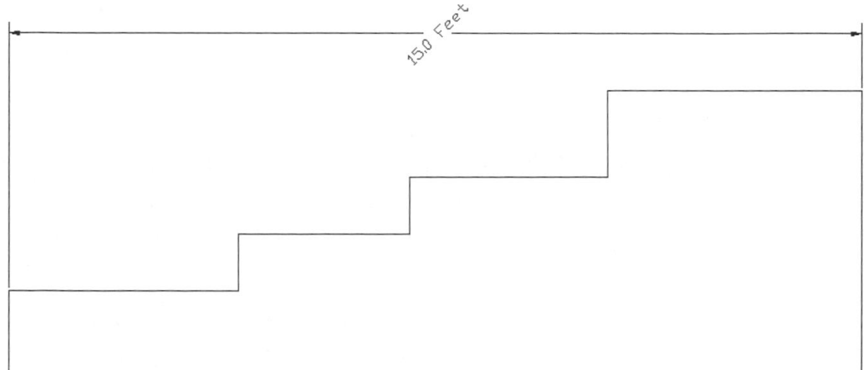

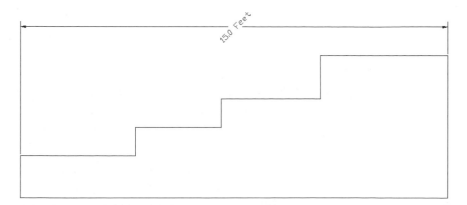

Figure 18–7 *The text of the dimension after you rotate it 45 degrees.*

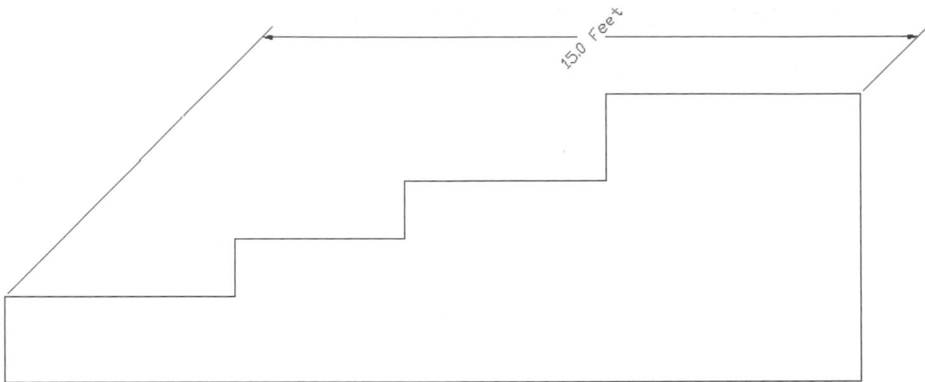

Figure 18–8 *The dimension after you rotate the text and oblique it 45 degrees.*

AutoCAD controls dimension styles through the use of dimension variables (DIMVARS). You can control these variables in two different ways. You can use the DDIM Dimension Style dialog box to access the variables using a graphical interface, or you can type the variable name at the command prompt and assign it a new value. There are approximately sixty-one dimension variables in AutoCAD 2002. Most of the time, adjusting the dimension variables through the Dimension Style Manager dialog box is the best way to go (see Figure 18–9). This dialog box is accessible through the Dimension toolbar or the Styles option on the Dimension menu.

The Dimension Style Manager enables you to set the current dimension style, create a new style, modify an existing style, override a part of the current style, or even compare two existing styles in the same drawing. To make matters even easier, the Dimension Style Manager gives you a graphical preview of what the currently selected dimension style will look like in use.

The Dimension Style Manager

Dimension styles provide you with a method for saving different sets of dimension variables for the various types of drawings you might create. There are many different options for defining how a dimension looks. To help you understand some of these options, Figure 18–10 shows a standard linear dimension with all the parts of the dimension labeled.

To edit a dimension style, you simply pull up the Dimension Style Manager, select the style you want to edit, and then click on the Modify button. To create a new style, you can choose New in the Dimension Style Manager. When you do, a dialog box pops up (see Figure 18–11); here you can name the style and base it on an existing style.

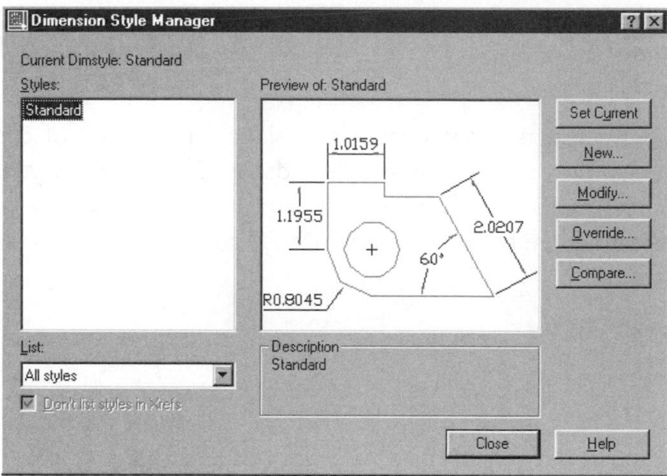

Figure 18–9 *The Dimension Style Manager dialog box. Here you can create, edit, or delete dimension styles.*

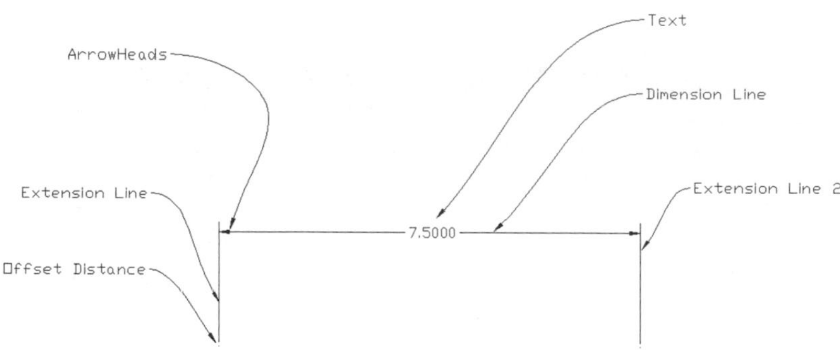

Figure 18–10 *The major parts of a linear dimension.*

Figure 18–11 *The Create New Dimension Style dialog box, where you can give the style a name, select the base style, and select how the dimension style will be used.*

Regardless of whether you choose to modify an existing style or create a new style, the Dimension Style dialog box (see Figure 18–12) will be displayed. Here you can edit the individual parts of the style.

The Dimension Style dialog box contains a folder for each set of controls for the style. A preview window accompanies each folder window to show you exactly what the dimension style looks like. This preview window enables you to make a change to the style and immediately see the effect of that change. The folders of the dialog box are as follows:

- Lines and Arrows
- Text
- Fit
- Primary Units
- Alternate Units
- Tolerances

Each of these folders is explained in the following sections.

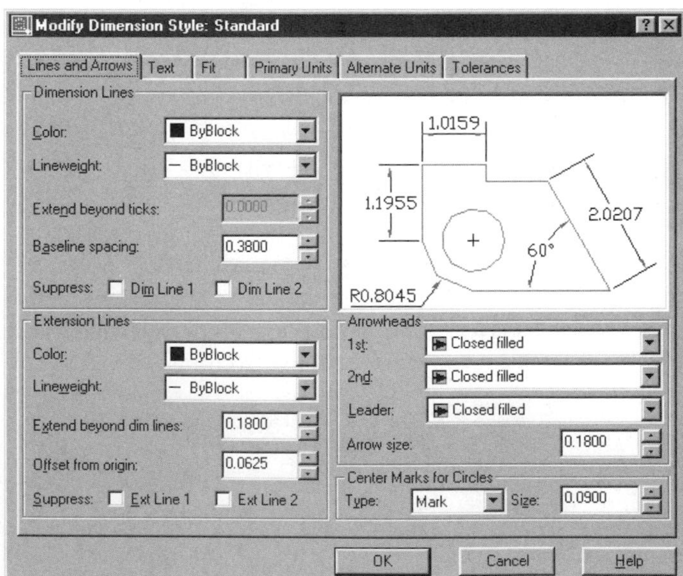

Figure 18–12 *The Dimension Style dialog box, where you can edit the parts of a dimension.*

Lines and Arrows Folder

The Lines and Arrows folder enables you to control all the dimension system variables related to the geometry of the dimension, except for the text. The folder is broken down into four distinct areas: Dimension Lines, Arrowheads, Extension Lines, and Center Marks for Circles.

The Dimension Lines section controls the appearance of the dimension line. In a linear dimension, this is the line beside or below the dimension text. In certain circumstances, you may want to create a dimension without the dimension line. For example, you may have a short dimension with large text. In this situation you can suppress the first or second dimension line, or both. This option takes effect only when the dimension text is centered inside of the dimension line. When text is above the dimension line, the suppression options have no effect. The location of the dimension text is controlled through the format options, which are covered in the next section.

The Extend Beyond Ticks option, which is grayed out by default, is used in conjunction with certain arrowhead types. In particular, the oblique and architectural tick arrowheads make use of this option. When one of these two arrowheads is active, you may adjust the dimension line's extension variable, which defines how far beyond the extension lines of the dimension the arrowhead goes.

Probably the most-used dimension line option is Color. The default color of the dimension line is BYBLOCK, which means the line will take on the color of the dimension as a whole. The only reason to change this color is if you want to have a different line width for the dimension line. For example, you could have a thinner line for the dimension line than for the extension lines.

In the Extension Lines area of the dialog box, the options perform the same function as those in the Dimension Lines area, with the notable exception of the Origin Offset option. When you create a dimension, such as a linear dimension, you select two points as the start point and endpoint. These points are considered the origin points and are drawn in as part of the dimension. The origin offset defines the distance above these points that the extension line is started.

The Arrowheads section of the Lines and Arrows folder provides you with complete control over the arrows. AutoCAD 2002 provides you with additional standard arrowheads, including architectural tick, open 30, dot blanked, box filled, and a few others. Even with all of these new arrowheads, you may want to create your own. To create your own arrowheads, from the 1st and 2nd drop-down lists, select the User Arrow option. This option enables you to select any block as any arrow, as long as that block is already defined in the current AutoCAD drawing. The arrowhead block should be created with an overall size of one unit in AutoCAD so that it will be correctly scaled when it is used in the dimension. It should also be created using the right end of the dimension. The block will be rotated for the left end.

When you create a custom arrowhead, you should save the arrowhead as a block in a template file so that it is available to all drawings based on that template. If you prefer, you can also use a block from an xref as well.

Figure 18–13 shows the dialog box in which you can enter the block name for the arrowhead.

The following exercise shows you how to create your own arrowheads.

EXERCISE: CREATING YOUR OWN ARROWHEADS

1. Start a new drawing from scratch.
2. Create a hexagon using the Polygon command. Make the radius of the polygon one unit long.
3. Make a block of the polygon with an insertion point at the center of the polygon. Name the block P1.
4. From the Dimension pull-down menu, choose Style.
5. In the Dimension Styles dialog box, select the Standard style and click on the Modify button.
6. In the Lines and Arrows folder, select User Arrow from the 1st drop-down list.
7. In the User Arrow dialog box, enter **p1** as the arrow name.
8. Click on OK to close the User Arrow dialog box.
9. Click on OK to close the Geometry dialog box and return to the Dimension Styles dialog box.
10. Choose Save to save the dimension style changes to Standard.
11. Click on OK.
12. Create a linear dimension.

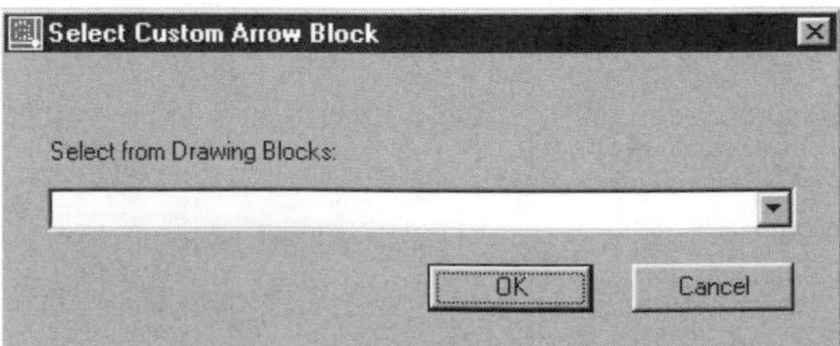

Figure 18–13 *The Select Custom Arrow Block dialog box, where you can select a predefined block for use as a custom arrowhead in your style.*

The last section of the Lines and Arrows folder controls center marks and scaling. These options define how center marks appear when they are used with radius and diameter dimensions.

Text Folder

The Text folder, shown in Figure 18–14, enables you to control how the text is displayed in your dimensions. In the Text Style drop-down list, you can select any previously defined text style. Additionally, you can create a style by clicking on the "..." button to the right of the Text Style drop-down list. This launches the Text Style dialog box, where you can immediately create the new style you want.

After you select a style, you can apply other properties, such as height and color, to the style. The Height option in particular depends on how the text style is defined. If the text style is defined with a fixed height, that height will be used. If the text style's height is zero, the height specified in the Annotation dialog box will be used instead.

In addition to selecting the text style, you can also control the placement of the text around the dimension line, as well as the alignment of the text in the drawing.

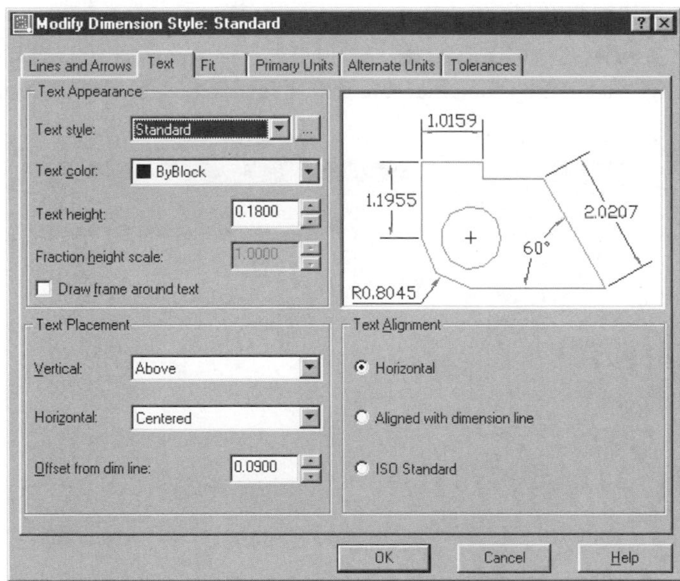

Figure 18–14 *The Text folder of the Dimension Styles dialog box. Here you can define how and where the text appears inside of your dimensions.*

Fit Folder

In the Fit Options section of the Fit folder (see Figure 18–15) you will find six different options, which are described as follows:

- Either the Text or the Arrows—When this option is selected, AutoCAD will try to determine the best method to use to create the most readable dimension. This is the default option.
- Arrows—When this option is selected, only the arrows will be forced inside of the extension lines. Text may appear outside of the extension lines when the distance between the lines is too small for the text to fit otherwise.
- Text—When this option is selected, only text will be forced inside the extension lines. Arrows may be pushed outside the extension lines when the distance between the lines is too small.
- Both Text and Arrows—When this option is selected, both the text and arrows will be forced inside the extension lines, even when there is not enough room for them.
- Always Keep Text between Ext. Lines—When this option is selected, the text will always be placed between the extension lines of the dimension, regardless of whether it will fit or not.
- Suppress Arrows—This option enables you to suppress the drawing of the arrows on a dimension when the text is forced inside of the extension lines.

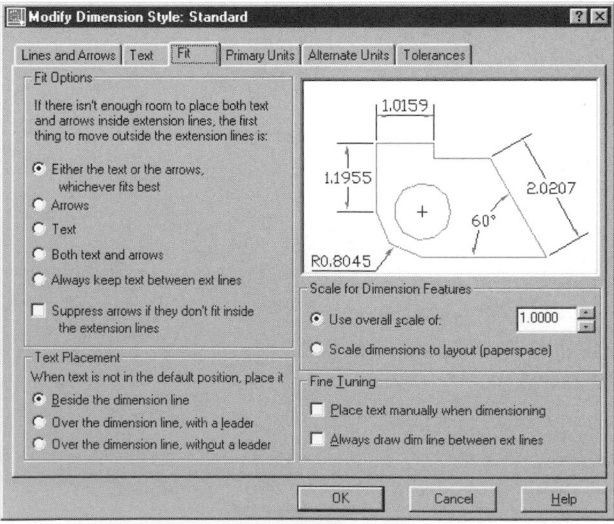

Figure 18–15 *The Fit folder of the Dimension Styles dialog box. Here you can control what happens to the text when there is no room for it in the dimension.*

The Text Placement area's options enable you to define where the text is placed in relation to the dimension line when the text is not in the default position. You can select one of three different ways to handle this: place the text beside the line, place the text over the line with a leader, or place the text over the line without a leader.

Probably the most important options in the Fit folder of the Dimension Styles dialog box, however, are the Scale area's options: Overall Scale and Scale to Layout. The Overall Scale option controls how large all the features of the dimension, such as arrowheads, will appear in the drawing. This scale is directly related to the final plot scale for the drawing. For example, if you are plotting at an architectural scale of 1/4 inch=1'0", your scale factor should be 48. This factor is obtained by taking the denominator of 1/4 inch and multiplying it by 12. For example, you can equate a scale such as 1:50 to 1/50=1'; so the overall scale factor is 600.

When you set the scale factor, you define a scale multiplier by which all dimension size variables are multiplied. For example, arrowheads default to a size of 0.18 units. If you have a scale factor of 48, you multiply the 0.18 by 48 to arrive at the current size, correctly scaled for plotting.

If you are going to work in layout mode, you can leave the Overall Scale factor set to 1, or you can turn on the Scale Dimensions to Layout option. The Scale Dimensions to Layout option sets the dimension variable DIMSCALE to zero. When you work in layout mode, a default value of 1.0 will be used. In this situation, when you are working in a model space viewport in layout mode, you can create a dimension in either space and it will be scaled correctly. This assumes that you have used the ZOOM command to scale the geometry correctly in the model space viewport.

Primary Units Folder

The Primary Units folder (see Figure 18–16) is used to define the units that dimensions will use in their text. Unfortunately, AutoCAD does not automatically set the dimension units to match your units setting in AutoCAD. So, you must correctly define the units for your dimensions separately.

In the Linear Dimensions section of the folder you can select the type of units, precision of the display, and even prefixes and suffixes. For example, if you select architectural units, you can select the precision with which the units will be displayed, how fractions are drawn (horizontal, slanted, or not stacked) and whether you want to append FT or IN to the dimension as a suffix. One important option in this section of the dialog box is the Round Off option, which enables you to define how the dimension measurements are rounded off.

The Measurement Scale option is used to adjust how the distance between the start point and endpoint of the dimension is measured. Most of the time this option is

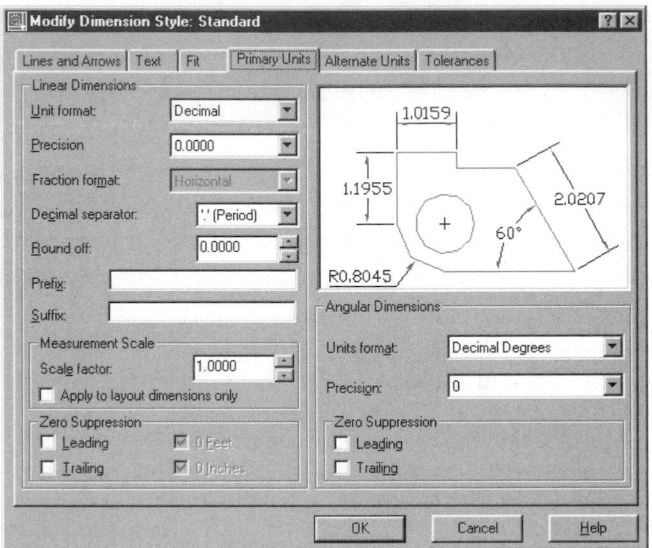

Figure 18–16 *The Primary Units folder, where you can control the type of units that are displayed and measured in your dimensions.*

reserved for layout mode. In layout mode, when you place a dimension, AutoCAD measures the dimension in layout units, not model space units.

In a viewport that is scaled to ¼"=1'0", for example, a 4' line will measure 1". This is because of the underlying scale factor difference between layout mode and model space units. Just as you have to adjust the overall scale factor of a dimension style for model space units, you must adjust the linear scale to match layout units. The linear scale factor is calculated the same way that the overall scale factor is calculated. In the preceding example, a linear scale of 48 is correct.

The Zero Suppression option is used to control when a zero appears in a dimension. For example, 6' is a valid dimension in architectural units. But 6' is easy to confuse with 6", especially if the blueprint of the drawing is not very good. In both cases, the leading or trailing zeros have been suppressed. These dimensions are much easier to read when they appear as 6'0" and 0'6". You can set this up by disabling zero suppression for feet and inches. You can also control zero suppression for leading and trailing zeros, as in 0.6 and 6.000.

Lastly, you can set the options for angular dimensions. Here you can define what unit format angles are read back and how precise the angle measurements are. Like linear dimensions, angles also have zero suppressions.

Alternate Units Folder

The Alternate Units folder (see Figure 18–17) is used when you want to display alternate units in your drawing. For example, you might have drawings with architectural units and metric units as alternate units. Such a drawing would be compatible in Europe as well as in the US.

Tolerances Folder

The Tolerance folder (see Figure 18–18) of the Dimension Style dialog box enables you to add tolerances to the end of the dimension text. These tolerances are different from the tolerance command discussed in Chapter 17, Dimensioning Basics. There are five different types of tolerance methods:

- **None**—No tolerances are used in the dimension.
- **Symmetrical**—The tolerance is applied with the a high and a low limit that are the same. For example, 1.00 ± 0.1 is a symmetrical tolerance.
- **Deviation**—The tolerance is applied with a high and a low limit that can be different. As long as the object is manufactured within the limit, it is acceptable.
- **Limits**—The tolerance completely replaces the dimension. As long as the object is manufactured within the tolerances, it is acceptable.
- **Basic**—No tolerance is used, but a box is drawn around the dimension to help emphasize it.

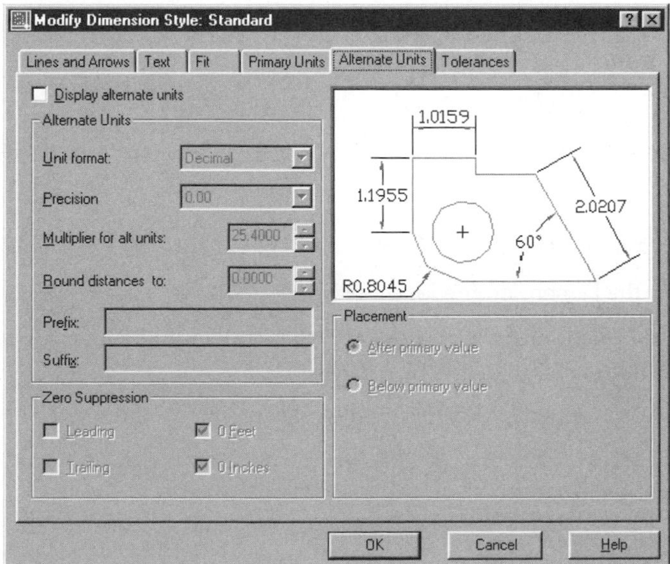

Figure 18–17 *The Alternate Units folder of the Dimension Style dialog box has the same controls as those found in the Primary Units folder.*

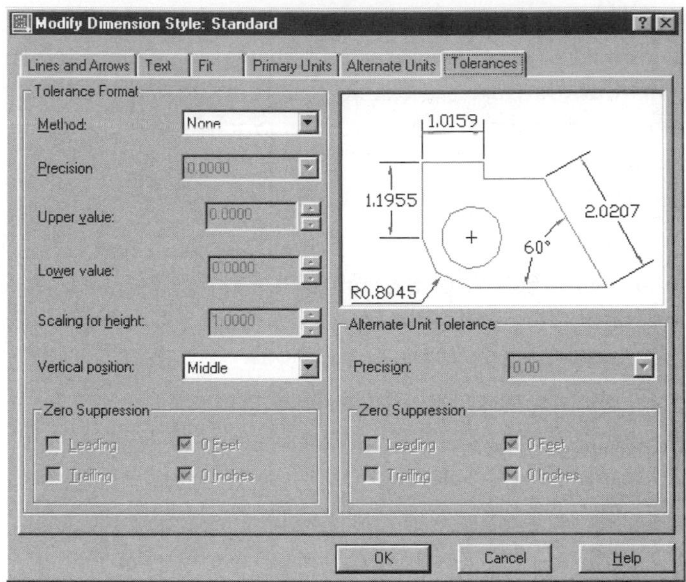

Figure 18–18 *The Tolerances folder of the Dimension Style dialog box, where you can define what tolerances are to be displayed with your dimensions and how.*

After you select the tolerance method, you may apply upper and lower values, as well as justification of the text in the dimension line.

The following exercise ties together all the information you have learned in this chapter so far. It shows you how to set up a complete dimension style quickly and easily for use in AutoCAD.

EXERCISE: CREATING A DIMENSION STYLE FOR AN ARCHITECTURAL DRAWING

1. Start a new drawing.
2. From the Dimension menu, choose Style.
3. In the Dimension Styles dialog box, click on New and give the style the name Architectural1.
4. Click on Continue. Then click on Modify to change the style.
5. In the Arrowheads area, select Architectural Tick from both the 1st and 2nd drop-down lists.
6. In the Extension Line area, select Red as the Color.
7. Click on the Text tab.
8. In the Text Placement section, set the Vertical option to Above.

9. In the Text Alignment area, turn on Aligned with Dimension Line.
10. Click on the Primary Units tab.
11. Set Unit Format to Architectural, and set Precision to 1/8".
12. In the Angular Dimensions area, set Precision to 0.
13. Turn off the 0 Feet and 0 Inches suppression so feet and inches will be displayed in the drawing.
14. Click on OK to close the Dimension Styles dialog box and return to AutoCAD.
15. Create a couple of dimensions in this style.

Now that the processes and steps for working with dimension styles have been covered, the following section introduces some important tips that will help you optimize your dimension styles.

Updating Dimensions

Another popular dimension editing task is updating an existing dimension to the current dimension style. This occurs frequently when users create drawings with many different dimension styles. In this situation, it is very easy to create dimensions in the wrong style by accident.

The three different ways to update the style of a dimension are as follows:

- Modify the style
- Update the dimension with DIMSTYLE
- Revise the dimension with the Dimnesion Update tool

If you modify a dimension style that is currently in use in the drawing, when you save the style and exit the Dimension Styles dialog box, all dimensions using that style will automatically be updated to the new version. In some instances, you may have to regenerate the screen to see the changes.

If you want to change a dimension to a different style, you must first set the new style as the current dimension style. You can do this in the Dimension Styles dialog box, or you can use the DIMSTYLE system variable. After the style is set to current, you can choose Dimension>Update or use the Dimension Update tool on the Dimension toolbar. Then select the dimension; it will be updated to match the new style.

The following exercise shows you how to update AutoCAD dimensions.

EXERCISE: UPDATING DIMENSIONS IN AUTOCAD

1. Open the file 18DWG03.DWG from the accompanying CD-ROM.
2. Choose Dimension>Style to open the Dimension Styles dialog box.
3. Set the style Architectural as the current style.

4. Click on Close to close the Dimension Styles dialog box.
5. Choose Dimension>Update.
6. Select the dimension in the drawing and then press Enter. The dimension is updated, as shown in Figure 18–19.

Tips for Creating Effective Dimension Styles

The following are a few tips and techniques concerning dimension styles:

- Create all of your necessary styles and then save them in a drawing template. That way, you never have to recreate the same styles.
- Try to create your styles in a somewhat generic fashion so that you can easily modify them.
- When naming your styles, give them names that make sense to you and to others. For example, a style named ARCH14 is easier to understand as an architectural dimension style for a ¼ drawing than a style named STYLE1.
- Make use of families when you need to have your dimension styles change slightly when you use different dimension types. This saves you from having to set dimension styles every time you change dimension types such as linear or angular.

Dimension Variables

A lesser-known method of modifying a dimension is to override a dimension variable. When you create a dimension in a specific style, it is possible to override one or more dimension variables in the style. This enables you to change the color of the dimension text for a couple of dimensions and then revert back to the original if you so desire.

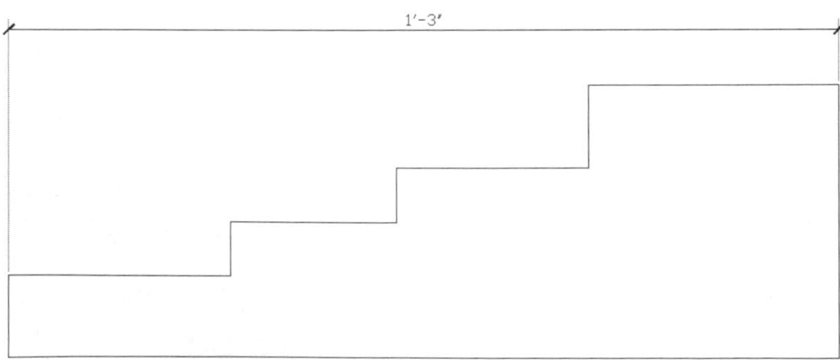

Figure 18–19 *The updated dimensions.*

There are several ways to implement dimension variable overriding. The easiest way is to override the dimension variable when you are creating the dimension. Unfortunately, to do this, you must know the name of the dimension variable you want to override. When you select a dimension command, such as DIMLINEAR, simply enter the name of the dimension variable you want to override. Give it the new value, and that value will be used until you clear the override. For example, DIMASZ controls the size of the arrowheads. You can override this variable with a larger or smaller value than that found in the dimension style.

To clear a dimension override, you must use the DIMOVERRIDE command, which is available on the Dimension pull-down menu as Override. At the command prompt, you will be asked for the dimension variable to override. If you type **Clear** at this prompt, you will clear all overrides and revert back to the original style definition. Alternatively, you can enter any dimension variable, override it, and then apply it to existing dimensions.

Overrides stay valid until the CLEAR command is executed, or a new style is chosen, or you change the override to another value.

To help you make use of the override command, Table 18-1 lists all the dimension variables along with a brief description of what each one does.

For most of the dimension variables, you may need to look up exactly what values you can use. Many are simply 1 or zero, and others accept text strings like DIMSTYLE does. If you are going to use overrides, though, you need to know which variables you want to override and how you want to override them.

Table 18–1: *Dimension Variables and Meanings*

Variable	Function
DIMADEC	Angle precision
DIMALT	Enables the use of alternate dimensions units
DIMALTD	Controls the decimal places used in alternate units
DIMALTF	Controls the alternate unit scale factor
DIMALTRND	Controls the alternate unit roundoff
DIMALTTD	Number of decimals in a tolerance in an alternate unit
DIMATTZ	Toggles suppression of zeros for tolerances
DIMALTU	Unit format for alternate units (except for angular dimensions)

Table 18–1: *Dimension Variables and Meanings (Continued)*

Variable	Function
DIMALTZ	Controls suppression of zeros for alternate units
DIMAPOST	The text prefix or suffix for alternate dimensions (except for angular dimesnions)
DIMASO	Enables associative dimensions
DIMASZ	Controls the arrowhead sizes
DIMATFIT	Fit for arrows and text
DIMAUNIT	Angle format for angular dimensions
DIMAZIN	Angle zero suppression
DIMBLK	Name of block to be drawn instead of regular arrowhead
DIMBLK1	User-defined arrowhead 1
DIMBLK2	User-defined arrowhead 2
DIMCEN	Enables use of center marks
DIMCLRD	Color of the dimension line
DIMCLRE	Color of the extension line
DIMCLRT	Color of the dimension text
DIMDEC	Number of decimal places for primary tolerances
DIMDLE	Controls extension of dimension line when it is oblique or architectural
DIMDLI	Dimension line spacing for baseline dimensions
DIMDSEP	Decimal separator
DIMEXE	Distance extension lines extend beyond the dimension line
DIMEXO	Extension line offset
DIMFIT	Placement of arrows and dimension lines inside of extension lines
DIMFRAC	Dimension fraction format
DIMGAP	Gap around dimension text
DIMJUST	Horizontal dimension text position
DIMLDRBLK	Leader arrow block

Table 18–1: *Dimension Variables and Meanings (Continued)*

Variable	Function
DIMLFAC	Global scale factor for linear measurements
DIMLIM	Generates dimension limits as default text
DIMLWD	Dimension line lineweight
DIMLWE	Dimension line extension lineweight
DIMPOST	Prefix or suffix for text
DIMRND	Dimension rounding value
DIMSAH	Enables use of user-defined arrowheads
DIMSCALE	Overall scale factor
DIMSD1	First dimension line suppression
DIMSD2	Second dimension line suppression
DIMSE1	First extension line suppression
DIMSE2	Second extension line suppression
DIMSOXD	Suppresses drawing of dimension lines outside extension lines
DIMSTYLE	Current dimension style
DIMTAD	Vertical position of text in relation to the dimension line
DIMTDEC	Number of decimals in a tolerance
DIMTFAC	Scale factor for text height in tolerances
DIMTIH	Position of text inside extension lines
DIMTIX	Draws text between extension lines
DIMTM	Lower tolerance limit
DIMTOFL	Forces drawing of dimension line
DIMTOH	Position of text outside of extension lines
DIMTOL	Appends tolerances to text
DIMTOLJ	Vertical justification of tolerances
DIMTMOVE	Fit for text movement
DIMTP	Upper tolerance limit

Table 18–1: *Dimension Variables and Meanings (Continued)*

Variable	Function
DIMTSZ	Size of oblique dimension arrowheads
DIMTVP	Vertical position of text
DIMTXTSTY	Text style for the dimension
DIMTZIN	Zero suppression of tolerance values
DIMUNIT	Unit format for dimensions except angular
DIMUPT	Cursor functionality for user-positioned text
DIMZIN	Suppression of primary unit value

SUMMARY

As you learned in this chapter, the ability to edit dimensions is very important, especially in the design phase of any project that may change drastically. Also, of equal importance is the ability to create, define, and use dimension styles. These abilities help to ensure that all drafts people in the office use the same styles and that your drawings have the same look and feel. These features combined with the base dimensioning.

Layouts and Paper Space

Paper space layouts provide the ability to plot a model space drawing without cluttering the drawing with objects needed only for plotting purposes, such as title blocks and sheet borders. You can create a standard-sized sheet border in a layout, insert a view of your model, and plot it at a 1:1 scale. By creating multiple viewports of the model space objects, you can view the objects from different angles. After the views are established, you can move and arrange the viewports in the layout to any necessary position inside the sheet border. All this can be accomplished, without compromising the purity of the model space drawing, by allowing the project's design model to exist separately from objects needed only for plotting sheets. By using layouts properly, you can quickly and easily design the sheets needed to plot model space objects.

This chapter discusses the following subjects:

- Using paper space layouts
- Viewports and layouts
- Tiled viewports versus floating viewports

USING PAPER SPACE LAYOUTS

AutoCAD 2000 introduced a feature called Layouts. This is probably one of the most significant enhancements to AutoCAD because it makes working in paper space easy to understand. Layouts provide a preview of what your plotted sheet will look like, and they make working with paper space intuitive. Layouts simulate the piece of paper that your model will be plotted on, and they accurately reflect the plotted sheet's scale factor, paper orientation, the current lineweight setting of objects, and the layout's current plot style settings.

Layouts help avoid the confusion experienced by many CAD technicians when they deal with model space and paper space in previous releases of AutoCAD. Because of the layout's WYSIWYG approach, it is easy to visualize what it is designed to do. Specifically, as shown in Figure 19–1, layouts make it easy to visualize how your model space objects will be plotted on a sheet of paper.

Note: The Paper Space feature was broadened in AutoCAD 2000 to allow not just one paper space in which to work, but as many as desired, named however you wish. The term *paper space* now refers to the multiple layouts that can be created and used. And *layout* refers to an individual "plot space" of the paper space for any given drawing.

CREATING LAYOUTS

When you start a new drawing from scratch, AutoCAD automatically creates a single Model folder and two Layout folders, as shown in Figure 19–2. The Model folder is where your model space drawing is created and edited. The Model folder itself cannot be renamed or deleted. Extra Layout folders beyond Layout1, however, are not a requirement and can be renamed or deleted entirely. They are available to allow you to assemble easily the paper sheets you use to plot your drawing.

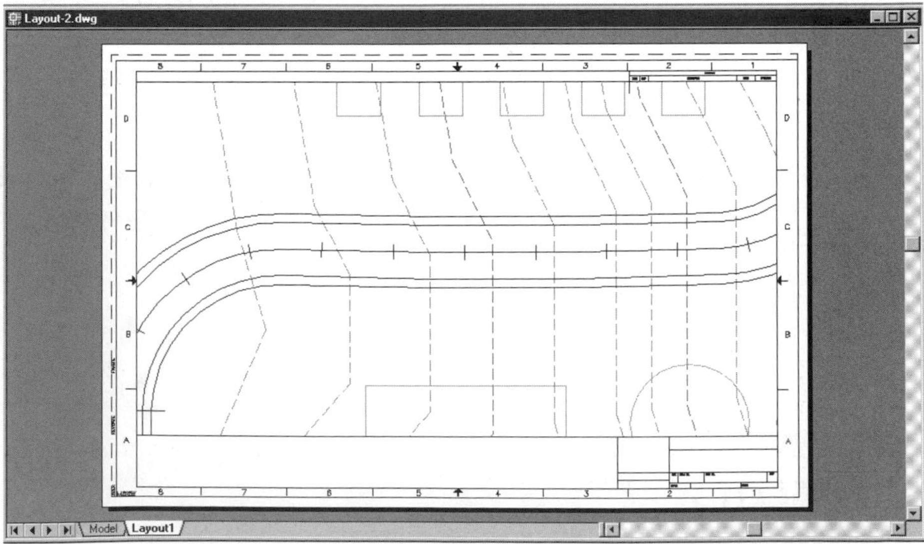

Figure 19–1 *Layouts make paper space easier to visualize.*

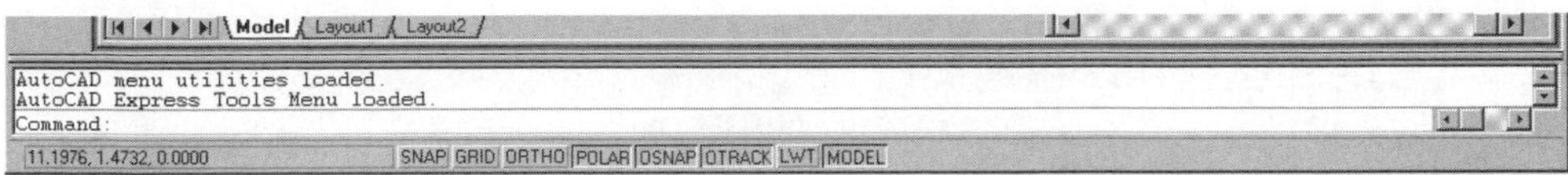

Figure 19–2 *When you start a new drawing from scratch, AutoCAD automatically creates a single Model folder and two Layout folders.*

Note: AutoCAD does not require you to plot a drawing from a layout. You can plot your drawing from model space, but some plot features are not supported. However, layouts make the process of creating a plot visually easier through its WYSIWYG display, and they are the preferred method for creating plots.

While AutoCAD automatically creates two Layout folders, your needs may require you to create more folders. AutoCAD provides three methods for creating layouts:

- Create layouts from scratch
- Create layouts from templates
- Create layouts through wizards

In the following exercise you will create a layout from scratch and then save it as a layout template.

EXERCISE: CREATING A LAYOUT FROM SCRATCH

1. Launch AutoCAD and start a new drawing from scratch. AutoCAD creates a new drawing with a single Model folder and two Layout folders.

   When you select a Layout folder for the first time, AutoCAD's default system settings instruct it to automatically prompt you for a page setup to apply to the layout, and to create a single viewport. For this exercise, you will disable these features.
2. From the Tools menu, choose Options, and then choose the Display tab.
3. In the Layout Elements area, clear the Show Page Setup Dialog for New Layouts option and the Create Viewport in New Layouts option, as shown in Figure 19–3.
4. Choose Apply, then choose OK.

   Next you will create three additional layouts.
5. Move the cursor over the Layout2 tab and right-click. AutoCAD displays the shortcut menu.
6. From the shortcut menu, choose New Layout. AutoCAD creates a new layout and automatically names it Layout3.
7. Repeat step 6 two more times until there are five Layout folders.

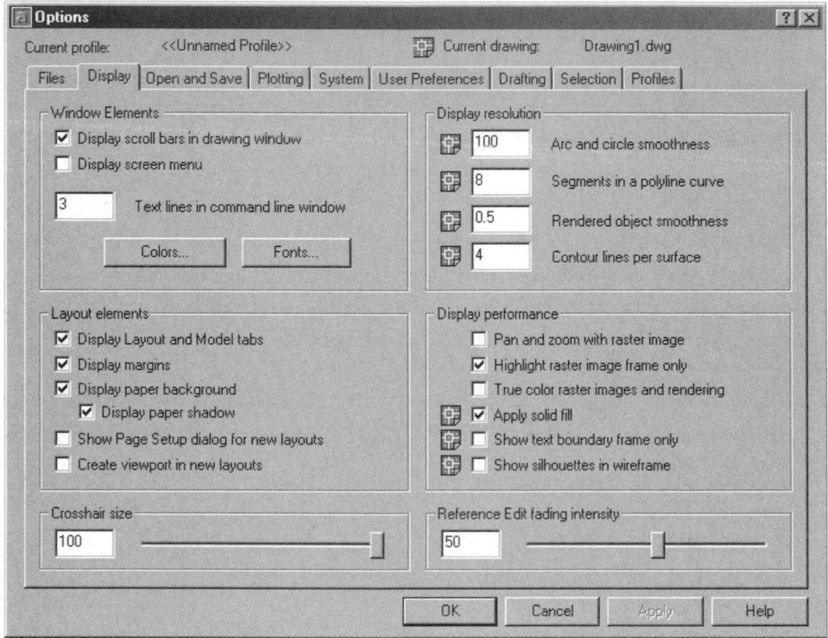

Figure 19–3 *You can stop AutoCAD from automatically creating a viewport in a layout and from prompting you for a page setup by clearing those features in the Options dialog box.*

Tip: Layouts can be created, imported from a template, deleted, renamed, moved, copied, setup, or plotted from the shortcut menu by a simple right-click of the mouse.

8. To rename a Layout folder, move the cursor over the Layout3 tab, and then right-click. AutoCAD displays the Layout shortcut menu.
9. From the shortcut menu, choose Rename. The Rename Layout dialog box is displayed.
10. Enter **First Layout** in the dialog box, then choose OK. AutoCAD renames the layout.
11. Repeat step 10 two more times, and rename Layout4 and Layout5 as Second Layout and Third Layout, respectively. AutoCAD renames the layouts as shown in Figure 19–4.

Note: You can name layouts using most keyboard characters (except for <>/\":;?*|.='), which can help provide very clear descriptions for your layouts.

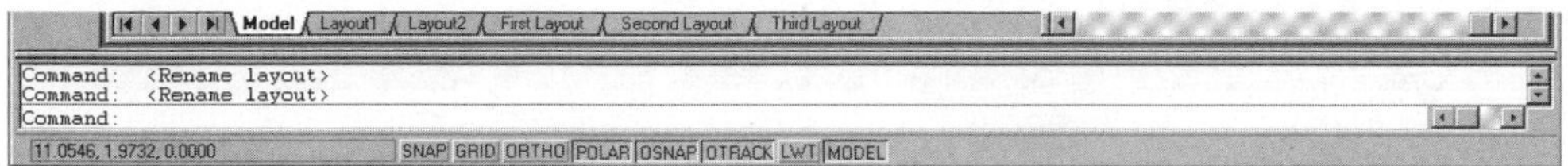

Figure 19–4 *The three last folders are renamed.*

12. To delete the two original Layout folders supplied by AutoCAD, hold the Shift key down, choose the Layout1 tab, and then choose the Layout2 tab. Both tabs are highlighted.
13. Move the cursor over either highlighted Layout tab and then right-click. AutoCAD displays the shortcut menu.
14. From the shortcut menu, choose Delete, as shown in Figure 19–5. AutoCAD issues a warning noting that the selected layouts will be deleted permanently.
15. Choose OK to permanently delete the two Layout folders. Only the three Layout folders you created and renamed remain.
16. Keep this DWG file open for the following exercise.

As you just experienced, creating layouts from scratch is simple. Once you have created a layout, you can save it in a template file and insert it into new drawings. This is a useful method for creating a set of predefined layouts and sharing them with others to insert into their drawings.

In the following exercise you will save the layouts you created in a template and then insert them into a new drawing.

EXERCISE: CREATING A LAYOUT FROM SCRATCH AND SAVING IT AS A TEMPLATE

1. Continuing from the previous exercise, open the File menu and choose Save As.
2. In the Save Drawing As dialog box, from the Save As Type drop-down list, choose AutoCAD Drawing Template File (*.dwt).
3. Name the drawing Layouts, then save it to the AutoCAD 2002/Template subdirectory, a shown in Figure 19–6. AutoCAD displays the Template Description dialog box.
4. In the Description text box, type **My Layouts**. Note that the Measurement setting can be set to English or Metric, but choose English, and then choose OK. AutoCAD saves the drawing as an AutoCAD template file.

   Next you will use the template file to insert a new layout from a template.
5. From the File menu, choose New to create a new drawing from scratch.
6. Move the cursor over the Layout2 tab and then right-click.

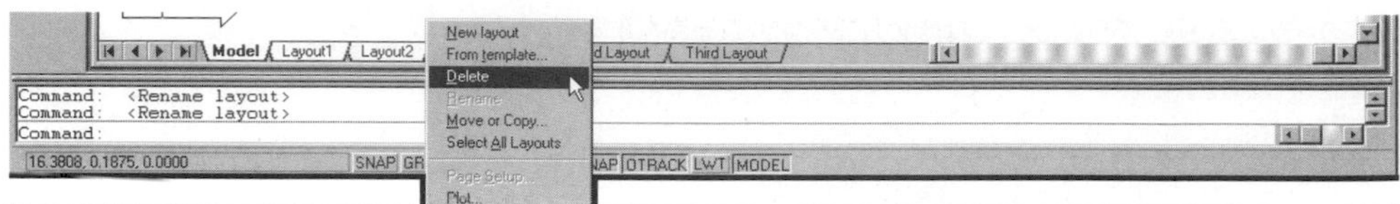

Figure 19–5 *The first two Layout folders are selected and then deleted.*

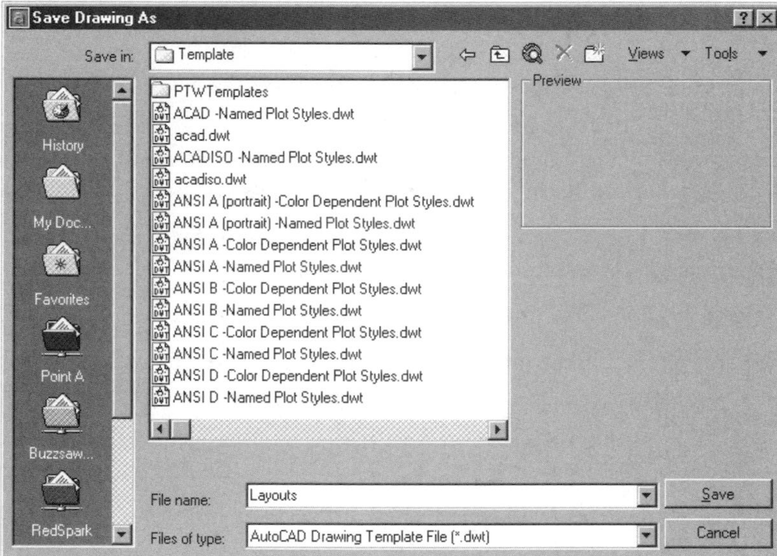

Figure 19–6 *Save the drawing with the new layouts as a template file for insertion into other drawings.*

7. From the shortcut menu, choose From Template. AutoCAD displays the Select File dialog box.
8. From the Select File dialog box, choose the Layouts template file you saved in step 3 from the AutoCAD 2002 Template folder, and then choose Open. AutoCAD displays the Insert Layout(s) dialog box.
9. From the Insert Layout(s) dialog box, choose Third Layout from the Layout names list, and then choose OK, as shown in Figure 19–7. AutoCAD inserts the Third Layout from the template created in the previous exercise.

Tip: You can use AutoCAD DesignCenter to browse drawing files and drag and drop layouts from these outside files directly into the current drawing. For more information on AutoCAD DesignCenter, refer to Chapter 11, Applying AutoCAD DesignCenter.

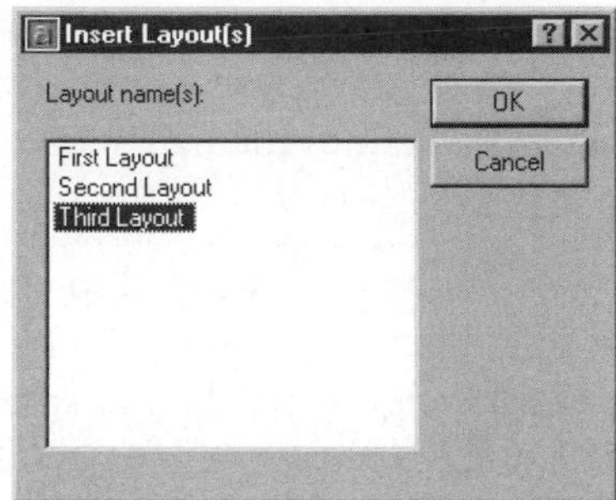

Figure 19–7 *You can insert predefined layouts from template files.*

Note: You can not replace a layout with another of the same name from a template. If you attempt this, AutoCAD will prefix the inserted layout with "Layout#-" when it is created. You should delete an existing layout before you insert a layout with the same name.

By using the technique demonstrated in this exercise, you can spend a little time at the beginning of a project to set up all the project's desired layouts, save them as a template file, and insert the layouts into new project drawings as they are needed.

UNDERSTANDING THE LIMITATIONS OF LAYOUTS

Understanding the limitations of working in a paper space layout is important. Because layouts are intended to make creating plots easier, certain commands that are available in model space do not work when you are in a Layout folder.

For example, the Layout folders are intended to display the contents of two-dimensional environments. In model space, however, not only can you create three-dimensional objects; you can also modify the model space view to look at these objects from different perspectives. Consequently, model space is where your project design work should be performed.

Remember that the Layout folder is intended as the environment in which you compose the plotted sheets of your model space project. It is not intended for modeling. Therefore, you should use model space to design your project, and use the Layout folder(s) to define your project's plots.

CONTROLLING OUTPUT THROUGH PAGE SETUP

The Page Setup feature is a powerful tool that controls certain paper and plotter configuration information and links it to the Model or Layout folder that is current. By using page setups, you can define certain characteristics of the sheet of paper on which you plot your drawing. You can also associate with your drawing a specific plotter configuration, which indicates the printer or plotter that plots your drawing. With page setups, you control values that define such things as a the plot device, paper size, scale factor, plot orientation, X and Y offset values, and more.

A single drawing may contain several layouts, and each layout can be assigned a page setup. In a single layout, you can switch between numerous page setups. When a page setup is selected, AutoCAD redefines the paper size, scale factor, and all other page setup properties.

In previous releases of AutoCAD, when you wanted to change your plot configuration either to plot a drawing to a different plotter or to plot a drawing at a different scale and still maintain proper lineweights, you had to spend time adjusting the plot parameters and/or loading PC2 files. However, if you select the appropriate page setup, you will not need to redefine the plot configuration file. For example, if you need to plot a full-scale drawing to a large format plotter, you can set up a unique page setup defined for the large format plotter. Then, in the same drawing, you can create a second page setup to plot the same drawing to a LaserJet printer on an 8 ½-by-11-inch sheet of paper. Because both page setups can exist in the same drawing simultaneously, you can quickly switch between the page setups to plot your drawing to the desired device. Additionally, when you choose the appropriate page setup, your drawing is plotted to the desired paper size, at the proper scale, using the appropriate lineweights, all without the need to redefine the plot parameters. This is a tremendous timesaver.

Note: Using page setups, you can assign plot styles to layouts, floating paper space viewports, and individual objects. For more information, refer to Chapter 20, Basic Plotting.

Not only does the ability to associate a unique page setup with a specific layout save time when you create different plots of your drawing, it also saves time by allowing you to save a page setup configuration and insert it into another layout. Just as you can save time when creating new drawings by inserting predefined layouts from a template file (as you learned in the previous exercise), you can also save time by inserting predefined page setups into the current drawing.

The Page Setup dialog box contains two folders, as shown in Figure 19–8. The Plot Device folder is where you select the plotter or printer device to which to send plots. It is also where you can assign a plot style table, which defines the pen assignments that control lineweights and line colors. The Layout Settings folder is where you define settings such as paper size, plot area, drawing orientation, and plot scale. If you are familiar with AutoCAD's Plot dialog box, you may notice that the Layout Settings folder is the same as the Plot dialog box's Plot Settings folder. This observation emphasizes that page setups allows you to predefine plot settings, which means you do not need to make any adjustments to the Plot Settings folder if you already selected the appropriate page setup. For detailed information on using page setups to control plot parameters, refer to Chapter 20, Productive Plotting.

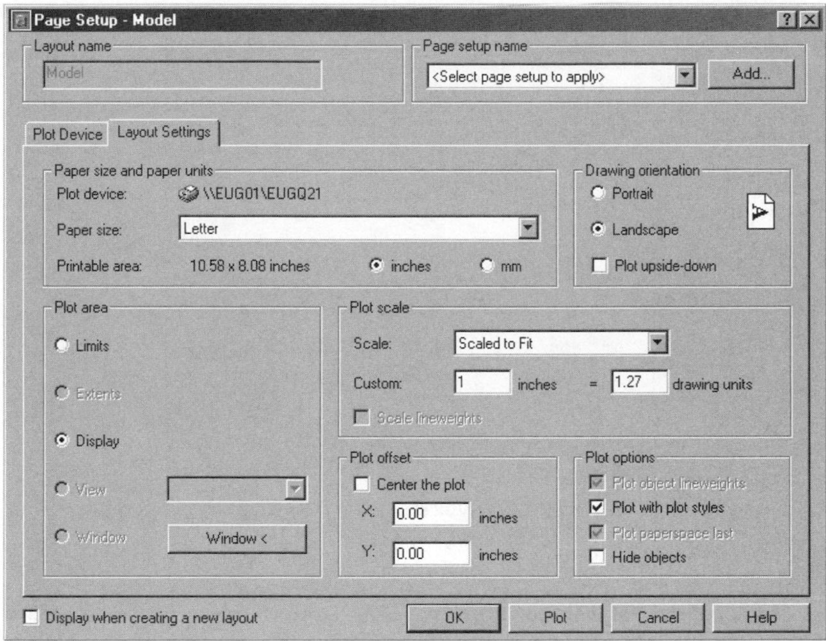

Figure 19–8 *The Page Setup dialog box allows you to define paper and plotter settings and link them to the current Layout folder.*

Notice at the top of the Page Setup dialog box the two areas that list the current layout's name and the page setup's name. The Layout Name area indicates the folder to which the current page setup values are linked. In Figure 19–8, the page setup is linked to the Model folder. The Page Setup Name area is where you select a predefined page setup from a drop-down list. Alternatively, by choosing the Add button, you can add the current page setup values to the drop-down list. When you choose the Add button, the User Defined Page Setups dialog box is displayed, as shown in Figure 19–9. In this dialog box you can name, rename, delete, or import page setups from other drawings.

Note: Saved page setups are saved only in the current drawing, and they cannot be saved to a unique file type. Therefore, if you wish to import a saved page setup, you must locate the drawing that contains the desired page setup and then use the User Defined Page Setups dialog box's Import option.

VIEWPORTS AND LAYOUTS

Viewports (Mviews) created in a layout have unique settings for controlling the appearance of your project's model. By controlling layer visibility, hidden line removal, and model-to-paper space scale, viewports allow your finished plot to display objects precisely as needed. More importantly, you can control these properties

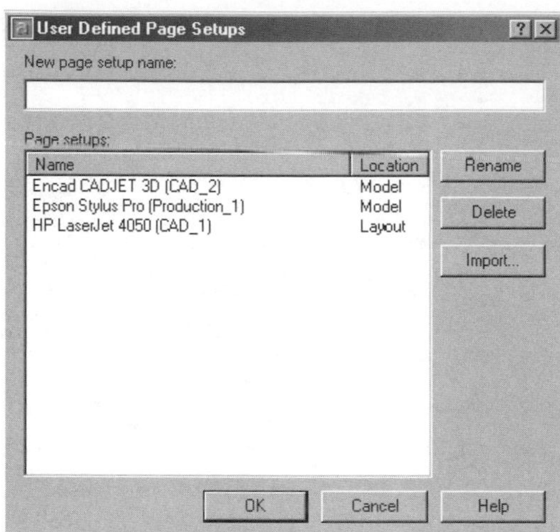

Figure 19–9 *The User Defined Page Setups dialog box allows you to import predefined page setups from other drawings.*

Note: The PSETUPIN command allows you to insert page setups from other drawings, templates, and DXF files, and it is executed at AutoCAD's command prompt. Also, when you call the command from AutoLISP, you can insert as many page setups as desired from outside files. For example, by entering the following AutoLISP code, you can find my-page-setups.dwg and read in saved page setup names Fullsize and Halfsize:

```
(command "psetupin" "my-page-setups" "Fullsize,Halfsize")
```

on a viewport-by-viewport basis, even when you have several viewports on a single sheet. The capability to control how your model space project appears through these viewports is a powerful feature.

AutoCAD 2000 introduced several features that greatly enhance the usefulness of floating viewports. For example, you can now create non-rectangular viewports. You can also clip a viewport and resize it to any shape using grips. Additionally, you can lock the display scale of a viewport so it is not accidentally modified when you zoom or pan in the viewport.

This section discusses the differences between tiled viewports and floating viewports, and it explores using floating viewports in layouts.

UNDERSTANDING TILED VIEWPORTS VERSUS FLOATING VIEWPORTS

Tiled viewports are created when you are working in model space. Untiled—or floating—viewports are created when you are working in paper space. When you create viewports, AutoCAD automatically determines the type of viewport to create based on the current space in which you are working. Therefore, tiled viewports are automatically created when the Model folder is active, and floating viewports are automatically created when Layout folders are active.

Tiled viewports, as the name implies, appear as tiles on the screen. They subdivide the original model space viewport (which is a single tiled viewport) into multiple viewports, as shown in Figure 19–10. Tiled viewports are fixed and cannot be moved. They never overlap, and their edges always lie adjacent to the surrounding viewports. Their usage is primarily for helping to view the model during its creation. The currently selected tile can be further divided into more tiles or joined with another tiled viewport to create a new larger one.

In contrast, floating viewports neither subdivide the screen nor remain fixed. Additionally, they can be copied, resized, and moved, just like any other AutoCAD object. They can even overlap each other, as shown in Figure 19–11.

THE VIEWPORTS TOOLBAR

The Viewports toolbar (see Figure 19–12), which was introduced with AutoCAD 2000, allows you to insert a single viewport, define a polygonal (or non-rectangular) viewport, and clip an existing viewport. Additionally, you can set the scale for model space objects displayed in a viewport, and you can display the new Viewports dialog box, which allows you to create multiple viewports by selecting the desired predefined viewport configuration.

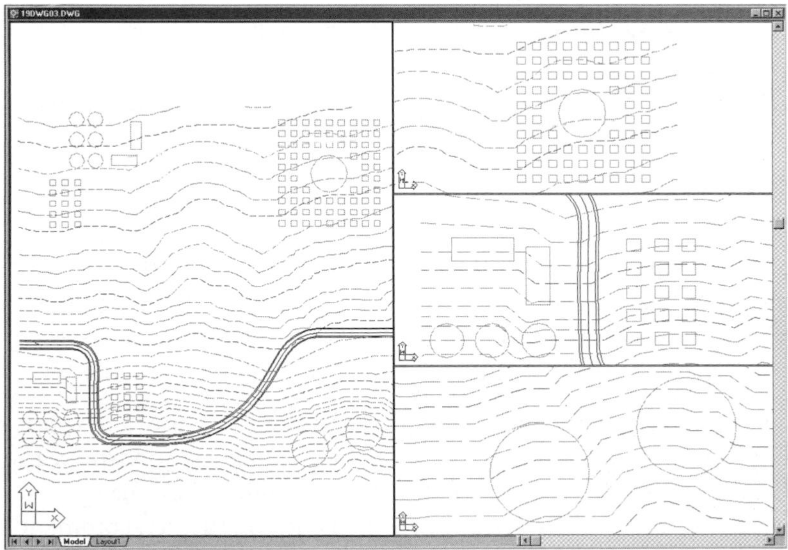

Figure 19–10 *Model space viewports subdivide the screen into smaller tiled viewports that cannot overlap.*

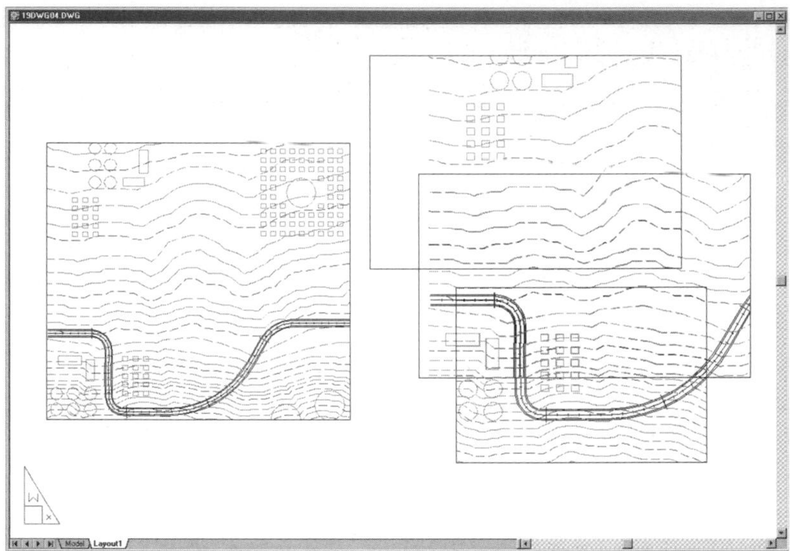

Figure 19–11 *Paper space viewports can be copied and resized, and they can overlap.*

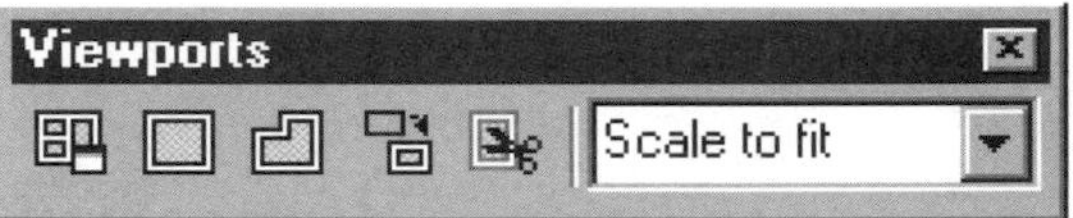

Figure 19–12 *The Viewports toolbar makes creating viewports very easy.*

Tip: To display the Viewports toolbar, right-click over any toolbar button. Then, from the shortcut menu, choose Viewports as shown in Figure 19–13.

Creating Non-rectangular Floating Viewports

For years, AutoCAD technicians have dreamed of the day that they would be able to create non-rectangular viewports. With the release of AutoCAD 2000, that day finally arrived. The Polygonal Viewport button allows you to create irregularly shaped viewports, as shown in Figure 19–14.

When you choose the Polygonal Viewport button from the Viewports toolbar, AutoCAD prompts you to specify the start point. Once you do, the feature works like the PLINE command, continuing to prompt you for additional points to define the polygon viewport's vertices. Also similar to using the PLINE command, you can switch between a line segment and an arc segment, as well as close the polygon or undo the last point selected. After you create the polygon viewport, you can even edit the pline with several options from the PEDIT command.

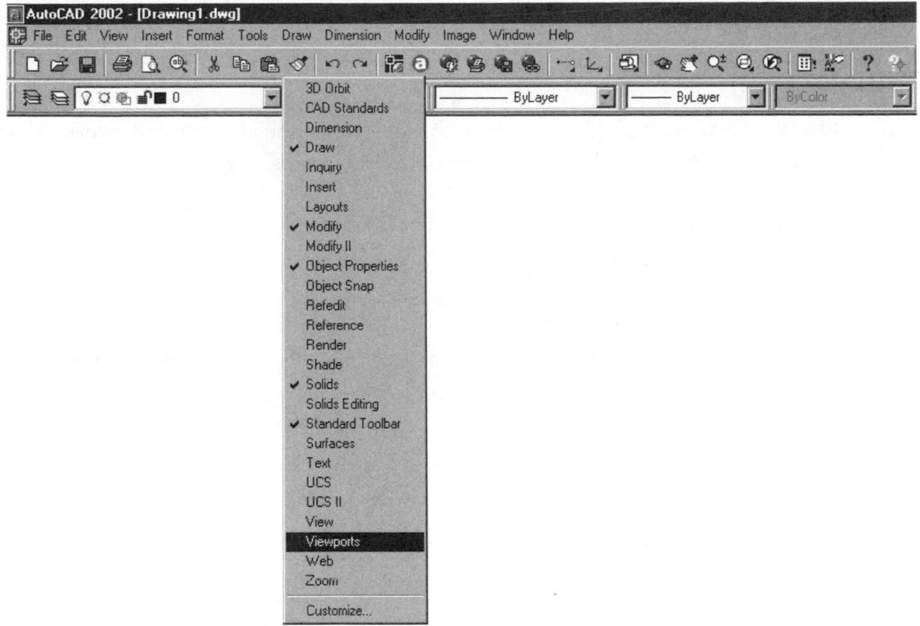

Figure 19–13 *You can display the Viewports toolbar by selecting it from the shortcut menu.*

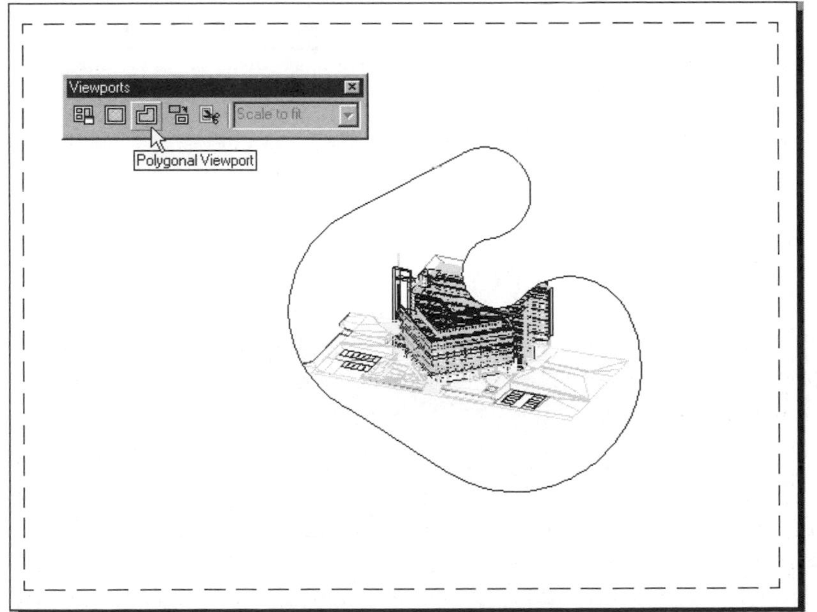

Figure 19–14 *The Polygonal Viewport button allows you to create irregularly shaped floating viewports.*

Tip: Once you have created a floating viewport, you can modify its shape by selecting the edge of the viewport to display its grips, and then by selecting the grips and moving them to new positions.

Note: If you wish to hide the display of a viewport's boundary, do not freeze the layer the viewport is; turn it off instead. If the viewport's layer is frozen, it does not display the model space objects as desired. However, the objects will be displayed properly if the viewport's layer is turned off. It is recommended that you create specific layers on which to place your viewports.

Converting Objects to Floating Viewports

Another handy feature is the ability to convert an existing AutoCAD object into a floating viewport. Any closed object such as a circle or a closed polyline can easily be converted to a floating viewport. For example, when you choose the Convert Object to Viewport button shown in Figure 19–15, the circle object can be selected and converted to a viewport, as shown in Figure 19–16.

Clipping Existing Floating Viewports

AutoCAD 2000 introduced a feature that seemingly clips an existing floating viewport. While the phrase "clip" may conjure visions of AutoCAD's TRIM command, this feature does not actually clip—or trim—an existing viewport. Instead, it replaces an existing viewport with a new clipped viewport.

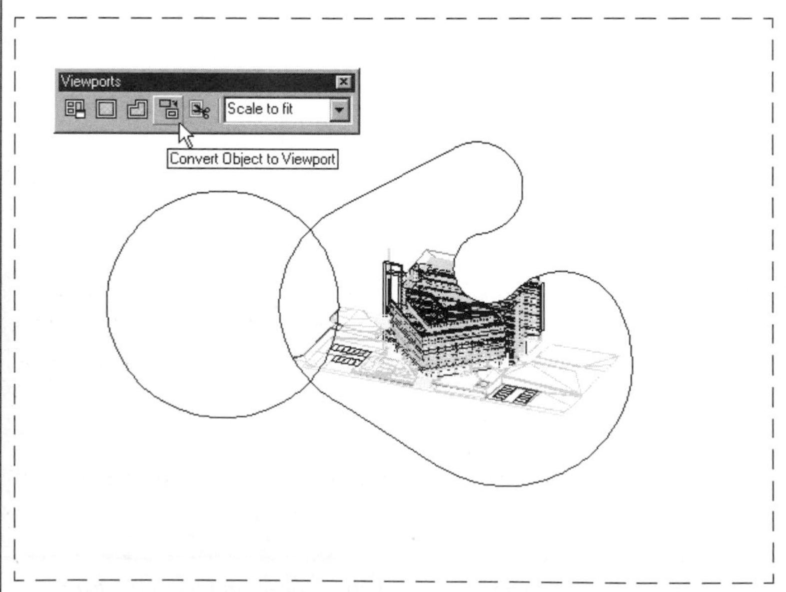

Figure 19–15 *The Convert Object to Viewport button allows you to select a closed object and convert it to a floating viewport.*

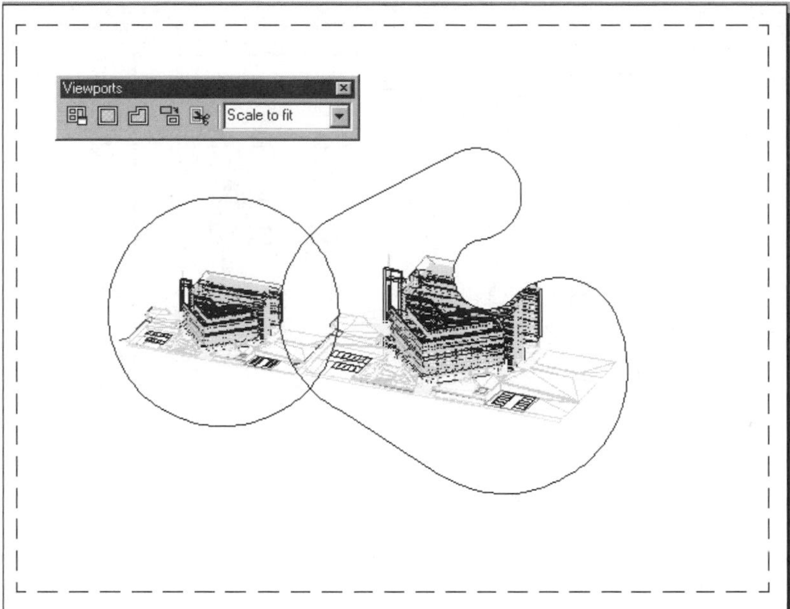

Figure 19–16 *The circle is converted to a floating viewport.*

Tip: You can convert a region object to a clipped viewport. By creating composite regions, you can define a region object with holes or voids, which can be used to blank out model space areas in the viewport, as shown in Figure 19–17. This is a great way to show notes on top of a viewport. Simply create the composite region in the shape desired and then convert it to a viewport.

What makes this feature useful is that you can replace an existing viewport with a new closed object that assumes the current properties of the existing viewport. So, to revise an existing viewport's shape quickly while retaining its properties, such as the model's view position and scale in the viewport, use the new viewport Clip feature.

To clip a viewport, choose the Clip Existing Viewport button on the Viewports toolbar. Once you do, AutoCAD prompts you to select the viewport to clip. Next AutoCAD prompts you to select the clipping object. After the clipping object is selected, AutoCAD converts the clipping object into a viewport and deletes the existing viewport.

The following exercise demonstrates how to create a composite region object and use it to clip an existing viewport.

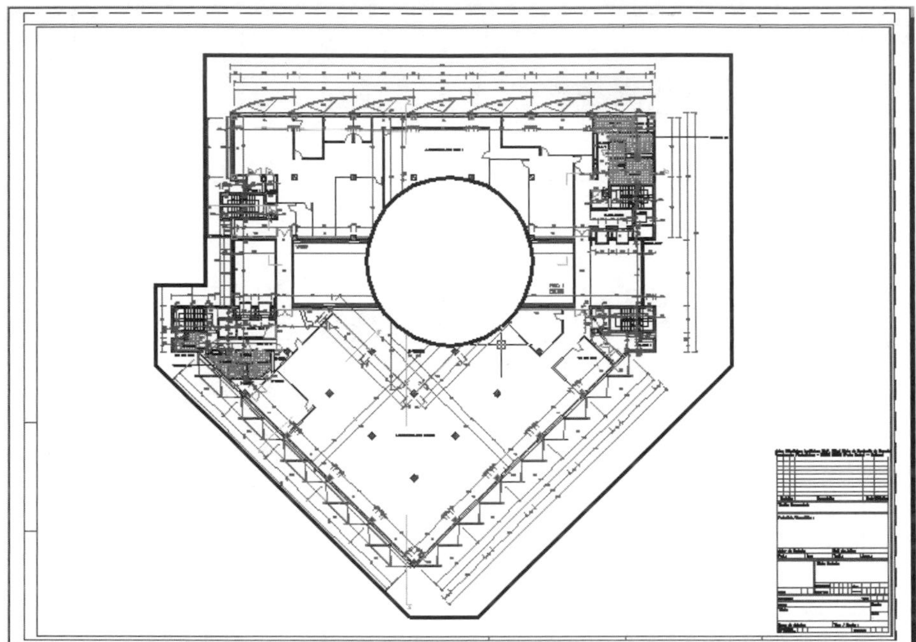

Figure 19–17 *The viewport, shown in bold, was created from a composite region object, which consists of the irregularly shaped polygon and the circle.*

Tip: You can copy floating viewport objects just like any other AutoCAD object. This is especially useful for duplicating properties such as the viewport's scale factor and display area.

EXERCISE: CLIPPING FLOATING VIEWPORTS

1. Open the 19DWG01.DWG drawing file found on the accompanying CD. When the drawing opens, it displays a single floating viewport, as shown in Figure 19–18. There is also a polyline and a circle object, both of which are in paper space.

   Next you will convert the polyline and the circle into regions.

2. From the Draw menu, choose Region. AutoCAD prompts you to select the objects to convert into regions.
3. Select the polyline and the circle, then press Enter. AutoCAD converts the two objects into region objects.

   Next you will convert the two region objects into a single composite region.

4. From the Modify menu, choose Solids Editing>Subtract. AutoCAD prompts you to select the regions from which to subtract.
5. Choose the polyline, then press Enter. AutoCAD prompts you to select the regions to subtract.

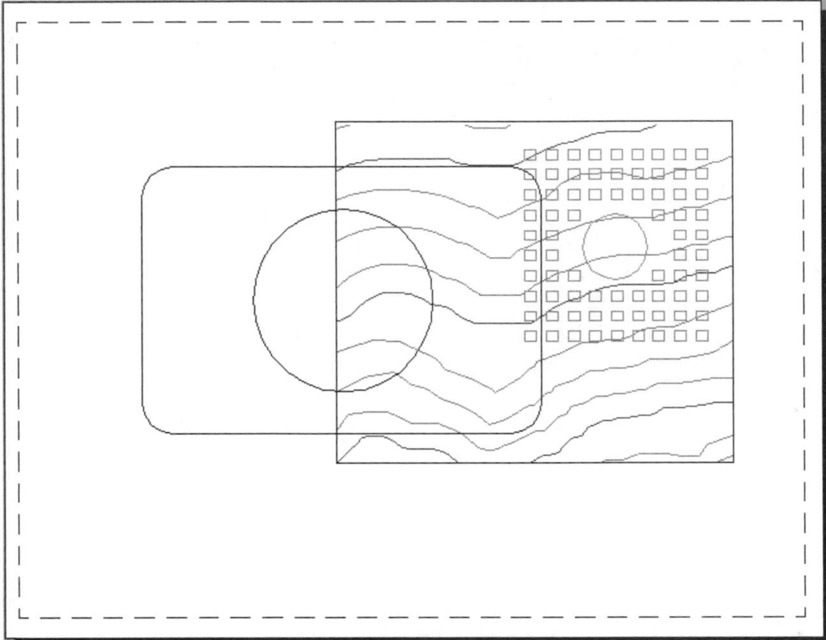

Figure 19–18 *The existing viewport, along with the polyline and circle objects.*

6. Choose the circle, then press Enter. AutoCAD subtracts the circle region from the polyline region and creates a single composite region in their place.

   Next you will use the composite region object to clip the existing viewport.
7. Right-click over any toolbar button to display the shortcut menu.
8. From the shortcut menu, choose Viewports to display the Viewports toolbar.
9. On the Viewports toolbar, choose the Clip Existing Viewport button. AutoCAD prompts you to pick the viewport to clip.
10. Choose the existing viewport. AutoCAD prompts you to select the clipping object.
11. Choose the composite region.

AutoCAD clips the existing boundary, replacing it with the composite region, as shown in Figure 19–19. Notice how the model space view is not displayed through the circle. This effect is achieved by using composite regions.

Controlling a Viewport's Scale

Another handy feature is the ability to scale your model easily in a floating viewport. In previous releases, to set the proper scale of your model, you had to calculate the

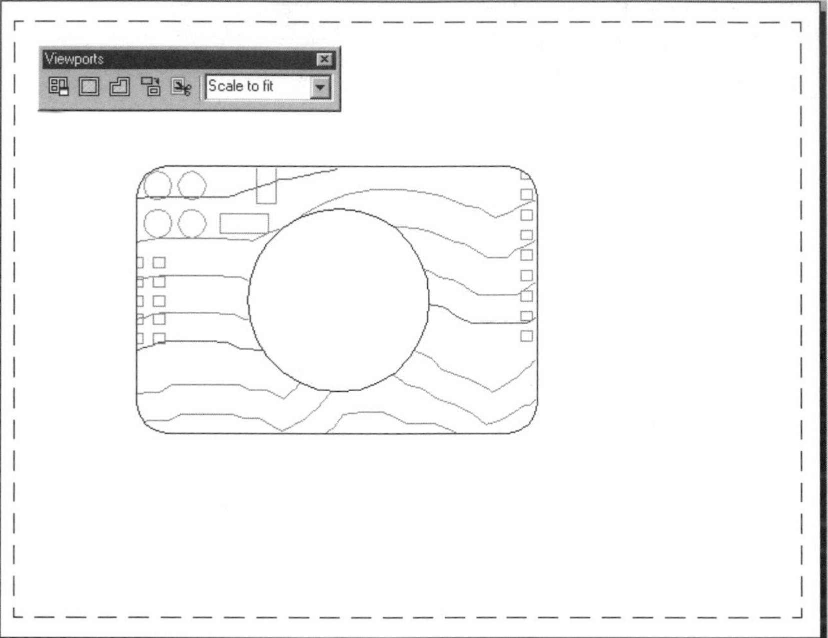

Figure 19–19 *The existing viewport is clipped by the composite region object.*

viewport's scale ratio and then use the ZOOM XP command. Now you can quickly select the desired scale ratio from the Viewports toolbar.

For example, suppose you were plotting your drawing from a layout, and you were displaying your model in a single floating viewport. If your plotted sheet's scale must be 1" = 40', then you would choose the 1:40 option from the Viewport Scale Control.

Tip: To set a floating viewport's scale ratio manually, double-click inside the viewport to switch to model space mode, and then execute the ZOOM command. At the command prompt, enter the scale ratio value followed by **XP**. For example, to set the viewport's scale ratio to 1" = 40' manually, after executing the ZOOM command, enter **1/40XP** at the command prompt.

The Viewport Scale Control comes with over thirty predefined scale ratios and includes the most common model-to-paper space ratios, as shown in Figure 19–20. To change an existing viewport's scale, select the viewport and then choose the desired scale ratio from the drop-down list. To set the desired ratio for new viewports as you create them, choose the desired ratio without selecting any objects.

You now can double-click inside or outside a viewport to switch between the layout environment and model space.

THE VIEWPORTS DIALOG BOX

In the previous section you learned about the various features of the Viewports toolbar. The only feature not discussed was the Display Viewports Dialog button shown in Figure 19–21. When you select this button, AutoCAD displays the Viewports dialog box, as shown in Figure 19–22.

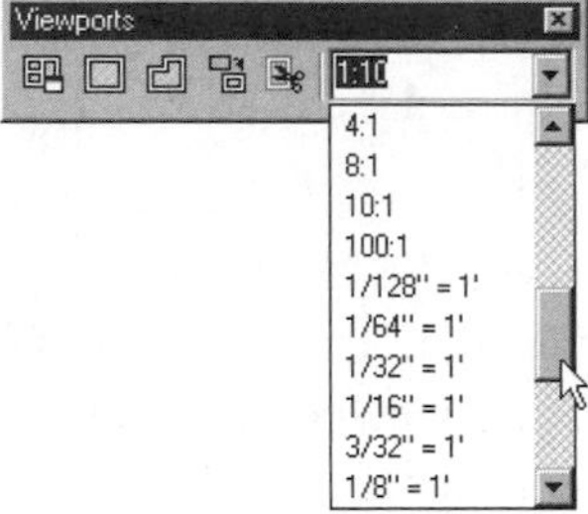

Figure 19–20 *The Viewport Scale Control allows you to set the scale for a floating viewport quickly.*

Figure 19–21 *The Display Viewports Dialog button displays the new Viewports dialog box.*

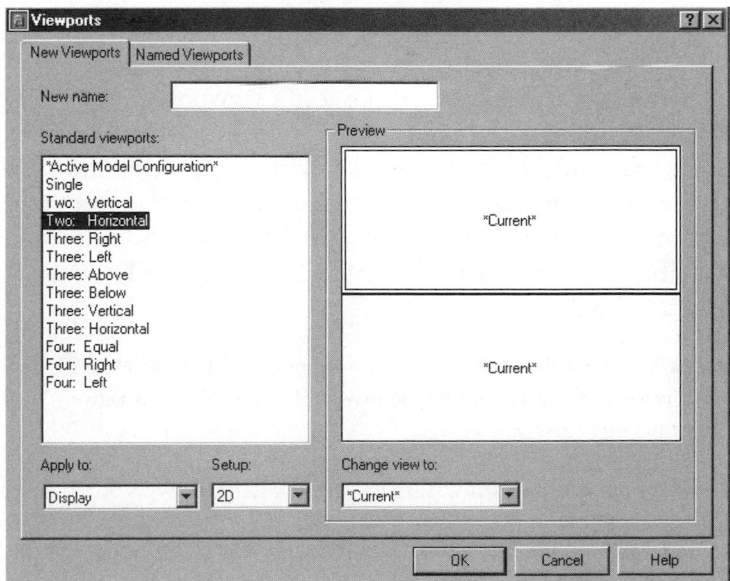

Figure 19–22 *The new Viewports dialog box.*

The Viewports dialog box allows you to create and edit both tiled and floating viewports. You can select from a list of standard viewport configurations, define new configurations, and assign saved views to each viewport in a configuration. If you are in the Model folder, you can name a defined configuration and save it for later use.

The Viewports dialog box is comprised of two folders: the New Viewports folder and the Named Viewports folder. The options displayed in each folder vary depending on whether the Model folder is current or a Layout folder is current. The New Viewports folder shown in Figure 19–22 shows the options available when the Model folder is current.

When the Model folder is current, the New Viewports folder allows you to select a viewport configuration from the Standard Viewports list and either apply it to the display or insert it into an existing viewport. When you apply the configuration to the display, the Model folder's current viewport configuration is replaced by the new configuration. When you insert the configuration into an existing viewport, the original viewport configuration is retained, and the new configuration is inserted into the Model folder's current viewport. You can also indicate whether the configuration is a 2D or a 3D setup. A 2D setup allows you to define each viewport's view by selecting defined views from the Change View To drop-down list. In contrast, when you select a 3D setup, AutoCAD allows you to define each viewport's view by selecting from a set of standard orthogonal 3D views from the Change View To drop-down list. Once you have created the desired viewport configuration, you can name the configuration and save it for later use.

In contrast, when a Layout folder is current, the options available in the New Viewports folder are slightly different than when the Model folder is current, as shown in Figure 19–23. For example, while you can select a viewport configuration from the Standard Viewports list, you can only apply it to the display. You cannot insert it into an existing viewport. However, unlike with tiled viewports created in the Model folder, you can indicate the viewport spacing, which defines the amount of space to apply between viewports when they are created. Additionally, when you insert a new configuration into a Layout folder, AutoCAD allows you to select the location at which to insert the viewports.

Note: You can create viewport configurations only when the Model folder is active. You cannot create viewports configurations when a Layout folder is active.

The Named Viewports folder shown in Figure 19–24 allows you to select a named viewport configuration to insert. While named viewports must be created in the Model folder, they can be inserted into either the Model folder or a Layouts folder. The options available in the Named Viewports folder are the same for both Model and Layout folders.

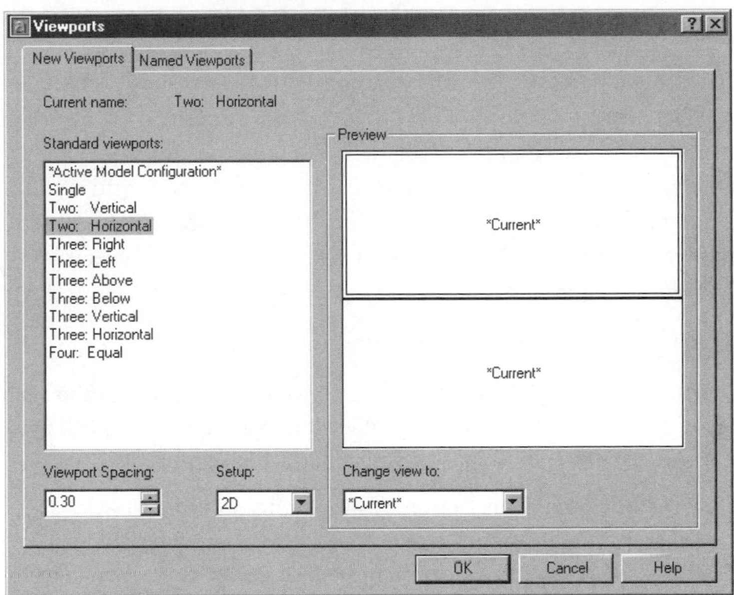

Figure 19–23 *The options available in the New Viewports folder when a Layout folder is current.*

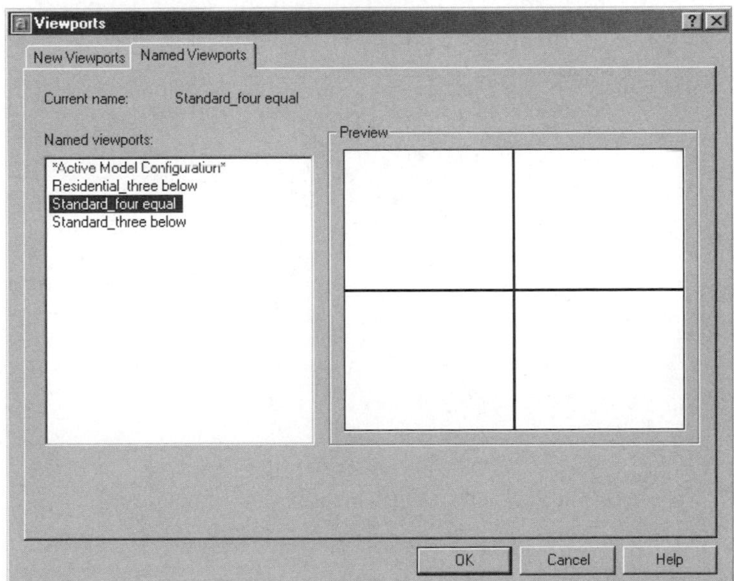

Figure 19–24 *The Named Viewports folder allows you to select a named viewport configuration to insert.*

ACCESSING VIEWPORT COMMANDS THROUGH THE SHORTCUT MENU

AutoCAD allows you to access many commands through the shortcut menu. These commands are accessed by a right-click of your mouse, and they are context-sensitive. The options that are displayed on the shortcut menu vary depending on the cursor's position in the drawing, whether objects are selected, and whether or not a command is currently in progress. Consequently, if you are in a Layout folder and a floating viewport is selected, a right-click of your mouse provides a wide selection of options, as shown in Figure 19–25.

Among these options are three related to floating viewports:

- **Display Viewport Objects**—This option allows you turn the visibility of objects in a viewport on or off. When this option is turned off, the objects in the viewport are not displayed and will not be plotted.
- **Display Locked**—This option allows you to lock the viewport so that its model space scale cannot be altered during zooms and pans. When you turn this option on, any zooms or pans occur in paper space, not in the viewport, thereby ensuring that the proper model space scale is maintained.
- **Hide Plot**—This option allows you to remove hidden lines from objects in floating viewports when drawings are plotted. Basically, hidden lines are the lines on the back side of 3D objects. When you view a 3D object such as a sphere in wireframe mode, AutoCAD allows you to see the lines in the back of

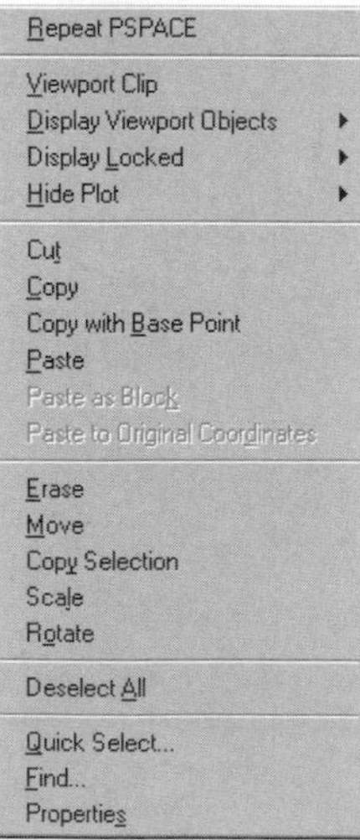

Figure 19–25 *When a floating viewport is selected, a right-click of your mouse displays many options, of which several are related to floating viewports.*

the object, as well as those in the front. In essence, the lines in the back of the object show through the object and are not hidden.

To remove the lines from view, turn the Hide Plot option on. When the 3D objects in the floating viewport are plotted from the layout, AutoCAD will hide any lines in the back of the object and not allow them to show through. In Figure 19–26, the left viewport has Hide Plot turned off, while the right viewport has Hide Plot turned on. Notice that the hidden lines show through the sphere in the viewport on the left, but they do not in the viewport on the right.

Note: AutoCAD provides another method of removing hidden lines. The Hide Objects option found in the Plot Options area of the Plot dialog box removes hidden lines from objects plotted from the Model folder and from objects in layouts that are not displayed in a floating viewport. (The Hide Plot feature removes hidden lines only from objects displayed in floating viewports.)

ALIGNING OBJECTS IN FLOATING VIEWPORTS

Floating viewports can be edited in several ways. You can use grips to scale, move, or resize viewports. Viewports can be copied or erased. You can even create an array of viewports.

Although creating multiple viewports is easy, aligning objects in different viewports can be difficult unless you take advantage of the MVSETUP command.

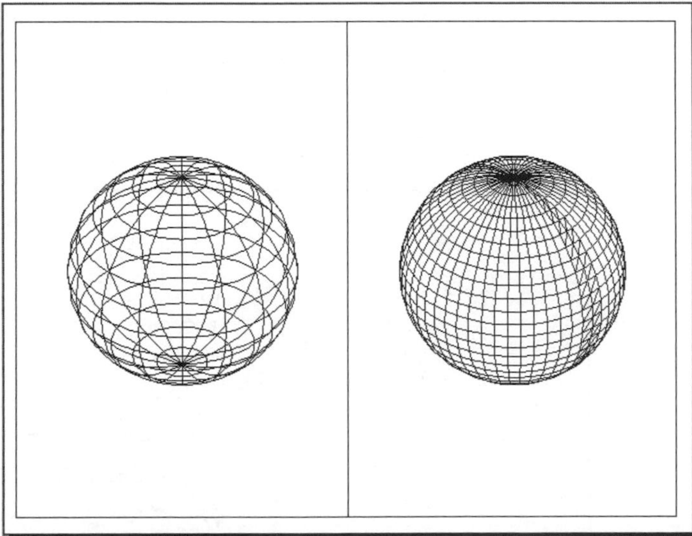

Figure 19–26 *When the Hide Plot option is turned on, hidden lines are removed and are not displayed, as shown in the sphere on the right.*

The following exercise demonstrates how to use the MVSETUP command to align objects in two different viewports.

EXERCISE: ALIGNING OBJECTS IN TWO DIFFERENT FLOATING VIEWPORTS

1. Open the 19DWG02.DWG drawing file found on the accompanying CD.

   When the drawing opens, it displays two floating viewports in a layout. Each viewport shows a different view of the same model space objects. It is important to note that both viewports have the same scale.

2. Type **MVSETUP** at the command prompt. AutoCAD initializes the MVSETUP routine.
3. Type **A** to start the Align feature.
4. Type **H** to start the Horizontal feature. AutoCAD prompts you for the base point. The view in the other viewport will be aligned with this point. If the lower-right viewport is not already highlighted, pick a point inside it to make it current.
5. With the lower-right viewport current, use endpoint snap to snap to the small rectangle as shown in Figure 19–27.

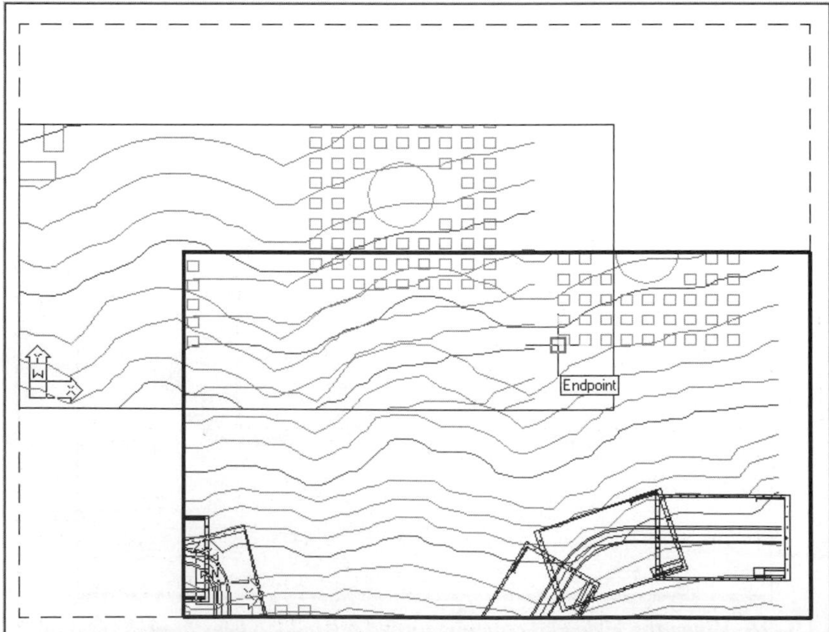

Figure 19–27 *To align objects in two different viewports using the MVSETUP command, first snap to the endpoint of an object that appears in both viewports.*

6. AutoCAD prompts you for the other point. Pick inside the upper-left viewport to make it current.
7. With the upper-left viewport current, use endpoint snap to snap the small rectangle as shown in Figure 19–28.

   AutoCAD moves the view in the upper-left viewport down and aligns the two small rectangles.
8. Type **V** to start the Vertical feature. Once again, AutoCAD prompts you for the base point. Pick a point inside the lower-right viewport to make it current.
9. With the lower-right viewport current, use endpoint snap to snap the same small rectangle shown in Figure 19–27.
10. AutoCAD prompts you for the other point. Pick a point inside the upper-left viewport to make it current.
11. With the upper-left viewport current, use endpoint snap to snap the same small rectangle shown in Figure 19–28.

    AutoCAD moves the view in the upper-left viewport to the right and aligns the two small rectangles, as shown in Figure 19–29.
12. Press Enter twice to end the command.
13. You may close the drawing without saving it.

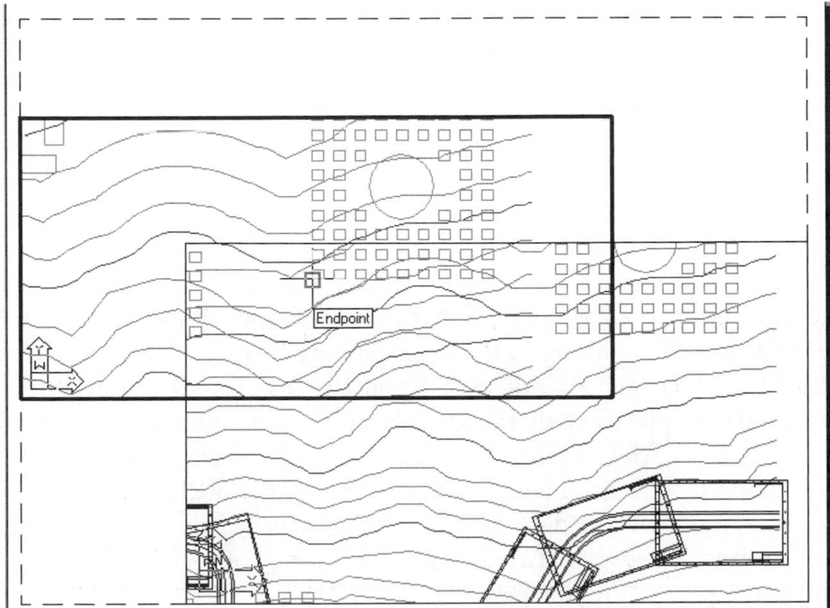

Figure 19–28 *Finish the MVSETUP command by snapping to the same endpoint of the same object in the other viewport.*

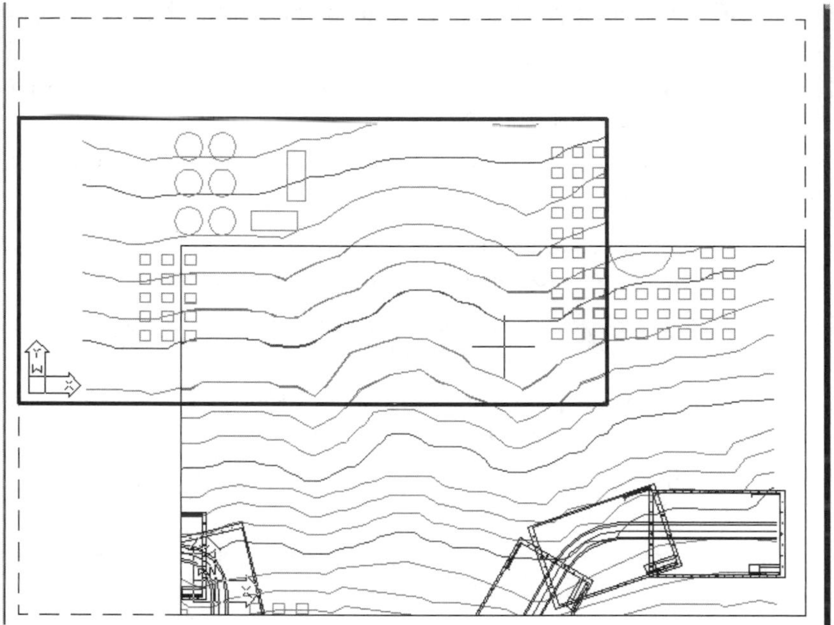

Figure 19–29 *The MVSETUP command aligns the objects in the two viewports.*

Notice that the MVSETUP command was started in paper space but exited in a model space viewport. Also note that the objects in these two viewports aligned perfectly because both viewports have the same scale.

CONTROLLING LAYER VISIBILITY IN FLOATING VIEWPORTS

Floating viewports provide the capability to freeze and thaw layers individually, independent of other viewports. This means that you can make a given object in model space invisible in one viewport by freezing its layer, and the object will remain visible in another viewport, where its layer is thawed. You can do this within the Layer Properties Manager or by using the VPLAYER command. The advantage of using the Layer Properties Manager is that you can simply choose the Freeze/Thaw in Current Viewport icon to toggle the layer visibility. The disadvantage of using the Layer Properties Manager for this task is that the settings affect only the current viewport. Consequently, to apply the same layer freeze/thaw properties to multiple viewports, you must select each viewport individually, make it current, and then choose the desired settings. The advantage of using the VPLAYER command is that you can apply the desired freeze/thaw settings to multiple viewports simultaneously. However, because VPLAYER does not have a dialog box interface, you must type in the layer names manually.

Note: Although you can control a layer's freeze/thaw property in the current viewport, the global freeze/thaw value can override a specified viewport's setting. If a particular layer is thawed in a viewport but frozen globally, for example, the model space objects on the frozen layer will not appear in any viewports.

The following exercise demonstrates the usefulness of the Layer Properties Manager and the VPLAYER command.

EXERCISE: CONTROLLING LAYER VISIBILITY IN FLOATING VIEWPORTS

1. Open the 19DWG03.DWG drawing file found on the accompanying CD. The drawing opens in Layout1 and displays two floating viewports. At this point it is obvious that there are layers that are not visible in the viewport on the right.

   Next you will determine which layers are frozen in the viewport on the right.

2. Choose the Layers button on the Object Properties toolbar. The Layer Properties Manager opens and displays the list of layers, as shown in Figure 19–30. Notice that the icons in the Current VP Freeze column indicate that all layers are thawed.

3. Choose OK to close the dialog box.

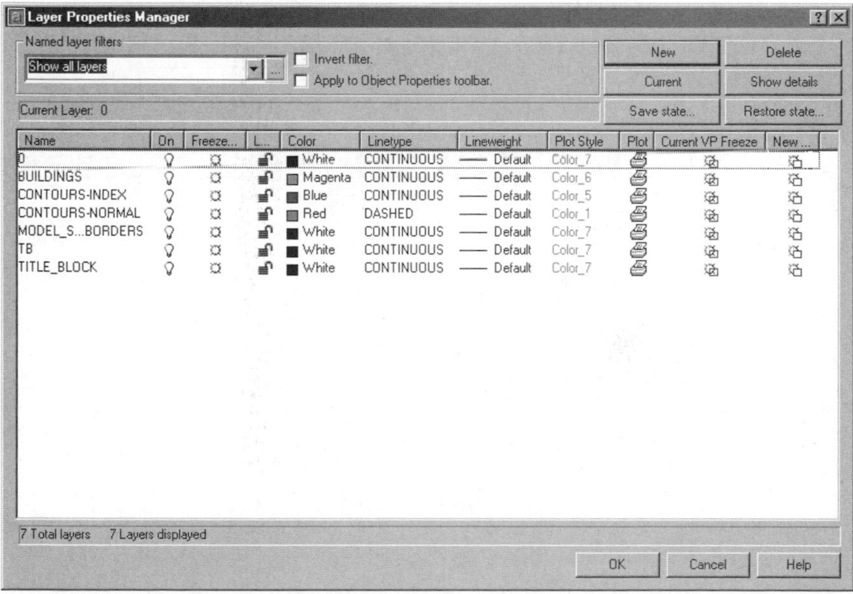

Figure 19–30 *The Current VP Freeze column indicates that all layers are thawed.*

4. Double-click in the right viewport. AutoCAD switches from paper space to model space, and the right viewport becomes active.
5. Choose the Layers button on the Object Properties toolbar.

   The Layer Properties Manager opens and displays the list of layers as shown in Figure 19–31. Notice that the icons in the Current VP Freeze column indicate that three layers are frozen in the current viewport.
6. Choose OK to close the dialog box.
7. Double-click outside of the floating viewports. AutoCAD switches from model space to paper space.

   The next part of this exercise uses the VPLAYER command to list the frozen layers in the two viewports.
8. Type **VPLAYER** at the command prompt.
9. Type **?** at the command prompt. AutoCAD prompts you to select a viewport.
10. Select the right viewport. AutoCAD lists the layers frozen in the selected viewport:

    Layers currently frozen in viewport 3:

    CONTOURS-INDEX

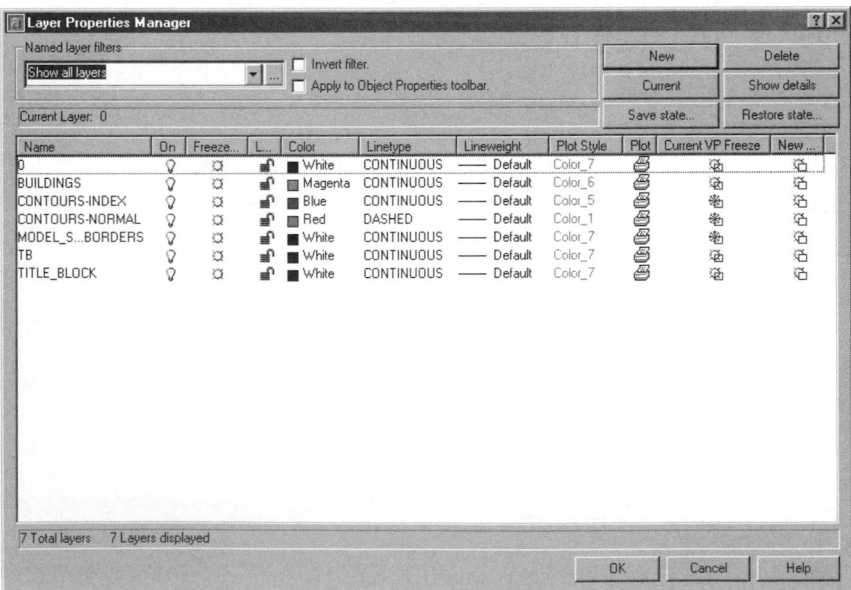

Figure 19–31 *The Current VP Freeze column indicates that three layers are frozen in the current viewport.*

CONTOURS-NORMAL

MODEL_SPACE_BORDERS

11. Type **?** at the command prompt.
12. Choose the left viewport. You will see the following information:

    Layers currently frozen in viewport 2:

    MODEL_SPACE_BORDERS

Note: Notice that the VPLAYER command listed the first viewport selected as viewport 3, and the second as viewport 2. Although only two paper space viewports appear in this drawing, the layout's view is considered viewport 1.

The next steps use the global freeze/thaw layer settings to set both viewports' current freeze/thaw layer settings.

13. Press Enter to end the VPLAYER command.
14. Choose the Layers button on the Object Properties toolbar.
15. Freeze all the layers except for layer 0 in the New VP Freeze column, as shown in Figure 19–32.
16. Choose OK to accept the changes and close the dialog box.

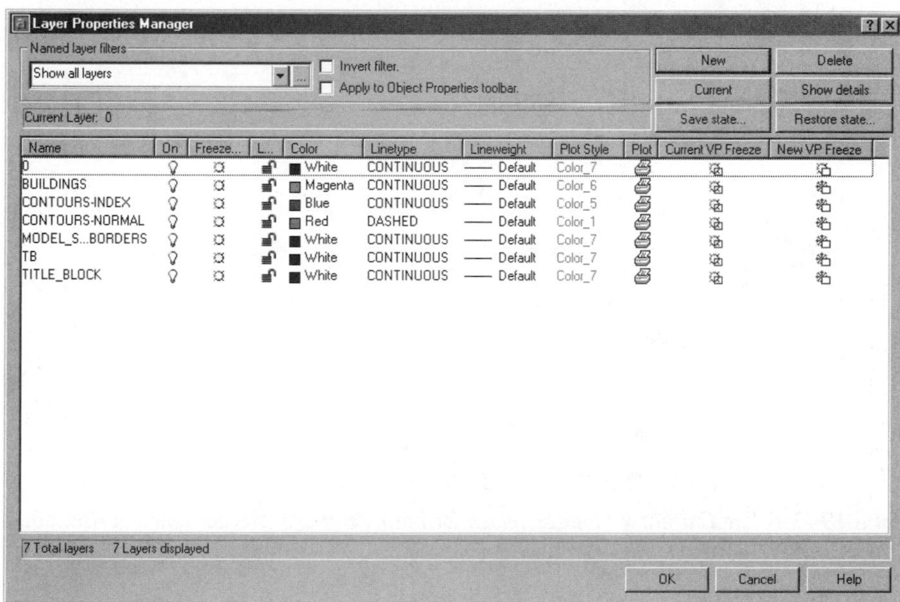

Figure 19–32 *All layers except layer 0 are frozen in the New VP Freeze column.*

Tip: To view a column's entire heading, click and drag the line separating column titles to the right until the heading is visible. To display more information, you can also stretch the dialog box to make it wider.

17. Type **VPLAYER** at the command prompt.
18. Type **R** for Reset.
19. Type * to reset all layers to the current values in the New VP Freeze column.
20. Type **S** for Select.
21. Select the two viewports, then press Enter to end object selection.
22. Press Enter to end the VPLAYER command.

The two viewports' Current VP Freeze values are set equal to the New VP Freeze column's current values. Consequently, only the road alignment is visible, as shown in Figure 19–33.

Tip: When you have multiple viewports in which you need the same Current VP Freeze values, use the technique of setting the New VP Freeze values to the desired values in the Layer Properties Manager. Then use the VPLAYER's Reset option to select the viewports and automatically update their Current VP Freeze values.

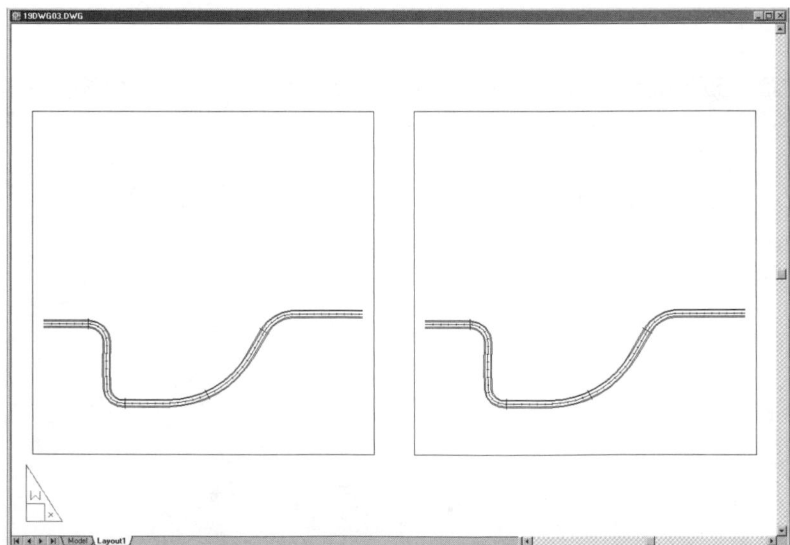

Figure 19–33 *The Current VP Freeze values for both viewports are automatically set equal to the New VP Freeze column's current values using the VPLAYER command's Reset option.*

Note: You might have experienced a problem with retaining changes you made to layer values for xref objects after you closed the drawing and reopened it. If you make changes to the layer values of xrefs, and you want those values to be saved with the drawing, set the VISRETAIN system variable to 1. This instructs AutoCAD to save any changes you make to xref-dependent layers with the drawing. You can also control this feature from the Options dialog box's Open and Save folder, in the External References area, by toggling the Retain Changes to Xref Layers check box.

SUMMARY

This chapter covered paper space layout basics and why you should use layouts to plot your drawings. You learned about the TILEMODE system variable and the difference between tiled and untiled, or "floating," viewports. Comparisons between model space and paper space were presented, as well as the differences between the Hide Plot and Hide Objects features. You learned how to use the MVSETUP command to align viewport objects relative to model space coordinates. Finally, you learned how to change the Active VP Freeze values in multiple viewports simultaneously.

This chapter has shown you how to use the layouts and paper space features in AutoCAD in productive ways. By using the techniques discussed, you can easily control the appearance of your final drawings and ease the time-consuming process of plotting your drawings, thereby increasing your productivity.

Basic Plotting

The ultimate goal of most AutoCAD drawings is the final plot, because plots are what your client uses to build the design you created inside AutoCAD. Consequently, when you begin a project, it is important to consider the form your final output must take to meet the needs of your client. For example, you should consider how dark and heavy or how light and thin objects must appear when they are plotted. By considering the form of the final output at the beginning of a project, you take an important step toward minimizing the amount of time spent reediting a nearly complete project just so the final output meets your client's needs.

AutoCAD 2002 includes features that make the process of creating output very easy. You can assign page setups to paper space layouts, which determines the printer or plotter to which the layout is sent, and how it will appear. You can control plot-specific information at the layout level, the layer level, or the object level by assigning the layout, layer, and object unique page setups.

AutoCAD 2002's features can be grouped into three main tasks:

- Configuring plotters
- Defining plot styles
- Creating page setups

The first two tasks employ wizards that make the process of creating plotter configurations and plot style tables easy. The third task, creating page setups, employs a dialog interface that lets you combine plotter configurations and plot style tables together to control how plotted drawings look. Using page setups you can mix and match a variety of plotters and plot styles on the fly, which, for example, lets you plot out full-scale black-and-white drawings one moment, then half-scale color plots the next.

Because the way that plots are generated using the new features in AutoCAD 2002 is so different from how they are generated in Release 14, the next several sections are dedicated to helping you get up to speed quickly. The first several topics explain creating plotter configurations, and then the next several discuss creating and editing plot styles. The final sections review using page setups. By understanding how plotter configurations, plot styles, and page setups work together, you can establish an easy-to-use yet flexible system for creating hardcopy output of your drawings.

CONFIGURING A PLOTTER

The first step in plotting drawings from AutoCAD is configuring the plot devices intended for plotting. By configuring the printers and plotters to which AutoCAD will plot drawings, you can predefine certain output properties, and then later refine them with plot styles and page setups.

AutoCAD supports many printers and plotters, and it is shipped with a variety of drivers. The drivers allow AutoCAD to communicate with the printers and plotters, including those that support raster and PostScript file formats. The device drivers support many plotting devices, including Hewlett-Packard, Xerox, and Océ plotters. Additionally, AutoCAD plots drawings to Windows system printers, which include any device that is listed in the Windows Printer folder, including Adobe Acrobat PDFWriter, which allows you to create PDF files.

Note: You access the Windows Printer folder from the Windows Start menu or by choosing Settings>Printers.

THE AUTODESK PLOTTER MANAGER

The Autodesk Plotter Manager allows you to configure non-system and Windows system plotter and printer devices easily. You can use it to configure AutoCAD to use local and network plotters and printers, and you can use it to predefine non-default output settings for Windows system devices.

AutoCAD 2002 stores information about media and plotting devices in plot configuration files called PC3 files. If you have used previous releases of AutoCAD, you are probably familiar with PCP and PC2 files. The PC3 files are similar to the earlier versions except they do not store any pen settings information.

Note: Pen settings information exists separately from PC3 files and is stored with the plot styles.

The Autodesk Plotter Manager takes you step by step through the process of creating PC3 files, as shown in the following exercise.

EXERCISE: CONFIGURING PRINTERS AND PLOTTERS

1. From the File menu, choose Plotter Manager. AutoCAD displays the Plotter Manager folder, as shown in Figure 20–1. This folder is where AutoCAD stores PC3 files and where you access the Add-A-Plotter wizard.

   You can display the Plotter Manager folder one of four ways:

   - From AutoCAD's File menu, choose Plotter Manager.
   - From AutoCAD's Options dialog box, in the Plotting folder, choose the Add or Configure Plotters button.
   - At AutoCAD's command line, enter **PLOTTERMANAGER**.
   - From the Windows Control Panel, double-click on the Autodesk Plotter Manager icon.

2. Double-click on the Add-A-Plotter Wizard icon. The wizard displays the Add Plotter - Introduction Page.
3. Choose the Next button. The wizard displays the Add Plotter - Begin page.

   The Begin page is where you indicate whether you want to use a local, network, or system printer. There are three choices:

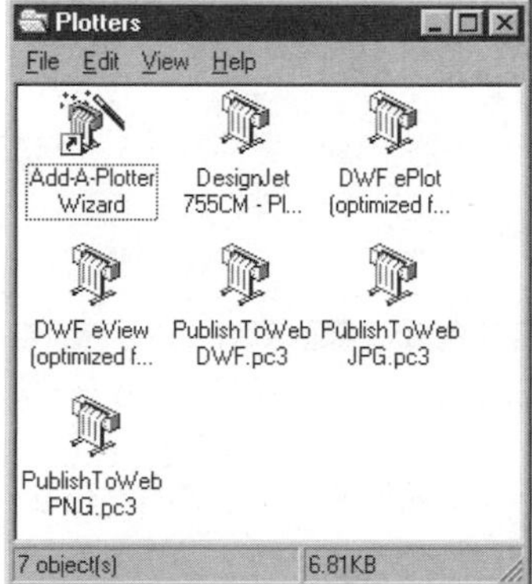

Figure 20–1 *The Autodesk Plotter Manager stores PC3 files, and it is where you access the Add-A-Plotter wizard.*

- **My Computer**—Configures plotter driver settings to be managed by your computer.
- **Network Plotter Server**—Configures plotter driver settings to be managed by the network plotter server.
- **System Printer**—Configures a Windows system driver that already resides in your computer's operating system. It allows you to set default printing/plotting parameters that apply only when the system printer is plotting from AutoCAD.

4. Select the My Computer radio button and then choose the Next button. The wizard displays the Add Plotter - Plotter Model page. The Plotter Model page is where you select the printer/plotter manufacturer and model type.
5. From the Manufacturers list, choose Hewlett-Packard.
6. From the Models list, choose DesignJet 755CM C3198A, as shown in Figure 20–2, and then choose the Next button.

   The wizard issues a warning if you have not already installed the HP DesignJet Windows system printer supplied on the AutoCAD 2002 installation CD. This driver is developed by Hewlett-Packard and is optimized for use with AutoCAD 2002. If the wizard issues a warning, choose Continue.

   The wizard displays the Add Plotter - Import Pcp or Pc2 page. This page allows you to import certain PCP and PC2 file information into the PC3 file.

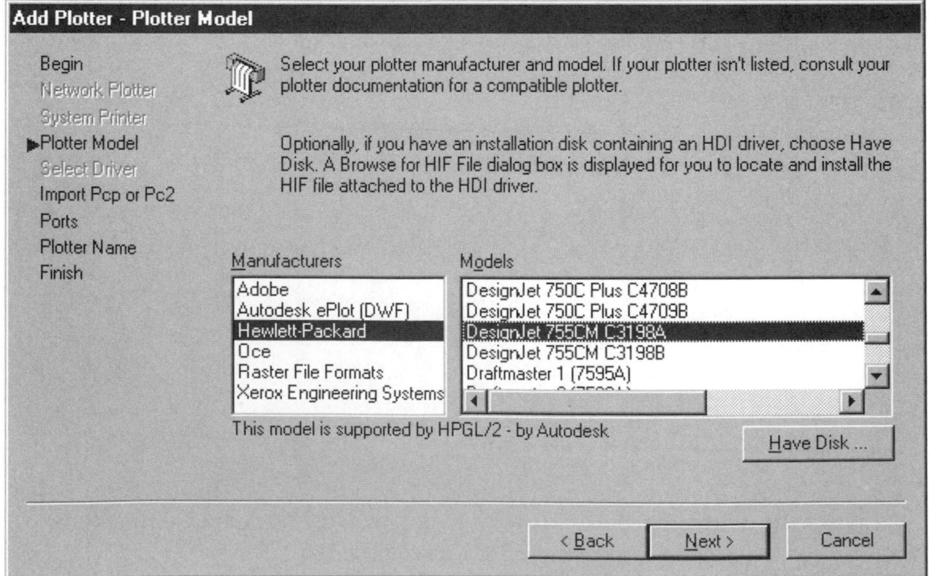

Figure 20–2 *The Plotter Model page is where you select your plotter's manufacturer and model.*

Note: AutoCAD allows PC3 files to import certain PCP and PC2 information, such as pen optimization, plot-to-file configurations, paper size and orientation, resolution, device name, and plot destination.

7. Choose the Next button. The wizard displays the Add Plotter - Ports page. This page is where you indicate whether you want your plot to be sent to a port (serial, parallel, or network), or to a file (PLT), or to AutoCAD's AutoSpool directory.
8. Choose the Plot to File radio button and then choose the Next button. The wizard displays the Add Plotter - Plotter Name page. This page is where you indicate the name to assign to the PC3 file.

Note: Although you can use special characters when naming a plotting device, it is recommended that you not use a space when doing so. Certain aspects of plotting with scripts are complicated when devices have spaces in their names.

9. In the Plotter Name text box, type **DesignJet 755CM - Plot to File** as shown in Figure 20–3, and then choose the Next button.

   The wizard displays the Add Plotter - Finish page. This page notes that the PC3 file is installed, and it allows you to edit the PC3 file settings and calibrate your file for the plotter.
10. Choose the Finish button. The PC3 file is created and saved in the Plotter Manager folder and is now available for use as a plotter configuration.

    Next you will create another PC3 file configured for a LaserJet printer.
11. If the Plotter Manager folder is not visible, then choose Plotter Manager from the File menu. AutoCAD displays the Plotter Manager folder.

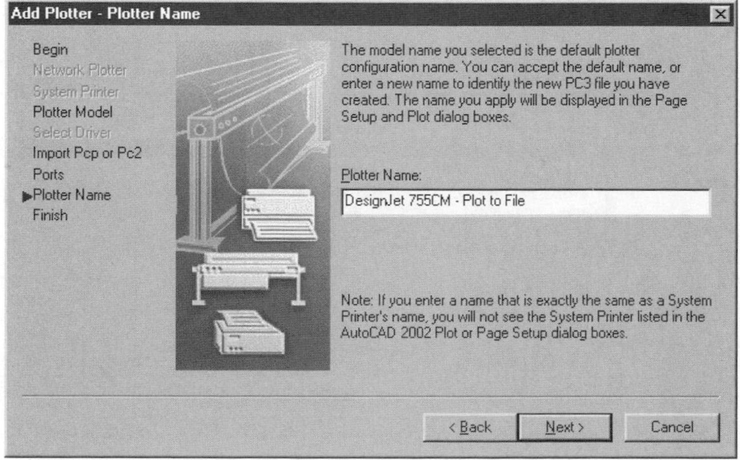

Figure 20–3 *The Plotter Name page is where you assign a name to the PC3 file.*

Note: The Plotter Manager folder's display is independent of AutoCAD. Consequently, you may leave it open on your desktop, even after you end your AutoCAD session.

12. Double-click on the Add-A-Plotter Wizard icon. The wizard displays the Add Plotter - Introduction Page.
13. Choose the Next button. The wizard displays the Add Plotter - Begin page.
14. Select the My Computer radio button, and then choose the Next button. The wizard displays the Add Plotter - Plotter Model page.
15. From the Manufacturers list, choose Hewlett-Packard.
16. From the Models list, choose LaserJet 4MV, as shown in Figure 20–4.
17. Choose the Next button. The wizard displays the Add Plotter - Import Pcp or Pc2 page.
18. Choose the Next button. The wizard displays the Add Plotter - Ports page.
19. Choose the Plot to File radio button, and then choose the Next button. The wizard displays the Add Plotter - Plotter Name page.
20. In the Plotter Name text box, type **LaserJet 4MV - Plot to File** as shown in Figure 20–5, and then choose the Next button. The wizard displays the Add Plotter - Finish page.

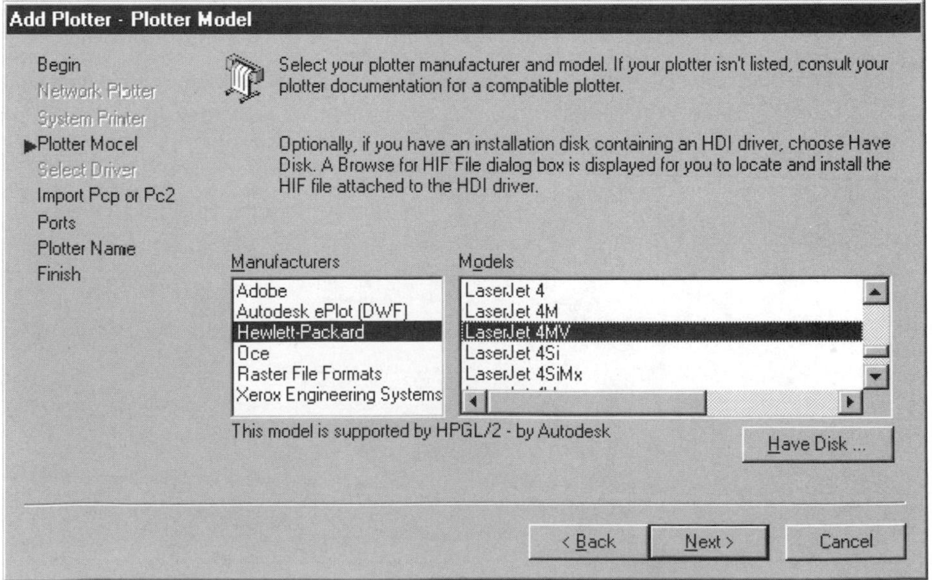

Figure 20–4 *The Hewlett-Packard LaserJet 4MV is selected.*

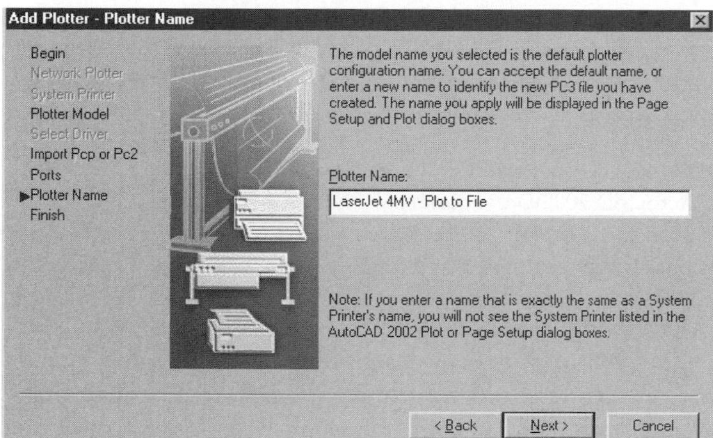

Figure 20–5 *The PC3 file is named LaserJet 4MV - Plot to File.*

21. Choose the Finish button.

    The second PC3 file is created and saved in the Plotter Manager folder and is now available for use as a plotter configuration.

22. You may close the Plotter Manager folder.

Note: The Add Plotter - Plotter Model page allows you to install drivers that are not found in the list. When you choose the Have Disk button, the wizard prompts you to identify the driver to install.

Tip: You can set a PC3 file as the default output device for new drawings and for earlier-version drawings first opened in AutoCAD 2002. From the Options dialog box, in the Plotting folder, choose the desired PC3 file from the Use as Default Output Device list.

Creating PC3 files is easy with the new Add-A-Plotter wizard. You can create as many PC3 files as you need and then share them with others. Later in this chapter you will use the PC3 files you created and apply them to page setups.

Next you will use the Plotter Configuration Editor to modify one of the PC3 files.

THE NEW PLOTTER CONFIGURATION EDITOR

You can create AutoCAD 2002 plotter configuration (PC3) files to predefine the print device to which your drawing is sent. You can create as many PC3 files as you need to meet all of your plotting conditions. By creating the PC3 files you need, you can quickly switch to the proper PC3 file to plot your drawing to the desired device.

It is not necessary to create PC3 files using the Add-A-Plotter wizard. For example, you can create a new PC3 file by copying an existing file and then modifying its settings. By copying an existing file and modifying its settings, you can quickly define a new PC3 file.

Tip: To create a copy of a PC3 file quickly, select Plotter Manager from the File menu. Then right-click on the desired PC3 file and drag and drop it to a blank area in the Plotter Manager folder. When prompted, choose Copy Here. Finally, right-click on the copy, choose Rename, and then rename the copy.

Note: When you copy PC3 plotter files, be sure to copy any attached PMP files from the AutoCAD 2002\Drv directory. A PMP file is where AutoCAD 2002 stores changes/additions to the paper sizes for a given plotter. A PMP file can be reattached to the PC3 files once it has been copied to the new location.

The new Plotter Configuration Editor allows you to modify your PC3 files. It has features that allow you to provide a description of the PC3 file, switch the port to which drawings are plotted, and control device and document settings such as the media source and custom paper sizes. By using the Plotter Configuration Editor, you can quickly edit your PC3 files, as shown in the following exercise.

This next exercise uses a PC3 file that was created in the previous exercise. Alternatively, you can copy the DesignJet 755CM - Plot to. File.pc3 file from the accompanying CD to the Plotter Manager folder. After you copy the file, be sure to right-click on the PC3 file, choose Properties, and then clear the Read-only attribute.

EXERCISE: EDITING AUTOCAD 2002 PLOTTER CONFIGURATION (PC3) FILES

1. From the File menu, choose Plotter Manager. AutoCAD displays the Plotter Manager folder.
2. Double-click on the DesignJet 755CM - Plot to File.pc3 file. The Plotter Configuration Editor is displayed.
3. From the General folder, in the Description box, type **Configured for Plots on Translucent Bond** as shown in Figure 20–6.

   The Ports folder allows you to select the port to which to send your plot. The DesignJet 755CM - Plot to File.pc3 file is currently set to Plot to File. This instructs AutoCAD to create a plot file of the drawing instead of sending the drawing to a plotter device.
4. Choose the Device and Document Settings tab.

   The Device and Document Settings folder allows you to define many of the PC3 file's settings. These include the paper source and size, custom properties settings for the device (if available), and plotter calibration files. You can also define custom paper sizes, adjust the paper's printable area, and identify the type of media to use, such as Opaque Bond or High-Gloss Photo.
5. In the tree view window, under the Media branch, select Media Type. The Media Type list is displayed.
6. From the Media Type list, choose Translucent Bond, as shown in Figure 20–7.

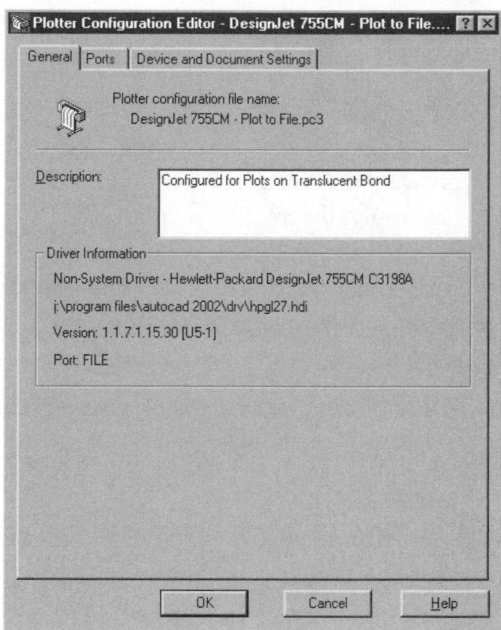

Figure 20–6 *The PC3 file's description is given.*

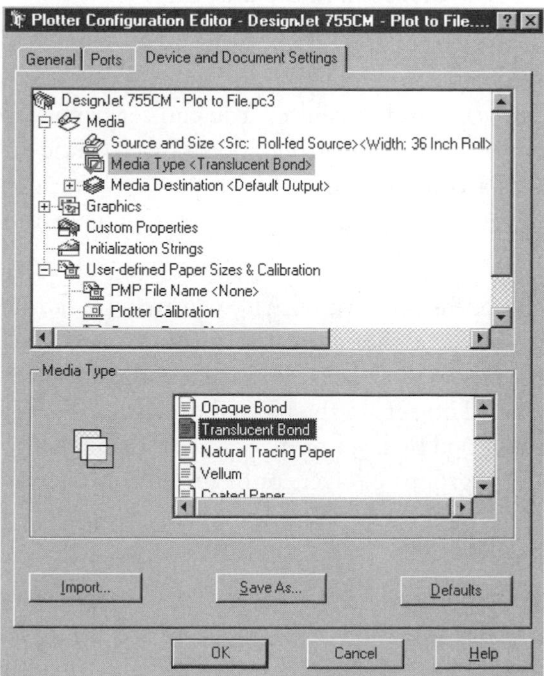

Figure 20–7 *The PC3 file's media type is set to Translucent Bond.*

7. Choose the OK button. The Plotter Configuration Editor modifies the PC3 file and saves the changes.
8. You may close the Plotter Manager folder.

The Plotter Configuration Editor makes modifying PC3 files very easy. This is especially useful when you want to create several nearly identical PC3 files whose settings vary slightly. For example, by creating a single PC3 file using the Autodesk Plotter Manager, and then copying the PC3 file, you can use the Plotter Configuration Editor to make minor changes to the PC3 file copy quickly. This saves time by duplicating all the settings from the original PC3 file, and it allows you to change the one or two settings necessary to customize the file.

In the next section you will continue the process of creating a plot by defining plot styles.

DEFINING PLOT STYLES

AutoCAD 2002 provides a feature called Plot Styles, which allow you to control how objects in drawings appear when they are plotted. Through a plot style you can tell AutoCAD to override certain object properties and plot the objects with a wide line, a different color, or even a different linetype than what is assigned to the object in the drawing. With plot styles you can easily vary the way objects look when they are plotted, no matter how they look in the drawing.

Plot styles allow you to control several object properties when you plot a drawing. You can control the object's color, linetype, and lineweight. You can also control values for dithering, gray scale, and screening. You can set pen assignments and define object fill styles. You can also control the appearance of line endpoints, including when two lines join at a common point. With plot styles you can create from the same drawing many plots that look different, without actually modifying any objects in the drawing.

If you have used prior releases of AutoCAD, you may be familiar with controlling linetypes and lineweights by assigning pen settings in the Print/Plot Configuration dialog box. The limitation with this feature is that you can only assign linetypes and lineweights globally, to all objects in the plotted drawing. With plot styles, however, the pen settings are assigned by layer or by object. This means you can control linetypes and lineweights by groups of layers or by groups of objects. More importantly, you can also control other properties, such as object screening and fill styles, as noted in the previous paragraph.

Plot styles are saved in a plot style table, and a plot style table can have as many plot styles as needed. In order to assign a plot style to a layer or an object, you must attach a plot style table to the Model or Layout folders. Once the plot style table is attached, you can assign its various plot styles to layers or to objects. Additionally, you can assign different plot style tables to the Model and Layout folders. The ability to assign different plot styles to layers and objects, and then to attach different plot style tables to the Model and Layout folders, provides you with a tremendous amount of flexibility in how your plotted drawings appear.

THE PLOT STYLE MANAGER

The Plot Style Manager provides a location at which to store your plot style tables. Similar in design to the Autodesk Plotter Manager, it also provides access to the Add-A-Plot Style Table wizard, which takes you step by step through the process of creating plot style tables, as demonstrated in the following exercise.

EXERCISE: CREATING A PLOT STYLE TABLE

1. From the File menu, choose Plot Style Manager.

   AutoCAD displays the Plot Style Manager folder, as shown in Figure 20–8. This folder is where AutoCAD stores color-dependent plot style tables and named plot style tables, and where you access the Add-A-Plot Style Table wizard.

2. Double-click on the Add-A-Plot Style Table Wizard icon. The wizard displays the Add Plot Style Table introduction page.

3. Choose the Next button. The wizard displays the Add Plot Style Table - Begin page.

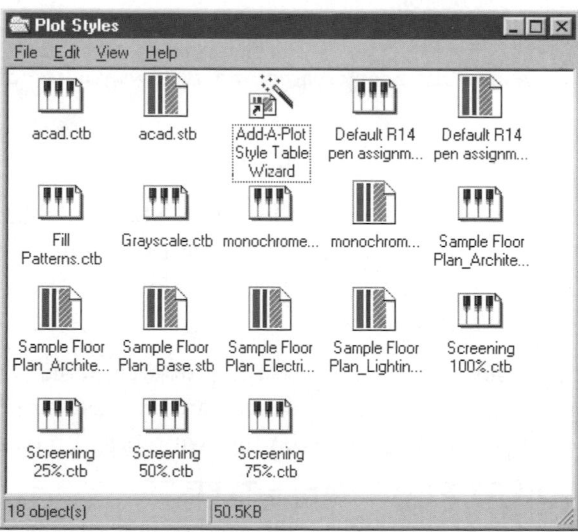

Figure 20–8 *The Plot Style Manager folder stores color-dependent and named plot style tables, and it is where you access the Add-A-Plot Style Table wizard.*

The Begin page is where you indicate whether you want to create a plot style table from scratch or from an existing file. There are four choices:

- **Start from Scratch**—Creates a new plot style table
- **Use an Existing Plot Style Table**—Creates a new plot style table from an existing plot style table, duplicating all the defined plot styles
- **Use My R14 Plotter Configuration (CFG)**—Creates a new plot style table by importing pen assignments stored in the acadr14.cfg file
- **Use a PCP or PC2 File**—Creates a new plot style table by importing pen assignments stored in a PCP or PC2 file

4. Select the Start from Scratch radio button, and then choose the Next button. The wizard displays the Add Plot Style Table - Pick Plot Style Table page.

   The Plot Style Table page is where you select the type of plot style table to create, either color-dependent or named. The two types of plot style tables are discussed in the section following this exercise.

5. Select the Named Plot Style Table radio button, and then choose the Next button. The wizard displays the Add Plot Style Table - File Name page. This page is where you indicate the name to assign to the plot style table file.
6. In the File Name text box, type **AP-2002** as shown in Figure 20–9, and then choose the Next button.
7. The wizard displays the Add Plot Style Table - Finish page. This page notes that the plot style table has been created, and it allows you to edit the file settings.
8. Choose Finish.
9. You may close the Plot Style Manager folder.

The plot style table is then saved in the Plot Style Manager folder. With the plot style table created, you can now define plot styles in the table. Then the plot style table can be attached to the Model or Layout folders, and its plot styles can be assigned to layers or objects in the drawing.

Before you proceed to define plot styles in the plot style table, you should understand the differences between color-dependent and named plot style tables. The next section reviews the differences between the two types of tables.

CHOOSING THE PLOT STYLE TABLE TYPE

AutoCAD provides two types of plot style tables—color-dependent and named—and both types behave similarly. For example, both types allow you to control the appearance of objects when they are plotted, including an object's color, linetype, and lineweight. Both types can be attached to the Model and Layout folders, and both can assign plot styles by layer or by object. So, if both types are so similar, then why have two?

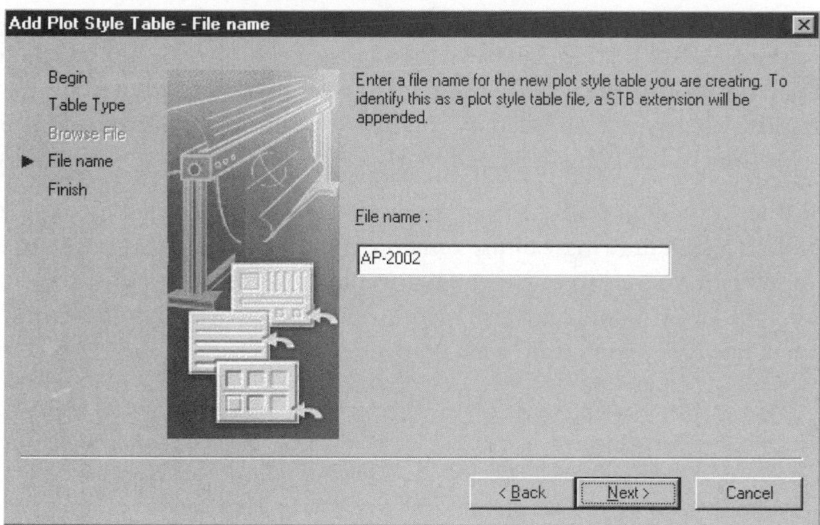

Figure 20–9 *The File Name page is where you assign a name to the plot style table file.*

While the two types are similar, there are two important differences between them. One difference is the number of plot styles each table type can hold. A color-dependent plot style table has a maximum of 255 color designations, which means you can define only 255 different plot styles per table. In contrast, a named plot style table can have an unlimited number of plot styles, because the plot styles in each table are assigned a unique name.

The other—and perhaps more important—difference is in how each type assigns plot styles to objects. As their name implies, color-dependent plot styles are based on the color of the object, or the color of the layer on which the object resides. Consequently, you must be careful when you select an object's color to ensure that it appears as desired when it is plotted.

If you have plotted drawings from prior releases of AutoCAD, then you are probably familiar with using colors to define an object's lineweight or linetype. If you have plotted drawings from prior releases of AutoCAD, then you have used color-dependent plot styles, whether you realized it or not.

Named plot styles are not based on object color. Named plot styles assign properties by name. Therefore, you can assign objects any color and then control properties like lineweight and linetype regardless of their color. This provides a new flexibility in creating objects, because you no longer need to use colors to control an object's linetype or lineweight. With named plot styles, color is just another independent property like lineweight and linetype. To demonstrate this, suppose that you have a drawing and all of its objects are red in color. You can override the red color of the objects by assigning each object a specific named plot style. When the drawing is plotted, the named plot styles override the color of the objects, applying the color listed in each named plot style. The same is true for objects whose color is set to ByLayer and is therefore controlled by the layer's color. You can override the layer's color at plot time by assigning a named plot style to each layer.

There is one very important feature to understand about color-dependent and named plot styles. A drawing can use only one type and cannot switch between the two types. When you create a drawing, the plot style table type is automatically assigned to the drawing and is permanent. Therefore, you must determine which type you will use *before* you create your drawing.

Note: The default plot style table type is set in the Options dialog box, which is accessed from the Plotting folder by choosing either the Use Color Dependent Plot Styles radio button or the Use Named Plot Styles radio button. Once the default plot style table type is changed, the new default plot style table type takes effect on new drawings.

Tip: While a drawing's plot style table type is permanently set to color-dependent or named, you can simulate changing a drawing from one type to the other. To do so, copy all the objects in the current drawing into a new, blank drawing whose plot style table is set to the type you wish to use.

In the next section you will use the Plot Style Table Editor to add two plot styles to an existing plot style table.

ADDING PLOT STYLES TO PLOT STYLE TABLES

Plot styles reside in plot style tables and provide the ability to control an object's appearance when it is plotted. You can control color, linetype, and lineweight, as well as other properties. By using plot styles, you can control how an object looks when it is plotted without changing the object's appearance in the drawing.

Once a named plot style table is created, you can add new plot styles, and you can edit or delete existing plot styles by using the Plot Style Table Editor. After you add plot styles to a plot style table, the table can be attached to the Model or Layout folders, and its plot styles can be assigned to layers or objects.

Note: Plot styles can be added or deleted only to named plot style tables. Color-dependent plot style tables have 255 predefined plot styles whose "names" are based on colors, and whose values can only be edited, not added or deleted.

The following exercise shows how to add and edit plot styles in a named plot style table. The example uses the named plot style table file created in the previous exercise. Alternatively, you can copy the AP-2002.stb file from the accompanying CD to the Plot Styles folder. After you copy the file, be sure to right-click on the STB file, choose Properties, and then clear the Read-only attribute.

EXERCISE: ADDING PLOT STYLES TO A NAMED PLOT STYLE TABLE

1. From the File menu, choose Plot Style Manager. AutoCAD displays the Plot Style Manager.
2. Double-click on the AP-2002.stb file. The Plot Style Table Editor is displayed.

   The Plot Style Table Editor has three folders. The General folder allows you to provide a description for the table. The Table View and Form View folders display property information for the selected plot style. The latter two display the same information in different formats.
3. In the General folder, in the Description box, type **Plot styles configured for D Size sheets** as shown in Figure 20–10.
4. Choose the Form View tab.

   The Form View folder displays the property information for the Normal plot style, as shown in Figure 20–11. Notice that only the Normal plot style is displayed in the Plot Styles list. The Normal plot style is automatically created as the default plot style in a table, and it cannot be edited or deleted.

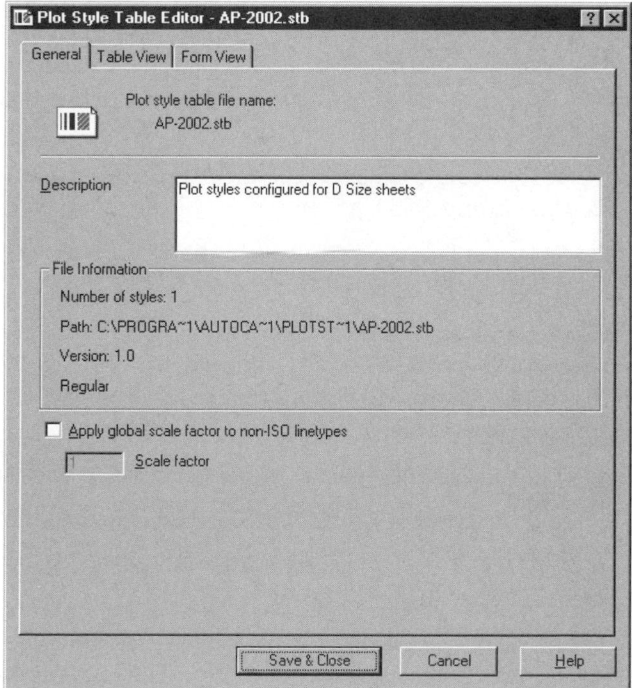

Figure 20–10 *The plot style table's description is set.*

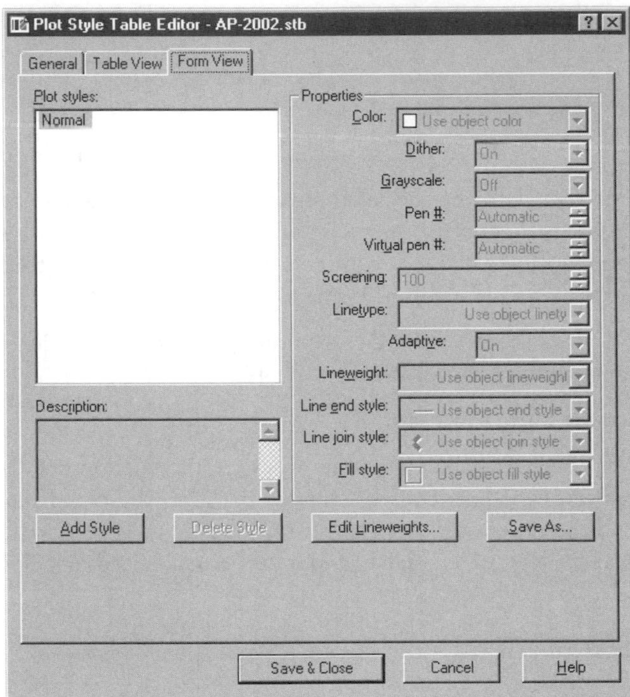

Figure 20–11 *Every named plot style table has the default Normal plot style.*

5. To add a plot style to the table, choose the Add Style button. The Add Plot Style dialog box is displayed.
6. In the Plot Style text box, type **Contours - Normal**, and then click OK. The Contours - Normal plot style is created and added to the Plot Styles list.

   The Properties area displays the properties you can edit for the selected plot style. When you assign the Contours - Normal plot style to an object's plot style property, the Contours - Normal plot style's property values will override the object's properties when it is plotted.

   Next you will edit several properties of the Contours - Normal plot style.
7. From the Color list, choose Black. The selected color is the color the plotter uses to draw the object.
8. In the Screening scroll box, type **30**.

   The screening value specifies a color intensity setting, which controls the amount of ink the plotter uses to draw a color. The lower the screening value, the less ink is used to draw the object, and the lighter—or less intense—the object appears when it is plotted. The range of screening values is zero through 100. Selecting 100 displays the color at its full intensity.

9. From the Lineweight list, choose 0.1000 mm.
10. In the Description text box, type **Objects are screened to 30%**. The new plot style definition is complete (see Figure 20–12).
11. Choose the Table View tab. Notice that the plot style properties are the same as those displayed in the Form View folder; they are simply arranged differently. Both forms allow you to control identical properties, and the form you use is up to you.
12. Choose the Add Style button. A new plot style column is created and assigned the default name Style 2.
13. In the Name row, type **Contours - Index** to rename the plot style.
14. Choose the Description box in the Contours - Index column. The description box is highlighted.
15. In the Description box, type **Objects are screened to 70%.**
16. From the Color list, choose Black.
17. In the Screening scroll box, type **70**.
18. From the Lineweight list, choose 0.2000 mm. The new plot style definition is complete (see Figure 20–13).

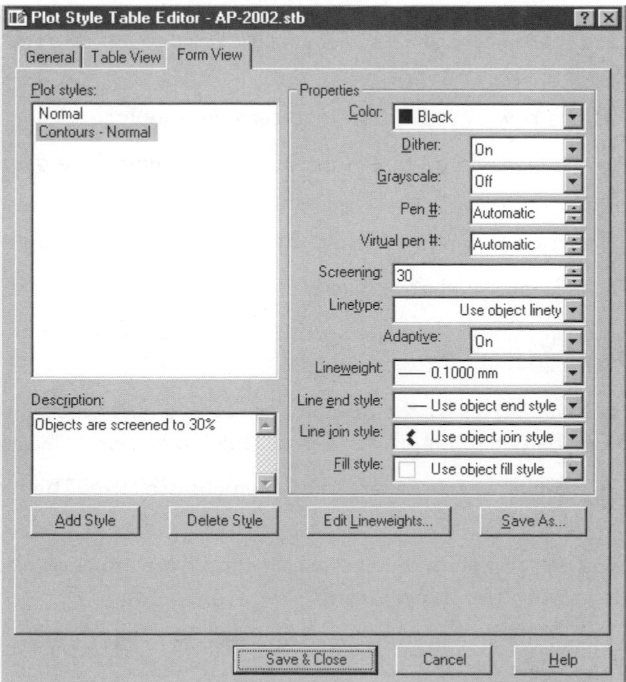

Figure 20–12 *The Contours - Normal plot style is defined.*

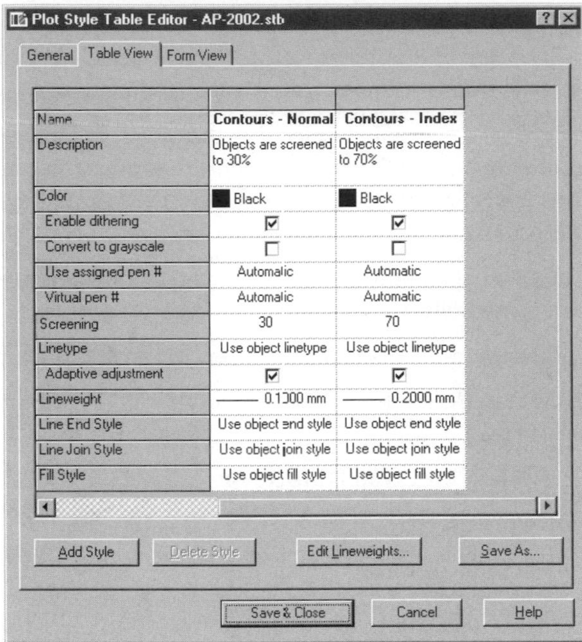

Figure 20–13 *The Contours - Index plot style is defined.*

19. Choose the Add Style button. A new plot style column is created and assigned the default name Style 3.
20. In the Name row, type **Buildings** to rename the plot style.
21. Choose the Description box in the Buildings column. The description box is highlighted.
22. In the Description box, type **New fill style applied to objects**.
23. From the Color list, choose Black.
24. From the Fill Style list, choose Solid. The new plot style definition is complete (see Figure 20–14).
25. Choose Save As to save the new plot styles to a new plot style table. The Save As dialog box is displayed.
26. Name the table "D Size Sheets," and then choose Save. The plot style table is saved using the new name.

    Next you will complete the exercise by modifying the current plot style table and then saving it as a new table.
27. In the General folder, in the Description box, type **Plot styles configured for A Size sheets**.
28. In the Form View folder, choose the Buildings plot style.

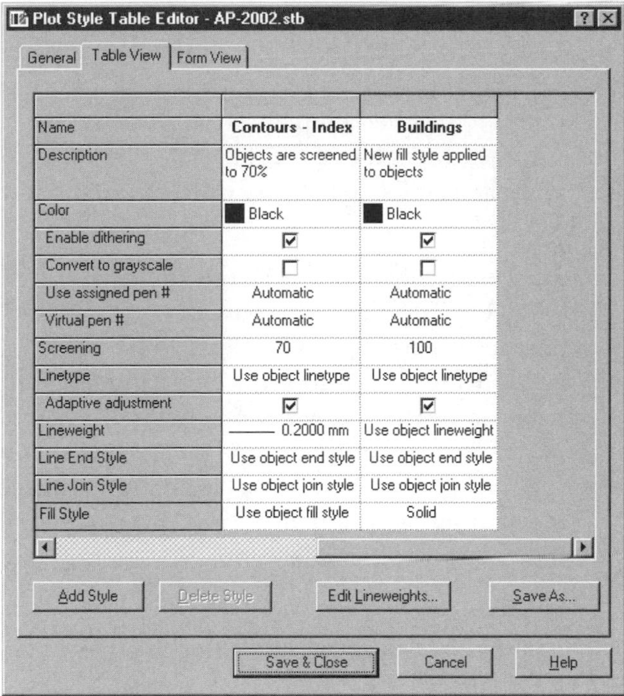

Figure 20–14 *The Buildings plot style is defined.*

29. From the Fill Style list, choose Horizontal Bars.
30. Choose Save As to save the plot styles to a new plot style table.
31. Name the table "A Size Sheets," and then choose Save. The plot style table is saved using the new name.
32. Choose the Save & Close button to save the new plot style table and close the Plot Style Table Editor.

The two new plot style tables are saved to the Plot Style Manager folder. The tables are now available for use with drawings and can be attached to the Model or Layout folders. Then their plot styles can be assigned to layers and objects.

As you just experienced, creating and editing plot styles is very easy. By using the various properties provided, you can alter the appearance of plotted objects, creating numerous versions of your drawing when they are plotted, without actually changing any object properties in the drawing.

In the next section you will create page setups that use the plot style table you created.

CREATING PAGE SETUPS

Page Setups are a feature of AutoCAD 2002 that provides the ability to control certain paper and plotter configuration information. Page setups work similarly to PC2 files in that they allow you to save certain plot settings and restore them as needed. You may save plot settings such as plot device, paper size, scale factor, and plot orientation, as well as the assigned plot style table, which contains plot styles used for the current drawing.

What makes page setups especially powerful is the ability to assign different page setups to the Model and Layout folders in your drawing. Consequently, a single drawing can have numerous page setups, with a particular page setup recalled and assigned to the current Model or Layout folder, which can then produce the desired results when the drawing is plotted. What this ultimately means is that you no longer have to worry about restoring prior plot and pen settings to duplicate a previous plot. All plot settings are saved with the drawing and instantly recalled by selecting the desired Model or Layout tab and then assigning the appropriate page setup.

The following exercise shows how to create page setups. This exercise uses the plotter configuration (PC3) files and named plot style tables created in previous exercises in this chapter. Alternatively, you can copy the following files from the accompanying CD to the designated folders:

- DesignJet 755CM - Plot to File.pc3 to the Plotter Manager folder
- LaserJet 4MV - Plot to File.pc3 to the Plotter Manager folder
- A Size Sheets.stb to the Plot Style Manager folder
- D Size Sheets.stb to the Plot Style Manager folder

After you copy the files to their folders, be sure to right-click on each one, choose Properties, and then clear the Read-only attribute.

EXERCISE: CREATING PAGE SETUPS

1. Open the 20DWG01.dwg drawing file found on the accompanying CD. The drawing opens, displaying a D Size Sheet border and a single viewport that displays model space objects.
2. Right-click over the Layout1 tab, and then choose Page Setup. The Page Setup dialog box is displayed.
3. In the Plot Device folder, in the Plotter Configuration area, select the DesignJet 755CM - Plot to File.pc3 file from the name list. This is the PC3 file that was created in a previous exercise in this chapter, and it indicates the device to which AutoCAD plots drawings.
4. In the Plot Style Table (Pen Assignments) area, from the Name list, choose D Size Sheets.stb, as shown in Figure 20–15. This is the plot style table that con-

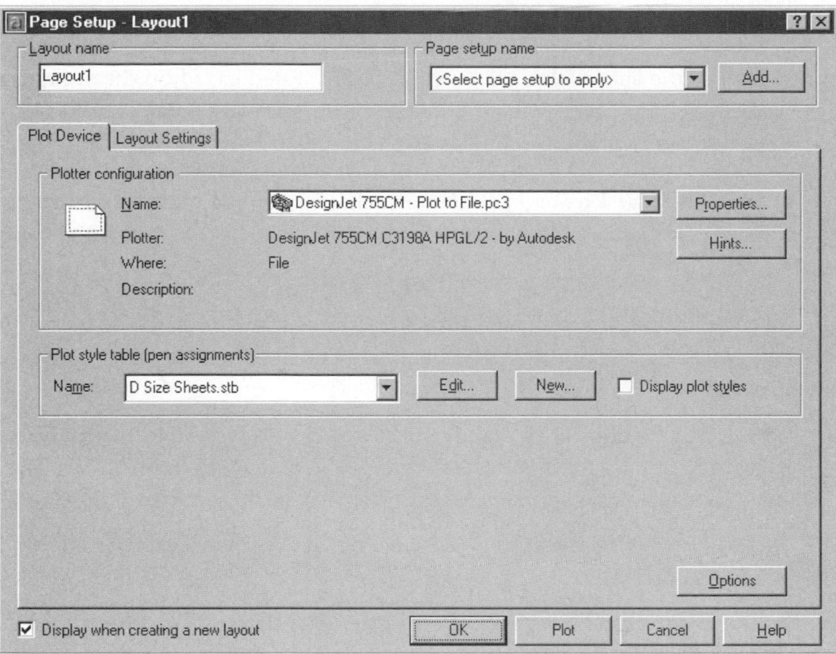

Figure 20–15 *The PC3 file and plot style table are selected for the current page setup.*

tains the plot styles created in a previous exercise in this chapter, and it controls the appearance of objects when they are plotted.

5. In the Layout Settings folder, in the Paper Size and Paper Units area, make sure the "ANSI expand D (34.00 x 22.00 Inches)" paper size is selected.
6. In the Drawing Orientation area, choose the Landscape radio button. This instructs AutoCAD to use the long edge of the paper as the top of the page.
7. In the Plot area, choose the Extents radio button. This instructs AutoCAD to calculate the plot area based on all objects in the current space—in this case, paper space. The extents does not include the layout's paper image and shadow, which are displayed in the background and provided only as visual aids for positioning the drawing on paper.
8. In the Plot Scale area, make sure Scale is set to 1:1. Typically, in the case of layouts, the paper background represents your drawing's paper sheet and is set to the sheet's actual size. In this case, the sheet size is 34.00 x 22.00 inches. Therefore, the scale is set to 1:1; one unit equals one inch.
9. In the Plot Scale area, make sure the Scale Lineweights check box is selected.

   Lineweights specify the line width of plotted objects, and they are normally plotted using the lineweight's value, regardless of the plot scale. This means that if an object's lineweight is set to 0.1000 inches, the object's lineweight will always be plotted at 0.1000 inches. This is true even if the plot scale is to 1:2. When you select Scale Lineweights, if the drawing's scale is set to a value other than 1:1, the lineweights are proportionally scaled based on the scale factor. In the case of a 1:2 plot scale, if the lineweight is set to 0.1000 inches, AutoCAD rescales the plotted object's lineweight to 0.0500 inches.

Note: The Scale Lineweights option is only available in a paper space layout. This is another valid reason to do all sheet plotting from paper space rather than model space.

10. In the Plot Offset area, make sure the Center the Plot check box is selected. This ensures that the plotted objects are centered on the sheet when they are plotted.
11. In the Plot Options area, make sure the Plot with Plot Styles check box is selected. This instructs AutoCAD to use the plot styles in the D Size Sheets.stb plot style table that you attached to the page setup in step 4.

    Next you will save the settings of the Plot Device and Layout Settings folders as a named page setup.
12. In the Page Setup Name area, choose the Add button. The User Defined Page Setups dialog box is displayed.
13. In the New Page Setup Name text box, type **D Size Plots**, and then choose OK. AutoCAD names the page setup as shown in Figure 20–16.

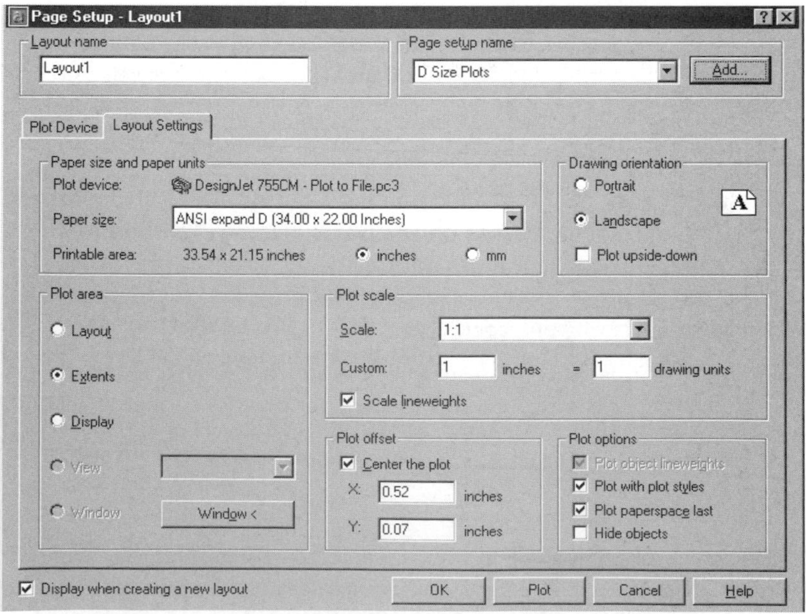

Figure 20–16 *The current settings are stored in the D Size Plots page setup.*

The current settings of the Plot Device and Layout Settings folders are saved as a named page setup called D Size Plots. The settings are saved in the current drawing, and you can recall them anytime by selecting D Size Plots from the Page Setup Name list.

Note: You can insert named page setups from other drawings by choosing the Import button from the User Defined Page Setups dialog box. You can also insert page setups using the command PSETUPIN.

Next you will define one more page setup.

14. In the Plot Device folder, in the Plotter Configuration area, select the LaserJet 4MV - Plot to File.pc3 file from the Name list. This is the second PC3 file that you created in a previous exercise in this chapter.

    AutoCAD displays a warning noting that the paper size in the layout is not supported by the selected plot device and that the layout will use the paper size specified by the plot device.

15. Choose OK to dismiss the warning message.
16. In the Plot Style Table (Pen Assignments) area, from the Name list, choose A Size Sheets.stb.

    This is the second plot style table that contains the plot styles that were created in a previous exercise in this chapter. The only difference between the two plot style tables is the Buildings plot style in A Size Sheets.stb displays fills using the Horizontal Bars fill style, while the Buildings plot style in D Size Sheets.stb displays fills using the Solid fill style.

17. In the Layout Settings folder, in the Paper Size and Paper Units area, make sure the ANSI A (8.50 x 11.00 Inches) paper size is selected.
18. In the Drawing Orientation area, choose the Landscape radio button.
19. In the Plot area, choose the Extents radio button.
20. In the Plot Scale area, from the Scale list, choose Scaled to Fit. (You may need to scroll to the top of the list.)

Note: You may not get the desired results when you use the Scaled to Fit plot scale size if the Layout option is selected in the Plot area. Therefore, when you use the Scaled to Fit plot scale size, choose the Extents option.

21. In the Plot Scale area, make sure the Scale Lineweights check box is selected.
22. In the Plot Offset area, make sure the Center the Plot check box is selected.
23. In the Plot Options area, make sure the Plot with Plot Styles check box is selected.

24. In the Page Setup Name area, choose Add. The User Defined Page Setups dialog box is displayed.

25. In the New Page Setup Name text box, enter **A Size Plots**, and then choose OK. AutoCAD names the page setup as shown in Figure 20–17.

26. Choose OK to accept the page setup changes.

AutoCAD saves the page setups to the current drawing, where they may be assigned to the Model or Layout folders, or to any new Layout folders. By right-clicking on the Model or Layout tab, then selecting Page Setup, you can assign the selected folder a saved page setup or create a new page setup.

Notice that when you chose OK in the Page Setup dialog box, the currently selected page setup—A Size Plots—redefined the drawing's paper size and scale. This is one of the big benefits of page setups. Without having to reset any scale values, you can instantly change the plot device, pen assignments, sheet size, *and scale* of the current folder by selecting the desired page setup.

Now that the plot style tables are assigned to the page setups, the only step remaining is to take advantage of the plot style pen settings by assigning them to layers in the drawing, as demonstrated in the next exercise.

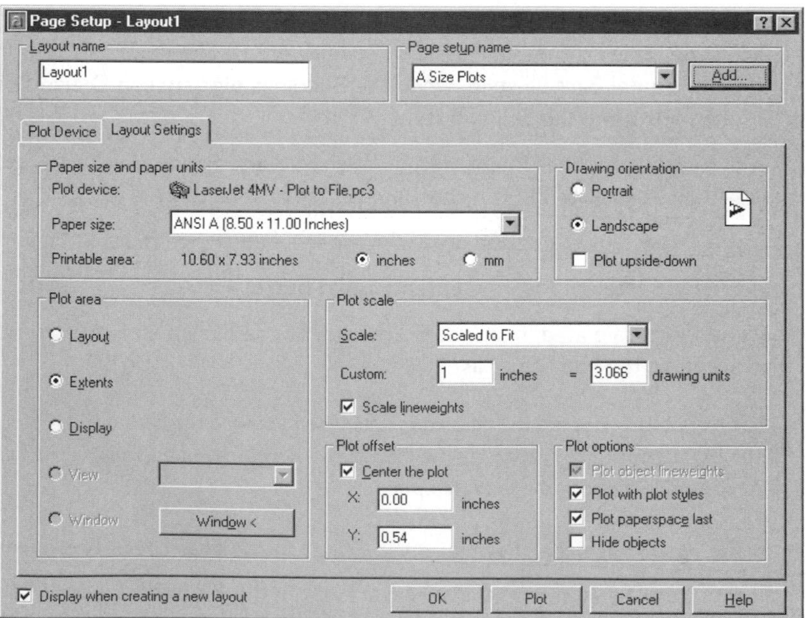

Figure 20–17 *The current settings are stored in the A Size Plots page setup.*

EXERCISE: ASSIGNING PLOT STYLES

1. Continue from the previous example.
2. Choose the Layers button on the Object Properties toolbar. AutoCAD displays the Layer Properties Manager.
3. Choose the Normal Plot Style from the BUILDINGS layer. AutoCAD displays the Select Plot Style dialog box.
4. Choose the Buildings plot style, and then choose OK. AutoCAD assigns the Buildings plot style to the BUILDINGS layer.
5. Choose the Normal Plot Style for the CONTOURS-INDEX layer. AutoCAD displays the Select Plot Style dialog box.
6. Choose the Contours - Index plot style, and then choose OK. AutoCAD assigns the Contours - Index plot style to the CONTOURS-INDEX layer.
7. Choose the Normal Plot Style for the CONTOURS-NORMAL layer. AutoCAD displays the Select Plot Style dialog box.
8. Choose the Contours - Normal plot style, and then choose OK. AutoCAD assigns the Contours - Normal plot style to the CONTOURS-NORMAL layer, as shown in Figure 20–18.

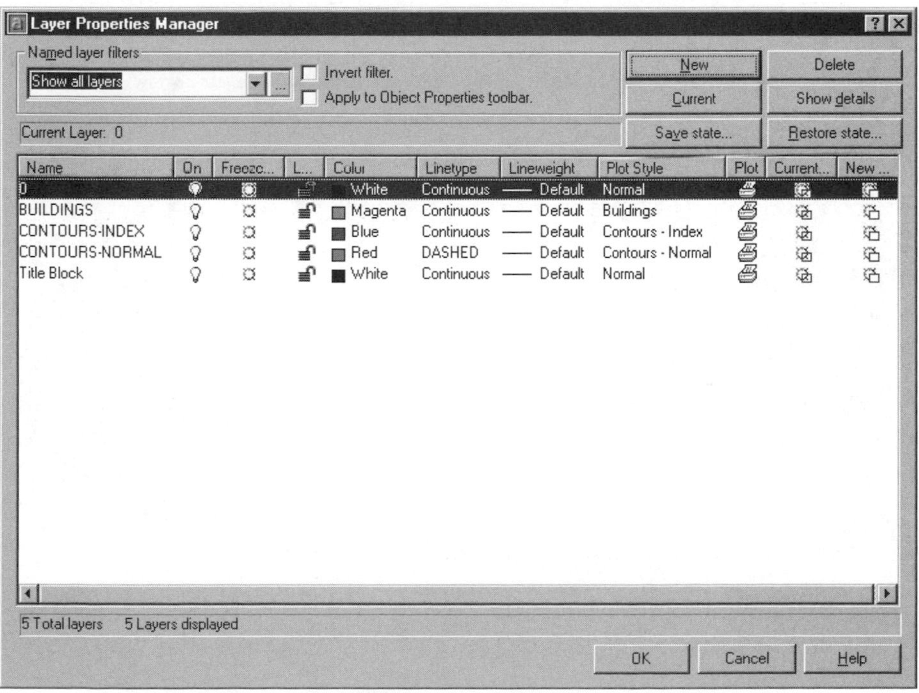

Figure 20–18 *The plot styles are assigned to the appropriate layers.*

9. Choose OK to close the dialog box and save the plot style assignments.

   With the plot styles now assigned to the appropriate layers, you can preview the effect that a page setup will have on a plotted drawing.

10. From the File menu, choose Plot Preview.

    The Plot Preview of the A Size Plots page setup is shown in Figure 20–19. Notice that the contours (the jagged polylines) are displayed in shades of gray, and the fill style of the buildings is horizontal bars. In the drawing, the colors used to depict the polylines are red and blue, and the fill is solid and magenta. This demonstrates the ability of plot styles to override object property settings.

    Next you will preview the D Size Plots page setup.

11. Right-click in the plot preview, and then choose Exit. AutoCAD closes the plot preview.

12. Right-click on the Layout1 tab, and then choose Page Setup. The Page Setup dialog box is displayed.

13. In the Page Setup dialog box, choose D Size Plots from the Page Setup Name list, and then choose OK. AutoCAD automatically switches to the D sheet size and rescales the drawing to the correct size.

14. From the File menu, choose Plot Preview. AutoCAD displays a preview of the drawing, as shown in Figure 20–20.

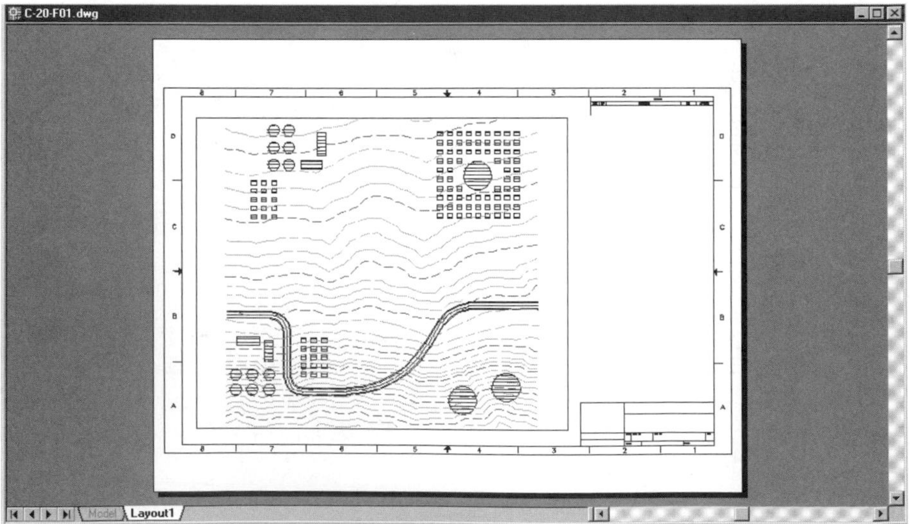

Figure 20–19 *The plot preview of the A Size Plots page setup demonstrates that plot styles override object property settings.*

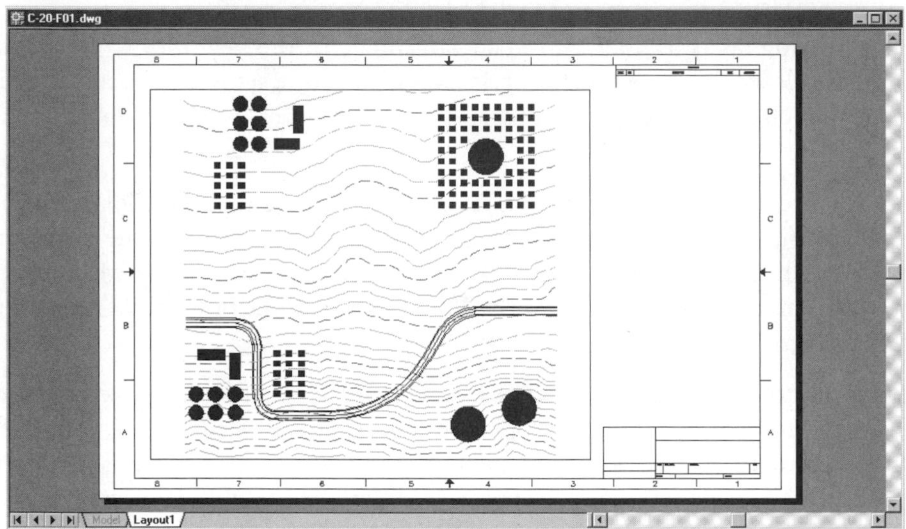

Figure 20–20 *The effect of the D Size Plots page setup is previewed.*

Notice that the contours are displayed in shades of gray just as they were in the A Size Plots preview. However, the fill style of the buildings is now solid. This occurs because the two plot style tables use the same plot style names. In other words, when the A Size Plots page setup is selected, it uses the A Size Sheets plot style table, whose Buildings plot style uses the Horizontal Bars fill style. In contrast, when the D Size Plots page setup is selected, it uses the D Size Sheets plot style table, whose Buildings plot style uses the Solid fill style.

15. You may exit plot preview and close the drawing.

By defining multiple page setups and assigning the desired PC3 files and plot style tables to each one, you can quickly switch a drawing's plot settings and achieve different output results when the drawing is plotted.

PLOTTING FROM AUTOCAD 2002

AutoCAD 2002 provides several methods for plotting drawings, with the most traditional method being the Plot dialog box. Additionally, AutoCAD provides a powerful batch plot utility that allows you to select many drawings, define different plot settings for each one, and then automatically plot all the files. The batch plot utility is discussed later in this chapter.

You can also produce and publish electronic drawing files to the Internet using the ePlot feature. The ePlot feature is discussed in Chapter 3, Expanded Reach.

These methods of producing plotted drawings, combined with the new Page Setup and Plot Style features, present a broad variety of options that ensure that you can produce output that meets your needs.

NEW FEATURES OF THE PLOT DIALOG BOX

The Plot dialog box was updated to take advantage of the new Plot Configuration (PC3) files, plot styles stored in plot style tables, and page setups, as discussed previously in this chapter. Additionally, there are several other improvements of which you should be aware.

When you execute the plot command, AutoCAD displays the new Plot dialog box, as shown in Figure 20–21. Notice that the Plot dialog box resembles the Page Setup dialog box, which is shown in Figure 20–15. The main difference between the two is the additional options available in the What to Plot and Plot to File areas of the Plot dialog box.

In the What to Plot area, you can choose to plot the current folder, selected folders, or all folders. When you choose to plot selected folders or all folders, AutoCAD automatically uses the page setup assigned to each folder. Therefore, it is good practice to define the page setup options for each folder when the folder is created.

Another very useful feature in the What to Plot area is the new Number of Copies control. This feature allows you to define the number of copies of the current or selected folder that AutoCAD should plot.

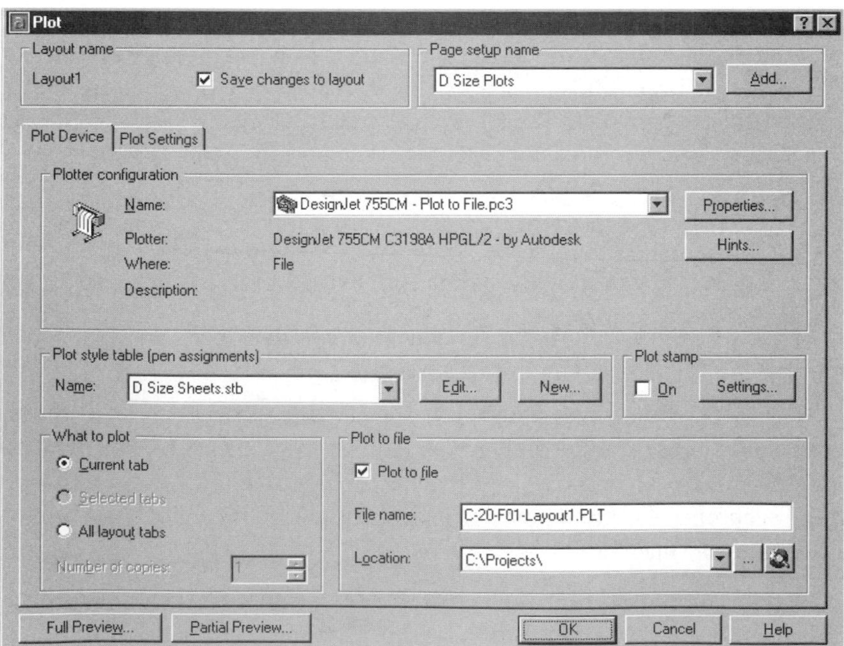

Figure 20–21 *The new Plot dialog box is similar to the Page Setup dialog box.*

One other new feature worth mentioning is found in the Plot Settings folder. The new Plot Paperspace Last option found in the Plot Options area allows you to force AutoCAD to define objects to be plotted in model space viewports first, and then to define paper space objects. This ensures that paper space objects are plotted "on top" of model space objects, and it avoids problems encountered in previous releases of AutoCAD, when model space objects were plotted on top of paper space objects, producing undesired effects.

PLOT STAMP

AutoCAD 2002 includes a Plot Stamp feature that lets you automatically add text to plotted drawings. You select the desired text from predefined fields, which include Drawing Name, Date and Time, and Plot Scale. When you plot a drawing, the selected fields are automatically inserted (stamped) into the corner of the paper sheet. This feature is very useful for quickly confirming useful and important information pertinent to the plotted drawing. By stamping the sheet, you can quickly determine which AutoCAD drawing was plotted, when it was plotted, and its scale, all by simply looking in the corner of the sheet.

How Plot Stamp Works

The data that Plot Stamp inserts onto a sheet is determined by settings saved in a plot stamp parameter (PSS) file. AutoCAD includes two predefined PSS files, Inches.pss and mm.pss, which are stored in AutoCAD's Support directory. You can edit these files, changing their settings, or you can create new files.

Note: If you wish to edit Inches.pss, you must first clear its Read-only attribute.

To insert a plot stamp, in the Plot dialog box's Plot Device folder, select the On check box in the Plot Stamp area, as shown in Figure 20–22. With the On check box selected, AutoCAD will stamp the sheet at plot time, using the current PSS file's settings.

Note: The Plot Stamp area is displayed only in the Plot Dialog box. Therefore, you cannot set the On option or indicate the PSS file from the Page Setup dialog box. Additionally, the selected values are stored in your systems registry, not with the drawing. Therefore, the values will not change from drawing to drawing, or from AutoCAD session to session, until you edit the values.

Tip: You can turn Plot Stamp on or off from the command line by typing **-PLOTSTAMP** and then choosing the desired value.

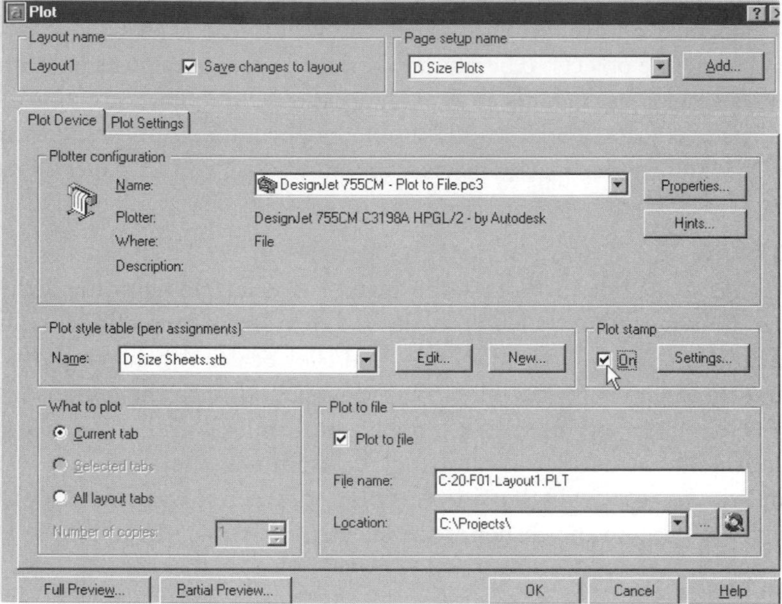

Figure 20–22 *To stamp a sheet at plot time, select the On check box in the Plot Stamp area of the Plot dialog box.*

Setting Plot Stamp Values

Plot Stamp includes a variety of predefined values that you can select. You can select the values by using the Plot Stamp dialog box or from AutoCAD's command line. Entering PLOTSTAMP at the command line opens the dialog box as shown in Figure 20–23; entering -PLOTSTAMP lets you select values from the command line. You can select from the following predefined fields:

- Drawing Name
- Layout Name
- Date and Time
- Login Name
- Device Name
- Paper Size
- Plot Scale

Selected fields are stamped onto the plotted sheet, and their values are determined at plot time. For example, the Date and Time value is based on the date and time that the plot is generated, i.e., the time you choose the OK button on AutoCAD's

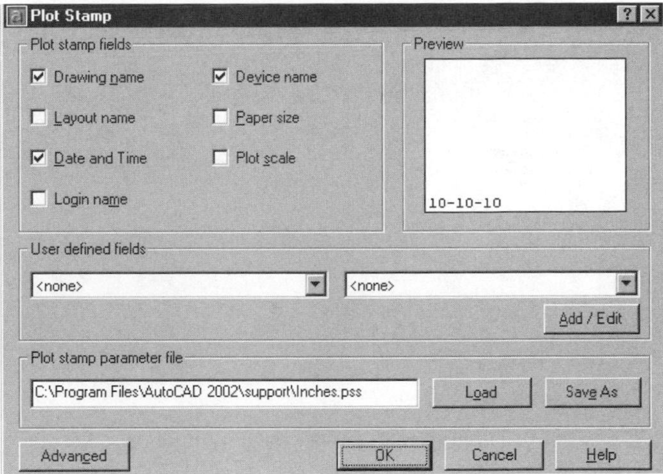

Figure 20–23 *The Plot Stamp dialog box lets you select the fields that AutoCAD stamps onto your plotted drawing.*

Plot dialog box. Therefore, the values for all predefined fields are dynamic, and they depend on your system's settings at the time you create the plot.

User-Defined Fields. In addition to the predefined fields, you can include user-defined fields. When you choose the Add/Edit button in the User Defined Fields area, AutoCAD displays the User Defined Fields dialog box, which lets you add, edit, or delete user-defined fields.

You can add as many user defined fields as you need, as shown in Figure 20–24. However, while there is no limit on the number of user-defined fields you can add to the User Defined Fields dialog box, you can select no more than two fields to stamp onto your plotted sheet.

Note: User-defined fields are static text only, which means that you cannot create dynamic user-defined fields. Consequently, if you use RTEXT, DIESEL code, or AutoLISP code to create a user-defined field, the code is not executed at plot time, and the syntax is therefore stamped onto your sheet.

Advanced Options

The Plot Stamp tool includes advanced options that let you control the position of the stamp on your plotted sheet, and that let you select the stamp's text font and size. When you choose the Advanced button on the Plot Stamp dialog box, AutoCAD displays the Advanced Options dialog box, as shown in Figure 20–25. The Advanced Options dialog box also lets you set the type of unit used to measure the stamp's XY offset and text height values (inches, millimeters, or pixels), and it lets you determine if the stamp's field values are stored in a log file.

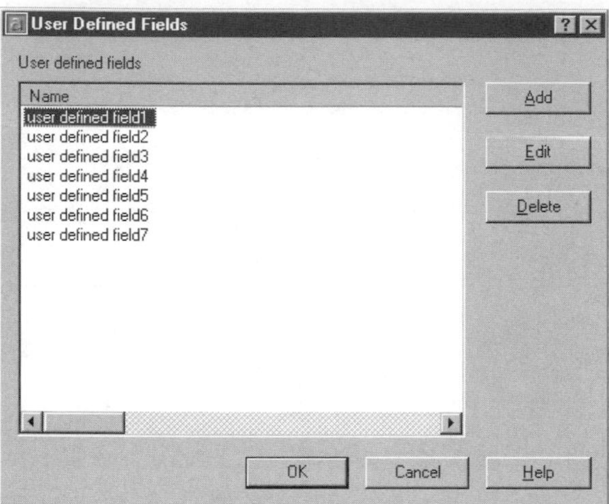

Figure 20–24 *You can create user-defined fields to stamp onto your drawings at plot time.*

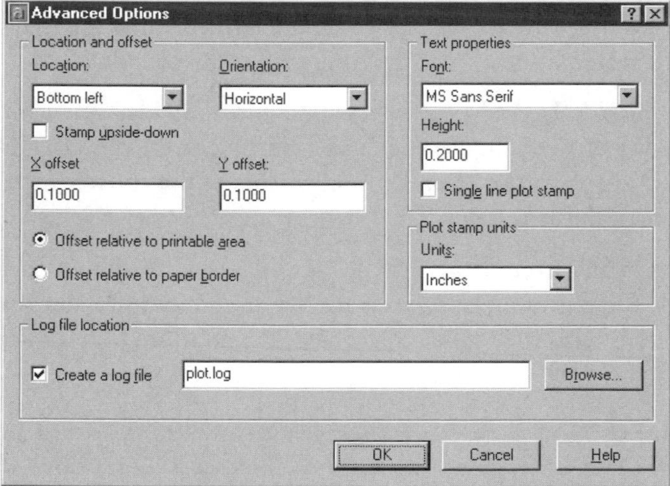

Figure 20–25 *Plot Stamp's Advanced Options dialog box lets you control the stamp's position and its text font and size.*

Note: As you adjust the stamp's position by setting values in the Advanced Options dialog box's Location and Offset area, AutoCAD updates the Preview area in the Plot Stamp dialog box (see Figure 20–23). This preview pane is designed to help you visualize the effects of your changes because the Plot dialog box's Full Preview feature does not display the plot stamp.

Controlling Location and Offset. The Advanced Options dialog box's Location and Offset area provides the following positional settings:

- **Location**—Options are Top Left, Bottom Left, Bottom Right, and Top Right.
- **Orientation**—Options are Horizontal and Vertical.
- **Stamp Upside Down**—Rotates the stamp 180 degrees.
- **X Offset**—Indicates the stamp's X offset value.
- **Y Offset**— Indicates the stamp's Y offset value
- **Offset Relative to Printable Area**—Calculates the offset values from the corner of the paper's printable area.
- **Offset Relative to Paper Border**—Calculates the offset values from the paper's corner.

Creating a Plot Log. The plot log lets you write the plot stamp data to a log file. After the first log file is created, AutoCAD appends each subsequent plot stamp to the log file, thereby maintaining a history of plot stamp data. Each drawing's plot stamp data is added as a single line of text.

Tip: The plot stamp log file can be placed on a network drive and shared by multiple users.

Creating a Single-Line Plot Stamp. One other Advanced Options setting to be aware of is the Single Line Plot Stamp option. If selected, this option forces the plot stamp data to be a single line of text. If this option is cleared, the plot stamp data can be up to two lines of text. If the option is selected and the plot stamp contains text that is longer than the printable area, the plot stamp text is truncated. If this option is cleared, the plot stamp text is wrapped after the third field.

MERGE CONTROL

The Merge Control feature provides the ability to merge overlapping plotted objects so that underlying objects are visible. This is not actually a new feature, but one that made a brief hiatus when AutoCAD 2000 was introduced. Originally available in AutoCAD Release 14 and accessed through plotter configuration interfaces such as HPCONFIG and OCECONFIG, merge control was reintroduced with AutoCAD 2000i.

Accessing Merge Control

In AutoCAD 14, entering **HPCONFIG** (or **OCECONFIG**) and then selecting the Pens button displayed the Pens dialog box. From this dialog box you selected the desired merge control option, either Lines Overwrite (Overlay) or Lines Merge (Merge). In AutoCAD 2002, merge control is a property of the plot device. If a particular device supports merge control, the property is displayed as an option under

Graphics in the Device and Document Settings folder, which is located in the Plotter Configuration Editor, as shown in Figure 20–26.

If you select Lines Overwrite, the more recently created objects in your drawing are plotted on top of earlier created objects. If Lines Merge is selected, earlier created objects show through the more recently created objects.

Note: You can access the Plotter Configuration Editor either from AutoCAD's menu by selecting File>Plotter Manager and then opening the desired plotter configuration (PC3) file, or by choosing the Properties button located on the Plot Device folder, which is found in both the Page Setups dialog box and the Plot dialog box.

TRUE COLOR PLOT STYLES

AutoCAD 2002 supports true color plot styles, which means it plots drawings using 24-bit color. A plot style that uses 24-bit color plots drawings using 16.7 million colors (assuming your plot device supports true color output.) In Release 2000, plot styles could only plot drawings using 8-bit color, or 256 colors. With AutoCAD 2002, you can specify 24-bit colors and assign them to objects at plot time using AutoCAD's plot styles.

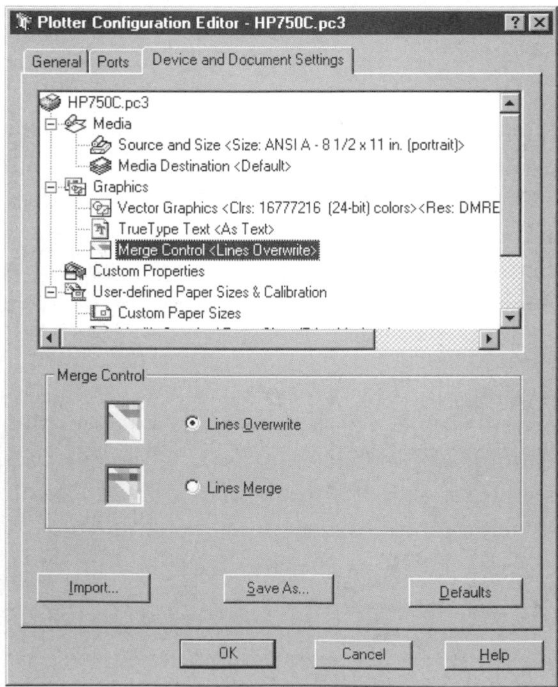

Figure 20–26 *You set the value for merge control under Graphics in the Plotter Configuration Editor.*

Specifying True Colors

To specify a 24-bit color, from either the Plot Style Table Editor's Table View or Form View folder, scroll to the bottom of the Color list and choose True Color, as shown in Figure 20–27. Once you do, AutoCAD displays the True Color Dialog box, which lets you specify a color from over 16.7 million possible colors.

You can specify a 24-bit color using one of three methods:

- **Hue/Value:** The Hue/Value area lets you specify color hue and value by clicking and dragging inside its rainbow-colored square, as shown in Figure 20–28. As you drag your cursor inside the Hue/Value area, a triangle along the top indicates the change in color hue, and the triangle along the left side indicates a change in color value.
- **Red, Green, Blue:** The Red, Green, and Blue values in the Color Values area let you specify the red, green, and blue values of the color. You can either enter a specific value in the text box or use the slider to specify a color by clicking or dragging.
- **Hue, Sat, Val:** The Hue, Sat, and Val values in the Color Values area let you specify the hue, saturation, and value of the color. You can either enter a specific value in the text box or use the slider to specify a color by clicking or dragging.

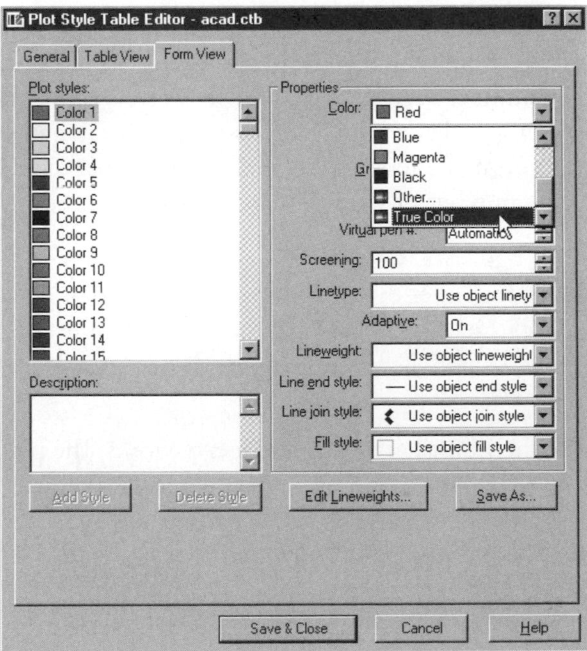

Figure 20–27 *You assign a 24-bit color by accessing the True Color option from the Plot Style Table Editor.*

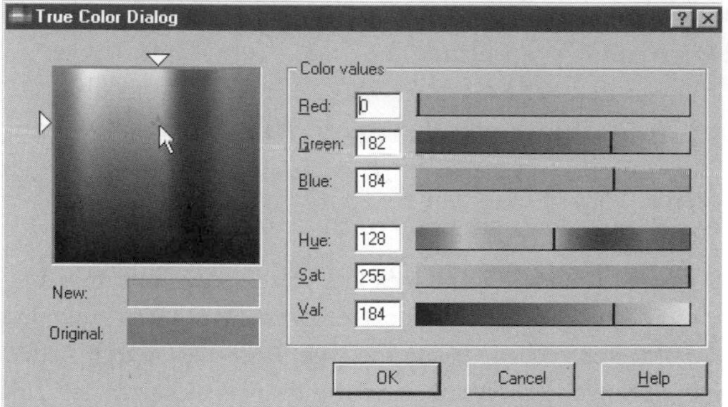

Figure 20–28 *The True Color Dialog box lets you specify a 24-bit color.*

As you adjust the color values using one of the three methods, the other two methods automatically adjust to their corresponding values. Also, as you adjust the color values, the new color is displayed in the New color box. For comparison purposes, the plot style's current color setting is displayed in the Original color box. When you specify a 24-bit color in the True Color Dialog box and then choose OK, AutoCAD displays the color's name in the Plot Style Table Editor's Color list as Custom Color.

FILTER PAPER SIZES

The Filter Paper Sizes feature lets you control the number of sizes that are displayed in the list of paper sizes from which you can choose in the either the Page Setup or Plot dialog box. The number of available paper sizes is based on the plot device and can produce a fairly lengthy list that is cumbersome to use when you try to select a particular paper size. To make identifying the desired paper size easier, AutoCAD 2002 lets you limit the number of sizes that are displayed in the paper sizes list.

Filtering Paper Sizes

In AutoCAD 2002, Filter Paper Sizes is a property of the plot device. It is displayed as an option under User-Defined Paper Sizes & Calibration in the Device and Document Settings folder, which is located in the Plotter Configuration Editor, as shown in Figure 20–29.

To demonstrate how the Filter Paper Sizes property works, the plot device shown in Figure 20–29 contains almost fifty different paper sizes. If you use only ANSI paper sizes that are rotated to landscape, you can limit the paper size list in the Page Setup or Plot dialog box to just those sizes. When you clear the check box next to each paper size you do not wish to display, you instruct AutoCAD to display only those paper sizes whose check boxes are selected in the Page Setup and Plot dialog boxes' Paper Size list, as shown in Figure 20–30.

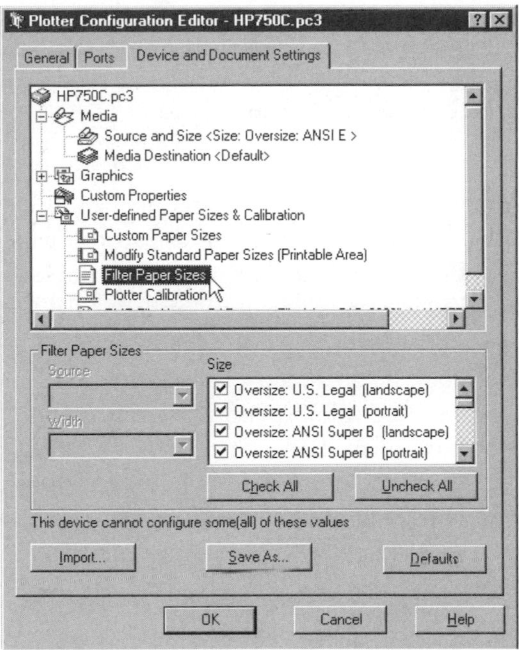

Figure 20–29 *The Filter Paper Sizes property is listed under User-Defined Paper Sizes & Calibration in the Plotter Configuration Editor.*

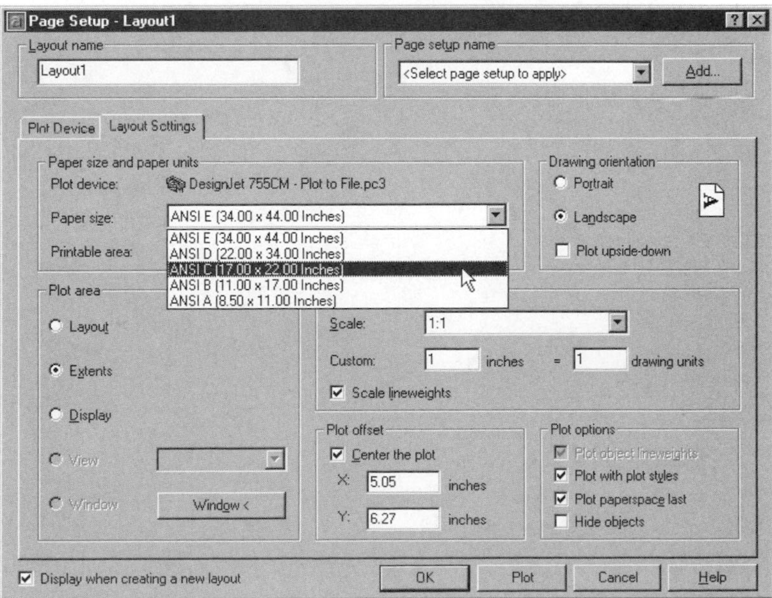

Figure 20–30 *The Paper Sizes list displays only those paper sizes that are checked in the plot device's Filter Paper Sizes property.*

Note: You can access the Plotter Configuration Editor either from AutoCAD's menu by selecting File>Plotter Manager and then opening the desired Plotter Configuration (PC3) file, or by choosing the Properties button located in the Plot Device folder, which is found in both the Page Setups and the Plot dialog boxes.

FILTER PRINTERS

The Filter Printers option lets you control the number of Windows system printers that are displayed in the list of plotter configurations from which you can choose in either the Page Setup or Plot dialog box. The number of available printers is based on the printers configured for your system using Windows' Add Printer wizard, which you access by choosing Start>Settings>Printer from the Windows taskbar and then opening the Add Printer tool. If you added a large number of printers to your system, they are all displayed in the list of plotter configurations in both the Page Setup and Plot dialog boxes. To make identifying the desired plotter configuration easier, AutoCAD 2002 lets you limit the number of Windows system printers that are displayed to only those for which you created a plotter configuration (PC3) file.

Filtering Printers

In AutoCAD 2002, to limit the display of system printers to only those that are associated with PC3 files, choose Options from AutoCAD's Tools menu, and then choose the Plotting tab. In the lower-left corner of the Plotting folder, select the Hide System Printers check box as shown in Figure 20–31, choose Apply, and then choose OK.

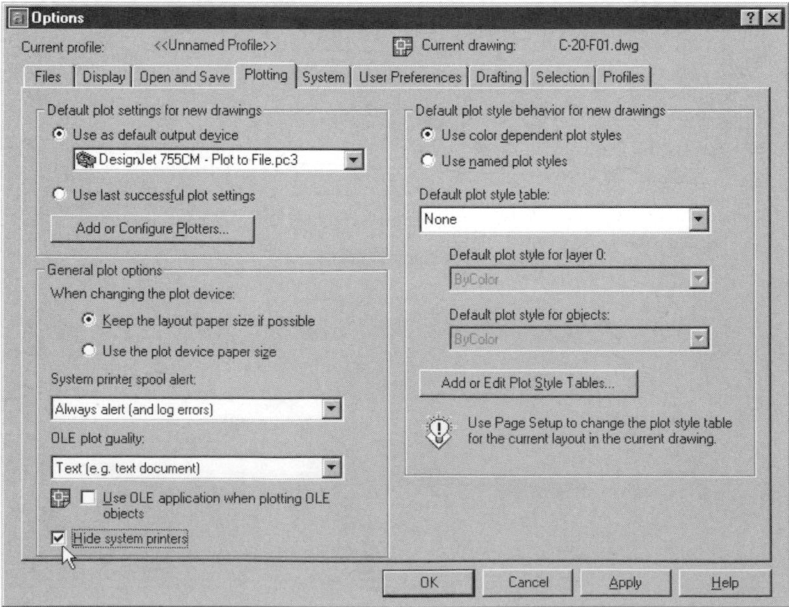

Figure 20–31 *The Hide System Printers check box limits the display of system printers.*

To demonstrate the effect of selecting the Hide System Printers check box, Figure 20–32 shows the list of available printers in AutoCAD. The list contains over twenty different print devices, most of which are not associated with a PC3 file. In contrast, Figure 20–33 shows that after you select the Hide System Printers check box, the list of available printers is reduced to just five printers, which represent the number of PC3 files in this system's Plotter Manager.

PLOT PREVIEW BACKGROUND COLOR CONTROL

The Plot Preview Background Color Control feature lets you control the color of the paper that is displayed during a plot preview. In AutoCAD 2000, while you could control the background color of the Model and Layout folders, when you selected the Full Preview button from the Plot dialog box, AutoCAD ignored the background color of the Model or Layout folder and displayed the Preview's background color as white. With AutoCAD 2002, you can now also control the background color during a full preview of your drawing.

To set the background color for plot previews, from AutoCAD's Tools menu, choose Options, and then select the Display tab. In the Windows Elements area of the Display folder, choose the Colors button to display the Color Options dialog box. From the Window Element list, choose Plot Preview Background, as shown in Figure 20–34. Next, from the Color list, choose the desired background color, and then choose Apply & Close to save your changes.

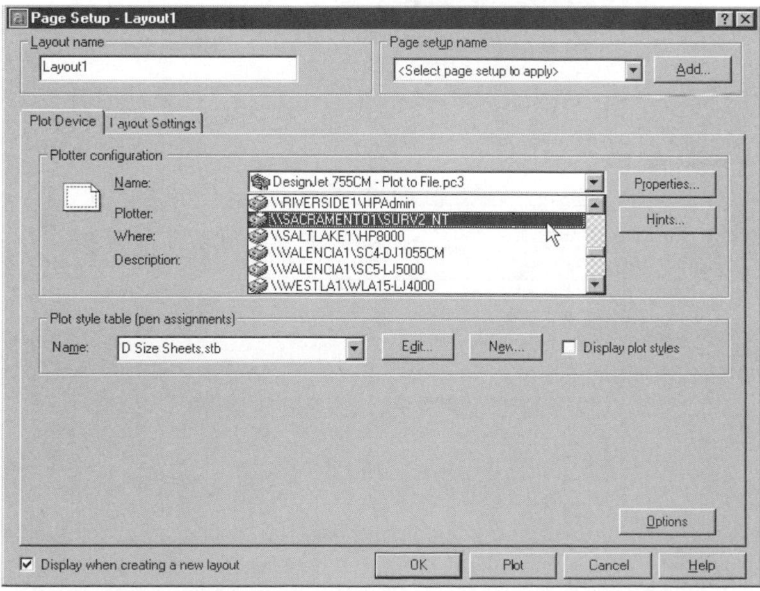

Figure 20–32 *With the Hide System Printers check box cleared, the list of available printers includes all Windows system printers that are installed on the computer.*

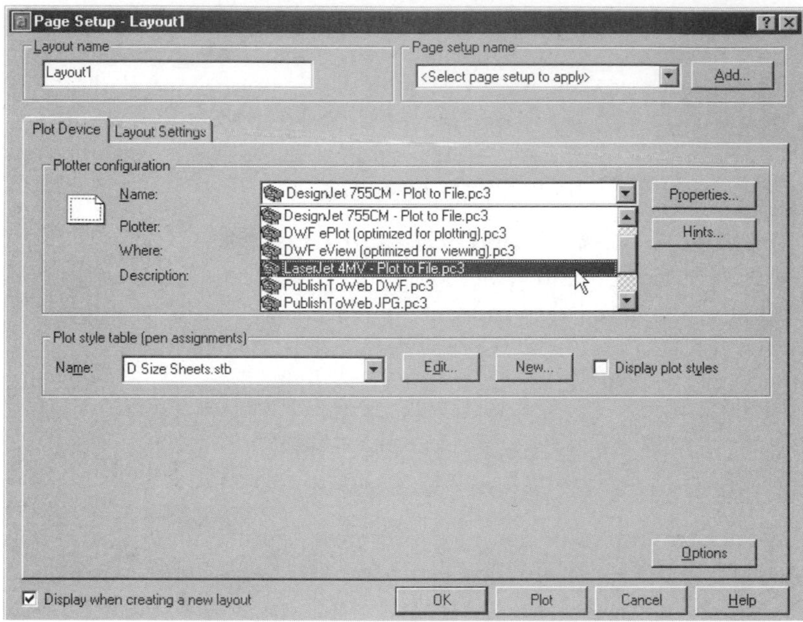

Figure 20–33 *With the Hide System Printers check box selected, the list of available printers is limited to only those printers associated with a PC3 file.*

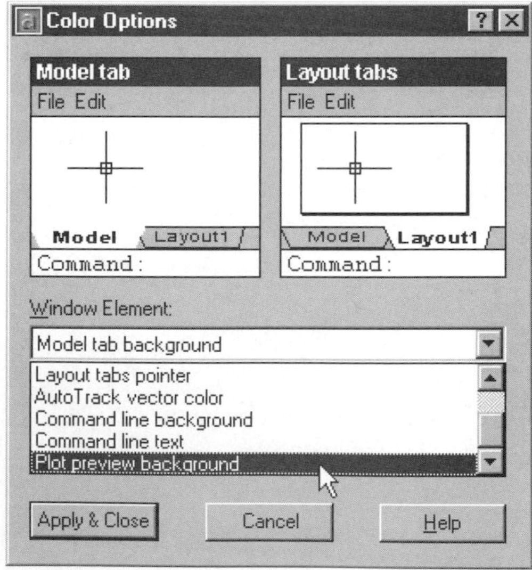

Figure 20–34 *The Plot Preview Background option lets you choose the background color that is displayed when you perform a full preview of drawings from the Plot dialog box.*

CONTROL WINDOWS METAFILE BACKGROUND COLOR

The Control Windows Metafile Background Color feature, while it is listed as a new system variable in AutoCAD 2002's "What's New in this Release" overview, was actually introduced with Release 2000. The WMFBKGND system variable was available in AutoCAD 2000, and it controls whether the background display of AutoCAD objects is transparent when a drawing is exported as a Windows Meta File (WMF). However, while controlling the transparency of the WMF file's background color is not new, the ability to control the foreground color is new. AutoCAD 2002 introduces the WMFFOREGND system variable, which controls the foreground color of AutoCAD objects exported as a WMF file.

Note: The WMFFOREGND system variable is active only if the WMFBKGND system variable is set to Off (0).

Note: References to the foreground color of AutoCAD objects actually refer to the color objects.

By default, the WMFBKGND system variable is set to On (1), and it produces a WMF file whose background color is the same as the Model or Layout folder's color. In Figure 20–35, a WMF file was exported from an AutoCAD drawing whose Layout folder's color was set to white. The WMF file was then overlaid on a gray image. Notice in Figure 20–35 that the WMF file's background is opaque and it covers the gray image. When the WMFBKGND system variable is set to Off (0), the WMF file's background is transparent and, if it is overlaid on the gray image, it lets the image's color show through the WMF file, as shown in Figure 20–36.

If the WMFBKGND system variable is set to Off (0), when AutoCAD exports the WMF file, it checks the WMFFOREGND system variable's setting to modify the color of the drawing's objects. When the WMFFOREGND system variable is set to Off (0), the color of AutoCAD objects is darkened as necessary to ensure that the objects are visible, as shown in Figure 20–36. When the WMFFOREGND system variable is set to On (1), the color of AutoCAD objects is lightened as necessary to ensure that the objects are visible, as shown in Figure 20–37. Notice that black lines in Figure 20–36 appear white in Figure 20–37.

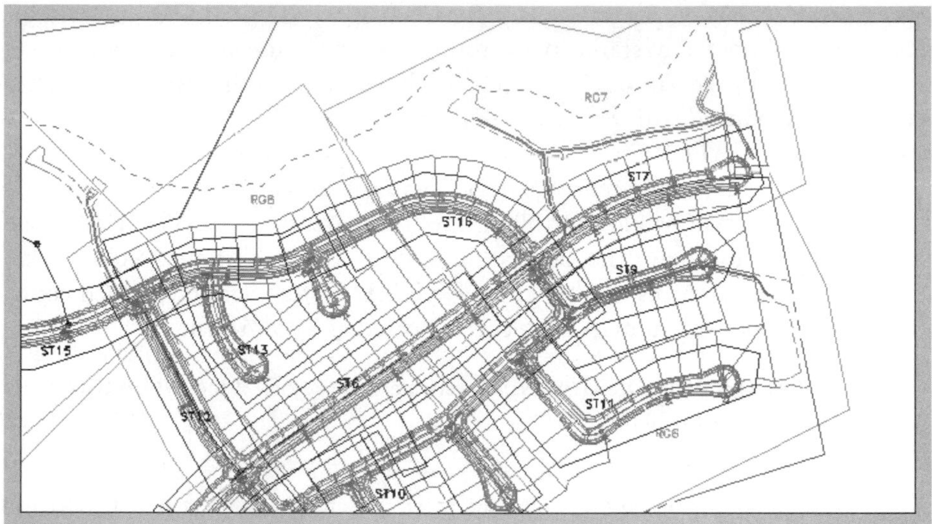

Figure 20–35 *With the WMFBKGND system variable set to On (1), the WMF file's background is opaque, and it covers the gray image.*

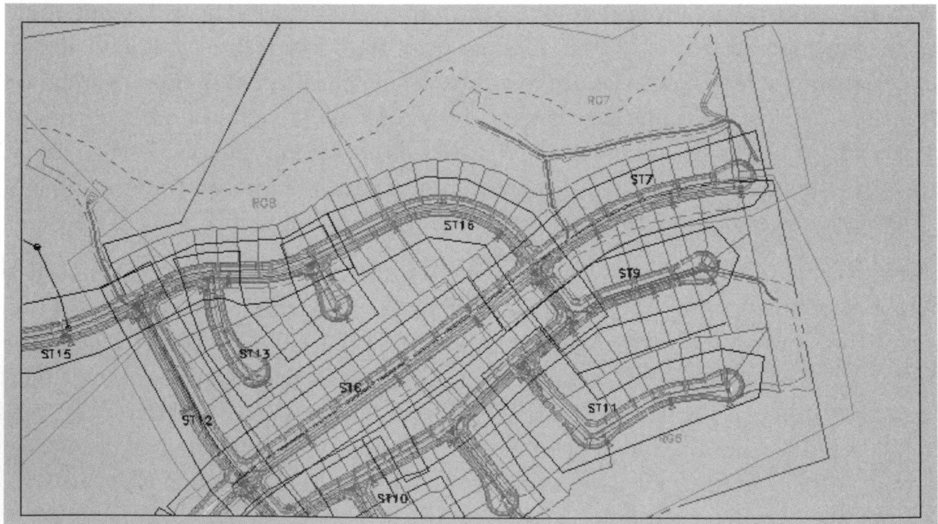

Figure 20–36 *With the WMFFOREGND system variable set to Off (0), the color of AutoCAD objects is darkened.*

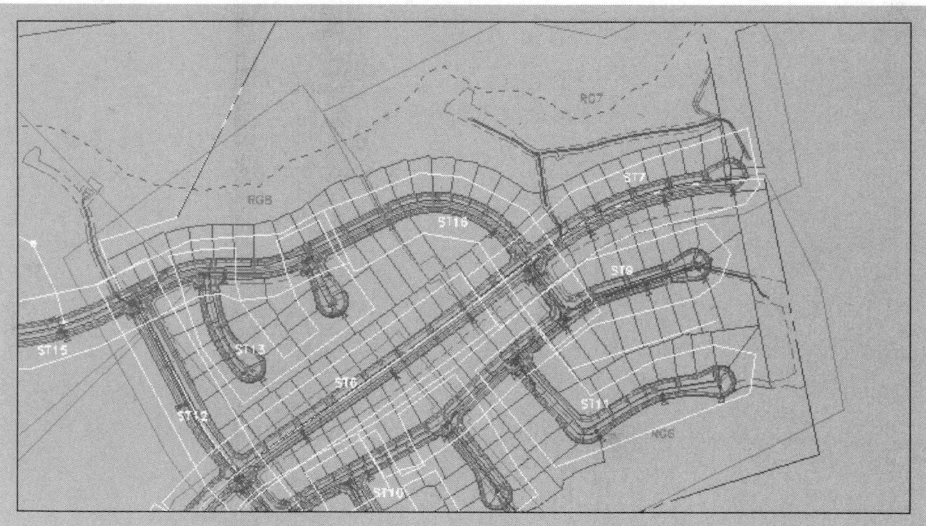

Figure 20–37 *With the WMFFOREGND system variable set to On (1), the color of AutoCAD objects is lightened.*

SUMMARY

In this chapter you learned about configuring a plotter, the new Autodesk Plotter Manager, and the new Plotter Configuration Editor. You created plot styles and saved them in plot style tables, and you reviewed the difference between color-dependent and named plot style tables. You learned how to create page setups. You also learned about AutoCAD's expanded plotting capabilities that make visualizing and producing plots easier and more flexible. Whether controlling the background color of layouts or Windows Meta Files, filtering for paper sizes or available printers, or stamping your drawings at plot time, AutoCAD 2002 continues to make working with CAD drawings easier.

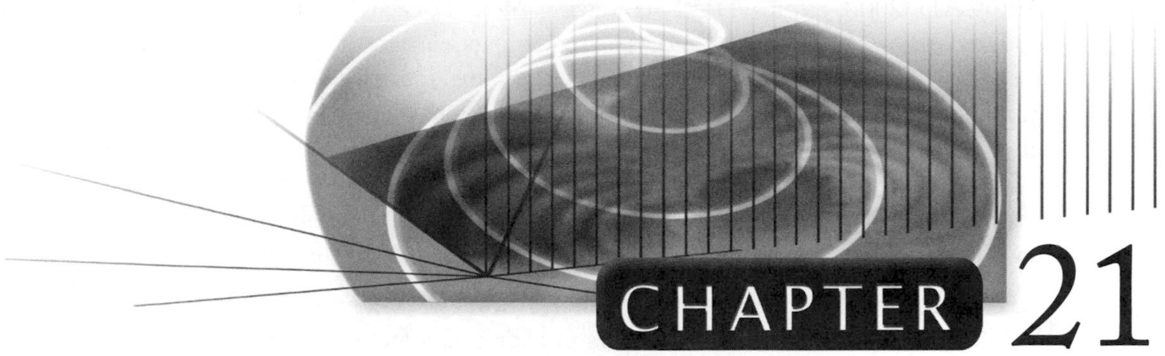

Advanced Plotting

AutoCAD provides three useful features that enhance your plotting ability. The first two are AutoCAD's ePlot and eView features, which let you create electronic (DWF) plots that are displayed over the Internet. The third feature is AutoCAD 2002's Batch Plot Utility, which lets you identify a group of drawings and then automatically plot them all using a single command. When you use these features, you can enhance your productivity by either sharing electronic drawings with clients and colleagues over the Internet, or by automatically plotting dozens, or even hundreds, of drawings with a single button click.

This chapter discusses the following subjects:

- AutoCAD's updated DWF format (creating ePlot and eView files)
- Autodesk's Volo View product (viewing ePlot and eView files)
- Creating batch plot (BP3) files
- Editing batch plot files
- Automatically plotting multiple drawings using a batch plot file

This chapter is divided into two major sections. The first discusses AutoCAD's updated DWF format. The second discusses AutoCAD's Batch Plot Utility.

UPDATED DWF FORMAT

AutoCAD 2002 creates DWF (Drawing Web Format) files using the recently updated ePlot/eView format. DWFs are compact electronic copies of your drawing files that produce high-quality images that are specifically designed for being quickly displayed over the Internet. While DWFs have been around for awhile, the updated version includes enhancements that enable you to produce DWF files optimized either for accurate plotting or for viewing drawings without the need for AutoCAD software. Though their file size is compact, DWFs contain the same level of detail and quality as a plot or view generated from AutoCAD itself.

To compliment the updated DWF format, Autodesk enhanced both Volo View and Volo View Express. These two viewers provide many features, including Zoom and Pan, layer on/off control, and a set of markup tools for redlining and adding comments to DWFs. Volo View Express is a free download, while Volo View includes several additional features and may be purchased online at www.autodesk.com.

CREATING DWFS OVERVIEW

AutoCAD 2002 installs two new plotter models onto your system, *DWF ePlot (optimized for plotting)* and *DWF eView (optimized for viewing)*. Using these new plotter models you can create plotter configuration (PC3) files that are then selected from the Plot Device folder in AutoCAD's Plot dialog box.

COMPARING EPLOT AND EVIEW

The ePlot and eView plotter models produce nearly identical PC3 files. Both allow you to control the media size and the color depth of vector graphics. Both allow you to include layer information as well as scale and measurement information with the DWF files. But while they are nearly identical, their subtle differences are important. ePlot PC3 files are optimized for creating DWFs used for plotting (i.e., for creating hardcopy output of your files), whereas eView PC3 files are optimized for creating DWFs used for online viewing, over either your company LAN/WAN or the Internet.

PC3 files created with ePlot are designed to produce high-quality plots of drawings that contain raster images. You can control the resolution of plotted drawings from the PC3 file's custom DWF Properties dialog box, as shown in Figure 21–1. When you create ePlot DWF files, the resolution should be set to no greater than the plotter's resolution. If your output device plots at 600 dpi, then you should set the DWF's resolution to no higher than 600 dpi. To keep DWF file sizes small, use the lowest resolution you can that produces legible, high-quality hardcopies.

PC3 files created with eView are designed to produce high-quality images for viewing online. Consequently, instead of focusing on producing high-quality copies of drawings that contain raster images, eView focuses on producing high-quality vector graphics that maintain high resolution when you zoom in close to objects. The

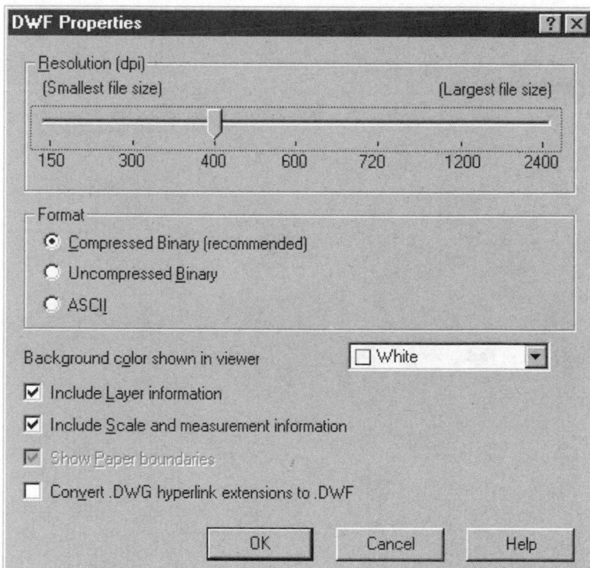

Figure 21–1 *ePlot's custom DWF Properties dialog box.*

resolution you select from eView's custom DWF Properties dialog box, shown in Figure 21–2, depends on the level of detail in your drawing. If your drawing contains a lot of detail, choose a higher resolution to ensure that the objects remain crisp and clear as users zoom in close to see the detail. But remember, as with ePlot's DWF files, the higher the resolution, the larger the file size.

Available Options

In addition to controlling the DWF's resolution, there are several other options you can control using the DWF Properties dialog box:

- **Background Color Shown in Viewer**—This option allows you to preset the background color of the DWF file. The most common color selected is white.
- **Include Layer Information**—If this option is selected, all layers that are turned on and thawed when the DWF file is created are included with the DWF. Consequently, someone viewing the DWF can turn the layers on or off and thereby control the display of objects in the DWF. If this option is cleared, no layers are included when the DWF is created.
- **Include Scale and Measurement Information**—If selected, this option includes scale and measurement information in DWF files. If the DWF file will be plotted, it is highly recommended that you leave this option enabled.

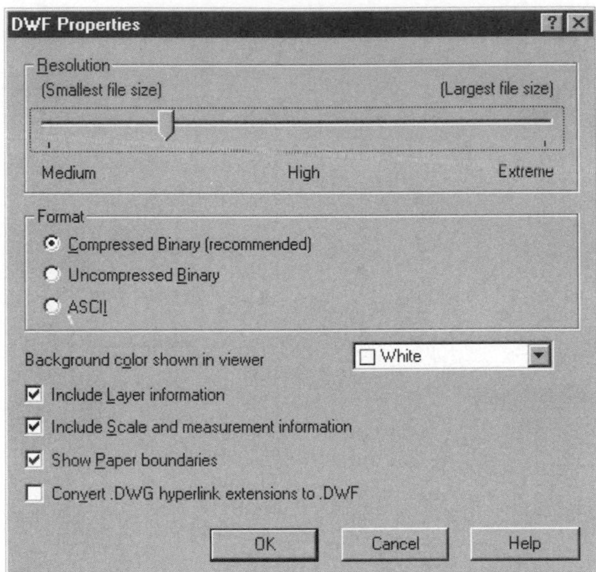

Figure 21–2 *eView's custom DWF Properties dialog box.*

- **Show Paper Boundaries**—This option ensures that the paper boundary is included in the DWF file. For ePlot, this option is set automatically and cannot be turned off.
- **Convert .DWG Hyperlink Extensions to .DWF**—If this option is selected and your drawing contains hyperlinks, the DWG ending in every hyperlink name is renamed as DWF. Select this option if you intend to convert the hyperlinked drawing files to DWF files.

ENHANCED PLOT FILE FEATURES

The updated DWF file format has enhanced raster image-handling capabilities, allowing it to optimize raster image resolution intelligently based on whether you create DWFs using ePlot (high resolution) or eView (lower resolution.) Additionally, raster images are stored in DWFs as PNG or G4 TIFF data, which are ideal formats for electronic data files.

DWFs also now support ISO linetypes and fill patterns, which let you control the appearance of objects using industry-standard formats.

Another enhancement is greater color depth control. Through this feature, you control the color depth of vector graphics. There are three levels of depth available for color output (8-bit, 24-bit, and 32-bit), and two levels for monochrome output (256 shades of gray, and black and white).

CREATING DWF PLOTTER CONFIGURATIONS

The ePlot and eView plotter models are accessed through AutoCAD's Plotter Manager folder, which you access from AutoCAD's File menu. When you open the Plotter Manager folder, you will notice that two predefined PC3 files already exist, "DWF eView (optimized for viewing).pc3" and "DWF ePlot (optimized for plotting).pc3." You can use thc predefined PC3 files, or you can create your own custom PC3 files using the Plotter Manager's Add-A-Plotter wizard. To do so, choose Autodesk ePlot (DWF) from the wizard's Manufacturers list and then select the desired DWF plotter model, as shown in Figure 21–3.

APPLYING DWF PLOTTER CONFIGURATIONS

After the new ePlot and eView PC3 files are added, you can plot DWF files by selecting the desired PC3 file from the Plot Device folder in AutoCAD's Plot dialog box, as shown in Figure 21–4. Once you select the PC3 file, you can edit its properties by choosing the Properties button located in the Plotter Configuration area of the Plot Device folder. Choosing the Properties button displays the Plotter Configuration Editor, allowing you to adjust the PC3 file's settings, as shown in Figure 21–5.

After you select and adjust the PC3 file, set the other desired parameters in the Plot dialog box as you normally do when you create plots. AutoCAD creates DWFs using all the Plot dialog box's parameters, including plot scale and plot style settings. Once the plot parameters are set, choose OK to create the plot. Doing so produces a DWF file that can then be viewed using the Volo View products.

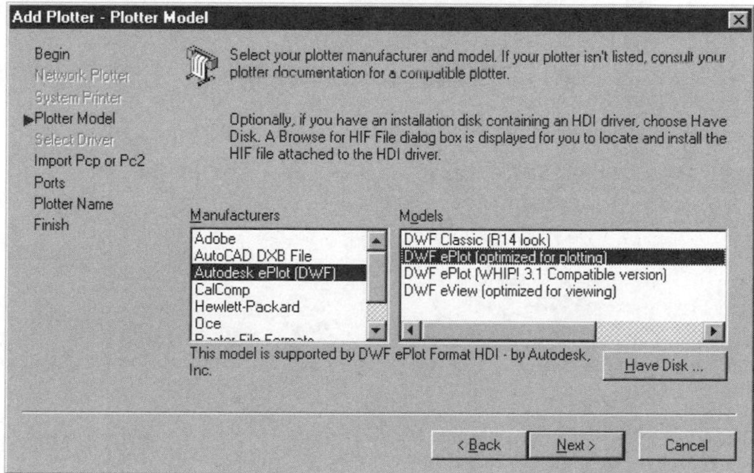

Figure 21–3 *The Add-A-Plotter wizard lets you select the desired DWF plotter model.*

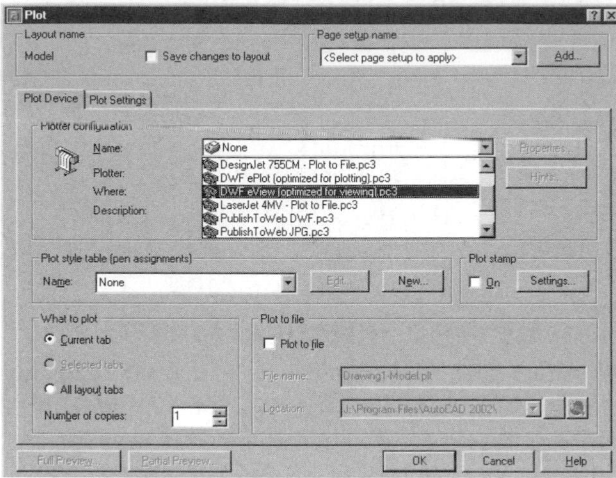

Figure 21–4 *You create DWFs by selecting the new ePlot and eView PC3 files from the Plot dialog box.*

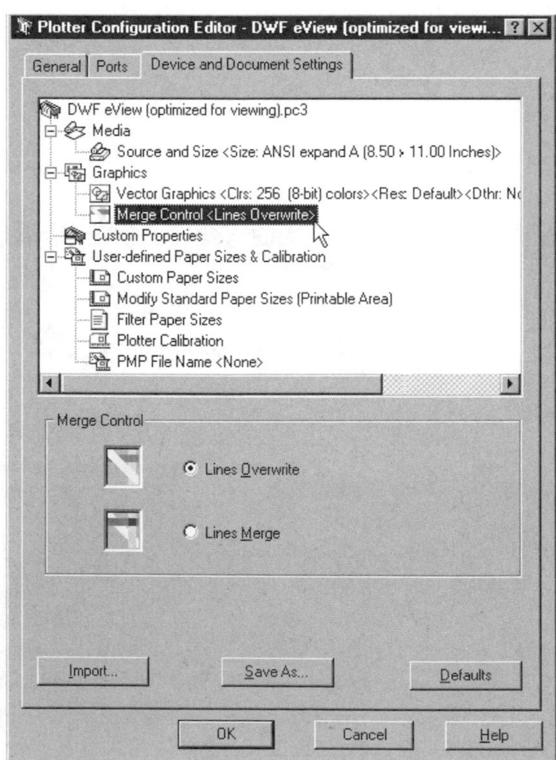

Figure 21–5 *You adjust a PC3 file's settings using the Plotter Configuration Editor.*

Note: You access ePlot's and eView's custom DWF Properties dialog boxes by choosing the Custom Properties option from the Plotter Configuration Editor's Device and Document Settings folder.

VOLO VIEW AND VOLO VIEW EXPRESS

Autodesk's Volo View and Volo View Express are standalone products that let you access AutoCAD's DWG, DXF, and DWF drawing files, and that support viewing the same raster formats that are supported by AutoCAD, including BMP, CLAS, GIF, JPG, PCX, PCT, TARGA, and TIF. They provide a variety of tools for working with drawings and collaborating with others. Because both are ActiveX controls, once Volo View or Volo View Express is installed, it can also view AutoCAD files within your Microsoft IE browser or any other ActiveX-enabled application.

Volo View

Volo View accesses the original native AutoCAD files. However, its tools are limited in functionality to ensure that accessed files cannot be edited, thereby ensuring the integrity of the original data. Because it uses the same plotting engine as AutoCAD, Volo View lets you produce hardcopy plots that are identical to those created using AutoCAD. Consequently, you can view and plot original drawing files, just like in AutoCAD. You can also print files to any Windows system printer.

Volo View provides object-based markup tools based on Autodesk's ActiveShapes technology. These tools allow you to mark up files by drawing lines, circles, and rectangles. Additionally, you can easily draw clouds, insert callouts, and sketch freehand. The markups can be exported and read directly into AutoCAD. Volo View also includes Autodesk's new ClearScale technology, which allows you to gray out the drawing's objects to improve visibility and readability of markups on the screen. Volo View's AutoSnap feature lets you accurately measure distances, areas, and paths of travel by snapping precisely to drawing geometry.

Volo View's expanded editing capabilities include real-time pan and zoom. Volo View is the only drawing viewer that includes a dynamic 3D Orbit rotation feature similar to that of AutoCAD 2002, allowing you to view 3D objects from any position. Volo View also lets you shade 3D models, and it displays AutoCAD's lineweights.

Volo View supports Autodesk's object enablers that were created for AutoCAD-based applications such as Mechanical Desktop and AutoCAD Architectural Desktop. Object enablers allow you to view custom objects instead of viewing only proxy graphics.

Volo View Express

In addition to Volo View, which is an excellent tool for viewing, redlining, and plotting AutoCAD drawings, Autodesk also offers Volo View Express (see Figure 21–6), a free product that can be redistributed to your entire design team. Volo View

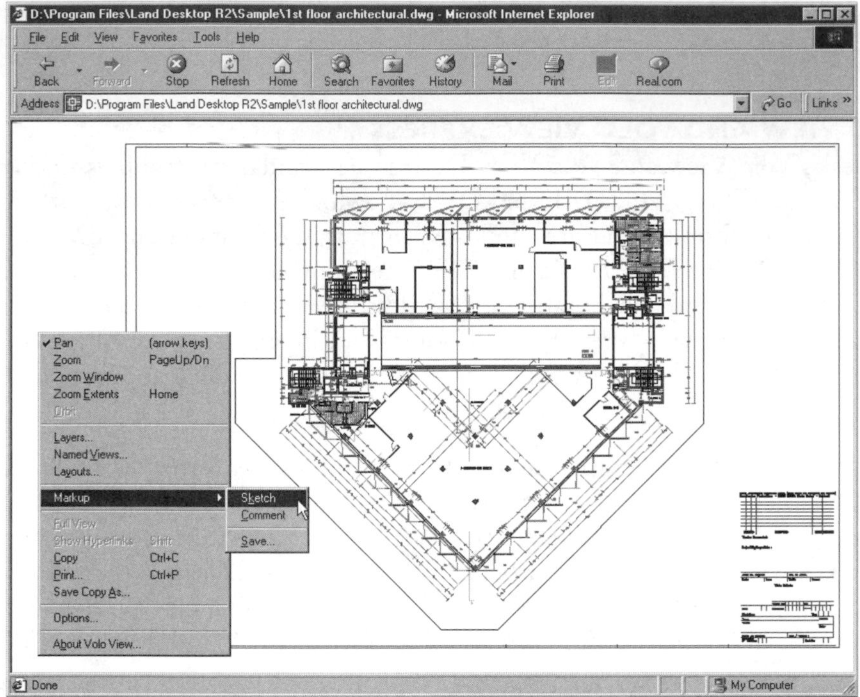

Figure 21–6 *Volo View Express is included with AutoCAD 2002.*

Express, which is included on the AutoCAD 2002 product CD, offers scaled-down Volo View functionality. Volo View Express allows you to open, view, do lightweight drawing markup on, and print AutoCAD drawings, including DWG, DXF, and DWF files. With Volo View Express you can view AutoCAD drawings and models either locally or over the Internet, with high fidelity and without risking changes to the original data. In contrast, Volo View (which may be purchased from Autodesk at www.autodesk.com) is more robust, providing richer viewing, more complete markup tools, precise measurement features, and high-quality plotting.

BATCH PLOT UTILITY

AutoCAD 2002 is designed to allow you to plot one drawing at a time. While plotting only one drawing at a time probably satisfies most of your needs, there will probably be occasions when you will want to plot multiple drawings by executing a single command. To meet the demands of plotting multiple drawings from a single command, AutoCAD provides a specially designed application called the AutoCAD Batch Plot Utility. When you use the Utility, you can plot an entire set of drawings by issuing a single command.

This section provides an overview of the AutoCAD Batch Plot Utility and explains how to use its features.

OVERVIEW

The AutoCAD Batch Plot Utility is a Visual Basic application that runs independently of AutoCAD. Using the Utility, you can create a list of drawings to plot and then plot all the drawings by picking a single button. Once you pick the button, the Utility takes control of AutoCAD 2002, loads each drawing in the Utility's list into AutoCAD, and then sends the drawing to the plotter engine using AutoCAD's Plot command. Using the AutoCAD Batch Plot Utility, you can specify and then plot an entire set of AutoCAD drawings repeatedly without having to load and plot each drawing manually from AutoCAD.

The AutoCAD Batch Plot Utility is designed so you can easily create a list of drawings to plot. As you add a drawing to the list, you can select which layers to plot for the drawing, and you can also associate a different layout and PC3 file to use when the drawing is plotted. Once you finish creating the list of drawings, you can save the list as a batch plot (BP3) file that can be recalled at a later time. When you create a BP3 file, you can plot entire sets of drawings by simply loading the list into the AutoCAD Batch Plot Utility and picking a single button.

Note: It is important that you do not interfere with the Utility once it takes control of AutoCAD. Doing so could cause the Utility to stop functioning.

The following section explains how to use the AutoCAD Batch Plot Utility to create a list of drawings to plot and then save the list as a BP3 file.

CREATING A BATCH PLOT FILE

The process of creating a batch plot (BP3) file is straightforward. From the Utility, you create a list by selecting the drawings you wish to plot. Then you set any plot parameters you desire, such as the page setup and plot device to use to plot each drawing. Next you save the list and its plot parameters as a BP3 file. Once the file is saved, it can be recalled at any time to plot all the drawings in the file's list. By using the AutoCAD Batch Plot Utility, you can easily create a BP3 file and use it to plot an entire set of drawings.

In the following exercise you will use the AutoCAD Batch Plot Utility to create and then save a BP3 file.

EXERCISE: CREATING AND SAVING A BATCH PLOT (BP3) FILE

1. From the Windows Taskbar, choose Start>Programs>AutoCAD 2002>Batch Plot Utility. The Batch Plot Utility launches, which in turn launches AutoCAD 2002.

   This exercise uses three AutoCAD drawings that you must copy from the accompanying CD to a folder on your PC. You must copy the drawings named 21DWG01.DWG, 21DWG02.DWG, and 21DWG03.DWG to a folder you select. After you copy the drawings, be sure to right-click on the drawings, choose Properties, and then clear the Read-only attribute.

2. From the AutoCAD Batch Plot Utility, choose the Add Drawing button. The Add Drawing File dialog box is displayed.
3. From the Add Drawing File dialog box, browse to the folder into which you copied the three drawings, select all three drawings, and then choose Open. The three drawing file names appear in the Batch Plot Utility list, as shown in Figure 21–7.

   Next you will indicate the layout to plot for the 21DWG01.DWG drawing.

4. From the list of drawings displayed in the Batch Plot Utility, choose 21DWG01.DWG, and then choose the Layouts button. The Layouts dialog box is displayed.

Tip: You can also display the Layouts dialog box by choosing the selected drawing's current layout value, which is displayed in the drawing list under the Layout column.

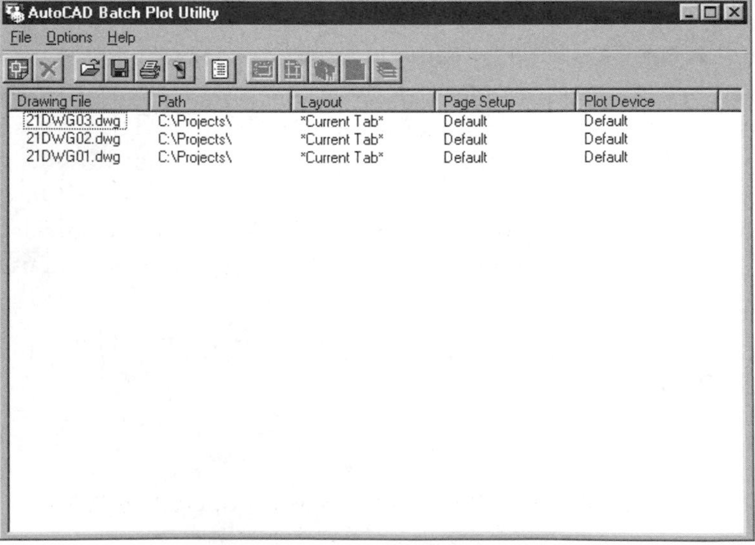

Figure 21–7 *Three drawings are added to the Batch Plot Utility list.*

5. From the Layouts dialog box, choose the Show All Layouts button. All layouts in the 21DWG01.DWG drawing are displayed.
6. From the list of layouts, choose layout Three as shown in Figure 21–8, and then choose OK. The Batch Plot Utility display is updated to indicate that layout Three will be plotted for drawing 21DWG01.DWG.

Tip: You can select multiple layouts to plot by holding down the Ctrl key as you select the layout names.

Next you will indicate the page setup to use to plot the 21DWG01.DWG drawing.

7. With the 21DWG01.DWG drawing still selected, choose the Page Setups button. The Page Setups dialog box is displayed.
8. From the list of page setups, choose page setup Drawing One as shown in Figure 21–9, and then choose OK. The Batch Plot Utility display is updated to indicate that page setup Drawing One will be used to plot drawing 21DWG01.DWG.

Next you will indicate the plot device to use for the 21DWG02.DWG drawing.

9. From the list of drawings displayed in the Batch Plot Utility, choose 21DWG02.DWG, and then choose the Plot Devices button. The Plot Devices dialog box is displayed.
10. From the list of plot devices, choose the plot device named "DWF ePlot (optimized for plotting).pc3" as shown in Figure 21–10, and then choose OK. The Batch Plot Utility display is updated to indicate that plot device DWF ePlot PC3 will be used to plot drawing 21DWG02.DWG, as shown in Figure 21–11.

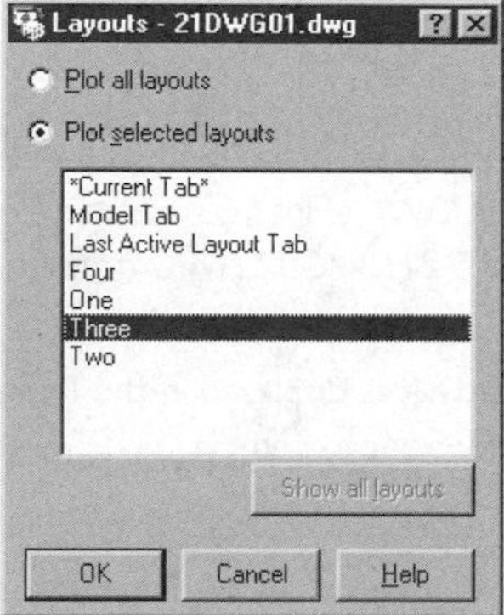

Figure 21–8 *The 21DWG01.DWG drawing's layout Three will be used when the drawing is plotted.*

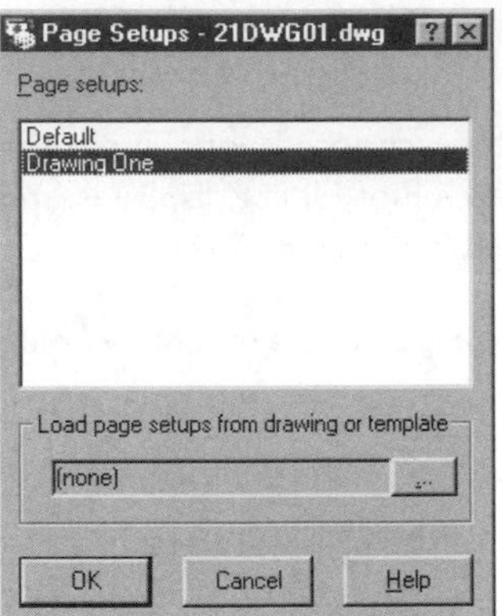

Figure 21–9 *The 21DWG01.DWG drawing's page setup Drawing One will be used to plot the drawing.*

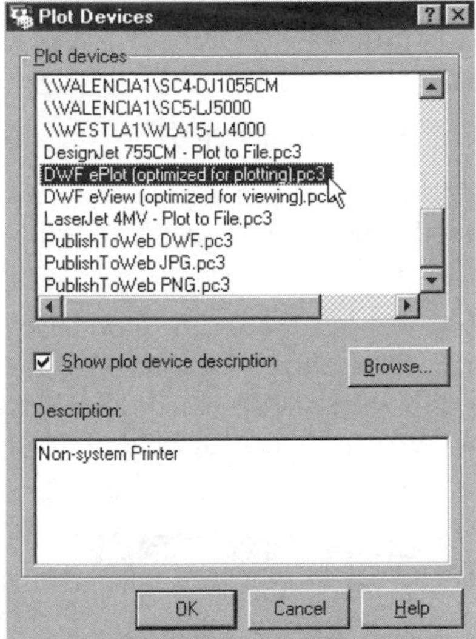

Figure 21–10 *The DWF ePlot PC3 plot device is selected to plot the 21DWG02.DWG drawing.*

Note: The list of plot devices displayed in the Plot Devices dialog box depends on the plot devices configured for your system.

Next you will save the current list of drawings and their settings as a BP3 file.

11. From the AutoCAD Batch Plot Utility, choose the Save List button. The Save Batch Plot List File dialog box is displayed.

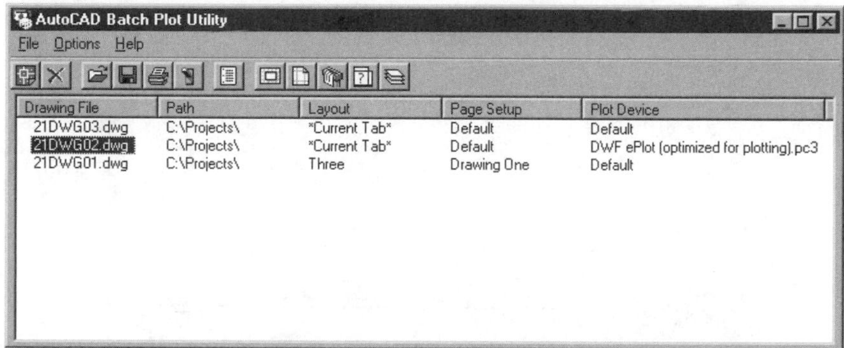

Figure 21–11 *The AutoCAD Batch Plot Utility drawing list, with its plotting parameters, is ready to be saved.*

12. From the Save Batch Plot List File dialog box, browse to the folder into which you copied the three drawings, save the BP3 file as List One (as shown in Figure 21–12), and then choose Save. The Batch Plot Utility creates the BP3 file.
13. You may quit the AutoCAD Batch Plot Utility by choosing Exit from the File menu. When you are prompted to quit the batch plot, choose OK. If you are prompted to save changes to the drawing, choose No.

The AutoCAD Batch Plot Utility lets you easily create a list of drawings to plot. Additionally, you can set plotting parameters to use to plot each drawing in the list, and thereby control which layout to plot, and which page setup and plot device to use. By using the AutoCAD Batch Plot Utility, you can easily create a list of drawings to plot, and you can control specific plotting parameters to use for each drawing.

The following section explains how to use the AutoCAD Batch Plot Utility to edit an existing BP3 file.

EDITING A BATCH PLOT FILE

The process of editing a batch plot (BP3) file is simple. Once you launch the AutoCAD Batch Plot Utility, you select the BP3 file you wish to edit. Then you modify the plot parameters you wish to change and save the updated BP3 file. Once it is saved, the BP3 file can then be used to plot all the drawings in the file's list using the new parameters. By using the AutoCAD Batch Plot Utility, you can easily edit an existing BP3 file, defining new plot parameters to use to plot the drawings in the list.

In the following exercise you will use the AutoCAD Batch Plot Utility to edit an existing BP3 file.

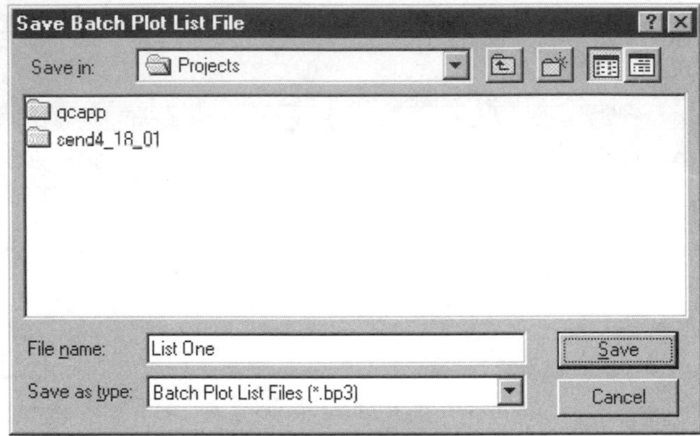

Figure 21–12 *The batch plot file is saved as List One.bp3.*

EXERCISE: EDITING AN EXISTING BATCH PLOT (BP3) FILE

1. From the Windows Taskbar, choose Start>Programs>AutoCAD 2002>Batch Plot Utility. The Batch Plot Utility launches, which in turn launches AutoCAD 2002.

   This exercise uses the List One.bp3 file created in the previous exercise.

2. From the AutoCAD Batch Plot Utility dialog box, choose the Open List button. The Open Batch Plot List File dialog box is displayed.
3. Browse to the folder where the List One.bp3 file was saved in the previous exercise, select the file as shown in Figure 21–13, and then choose Open. The List One.bp3 file is opened, and its settings are displayed in the AutoCAD Batch Plot Utility.

   Next you will modify the list of drawings by removing two of the drawings from the list.

4. In the list of drawings displayed in the AutoCAD Batch Plot Utility dialog box, while you hold down the Ctrl key, select the 21DWG01.DWG and 21DWG02.DWG drawing file names, and then choose the Remove button. The Utility prompts you to verify that you wish to remove the selected files.
5. Choose Yes. The Utility removes the selected drawing file names from the list of drawings, leaving only the 21DWG03.DWG drawing.

   Next you will modify the layout to use to plot the 21DWG03.DWG drawing.

6. Select the 21DWG03.DWG drawing file name, and then choose the Layouts button. The Layouts dialog box is displayed.
7. Choose the Plot Selected Layouts radio button, then choose the Model Tab layout (as shown in Figure 21–14), and then choose OK. The Utility changes the layout so it is plotted to the Model Tab layout.

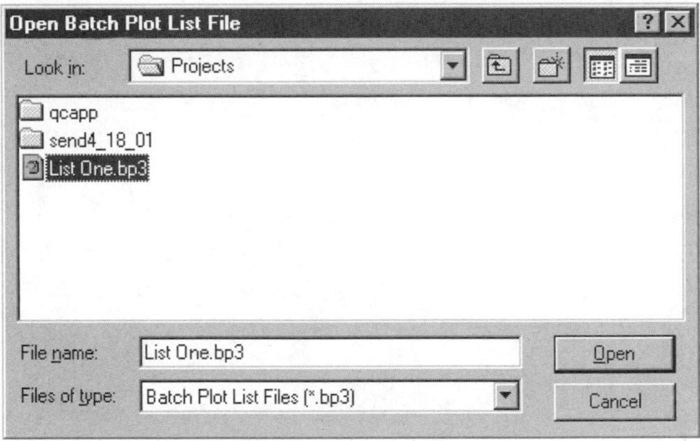

Figure 21–13 *The List One.bp3 file is opened for editing using the Open Batch Plot List File dialog box.*

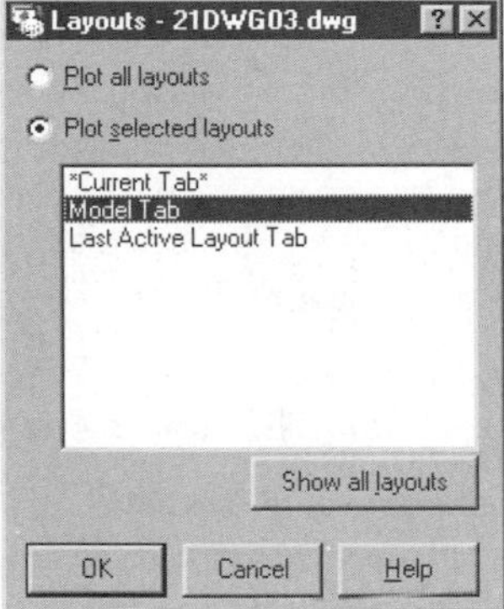

Figure 21–14 *The layout to plot is changed to Model Tab.*

Next you will modify the plot settings to use to plot the 21DWG03.DWG drawing.

8. With the 21DWG03.DWG drawing file name still selected, choose the Plot Settings button. The Plot Settings dialog box is displayed.

   The Plot Settings dialog box contains two folders: the Plot Settings folder and the Layers folder. The Plot Settings folder allows you to set the area to plot, the plot scale to use, and the file name and location to use when you plot drawings to a file. The Layers folder allows you to turn layers off and on, thereby controlling which layers to plot.

9. From the Plot Settings folder, in the Plot Area section, choose the Extents radio button as shown in Figure 21–15. This sets the drawing extents as the area to plot.

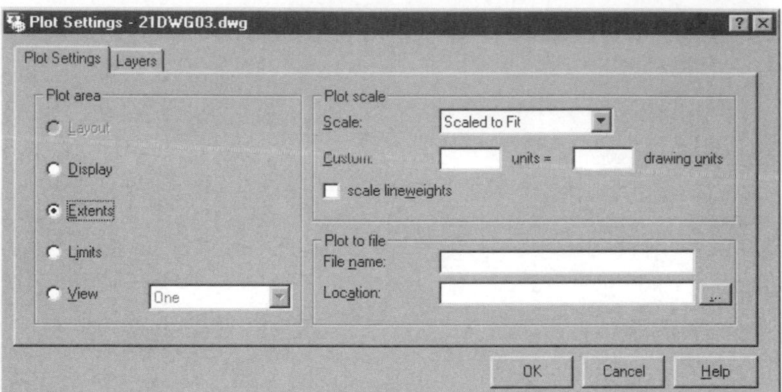

Figure 21–15 *The drawing extents is selected as the area to plot.*

Tip: While it may be useful at times to define the area to plot using the Plot Settings dialog box, it is not the best way to plot the area you desire to plot. To ensure that the area you desire is plotted, prior to using the AutoCAD Batch Plot Utility, you should open each drawing, zoom to the view you wish to plot, and then save the drawing. Then, as the Utility loads each drawing, it will plot the last view displayed when you saved the drawing file.

10. In the Plot Settings folder, in the Plot Scale section, choose 1:1 from the Scale drop-down list, as shown in Figure 21–16. The plot scale is set to 1:1.
11. In the Plot Settings folder, in the Plot to File section, choose the "..." button next to the Location text box. The Plot-To-File Filename dialog box is displayed.
12. In the Plot-To-File Filename dialog box, in the File Name text box, enter **Drawing Three** as shown in Figure 21–17, and then choose Save. The file name and location are set as shown in Figure 21–18.

    Next you will modify the layers to plot.
13. Choose the Layers tab. The 21DWG03.DWG drawing file's layers are displayed.
14. In the Layers folder, hold the Ctrl key down and select the layer names Red and Blue, then choose the Off button as shown in Figure 21–19, and then choose OK. The Plot Settings dialog box is dismissed, and its settings are assigned to 21DWG03.DWG.

    Next you will indicate the drawing's plot device.
15. From the AutoCAD Batch Plot Utility, with the 21DWG03.DWG drawing name still highlighted, choose the Plot Devices button. The Plot Devices dialog box is displayed.
16. From the list of plot devices, choose the plot device named "DWF ePlot (optimized for plotting).pc3" as shown in Figure 21–20, and then choose OK. The Batch Plot Utility display is updated to indicate that plot device DWF ePlot.pc3 will be used to plot drawing 21DWG03.DWG.

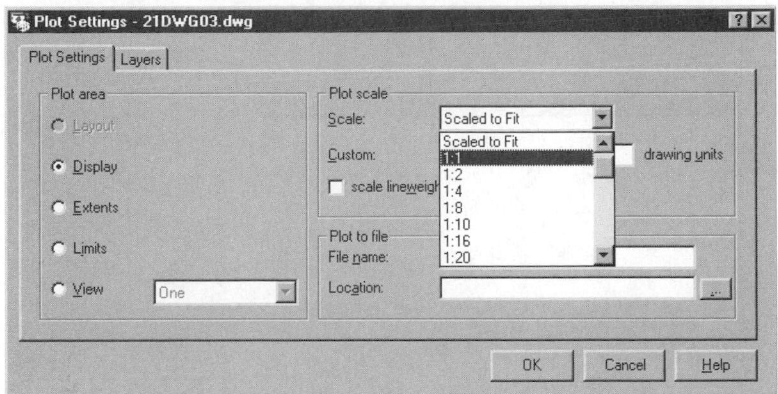

Figure 21–16 *The drawing's plot scale is set to 1:1.*

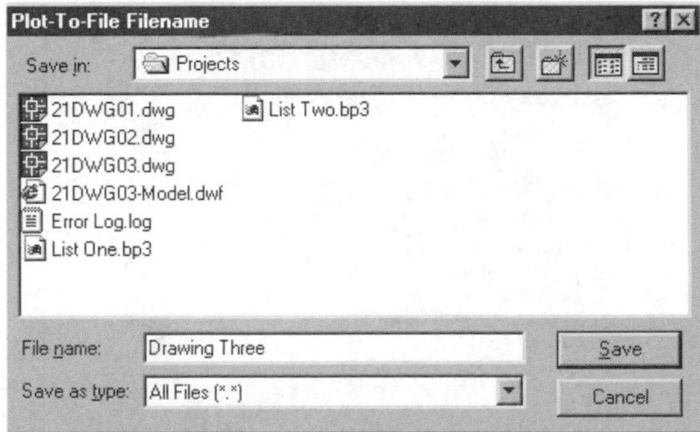

Figure 21–17 *The plotted drawing's file name is set to Drawing Three.*

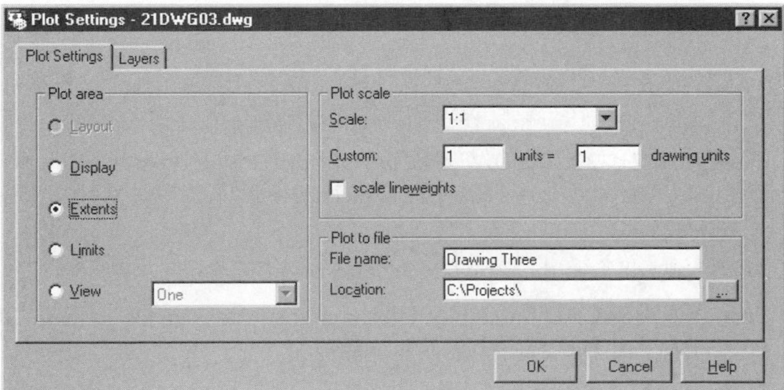

Figure 21–18 *If the plotted drawing is saved to a file, its file name and location are as shown in the Plot to File section.*

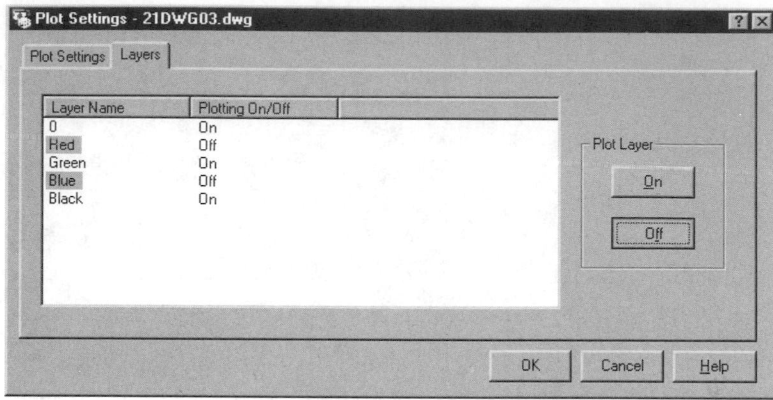

Figure 21–19 *The layers named Red and Blue are turned off; therefore they will not be plotted.*

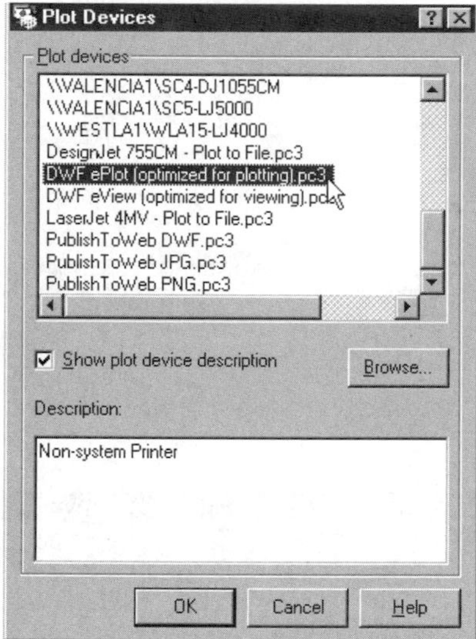

Figure 21–20 *The DWF ePlot.pc3 plot device is selected to plot the 21DWG03.DWG drawing.*

Next you will save the modified list as a new BP3 file.

17. From the AutoCAD Batch Plot Utility, choose the Save List button. The Save Batch Plot List File dialog box is displayed.
18. In the File Name text box, enter List Two as shown in Figure 21–21, and then choose Save. The Batch Plot Utility creates the new BP3 file.

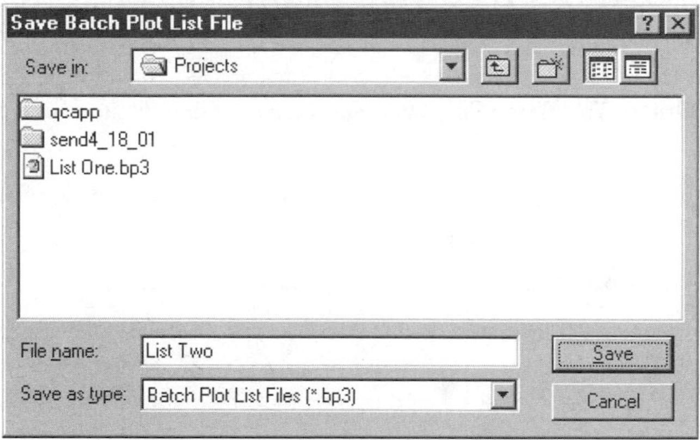

Figure 21–21 *The edited BP3 file is saved as a new file.*

19. You may quit the AutoCAD Batch Plot Utility by choosing Exit from the File menu. When you are prompted to quit the batch plot, choose OK. If you are prompted to save changes to the drawing, choose No.

The AutoCAD Batch Plot Utility allows you to edit an existing BP3 file quickly. You can modify the plotting parameters used to plot each drawing in the list, and you can add drawings to or remove drawings from the list. Additionally, as demonstrated in the last exercise, the AutoCAD Batch Plot Utility allows you to control a drawing's plot area and plot scale, as well as which layers of the drawing to plot. By using the AutoCAD Batch Plot Utility, you can easily edit an existing BP3 file, and thereby modify its list of drawings and the plotting parameters to use for each drawing.

The following section explains how to use the AutoCAD Batch Plot Utility to plot drawings from an existing BP3 file.

PLOTTING DRAWINGS USING A BATCH PLOT FILE

The ultimate purpose of creating a batch plot (BP3) file is to recall a list of drawings and then plot the drawings by picking a single button. Once you launch the AutoCAD Batch Plot Utility, you can load an existing BP3 file, which contains a list of drawings and their plotting parameters. Then you can plot the drawings listed in the BP3 file by simply choosing the Plot button.

In the following exercise you will use the AutoCAD Batch Plot Utility to plot an existing BP3 file.

EXERCISE: PLOTTING AN EXISTING BATCH PLOT (BP3) FILE

1. From the Windows Taskbar, choose Start>Programs>AutoCAD 2002>Batch Plot Utility. The Batch Plot Utility launches, which in turn launches AutoCAD 2002.

   This exercise uses the List Two.bp3 file created in the previous exercise.

2. From the AutoCAD Batch Plot Utility dialog box, choose the Open List button. The Open Batch Plot List File dialog box is displayed.

3. Browse to the folder where the List Two.bp3 file was saved in the previous exercise, select the file as shown in Figure 21–22, and then choose Open. The List Two.bp3 file is opened, and its settings are displayed in the AutoCAD Batch Plot Utility.

   The AutoCAD Batch Plot Utility provides two useful features which you can take advantage of when you prepare to plot the drawings listed in a BP3 file. The first feature is the Logging feature, and the second feature is the Plot Test feature. By using these two features, you can track and record important information about the plotting process, and you can ensure that AutoCAD can find all the files necessary to plot the drawings successfully.

   Next you will use the Logging feature.

4. From the AutoCAD Batch Plot Utility dialog box, choose the Logging button. The Logging dialog box is displayed.

   The Logging feature allows you to create two different types of logs. One log is a plot journal log, which records who plotted each drawing and when. The other log is an error log, which records any errors encountered during plotting.

   Next you will enable the Journal Logging feature.

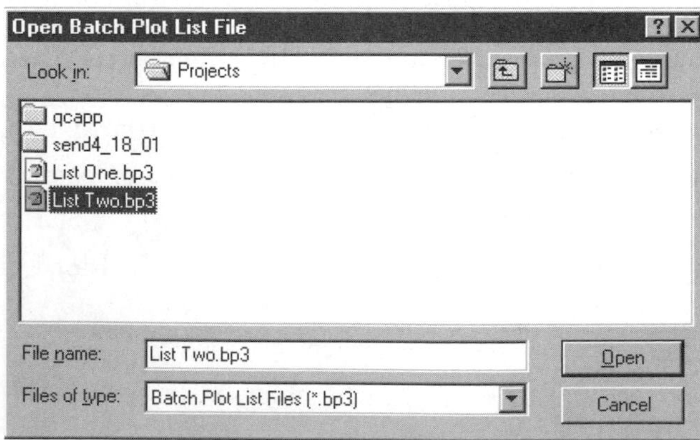

Figure 21–22 *The List Two.bp3 file is opened using the Open Batch Plot List File dialog box.*

5. From the Logging dialog box, in the Plot Journal section, make sure the Enable Journal Logging option is selected.
6. From the Logging dialog box, in the Plot Journal section, select the Browse button, which is next to the File Name text box. The Journal Filename dialog box is displayed.
7. Browse to the folder where the List Two.bp3 file was opened, enter **Journal Log** as shown in Figure 21–23, and then choose Save. The Journal Log.log file is ready to be created.
8. From the Logging dialog box, in the Plot Journal section, select the Append radio button. The plot journal log data will be appended to the designated file every time there is an entry, instead of overwriting the previous file.

Note: The Header and Comments text boxes allow you to include text that is inserted at the beginning of each log entry. You can leave these boxes blank.

Next you will enable the Error Logging feature.

9. From the Logging dialog box, in the Error Log section, make sure the Enable Error Logging option is selected.
10. From the Logging dialog box, in the Error Log section, select the Browse button, which is next to the File Name text box. The Error Log Filename dialog box is displayed.
11. Browse to the folder where the List Two.bp3 file was opened, then enter **Error Log** as shown in Figure 21–24, and then choose Save. The Error Log.log file is ready to be created.

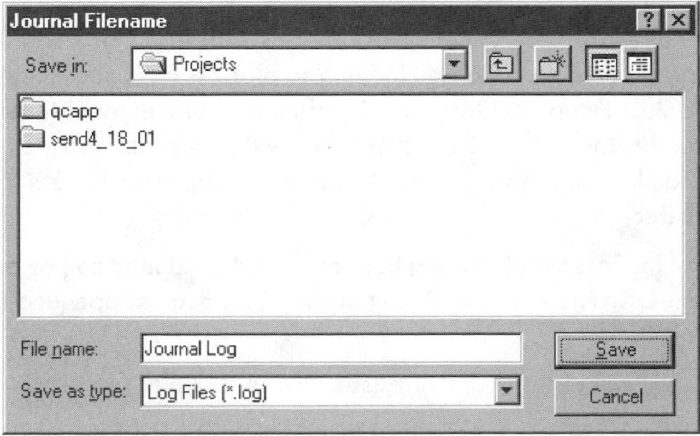

Figure 21–23 *The Journal Log.log file is ready to be created.*

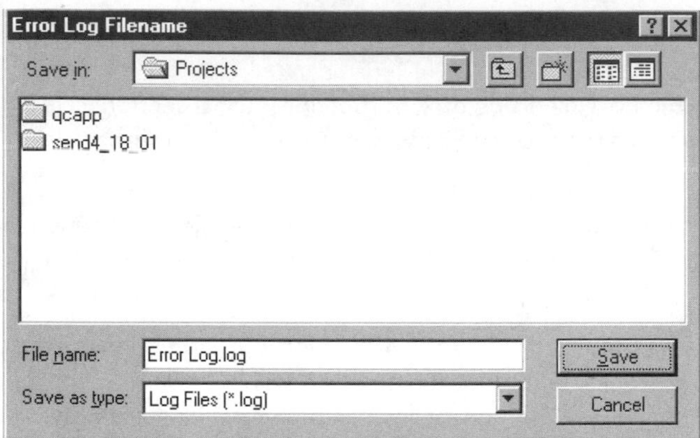

Figure 21–24 *The Error Log.log file is ready to be created.*

12. In the Logging dialog box, in the Error Log section, select the Append radio button. The error log data will be appended to the designated file every time there is an entry, instead of overwriting the previous file. The Error Log.log file is ready to be created, as shown in Figure 21–25.

13. In the Logging dialog box, choose OK. The two log files are created and saved.

    Next you will use the Plot Test feature.

14. In the AutoCAD Batch Plot Utility dialog box, choose the Plot Test button. The Plot Test Results dialog box is displayed as shown in Figure 21–26, and it lists any errors.

    The Plot Test feature causes the Utility to load the drawings in the list into AutoCAD 2002 and check for missing xrefs, fonts, or shape files. The Plot Test feature only loads the drawings; it does not plot them.

    In this particular exercise, because there are no xrefs or shapes in the drawing, and because the font file was found, the plot test was successful. However, if you look at Figure 21–26, you will notice that one error is listed as a warning. This error does not indicate that there will be a problem when the BP3 file is used to plot the drawing; it simply notes that the optional xref log file could not be located.

15. In the Plot Test Results dialog box, choose the Append to Log button, and then choose OK. The Plot Test Results dialog box data is appended to the Journal Log.log file.

    Next you will plot the drawing listed in the BP3 file.

16. From the AutoCAD Batch Plot Utility dialog box, choose the Plot button. The Utility loads the 21DWG03.DWG drawing into AutoCAD 2002 and creates a plot file. Additionally, the plotting information is generated and appended to the Journal Log.log file.

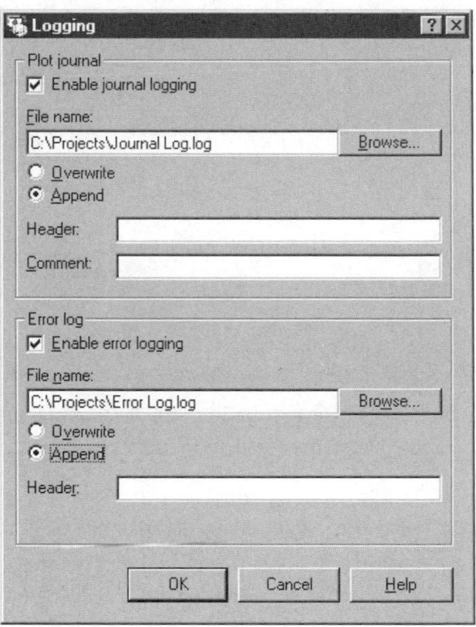

Figure 21–25 *The log files are enabled in the Logging dialog box.*

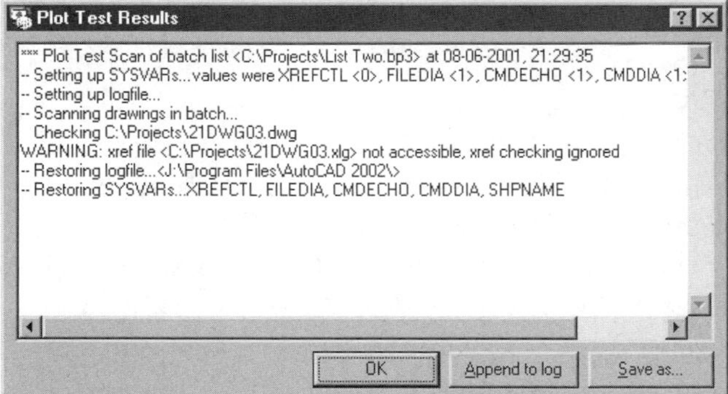

Figure 21–26 *The Plot Test Results dialog box displays any errors encountered while the drawings were loaded into AutoCAD 2002.*

17. You may quit the AutoCAD Batch Plot Utility by choosing Exit from the File menu. When you are prompted to quit the batch plot, choose OK. If you are prompted to save changes to the BP3 file or to the drawing file, choose No.

By using AutoCAD's Batch Plot Utility, you can create and save predefined lists of drawings in BP3 files and automatically plot large sets of drawings without having to attend the plotting process. You can plot many drawings overnight, or over the weekend, and then retrieve the finished plots the following morning.

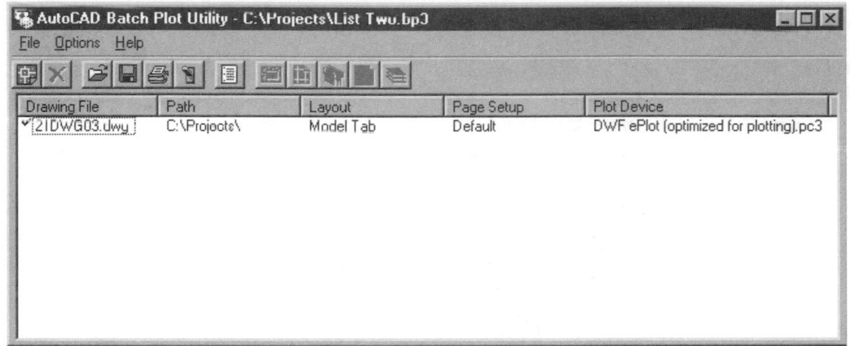

Figure 21–27 *The small check mark shown to the left of the drawing's file name indicates that the plot was generated successfully.*

Note: If the plot is created successfully, a small check mark appears to the left of the drawing's file name, as shown in Figure 21–27. However, if an error is encountered during the plotting process, an X appears to the left of the file name instead.

Tip: You should customize the Batch Plot Utility icon that you use to start the program so that it starts in a directory where paths to the xrefs and fonts will be found.

SUMMARY

In this chapter you learned about using AutoCAD's ePlot and eView features to publish electronic copies of your drawings on the Internet. You also learned how to configure a Drawing Web Format (DWF) plot device, and about creating a DWF plot file using the new ePlot and eView features. Additionally, you reviewed the AutoCAD Batch Plot Utility and learned how to create and edit batch plot (BP3) files. Finally, you learned about using BP3 files to plot multiple drawings by picking a single button, and about how to enable journal logs and error logs and use them to track the Batch Plot Utility's plotting progress.

By using AutoCAD's updated DWF and Batch Plot Utility features, you can enhance your ability to create plots quickly and share them with colleagues and clients, ultimately increasing your productivity.

Working in 3D

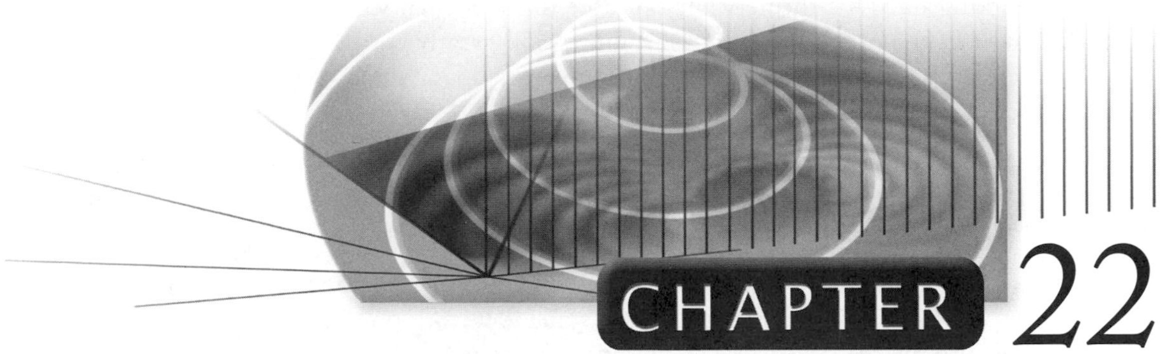

3D Fundamentals

Up to this point in this book you have been learning about drawing 2D representations of real-world, 3D objects. This method of designing and drafting is limited because its end result is a 2D drawing that must be visualized as a 3D real-world object. Drawing 3D objects as 3D objects largely eliminates the need to mentally visualize objects because the 3D information is included in the design drawing. Drawing and designing in 3D offers other advantages as well: the viewpoint can be changed to help define the form of the object; shaded and rendered presentations—and even animations—are possible (see Figure 22–1); and manufacturing information and other information such as Finite Element Analysis (FEA) data can be extracted. Having the ability to use the 3D capabilities of AutoCAD can be a valuable addition to your design skills. This chapter introduces you to the following topics:

- Understanding 3D coordinate systems
- Defining a User Coordinate System (UCS) in 3D space
- Using viewports
- Interactive viewing in 3D
- Shading a model

UNDERSTANDING 3D COORDINATE SYSTEMS

Working in 3D is theoretically no more difficult than working in 2D. In practice, however, the presence of the third dimension greatly complicates your task. In 3D there are an infinite number of drawing planes upon which you can work—not just the XY plane of 2D drawings.

Figure 22–2 shows the three axes that define AutoCAD's 3D world. The three axes shown can be aligned with any number of working planes conforming to the geometry of the 3D model.

Figure 22–3 shows a simple 3D model with five of the many possible coordinates systems that you might define for use while you work on the model. These coordinate systems are called User Coordinate Systems, or UCSs. There are, of course, many more UCSs possible for just this one model. It is the flexibility of the UCS that makes constructing and working with 3D models possible.

Figure 22–1 *A 3D solid model that has been shaded using AutoCAD's SHADE command.*

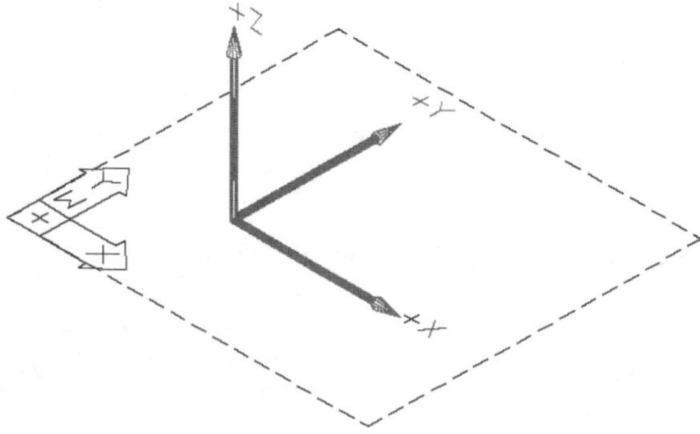

Figure 22–2 *Working in 3D means working with an additional axis, the Z-axis.*

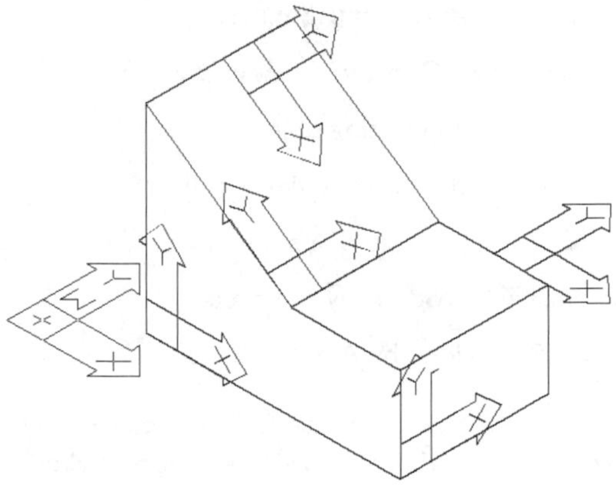

Figure 22–3 *You can have an unlimited number of User Coordinate Systems defined in a 3D drawing.*

Note: Figure 22–3 is for illustration purposes only. AutoCAD allows only one UCS per view.

Understanding how to use 3D coordinates systems is the key to creating 3D models in AutoCAD; and the key to understanding 3D coordinates is an understanding of AutoCAD's UCS.

DEFINING A USER COORDINATE SYSTEM IN 3D SPACE

The UCS provides the means to change the location of the 0,0,0 origin point, as well as the orientation of the XY plane and Z-axis. Any plane or point in 3D space can be referenced, saved, and recalled, and you can define as many UCSs as you require. Coordinate input and display are relative to the current UCS. If multiple viewports are active, you can assign a different UCS to each.

Note: Prior to AutoCAD 2000, if multiple viewports were active, they shared the same UCS. Now, to facilitate editing objects in different views, you can define a different UCS for each view. Each time you make a viewpoint current, you can begin drawing using the same UCS you used the last time that viewport was current.

Usually it is easier to align the coordinate system with existing geometry than to determine the exact placement of a 3D point.

SPECIFYING A NEW UCS

You can define a UCS in any of the following ways:

- Specify a new origin, new XY plane, or new Z-axis
- Align the new UCS with an existing object
- Align the new UCS with the current viewing direction
- Align the new UCS with the face of a solid object
- Rotate the current UCS around any of its axes
- Select a preset UCS provided by AutoCAD

Note: The ability to align a UCS with the face of a solid was introduced in AutoCAD 2000. While you can align a UCS with a solid face using several other options of the UCS command, the Face option is usually faster.

In the following exercise you will use UCS command options that you might not be familiar with from working in 2D drawings to define UCSs on 3D objects.

Note: Before you explore the methods of establishing 3D coordinate systems, you may want to review the basics of the World Coordinate System (WCS) in Chapter 6, Precision Drawing.

EXERCISE: SPECIFYING A NEW UCS BY USING THE Z-AXIS, 3POINT, OBJECT, VIEW, FACE, AND PRESET OPTIONS

1. Open the 22DWG01.DWG drawing file found on the accompanying CD-ROM. Your drawing should resemble Figure 22–4. If it does not, use the VIEW command to set the view VIEW1 current. Note the position of the WCS icon to the lower-left of the model. Although it is not necessary, you may want to set a running Endpoint Osnap.

   In the following step you will use the Z-Axis option to define a new UCS. The Z-Axis option specifies a new UCS origin and a point that lies on the positive Z-axis.

2. Start the UCS command by typing **UCS** and pressing Enter. Specify the New option.
3. Choose the Z-Axis option by typing **ZA** and pressing Enter. Specify the new origin point by snapping to the endpoint at ① in Figure 22–4. Specify a point on the positive portion of the new Z-axis by picking at ②. The UCS should appear as it does in the upper-left viewport in Figure 22–5.

   In the following step you will establish a new UCS using the 3point option. The 3point option allows you to specify the new UCS origin and the direction of its positive X- and Y-axes.

4. Press Enter to restart the UCS command. Specify the New option.

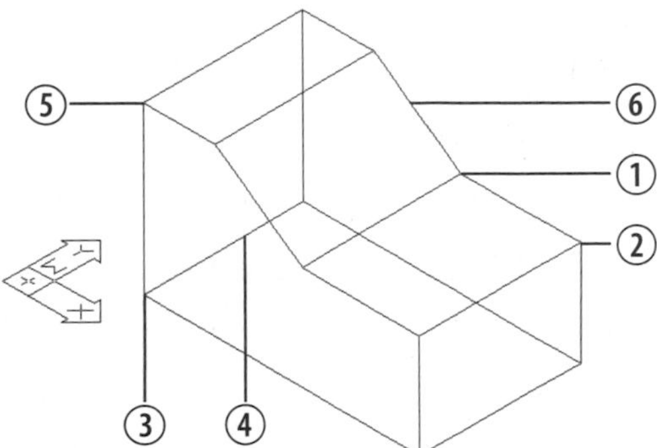

Figure 22–4 *The 22DWG01.DWG file's 3D object with pick points for new UCSs.*

5. Choose the 3point option and specify the new origin by picking the endpoint at ③ in Figure 22–4. Specify a point on the positive portion of the new X-axis by using a Midpoint snap and picking at ④. Specify a point on the positive Y portion of the XY plane by picking the endpoint at ⑤. The new UCS should be placed as shown in the upper-right viewport in Figure 22–5.

   In the next step you will use the Object option of the UCS command to establish a new UCS. The Object option defines the UCS with the same extrusion (positive Z) direction as the selected object. The origin and orientation of the new XY plane depend upon the object chosen and are somewhat arbitrary.

6. Restart the UCS command and choose New. Type **OB** and press Enter to select the Object option.
7. At the "Select object to align UCS:" prompt, pick near ⑥ in Figure 22–4. The new UCS should look like that shown in the lower-left viewport in Figure 22–5.

   In the following step you will set a new UCS using the View option. The View option establishes a new UCS with the XY plane perpendicular to the viewing direction (parallel to the display screen). The origin is unchanged.

8. Start the UCS command and choose New. Select the View option. The UCS aligns the XY plane with the screen without changing the UCS origin, as shown in the lower-right viewport in Figure 22–5.

Tip: The View option is useful when you wish to add text to a non-orthographic view. The text is automatically placed parallel to the plane of the display screen.

In the next step you will establish a preset UCS using the Orthographic option of the UCS command. An orthographic, preset UCS includes one of the six orthographic UCSs provided with AutoCAD.

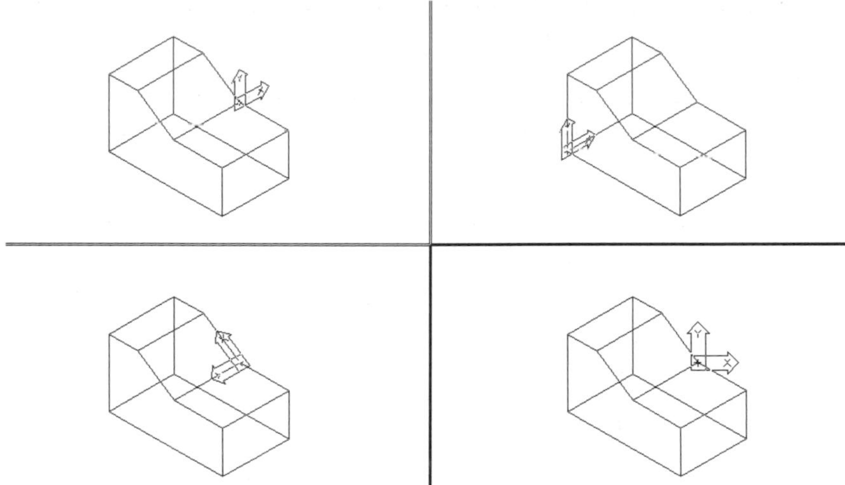

Figure 22–5 *Four possible UCSs for the 22DWG01.DWG model.*

9. Start the UCS command and choose the orthoGraphic option by typing **G** and pressing Enter. Options for the six standard orthographic UCS orientations are presented. Choose the Front option by typing **F** and pressing Enter. The UCS switches to the standard Front orientation, as shown in the upper-left viewport in Figure 22–6. Note that the Front UCS shares the WCS's origin.

   In the next step you will use the X,Y,Z rotation option to establish a new XY plane. The X,Y,Z option rotates the current UCS about a specified axis.

10. Start the UCS command, select New, and then type **Y** to indicate that you want to rotate the current UCS about the X-axis. When you are prompted to specify the rotation about the X-axis, enter **90**. The new UCS retains the same origin, but the XY plane is rotated about the X-axis 90 degrees, as shown in the upper-right viewport in Figure 22–6. (See the section titled Right-Hand Rule later in this chapter for rules governing rotations about a 3D axis.)

Tip: When you establish a new UCS by rotating the current UCS about the X-,Y-, or Z-axis, you can either specify a rotation angle or accept the default angle. This default angle can be changed using the UCSAXISANG system variable. Changing the angle to one you frequently use can save an input step when you use the X-,Y-, or Z-axis rotation options of the UCS command.

The Face option can be used to align a new UCS quickly with the face of a 3D solid. This option is shown in the next step.

11. Start the UCS command, choose New, and then type **F**. At the "Select face of solid object:" prompt, pick near ① in the lower-left viewport in Figure 22–6. Type **X** and press Enter several times to flip the UCS 180 degrees about the X-

axis. Type **Y** and press Enter to flip the UCS about the Y-axis. Finally, press Enter to accept the current orientation.

Note: When you use the Face option, the X-axis is aligned with the nearest edge of the first face found. The Xflip and Yflip options allow you to reorient the positive directions of each axis.

In the next step you will use the Move option of the UCS command. Just as the X,Y,Z option changes the XY plane without changing the origin, the Move option changes the origin without changing the XY plane orientation.

12. Start the UCS command and choose the New option. When you are prompted for the new origin point, snap to the endpoint at ② in the lower-right viewport in Figure 22–6. The XY plane orientation may differ from that shown in Figure 22–6.
13. If you plan to continue with the next exercise now, leave this drawing open. Otherwise, close the drawing without saving changes.

RIGHT-HAND RULE

In the previous exercise you established a new UCS by rotating the current UCS about one of its axes. The convention for defining the positive direction of rotation about an axis is summarized in the so-called "Right-Hand Rule." To determine the positive rotation direction about an axis, point your right thumb in the positive direction of the axis, and then curl your fingers in a fist around the axis. The "curl" of your fingers indicates the positive rotation direction of the axis. This sense of the direction of rotation about an axis is used consistently throughout AutoCAD's 3D commands that involve axes—such as the REVSURF command in surface modeling, and the REVOLVE command used with solids. The positive rotation direction about the three axes is shown in Figure 22–7.

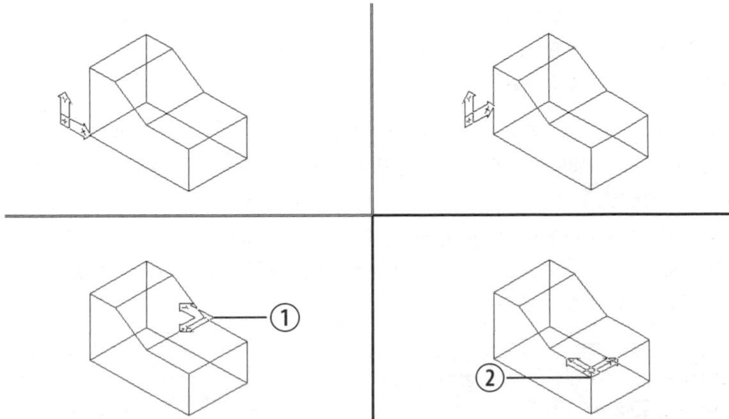

Figure 22–6 *Moving the UCS around in a 3D model.*

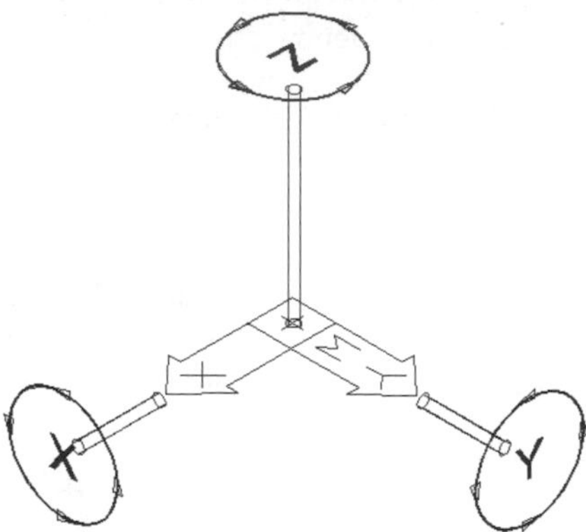

Figure 22–7 *The direction of positive rotation about the X-, Y-, and Z-axes.*

USING THE UCS MANAGER

Although the UCS command provides you maximum flexibility in setting, naming, restoring, and deleting a UCS, the UCSMAN (UCS Manager) command displays a multi-folder dialog box that presents a convenient, graphical method of restoring saved UCSs, establishing orthographic UCSs, and specifying UCS icon and UCS settings for viewports.

In the following exercise you will use the Orthographic UCSs folder of the UCS dialog box to establish and position a preset, orthographic UCS.

EXERCISE: SPECIFYING AND PLACING A PRESET UCS WITH THE UCS MANAGER

1. Continue in, or open 22DWG01.DWG from the accompanying CD. If necessary, restore a WCS by using the World option of the UCS command. (Type **UCS** and press Enter twice.) Your drawing should resemble Figure 22–4 from the previous exercise.
2. From the Tools menu, choose Orthographic UCS> Preset. AutoCAD displays the UCS dialog box with the Orthographic UCSs folder current (as shown in Figure 22–8).
3. Choose the Right orthographic UCS. Select Set Current and OK. AutoCAD establishes a standard right orthographic UCS. Note that this UCS shares the 0,0,0 origin of the WCS.

   In the next step you will change the origin of the current UCS using the Depth feature of the UCS dialog box. This allows you to move the origin along the current Z-axis.

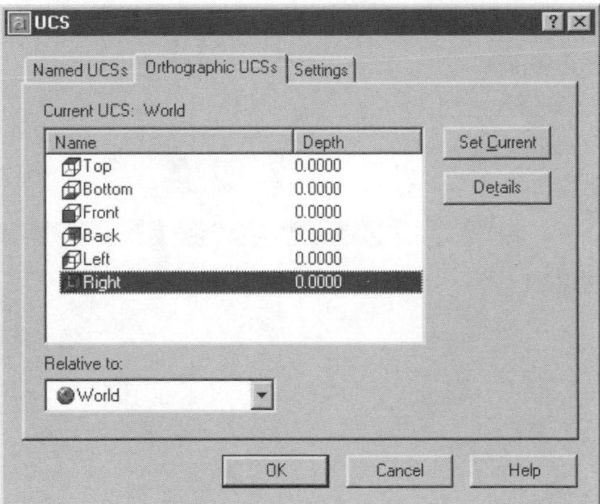

Figure 22–8 *The UCS Manager's UCS dialog box—Orthographic UCSs folder.*

4. Press Enter to redisplay the UCS dialog box. With the Right orthographic UCS chosen, right-click and choose Depth to display the Orthographic UCS Depth dialog box, as shown in Figure 22–9.
5. Choose the Select New Origin button and, in the drawing, use an Endpoint Osnap to pick at ① in the upper-left viewport shown in Figure 22–10. In the Orthographic UCS Depth dialog box, select OK, and then select OK in the UCS dialog box. AutoCAD moves the origin of the UCS along its Z-axis so it equals the Z-axis value of the point at ①.

   Now that you have established a new, non-preset UCS, in the next step you will name it and save it.
6. Start the UCS command by typing **UCS** and pressing Enter. Choose the Save option. When you are prompted for the name to save the current UCS with, type **MIDWAY** and press Enter. The UCS is saved with the name MIDWAY.

Note: You can restore named UCSs by using the Restore option of the UCS command or through the Named UCSs folder of the UCS dialog box. You can access the Named UCSs folder from the Tools menu by choosing Named UCS.

   Once you have named and saved a non-preset USC in steps 5 and 6, you can establish a new UCS relative to the named UCS, as shown in the next step.
7. From the Tools menu, choose Orthographic UCS>Preset. In the UCS dialog box's Orthographic UCSs folder, choose (highlight) the Left preset orthographic UCS. From the Relative To drop-down list, choose Midway. Choose Set Current and OK. The new UCS is changed to a "back" orientation relative to the previously named MIDWAY UCS, as shown in the upper-right viewport in Figure 22–10.

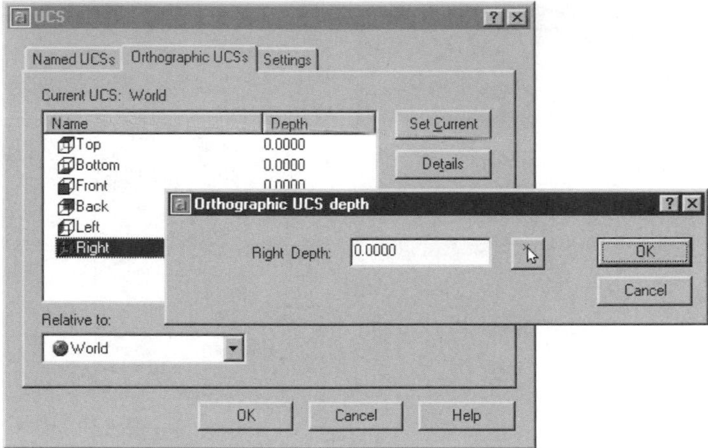

Figure 22–9 *Orthographic UCS Depth dialog box.*

Note: In the previous step you could have also simultaneously changed the depth of the new UCS by using the Depth feature of the UCS dialog box. as you did in steps 4 and 5.

AutoCAD keeps track of the last ten coordinate systems created in both model and paper space. Repeating the Previous option of the UCS command takes you back through the list. In the following step you will use the Previous option of the UCS command to return to the previous UCS.

8. Start the UCS command and select the Previous option. AutoCAD restores the most recent UCS (see the lower-left viewport in Figure 22–10).

   In the next step of this exercise you will reestablish the WCS for the model.

9. Start the UCS command and accept the default World option by pressing Enter. AutoCAD restores the WCS as shown in the lower-right viewport in Figure 22–10.

10. Quit AutoCAD or close the drawing without saving it.

UCSS AND VIEWPORTS

The ability to have multiple viewports is helpful when you work in 3D. Multiple viewports provide different views of your model and facilitate editing and visualizing your work. Each viewport can offer a different view, and each can be assigned a different UCS. In addition, the UCS in one viewport can be transferred, or "copied," to any number of other viewports.

The UCS icon is normally visible in the lower-left corner of the viewport. It can, however, be either turned off or set to be displayed at the origin of the current coordinate system on a per-viewport basis. You can also configure the UCS to change to the associated UCS automatically whenever an orthographic view is restored in a

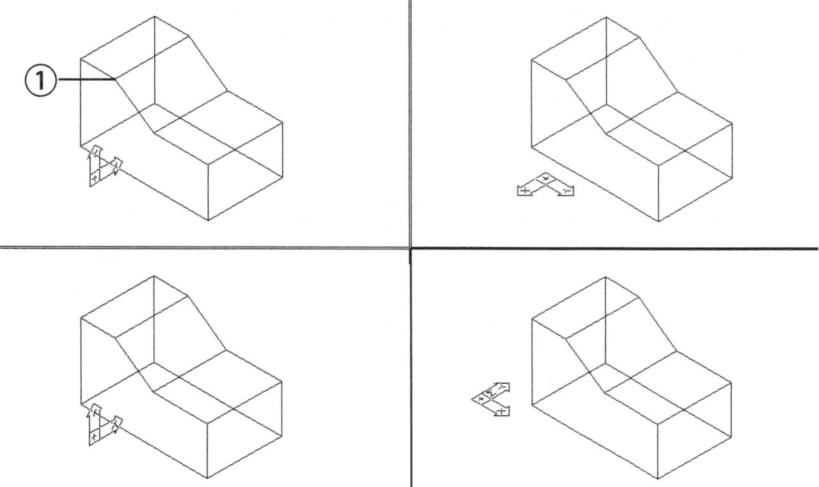

Figure 22–10 *Moving the UCS around your model.*

Tip: When you work in the Orthographic UCSs folder of the UCS dialog box, make sure that the desired UCS is displayed in the Relative To list before you change to any of the six preset UCSs. Orienting a new UCS relative to the WCS versus to a UCS you have named can lead to very different results.

viewport. Lastly, you can set each viewport to generate a plan view whenever you change coordinate systems. Two of these viewport features are demonstrated in the following exercise.

EXERCISE: SETTING DIFFERENT COORDINATE SYSTEMS IN VIEWPORTS

1. Open 22DWG01A.DWG from the accompanying CD. This drawing shows a solid model of a bracket. Make sure that the upper-left viewport is active. Your drawing will resemble Figure 22–11. The UCS in each viewport is controlled by the UCSVP system variable. When UCSVP is set to 1 in a viewport, the last used UCS in that viewport is saved, and then restored when the viewport is once again made current. If UCSVP is set to zero in a viewport, the UCS for that viewport always reflects the UCS in the current viewport.

   The three viewports of this drawing show a top view, a front view, and an isometric view. The UCSVP system variable in all three viewports is set to 1, allowing each viewport to have a separate UCS.

   In the next step you will set the UCSVP system variable for the right isometric viewport to zero.

2. Pick in the bottom-left viewport to set it current. Note that the UCSs in all three viewports remain unchanged. Now pick in the right, isometric viewport

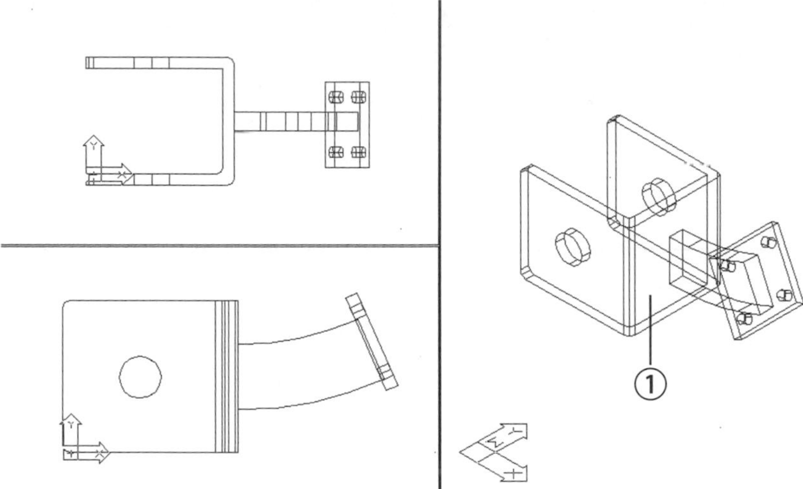

Figure 22–11 *The Orthographic UCS Depth dialog box of the UCS dialog box.*

to set it current. Type **UCSVP** and press Enter. When you are prompted, set the UCSVP variable to zero.

3. Now pick in the upper-left viewport to make it current. Note that the UCS in the isometric viewport changes to match the current viewport.

4. Pick in the lower-left viewport. Note that the UCS in the isometric viewport again changes to match the current viewport. Note that as you alternately pick in the upper- and lower-left viewports, the UCS in the orthographic viewport changes to reflect the UCS of the current viewport.

   In the next two steps you will establish a new UCS in the orthographic view and then transfer the UCS to the top view's viewport.

5. Pick in the orthographic viewport to make it current. Start the UCS command and then choose the New option. Select the Face option. When you are prompted to select the face of the solid object, pick on the face at ① in Figure 22–11, and then press Enter. A new UCS is established on the side face of the bracket.

6. Press Enter to restart the UCS command, and then choose the Accept option. When you are prompted to pick the viewport to apply to the current UCS, pick in the upper-left viewport, and then press Enter.

7. With the top-left viewport still current, type **PLAN** and press Enter to start the PLAN command. Press Enter to accept the default Current UCS. The upper-left viewport now shows a plan view of the new UCS.

8. Once again, alternately pick in the top-left and bottom-left viewports. Note that the UCS in the orthographic view changes to reflect the UCS in the current viewport. Your drawing should now resemble Figure 22–12.

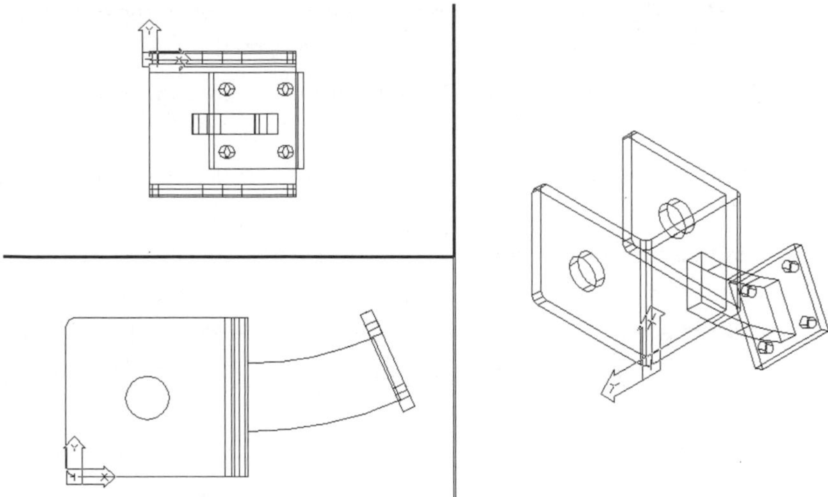

Figure 22–12 *Transferring a UCS from one viewport to another.*

Note: Because you used the Face option of the UCS command, the orientation of the X- and Y-axes that you obtained in step 5 may differ from that shown in Figure 22–12. As explained earlier in this chapter, using the Face and Object options of the UCS command yields X-axis and Y-axis orientations that depend upon the geometry of the object or solid face. If axis orientation is important for your editing, you can reorient the X- and Y-axes using the Xflip and Yflip options of the Face option or the Z-axis rotation option of the UCS command once the basic XY plane has been established.

Being able to maintain separate UCSs on a per-viewport basis and being able to transfer a UCS from one viewport to other viewports are effective tools that help you quickly navigate to various planes and orientations in your model. In the next step you will turn the UCS icon off in the active viewport.

9. Pick in the orthographic viewport to make it active. From the View menu, choose Display, and then choose the UCS Icon. Select ON to remove the check mark. The UCS icon in the active viewport is turned off.
10. Close this drawing without saving the changes you have made.

The ability to establish new coordinate systems easily, as well as the ability to move, orient, name, recall, and associate UCSs with individual views, are among the most important tools you have for working in 3D. The following list summarizes the many AutoCAD system variables that create or control UCSs.

- **UCSAXISANG**—Stores the default angle when you rotate the UCS around one of its axes using the X, Y, or Z options of the UCS command
- **UCSBASE**—Stores the name of the UCS that defines the origin and orientation of orthographic UCS settings

- **UCSFOLLOW**—Generates a plan view whenever you change from one UCS to another (also controlled from the USC dialog box)
- **UCSICON**—Displays the USC icon for the current viewport (also controlled from the USC dialog box)
- **UCSNAME**—Stores the name of the current coordinate system for the current space
- **UCSORG**—Stores the origin point of the current coordinate system for the current space
- **UCSORTHO**—Determines whether the related orthographic UCS setting is automatically restored when an orthographic view is restored (also controlled from the View dialog box)
- **UCSVIEW**—Determines whether the current UCS is saved with a named view
- **UCSVP**—Determines whether the UCS in active viewports remains fixed or changes to reflect the UCS of the currently active viewport (also controlled from the UCS dialog box)
- **UCSXDIR**—Stores the X direction of the current UCS for the current space
- **UCSYDIR**—Stores the Y direction of the current UCS for the current space
- **VIEWUCS**—Determines whether the current UCS is saved with a named view

USING VIEWPORTS

Another important tool that helps you with your work in 3D is AutoCAD's ability to display more than one model space viewport simultaneously. In model space, AutoCAD usually displays a single viewport that fills the entire drawing area. You can, however, divide the drawing area into several viewports. In model space these viewports are "tiled" or fit together as adjacent, rectangular regions, much like tiles on a floor. Unlike the viewports of paper space that were discussed in Chapter 18, Paper Space Layouts, model space viewports cover the entire screen and do not behave as sizable objects. A 3D drawing with three viewports defined is shown in Figure 22–13.

Multiple viewports are especially useful when you work in 3D, because you can set up a top, or plan, view in one viewport; a front, or elevation, view in another viewport; and have yet a third viewport show an isometric view of your model. You can see the effects of changes in one view reflected in the other views. As you learned in an earlier section of this chapter, you can also set the UCS in one viewport so that it is always the same as the UCS in the active, or current, viewport. Each viewport is largely independent of other viewports, giving you a great amount of flexibility in

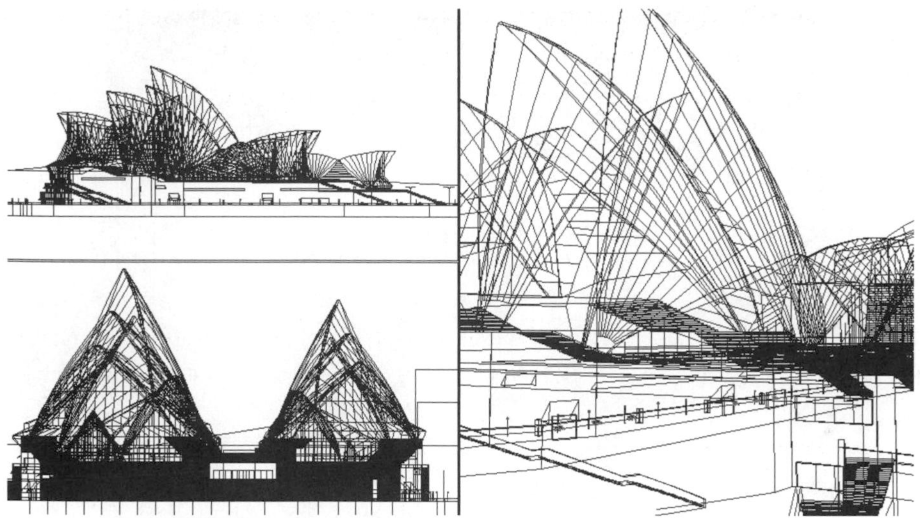

Figure 22–13 *A 3D drawing displayed in three viewports.*

viewing and editing your model. In each viewport, for example, you can independently do the following:

- Pan and zoom
- Set grid and snap distances
- Control the visibility and placement of the UCS icon
- Set coordinate systems and restore named views

Most of the operations controlling tiled viewports can be carried out from the two folder of the Viewports dialog box. You display this dialog box, which is shown in Figure 22–14, by selecting the Display Viewports Dialog icon on the Standard toolbar, or by choosing Viewports from the View menu, or by entering **vports** at the command prompt.

The functionality of the New Viewports folder of the Viewports dialog box is summarized as follows:

- **New Name**—If you want to save the viewport configuration you are creating, enter a name here. A list of named viewport configurations is displayed in the Named Viewports folder.
- **Standard Viewports**—The standard viewport configurations are listed here by name.
- **Preview**—This window displays a preview of the viewport configuration you select from the Standard Viewports box. The default views assigned to individual viewports in each configuration are displayed.

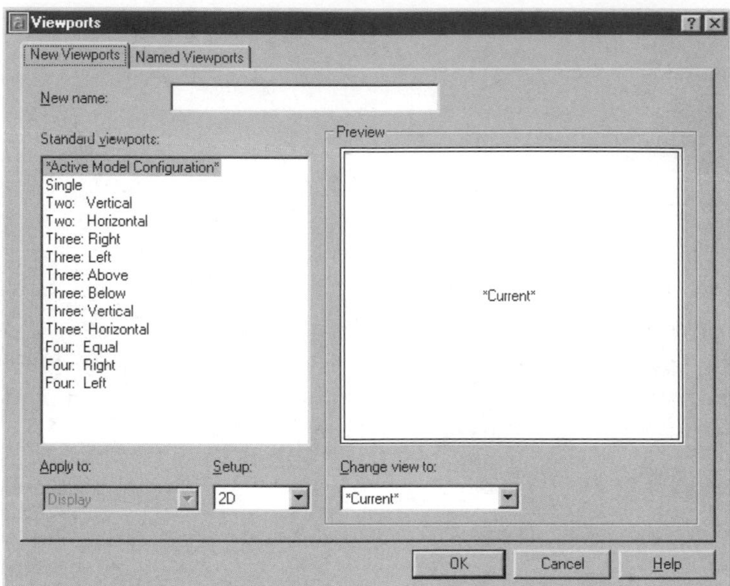

Figure 22–14 *The Viewports dialog box controls most viewport operations.*

- **Apply To**—This function allows you to assign the configuration to either the display or the current viewport. Assigning the configuration to the display assigns the configuration to the entire display area. Assigning the configuration to the current viewport applies the configuration to the current viewport only.
- **Setup**—You can choose either a 2D or 3D setup. When you select 2D, the new viewport configuration is initially created with the current view in all the viewports. When you select 3D, a set of standard orthogonal 3D views is applied to the viewports in the configuration.
- **Change View To**—This function replaces the selected viewport configuration with the viewport configuration you select from the list. You can choose a named viewport configuration, or, if you have selected 3D setup, you can select from the list of standard viewport configurations. Use the Preview area to view the viewport configuration choices.

In the following exercise you will learn about setting up multiple viewports configurations for working in 3D.

EXERCISE: CONFIGURING VIEWPORTS FOR WORKING IN 3D

1. Open 22DWG02.DWG from the accompanying CD. This drawing represents a model of a sliding glass door's locking mechanism. Your drawing should resemble Figure 22–15.

   As mention earlier in this chapter, the UCSORTHO system variable determines whether or not the related orthographic UCS setting is restored automatically when a preset orthographic view is established in a viewport. In the next step you will ensure that this system variable is turned on.

2. Type **View**, or, from the Standard toolbar, choose the Named View icon. In the View dialog box, choose the Orthographic & Isometric Views tab and, if necessary, select the Restore Orthographic UCS with View option to placed a mark in the check box. Select OK to close the dialog box.

Note: You can also display the View dialog box by choosing View from the menu and then choosing Named Views. You can also set the UCSORTHO system variable by entering **UCSORTHO** at the command prompt.

In the next three steps you will establish and name a standard 3D three-viewport configuration.

3. From the Standard toolbar, choose Display Viewports Dialog. If necessary, select the New Viewports tab in the Viewports dialog box. In the New Name input box, type **INITIAL-3**.

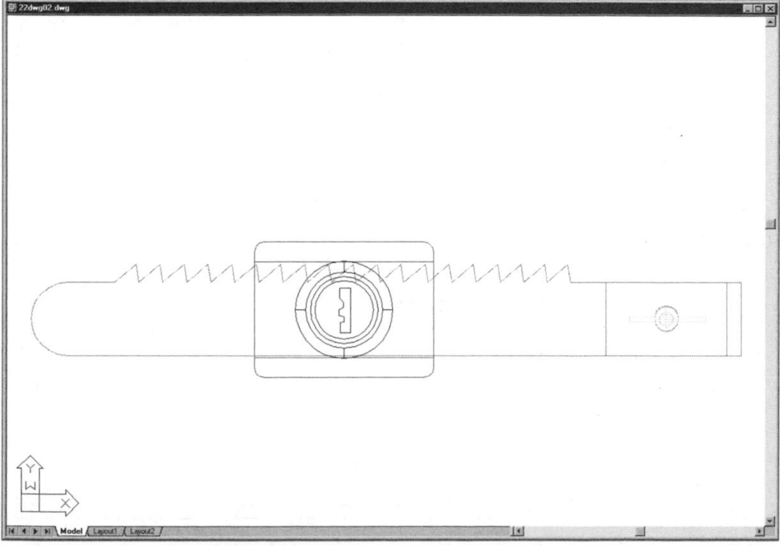

Figure 22–15 *The initial single viewport view of 22DWG02.DWG.*

4. In the Standard Viewports selection box, select Three Right. Note that the Three Right viewport configuration is shown in the Preview window. The word "Current" indicates that the current screen view of your model will appear in each viewport.

5. To establish more meaningful views for 3D work in the three viewports, select #D from the Setup drop-down list. Note that the viewports in the Preview window now have the standard top, front, and SE isometric views assigned. Make sure that "Display" is showing in the Apply To selection box, and then choose OK to exit the dialog box. Your drawing should now resemble Figure 22–16.

   In the next step you will establish a right view in the lower-left viewport.

6. Press Enter to redisplay the Viewports dialog box. In the Preview window, pick in the lower-left viewport to set it current. In the Change View To drop-down list, scroll to find the Right view and then select it. The viewport should now display "Right." Select OK to close the dialog box. Your display should have a right view in the lower-left viewport.

   As you follow the next step, note that the locking screw's position is dynamically updated in all three viewports as you drag it to a new position.

7. Set the upper-left viewport current. Start the MOVE command, select the green locking screw, and press Enter. Drag the locking screw downward until it is approximately even with the screw hole at the right end of the main locking bar. Leave it in this new position.

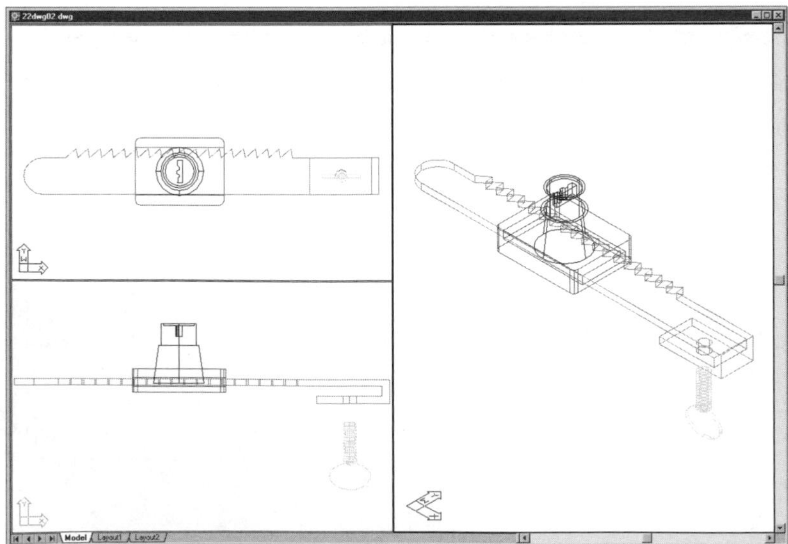

Figure 22–16 *You can easily establish standard top, front, and isometric views.*

You can further divide any viewport into smaller viewports. This is shown in the next step.

8. Pick in the right, isometric viewport to set it current. Redisplay the Viewports dialog box. In the Standard Viewports selection box, choose Two: Horizontal. Note the names of the proposed views in the Preview window. Select the top or upper viewport in the Preview window. From the Change View To drop-down list, select SW Isometric. Note that the top viewport reflects the change. In the Apply To box, make sure that Current Viewport is selected. Close the dialog box. Your drawing should now resemble Figure 22–17.

   In the next step you will name the current viewport configuration and then return to a single viewport configuration.

9. In the Viewports dialog box, type **4-SE-SW** in the New Name input box. Select OK to close the dialog box. Select the upper-left viewport to set it current. Then redisplay the Viewports dialog box. In the Standard Viewports window, select Single. Make sure that Display appears in the Apply To box, and then close the dialog box. The previous "current" viewport is now the single viewport (as shown in Figure 22–15).

   To restore a named viewport configuration, perform the following step.

10. Display the Viewports dialog box. Select the Named Viewports tab. From the Named Viewports drop-down list, select 4-SE-SW. The configuration is shown in the Preview window. Select OK to close the dialog box and restore the 4-SE-SW configuration (as shown in Figure 22–17).

11. Close this drawing without saving changes.

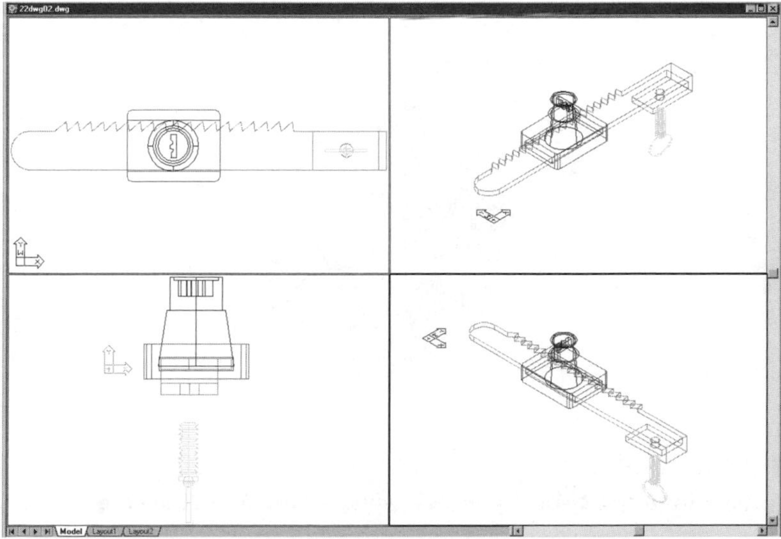

Figure 22–17 *You can quickly restore a named viewport configuration.*

User-definable coordinate systems (UCSs) and the ability to establish, name, and recall multiple viewport configurations quickly serve as two of the most useful tools for use in 3D modeling. Just as establishing a new UCS allows you to move the drawing's origin and XY plane to the position and orientation that is best suited for work on various parts or aspects of the model, multiple viewports allow you to see the model from the most advantageous viewpoint.

INTERACTIVE VIEWING IN 3D

While multiple viewports provide useful static views of a 3D model, the 3D Orbit command allows you to view 3D models interactively. Because the 3D Orbit command is transparent, you can call it in the middle of other commands, such as Move or Copy. When the 3D Orbit command is active, you manipulate the view with a screen pointing device, such as the mouse, and you can view the model from any point in 3D space.

The 3D orbit view overlays an "arcball" on the current view of the model. The arcball consists of a circle divided into four quadrants by smaller circles, as shown in Figure 22–18. The center of the arcball represents the target, which remains stationary. As you manipulate the view, the camera, or viewpoint, moves around the target, depending upon where you place the pointing device. You click and drag the cursor to rotate the view.

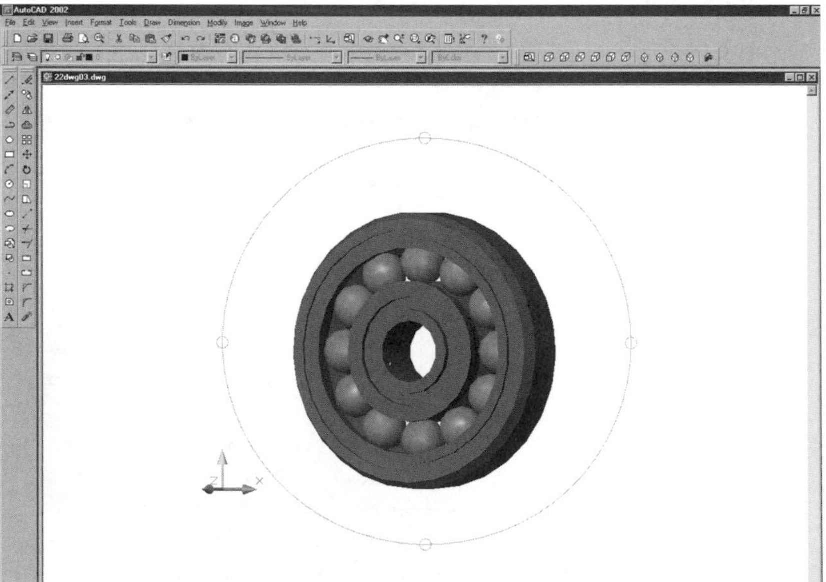

Figure 22–18 *The arcball symbol surrounds a 3D model.*

Different cursor icon symbols appear depending upon the position of the cursor with respect to the arcball. Four basic movements are possible. These movements and their associated cursor icons are as follows:

- When the cursor is moved inside the arcball, a small sphere encircled by two arrowhead lines is displayed. You can manipulate the view freely in all directions when this icon is active.

- When the cursor is moved outside the arcball, a circular arrow surrounding a sphere is displayed. When this cursor is active, clicking and dragging the cursor in a vertical motion causes the view to move around an axis that extends through the center of the arcball and perpendicular to the screen. If you move the cursor inside the arcball, it changes to the sphere encircled by two ellipses, and the view moves freely.

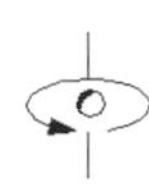

- When you move the cursor over one of the small circles on either side of the arcball, a horizontal ellipse encircling a sphere appears. Clicking and dragging the cursor when this icon is active rotates the view around the vertical axis, or Y-axis, that extends through the center of the arcball. This axis is represented by the vertical line in the cursor icon.

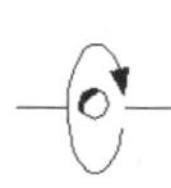

- When you move the cursor over one of the circles at the top or bottom of the arcball, a vertical ellipse encircling a sphere appears. Clicking and dragging the cursor when this icon is active rotates the view around the horizontal axis, or X- axis, that extends through the center of the arcball. This axis is represented by the horizontal line in the cursor icon.

With these four modes of movement you can view the model from any position. Using any mode except the free movement mode that is available when the cursor is inside the arcball constrains motion to one axis.

EXERCISE: VIEWING A MODEL WITH 3D ORBIT

1. Open 22DWG03.DWG from the accompanying CD. This model shows a ball bearing assembly from a side view (see Figure 22–19).

   In the following steps note the shape and orientation of the 3D Orbit UCS icon as you manipulate the view of the model.

2. From the View menu, choose 3D Orbit to start the 3D Orbit command. Move the screen cursor to a position outside the large arcball circle, and note the circular arrow icon. Remaining outside the arcball circle, click and drag the cursor in up and down motions. Observe that the model view rotates about an axis that extends through the center of the arcball and perpendicular to the plane of the screen. Releasing the left mouse button "fixes" the view in a new orientation.

3. Right-click to display the 3D Orbit shortcut menu. Select Reset View to restore the original view of the model.

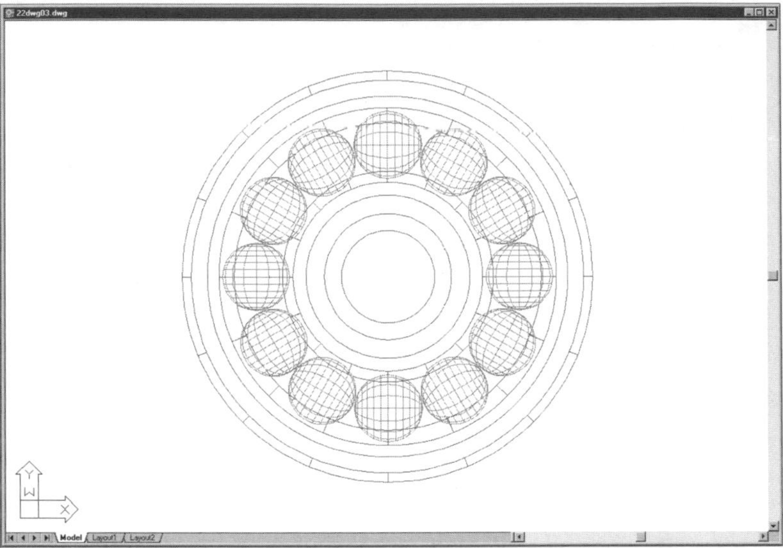

Figure 22–19 *The 2D side view of a ball bearing assembly.*

4. Move the screen cursor over either of the small circles at the left or right edge of the arcball circle. Note the horizontal elliptical arrow icon. Click and drag the cursor in horizontal motions. Observe the view rotating about an axis that extends vertically through the arcball. Note that dragging the cursor outside the arcball has no effect on the axis of rotation or on the icon shape. Releasing the left mouse button "fixes" the view in a new orientation.
5. Right-click to display the 3D Orbit shortcut menu. Choose Shading Modes and select one of the five remaining shade modes.
6. Repeat the previous step four times, choosing one of the other remaining shade modes each time.
7. With the model in a shade mode other than wireframe, move the screen cursor over either of the small circles at the top or bottom edge of the arcball. Note the vertical elliptical arrow icon. Click and drag the cursor in a series of vertical, up and down, motions. Observe the view rotating about an axis that extends horizontally through the center of the arcball. Note that the model remains shaded as the view changes. Releasing the left mouse button "fixes" the view in a new orientation.
8. Right-click to display the 3D Orbit shortcut menu. Select Reset View to restore the original view of the model. Now move the icon anywhere inside the arcball. Note the double elliptical icon that has both a horizontal and a vertical arrow. Click and drag the cursor in a series of horizontal, vertical, and diagonal motions. Observe the view rotating freely. Releasing the left mouse button "fixes" the view in a new orientation.

9. Right-click to display the 3D Orbit shortcut menu, and then choose Reset View.
10. Close 22DWG03.DWG without saving changes.

The 3D Orbit command allows you not only to establish a shaded, static view of a model from virtually any vantage point; it allows you to view the model dynamically as you change the view. Shaded or hidden line removal modes persist even while the viewpoint is being dynamically and interactively modified. This capability allows you to gain new insights into the geometric relationships among the various aspects of even a simple model.

You should be aware of several other features, and a few limitations, of the 3D Orbit command:

- The Orbit Maintains Z option, which you access from the context menu, locks the X-axis in its current orientation as you drag the cursor horizontally within the arcball circle or from the small circles on the left and right sides of the arcball. This prevents models from tumbling end over end, and it is useful for avoiding inappropriate views of architectural models, such as a house in an upside-down orientation.
- You can control the view height and view width of 3D Orbit from the Properties dialog box. If you open the Properties dialog box and then call the 3D Orbit command, you will find a View Height and View Width parameter listing. Click the parameter you want and enter a new value. This capability allows you to enter specific values for these parameters rather than use the less-precise pointer-drag method offered by the 3D Orbit command's context menu.
- Depending upon the complexity of the model and the efficiency of your particular CPU/graphics hardware, the components of the model may be reduced and displayed as simple 3D boxes during real-time manipulation of the view. These "bounding box" surrogates accurately represent the volumetric extents of the original elements. Once dynamic motion input is stopped, the model reverts to its original geometric forms. Figure 22–20 shows this effect with 22DWG03.DWG.
- If you choose only individual elements of the model before you enter the 3D Orbit command, only those selected elements will participate in the dynamic viewing and any subsequent presentations. If your model is large or complex and the speed of your CPU/graphics hardware is a limiting factor, choosing only a few key elements may allow a smoother, more accurate dynamic presentation.
- As described in the following section, placing your model in a perspective presentation is helpful in establishing spatial relationships.
- You cannot enter commands at the command line while 3D Orbit is active. Commands such as ZOOM and PAN are available from the 3D Orbit right-click shortcut menu, however.

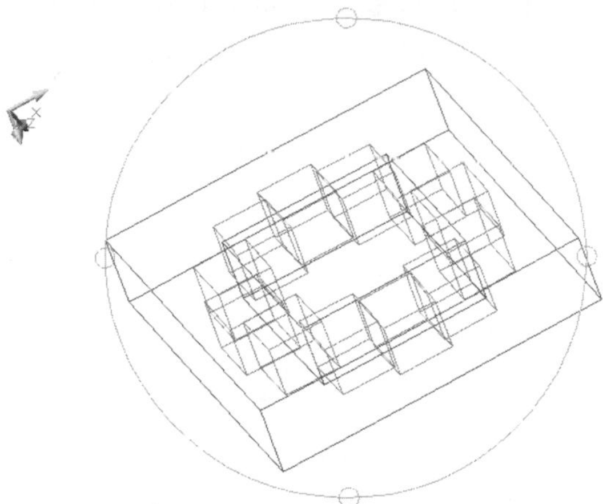

Figure 22–20 *Bounding boxes may be used as substitutes for more complex shapes in slower systems.*

- Keep in mind that 3D Orbit alters only the view of the model; the model remains stationary during dynamic viewing.
- If the view of the model becomes ambiguous, the Reset View option available on the 3D Orbit right-click shortcut menu will restore the view that was in effect when 3D Orbit was last entered.

In addition to the modes of view manipulation and the shading options demonstrated in the previous exercise, the right-click shortcut menu of the 3D Orbit command presents several other features and options. You can pan and zoom; turn two additional viewing aids on and off; establish standard, preset views; adjust camera distance and swivel the camera; and set either a parallel or perspective view. An ability to establish clipping planes is also available. Several of these features are demonstrated in the following exercise. The clipping planes feature will be demonstrated in a later exercise.

EXERCISE: ADDITIONAL 3D ORBIT VIEWING OPTIONS

1. Open 22DWG04.DWG from this book's accompanying CD. This drawing represents a model of a battery canister. The initial view is shown in Figure 22–21.
2. From the View menu, choose 3D Orbit. Note that the 3D Orbit arcball and UCS icon are displayed.
3. Right-click to display the 3D Orbit shortcut menu. Choose Visual Aids and then select Grid. Note the 3D Orbit Grid display.

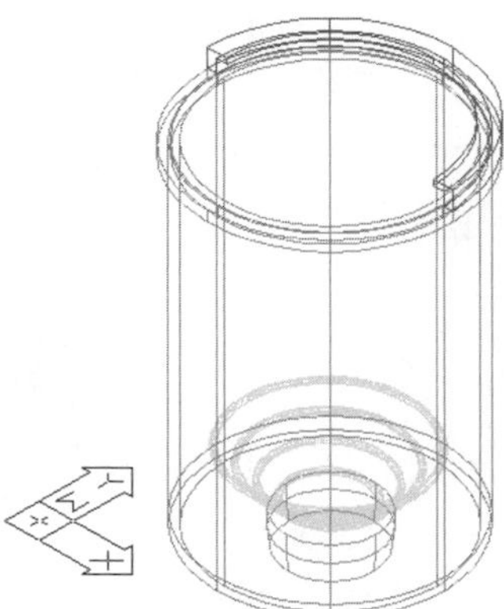

Figure 22–21 *An isometric view of a battery canister.*

4. Repeat step 3 but this time select the Compass visual aid. While you manipulate the view as you did in the preceding exercise, note the appearance of the grid and compass visual aids.
5. Use the right-click shortcut menu to turn both the grid and compass visual aids off.
6. Manipulate the view so it resembles the view shown in the right viewport in Figure 22–22.
7. Right-click to display the 3D Orbit shortcut menu. Choose Projection and then select Perspective. Your view should now resemble the model in the left viewport in Figure 22–22.
8. From the 3D Orbit shortcut menu, choose Preset Views and then select Front.
9. Leave this drawing open. You will continue from this view in the next exercise.

ADJUSTING CLIPPING PLANES IN 3D ORBIT VIEW

You can establish clipping planes in 3D orbit view. With clipping planes, objects or portions of objects that move beyond a clipping plane become invisible in the view. Adjustments to clipping planes are made interactively in the Adjust Clipping Planes window shown in Figure 22–23. You display this window by choosing More>Adjust Clipping Planes from the 3D Orbit shortcut menu.

The Adjust Clipping Planes window shows two clipping planes: front and back. These planes are represented by two horizontal lines, which initially may be superimposed upon each other. You use buttons on the toolbar in the Adjust Clipping Planes window, or the options available from the right-click shortcut menu shown in Figure 22–23, to choose the clipping plane that you want to adjust and to turn clipping planes on and off. The options available are as follows:

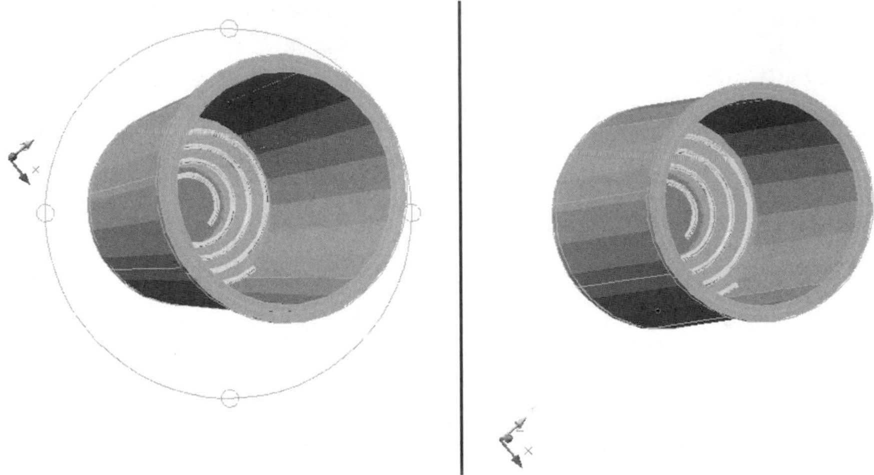

Figure 22–22 *A perspective view (left) versus a parallel view (right).*

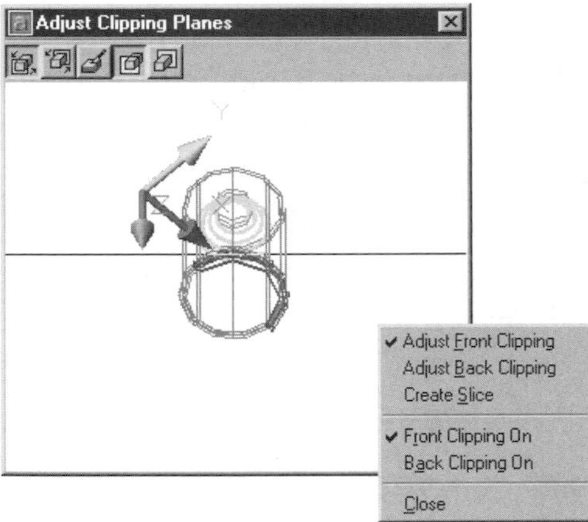

Figure 22–23 *You adjust clipping planes in the Adjust Clipping Planes window.*

- **Front Clipping On**—Toggles the front clipping plane on and off. When the front clipping plane is on, the results of moving this plane are shown interactively in the main 3D orbit display.
- **Back Clipping On**—Toggles the back clipping plane on and off. When the back clipping planes is on, the results of moving this plane are shown interactively in the main 3D orbit display.
- **Adjust Front Clipping**—Adjusts the front clipping plane. The line nearest the bottom of the window shows the front plane. If Front Clipping On is active, you see the results of moving the plane up or down in the main 3D orbit display.

- **Adjust Back Clipping**—Adjusts the back clipping plane. The line nearest the top of the window shows the back plane. If Back Clipping On is active, you see the results of moving the plane up or down in the main 3D orbit display.
- **Create Slice**—When this feature is active, the front and back clipping planes move together at their current separation distance, creating a "slice" of the objects contained between the two planes. If both Front Clipping On and Back Clipping On are active, the slice is displayed in the main 3D orbit display.

Note: Except when you use the Create Slice option, you can adjust only one clipping plane at a time. If the Create Slice button is depressed, you adjust both planes simultaneously. On the toolbar, the button for the plane you are adjusting appears depressed.

Tip: Although you can establish clipping planes from any 3D orbit view, standard orthogonal views such as front or side usually yield the most effective results.

The use of front and back clipping planes and slices is demonstrated in the following exercise.

EXERCISE: ESTABLISHING CLIPPING PLANES

1. Continue from the previous exercise with a front view of the model in 22DWG04.DWG. From the 3D Orbit shortcut menu, choose More>Adjust Clipping Planes. Note that the view is rotated 90 degrees in the Adjust Clipping Planes window.
2. Right-click in the Adjust Clipping Planes window and then make sure that a check appears beside the Adjust Front Clipping and Front Clipping On options. Click in the Adjust Clipping Planes window to close the shortcut menu.

   In the Adjust Clipping Planes window, the bottom line represents the position of the front clipping plane, while the top line represents the position of the back clipping plane. These two lines are initially superimposed and pass through the middle of the model.
3. Click and drag the special clipping plane icon downward and then observe the view of the model in the main 3D orbit display. As the clipping plane approaches the "front" of the model, more of the model becomes visible because less of the model is "clipped." In this model, this effect is most evident in the spring element.
4. With the front clipping plane line near the "front" of the model, select the Back Clipping On option from the right-click shortcut menu. (Alternatively, you can also activate the Back Clipping On/Off button on the Adjust Clipping Planes window toolbar.) The model is displayed as a slice, with only those portions falling between the current front and back clipping plane lines visible in the main display.

5. Activate the Create Slice button or select Create Slice from the shortcut menu. Click and drag the icon to vary the position of the slice. Note that the front and back planes appear more together when Create Slice is active. Figure 22–24 shows a typical front-view slice of the model.
6. Close the Adjust Clipping Planes window by selecting Close from the shortcut menu. Manipulate the current view to observe the slice from different views. Restore the previous view by selecting Reset View from the shortcut menu.
7. From the 3D Orbit shortcut menu, choose More and then select Back Clipping On to turn the back plane off. Notice that the display now shows all portions of the model behind the front clipping plane.
8. Leave this drawing open. You will continue from this view in the next exercise.

   With clipping planes and the use of 3D Orbit's shading modes, you can establish new and informative views of your model. These clipped views remain when you exit 3D Orbit.

Note: Clipped views are especially helpful when they are combined with shaded views in 3D Orbit. They may not be as effective when you exit 3D Orbit and return to standard 3D and 2D views. The 3D Orbit shortcut menu offers a quick way to disable the front and back clipping planes prior to exiting the 3D Orbit command.

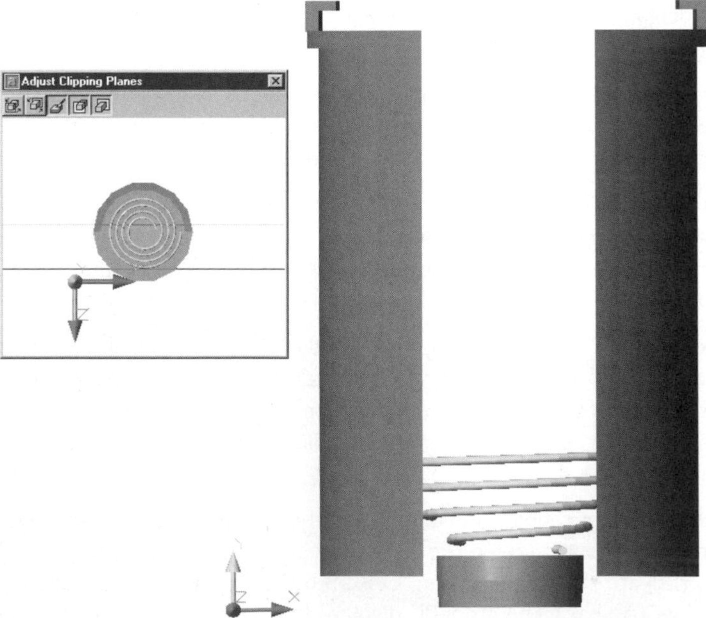

Figure 22–24 *A slice view through the middle of the canister model.*

USING A CONTINUOUS ORBIT

While 3D Orbit is active, you can choose to establish a continuous motion of the view around your model. Such continuous motion studies can yield information about the structure and geometric relationships in the model that are less apparent in static views. A continuous orbit is demonstrated in the following exercise.

EXERCISE: ESTABLISHING A CONTINUOUS ORBIT

1. Continue from the previous exercise with a front, shaded view of the model in 22DWG04.DWG. Start 3D Orbit. From the shortcut menu, choose Shading Modes and then select Flat Shaded.
2. From the shortcut menu, choose More and then select Continuous Orbit. Note that the cursor changes to a small sphere encircled by two lines.
3. To start the continuous orbit motion, click and drag in the direction in which you want the continuous orbit to move and then release the pick button. Observe that the model continues to move in the direction of the motion of the pointing device.
4. To change the direction of the continuous orbit motion, click and drag in a new direction and then release the pick button.
5. To stop the motion at any point without leaving 3D Orbit, left-click while keeping the pointing device stationary.
6. To start a new continuous orbit motion, click, drag, and release the pick button again.
7. To change the projection type while the model is in continuous orbit, right-click to display the shortcut menu, choose Projection, and then toggle the projection mode.
8. To stop the motion at any point and exit 3D Orbit, press Esc.
9. You are finished with this model. Close the drawing without saving changes.

While the continuous orbit is active, you can modify the model view by right-clicking to display the shortcut menu and then choosing Projection, Shading Modes, Visual Aids, Reset View, or Preset Views. You can also turn the front and back clipping planes on and off. You cannot, however, adjust the clipping planes while continuous orbit is active. Choosing Pan, Zoom, Orbit, or Adjust Clipping Planes from the shortcut menu will end the continuous orbit

Note: The speed with which you drag the pointing device to establish a continuous orbit determines the speed of the orbit motion. If your model is complex or if your CPU/hardware is less efficient, you may want to make your orbit motions slower to yield smoother, less jerky, results.

Tip: Continuous orbits are frequently more realistic with the projection mode set to perspective.

SHADING A MODEL

AutoCAD's SHADEMODE command provides you with the same hidden line removal and enhanced shading options that are available with the 3D Orbit command. Like the 3D Orbit shade modes, the shading provided by SHADEMODE uses a fixed light source coming from behind and over your left shoulder. The various SHADEMODE options are shown in Figure 22–25. The options are described as follows:

- **2D Wireframe**—Displays the model objects using lines and curves to represent boundaries. This is AutoCAD's default display mode for 2D and 3D objects.
- **3D Wireframe**—Same as 2D Wireframe but also displays a shaded 3D UCS icon.
- **Hidden**—Same as 3D Wireframe but hides lines representing back faces. Its results are similar to the results obtained with the HIDE command.
- **Flat Shaded**—Shades the objects between polygon faces. Objects appear flatter and less smooth than Gouraud-shaded objects.
- **Gouraud Shaded**—Shades the objects and smooths the edges between faces; yields a smoother, more realistic appearance.
- **Flat Shaded, Edges On**—Combines the Flat Shaded and Wireframe options. Objects appear flat-shaded with wireframe showing.
- **Gouraud Shaded, Edges On**—Combines the Gouraud Shaded and Wireframe options. Objects appear Gouraud-shaded with wireframe showing.

The options offered by the SHADEMODE command are available from the View menu. Choose Shade and select one of the six shading modes. You can also return your display to standard 2D wireframe mode from this menu. SHADEMODE is also available from the command line.

Several enhancements were introduced in AutoCAD 2000 to make working with shaded models easier. These improvements include the following:

- **3D Grid:** When any of the shade modes (including 3D wireframe) are active, a distinctive "3D" grid is displayed if the Grid feature is enabled. The grid features major (i.e., heavier) grid lines with ten horizontal and vertical lines drawn between the major lines. The number of major lines corresponds to the value that is set using the grid spacing option of the GRID command and stored in the

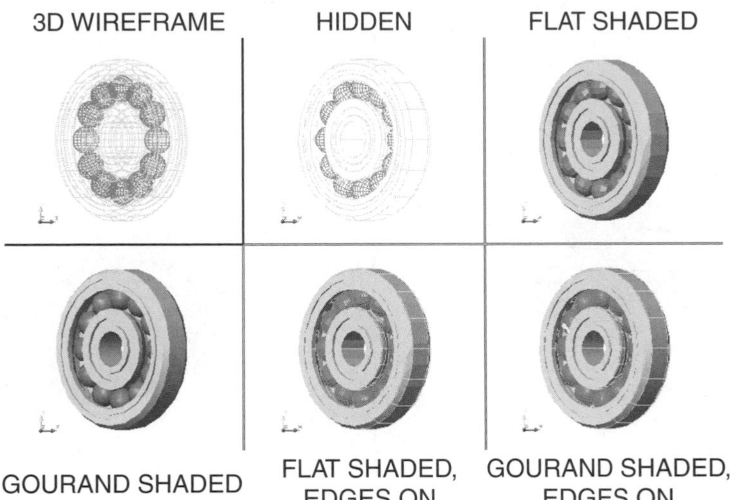

Figure 22–25 *Shading and hidden-line removal options available with SHADEMODE.*

GRIDUNIT system variable. The 3D grid is displayed coincident with the current XY plane, and it "tracks" UCS changes. The 3D grid is shown in Figure 22–26.

- **3D UCS Icon:** When any of the shade modes are active, a distinctive 3D USC icon is displayed whenever the UCS icon is enabled, as shown in Figure 22–26. This icon is shaded and has the X-, Y-, and Z-axes labeled. The X-axis is red, the Y-axis is green. and the Z-axis is blue.
- **3D Compass:** When any of the shade modes are active, a Compass feature can be activated. This feature superimposes on the model a 3D sphere that is composed of three lines representing the X-, Y-, and Z-axes. The apparent center of the sphere coincides with the center of the viewport. The display of the 3D compass, shown in Figure 22–26, is controlled by the COMPASS system variable.
- **Regeneration:** Regenerating the drawing does not affect shading.
- **Editing:** You can edit shaded models by selecting them in the normal manner. If a shaded object is selected, the wireframe and grips appear on top of the shading.
- **Saving:** You can save a drawing with objects shaded, and then open it again with objects still shaded.

Note: If you exit 3D Orbit with a shade mode active, the only means of changing to a different shade mode or returning to standard 2D wireframe mode is through the SHADEMODE command or the Shade options of the View menu.

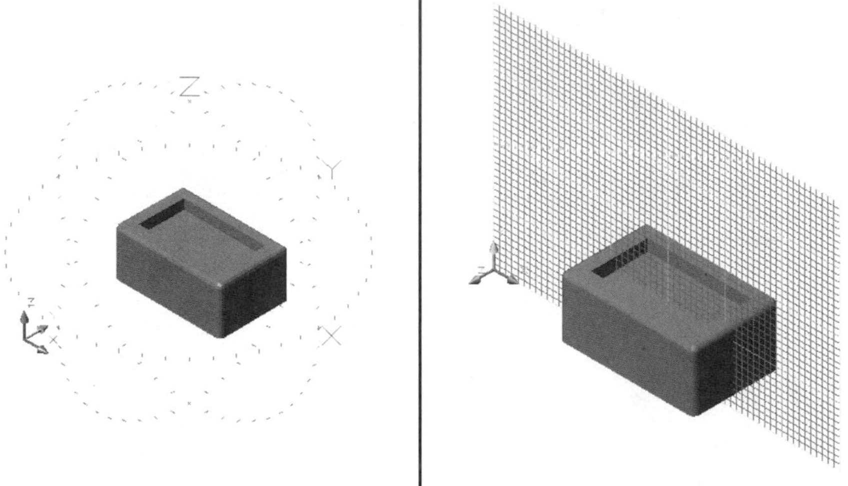

Figure 22–26 *The 3D Icon, Compass, and Grid features are available in 3D shade modes.*

SUMMARY

Understanding the orientation of AutoCAD's 3D axes is the first step toward working comfortably in 3D space. The UCS command sets the orientation of the UCS in 3D space. Using multiple viewports allows you to orient yourself in 3D space easily and provides an easy means of switching views of your model. AutoCAD's 3D Orbit command lets you move around and view a shaded model in real time. Together these 3D tools make working in 3D space more efficient.

Drawing In 3D

As you work with 3D models inside AutoCAD, you will soon find that there are two skills that will quickly enable you to become as comfortable working in 3D as you are working in 2D. One of these skills is the ability to move around inside the 3D model. The tools needed to develop this ability were discussed in the preceding chapter, 3D Fundamentals. You learned the tools and AutoCAD commands needed to establish UCSs, the various viewpoints from which to view your 3D geometry, and the usefulness of working with more than one viewport.

In this chapter you will learn about using AutoCAD commands that are already familiar to you from your work with 2D drawings in new and different ways as you apply these commands to 3D work. Familiar commands such as LINE and PLINE can be used in 3D work as well—you just use them a little differently. Likewise, editing commands such as ROTATE, MIRROR, and ARRAY have their 3D counterparts in ROTATE3D, MIRROR3D, and 3DARRAY. You will also see that tools such as Point Filters and Object Snaps are even more important and necessary as you work with 3D space.

Lastly, you will also learn how to create a special AutoCAD object called a Region, which acts like a 2D object in some ways and like a 3D object in others.

The following topics will be discussed and demonstrated in this chapter:

- Starting with lines, polylines, and 3D polylines
- Taking advantage of object snaps and point filters
- 3D editing commands
- Using EXTEND and TRIM in 3D
- Creating regions

STARTING WITH LINES, POLYLINES, AND 3D POLYLINES

A standard line in AutoCAD is usually drawn in 2D and exists on the WCS or on the current UCS. Each endpoint of this line is defined by a set of coordinates: X, Y, and Z. If you have spent most of your time drawing in 2D, then you have probably always left the Z value set to zero. When you adjust the Z value of either endpoint of the line, the line becomes a line used in all three dimensions.

The fact that the line is now in three dimensions, however, becomes apparent only when you observe it from the correct view. For example, a line may appear to be 2D when it is viewed from the top, but 3D when it is viewed from other angles (see Figure 23–1).

The following exercise shows you how to draw a wireframe version of a box in 3D by using the LINE command. In the last chapter you created a similar object, but you changed the UCS frequently. In this exercise you will create the object without changing the UCS, which is significantly faster.

EXERCISE: USING THE LINE COMMAND TO DRAW A WIREFRAME BOX

1. Load AutoCAD if it is not already loaded, and start a new drawing.
2. Select the LINE command from the Draw toolbar.
3. Enter the following coordinates at the command prompt:

   0,0

   5,0

   5,5

   0,5

   c

   Press Enter to end the command.
4. From the View menu, choose 3D Views>SW Isometric to switch to an isometric view of the drawing.
5. Select the COPY command from the Modify toolbar. At the "Select objects:" prompt, select the four lines and press Enter. Specify a displacement by typing **0,0,5** and pressing Enter twice. The four lines are copied "up" five units on the

Z-axis. Perform a Zoom Extents to view the newly copied lines. Figure 23–2 shows the drawing at this point.

6. Start the LINE command again. Enter the following coordinates:

   0,0,0

   0,0,5

   Press Enter to end the command.

7. Repeat step 6 three more times using the following coordinate pairs for the From and To points:

   0,5,0 & **0,5,5**

   5,5,0 & **5,5,5**

   5,0,0 & **5,0,5**

   Your drawing should now resemble Figure 23–3.

8. Exit this drawing without saving changes.

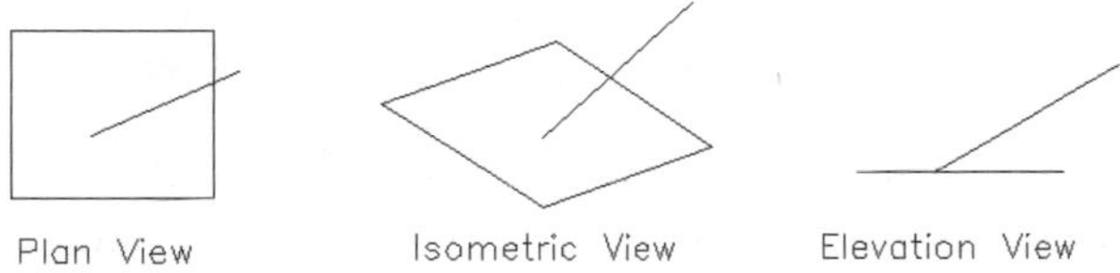

Figure 23–1 *Three views of the same line, showing the importance of correctly viewing a 3D model.*

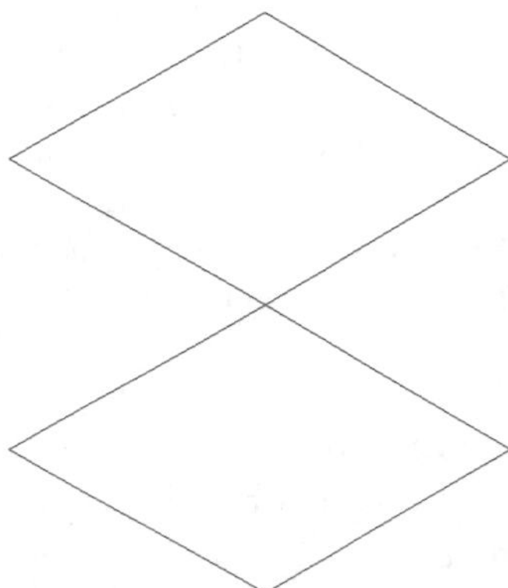

Figure 23–2 *The "top" and "bottom" of the 3D box from a SW isometric viewpoint.*

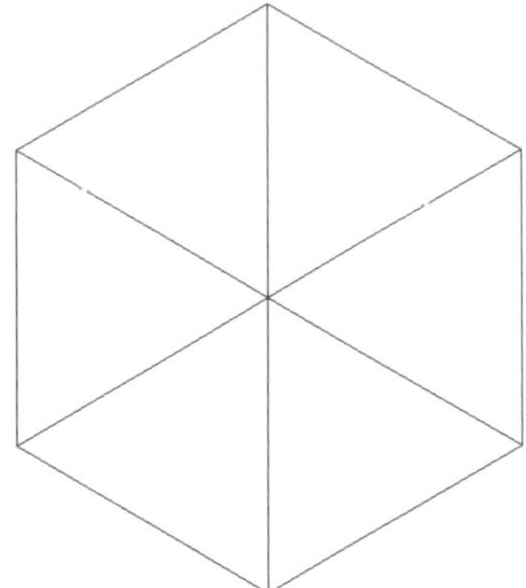

Figure 23–3 *The completed wireframe box in a SW isometric view.*

Note: During the last exercise you created what is called a wireframe model of a box. A wireframe model does not have any 3D surfaces; it represents only the edges of the object you have drawn. These edges, however, can have surfaces applied later.

You can also give a line or polyline a thickness (similar to an extrusion) by changing the properties of the line with the PROPERTIES or CHANGE command.

As you can see from this exercise, drawing in 3D with the LINE command is relatively easy; you just type in the coordinates. But it is also very tedious and potentially inaccurate. As you will see later in this chapter, the use of drawing aids can dramatically increase your 3D drafting speed.

A standard polyline is not quite as flexible as a regular line in 3D work. When you create a standard polyline, the vertices must lie on the same 3D plane. So if you start a polyline at three units above the coordinate system, all the vertices of the polyline would be at that height, regardless of where you select the points in the drawing. This makes polylines very useful for creating outlines or other planar construction elements that can be surfaced later with a variety of commands in AutoCAD.

Note: Surfacing is the process of creating continuous surfaces between boundary edges, such as lines. These surfaces can then be used to hide objects or lines that exist behind them.

Because of the planar limitation of a polyline, AutoCAD also includes a 3D polyline object. This polyline differs from a standard polyline in that you can place each vertex of the polyline at a different point in 3D space. Several of the polyline editing commands, however, are not available with the 3D polyline. For example, although you can still use PEDIT to create a spline curve from a 3D polyline, you cannot create a fit curve by using PEDIT. Figure 23–4 shows a 3D polyline that has been

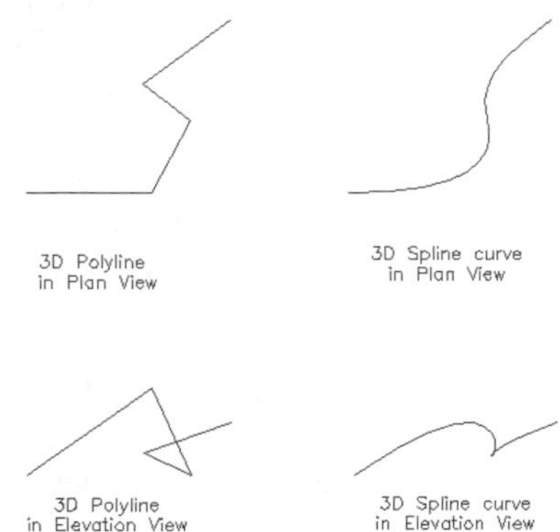

Figure 23–4 *A 3D polyline changed into a splined curve. The curves are smoothly translated in both 2D and 3D space.*

changed into a spline curve by using PEDIT. As mentioned earlier, lines and polylines are frequently used in 3D work as construction elements upon which more complex surfaces are created. The actual choice of which type of line to use to create the construction elements will depend on your requirements and specific drawing conditions.

Now that you have an idea of how lines and polylines can be used to create 3D drawings in AutoCAD, it is time to look at how those lines and polylines can be created more quickly and accurately using snaps and point filters.

TAKING ADVANTAGE OF OBJECT SNAPS AND POINT FILTERS

Drawing a line in 3D is relatively easy if you type in the coordinates for the endpoints of the line, instead of picking those points with a mouse. However, typing in points is not overly intuitive, and it can become tedious very quickly. To get around this, you need to make use of AutoCAD's drawing aids—in particular, object snaps and point filters.

OBJECT SNAPS

As you already know, an object snap is used to input a point accurately from an existing piece of geometry. Using object snap modes makes it easy to draw a line from the endpoint of one line to the endpoint of another. When you consider this concept from a 3D perspective, object snaps become almost critical to the drawing process.

For example, if you draw a box in 3D, you can draw only so much of the box in a 2D plan view. To create the rest of the box, you must switch to a 3D view, such as a SW isometric view, to see the model in its 3D form. You must make this switch because any lines that extend along the Z-axis appear as dots in a plan view. Only from a 3D view, such as an isometric view, are you truly able to see the model. Because this is an isometric view, however, what you see in the view might not be apparent. Figures 23-5 and 23-6 illustrate this problem.

As you can see in Figures 23-5 and 23-6, it is very easy to draw lines that appear to be 3D but that really exist only in the flat plane of the current UCS instead of in 3D space. This is where object snaps come in handy. When you snap to geometry that exists in 3D space, you ensure that you will not run into this problem.

EXERCISE: USING OBJECT SNAP MODES TO DRAW A 3D BOX

1. Start AutoCAD if it is not loaded, and begin a new drawing.
2. Select the LINE command from the Draw toolbar.

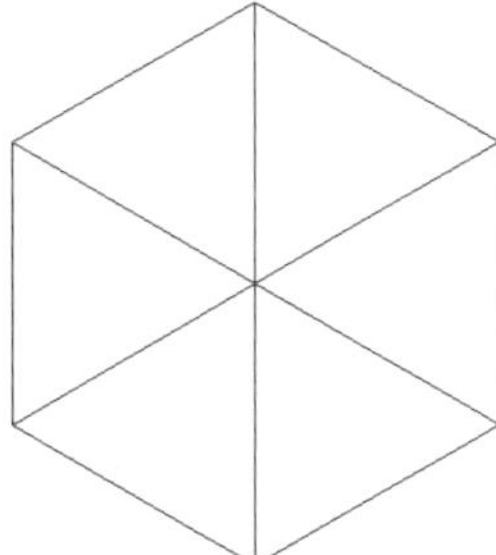

Figure 23–5 *A drawing that looks like a 3D box when it is viewed from a SW isometric viewpoint.*

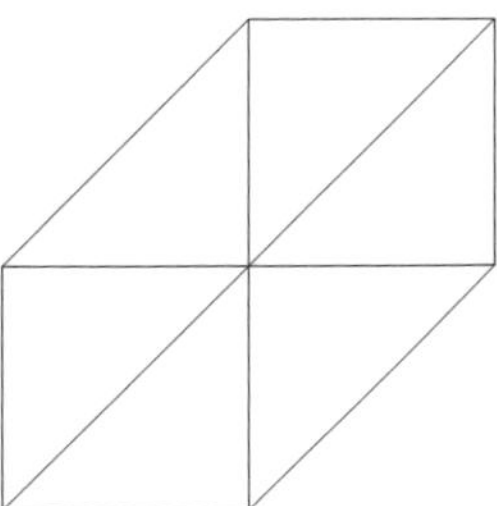

Figure 23–6 *The same box in plan view. The box was never actually drawn in 3D.*

3. Create a line with the following coordinates:

   0,0

   5,0

   5,5

   0,5

   c

   Press Enter to end the command.

4. From the View menu, select 3D Views>SW Isometric to switch to a 3D isometric view.

5. Copy the four lines "up" three units on the Z-axis using the COPY command. Specify a displacement of **0,0,3** and press Enter twice to complete the copy. Perform a Zoom Extents. Figure 23–7 shows the drawing at this point.

   Now that you have built the "top" and the "bottom" of the boxes, it is time to use object snap modes to create the lines in between.

6. From the Tools menu, select Drafting Settings to display the Drafting Settings dialog box. Select the Object Snap tab to make the Object Snap folder current, as shown in Figure 23–8.

7. Use the Clear All button to remove any checks from snap modes, and then place a check next to the Endpoint Osnap mode only. This enables endpoint snapping. Choose OK to exit the dialog box.

8. Select the LINE command from the Draw toolbar.

9. Draw a vertical line from the endpoint of each corner on the lower portion of the box to the corresponding upper portion, as shown in Figure 23–9.

10. Exit this drawing without saving changes.

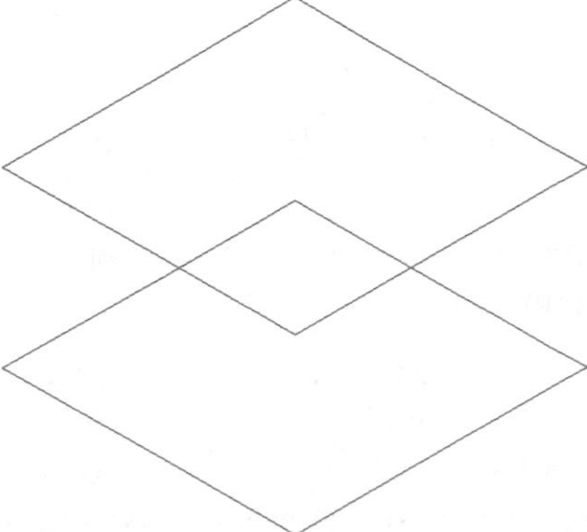

Figure 23–7 *The box with a completed "top" and "bottom."*

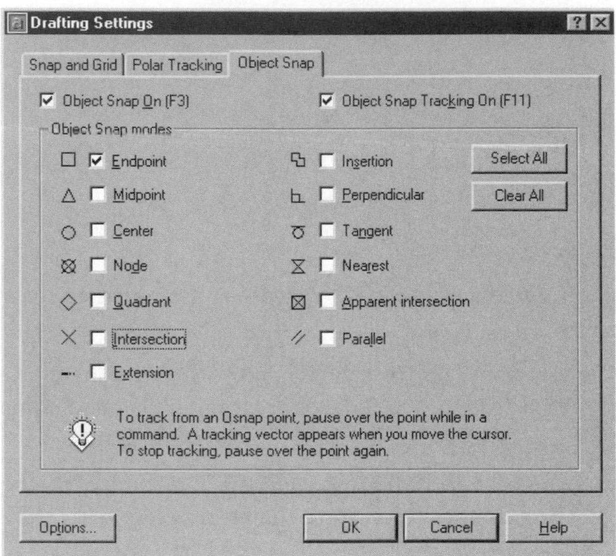

Figure 23–8 *Drafting Settings dialog box—Object Snap folder.*

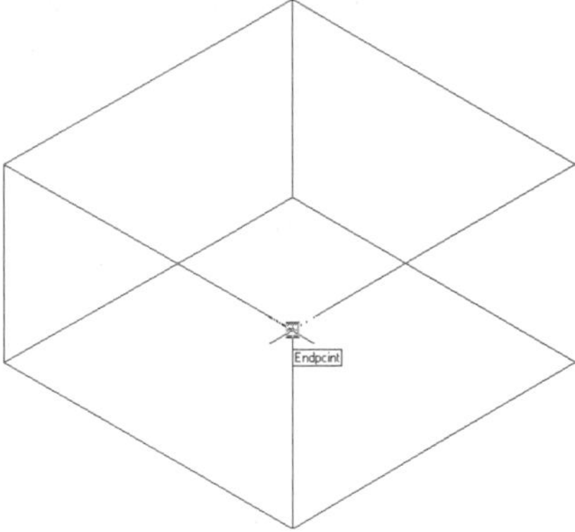

Figure 23–9 *Constructing the vertical lines of the box using endpoint Osnaps.*

As you can see in this exercise, the use of object snap modes almost immediately gives you a performance gain. It becomes quite easy to "snap" to a point in 3D.

However, even object snaps can be troublesome from certain 3D viewpoints. Consider the box example again. Depending upon the dimensions of the sides of the box, when you view the box from an isometric view, you can have overlapping endpoints, as shown in Figure 23–10. To get around this problem, you can change the view of the 3D model so that the endpoints do not overlap.

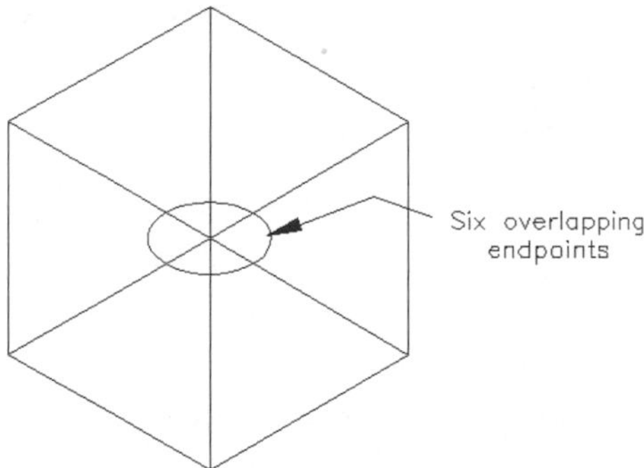

Figure 23–10 *A view of a wireframe box showing how endpoints can overlap in certain views.*

Tip: 3D views of wireframe models are notoriously ambiguous. When you work in 3D, it is always a good idea to check your model from different views to make sure you have drawn it correctly. What looks correct in one view might be incorrect, because the view hides the problem. Because of this, even experienced AutoCAD users occasionally draw something incorrectly in 3D.

POINT FILTERS

Point filters are a handy method of drawing 3D lines without having to enter all three coordinates. A point filter, such as the .XY (pronounced "point X-Y") filter for example, is used to filter out one or more of the three coordinates that your mouse selections, or "picks," make. For example, when you use a .XY filter, AutoCAD receives only the X and Y coordinate values from the point you select. You then type or use a .Z point filter to supply the Z coordinate. Because of the way they work, point filters—and especially the .XY filter—are extremely useful in 3D work.

Up to this point, you have drawn lines in 3D by relying upon existing geometry or by typing in all three coordinates. By using point filters, however, you can create 3D lines even more quickly and easily.

EXERCISE: CONSTRUCTING THE VERTICAL LINES OF A BOX WITH A .XY FILTER

1. Start a new drawing in AutoCAD.
2. From the Tools menu, choose Drafting Settings. In the Drafting Settings dialog box's Snap and Grid folder, turn on both your grid and snap, with both set to one unit. Set a running endpoint Osnap as you did in step 7 of the previous exercise.
3. From the View menu, choose 3D Views>SW Isometric.

4. Choose the RECTANGLE command from the Draw toolbar, and draw a rectangle by dragging the mouse. The rectangle should resemble Figure 23–11.
5. Choose the LINE command.
6. Select a corner of the box to set the start point of the line.
7. At the "To Point:" prompt, type in **.XY** and press Enter. AutoCAD then prompts you with "OF," meaning ".XY of what point?".
8. At the "OF" prompt, select the same point that you picked in step 6. AutoCAD then prompts you for a Z value.
9. Enter a value of **3**. Figure 23–12 shows the box at this point.
10. Repeat steps 5 through 9 for the remaining three corners, or use the COPY command to copy the vertical line to each corner.
11. Use the LINE command with the endpoint Osnap mode to complete the top of the box.
12. Exit this drawing without saving changes.

Point filters are always relative to the current UCS; so make sure that you are aware of your current UCS orientation before you use them, or you may get results you did not expect. For example, if your current UCS is aligned with the WCS X-axis, when you use a .XY filter, the Z value you type in is Z of that UCS, not of the WCS. Refer to Chapter 22, 3D Fundamentals, for a description of the UCS command.

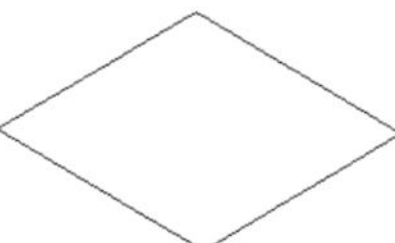

Figure 23–11 *The "base" of the box before you use point filters to construct the vertical elements.*

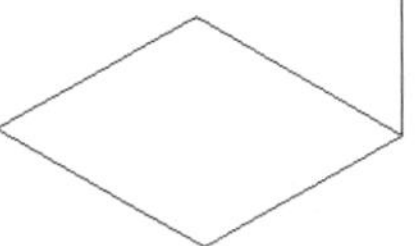

Figure 23–12 *The box with one vertical line segment drawn using a .XY filter.*

Now that you have explored some of the basic commands used to draw lines and polylines in 3D, it is time to look at how AutoCAD handles editing in 3D.

3D EDITING COMMANDS

Like the line and polyline commands, you can utilize most AutoCAD editing commands for work in 3D simply by using them differently. In addition to the standard editing commands, AutoCAD also provides you with several 3D-specific editing commands. We will first look at the basic 2D editing commands to see how they are utilized in 3D.

USING MOVE, COPY, SCALE, AND ROTATE IN 3D

The basic 2D editing commands in AutoCAD include MOVE, COPY, SCALE, and ROTATE. Depending upon your view of the 3D model, you can easily use each of these commands in 3D as well. For example, consider the MOVE command. If you are working in an isometric view, you can use the MOVE command in combination with object snap modes to move and place objects in 3D space. This works well when you have existing geometry to use; i.e., when you have a 3D point to move from and to.

Suppose, however, you have drawn the outline of a table top and you want to move it into the correct position in 3D space but you have no existing geometry to work with. In this situation, you have two options. First, you can combine the MOVE command with a .XY point filter to move the tabletop correctly. A second method involves using a shortcut in the MOVE command. When you select the MOVE command in AutoCAD, you are first prompted to select the object(s) you want to move. Then AutoCAD prompts you for the base point from which the move occurs. Instead of supplying a base point, you can supply a displacement. If, for example, you type in a displacement of **0,0,10** and press Enter twice, the selected objects move zero units along the X- and Y-axes but ten units along the Z-axis. The COPY command offers the same option at its "Specify base point:" prompt. Instead of specifying a base point, you can enter a 3D displacement vector. When you then press Enter twice, the copy appears at the displaced distance. In some instances, even when reference geometry is present in the model, it is more convenient to supply displacements when you move or copy objects.

In the following exercise you will use the displacement option of the COPY and MOVE commands with 3D objects.

EXERCISE: WORKING WITH MOVE IN 3D

1. Load the 23DWG01.DWG file found on the accompanying CD. The model consists of a box within a box, as shown in Figure 23–13.
2. Choose Move from the Modify toolbar.

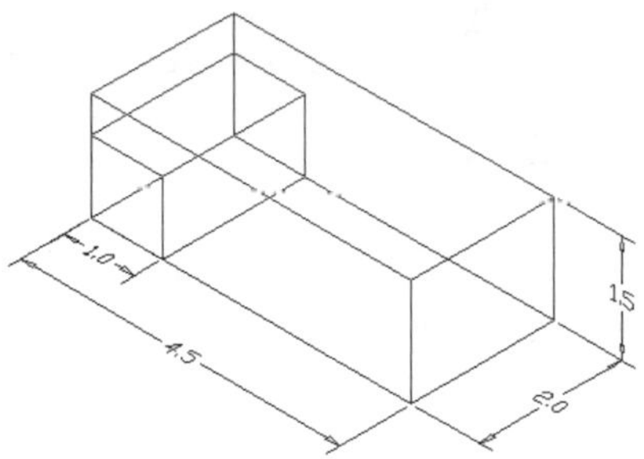

Figure 23–13 *A 3D model showing a box within a box.*

3. Select an edge of the smaller box and press Enter.
4. When you are prompted for the base point or displacement, type in **0,0,1.5** and press Enter twice. The smaller box is displaced 1.5 units in the current Z direction, as shown in Figure 23–14.

   In the next steps you will displace a copy of the box that you just moved 3.5 units along the X-axis.
5. Choose Copy from the Modify toolbar.
6. When you are prompted to select objects, type **P** for previous and then press Enter.
7. When you are prompted for the base point or displacement, enter **3.5,0,0** and press Enter a second time. A copy of the box is placed at the specified displacement, as shown in Figure 23–14.

As you can see in this exercise, using the MOVE and COPY commands with the displacement option is a fast and easy way to specify displacements of objects in 3D space. There are situations in which the displacement method is not practical, and in these situations the most efficient method is using object snaps, perhaps with point filters, and existing geometry reference points.

The SCALE and ROTATE commands can also be used in 3D, and they work in 3D essentially the same as they do in a "flat" 2D model. Keep in mind that when you edit 3D models, it is essential that you move away from plan views and utilize isometric views so that the 3D geometry can be fully seen.

In addition to the standard 2D editing commands, AutoCAD provides several editing commands specifically intended for use in 3D modeling. These include MIRROR3D, ROTATE3D, 3DARRAY, and ALIGN. We will examine these specialized 3D editing commands in the following sections.

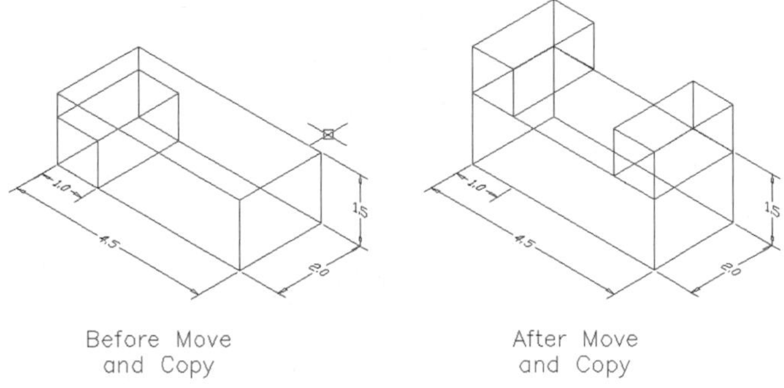

Figure 23–14 *The 3D model after a MOVE and COPY displacement.*

MIRROR3D

The MIRROR3D command is a modified version of the standard MIRROR command. In the standard 2D command, the mirroring plane is always the XY plane of the current UCS. In the 3D version, provision is made to specify any plane in 3D space.

After you select the objects you want to mirror, the following prompt appears:

Specify first point of mirror plane (3 points) or

[Object/Last/Zaxis/View/XY/YZ/ZX/3points] <3points>:

You have several methods of specifying the 3D mirroring plane:

- **Object:** Use the plane of a current planar object as the mirroring plane. Qualifying objects include a circle, arc, and 2D polyline.
- **Last:** Use the last specified mirroring plane.
- **Z-axis:** Define the plane with a point on the plane and a point normal to the plane.
- **View:** Use the current viewing plane as the mirroring plane.
- **XY/YZ/ZX:** Use one of the standard planes through a specified point.
- **3 points:** Define the plane by specifying three points on the plane.

Often, two or more of these methods may work. The three-points method is the default and it can usually be used when sufficient geometry is available.

EXERCISE: USING THE MIRROR3D COMMAND

1. Load the 23DWG02.DWG file found on the accompanying CD. This drawing is shown in Figure 23–15.
2. From the Modify menu, choose 3D Operations>Mirror 3D.

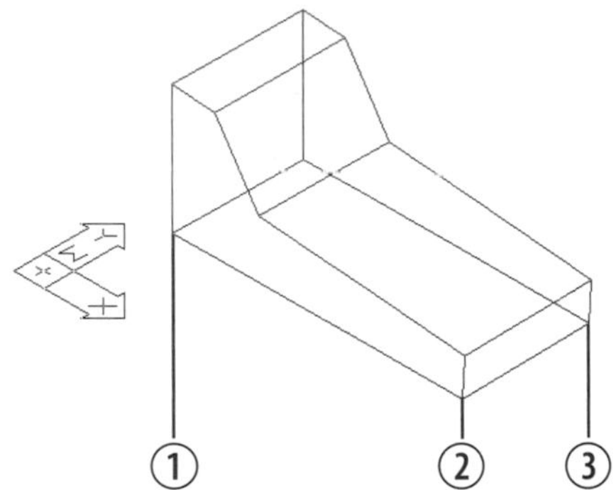

Figure 23–15 *A 3D model for use with the MIRROR3D command.*

3. Select anywhere on the model and then press Enter. You are prompted with the Mirror 3D options.

   Use the default 3 Points option to define the mirroring plane. Because 3 Point is the default method, you can immediately identify the first point.

4. Use an endpoint Osnap to pick at ①, ②, and ③ in Figure 23–15.
5. When you are prompted to delete the old objects, press Enter to accept the default NO. Figure 23–16 shows the mirrored result.
6. Type **U** and press Enter to undo the last command.

   Next you will use the XY plane option to identify the mirroring plane.

7. Repeat steps 2 and 3, but specify the XY option instead of the default.
8. When you are prompted for the point on the XY plane, use an endpoint Osnap and pick at ① in Figure 23–15. Press Enter to accept the default NO when you are asked if you wish to delete source objects. Again, Figure 23–16 shows the same mirrored result.
9. If you are not continuing with the next exercise, close this drawing without saving changes.

As you can see in this exercise, the MIRROR3D command enables you to mirror one or more objects around any planar axis in 3D space, as long as you have three points on that axis to select.

ROTATE3D

The ROTATE3D command is very much like the MIRROR3D command. It, too, differs from its 2D counterpart by enabling you to rotate an object around any axis, not just the current Z-axis.

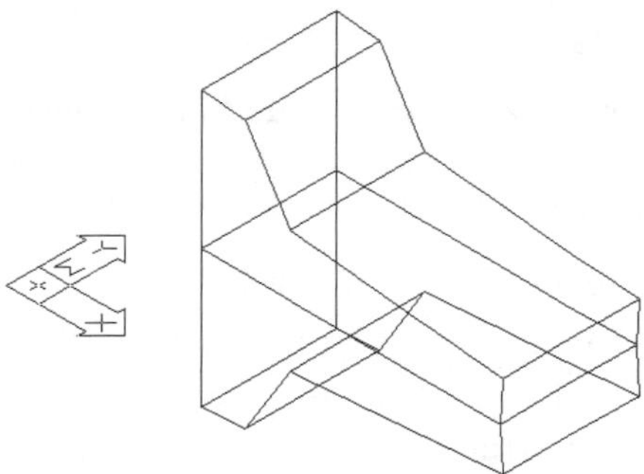

Figure 23–16 *The model after you use the MIRROR3D command.*

After you select the objects you want to mirror, the following prompt appears:

Specify first point on axis or define axis by

[Object/Last/View/Xaxis/Yaxis/Zaxis/2points]:

You have several methods of specifying the 3D mirroring plane:

- **Object:** Align the axis of rotation with an existing object. Qualifying objects include a line, circle, arc, and 2D polyline.
- **Last:** Use the last specified axis of rotation.
- **View:** Align the axis of rotation with the current viewing direction.
- **X-/Y-/Z-axis:** Align the axis of rotation with one of the standard axes that pass through the selected point.
- **2 points:** Define the axis of rotation by specifying two points.

You access the ROTATE3D command either from the Modify menu by choosing 3D Operations>3D Rotate, or by typing **ROTATE3D** at the command prompt. In the following exercise you will use the X-/Y-/Z-axis method of specifying an axis of rotation for the ROTATE3D command.

EXERCISE: USING THE ROTATE3D COMMAND

1. Either continue from the last exercise or open the drawing 23DWG03.DWG. This drawing, shown in Figure 23–16, is from the preceding exercise.
2. From the Modify menu, choose 3D Operation>Rotate 3D.
3. When you are prompted, select both halves of the mirrored model and then press Enter.
4. Type in **Y** and press Enter to select the X-axis, around which the rotation will be performed.

5. When you are prompted for a point on the Y-axis, use an endpoint Osnap to pick ① in Figure 23–15.
6. When you are prompted for the rotation angle, type in **90** and press Enter. The model is rotated 90 degrees about an axis that is parallel to the current Y-axis and that passes through the specified point, as shown in Figure 23–17.
7. Close this drawing without saving changes.

In this exercise you could have performed the same rotation of the model by first changing the UCS and then using the standard 2D ROTATE command. This would have involved aligning the new UCS so that its Z-axis aligned with the desired axis. Although this would not have been a difficult UCS realignment, leaving the UCS intact and using an option of the ROTATE3D command is quicker and simpler. There are also circumstances when using the ROTATE3D command options provides the only convenient means of identifying the rotation axis.

ALIGN

The ALIGN command is used to cause one object to line up, or "align," with another object based on common points shared by the two objects. The two objects do not necessarily need to line up exactly; the ALIGN command will either make the best alignment possible with the point you supply or, alternately, it will scale the object being aligned to the object to which it is being aligned. The ALIGN command is a very powerful 3D tool that is probably under-used.

You access the ALIGN command either from the Modify menu by choosing 3D Operations>Align, or by typing **ALIGN** at the command prompt. The command prompts you for the object(s) you want to align. You then pick source points on the source object and destination points on the destination object. The use of object

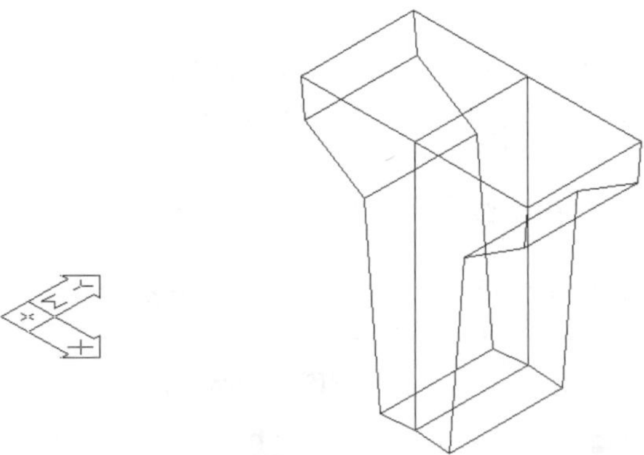

Figure 23–17 *The model after you use the ROTATE3D command.*

snaps is essential. These points are then lined up, and the object is moved and rotated as necessary to make it line up with the destination object. In the following exercise you will use the ALIGN command to align two 3D objects.

EXERCISE: USING THE ALIGN COMMAND

1. Open the 23DWG04.DWG file found on the accompanying CD. The model consists of a 3D wedge and box (see Figure 23–18).

   Because it is difficult, if not impossible, to use the ALIGN command effectively without object snaps, you will first set a running endpoint Osnap.
2. From the Tools menu, choose Drafting Settings. In the Drafting Settings dialog box's Object Snap folder, make sure that a check appears beside Object Snap On and the Endpoint mode. Choose OK to close the dialog box.
3. From the Modify menu, choose 3D Operation>Align.
4. When you are prompted to select an object, select the box.
5. You are then prompted for the first source point and first destination point. Create these points as shown in Figure 23–18.
6. Create the second pair of source and destination points, as shown in Figure 23–19.Create the third pair of source and destination points, as shown in Figure 23–20.
7. Once you select the third destination point, the box is aligned, as shown in Figure 23–21. You can exit this drawing without saving changes.

The first destination point determines the final position of the source object. The other source/destination points determine the alignment. In the preceding exercise, if you were to switch the first and third pairs of points, the box would be positioned at the bottom of the wedge rather than at the top.

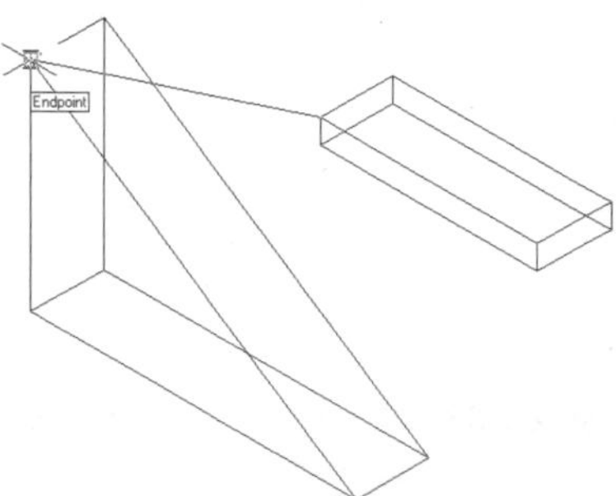

Figure 23–18 *The first pair of source and destination points.*

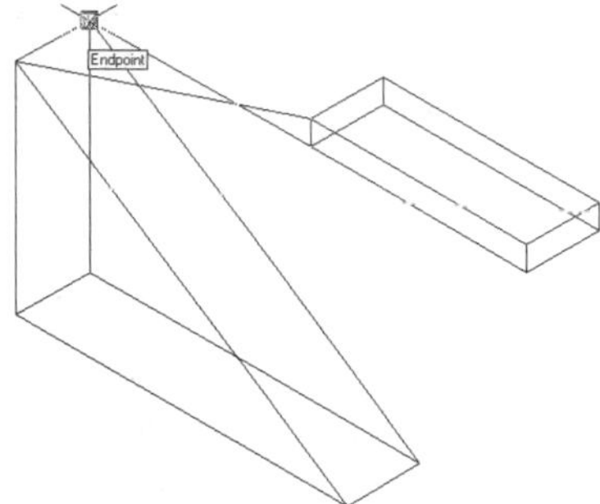

Figure 23–19 *The second pair of source and destination points.*

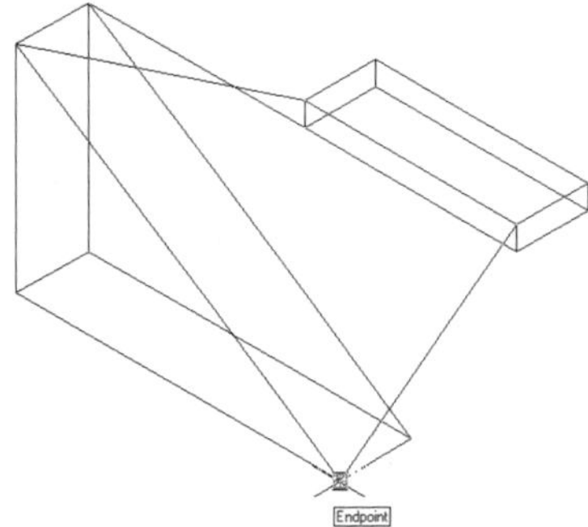

Figure 23–20 *The third pair of source and destination points.*

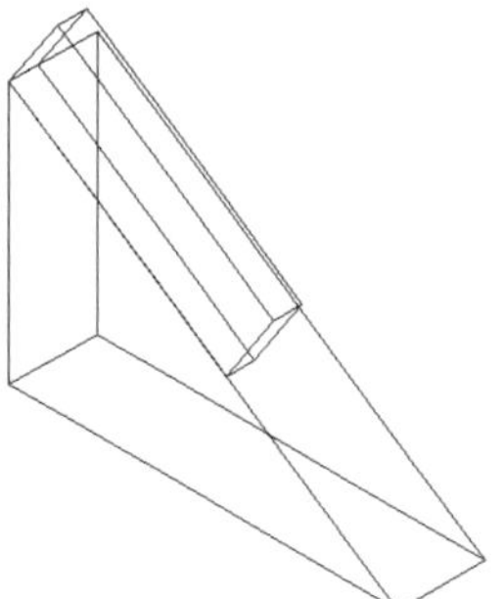

Figure 23–21 *The newly aligned box.*

If you supply only two sets of points to the ALIGN command, you can cause the source object to be scaled to "fit" the destination object, with the two sets of points acting as a "reference" for the scaling operation. This may not always lead to the alignment you want, but subsequent use of the ALIGN, MOVE, or ROTATE3D command can achieve the proper alignment.

3DARRAY

The last 3D editing command to look at is the 3DARRAY command. Like the other 3D editing commands, you can access this command either from the Modify pull-down menu or by typing **3DARRAY** at the command prompt. This command is a modified version of the standard array, but it can create objects quickly in the third dimension. 3DARRAY accomplishes this in the rectangular array by adding levels to the rows and columns.

EXERCISE: USING THE 3DARRAY COMMAND

1. Load the 23DWG05.DWG file found on the accompanying CD. The file contains a three-element cell, as shown in Figure 23–22.
2. From the Modify menu, choose 3D Operation>3D Array.
3. Select the three cylinders and then press Enter.
4. Select R (for rectangular) as the type of array you want to create.
5. Set the number of rows, columns, and levels to 4.
6. Specify the distance between rows, levels, and columns as 1. When you complete this step, AutoCAD begins arraying the objects. Depending on the speed of your machine, this may take several seconds. You many need to perform a Zoom Extents to view the entire array. Figure 23–23 shows the array after you perform a HIDE command.
7. Close this drawing without saving changes.

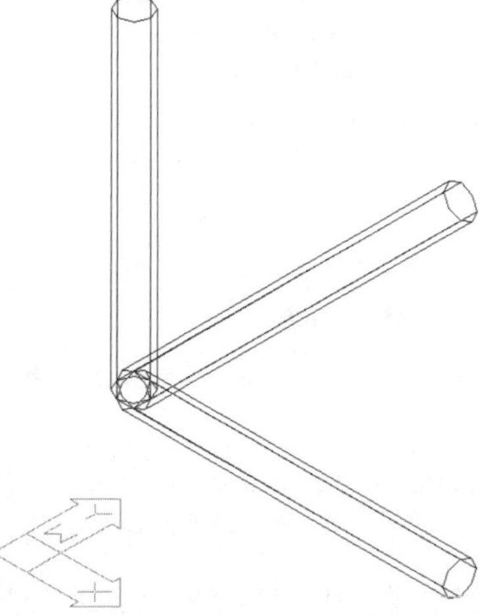

Figure 23–22 *The basic cell used for the 3DARRAY command.*

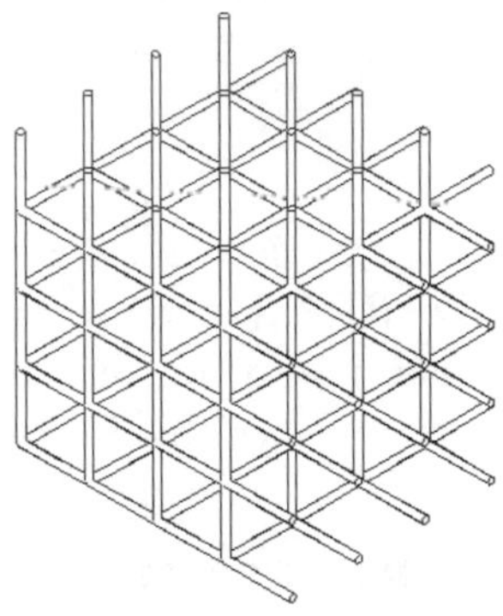

Figure 23–23 *The completed sixty-four-unit 3D array.*

The 3DARRAY command also supports a Polar option. As with the other specialized 3D editing commands, the ability to specify an axis or plane in space provides the command's unique 3D functionality.

The special 3D editing commands—MIRROR3D, ROTATE3D, and 3DARRAY—combined with the ALIGN command allow you to edit objects easily in 3D space. Two other editing commands—TRIM and EXTEND—also have functionality that makes them more useful in 3D work.

USING EXTEND AND TRIM IN 3D

AutoCAD's TRIM and EXTEND commands both have a Project option that allows you some degree of flexibility when you use these commands in 3D space. As shown in Figure 23–24, for example, the EXTEND command can extend a line lying on one 3D plane to another line lying on a different plane. In this case, the boundary line can be "projected" onto the plane of the line being extended. Alternately, you can cause the extension to be to a plane perpendicular to the current viewing plane. Using the Project modes, you can have TRIM and EXTEND behave in the following ways:

- Using no projection causes TRIM and EXTEND to trim or extend to objects that actually intersect in 3D space.
- Using the USC projection mode causes objects to be projected onto the current XY plane.
- Using the View projection mode causes the objects to be projected onto a plane perpendicular to the current viewing plane.

The following exercise demonstrate the USC and View modes of the Project option of the EXTEND command.

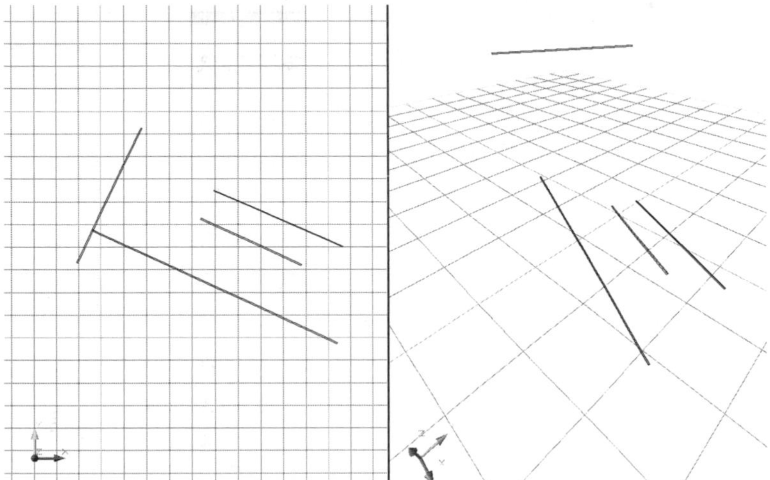

Figure 23–24 *You can extend a line to a line on a different 3D plane.*

EXERCISE: USING THE PROJECT MODES OF THE EXTEND COMMAND

1. Open the 23DWG06.DWG file found on the accompanying CD. The file contains three polylines with thickness. The longest line is drawn at a higher elevation, as shown in Figure 23–25.
2. Click in the large, right viewport, if necessary, to make it active.
3. From the Modify menu, choose Extend. Note that the current setting for the projection mode is reported on the command line.
4. When you are prompted to select boundary edges, select the longest line near the top of the viewport and then press Enter.
5. When you are prompted to select objects to extend, type **P** and press Enter. At the next prompt, make sure that the projection mode is set to UCS, and then press Enter.
6. When you are prompted again for objects to extend, select the shorter line at ① in Figure 23–25. Note that the line is extended to the USC projection of the boundary line, down onto the current XY plane.
7. When you are prompted again for objects to extend, type **P**. Then change the projection mode to View by typing **V** and pressing Enter.
8. When you are prompted for objects to extend, choose the shorter line at ② and then press Enter to end the command. Note that the line is projected so it appears to end at the current view of the boundary line. Your drawing should resemble Figure 23–26.
9. Close the drawing without saving changes.

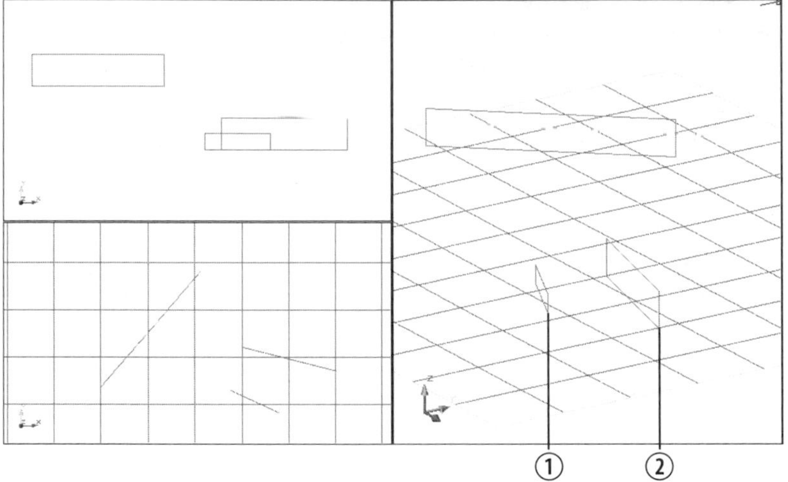

Figure 23–25 *The views of the boundary line and the extension lines.*

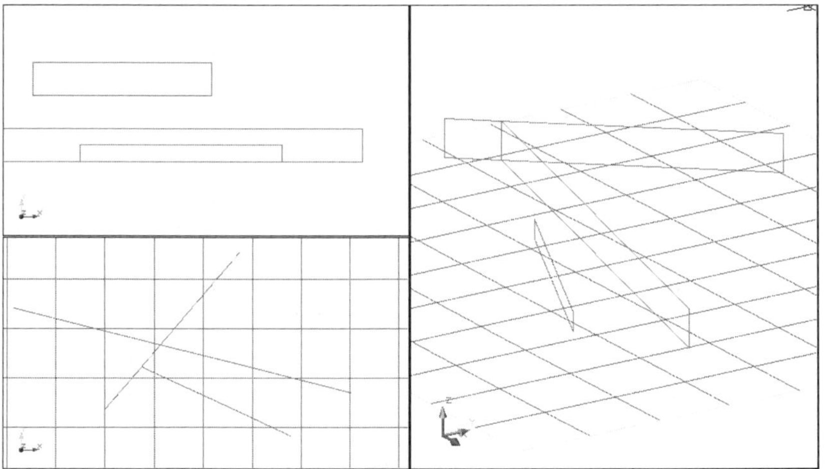

Figure 23–26 *The views of the boundary line and the extension lines.*

When you work with the TRIM and EXTEND commands in 3D space, you may want to have the Project option of these commands set to UCS to ensure accuracy no matter what viewpoint you are using to view the 3D geometry.

Before we look at actual 3D objects, such as objects constructed with surfaces and objects composed of solids, in the following two chapters, we will examine a special 2D object that behaves like a 3D object.

CREATING REGIONS

Regions are special AutoCAD objects. In many ways, regions act like 2D planar objects; in other ways, regions exhibit many of the properties of 3D objects. Like 3D surfaces, for example, regions can hide objects "behind" them and have materials applied to them for rendering purposes. Like 2D objects, regions have no third dimension, Z-axis information. In many ways, regions can be considered infinitely thin solids.

Technically regions are enclosed 2D areas. You create regions from closed shapes. Closed shapes consist of a curve or a sequence of curves that define an area on a single plane with a non-self-intersecting boundary. Closed curves can be combinations of lines, polylines, circles, ellipses, elliptical arcs, splines, 3D faces, and solids. The objects that compose a closed curve must either be closed themselves or form a closed area by sharing endpoints with other objects.

As you can see, a large number of AutoCAD objects can join together to form a closed curve from which a region can be created. The only restrictions are that the object must be coplanar, or rest on the same plane, and sequential elements must share endpoints. Assuming that the arcs and lines shown in Figure 23–27 are coplanar, the set of curves on the left could not be converted into a region, while the curves on the right could.

To convert a set of qualifying curves into a region, you use the REGION command. In the following exercise you will create a region from a set of straight polyline "curves" using the RECTANGLE command.

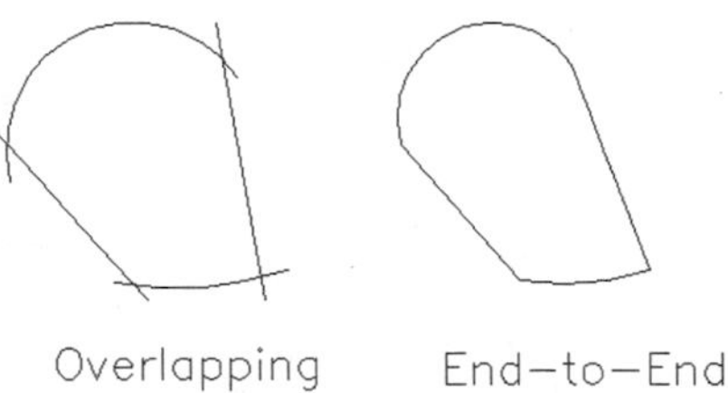

Figure 23–27 *Sets of overlapping and end-to-end curves. Only end-to-end curves can be used to create a region.*

EXERCISE: CREATING REGIONS WITH THE REGION COMMAND

1. Open the 23DWG07.DWG file found on the accompanying CD. This drawing is composed of only a circle and a polyline drawn at an elevation of -1.0 units.
2. From the Draw menu, choose Rectangle. When you are prompted for the first corner of the rectangle, enter **4,2**. At the "Specify other corner:" prompt, enter **@3.5,5.5**. Your drawing should now resemble Figure 23–28.
3. From the Draw menu, choose Region. At the "Select objects:" prompt, select the rectangle you just drew and then press Enter. AutoCAD reports "1 Region created" at the command line. The polyline used to create the rectangle has now been converted to a region.
4. Type **HI** and press Enter to invoke the HIDE command. Note that the newly created region acts as a planar surface, hiding the circle and line objects behind it, as shown in Figure 23–29.

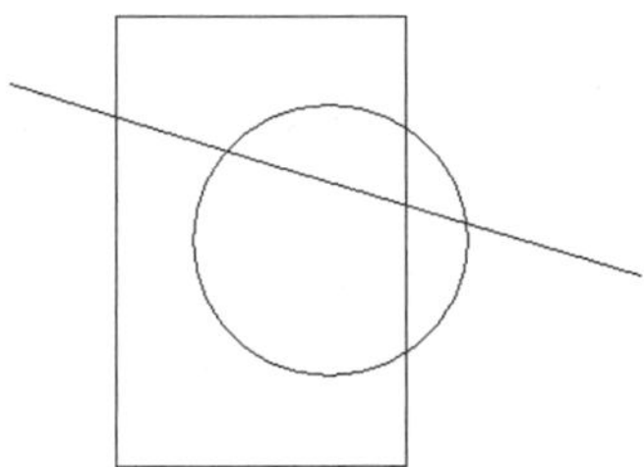

Figure 23–28 *A closed polyline curve drawn over a circle and a line.*

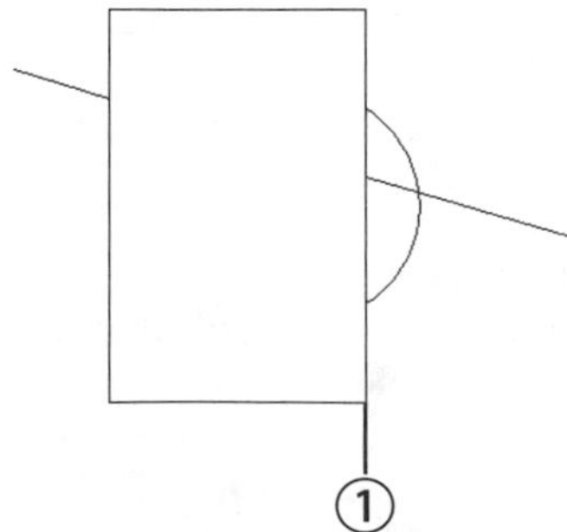

Figure 23–29 *A region created with the REGION command. Regions hide objects behind them.*

In the following step you will rotate the region in 3D space.

5. From the Modify menu, choose 3D Operation>Rotate 3D. When you are prompted to select objects, select the region and then press Enter.
6. At the next prompt, type **Y** to indicate you want to rotate the region about its Y-axis.
7. When you are prompted to specify a point on the Y-axis, use an endpoint Osnap and pick at ① in Figure 23–29. To specify an angle of rotation, type **90** and press Enter.
8. From the View menu, choose 3D Views>SE Isometric. Your drawing should resemble Figure 23–30.
9. Again type **HI** to invoke the HIDE command. Note that the region is rotated in 3D space and it hides objects behind it, as shown in Figure 23–31.
10. You can close this drawing without saving changes.

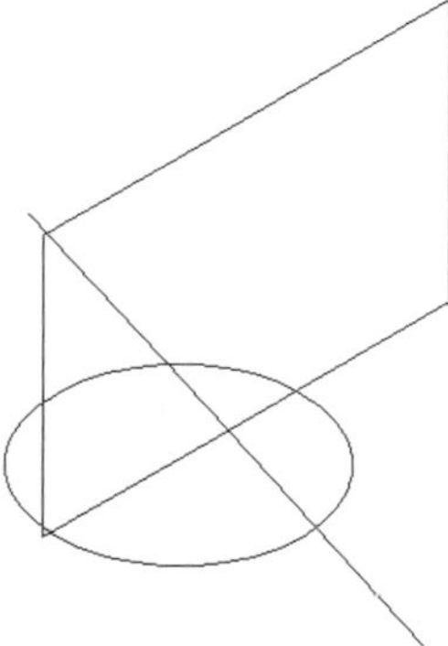

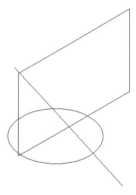

Figure 23–30 *Regions can be moved and positioned in 3D space.*

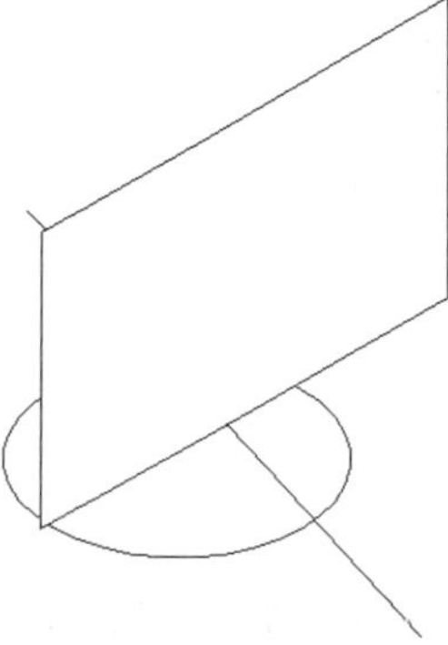

Figure 23–31 *Regions behave like solids and hide objects behind them.*

As you can see in this exercise, regions have many of the properties of 3D surfaces; they hide objects behind them, and they can be easily moved in 3D space. One of the most interesting and useful properties of regions is their ability to undergo Boolean operations—a trait they share with AutoCAD solids. Boolean operations will be discussed and demonstrated later in Chapter 25, Modeling in 3D.

SUMMARY

In this chapter you learned some of the basic principles of drawing in three dimensions. Starting with simple, basic lines and polylines, you learned the importance of using object snaps and point filters to ensure accuracy in 3D space. You learned that many of the editing operations you use in 2D, such as move and rotate, have equivalent 3D commands. Finally, you learned about a special 3D object called a region.

Creating 3D Surfaces

In the previous two chapters you learned how to create simple 3D objects in AutoCAD. By creating and manipulating these 3D primitives, you can create a wide variety of more complex objects such as chairs, tables, and so on. There are, however, many instances when you need more flexibility in modeling complex objects such as irregularly curved surfaces.

Fortunately, AutoCAD provides a set of surfacing commands with which you can create the outlines of these types of objects in profile, using lines or polylines, and then create surfaces between these lines. This chapter takes a look at the various methods and techniques AutoCAD provides for creating both simple and complex 3D surfaces. This chapter covers the following topics:

- Basic surfacing commands
- Advanced surfacing commands
- 3D meshes
- Editing mesh surfaces

BASIC SURFACING COMMANDS

In the previous chapter the techniques used to create 3D lines and polylines were discussed. Objects modeled with such lines are called wireframe models. To create a true 3D object, however, surfaces must be applied across or around the edges of the wireframe. Such surfaces can then be used to hide lines and other surfaces that lie "behind" other surfaces in any particular view. Surfaces can also have hatch patterns and even materials applied to them to yield more realistic-looking objects.

In some cases you need to create the wireframe "skeleton" of an object before you create the surfaces, using the wireframe as a starting point from which to apply the surfaces. This method employs two relatively simple 3D objects: 3D faces and polyfaces.

Note: All of the surfacing commands discussed in this chapter are available from the pull-down menus. Most are also available on the Surfaces toolbar. To access the toolbar, right-click on any toolbar in AutoCAD; the Toolbars menu appears. Scroll down the list and select Surfaces. In the exercises in this chapter, reference to the menu commands is usually made. You can also use the corresponding commands from the Surface toolbar, and you can access the basic commands by typing them at the command prompt.

USING 3D FACES AND POLYFACE MESHES

A 3D face is a surface defined by either three or four sides, forming either a triangular or rectangular surface, respectively. In most instances, a triangular, three-sided face is preferable because a 3D face is a planar element. A rectangular 3D face can be either planar or non-planar. Figure 24–1 show a triangular face and a rectangular 3D face, with the upper-right corner of the rectangular face lying at a different elevation than the other three corners. Four-sided faces are more difficult to visualize.

You construct 3D faces with the 3DFACE command, which you access by choosing Draw>Surfaces>3D Face. Specifying the corner points creates the face. The 3DFACE command enables you to continue creating 3D faces by selecting more points. When you have specified the last corner point, right-click to end the command.

Note: 3D faces can be difficult to visualize. To be able to see if the 3D faces you created are correct, use the HIDE and SHADEMODE commands to check your drawing. Also make sure that you check the drawing from different viewpoints to ensure that the faces are drawn correctly.

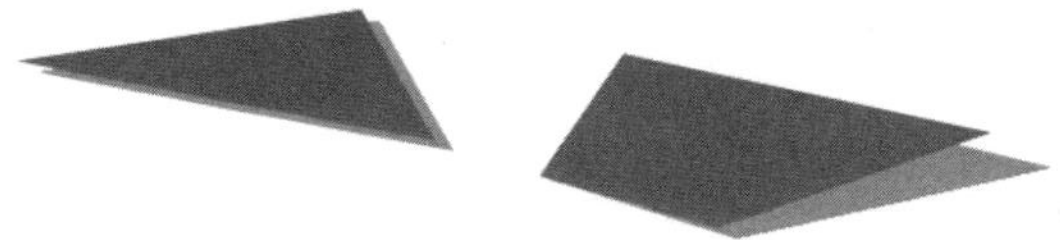

Figure 24–1 *A three-sided (left) and a four-sided (right) 3D face. Three-sided faces must be planar, while four-sided faced may have vertices at different elevations.*

When you create your 3D face, always work your way around the perimeter of the face. This is especially important if you decide to create a four-sided face. Choosing points in an X-wise fashion will result in a bow-tie effect, as shown in Figure 24–2.

Although you can construct a surfaced box with the AI_BOX command, there are instances when you may need to place 3D faces on a wireframe, as shown in the following exercise.

EXERCISE: USING THE 3DFACE COMMAND TO SURFACE A BOX

1. Load the 24DWG01.DWG drawing file found on the accompanying CD. The file contains a wireframe box, as shown in Figure 24–3. This box consists of twelve individual polylines arranged in such a way as to represent a 3D box.
2. At the command prompt, type **HIDE** and then press Enter. Because there are no surfaces in this collection of polylines, the HIDE command has no visual effect.

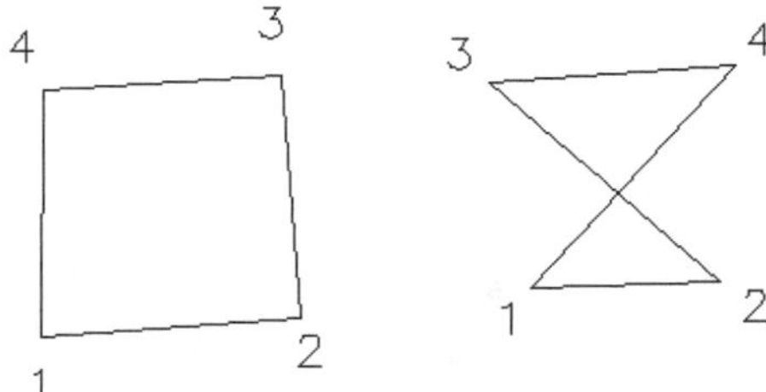

Figure 24–2 *Always specify the corners of a 3D face in a clockwise or counterclockwise sequence.*

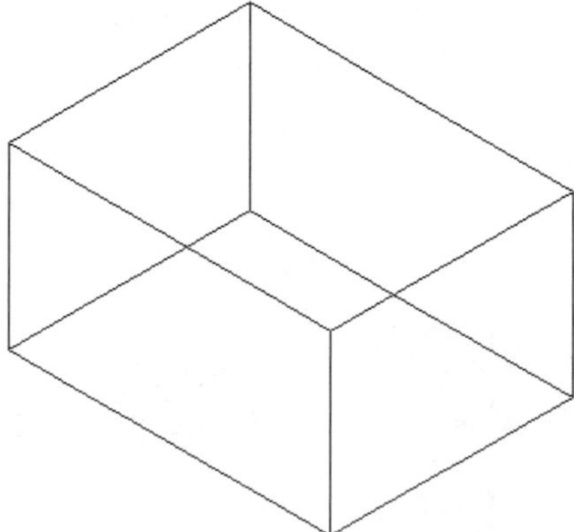

Figure 24–3 *A wireframe box that you will surface with the 3DFACE command.*

3. From the Tools menu, choose Drafting Settings. In the Drafting Settings dialog box's Object Snap folder, make sure that the Object Snap On and Endpoint Object Snap options are checked. Choose OK to close the dialog box.
4. From the Draw menu, choose Surfaces>3D Face.
5. Select the seven corners of the box shown in ① through ⑦ in Figure 24–4. Make sure you select the corners in numerical sequence.
6. When you have selected the corner at ⑦, press Enter twice to end the 3DFACE command.
7. Type **HIDE** and press Enter. Figure 24–5 shows the resulting box.
8. Leave this drawing open for use in the next exercise.

In this exercise you constructed only two of the six 3D faces required to cover the frame of the box completely. The other four sides could be applied in a similar manner. The second face actually consists of two three-sided faces, with the diagonal line appearing between corners 5 and 7 representing the shared edge. Because of the man-

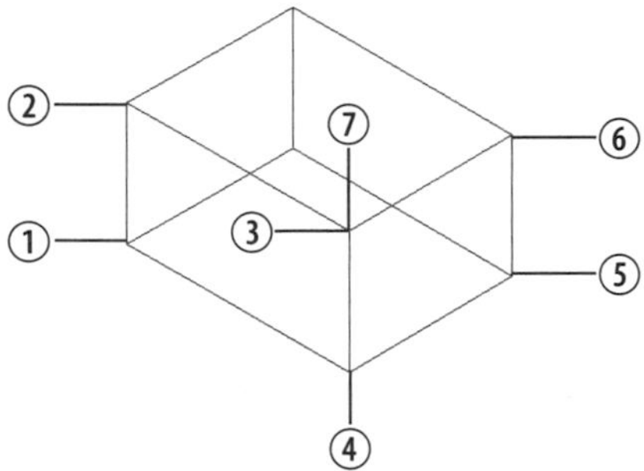

Figure 24–4 *Choose the corners in sequence to apply 3D faces.*

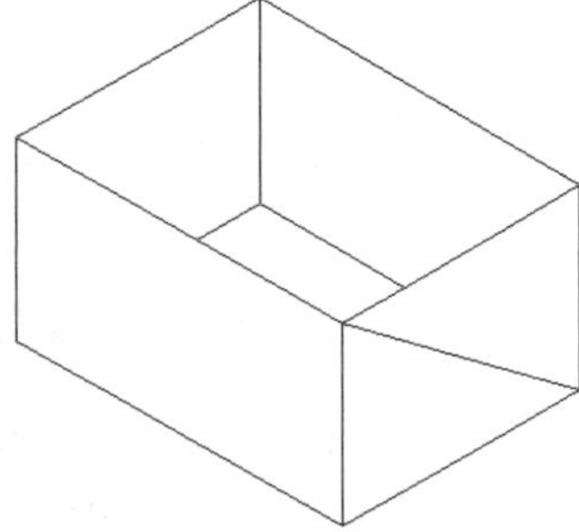

Figure 24–5 *The box with faces applied to two sides.*

ner in which the 3DFACE command works, four-sided faces are usually drawn this way because AutoCAD attempts to construct three-sided faces wherever possible.

AutoCAD provides two means of hiding face edges. You can hide an edge during the construction of a face by entering an **i** just prior to specifying the point that will create the edge. You can also make a 3D face edge invisible after it is formed by using the EDGE command.

The following exercise shows you how to use the EDGE command to hide a face edge once it has been drawn.

EXERCISE: USING THE EDGE COMMAND

1. Continue in 24DWG01.DWG from the preceding exercise.
2. From the Draw menu, choose Surfaces>Edge.
3. When you are prompted, select the diagonal edge appearing between corners ⑤ and ⑦ in Figure 24–4. Notice that AutoCAD switches you to midpoint object snap mode automatically.
4. Press Enter. The edge is hidden.

   After an edge is hidden, you can no longer select it. To unhide an edge that is already hidden, you can use the Display option of the EDGE command.
5. Press Enter to restart the EDGE command.
6. Select an edge of the face containing the hidden edge. Select between ④ and ⑤, for example.
7. Type **D** for Display and then press Enter. You are then prompted for the Display options.
8. When you are prompted for the Display options, type **A** for All and then press Enter. All the edges of the face become visible and highlighted.
9. Select the edge you want to make visible and press Enter. The visibility of the edge is restored.
10. Exit this drawing without saving changes.

Note: The AutoCAD system variable SPLFRAME can be used to control the display of the invisible edges of 3D faces. If you are having trouble using the EDGE command to change the visibility of an edge, change SPLFRAME to 1 to display all edges. Then you can more easily use the EDGE command to make edges visible or invisible.

3D faces are limited as surfacing tools, principally because they are limited to three or four sides, which do not serve well in surfacing curved objects. Applying 3D faces to define a large surface would be tedious work and likely to yield less than satisfactory results. To get around this limitation, you can make use of the AutoCAD

PFACE (polyface) command, which enables you to create larger faces composed of many more individual faces. The PFACE command works by defining multiple faces that define how the surface is created between points. Such a large non-planar surface is termed a polyface mesh. Here again, to construct a large polyface mesh manually would be prohibitively tedious and time-consuming, and the PFACE command is therefore usually used by another application. The application "communicates" with the PFACE command through an application programming interface, such as AutoLISP, to construct the mesh automatically.

ADVANCED SURFACING COMMANDS

AutoCAD provides five true surfacing commands. These commands make use of existing geometry—usually polylines—to create surfaces. These surfacing tools include the following commands:

- EDGESURF
- RULESURF
- TABSURF
- REVSURF
- 3DMESH

EDGESURF

The EDGESURF command is used to create a 3D surface between four connected lines or polylines. Figure 24–6 shows an example of a surface created using EDGESURF.

You can access the EDGESURF command either from the Draw menu, by choosing Surfaces>Edge Surface, or by typing **EDGESURF** at the command prompt.

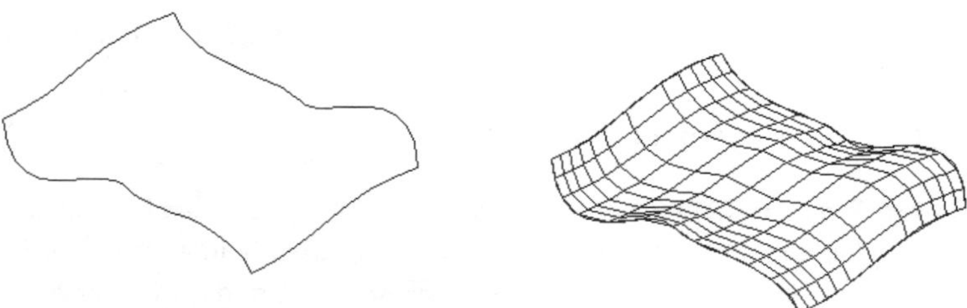

Figure 24–6 *Four splined curves before (left) and after (right) you apply the EDGESURF command.*

The resolution of the 3D surface generated by EDGESURF, as well as by the other surfacing commands, is controlled by two system variables: SURFTAB1 and SURFTAB2. These two variables represent the number of mesh faces in the M and N directions, respectively. The letters M and N are used to reduce the confusion with the standard X- and Y-axes. The M direction, however, is generally considered to coincide with the X-axis direction, and the N direction with the Y-axis direction. The default value of both SURFTAB1 and SURFTAB2 is 6. By increasing these default values, you get more accurate surfaces, because more individual faces are generated between the bounding edges. Figure 24–7 shows the difference between a surface with both variables set to 6, and the same surface with both variables set to 24.

Tip: The SURFTAB variables do not always have to be set to the same values. For example, suppose you have a set of four lines that you want to surface, but in one direction the lines are more complex than in the other direction. You might want to increase the SURFTAB value in the more complex direction only. This would result in a more accurate mesh without creating an overly complex mesh.

The following exercise shows you how to use EDGESURF to create the canopy of an airplane.

EXERCISE: USING EDGESURF TO CREATE THE CANOPY

1. Load the 24DWG02.DWG file found on the accompanying CD. The file contains the outline of an airplane canopy, as shown in Figure 24–8.
2. At the command prompt, type **SURFTAB1**. Set the value to **24**.
3. Repeat step 2, setting SURFTAB2 to a value to **24**.
4. From the Draw menu, choose Surfaces>Edge Surface.
5. Select the four lines in the drawing. The surface is then created, as shown in Figure 24–9.
6. Close this drawing without saving changes.

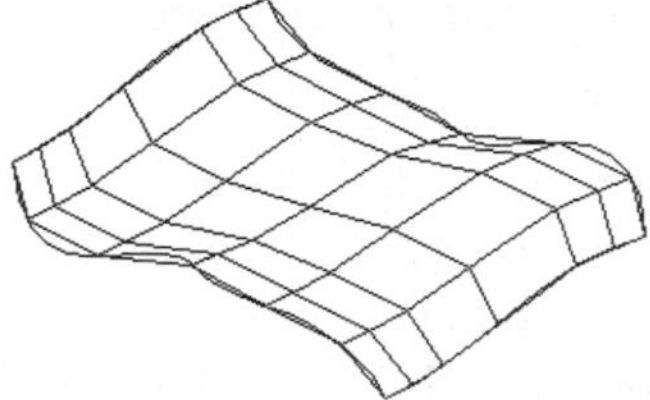

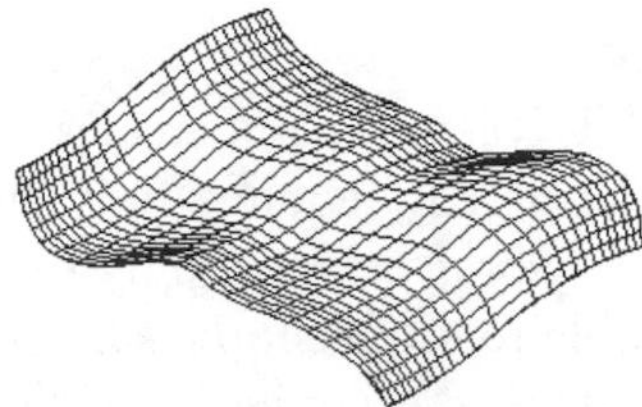

Figure 24–7 *Two surfaces created with the EDGESURF command. The surface on the left was created with SURFTAB settings of 6; the surface on the right has SURFTAB values of 24.*

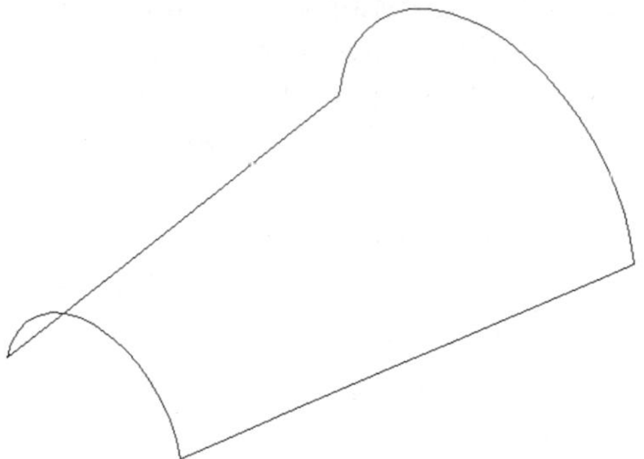

Figure 24–8 *The outline drawing of an airplane canopy that is ready for the EDGESURF command.*

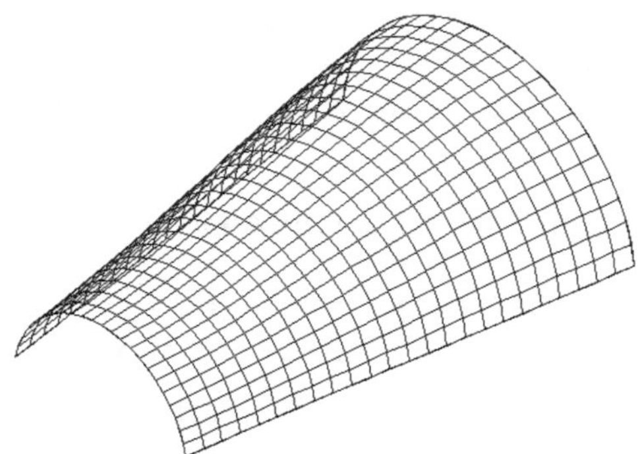

Figure 24–9 *The canopy with the surface applied.*

In this exercise you set SURFTAB1 and SURFTAB2 to relatively high values. This resulted in a smoother surface at the expense of longer regeneration times and a larger drawing. If you change SURFTAB1 and SURFTAB2 to either higher or lower values, you must erase or undo the previous mesh and perform the surfacing operation again. Also note that the surface is generated on the current layer.

RULESURF

The RULESURF command creates a ruled surface. Unlike the EDGESURF command, RULESURF requires only two defining edges instead of four. Because RULESURF works between only two lines, it only makes use of the SURFTAB1 system variable to establish the surface mesh density. Figure 24–10 shows examples of typical surfaces created with the RULESURF command. The defining edges can be points, lines, splines, circles, arcs, or polylines. If one of the edge objects is closed—a circle, for example—then the other object must be closed.

If you generate a surface between two open boundary edges, you have to be careful about the points you use to select the boundaries. RULESURF starts generating the mesh at each boundary by dividing the boundary curve into a number of segments equal to the current setting of SURFTAB1, starting from the endpoint nearest the pick point. If you use pick points on opposite sides of the two boundary curves, the resulting curve will be self-intersecting, as shown in Figure 24–11.

The following exercise shows you how to use RULESURF to surface the contours of a site.

EXERCISE: USING RULESURF

1. Load the 24DWG03.DWG file found on the accompanying CD. The file contains four contours, as shown in Figure 24–12.
2. From the Draw menu, select Surfaces>Ruled Surface.
3. Select the top two lines in order—① then ② in Figure 24–12. The ruled surface is generated, as shown in Figure 24–13.

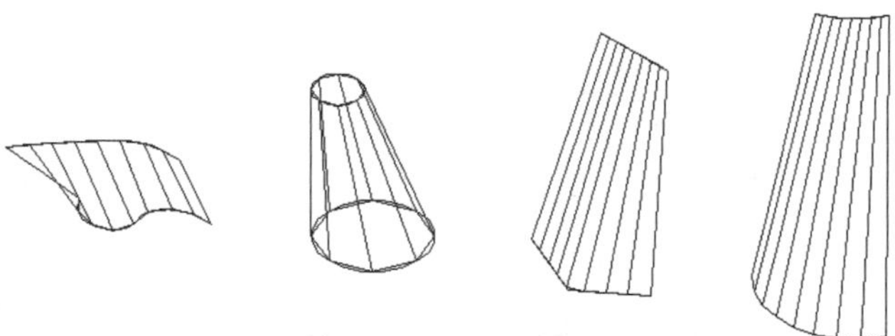

Figure 24–10 *Typical surface meshes created with the RULESURF command.*

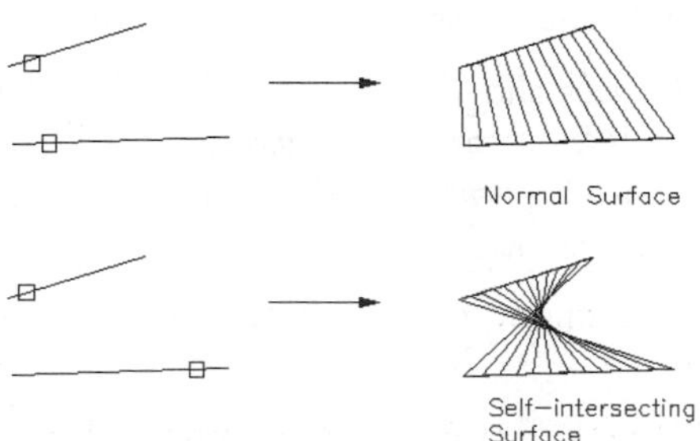

Figure 24–11 *With open boundary edges, selecting objects at opposite ends creates a self-intersecting polygon mesh.*

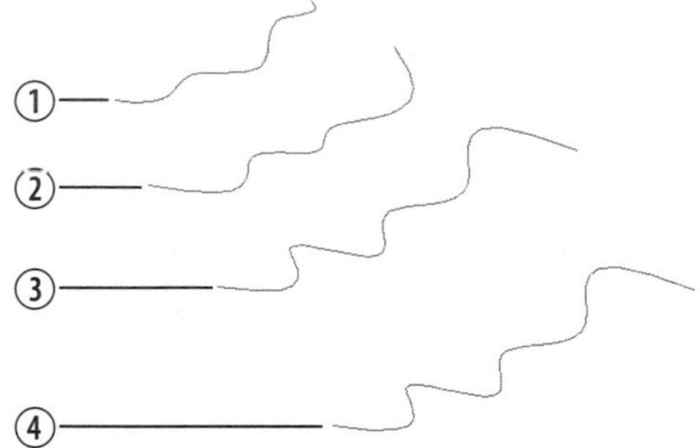

Figure 24–12 *Four contour lines that you will surface with RULESURF.*

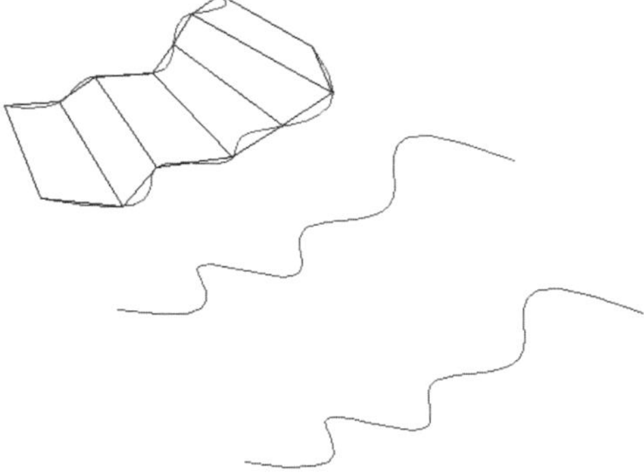

Figure 24–13 *The first surface generated between the first two contour lines.*

4. The smoothness of this curve is inadequate. Type **SURFTAB1** and set the value to **24**.
5. Use the ERASE command to erase the surface. Repeat step 3 with the new SURFTAB1 setting. The surface should now resemble Figure 24–14.
6. Select the new surface and use the PROPERTIES command to change its layer from 0 to Surface. Note that the Surface layer is currently frozen. Press OK in the frozen layer warning dialog box. The surface disappears when it is transferred to the frozen layer.
7. Repeat step 3, selecting at ② and ③ in Figure 24–12. RULESURF generates a second surface.
8. Use the same procedure as in step 6 to transfer the second surface to the Surface layer.
9. Use RULESURF and the boundaries at ③ and ④ to generate the third surface.

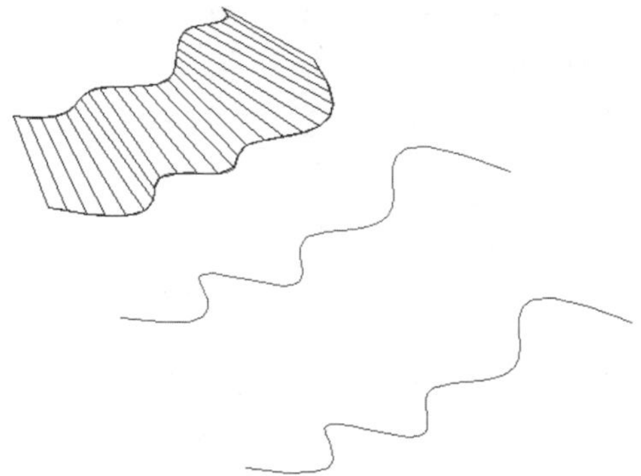

Figure 24–14 *The first surface generated between the first two contour lines at higher density.*

10. Transfer the third surface to the Surface layer and thaw the Surface layer. Your drawing should resemble Figure 24–15.
11. Close this drawing without saving changes.

This exercise demonstrates that the amount of detail necessary in a RULESURF mesh often depends upon the boundary object(s). With curved boundaries, you may need to set SURFTAB1 to a higher value to cause the surface to follow the boundary edge more closely. With sequential boundaries such as in this exercise, it is a good idea to create the surfaces on a separate layer, and then transfer them to a frozen layer. This allows the boundary edges to be more available for selection.

TABSURF

The TABSURF command creates a tabulated surface, or a surface that is extruded along a linear path. You access TABSURF by choosing Draw>Surfaces>Tabulated Surface, or by typing **TABSURF** at the command prompt.

To create a surface with TABSURF, you must have two elements: an outline, or curve, to be extruded; and a direction vector indicating the direction and distance that the curve is to be extended. Figure 24–16 shows several examples of tabulated surfaces.

The path curve can consist of a line, arc, circle, ellipse or a 2D or 3D polyline. TABSURF draws the surface starting at the point on the path curve closest to the selection point.

If the direction vector is a polyline, TABSURF considers only the first and last vertices of the line in determining the length and direction of the vector. In other words, TABSURF will extrude only a straight line. The end of the vector line chosen determines the direction of the extrusion. As with RULESURF, only SURFTAB1 has meaning with TABSURF.

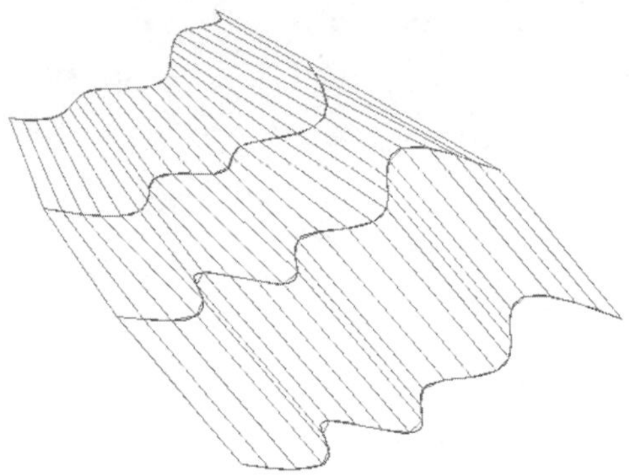

Figure 24–15 *The completed contour surfaces.*

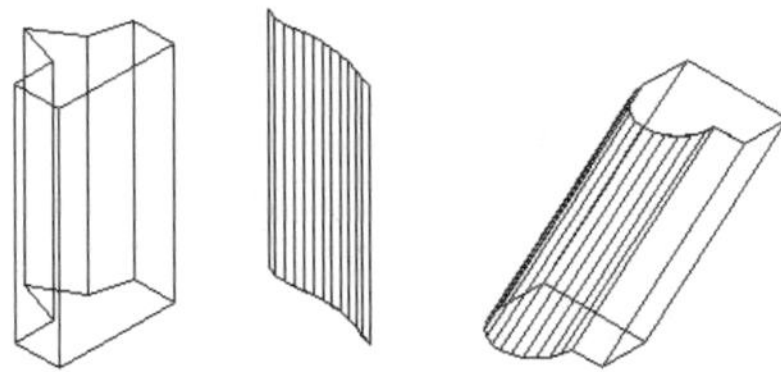

Figure 24–16 *Typical surfaces constructed with RULESURF.*

The following exercise shows you how to use the TABSURF command to create a stair railing.

EXERCISE: USING TABSURF

1. Load the 24DWG04.DWG file found on the accompanying CD. This drawing is shown in Figure 24–17.
2. From the Draw menu, choose Surfaces>Tabulated Surface.
3. Select the outline of the hand rail.
4. Select the vertical line near ① in Figure 24–17. The curve is extruded, as shown in Figure 24–18.
5. Close this drawing without saving changes.

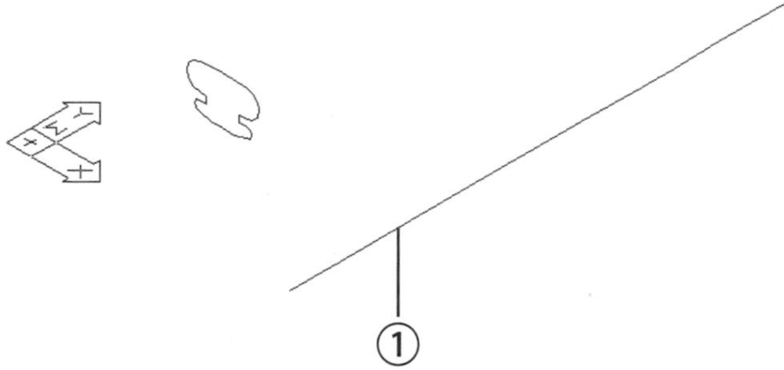

Figure 24–17 *The path curve and direction vector for use in constructing a stair railing.*

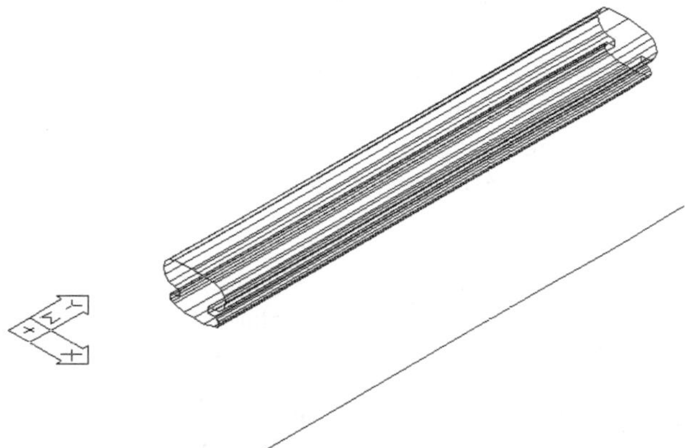

Figure 24–18 *The completed railing.*

REVSURF

REVSURF is perhaps the most useful of the 3D surfacing commands. REVSURF generates a 3D mesh object in the form of a surface of revolution by taking an outline—the path curve—and revolving it about an axis of revolution, as shown in Figure 24–19. You access the REVSURF command either from the Draw menu, by choosing Surfaces>Revolved Surface, or by typing **REVSURF** at the command prompt.

The path curve is the outline that will be revolved. It must be a single object—a line, arc, circle, ellipse, elliptical arc, polyline, polygon, spline, or donut.

The axis of revolution is the axis about which the path curve is revolved. The axis can be a line or an open 2D or 3D polyline. If a polyline is selected, the axis is assumed to be a line running through the first and last vertices.

REVSURF is made even more powerful by the fact that it can revolve the path curve through an included angle that can range from zero to a full 360 degrees. The

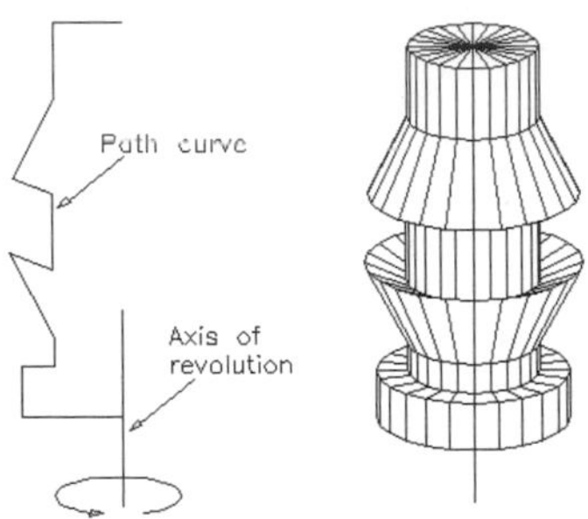

Figure 24–19 *REVSURF revolves an outline around an axis.*

default angle is a full circle, which results in the generation of a closed surface of revolution such as that shown in Figure 24–19. If you specify an angle less than 360 degrees, then the surface is generated in a counterclockwise direction. If you specify a negative angle smaller than -360 degrees, then the surface is generated in a clockwise direction.

You can also specify the start angle, which is the angular offset from the path curve at which the surface of revolution begins. The default value, zero, indicates that the surface of revolution will begin at the location of the path curve.

Determining the Positive Direction of Rotation

To have a start angle other than zero degrees, or to have an included angle other than a full circle, you must be able to determine the positive direction of rotation. REVSURF follows the following conventions: a negative value dictates an angular distance in the clockwise direction; a positive value dictates an angular distance in a counterclockwise direction. You can determine the direction of rotation by applying the so-called "Right-Hand Rule."

According to the Right-Hand Rule, if you point your right thumb in the positive direction of the axis about which you are rotating, and then wrap the fingers of your right hand around the axis, the curl of your fingers indicates the direction of positive rotation. But how do you determine the "positive direction of the axis"?

The positive direction along the axis of rotation runs from the endpoint of the object nearest the pick point used to select the object to the other endpoint. For example, in Figure 24–20, if you select the line at point ①, then the positive direction of the axis runs from ① to ②. If you select the line near ②, then the positive direction of the axis runs from ② to ①.

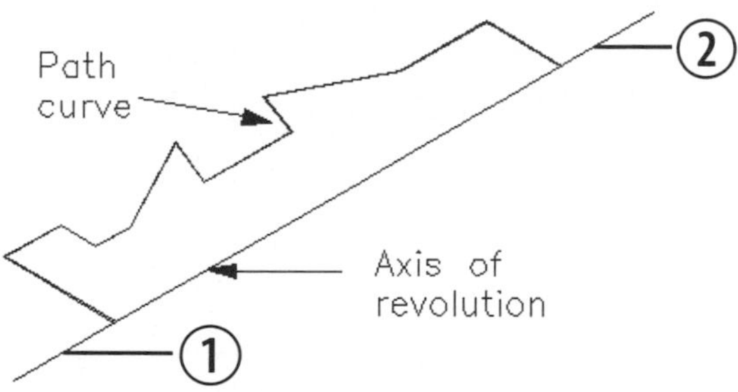

Figure 24–20 *Where you select the axis of revolution determines the positive direction of revolution.*

Unlike the surfaces generated by TABSURF and RULESURF, but like the surfaces generated by EDGESURF, REVSURF generate a 2D mesh. Therefore, both SURFTAB1 and SURFTAB2 influence the density of a SURFTAB-generated mesh. Again, keep both SURFTAB values as low as is consistent with a mesh that meets your needs.

Tip: The original objects used to define the profile (path curve) and the axis are left untouched by REVSURF; they are not incorporated into the resulting mesh. They are often difficult to distinguish, however, because the mesh can obscure them. It is, therefore, a good habit to generate the surface mesh on a separate layer. This will allow you to freeze or turn off the two layers and isolate the path curve and axis object should you want to use them again. This tip applies equally to the TABSURF, EDGESURF, and RULESURF commands.

The following exercise shows you how to use REVSURF to create a rivet.

EXERCISE: USING REVSURF

1. Load the 24DWG05.DWG file found on the accompanying CD. The file contains an outline of the piston and its axis of rotation, as shown in Figure 24–21.
2. At the command prompt, type **Surftab1**. Set the value to **24**. Type **Surftab2**. Set its value to **24**.
3. Change the current layer to Mesh.
4. From the Draw menu, choose Surfaces>Revolved Surface.
5. When you are prompted for the object to revolve, pick the piston outline.
6. When you are prompted to select the object that defines the axis of revolution, pick the vertical line next to the outline.
7. Press Enter to accept the default starting angle of zero degrees.

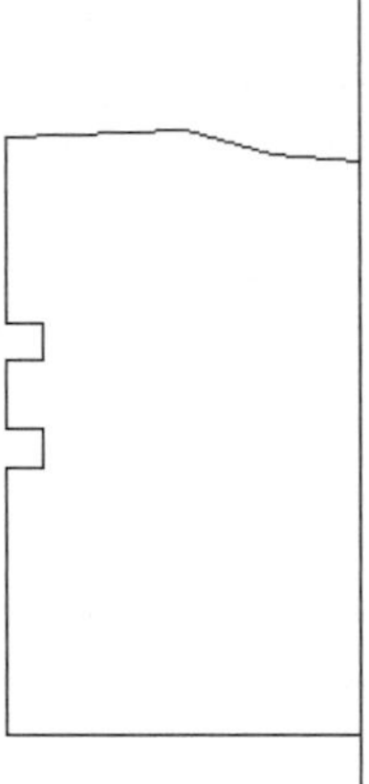

Figure 24–21 *The outline of a piston and the axis of revolution line.*

8. When you are prompted for the included angle, press Enter to accept the default of 360 degrees. REVSURF creates the revolved surface. Your drawing should now resemble Figure 24–22.
9. From the Modify menu, choose 3D Operation>Rotate 3D.
10. Type **All** and press Enter to select all objects for rotation. Press Enter again.
11. Type **X** and press Enter to specify the X-axis as the axis of rotation.
12. When you are prompted for a point on the X-axis, use a midpoint Osnap to pick the midpoint of the line representing the axis of rotation.
13. When you are prompted for the rotation angle, type **90**.
14. The piston, its outline, and the axis line are rotated 90 degrees around the X-axis.
15. From the View menu, choose 3D Views>SE Isometric. AutoCAD switches to an isometric view. Notice that the original piston outline and the axis line are visible.
16. From the View menu, select Shade>Flat Shaded. Your model should now resemble Figure 24–23.
17. Close this drawing without saving changes.

REVSURF offers more options and parameters than the other commands presented in this chapter. Although it is a somewhat complicated surfacing command, it is one of the most flexible and useful 3D tools available.

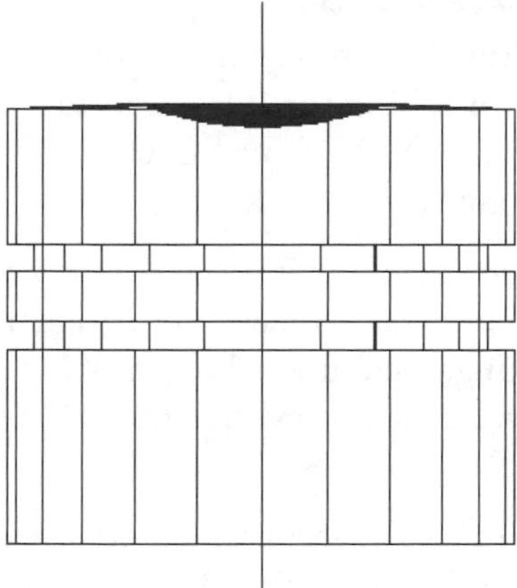

Figure 24–22 *The revolved piston surface.*

Figure 24–23 *The revolved piston surface in a shaded rendering.*

3DMESH

Like the other 3D surfacing commands discussed in this chapter, the 3DMESH command creates a mesh of contiguous 3D faces. An M×N matrix—where M is generally associated with the X-axis, and N with the Y-axis—defines the size of the mesh. Constructing even a relatively simple mesh of this type requires a large amount of input and, if it is done manually, is quite tedious and error-prone. The 3DMESH command, therefore, is intended primarily as the avenue of input for external programs. In this respect, 3DMESH is similar to the PFACE command that was discussed earlier in this chapter. Programs written in programming languages such as AutoLISP are adept in supplying the vertex information required by the 3DMESH command.

EDITING MESH SURFACES

After a surface has been created using one of the commands described in this chapter, there are two methods for editing that surface in addition to the standard editing commands, such as MOVE, ROTATE, or SCALE. Those two methods are Grip Editing and the PEDIT command.

Grip editing works on a surface just as it does on any other AutoCAD object. The one difference is that surfaces generally have more grips that you can easily manipulate. All forms of grip editing, including MOVE, SCALE, ROTATE, MIRROR, and so on, will work with surfaces, however. Figure 24–24 show the grips density of a typical 3D surface.

The PEDIT command provides the second method of editing any 3D mesh object. Much like you can edit a polyline with PEDIT, you can manipulate a mesh surface in certain ways. PEDIT can be used to perform the following functions on a 3D mesh surface:

- **Vertex Editing**—This includes adding, moving, and deleting vertices from the mesh.
- **Smooth Surfaces**—This is generally used with 3D meshes or other surfaces that are not already smoothed. This is very similar to the PEDIT SPLINE command for polylines, but it works in three dimensions instead of two.
- **Desmooth Surfaces**—This removes any smoothing from surfaces that have been smoothed with the PEDIT SMOOTH command.
- **MOPEN, NOPEN, MCLOSE, and NCLOSE**—Basically, surfaces are created as polylines with 3D faces between them. These commands either open or close the polylines in the M or N directions. For example, if you create a sur-

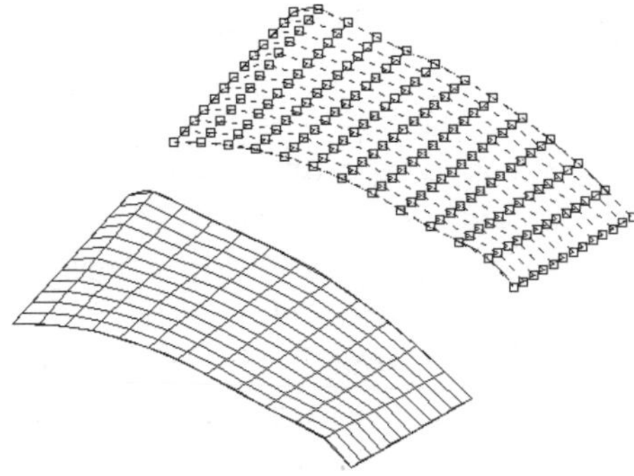

Figure 24–24 *A typical 3D mesh surface with (right) and without (left) grips displayed.*

face that forms a dome, using MCLOSE and NCLOSE closes all the polylines in the M direction and forms a floor. An example of closing a 3D mesh in the M and N directions is shown in Figure 24–25.

Editing a 3D mesh surface is not easy. If you need to change a surface in AutoCAD, it is frequently more efficient to change the construction edges defining the surface rather than to use grips or the PEDIT command to perform the editing.

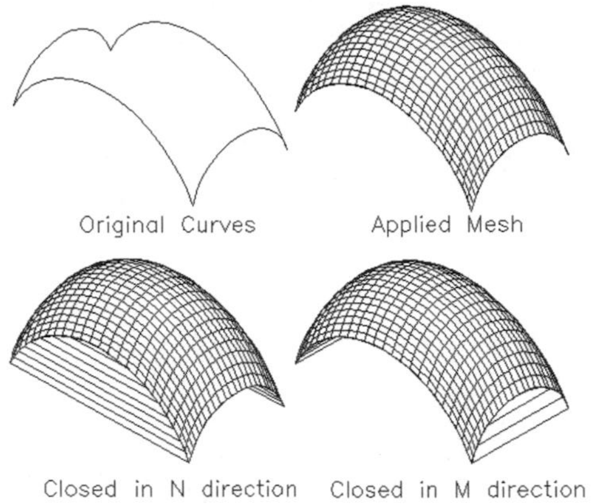

Figure 24–25 *Closing a 3D mesh in the M and N directions with the PEDIT command.*

SUMMARY

In this chapter you learned how to create 3D surfaces based on existing geometry in your model by using a variety of 3D surfacing commands. You learned that adjusting the SURFTRAB1 and SURFTAB2 system variables controls the resolution of these surfaces. You also discovered how you can edit surfaces by using PEDIT. The 3DMESH and PFACE commands are used primarily by outside programs that provide the high amount of input required by these commands.

In the next chapter you will explore how to use AutoCAD's solid modeling commands to create many types of 3D solid objects that would be difficult, if not impossible, to create using the surfacing commands covered in this chapter.

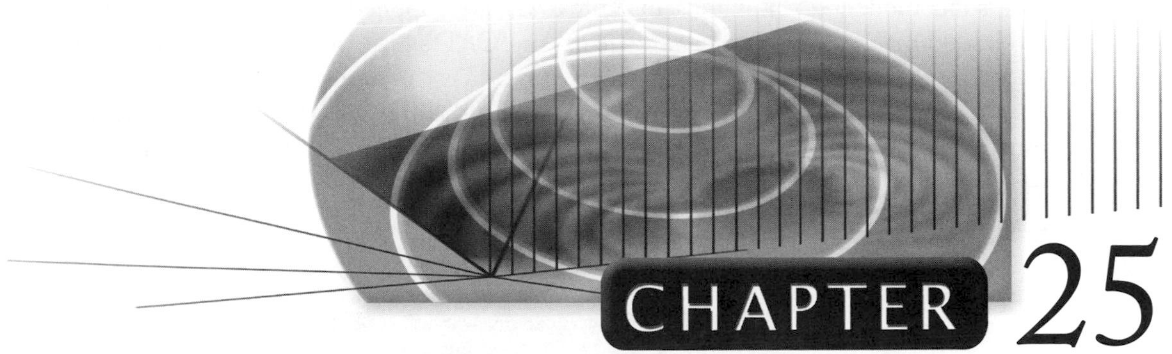

Modeling in 3D

Solid modeling is a relatively new method of modeling that has been around for only a few years and is now becoming very popular. Solid modeling consists of creating primitives and combining and manipulating those primitives into more complex objects. It is called solid modeling because you can attach material information to the model so that you can find information such as the object's center of gravity, how much it weighs, and so on.

This chapter focuses on how the ACIS solid modeler in AutoCAD works as an integrated modeling tool. Specifically, this chapter explores the following concepts:

- The ACIS solid modeler
- Creating primitives solids
- The EXTRUDE and REVOLVE commands
- Using Boolean operations in 3D
- Using FILLET and CHAMFER in 3D
- Controlling surface resolution
- Advanced solid modeling commands
- Editing solid objects

THE ACIS SOLID MODELER

The solid modeling commands in AutoCAD are supplied by the ACIS 4.0 solid modeler. This modeler takes advantage of AutoCAD's load-on-demand architecture, so the modeler program tools are loaded only when you access the commands. As a result, when you access any one of the various solid commands for the first time in a modeling session, there may be a slight delay as the set of modeler functions loads.

The solid modeler commands are located on their own toolbar, called the Solids toolbar, as shown in the Figure 25–1.

Figure 25–1 *The Solids toolbar.*

Note: You can also access the solid modeling commands by selecting Solids from the Draw pull-down menu or by typing in the individual commands at the command prompt.

CREATING PRIMITIVE SOLIDS

AutoCAD supports six different primitives:

- Box
- Sphere
- Cylinder
- Cone
- Wedge
- Torus

These rather simple solid geometric forms are termed "primitives" because they can be used in various combinations to produce hundreds of other, more complex solid geometric shapes. Primitives themselves are easy to construct.

Creating a solid box, for example, is as simple as selecting the BOX command from the Solids toolbar and then specifying the two points that define the opposite corners of the box and the height. The resulting solid box appears similar to a 3D box created with surfaces. The display of solid objects, however, is quite different in some cases. Consider, for example, a surface and a solid sphere. The surface sphere looks quite different than the solid version, as shown in Figure 25–2.

The displays differ because the solid modeler works with boundaries, whereas the surface modeler works only with faces joined together to form surfaces. When you

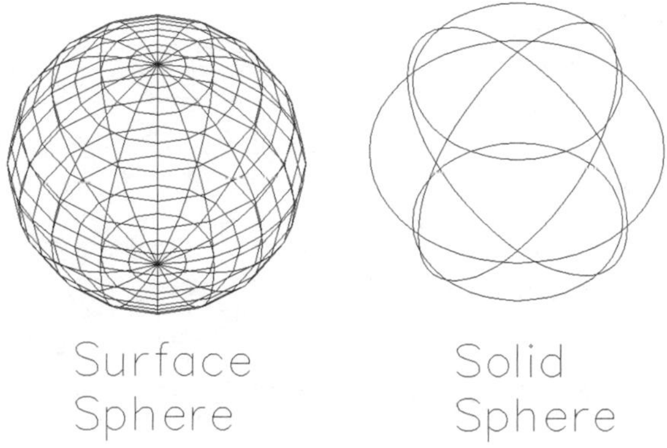

Figure 25–2 *A surface sphere versus a solid sphere.*

hide the solid sphere or render it, it is converted to a surface-like model for that purpose only. Figure 25–3 shows the spheres represented in Figure 25–2 after the HIDE command is used.

Note: The lines used to display AutoCAD" solids are called isolines. Usually, four isolines per solid object adequately represent the shape and form of a solid, and the default value for isolines is 4. The system variable Isolines controls the number of isolines used to show solids. For more complex solids or in views containing a large number of solid objects, you may wish to increase the number of isolines. A value of 8 or 16 is usually sufficient. Values significantly higher than 16 may slow down operations such as regenerations. If you alter the number of isolines, you must perform a regeneration to have the change take effect on solids that are already drawn.

The following exercise shows you how to use solid primitives to create a simple chair model.

EXERCISE: USING THE SOLID PRIMITIVES TO CREATE A SIMPLE CHAIR

1. Open the 25DWG01.DWG file found on the accompanying CD.
2. At the command prompt, type **BOX**, or select Box from the Solids toolbar.
3. Specify the corners at **0,0** and **1,1**. Specify a height of **24**.
4. Copy this box eighteen units along the X-axis.
5. Copy both existing boxes eighteen units along the Y-axis.

   Figure 25–4 shows the drawing at this point.

   Next you will create the seat of the chair.

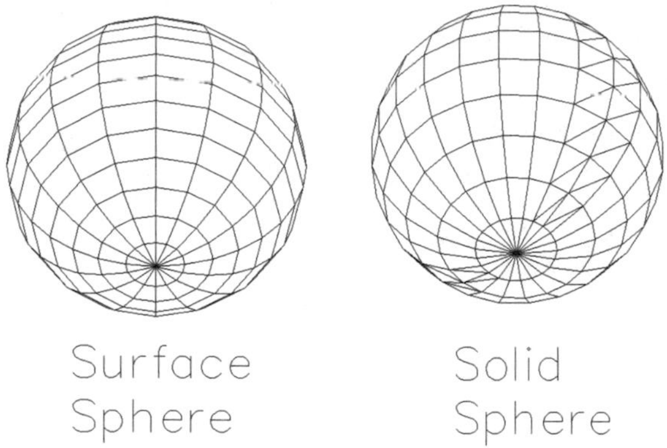

Figure 25–3 *A surface sphere and a solid sphere after the HIDE command is used.*

Figure 25–4 *The chair after you create the four legs.*

6. Create another box with corners at -0.5,-0.5 and 19.5,19.5 and a height of 1.
7. Choose Move from the Modify toolbar. Select the box and move the box twenty-four units along the Z-axis by typing in **0,0,24**. Then press Enter twice.
8. Create three boxes to form the backrest of the chair. You will create these boxes "on the ground," and then move them into position on top of the chair.
9. From the View pull-down menu, choose 3D Views>SW Isometric.
10. Create a box with corners at 18,0 and 19,1 and a height of 18. This box should appear "inside" the lower-right leg.

11. Choose Move from the Modify toolbar. Type in **L** for Last at the "Select Objects" prompt and then press Enter. Specify a displacement of **0,0,25** and then press Enter twice to move this box twenty-five units along the Z-axis. Perform a Zoom Extents to show the objects you have modeled so far.
12. Choose Copy from the Modify toolbar. Make a copy of the box and place it over the upper-right leg of the chair by specifying a displacement of **0,18**.
13. Make a second copy of the box and place it between the two other backrest supports by specifying a displacement of **0,9**.
14. Draw the back's top crossbar using the other options of the BOX command. Specify the first corner at **18,0,43**. Now type **L** to indicate that you want to specify a length next. Enter a value of **1**. At the width prompt, type **19**. Specify a height of **1**.

    Figure 25–5 shows the drawing at this point. Now you can add a little decoration to the chair.
15. At the command prompt, type **PLAN** and press Enter twice. This switches you back to a plan view. Perform a Zoom 0.8X to generate a slightly zoomed-out view.
16. Choose Sphere from the Solids toolbar.
17. Create a sphere with its center at 18.5,0.5,44.5 and a radius of 0.5.
18. Choose Copy from the Modify toolbar. Copy the sphere to the other side of the back of the chair.
19. From the View pull-down menu, choose 3D Views>SW Isometric.

    Figure 25–6 shows the final chair model.
20. Save the drawing.

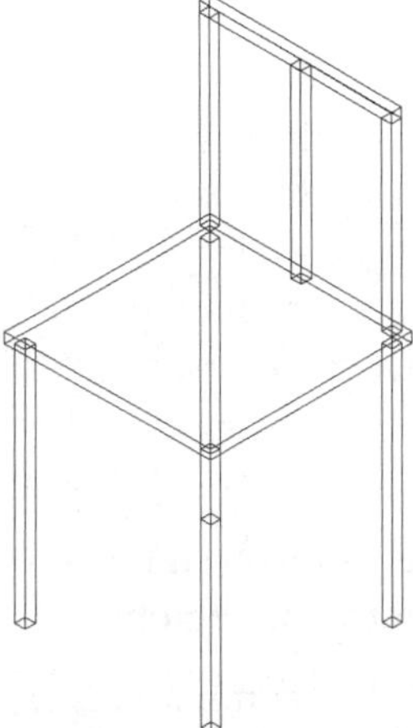

Figure 25–5 *The chair after you create the seat and backrest.*

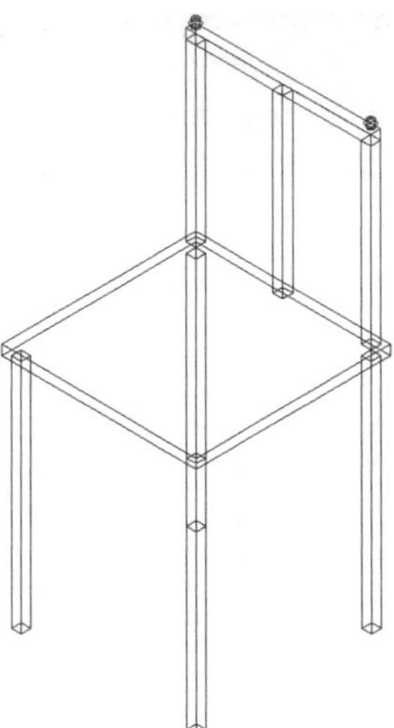

Figure 25–6 *The completed chair model in isometric view.*

As you can see in this exercise, modeling with primitives is not that different from modeling with surface primitives. As always, when you work with 3D models, you need to change your view direction frequently in order to keep the spatial relationships of the model's components in view.

THE EXTRUDE AND REVOLVE COMMANDS

As mentioned before, solid primitives such as box, sphere, cone, etc. are frequently used in combinations to produce new solid shapes. In addition to primitives, AutoCAD provides two powerful commands that produce non-primitive, often-complex shapes from closed 2D curves.

EXTRUDE

With the EXTRUDE command you can extrude, or give thickness to, certain 2D objects. You can extrude along a path or you can specify a height and taper angle. Objects must be "closed" to qualify for extrusion. Such objects can include planar 3D faces, closed polylines, polygons, circles, ellipses, closed splines, donuts, and regions. 2D objects such as these are termed profiles. EXTRUDE is especially useful for objects that contain fillets, chamfers, and other details that might otherwise be difficult to reproduce except in a profile.

The REVOLVE command creates a solid by revolving a 2D object (profile) about an axis. 2D objects capable of being revolved include closed polylines, polygons, circles, ellipses, closed splines, donuts, and regions.

A common usage of the EXTRUDE command is the creation of 3D walls. The following exercise shows you how to create a wall by using the EXTRUDE command.

EXERCISE: USING THE EXTRUDE COMMAND TO CREATE A WALL

1. Load the 25DWG02.DWG file found on the accompanying CD. The file contains the outline of a wall, as shown in Figure 25–7.
2. From the Solids toolbar, choose Extrude, or type in **EXTRUDE** at the command prompt.
3. Select the three polylines in the scene (the wall and two window openings).
4. When you are prompted for an extrusion height, type **0.5**.
5. Accept the default taper angle of zero by pressing Enter. The objects are extruded.
6. From the View pull-down menu, choose 3D Views>SW Isometric.
7. At the command prompt, type **HIDE**. Figure 25–8 shows the wall parts after the extrusion process.
8. Save the file as 25DWG02A.DWG for use in a later exercise.

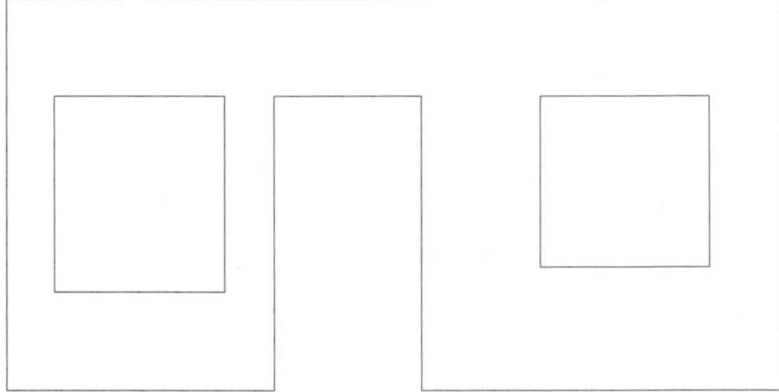

Figure 25–7 *The outline of a wall with openings ready to be extruded.*

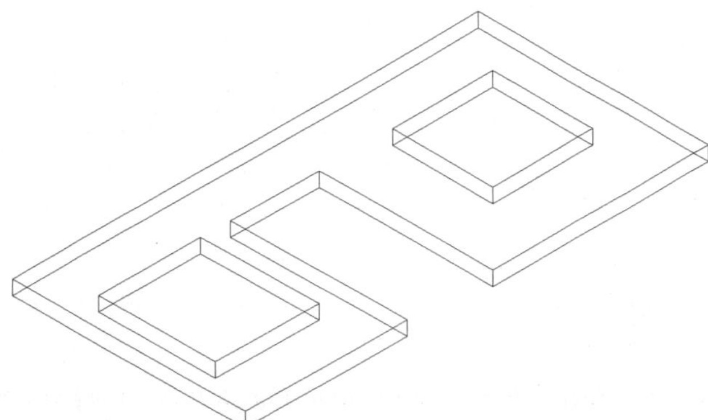

Figure 25–8 *The wall parts after extrusion. The windows are extruded as separate objects.*

As you can see, using the EXTRUDE command to extrude a profile in the object's Z-axis is simple and straightforward. If you extrude a closed polyline that is itself enclosed within the boundary of another extruded closed polyline, as in the preceding exercise, you end up with multiple objects. This is a common situation when you extrude walls with window openings. The windows can be subtracted using the SUBTRACT command. You will do this in a later exercise in the section titled Working with 3D Boolean Operations.

In addition to being able to extrude straight up or down with or without a taper, you can extrude a closed 2D object along a path. The following exercise shows you how to create a pipe by using the "extrude along a path" option. This exercise shows you some of the flexibility of the EXTRUDE command and how it can be useful for modeling many different objects.

EXERCISE: EXTRUDING ALONG A PATH

1. Load the 25DWG03.DWG file found on the accompanying CD. The file contains a circle and a polyline path, as shown in Figure 25–9.
2. From the Solids toolbar, choose Extrude, or type in **EXTRUDE** at the command prompt.
3. Select the circle as the object to extrude.
4. When you are prompted, type in **P** for Path.
5. Select the polyline. The object is then extruded along the path. Note that the path is moved by AutoCAD to match the center of the shape.
6. Type **HIDE** at the command prompt. The circle extrudes along the path, as shown in Figure 25–10.

Figure 25–9 *A circle and the polyline path along which it will be extruded.*

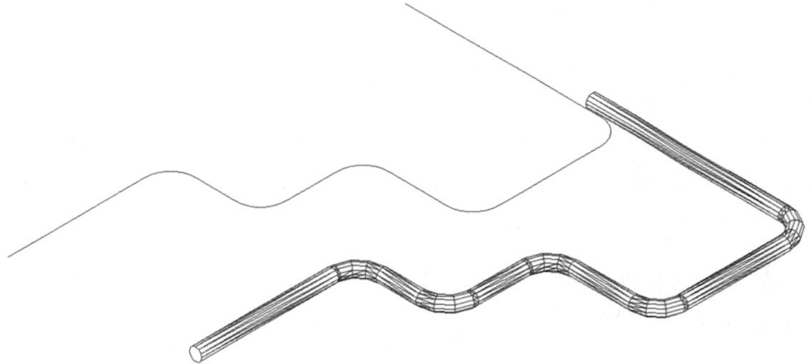

Figure 25–10 *The circle after it has been extruded along the polyline path.*

As you can see in this exercise, extruding shapes along a path is relatively easy. Just make sure that the shape you want to extrude is perpendicular to the path, as it was in this exercise. If it is not, you will not be able to extrude the object.

Tip: Extruding shapes along a path works well with polylines and splines as long as they are not closed. When a closed polyline or spline is used as a path, the extrusion may not always work.

REVOLVE

The REVOLVE command produces solid objects from 2D profiles much like the EXTRUDE command does. With REVOLVE, however, an axis of revolution is specified. 3D objects produced with the REVOLVE command therefore have a radial axis of symmetry. You can revolve any number of degrees about the axis of revolution, although full 360-degree revolutions are most common. In the following exercise you will use two different degrees of rotation to produce two models of a piston.

EXERCISE: USING REVOLVE TO PRODUCE A PISTON

1. Load the 25DWG04.DWG file found on the accompanying CD. The file contains two identical profiles of a piston. The axis of revolution is indicated for both profiles, as shown in Figure 25–11.
2. Choose Revolve from the Solids toolbar or type in **REVOLVE** at the command prompt.
3. Select the left piston outline and press Enter.
4. When you are prompted, type in **O** for Object.
5. Select the black vertical line associated with the profile. This line will serve as the axis of revolution.

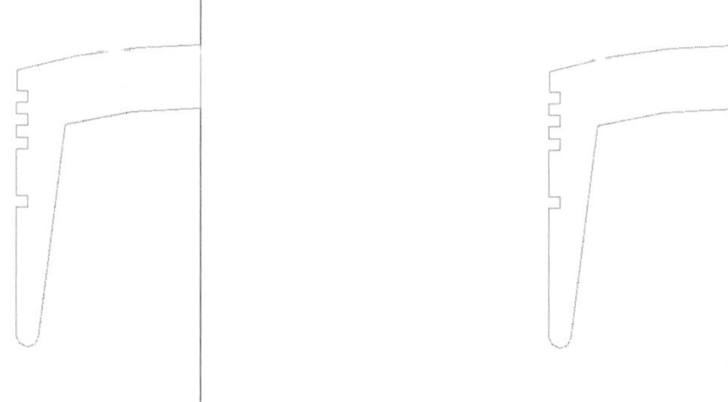

Figure 25–11 *Two profiles of pistons that are ready to revolve.*

6. Press Enter to accept a full 360 degrees of revolution.
7. Restart the REVOLVE command by pressing Enter. Repeat steps 3–5 with the right piston profile.
8. When you are prompted to specify the angle of revolution, enter **230**.
9. Click on the Orthographic & Isometric Views tab and set the SW isometric view current.
10. Type **HIDE** at the command prompt. Figure 25–12 shows the resulting solid objects.
11. Save the file as 25DWG04A.DWG for use in a later exercise.

As you can see, the REVOLVE command can be a flexible 3D design tool. Similar to EXTRUDE, REVOLVE is useful for objects that contain fillets or other details that would otherwise be difficult to reproduce in a common profile.

USING BOOLEAN OPERATIONS IN 3D

Much of the power of solid modeling comes from the ability to use Boolean operations as modeling tools. In Chapter 23, Drawing in 3D, you were introduced to Boolean operations while working with regions. These same Boolean commands work at the 3D level to enable you to create complex objects quickly and easily from simple primitives.

There are three Boolean operation types that you can create: union, intersection, and subtraction. Figure 25–13 shows two primitives and the object that results when you apply each of the three Boolean operations to the primitives.

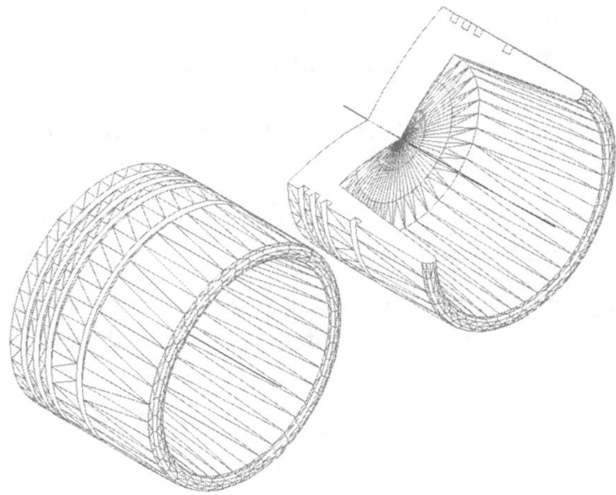

Figure 25–12 *Piston profiles fully (left) and partially (right) revolved about an axis.*

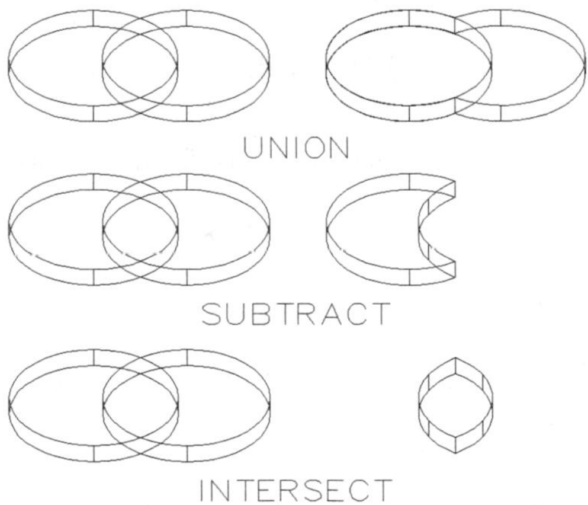

Figure 25–13 *Examples of each of the three Boolean operations.*

Tip: If you create a profile using lines or arcs that meet a polyline, use the PEDIT Join option to convert them to a single polyline object before you use REVOLVE.

The following exercise shows you how to use a Boolean Subtract operation to create holes in the wall that you built earlier in this chapter.

EXERCISE: USING SUBTRACT TO CREATE HOLES IN A WALL

1. Load the file 25DWG02A.DWG from the earlier exercise. If you did not complete the exercise, load the file from the CD-ROM.
2. At the command prompt, type **Subtract**, or, from the Modify pull-down menu, choose Solids Editing>Subtract.
3. Select the large wall object as the object to subtract from and then press Enter. Select the two boxes representing the windows as the objects to subtract and then press Enter.
4. To show the result of the subtraction, type **HIDE** and press Enter. The wall with window openings is shown in Figure 25–14.

As you can see in this exercise, it is easy to carry out Boolean operations. In some instances, performing a Boolean Intersect or Subtract operation is virtually the only way to obtain the solid object you want. The following exercise shows you a more complex example of a Subtract operation.

EXERCISE: CREATING A GROOVE WITH SUBTRACT

1. Load the 25DWG05.DWG file found on the accompanying CD. The file contains a box and a pipe, as shown in Figure 25–15.
2. From the pull-down menu, select Modify>Solids Editing>Subtract.

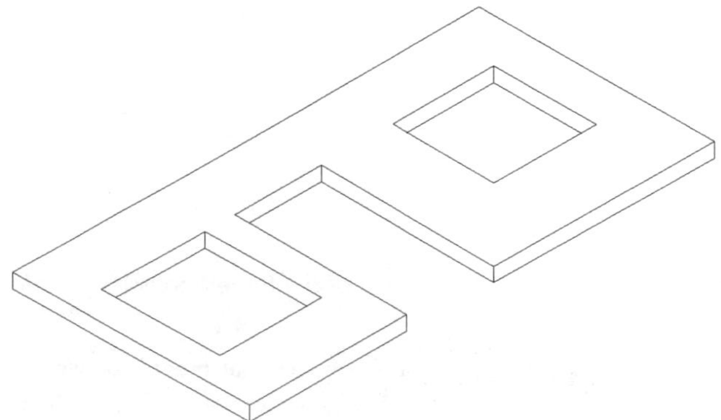

Figure 25–14 *The wall after you subtract the window openings.*

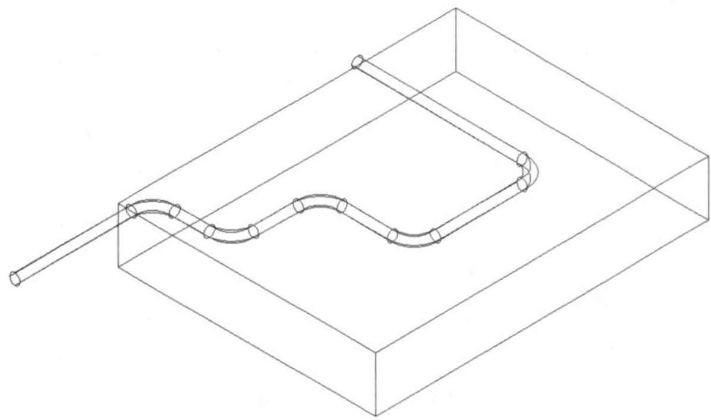

Figure 25–15 *A solid box and the pipe that will form a groove when it is subtracted.*

3. Select the box as the object to subtract from and then press Enter.
4. Select the pipe as the object to subtract and then press Enter. The subtraction is executed.
5. Hide the drawing to achieve the result shown in Figure 25–16.

Imagine trying to create the 3D model in this exercise without the Boolean Subtract operation. This would be difficult if not impossible to do using simple surfaces.

USING FILLET AND CHAMFER IN 3D

Although true solids editing capabilities were introduced for the first time in AutoCAD 2000, AutoCAD has allowed the adding of fillets and chamfers to solids since Release 13 using the standard "2D" FILLET and CHAMFER commands. To fillet or chamfer the edges of an AutoCAD solid, you simply execute the FILLET and CHAMFER commands, respectively.

The following exercise shows you how to use the FILLET command to round the edges of a solid model of a phone handset.

EXERCISE: USING THE FILLET COMMAND

1. Load the 25DWG06.DWG file found on the accompanying CD. The file consists of three "unioned" boxes, as shown in Figure 25–17.

   You will use the FILLET command to round the edges and refine the model.
2. Choose Fillet from the Modify toolbar, or simply type **FILLET** and press Enter.
3. When you are prompted to select the first object, select the vertical inner edge of the phone, as shown at ① in Figure 25–18.

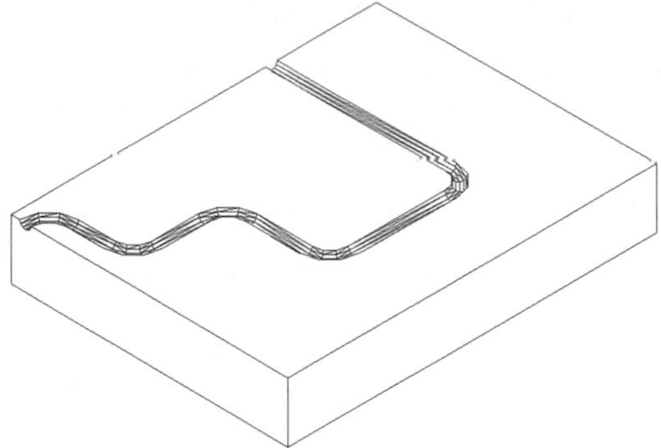

Figure 25–16 *The box with the pipe subtracted to create a groove.*

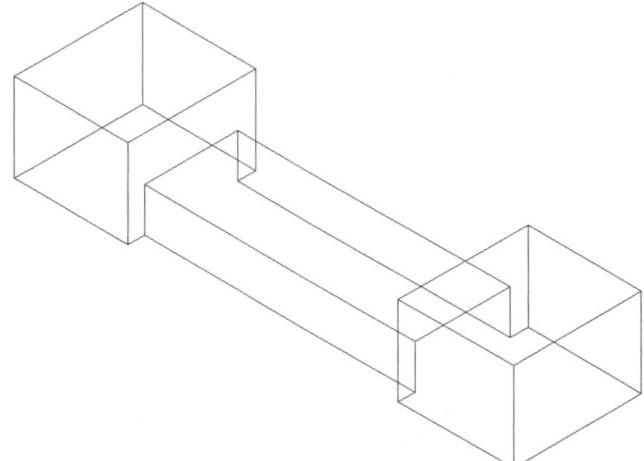

Figure 25–17 *A set of three solid boxes representing a basic telephone handset.*

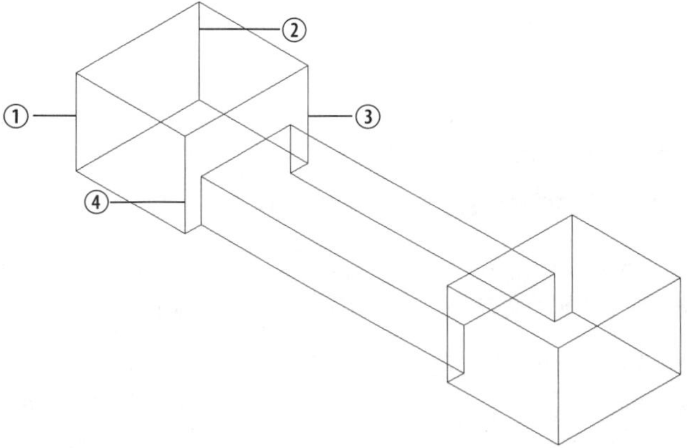

Figure 25–18 *Preparing to fillet four vertical edges.*

4. Specify a fillet radius of 0.25 and press Enter.
5. At the "Select an edge or [Chain/radius]:" prompt, select the other three vertical edges shown at ② through ④ in Figure 25–18, and then press Enter. AutoCAD fillets the four edges.
6. Choose FILLET again by pressing either the Spacebar or Enter.
7. Select the upper-outer edge of the phone, as shown at ① in Figure 25–19.
8. Set the radius to **0.2** and press Enter.
9. At the "Select an edge or [Chain/radius]:" prompt, type **C** for Chain and then select any of the other seven top edges that make up the top surface of the receiver. ② of Figure 25–19 is a valid choice. Note that selecting any edge selects all the connected edges in the chain. Press Enter to complete the operation. At this point your model should resemble Figure 25–20.
10. Repeat the FILLET command; pick the edge shown at ① in Figure 25–20 when you are prompted for the first object.
11. Enter a radius of **0.20**.
12. Select the edges shown at ② through ⑧ in Figure 25–20. Press Enter to complete the fillet option. Your model should resemble Figure 25–21.

    In the following steps you will use the CHAMFER command to mold the top edges of the transmitter portion of the model.

13. From the Modify menu, choose Chamfer.
14. When you are prompted for the first line, pick ① in Figure 25–21. If necessary, type **N** and press Enter until the top surface of the transmitter portion is highlighted. Then press Enter to OK the selection.

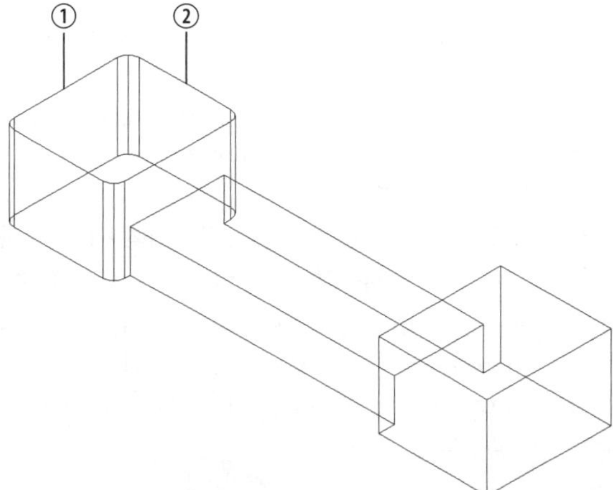

Figure 25–19 *Preparing to fillet a chain of connected edges.*

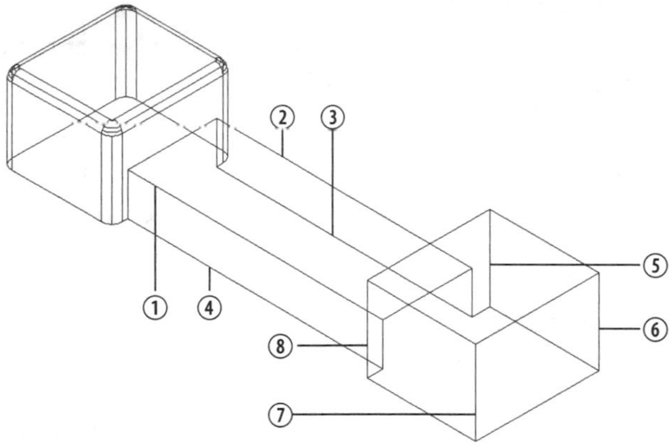

Figure 25–20 *A total of eight edges are filleted at one time.*

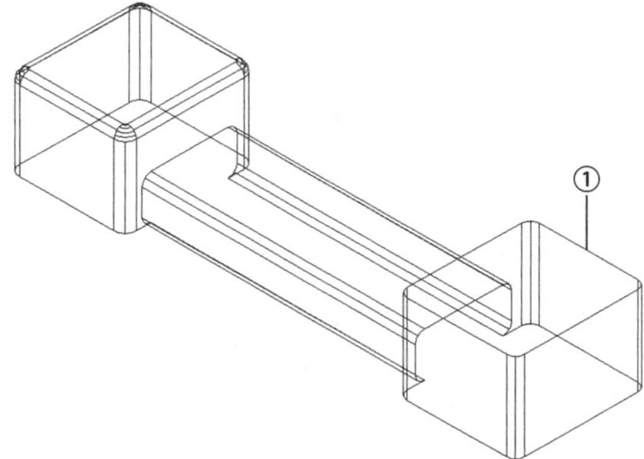

Figure 25–21 *Preparing to chamfer the handle and transmitter portions.*

15. When you are prompted, specify a base surface distance of **0.10** and specify the other surface distance as **0.10**. At the next prompt, type **L** for the Loop option and pick ① again to select an edge loop.
16. Press Enter to chamfer the top surface. Issue the HIDE command by typing **HI** and pressing Enter. Your model should resemble Figure 25–22.

As you can see in this exercise, adding a third dimension presents many more edges that can be filleted and chamfered. Also, you can see that the Chain option of the FILLET command and Loop option of the CHAMFER command make it easy to edit a continuous string of edges.

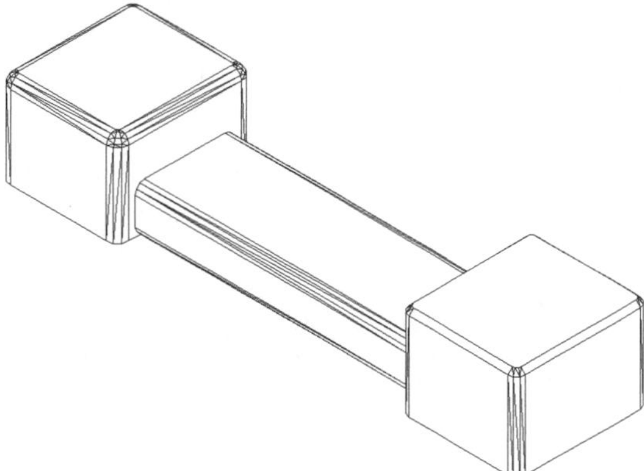

Figure 25–22 *The finished handset model, with fillets and chamfers applied and hidden lines removed.*

CONTROLLING SURFACE RESOLUTION

As was mentioned earlier in this chapter, AutoCAD displays solid objects on the screen using boundary representations of the objects (see Figure 25–23). Solid objects are converted to surfaces only when you hide, shade, render, plot, or export the solid geometry.

The process of converting a boundary representation into a surface representation is called *tessellation*. Tessellation places a series of contiguous three- and four-sided "tiles" on the solid's boundaries, resulting in a "surface" mesh that can be shaded, rendered, and used in a Hide operation. Figure 25–24 show the same solids as those in the previous figure but with tessellated surfaces applied.

The density of the surface mesh applied during tessellation is controlled by the system variable FACETRES (Facet Resolution). The default value for FACETRES is 0.5, but the variable can be set to any value from 0.01 to 10. Higher values result in finer meshes being generated from the solid objects, but higher-resolution meshes also take much longer to hide, render, or export to other programs. Figure 25–24 has FACETRES set to a value of 0.5. Compare Figure 25–24 to Figure 25–25, in which FACETRES is set to a value of 2.0.

ADVANCED SOLID MODELING COMMANDS

AutoCAD also provides you with three advanced solid modeling commands that can be quite useful:

- SLICE
- SECTION
- INTERFERENCE

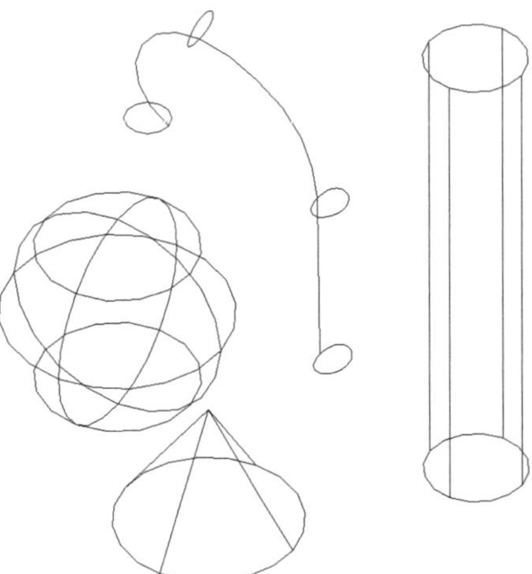

Figure 25–23 *Solids are represented on the screen by lines and curves denoting boundaries.*

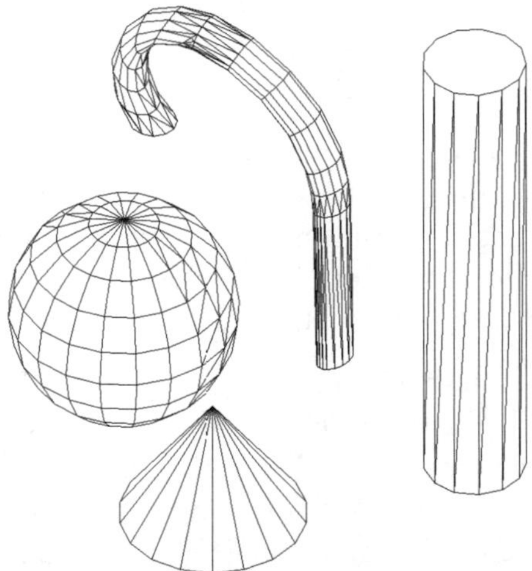

Figure 25–24 *Low-resolution tessellation of solid objects.*

The following sections define these advanced commands and take you through exercises designed to show you how you can use each to help build your solid models.

THE SLICE COMMAND

The SLICE command is used to divide solids objects on either side of a plane. If, for example, you create a complex model and want to cut it in half, the SLICE command will accomplish the task.

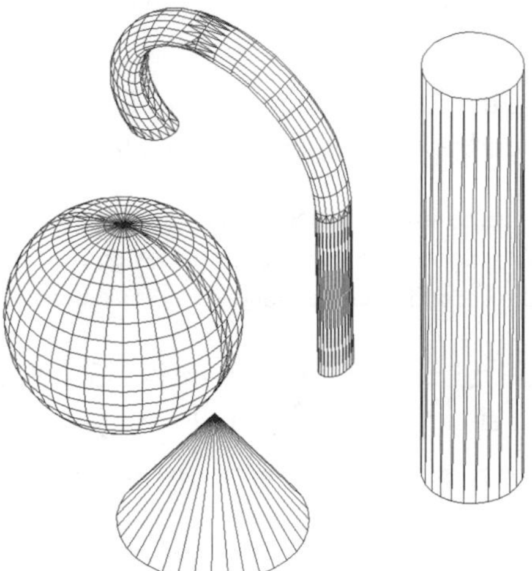

Figure 25–25 *Increasing the value of FACETRES increases the resolution of the surface representation of solids.*

There are five methods you can use to define the slicing, or cutting, plane:

- **3 points:** Defines the slicing plane using three points. If the desired plane is not parallel with the current XY, YZ, or ZX planes, this method is usually the best choice.
- **Object:** Aligns the cutting plane with a circle, an ellipse. a circular or elliptical arc, a 2D spline, or a 2D polyline. The object does not need to be separated from the volume of the solid.
- **View:** Aligns the cutting plane with the current viewport's viewing plane. Specifying a point defines the location of the cutting plane along the Z-axis of the view plane.
- **Z-axis:** Defines the cutting plane by specifying a point on the plane and another point on the Z-axis (normal) of the plane.
- **XY, YZ, and ZX:** Aligns the cutting plane with the XY, YZ, or ZX plane of the current UCS. Specifying a point defines the location of the cutting plane.

Note that the three-points method defines the plane immediately; no further point is required. The other four methods first align the plane; you must supply an additional point to place the plane relative to the model.

After the slice has been carried out, AutoCAD presents the "Point on the Desired Side of the Plane:" prompt. You must specify a point to determine which side of the sliced solids your drawing retains. This point cannot lie on the cutting plane. You can also choose the "Keep Both Sides" option. This option retains both sides of the sliced solids. Slicing a single solid into two pieces creates two solids from the pieces on either side of the plane.

Often you can use more than one method to accomplish the same slice. The method you choose will often depend upon the geometry available to you.

Tip: Do not hesitate to use the three-points method merely because a convenient third point is not available. Often you can use a relative coordinate designation (e.g., @0,0,1) after you specify the second point to identify a point coplanar with the first two.

All you have to do to make SLICE work is create the 3D ACIS model you want to slice and then determine a slicing plane. The following exercise illustrates this.

EXERCISE: USING SLICE TO CUT A COMPLEX MODEL IN HALF

1. Load the file 25DWG07.DWG. This is a model of a piston that is similar to the one you created earlier in this chapter. Note that the line used to define the axis of revolution in forming the piston from its profile is also visible. The piston is sitting on a plane coincident with the XY plane (see Figure 25–26).
2. From the Draw menu, choose Solids>Slice.
3. When you are prompted, select the piston in the upper-left viewport and then press Enter.
4. Type **ZX** and press Enter to indicate that you want to define the slicing plane parallel to the current ZX plane.
5. When you are prompted with "Specify a point on the ZX plane:," snap to either endpoint of the line defining the axis of revolution, ① or ② in Figure 25–26.

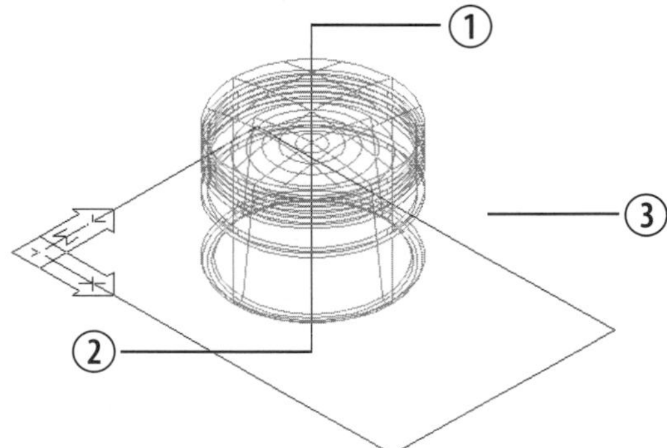

Figure 25–26 *The three points you use to specify the exercise's slice plane.*

6. When you are prompted with "Specify a point on the desired side of the plane:," pick a point near ③ in the NE corner of the ground plane rectangle and click. AutoCAD completes the slice.
7. Before closing this model, you may want to practice using other methods of specifying a slicing plane. Note that once a slice has been completed, you can use the Undo option of the SLICE command to restore the solid to its original state.
8. Close the drawing with or without saving changes.

This exercise gave you practice in visualizing and specifying "invisible" slicing planes. With practice you will be able to slice solids quickly and precisely.

THE SECTION COMMAND

The SECTION command works almost exactly the same as the SLICE command, with only one major difference. Instead of slicing the object, the SECTION command generates a region that is representative of a section of the selected ACIS solid object on the chosen plane. Suppose, for example, you model this great 3D part and want to draw a section of it for manufacturing purposes. The SECTION command can automatically generate most of the 2D drawing from your 3D model, with very little effort.

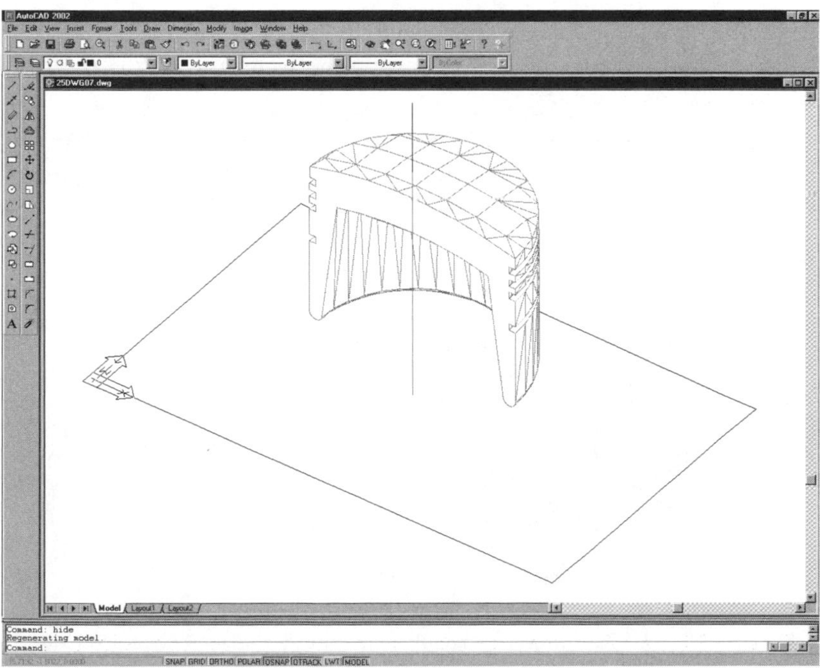

Figure 25–27 *The hidden line display of the sliced piston.*

The following exercise shows you how to create a section of a solid object.

EXERCISE: CREATING A SECTION OF A SOLID OBJECT

1. Load the file 25DWG06A.DWG. This is a completed telephone handset similar to the one you built earlier in this chapter. If you did not complete that exercise, load the file from the accompanying CD. The model will resemble Figure 25–28.
2. Select Section from the Solids toolbar, or type **SECTION** at the command prompt.
3. Select the handset and press Enter.
4. Press Enter again to define the section plane using three points.
5. When you are prompted for the first point, choose Midpoint and select the middle point of the end of the receiver at ① in Figure 25–28.
6. Select the next two points by using midpoint, as shown at ② and ③.
7. Choose Move from the Modify toolbar.
8. When you are prompted to select objects, type in **L** for Last and then press Enter.
9. Move the section so it is outside of the 3D object. Figure 25–29 shows the resulting 3D object and section.

As this exercise has shown, the ways in which the SLICE and SECTION commands are used are very similar. The difference between the two is the result. The

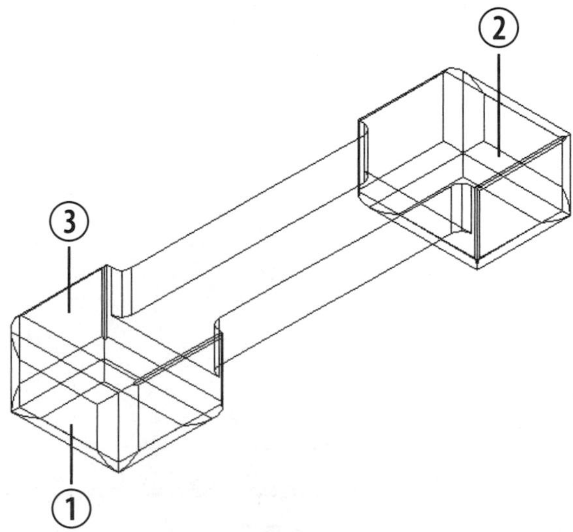

Figure 25–28 *The three points you use to specify the exercise's section plane.*

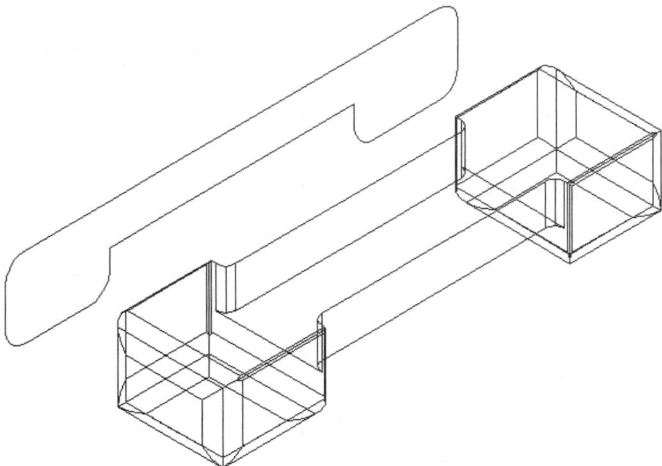

Figure 25–29 *After you create a section of the model, use the MOVE command to relocate the section line.*

SLICE command enables you to cut a model into pieces. When you use the SECTION command, the result is a region. You can convert this region to a series of lines by exploding it, if you need to edit the section further. You can also use it to develop new solids by extruding it.

THE INTERFERENCE COMMAND

The last advanced command that will be looked at here is the INTERFERE command. This command is used to determine whether two solids overlap or interfere with each other. If they do, you can generate a new solid that is the volume of the area where the two solids interfere. This enables you to calculate quickly how much volume is interfering between the two solids. The INTERFERE command uses a Boolean Intersect operation to generate any interference information. No exercise will be provided for this command.

EDITING SOLID OBJECTS

Just as with all other AutoCAD objects, once you create solid objects you may want to edit them. Prior to AutoCAD 2000, editing solids inside AutoCAD was, at best, moderately difficult and often possible. If, for example, in Release 13 or 14 you needed to decrease the diameter of a circular bore into a solid, you needed to first "fill in" the current bore and then "rebore" by subtracting another cylinder with the smaller diameter. This was hardly solid editing; it was just redoing your work.

AutoCAD 2000 introduced a solid editing facility that allows you to edit, or modify, the size and geometry of AutoCAD solids easily. Through the SOLIDEDIT command you can move, rotate, taper, resize, or even remove features of a solid. Blends created with the FILLET command, for example, can be removed. Through SOLIDEDIT you can also copy a face or edge from of a solid as a body or region (in the case of faces), or as a line, arc, circle, ellipse, or spline object (in the case of edges). You cannot, however, copy a feature, such as a hole, in a solid to make a second hole. This is technically not an edit operation.

SOLID EDIT OPERATIONS

The following operations are performed by the SOLIDEDIT command. Two SOLIDEDIT operations, Color and Copy, are common to both faces and edges. The Imprint, Shell, Separate, and Clean operations are performed on entire 3D solid bodies (See Figure 25–30).

- **Extrude**—Extrudes selected planar faces of a 3D solid object to a specified height or along a path. You can select multiple faces at one time.
- **Move**—Moves selected faces on a 3D solid object to a specified height or distance. You can select multiple faces at one time.
- **Rotate**—Rotates one or more faces or a collection of features on a solid about a specified axis.
- **Offset**—Offsets faces equally by a specified distance or through a specified point. A positive value increases the size or volume of the solid; a negative value decreases the size or volume of the solid.
- **Taper**—Tapers faces with an angle. The rotation of the taper angle is determined by the selection sequence of the base point and the second point along the selected vector.
- **Delete**—Deletes or removes faces, including fillets and chamfers.
- **Copy**—With Faces: Copies faces as a region or a body. If you specify two points, AutoCAD uses the first point as a base point and places a single copy relative to the base point. If you specify a single point (usually entered as a coordinate) and then press Enter, AutoCAD uses the coordinate as the new location. With Edges: Copies 3D edges. All 3D solid edges are copied as a line, arc, ellipse, or spline.
- **Color**—With Faces: Changes the color of faces. With Edges: Changes the color of edges.
- **Imprint**—Imprints an object on the selected solid. The object to be imprinted must intersect one or more faces on the selected solid in order for imprinting to be successful. Imprinting is limited to the following objects: arcs, circles, lines, 2D and 3D polylines, ellipses, splines, regions, bodies, and 3D solids.
- **Separate**—Separates 3D solid objects with disjointed volumes into independent 3D solid objects.

- **Shell**—Creates a hollow, thin wall with a specified thickness. You can specify a constant wall thickness for all the faces. You can also exclude faces from the shell by selecting them. A 3D solid can have only one shell. AutoCAD creates new faces by offsetting existing ones outside their original positions. Specifying a positive value creates a shell from the outside of the perimeter; specifying a negative value creates a shell from the inside.
- **Clean**—Removes shared edges or vertices having the same surface or curve definition on either side of the edge or vertex. Removes all redundant edges and vertices, imprinted as well as unused geometry.
- **Check**—Validates the 3D solid object as a valid ACIS solid, independently of the SOLIDCHECK system variable setting.

Picking Edges and Faces with SOLIDEDIT

The SOLIDEDIT command is the only AutoCAD command that requests you to select faces on 3D solid objects to perform some of its operations. Selecting individual faces while in the SOLIDEDIT command, however, can be tricky. If you pick on an edge at a "Select faces:" prompt, the two faces that share the picked edge will be highlighted, and you must remove the unwanted face from the selection set. To alleviate this inconvenience, an additional method of selecting, called Boundary Sets, is automatically available during SOLIDEDIT operations involving faces. With boundary set selection, you can pick a face inside of the edges that define the face. While elsewhere in AutoCAD, picking on "empty space" will, at best, establish one corner of a selection window (if the system variable PICKAUTO is set to its default value of 1), in editing operations involving 3D solid faces, you can successfully pick within the boundary of a face.

In the following exercises the Boundary Set pick method will be demonstrated as some of the options of the SOLIDEDIT command are examined.

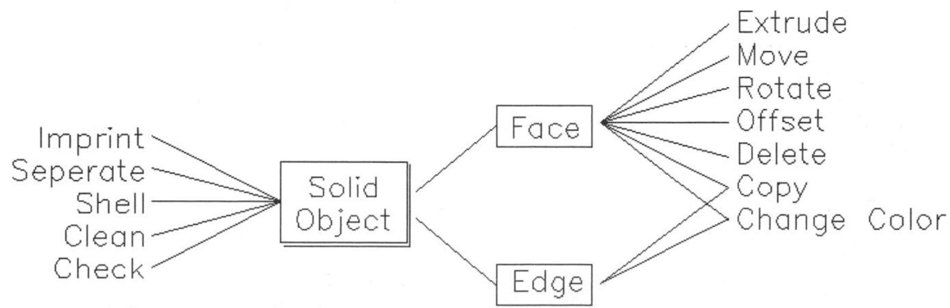

Figure 25–30 *An overview of the SOLIDEDIT operations.*

Note: Even with the convenience of the Boundary Set selection method, if there is any ambiguity due to the view, picking on the two faces will select the face that is "closest" visually. Picking again will select the second, more "distant" face. You must then remove the "top" face from the boundary selection set.

EXERCISE: USING BOUNDARY SELECTION TO EDIT 3D FACES

1. Open the drawing 25DWG08.DWG. The layer EXTRUDE should be current. Your screen should resemble Figure 25–31.

   You will first move the "front" face of the block inward twenty units by performing a negative extrusion.

2. From the Modify menu, choose Solids Editing>Extrude Faces. When you are prompted to select faces, select the front face. Use a single pick boundary selection by picking at ①. Notice that using a boundary selection allows you to select the face by picking inside the face boundary.

3. Press Enter to close the selection set. At the "Specify height of extrusion:" prompt, type **-20**. When you are prompted to specify the angle of taper for extrusion, press Enter to accept the default value of zero. The front face is extruded in a negative direction by twenty units with a zero-degree taper angle.

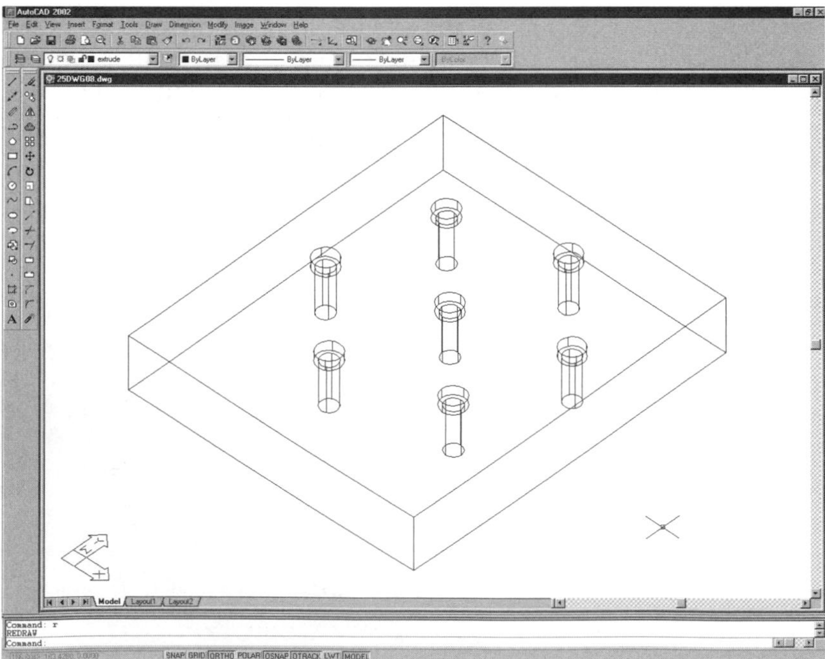

Figure 25–31 *Using boundary sets to select faces.*

Next you will extrude the top of the block and all the countersink holes upward by twenty units. Because "auto-windowing" is disabled during solid face editing to allow boundary set selection, you must manually start a crossing window.

4. From the Modify menu, choose Solids Editing>Extrude Faces. When you are prompted to select faces, start a crossing window by typing **C** and pressing Enter. Specify a crossing window that includes all faces and all countersink holes (see Figure 25–32).

   The previous step selects all the faces of the model. Because you only want to extrude the seven countersink holes and the top face, you must remove the side and bottom faces. Refer to Figure 25–33 during the next step.

5. While you are still being prompted to select faces, choose the Remove option by typing **R** and pressing Enter. When you are prompted to remove faces, pick the edges at ①, ②, ③, ④ (see Figure 25–33). Press Enter to end face selection. Notice that the first edge you select removes two faces: the side face and the bottom face.

6. When you are prompted to specify the height of extrusion, type **20**, and then press Enter to accept the default zero-degree angle for taper of extrusion. The top face and the seven countersink holes are extruded by twenty units. Press Enter twice to exit the SOLIDEDIT command.

   In the next steps you will rotate the array of countersink holes to align them better with the sides of the block.

7. Using the VIEW command, restore the view Rotate. From the Modify menu, choose Solids Editing>Rotate Faces. When you are prompted to select faces, initiate a crossing window by typing **C** and pressing Enter. Specify a window enclosing all the countersink holes.

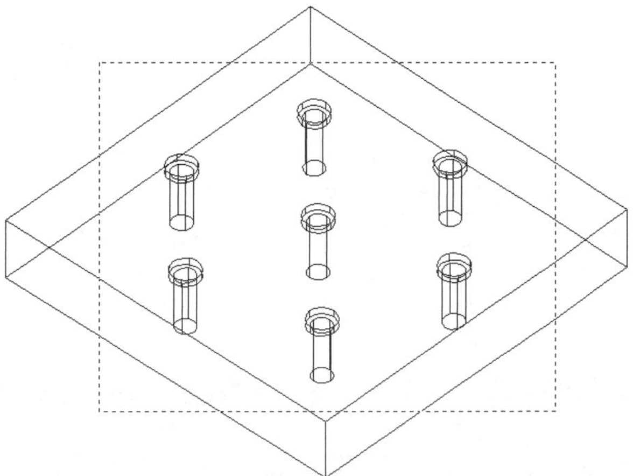

Figure 25–32 *Choosing faces with a crossing window.*

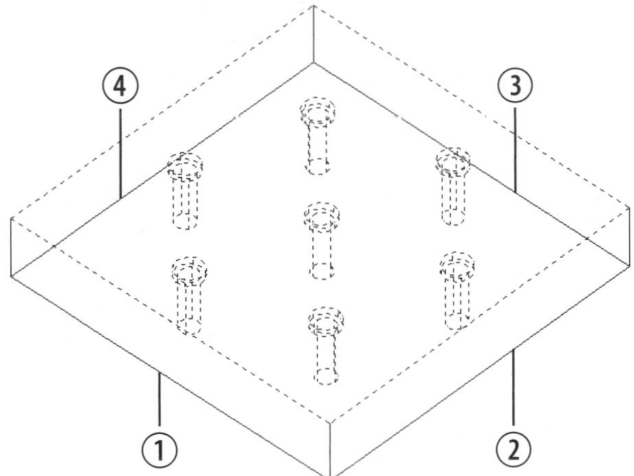

Figure 25–33 *Removing faces by selecting common edges.*

In the next step you will take advantage of Boundary Set selection to remove faces from the current set.

8. While you are still being prompted to select faces, choose the Remove option. When you are prompted to remove faces, pick on the sides, top, and bottom faces by picking at ①, ②, ③, ④, ⑤, ⑥ in Figure 25–34. Note that selecting at ① chooses the top face because it is "closest" in this view; picking at ② then removes the bottom face. Press Enter to end selection.

   In the next step you will specify the axis of rotation by picking the center of the array and then specifying a vector.

9. Answer the prompt to specify a rotation axis point by specifying a center Osnap and then picking the at the center of the central hole. Specify the second point on the rotation axis by typing the relative coordinate **@0,0,1** and pressing Enter. Specify a rotation angle by typing **-15** and pressing Enter. The array of holes rotates 15 degrees clockwise.

   In the next steps you will use the Offset option of SOLIDEDIT to edit the diameter of the central countersink hole.

10. Choose Offset from the face edit choices by typing **O**. Select faces by typing **C,** pressing Enter, and specifying a crossing window that encloses only the center countersink hole. Remove the top and bottom faces as in step 8 and press Enter to end selection.

11. Specify the offset distance by typing **-1.5**. Notice that supplying a negative offset distance decreases the volume of the solid by increasing the diameter of the countersink hole. End the command by pressing Enter twice.

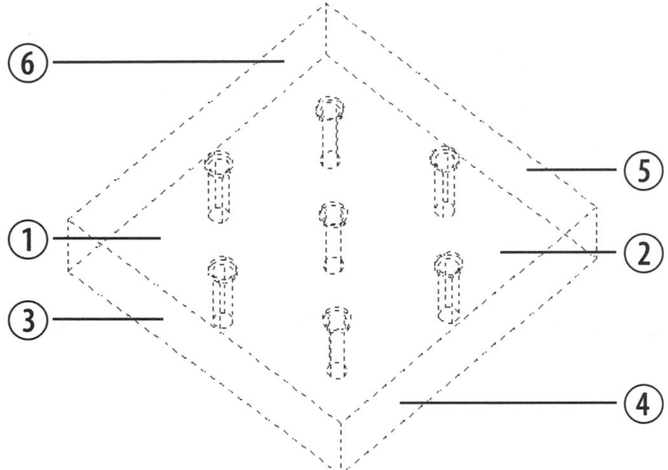

Figure 25–34 *Using Boundary Set selection to remove faces.*

Distinguishing between Extrude and Offset

Although similar in their effect, the Offset and Extrude options of the SOLIDEDIT command can yield quite different results. In Figure 25–35, the faces highlighted on the center solid are extruded and offset by the same amount in the solids on the left and right. The solid on the left underwent an extrusion, while the solid on the right was offset.

In the following exercise you will investigate the Move, Delete, and Taper options of the SOLIDEDIT command.

EXERCISE: USING THE MOVE, DELETE, AND TAPER OPTIONS OF SOLIDEDIT

1. If necessary, open 25DWG08.DWG. Thaw the layer DELETE and set it current. Freeze the layer EXTRUDE. Use the VIEW command to set the view EXTRUDE current. The drawing should resemble Figure 25–36.
2. To move the filleted structure on top of the block, from the Modify menu, choose Solids Editing>Move Faces. Pick the filleted face at ① and the cylinder at ② in Figure 25–36. Press Enter to end selection. Specify a displacement by typing **25,10** and pressing Enter twice.
3. Specify the Delete option by entering **D**. Select the filleted face again and press Enter to end selection and remove the face.
4. To taper the cylinder, choose the Taper option. Pick the cylinder face using Boundary Set selection and then press Enter. Use a center Osnap and pick the lower or upper center of the cylinder. Specify another point along the axis of

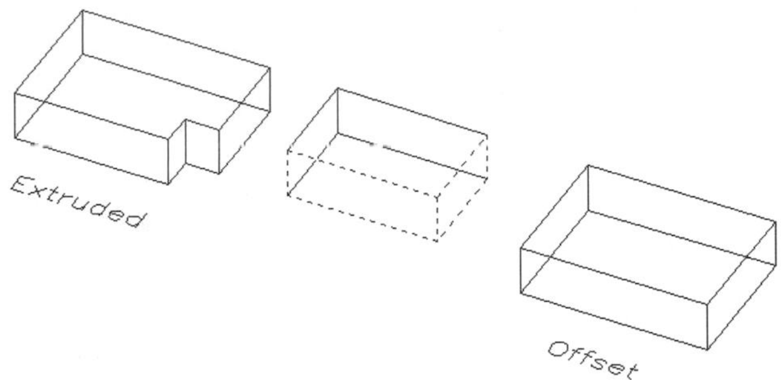

Figure 25–35 *The Extrude and Offset options yield different results.*

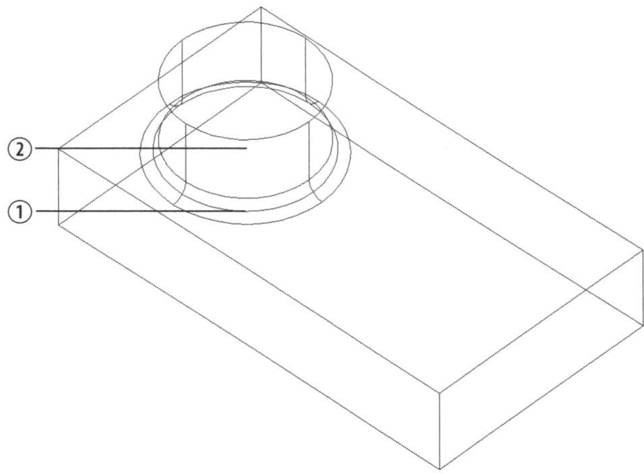

Figure 25–36 *Using the Move and Delete options of SOLIDEDIT.*

tapering by typing **@0,0,1**. When you are prompted for the taper angle, enter **6**. Your drawing should now resemble Figure 25–37.

In the next steps you will taper the four side faces in the drawing.

5. Choose the Taper option again. Referring to Figure 25–37, pick the four side faces of the model by picking at ①, ②; then pick at ③ twice to select the top and then back face; and then pick at ④. To remove the top and bottom faces, choose the Remove option and pick at ⑤ twice. Press Enter to end face selection.
6. Specify the taper base point by using an endpoint Osnap and picking at ⑥. Specify another point along the taper axis by using an endpoint Osnap and picking at ⑦.
7. Specify the taper angle by typing **-12**. This completes your work on this layer. Your drawing should now resemble Figure 25–38.

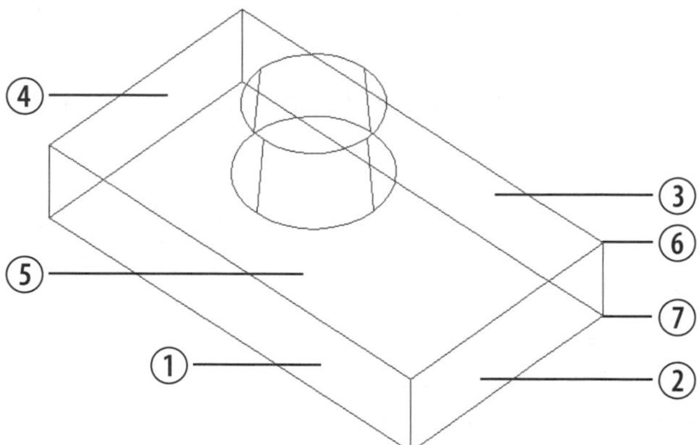

Figure 25–37 *Using the Taper option of SOLIDEDIT.*

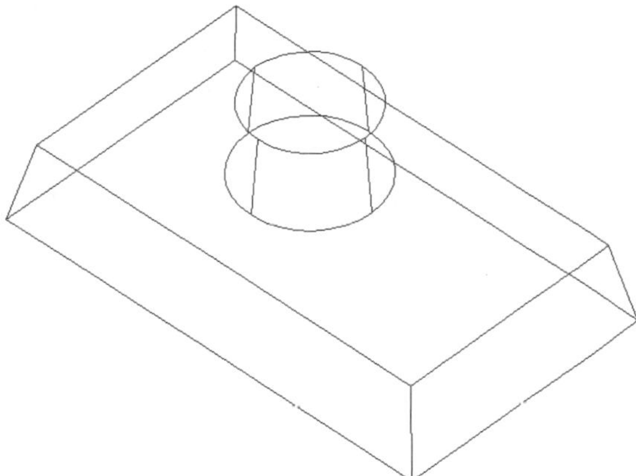

Figure 25–38 *The finished model with faces moved, deleted, and tapered.*

Note that the rotation of the taper angle is determined by the selection sequence of the base point and the second point along the taper axis. Tapering the selected face with a positive angle tapers the face in, and using a negative angle tapers the face out. If the pick sequence of ⑥ and ⑦ in step 6 had been reversed, for example, the specified tapering angle of -12 degrees would have resulted in an "inward" taper.

UNDERSTANDING THE IMPRINT AND SHELL OPTIONS

The Imprint and Shell options of SOLIDEDIT, although they are not technically editing, provide ways to alter a solid object. You can, for example, create new faces or solids by imprinting arcs, lines. circles, etc. onto existing faces. Once these objects are imprinted, you can use the imprinted face as the basis for extrusions. With the shell option, you can "hollow out" a solid forming a thin wall of specified thickness.

In the following exercise you will use the Imprint and Shell options of the SOLID-EDIT command to develop the object further in 25DWG08.DWG.

EXERCISE: USING THE IMPRINT AND SHELL OPTIONS OF SOLIDEDIT

1. In 25WDG08.DWG, thaw and set the IMPRINT layer current and freeze the DELETE layer. Use the VIEW command to set the view IMPRINT current. Then use the UCS command to restore the IMPRINT UCS. Your drawing should resemble Figure 25–39.

   In the next steps you will draw a rectangle on the face of the box.

2. Start the RECTANGLE command by typing **RECTANG** and pressing Enter. Type **F** to choose the Fillet option, and then specify a fillet radius of five units. When you are prompted for the first corner of the rectangle, type **8,40**. Enter coordinates of **46,10** for the opposite corner.

3. To imprint the rectangle on the face, from the Modify menu, choose Solids Editing>Imprint. Pick anywhere on the solid. When you are prompted to select an object to imprint, select the rectangle. Press Enter to retain the source object. Press Enter to end the select objects prompt. Your drawing should resemble Figure 25–40.

   Once objects are imprinted on a solid, they can form the basis for both positive and negative embossments.

4. Continuing from the previous step, press Enter to display the parent solid edit command line prompts. Choose the Face option by typing **F**, then choose the Extrude option.

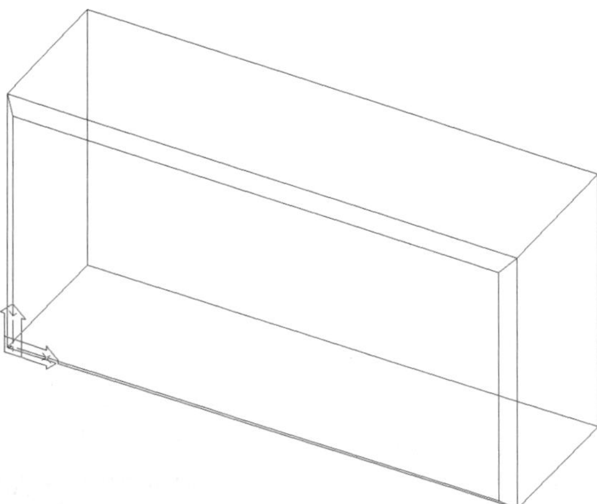

Figure 25–39 *Preparing to add imprinted figures.*

5. Select the rectangular imprint by picking at ① within the boundary of the rectangle and pressing Enter (see Figure 25–40). Press Enter again to end selection of faces. Specify a height of extrusion of 2.5 and press Enter to accept the default of a zero-degree angle of extrusion. Press Enter twice to end the SOLIDEDIT command. Your drawing should resemble Figure 25–41.

   In the following steps you will use the Shell option to "hollow out" the solid.

6. From the main menu, choose Modify>Solids Edits>Shell. Pick anywhere on the solid and then press Enter because you do not want to remove any faces.

7. Enter a shell offset distance of **2** and then press Enter twice to end the SOLIDEDIT command.

   In the next steps you will use the SLICE command to examine the effects of the previous Shell operation.

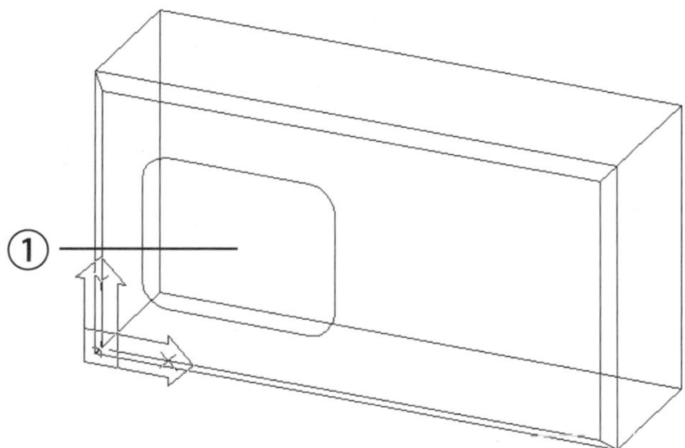

Figure 25–40 *Imprinted figures on a solid face.*

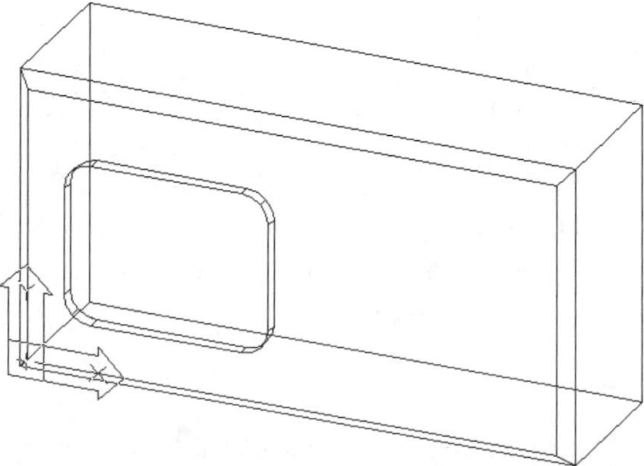

Figure 25–41 *Adding an embossed surface with Imprint.*

8. Type **UCS** and press Enter twice to change the UCS to the WCS. From the main menu, choose Draw>Solids>Slice, and then pick anywhere on the solid and press Enter.
9. Enter the following three sets of coordinates to define a slicing plane. After typing each set of coordinates, press Enter: **48,70,0,** then **48,140,0,** then **48,100,1**. When you are prompted to specify a point on the desired side of the slicing plane, pick near ① in Figure 25–42.
10. Use the HIDE command (type **HI** and press Enter) to remove hidden lines from the view. Your drawing should resemble Figure 25–42. Note that the original rectangle used to generate the imprint remains at ②.

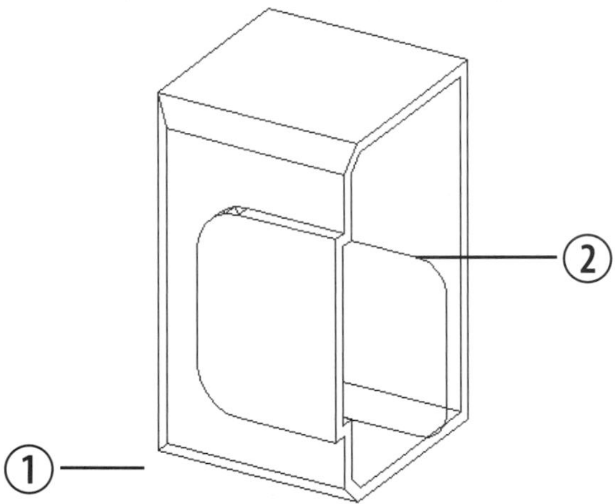

Figure 25–42 *The result of using the Shell option.*

SUMMARY

Solid modeling is a powerful method of creating complex models by combining and editing solid primitives and extruded 2D shapes. To make these models, you can use the Boolean operations of Union, Subtract, and Intersect, as well as the EXTRUDE, SLICE, REVOLVE, and SECTION solids commands. The SOLIDEDIT command allows you to copy and modify solid edges and faces once they are created. You can also imprint shapes onto solid objects and create shells from parent solids.

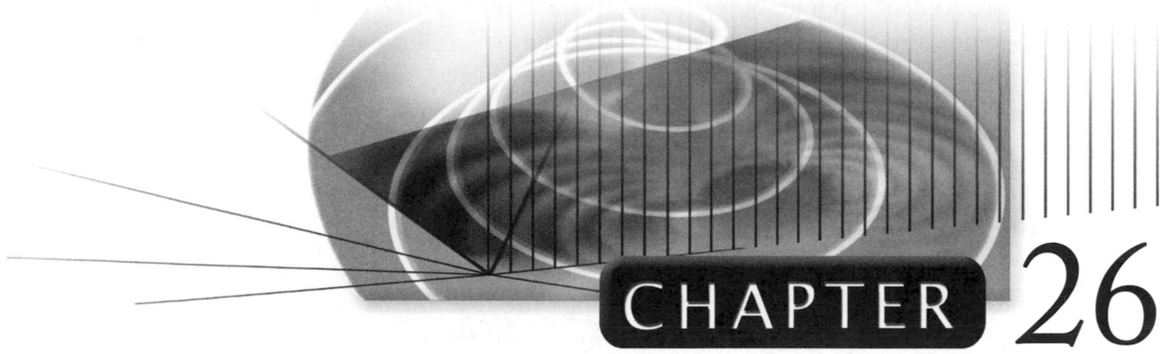

Rendering Your Model

Up to this point you have explored several different methods of modeling in 3D in AutoCAD. To produce a "picture" of the model, you will probably want to render it. Rendering is a process in which you attach materials to the surfaces of your 3D objects, create lights to illuminate the objects, establish a viewpoint from which to view the model, and then transform the scene into a realistic rendering, or picture.

This process is generally straightforward, but it can require a bit of trial and error, as well as an understanding of artistic and photographic principles, to produce good results. This chapter focuses on the basics of rendering and the rendering tools found in AutoCAD. It is beyond the scope of this chapter to cover all aspects of rendering; you are encouraged to investigate many of AutoCAD's more advanced rendering capabilities on your own. The topics of this chapter will serve to introduce the basic rendering process, including how to set up and create simple renderings inside AutoCAD. In particular, this chapter addresses the following topics.

- Types of renderings supported by AutoCAD
- Creating a view
- Creating and assigning materials
- Creating lights
- Rendering the model
- Generating output

TYPES OF RENDERING SUPPORTED BY AUTOCAD

AutoCAD incorporates a full-featured rendering engine that is capable of producing renderings ranging from simple shaded models all the way to full photo-realistic renderings. Generally, there are three types, or levels, of renderings offered by the rendering facility inside AutoCAD.

- Render
- Photo Real
- Photo Raytrace

These three different types of renderings produce increasingly more realistic outputs, with Photo Raytrace being the most realistic of the three. A price is paid for realism, however, because with increased realism comes increased rendering times.

The line of demarcation between shading and rendering is somewhat arbitrary. The shading modes available with the SHADEMODE command (which is discussed in Chapter 22, 3D Fundamentals) could be considered simple rendering because the surfaces of a model are depicted as they appear with a single light source. AutoCAD's RENDER command, however, takes up where simple shading leaves off in providing the abilities to apply materials to surfaces, to supply various types of lighting, and to cast realistic shadows. The three levels of rendering offered with the RENDER command are summarized as follows:

- **Render:** Basic Render is the next logical step beyond the Gouraud-shaded mode offered by the SHADEMODE and 3DORBIT commands. Render mode has the added advantage of allowing material assignment to surfaces. Materials—such as brick, chrome, or wood—can be assigned to objects in the model on a per-layer, per-color, or per-object basis. In addition, three types of light sources can be added to the scene. In the basic Render mode, however, light sources are not capable of casting shadows. Other enhancements, such as backgrounds and plants, are also supported. A typical rendering using the Render mode is shown in Figure 26–1.
- **Photo Real:** The Photo Real mode shares all the features of Render mode, plus it adds the ability to cast shadows and use bitmaps for materials. Figure 26–2 shows a rendering using the Photo Real mode.
- **Photo Raytrace:** The Photo Raytrace mode provides the most realistic renderings. It adds the ability to generate reflections, refraction effects, and true, detailed shadows. Figure 26–3 shows a rendering using the Photo Raytrace mode.

THE RENDERING PROCESS

Rendering takes place after you have constructed a 3D model. Hopefully the preceding chapters have given you the knowledge to allow you to build models that you want to render.

Figure 26–1 *A typical rendering using the Render mode of the RENDER command. Note the absence of shadows.*

Figure 26–2 *A typical rendering using the Photo Real mode of the RENDER command. Shadows and materials are present.*

Figure 26–3 *A typical rendering using the Photo Raytrace mode of the RENDER command. Note the presence of shadows and reflections.*

Now it is time to take a look at how to set up your scenes for rendering. In general, you can follow a simple process to render your scenes:

1. Create a view of the scene. Most of the views in AutoCAD are orthogonal, such as a SW isometric. For true realism, you must create a perspective view by using DVIEW.
2. After the perspective is set up, create and assign the materials in the scene. A material is a set of surface attributes that describe how that surface looks at rendering time. You must define these attributes and assign them to the appropriate surfaces.
3. After you create your materials, create lights for the scene. Without lights, you do not have any illumination or shadows in the scene. Correct placement of lights adds to the realism.
4. Once you have set up the scene, you can begin to create test renderings of it. Make sure that the materials, lights, and geometry are correct. Most of the time you will create your test renderings with the Photo Real method. You might end up creating dozens of test renderings before achieving the look you want.
5. Set up the final Photo Raytrace rendering and save the rendering to a bitmap file for printing or for use outside of AutoCAD.

These are the basic steps necessary to create a rendering. Now it is time to take a closer look at each step, starting with creating a view.

CREATING A VIEW

Establishing a compelling view of the model is one of the most important steps in creating an effective rendering. No matter how much effort and time are devoted to selecting and applying materials and establishing realistic lighting, a rendering of the model from an uninteresting point of view will detract from your efforts. Put another way, an interesting or even dramatic view of the model can make the difference between an average-looking rendering and a memorable one.

During the construction of the model, the 3DORBIT command can be effectively used to view and study the various parts of the model. 3DORBIT's ability to change the viewpoint dynamically in real time is useful to gaining a sense of the model's spatial relationships. For the final rendering, however, you generally will want more control of the viewpoint. Such flexibility and control are offered by the DVIEW command. Like 3DORBIT, DVIEW allows you to view the model from a perspective projection, which is almost always preferable because it is the way we view objects in the real world. In addition, DVIEW allows you to set and fine-tune such factors as the camera-to-subject distance and the field of view.

To use DVIEW to establish a perspective view of your model, you must know two things:

- **Camera point:** The location from which you want to be looking in the model
- **Target point:** The location that you want to look at in the model

As soon as you establish these points, you can adjust the perspective until you are happy with it. If you have any experience with 35 mm photography, many of the terms—such as *focal length*, which is used to adjust viewing angle of the camera—will be familiar to you.

After you set up the view by using DVIEW, you use the VIEW command to save the view so that you do not have to recreate it later.

The following exercise shows you how to set up a view using the DVIEW command. You will then save the view with the VIEW command.

EXERCISE: SETTING UP A VIEW WITH DVIEW

1. Load the 26DWG01.DWG file found on the accompanying CD. The file contains a model of a twelve-story office building, as shown in Figure 26–4. For reference, the roof of this building has an elevation of approximately 154 feet above ground.

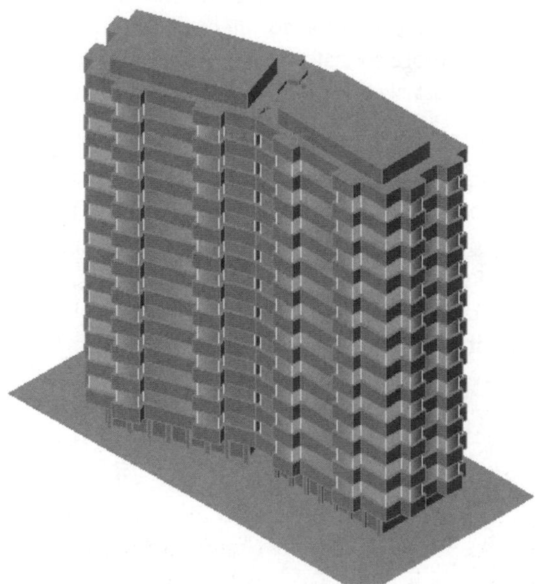

Figure 26–4 *A model of a twelve-story office building shown in an isometric, Gouraud-shaded view.*

2. From the View menu, choose Shade>2D Wireframe. Then, from the View menu, choose Named Views. In the View dialog box, right-click on view AAA and then choose Set Current from the shortcut menu. Choose OK to exit the dialog box. The model should now resemble Figure 26–5.

   In the next three steps you will draw the line of site, or sightline, that will establish the viewpoint for the rendering. You will place a line from the intended camera point to the target point.

3. From the Draw menu, choose Line. When you are prompted to specify the start point, type **.XY** and press Enter to indicate that you will specify the X and Y coordinates on the screen. Pick a point near ① in Figure 26–6. (This point should be near 163', -174'.)

4. When you are prompted for the Z value, type **8'-5"**. This establishes the camera position for the view.

5. When you are prompted for the next point, again type **.XY** and pick a point near ②. (This should be near 24', 33'.) When you are prompted for the Z value, type **83'**. Press Enter to end the LINE command. The completed sightline should resemble that shown in Figure 26–6. This line starts 8'5" off the ground and aims at a point a little more than halfway up the building.

   In the next steps you will use the Points option of the DVIEW command to establish a view along the sightline.

6. To start the DVIEW command, at the command line, type **dview** and press Enter. When you are prompted to select objects, pick only the sightline, and then press Enter. Note that all of the model's objects except the line you drew in steps 3-5 temporarily disappear.

7. Type **PO** and press Enter to select the Points option of the DVIEW command.

8. When you are prompted for the target point, use an endpoint Osnap and pick at ② in Figure 26–6 to specify the target end of the line. When you are prompted for the camera point, again use an endpoint Osnap to pick at ①. Although all of the model's objects seem to have disappeared at this point, DVIEW has established a view looking down the sightline from the camera to the target point.

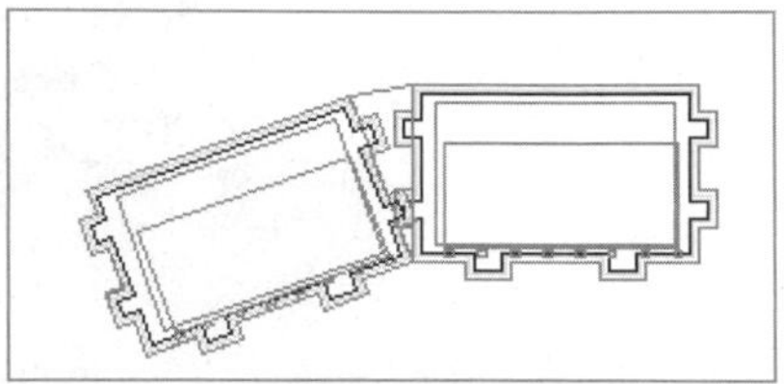

Figure 26–5 *A model of a twelve-story office building shown in plan view.*

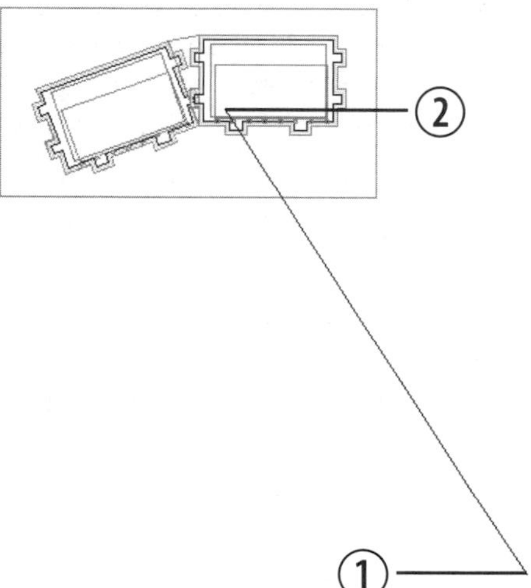

Figure 26–6 *The first step in establishing a view is drawing the sightline from the camera to the target.*

9. Type **D** and press Enter to choose the Distance option of the DVIEW command. Press Enter to accept the default distance—the length of the sightline.
10. At the next prompt, type **Z** and press Enter to select the Zoom option, and then specify a lens length of 33 mm and press Enter. Press Enter again to end the DVIEW command. The resulting view of the model should resemble Figure 26–7.
11. To generate a better view of the building, from the View menu, choose Shade>Hidden. Figure 26–8 shows a hidden-line view.
12. Use the VIEW command to save this view under the name Camera1.
13. Save this drawing as 26DWG01A.DWG. You will use it in a later exercise.

Note: When you set up a view using the sightline method, be careful when you adjust the camera-target distance. If you have objects representing a ground plane, increasing the distance using the Distance option of the DVIEW command may place the camera point below this plane. This will result in a rendering in which all of the model's geometry is hidden.

The advantage to using the sightline method with the DVIEW command is that you have complete control of the resulting viewpoint. Fine-tuning of distance, field of view, and target point and direction make establishing the exact view that you have in mind a straightforward matter of adjusting the sightline and then using the DVIEW command and the method in the preceding exercise to generate the view.

Figure 26–7 *The view resulting from looking "down" the sightline.*

Figure 26–8 *A view of the model with hidden lines removed.*

CREATING AND ASSIGNING MATERIALS

Once you have established a view of the model, you can begin assigning materials to the model's surfaces. A material is a set of attributes assigned to a surface. Attributes include such qualities as color, smoothness, reflectivity, texture, and transparency. When a model is rendered, the rendering engine takes into account those attributes of the image accordingly.

Materials are handled through the Materials and Materials Library dialog boxes in AutoCAD's built-in rendering engine. The Materials dialog box is used to assign, create, or modify materials and their associated set of attributes. The Materials Library dialog box is used to store a group of predefined materials or materials that you create.

You use the Materials Library dialog box to choose materials that you want to import into your model for assignment to surfaces or for use as the basis for creating a modified or new material. You access this dialog box by choosing Materials Library from the Render toolbar, or by selecting Render>Materials Library from the View menu, or by typing in **MATLIB** at the command prompt. Figure 26–9 shows the Materials Library dialog box.

The Materials Library dialog box is divided into three main sections:

- **Current Drawing List**—A list of all the materials currently loaded for use or that are assigned in the current drawing.
    - **Purge**—Deletes all unassigned materials from the Current Drawing list
    - **Save As**—Allows you to save the current drawing list to a materials library (MLI) file
- **Preview window**—This is a small window that gives you a preview of what a material would look like if you applied it to a sphere or cube object.
    - **Import**—Adds materials selected in the Current Library list to the Current Drawing list

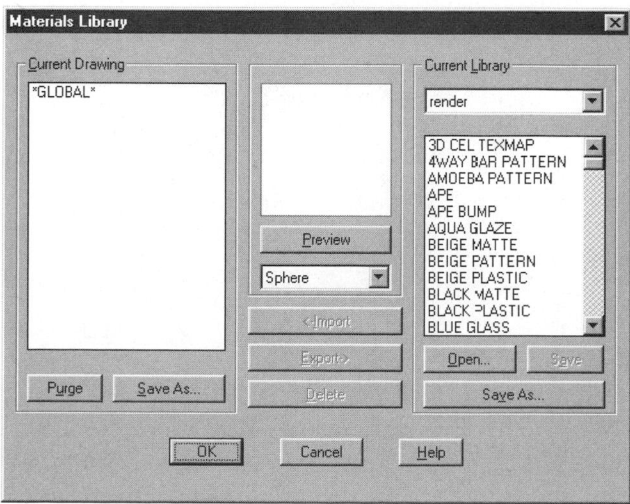

Figure 26–9 *Materials Library dialog box.*

- **Export**—Adds materials selected in the Current Drawing list to the Current Library list
- **Delete**—Deletes materials selected in the Current Drawing list or the Current Library list

- **Current Library list**—A list of all the materials contained in the current library. All material libraries for AutoCAD have an "MLI" extension. By default, AutoCAD is shipped with one library file—the Render.MLI library.
    - **Open**—Displays a standard file selection dialog box listing MLI files
    - **Save**—Saves the changes to the current MLI file in the current folder
    - **Save As**—Displays a standard file selection dialog box, where you can specify the name of the materials library (MLI) file in which AutoCAD saves the Current Library list.

Tip: The small Preview window in the Materials Library dialog box is a 256-color display. This means that the final rendering will invariably look better than this simple preview window. Keep this in mind when you select materials. If you are not sure about how a material will look, use AutoCAD's Multiple Drawing Environment to create a new drawing with a simple object in it. Use this second drawing to test how a material will appear in the larger model.

You use the Materials dialog box (which is shown in Figure 26–10) to manage the materials selected for use in your model. You access the Materials dialog box by choosing Materials from the Render toolbar, or by selecting Render>Materials from the View menu, or by typing in RMAT at the command prompt The Materials dialog box provides the following functions:

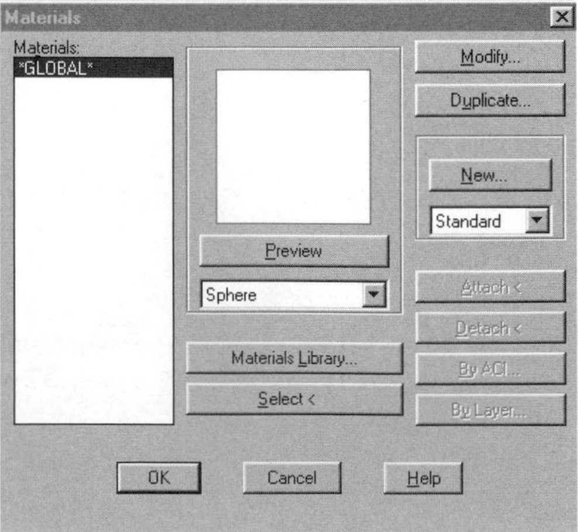

Figure 26–10 *Materials dialog box.*

- **Materials**—Lists the available materials. The default for objects with no other material attached is GLOBAL.
- **Preview**—Displays a selected material on either a sphere or a cube.
- **Materials Library**—Displays the Materials Library dialog box, from which you can select a material.
- **Select**—Closes the dialog box temporarily so that you can select an object with the pointing device and display the attached material. After you select the object, the Materials dialog box is redisplayed with the method of attachment specified at the bottom of the dialog box.
- **Modify**—Displays one of four dialog boxes, depending on which material type is selected in the list under the New button (Standard, Marble, Granite, or Wood). Use the dialog box to edit an existing material.
- **Duplicate**—Duplicates a material and displays one of four dialog boxes, depending on which material type is selected in the list under the New button (Standard, Marble, Granite, or Wood). Use the dialog box to name the new material and define attributes.
- **New**—Displays one of four dialog boxes, depending on which material type is selected in the list under the New button.
- **Attach**—Closes the dialog box temporarily so that you can select an object and attach the current material to it.
- **Detach**—Closes the dialog box temporarily so that you can select an object and detach the material from it.
- **By ACI**—Displays the Attach by AutoCAD Color Index dialog box, where you can select an ACI to which to attach a material.
- **By Layer**—Displays the Attach by Layer dialog box, where you can select a layer to which to attach a material.

The following exercise shows you how to load a few materials from the materials library and assign them to objects in a scene.

EXERCISE: ASSIGNING MATERIALS FROM THE MATERIALS LIBRARY TO A MODEL

1. Load the file 26DWG01A.DWG, which you worked on in the last exercise. If you did not complete the last exercise, load the file from the accompanying CD. Display the Render toolbar by right-clicking on any displayed toolbar and selecting Render from the toolbar list.
2. From the Render toolbar, choose Materials, or type in **RMAT** at the command prompt. AutoCAD displays the Materials dialog box, as shown in Figure 26–10.

3. Select the Materials Library button to launch the Materials Library dialog box (see Figure 26–9).
4. Select the material named Blue Glass on the right.
5. Choose the Import button. This places the Blue Glass material on the list of materials available for use in the current drawing.
6. Repeat steps 4 and 5 for the Dark Brown Matte material.
7. Choose OK to return to the Materials dialog box, which should appear as shown in Figure 26–11. The materials you just imported now appear in the Materials list.
8. Select the Blue Glass material from the list.
9. Choose By Layer and, in the Attach by Layer dialog box, select the glass layer from the Select Layer list. Select Attach to attach the Blue Glass material to objects on the glass layer.
10. Repeat steps 8 and 9 to attach the Dark Brown Matte material to the concrete layer. The Attach by Layer dialog box should resemble Figure 26–12. Choose OK to exit the Attach by Layer dialog box. Choose OK again to exit the Materials dialog box.
11. Choose Render from the Render toolbar to display the Render dialog box. Under Rendering Type, select Photo Raytrace. Under Rendering Options check that the Smooth Shade, Apply Materials, and Shadows option are all selected. The setting for this rendering are shown in Figure 26–13.
12. Choose Render. The scene is rendered as shown in Figure 26–14. Depending on your equipment, the rendering process may take a minute or two.
13. Save the file as 26DWG01B.DWG for use later.

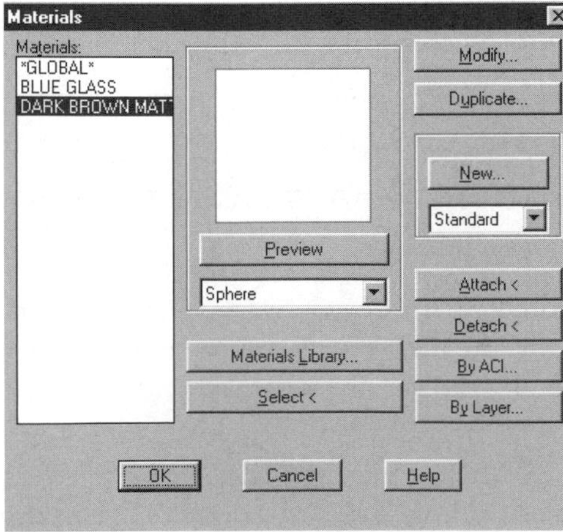

Figure 26–11 *The Materials dialog box after you select materials for use in the model.*

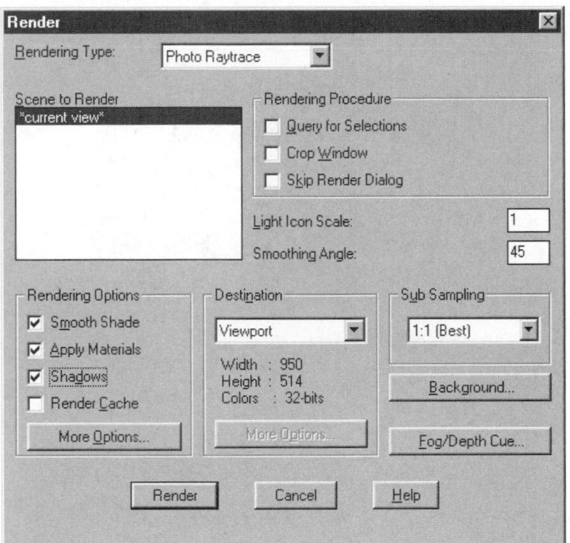

Figure 26–12 *The Attach by Layer dialog box shows current layer/material attachments.*

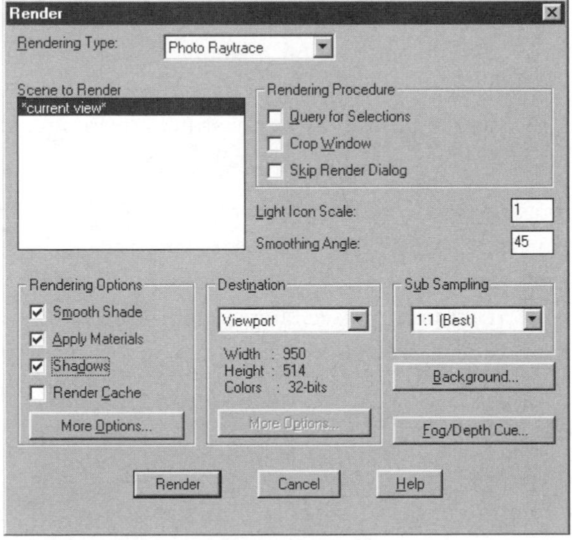

Figure 26–13 *The Render dialog box settings.*

This exercise shows that even with simple "off-the-shelf" materials and no lighting effects, effective renderings are possible. The key to such renderings is the viewpoint. In this example, a perspective viewpoint from near the ground looking upward at the building yields a life-like view—one you might see if you were standing in front of the building.

Figure 26–14 *The rendered model with materials applied.*

In this exercise you assigned materials on a "by-layer" basis. You can also assign materials on a "by-color" or "by-object" basis. The advantage to assigning materials by layer is that it results in a more orderly material assignment process. Just as layers in 2D drafting help to organize a drawing, placing objects in your model on layers that share a rendered material makes material assignment much more orderly and easier. Because the "by-layer" method offers a distinct advantage, you should adopt a layer-conscious scheme as you construct a model.

Materials can be roughly divided into two categories: basic materials and mapped materials. Mapped materials make use of a bitmap, or image, to represent the color or some other attribute of a material. A brick material, for example, may have a bitmap of the brick pattern superimposed along with the other material attributes. Basic materials achieve their appearance without the use of bitmaps. Mapped materials may require more effort to define, but they can yield more realistic results with materials that exhibit prominent textures or patterns.

Basic Materials

A basic material is a simple material that does not use any sort of bitmap. Generally these are materials that have no prominent surface texture or significant surface pattern. Examples of such materials include metals, paints, plastics, or glass.

Looking at the Modify Standard Material dialog box shown in Figure 26–15, you will see the attributes that you can modify. Each is briefly described as follows:

- **Color/Pattern**—This is the general color of the material. It is defined in either the RGB (Red, Green, Blue) color system or the HSV (Hue, Saturation, Value) color system. The color can also be derived from the ACI color value of the object to which it is assigned. The Value slider is used to control the overall intensity of the color.

- **Ambient**—This is the color of the material when it has a shadow cast on it. Generally speaking, this is simply a darker version of the Color/Pattern color.
- **Reflection**—This attribute determines the amount of reflection the material has. The Value slider determines the strength of that reflection. In general, most reflections are subtle and have a value of 0.20 or less. The Lock check box can be used to lock all the colors together, and the Mirror check box can be used to turn the material into a mirror, based on the color. For example, a true mirror material would have Lock and Mirror turned on and have a white Color/Pattern value.
- **Roughness**—This attribute adds roughness to the material. The adjustment you make here controls the value of the roughness. The higher the value, the rougher the material appears.
- **Transparency**—This attribute is used to determine the amount of light that passes through an object. For example, glass is highly transparent, whereas concrete is not. The amount of transparency is controlled with the Value slider.
- **Refraction**—Refraction is the bending of light as it passes through an object. For example, if you look at a pencil in a glass of water, from the side the pencil appears bent or broken. You can set the Value slider to determine the amount of refraction.
- **Bump Map**—This is the only attribute that makes a reference to a bitmap. A bump map is used to make the surface of an object, such as mortar joints, appear to have more detail or texture without you having to model the joints themselves. This detail and appearance of texture come from a bitmap, such as one created from a photograph of a set of bricks.

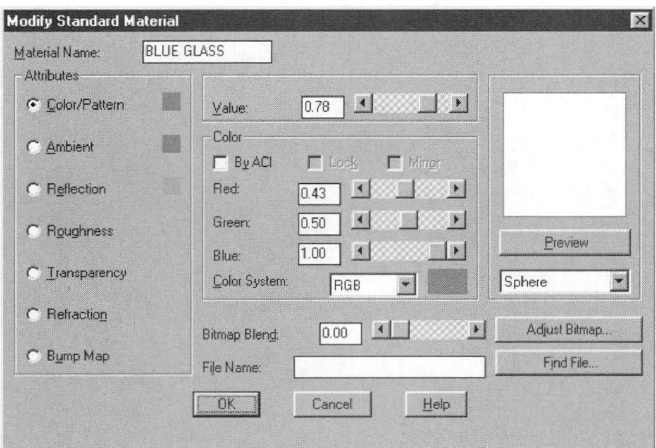

Figure 26–15 *The Modify Standard Material dialog box is used to assign attributes to a new material.*

By setting one or more of these attributes, you can create just about any material you want. In the following exercise you will create a new glass material for use in the model in 26EX01B.DWG from the preceding exercise. Because the new material is a modification of an existing material, most of the material attributes will remain unchanged. Other attributes will be changed slightly to yield a more appropriate material.

EXERCISE: CREATING A SIMPLE MATERIAL

1. Continue in or open 26DWG01B.DWG from the preceding exercise. If necessary, display the Render toolbar. From the Render toolbar, choose Materials, or type in **RMAT** at the command prompt.
2. From the Materials list, select the Blue Glass material.
3. Select Duplicate. AutoCAD displays the New Standard Material dialog box.
4. In the Material Name input box, type the name **Dark Blue Glass**.
5. Make sure that the Color/Pattern radio button is active. Verify that the ACI check box is off. In the Color section of the dialog box, select the color swatch. This launches the Color dialog box, as shown in Figure 26–16.
6. Set the Red, Green, and Blue values to **0**, **20**, and **160**, respectively, which creates a dark blue.
7. Choose OK to return to the New Standard Material dialog box. With Color/Pattern still selected, set the Value slider to 0.15.
8. Select the Ambient attribute radio button, and then select the Lock check box to place a check in it.
9. Select the Reflection radio button and set the Value slider to 0.10.

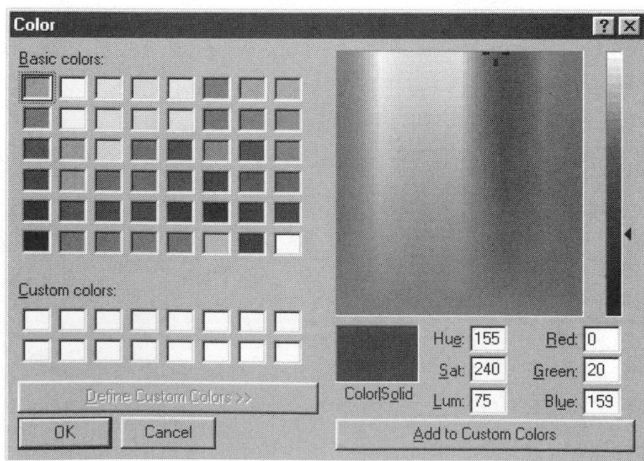

Figure 26–16 *The Color dialog box is used to assign colors to materials.*

10. With the Reflection attribute still selected, select the color swatch in the Color section to display the Color dialog box as you did in step 4. Set the Red, Green, and Blue values to **0**, **7**, and **77**, respectively. Choose OK to close the Color dialog box.

    With the Reflection attribute still selected, set the Value slider to 0.10. The New Standard Material dialog box should now resemble Figure 26–17.

11. Choose OK to return to the Materials dialog box. The Dark Blue Glass material now appears in the Materials list.
12. With Dark Blue Glass selected in the Materials list, select the By Layer button. The Attach by Layer dialog box is displayed.
13. From the Select Layer list, choose the glass layer, and then select the Attach button to assign Dark Blue Glass to all objects on the glass layer.
14. Choose OK to close the Attach by Layer dialog box. Choose OK again to close the Materials dialog box.
15. From the Render toolbar, choose Render to display the Render dialog box. Check that the settings duplicate those shown in Figure 26–13.
16. Choose the Render button. AutoCAD renders the model. The model should now resemble Figure 26–18.
17. Leave this drawing open. You will use it in the next exercise.

As you can see in this exercise, creating a new basic material is not difficult. As you did in this exercise, you can frequently use a currently defined material and modify one or more attributes to create the new material. The existing material serves as a template for the new material.

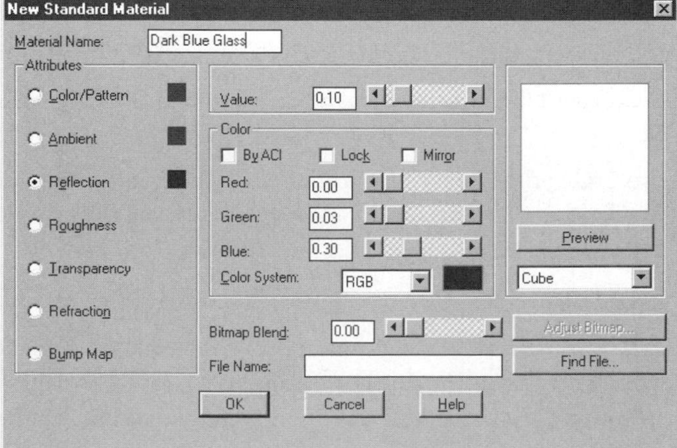

Figure 26–17 *The New Standard Material dialog box after you define the new Dark Blue Glass material.*

Figure 26–18 *The office building model with the new Dark Blue Glass material applied to the glass layer.*

Once you have created a new material, you may want to save it for use in other models. The following exercise demonstrates how to add a newly created material to the materials library.

EXERCISE: SAVING A NEW MATERIAL

1. Continue in 26DWG01B.DWG from the preceding exercise. From the Render toolbar, choose Materials, or type in **RMAT** at the command prompt.
2. Select the Materials Library button.
3. In the Materials Library dialog box, select Dark Blue Glass from the Materials list.
4. Select the Export button to add the selected material to the Current Library list.
5. In the Current Library list box, scroll down to find Dark Blue Glass added to the list.
6. Choose OK. In the Library Notification dialog box, select Save Changes to save the modification. Choose OK to close the Materials dialog box.
7. You can save and close this drawing.

Mapped Materials

A mapped material varies from a basic material in that you can replace one or more of the attributes with a bitmap image. For example, by replacing the Color/Pattern attribute with a photograph image of a wood pattern, you can make the surface of an object appear to have that wood pattern.

Mapped materials are slightly more complicated to work with in AutoCAD because you also need to supply mapping coordinates. Mapping coordinates tell the rendering engine where and how to place the map on the surface of the object. Without correct mapping coordinates, the texture may not appear at the desired orientation or it may be out of scale with the rest of the scene.

When you look at the Modify Standard Material dialog box in Figure 26–19, you will see the bitmap controls below and to the right of the Color area. When you choose the Find File button, you can navigate to any directory on your system and select a bitmap file for use in the material. You can use any bitmap file format supported by AutoCAD. These include the BMP, JPG, PNG, TIF, TGA, PCX, and GIF formats.

Note: In order for you to use a particular bitmap in your scene, that bitmap should reside either in the same directory as the drawing or one of the directories listed under Texture Maps Search Paths in the AutoCAD options. Otherwise, AutoCAD may not be able to find the file.

After you select the bitmap, you can choose the Adjust Bitmap button to crop and trim the bitmap as you desire. Figure 26–20 shows the dialog box that appears when you select this button.

Although in many instances you will not need to adjust the bitmap, there are circumstances in which bitmap adjustment is necessary to obtain the effect you want.

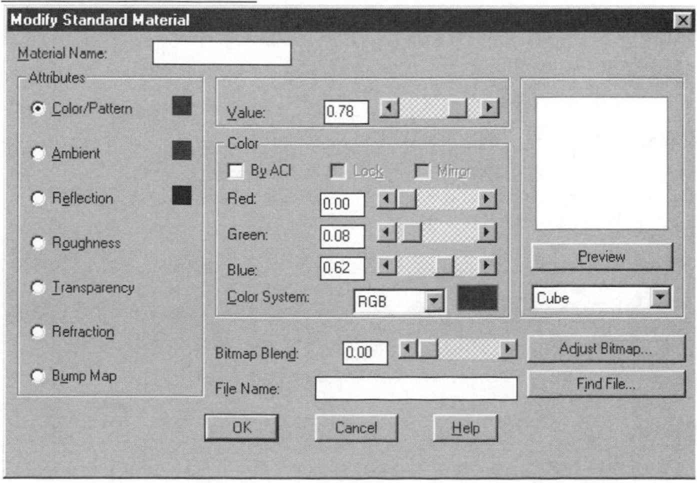

Figure 26–19 *The Modify Standard Material dialog box's bitmap controls appear near the bottom-right corner.*

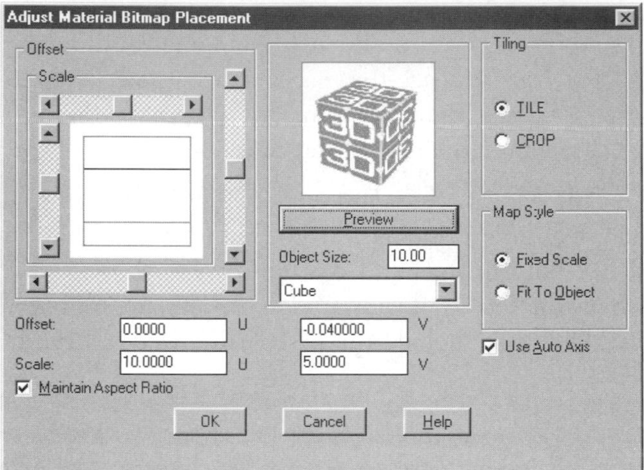

Figure 26–20 *Adjust Material Bitmap Placement dialog box.*

If, for example, you need to create a decal" map—such as the label on a bottle of wine—you can set the tiling to crop instead of tile, and only one copy of the image will appear on the object. Figure 26–21 shows the difference between a tiled material and a cropped material. You will learn more about mapping materials to objects later in this chapter.

The following exercise shows you how to create a wood grain material with a bump map.

EXERCISE: USING A BUMP MAP TO CREATE A WOOD GRAIN

1. Start a new drawing.
2. Load the Materials dialog box by choosing Materials from the Render toolbar or by typing in **RMAT** at the command prompt.
3. Choose New to create a new material.
4. Name the material by typing **Wood2** in the Material Name input box.
5. With the Color/Pattern radio button selected, choose Find File in the lower-right corner. This will display the Bitmap File dialog box. Make sure that the Files of Type input box is set to display *.tga files.
6. Navigate to the Textures directory in your AutoCAD installation.
7. Choose Teak.tga and select Open to open the file. At this point you are returned to the New Standard Material dialog box.
8. Set the Value slider to 1.00 and choose Preview to view the teak bitmap. Figure 26–22 shows the dialog box at this point.

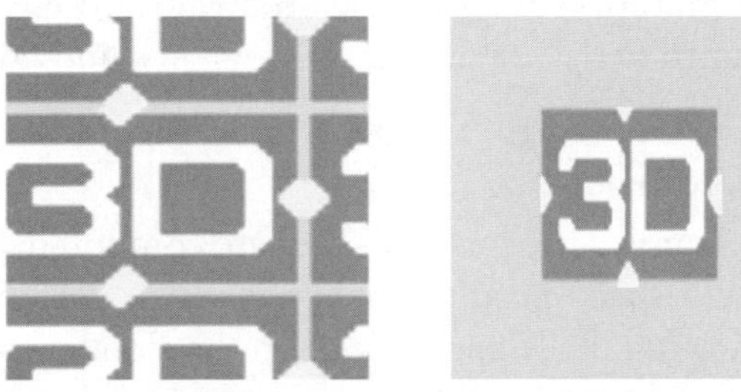

Figure 26–21 *A tiled material (left) and a cropped material (right).*

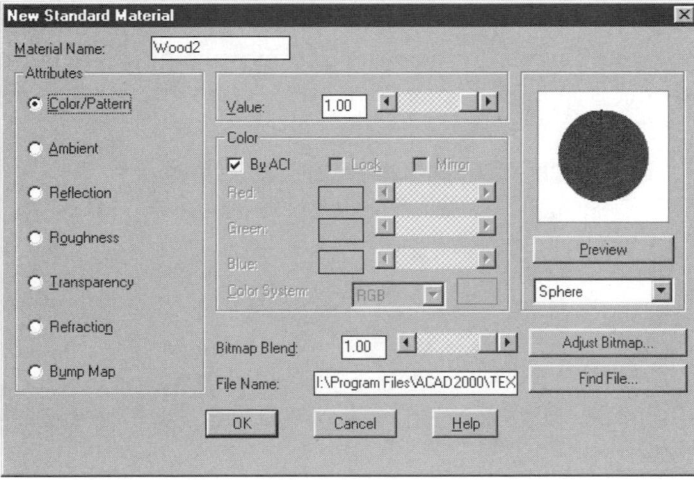

Figure 26–22 *The New Standard Material dialog box displaying the new Wood2 material.*

9. Select the Bump Map radio button.
10. Select OK to accept the new material and to return to the Materials dialog box.
11. The new Wood2 material now appears in the Materials list.
12. To save the new material and add it to the materials library, first select the Materials Library button.
13. In the Materials Library dialog box, select Wood2 from the Current Drawing list, and then select the Export button to add the material to the current library. Save the changes to the library by selecting the Save button.
14. Select OK to close the Materials Library dialog box and then close the Materials dialog box. Close the drawing without saving changes.

This exercise demonstrates that creating a mapped material is easy as long as you have the bitmap file and know its location. You can also supply your own bitmaps for use in defining new materials. Several libraries of third-party bitmaps suitable for use in making new materials are available. You can also create bitmap files by scanning photographs of materials such as metals, woods, and construction materials, to mention just a few.

Applying Mapping Coordinates

As mentioned earlier, when you use materials that are composed wholly or in part of bitmaps images, you will usually need to apply mapping coordinates to the object. This is accomplished through the Mapping icon on the Render toolbar, or by typing in **SETUV** at the command prompt. You are then prompted to select the objects to which you want to apply the mapping coordinates. Once the objects are selected, the Mapping dialog box appears (see Figure 26–23).

AutoCAD supports four types of mapping: planar, spherical, cylindrical, and solid. The first three generally refer to how the bitmap is wrapped around the object to which the mapping is applied. Figure 26–24 shows the various mapping types.

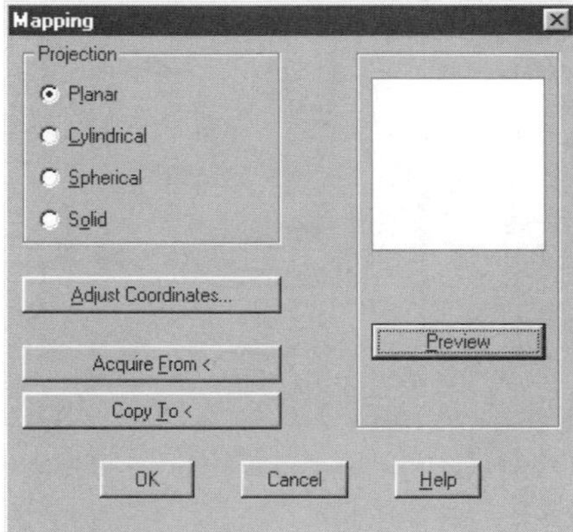

Figure 26–23 *The Mapping dialog box, where you select the type of mapping to apply to an object.*

Figure 26–24 *The four mapping types in AutoCAD: planar, cylindrical, spherical, and solid.*

The last mapping type is the solid mapping type. This type of mapping is intended for use with procedural materials such as wood, marble, and granite. These materials are generated based on mathematical formulas and do not make use of bitmaps. The solid mapping coordinates are used when you have a procedural material assigned to an object with highly varying geometry, such as a box with a sphere subtracted out of it. The curved inner portion of the subtracted area stretches the bitmap, due to the mapping coordinates, but a solid procedural material always appears correct.

After you select the type of mapping you want to use, you can select the Adjust Coordinates button to fine-tune the mapping. The specific dialog box presented will depend upon whether you choose a planar, cylindrical, spherical, or solid mapped object. As an example of the scope of adjustments available for mapping and adjusting coordinates, Figure 26–25 shows the Adjust Planar Coordinates dialog box.

In this dialog box you can select the WCS plane along which the mapping coordinates are aligned. You can also pick your own 3D plane by picking three points on that plane, much like you can define your own UCS.

In the Center Position section of the dialog box, you will actually see a wireframe representation of the selected object. A light-blue outline called the mapping icon appears around this box as well.

This blue outline represents the mapping coordinates and indicates the size of one copy of the bitmap on the surface of the object. By adjusting the slider to the right and bottom of the preview, you can control the position of the bitmap on the surface of the object. At the bottom of the dialog box, you can control the offset and rotation

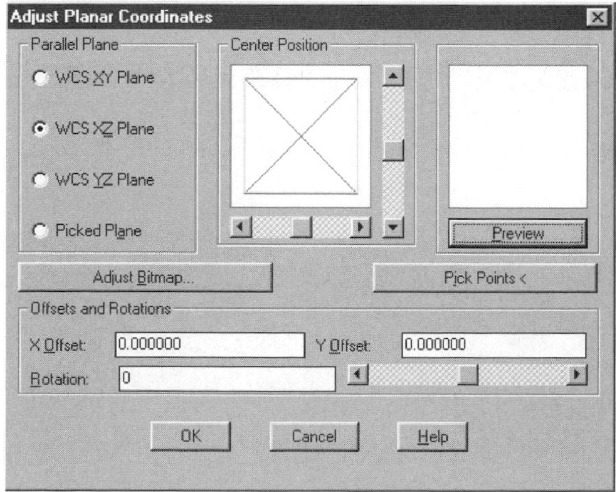

Figure 26–25 *The Adjust Planar Coordinates dialog box, which is used to adjust mapping settings on planar objects.*

of the mapping icon. If you want to control the scale of the bitmap, select the Adjust Bitmap button.

The combinations of geometric shapes and mapping planes make adjusting bitmapped materials a somewhat complex operation, due principally to the number of choices available. The procedures are relatively straightforward, however, and by adjusting the offset, rotation, scale, and coordinates of bitmapped materials, you can exert a great deal of control over how materials based on bitmaps will appear in the rendered model.

CREATING LIGHTS

After you select and perhaps create and then apply materials to the objects in your model, the next logical step is to add sources of light. Lighting is one of the most important aspects of creating an effective rendering. Without realistic lighting, the surfaces and materials of an otherwise realistic model will appear "flat." Lighting also provides the means to provide realistic shadows to your model.

You create lights by selecting the Lights button from the Render toolbar or by typing **LIGHT** at the command prompt. Either method displays the Lights dialog box, as shown in Figure 26–26.

In the Lights dialog box you will see a list of all the current lights in the scene, as well as the Ambient Light controls, the Light creation controls, and the North Location button.

The ambient light is the overall brightness of the scene. By adjusting the ambient light, you can set the overall brightness of the scene. Higher ambient values are good for outside light, whereas lower values are good for interior or night scenes. Strive to

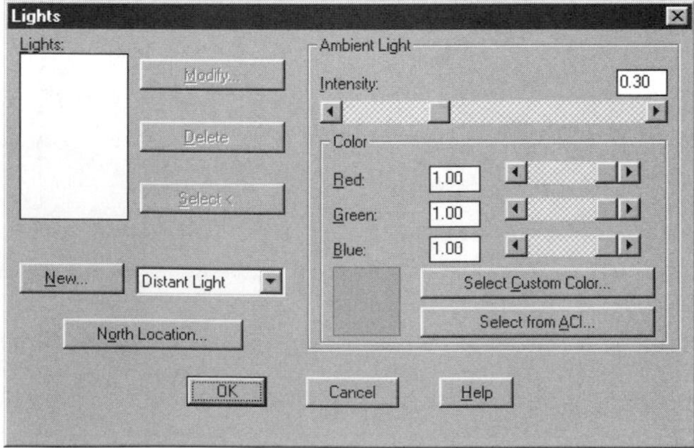

Figure 26–26 *From the Lights dialog box you can create, adjust, and manage all the lights in a model.*

avoid extreme settings of ambient light. Usually values between 30 and 70 yield suitable amounts of ambient light. You can also adjust the ambient light color to provide basic color tinting to the scene. Keep ambient light color variations very subtle for the most realistic effects. Ambient light exhibits no source and therefore is not capable of casting shadows.

The North Location button allows you to set the north direction in your models. By default, north coincides with the positive Y direction. You may want to alter the north direction in architectural models.

Types of Lights

AutoCAD supports three types of light:

- **Point light:** This type of light is similar to a single light bulb. Light is radiated from a single point in all directions. You can specify no attenuation, or attenuation that is inverse linear or inverse square.
- **Spotlight:** This type of light is similar to the light emitted from a flashlight. The spotlight has a source and a target location. Light from a spotlight is cast in a cone fashion, and you can define the angle of the cone. You can specify no attenuation, or attenuation that is inverse linear or inverse square.
- **Distant:** A distant light emits parallel light beams in one direction. Distant light sources cannot exhibit any attenuation; the light intensity remains constant regardless of its distance. Distant lights are often used to simulate sunlight.

Generally, point lights are used for general illumination in the scene; spotlights are used to add special highlighting in a relatively small area; and distant lights are usually used to simulate sunlight.

Creating a Light

Referring to the Lights dialog box shown in Figure 26–26, you can create a light simply by choosing the type of light from the drop-down list next to the New button and then selecting the New button. Each of the three light types has its own dialog box for specifying the attributes of the light. The dialog box for creating a spotlight is shown in Figure 26–27.

All lights that you create in an AutoCAD model must have a name. You enter the name in the Light Name input box. Choose a name that is descriptive, such as Spot1 or Point-main.

With point lights and spotlights, you need to position the light in the model. Point lights require only the location of the light source. Because spotlights are directional, they require the position of both the source and the target. You specify these locations by choosing the Modify button in the Position section of the New Spotlight dialog box. In the case of spotlights, you are then prompted to pick the target of the

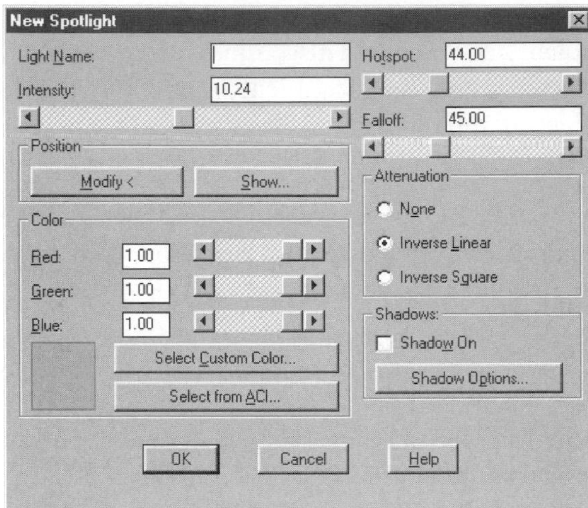

Figure 26–27 *The New Spotlight dialog box, where you create and set the attributes of a spotlight.*

light, followed by the location of the light. AutoCAD temporarily hides the dialog box to allow you to pick these points in the model. Using point filters, as described in Chapter 23, Drawing in 3D, makes specifying light locations easier. With distant lights, you need to specify the source direction as well as its height above the horizon.

After you position a light, you are returned to the New Spotlight dialog box, where you set the remaining parameters associated with the light. These parameters are explained as follows:

- **Intensity:** Sets the intensity or brightness of the light. Entering zero turns off a light. The maximum point light intensity depends on the attenuation setting and the extents of the drawing. If attenuation is none, maximum intensity is 1. If attenuation is inverse linear, maximum intensity is the value of twice the extents distance. If attenuation is inverse square, maximum intensity is twice the square of the extents distance.
- **Color:** Each light in the scene can have a different color assigned to it. For example, florescent lighting often has a slight blue cast, while typical incandescent lighting tends to be slightly yellow. For special effects such as sunrise or sunset, you may want to give a distant light a slightly red color. Keep these color assignments minimal for the best results.
- **Attenuation:** In real life, light fades, or grows weaker, as the distance between source and objects is increased. In AutoCAD, lights can remain the same strength over distance unless you specify an attenuation factor. There are two types of attenuation: inverse linear and inverse square. Inverse square lights drop off much faster than inverse linear lights. For point lights and spotlights,

inverse linear is the default attenuation factor, and it generally yields the most effective results. Direct lights exhibit no attenuation.

- **Shadows:** All lights in AutoCAD are capable of casting shadows. The type of shadow depends on the current rendering type (Photo Real or Photo Raytrace) and the settings of the Shadow Options dialog box. The shadow-casting capability of any light can be turned on or off through the Shadow On check box in the Shadows section of the New Spotlight dialog box.

Beyond these basic controls, which are common to all light types, the New Spotlight dialog box also allows you to adjust the angle of the cone and its fall-off rate, and the New Distant Light dialog box provides settings for sun position.

The following exercise shows you how to create a spotlight for use in a model.

EXERCISE: CREATING A SPOTLIGHT

1. Load the 26DWG02.DWG file found on the accompanying CD. The drawing represents a model of a telephone handset unit, as shown in Figure 26–28.
2. From the View menu, choose 3D Views>Plan View>World UCS. This switches you to a plan view. From this view you will add a spotlight.
3. Select Lights from the Render toolbar, or type in **LIGHT** at the command prompt.
4. In the Lights dialog box, select Spotlight from the drop-down list and select the New button.
5. In the New Spotlight dialog box, type the name **Spot1** in the Light Name input box.
6. In the Position section, select the Modify button. Autocad hides the dialog box and returns you to the drawing.

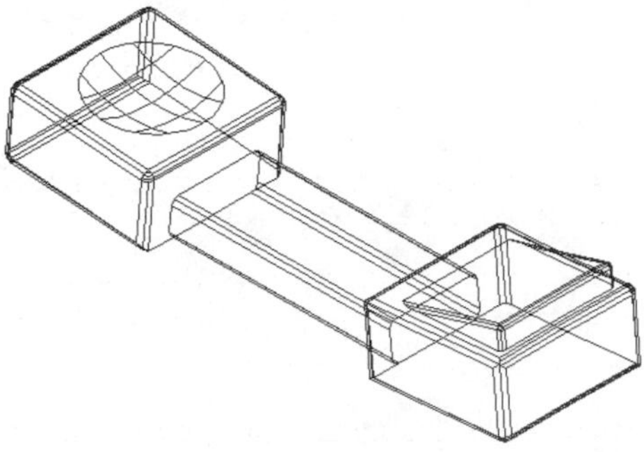

Figure 26–28 *A model of a telephone handset before rendering.*

7. At the "Enter light target:" prompt, type the coordinates **8.5,7.0**. At the "Enter light location:" prompt, type **.xy** to indicate that next you will provide the X-Y location of the light.
8. At the "OF:" prompt, type the coordinates **-2,14**. At the "(need Z):" prompt, type **25**. This places the light location twenty-five units above the model. AutoCAD returns to the New Spotlight dialog box.
9. Configure the remaining settings in the New Spotlight dialog box as shown in Figure 26–29.
10. In the Shadows section of the New Spotlight dialog box, select Shadow Options. In the Shadow Options dialog box displayed, check the Shadow Volume/Ray Traced Shadows check box. Choose OK to close the Shadow Options dialog box.
11. In the Lights dialog box, set the Ambient Light Intensity to **0.5**. Choose OK to accept the settings and close the dialog box.
12. From the View menu, choose Named Views. In the View dialog box, select View1, then Set Current. Choose OK to close the dialog box and restore the opening view.
13. From the Render toolbar, choose Render, or type **Render** at the command prompt.
14. Configure the settings in the Render dialog box as shown in Figure 26–30.
15. In the Render dialog box, choose Render. After a few moments, the rendered model appears. The rendering should resemble Figure 26–31.
16. You may want to render this model again after you adjust the Hotspot and Falloff angles (in the New Spotlight dialog box) as well as the intensities of the spotlight and ambient light. You can then close this drawing without saving changes.

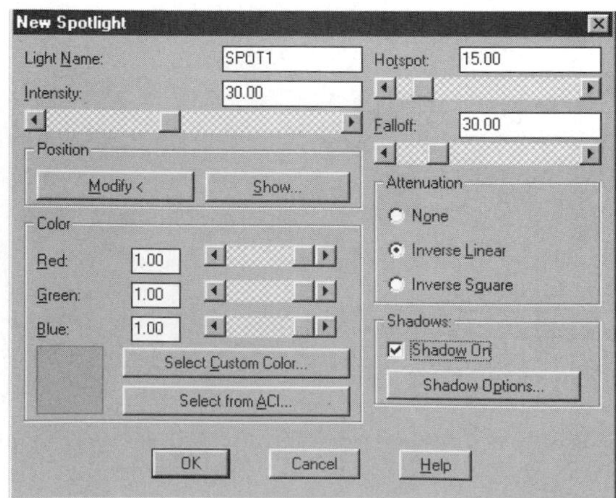

Figure 26–29 *The settings of the New Spotlight dialog box for the model's new spotlight.*

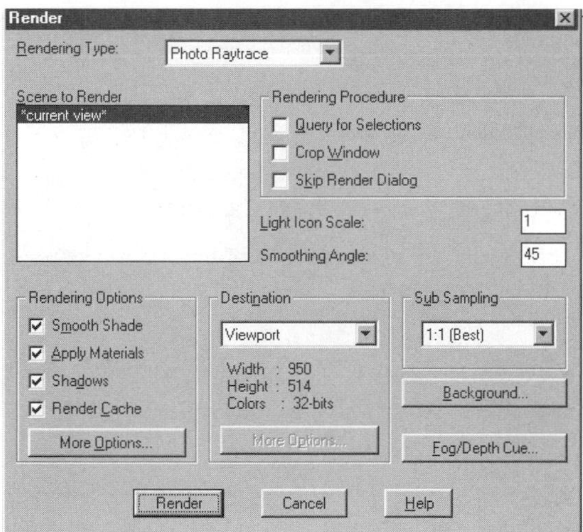

Figure 26–30 *Render dialog box settings before the model is rendered.*

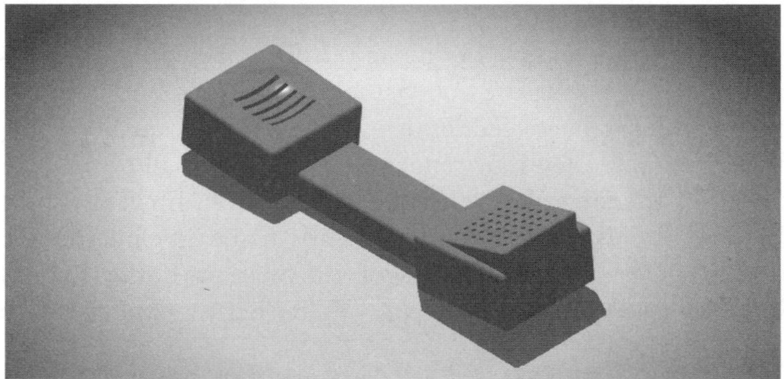

Figure 26–31 *The rendered model illuminated with the new spotlight.*

As was shown in this exercise, creating lights for a model is not difficult. Proper lighting of a model, however, is an art form in itself. Good results can be produced quickly, but great results take time and a lot of testing and adjusting of the lights in the scene. By testing different light types, positions, and intensities, you can speed up your test renderings in several ways. First, you do not need to render the complete model or view. Often, rendering a smaller portion of the view will give the information you need to decide if further adjustments are needed. In the Render dialog box shown in Figure 26–32, the Rendering Procedure area's Crop Window option allows you to window a portion of the model for rendering. In a similar way, the Query for Selections option allows you to select portions of the model for rendering on an object-by-object basis. Using one or both of these options can greatly speed up your rendering of portions of the model as you make lighting adjustments.

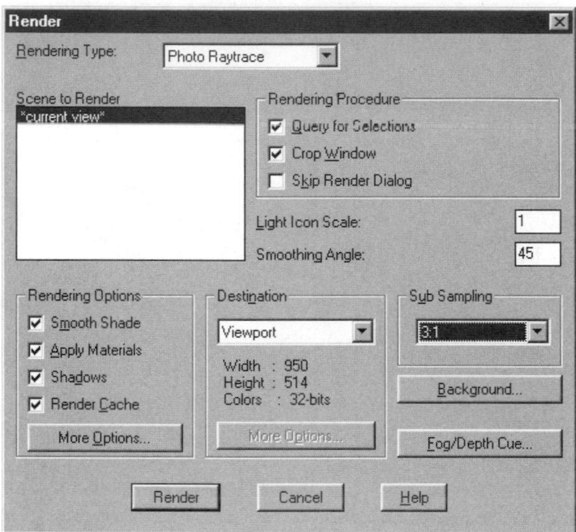

Figure 26–32 *The Render dialog box showing ways to speed test renderings.*

If you need to render the entire model view while making adjustments, you can make "course" renderings using the Sub-Sampling options found in the Render dialog box. Sub-sampling reduces rendering time and image quality, without losing lighting effects such as shadows, by rendering a fraction of all of the pixels. A sub-sampling ratio of 1.1 renders all pixels and takes the most time. A sub-sampling of 3.1, however, renders only every third pixel, greatly reducing rendering times. Shadows also increase rendering times. Turning off the Shadows option in the Rendering Options section of the Render dialog box is yet another way to make successive test renderings more attractive.

Creating Sunlight

Sunlight—in both exterior and interior views of your model—can be the most important light you use. You can manually set a distant light to simulate sunlight from any direction to yield any effect you want. You can also use AutoCAD's built-in Sun Angle Calculator to calculate the exact position of the sun for any point on earth on any day of the year at any time of day. Being able to place the sun so precisely is often important in renderings of exterior architectural models, where the exact position and extent of shadows cast by buildings or other structures may be important. In the following exercise you will learn how to create precise sun shadows using the Sun Angle Calculator.

EXERCISE: USING THE SUNLIGHT SYSTEM TO CREATE A DISTANT LIGHT

1. Load the 26DWG03.DWG file found on the accompanying CD. This drawing contains a model of an apartment building with an adjoining urban park.
2. Choose Lights from the Render toolbar.
3. In the New Distant Light dialog box, create a new distant light. Name this light Sun. Set its intensity to **1.00**.
4. In the Shadows section, select Shadow On. Select Shadow Options and, in the Shadow Options dialog box, select Shadow Volume/Ray Traced Shadows. Select OK to close the dialog box.
5. In the New Distant Light dialog box, select Sun Angle Calculator. AutoCAD displays the Sun Angle Calculator dialog box, as shown in Figure 26–33.
6. Select Geographic Location to display the Geographic Location dialog box. Select San Francisco from the City list box and then choose OK to close the dialog box.
7. In the Sun Angle Calculator dialog box, set the Date and Clock Time as shown in Figure 26–33, and then select OK to return to the New Distant Light dialog box. Select OK to close this dialog box.
8. In the Lights dialog box, check that the Ambient Light intensity is set to **0.90**. Choose OK to close this dialog box.
9. Select Render from the Render toolbar, or type **Render** at the command prompt.
10. In the Render dialog box, set the Rendering Type to Photo Raytrace. In the Rendering Options section, make sure that the Shadows option is checked.
11. Select the Render button to begin rendering the model. After a few moments, your rendered model should resemble Figure 26–34. Close this drawing without saving changes.

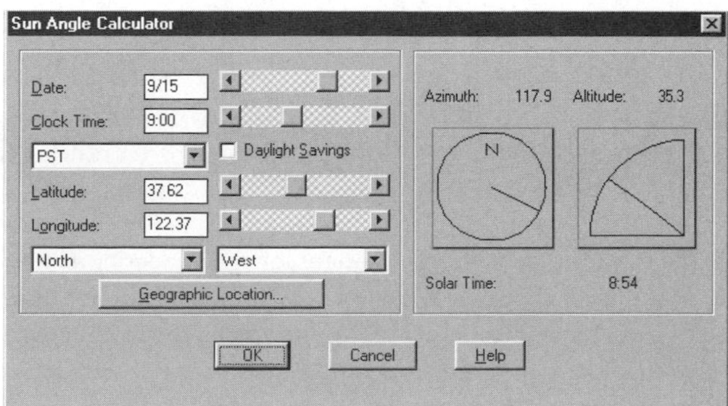

Figure 26–33 *The Sun Angle Calculator dialog box, where you set the geographic location, date, and time.*

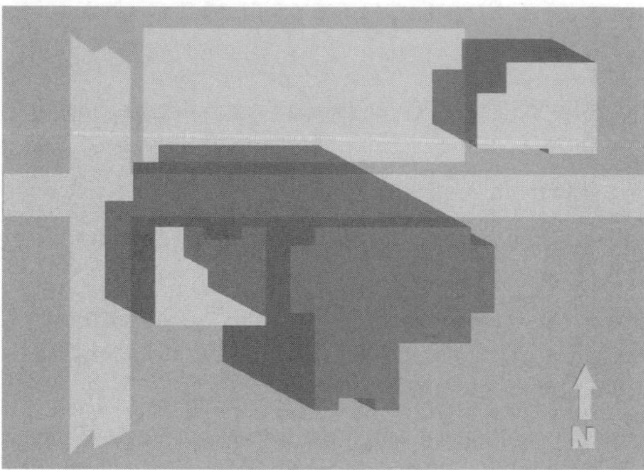

Figure 26–34 *The rendered building model showing the cast shadows on a specific date and time.*

Note: The Sun Angle Calculator takes its north as the current AutoCAD north direction. By default this is the positive Y-axis direction of the current coordinate direction. You can change the north direction in the North Direction dialog box, which is accessible from the Light dialog box.

This exercise demonstrates that the Sun Angle Calculator can be effectively used to place a distant light capable of producing accurate object shadows. Of course, a distant light can also be used more conventionally as a directional light source.

GENERATING AN OUTPUT

The default rendering method in AutoCAD renders the output to the current viewport. This is the method used in the exercises in this chapter. To print the image or use it in other programs, however, you must be able to save the rendering to a file. You accomplish this by setting the Destination drop-down list in the Render dialog box to File. Once File is selected, choose the More Options button to define the file type.

AutoCAD supports rendering the image out to any of five different file types: TARGA, TIF, Postscript, BMP, and PCX file formats. Below the file type drop-down list, you can select the resolution to which you want to render. Higher-resolution images, of course, take longer to render.

Note: There are other advanced features of the rendering system in AutoCAD to explore. These include fog, backgrounds, vegetation, and several others. This chapter is intended to give you an introduction to the principles of rendering and the basic capabilities of the rendering facility inside AutoCAD. Rendering is as much an art as a science, and you use this chapter's information as a basis for individual exploration of rendering inside AutoCAD.

SUMMARY

Once you have constructed a 3D model using the principles discussed in the 3D chapters of this book, you can use AutoCAD to render the model. Rendering consists of applying materials to the objects in the model and setting lights to illuminate the model. One of the most important steps in carrying out an interesting and informative rendering is establishing an effective viewpoint. Once these basic preparation steps are carried out, rendering is often a matter of trial and error and fine-tuning.

Advanced Concepts

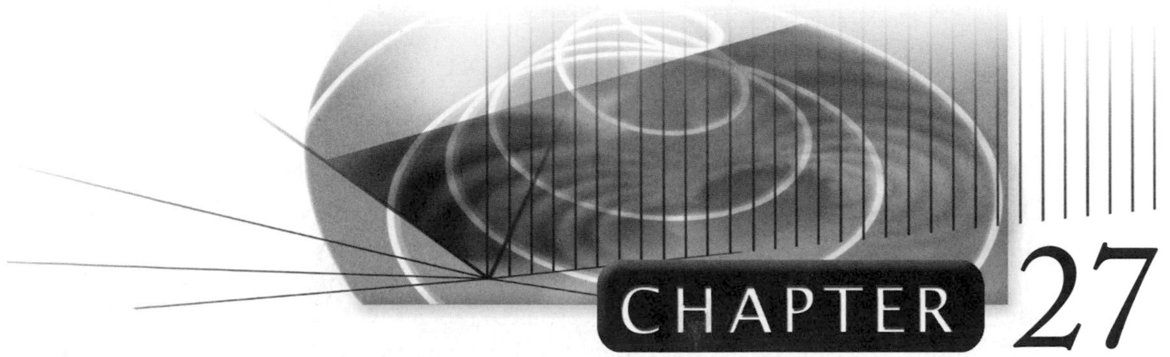

Taking Advantage of OLE Objects in AutoCAD

When you work on a set of drawings, you are typically working on one element of a project. Other elements may include word documents, spreadsheet data, and graphics created in programs other than AutoCAD. All of these elements combined are required to complete the project's deliverables and to meet your clients' needs.

Quite often, data created in other applications must be duplicated in your AutoCAD drawing. Elements such as General Notes created in a word processing program, or a Bill of Materials created in a spreadsheet application, must be duplicated in your drawing to satisfy the project's final delivery requirements. By adding this data to your drawing, you make the drawing a complete project deliverable.

Developing compound documents using Object Linking and Embedding (OLE) is a powerful, simple way to create the final documents required to satisfy your clients' needs. When you insert documents created in other applications into your AutoCAD drawing, you create a compound document such that you can take advantage of data already created in other applications. By simply dragging existing files into your drawing, you can insert data created in word processing applications such as Word or WordPerfect, as well as tabular data from spreadsheet programs such as Excel or Lotus, directly into your AutoCAD drawing. By using OLE, you make the process of completing a set of drawings easier by using existing data in its native format.

This chapter reviews AutoCAD OLE capabilities and covers the following subjects:

- Understanding OLE
- Importing objects into AutoCAD using OLE
- Exporting AutoCAD objects using OLE

UNDERSTANDING OBJECT LINKING AND EMBEDDING

Object Linking and Embedding (OLE) is a feature provided by the Windows operating system. Whether or not an application takes advantage of OLE is up to its program developers. In the case of AutoCAD, the application is designed to take advantage of OLE technology, allowing you to interact with other OLE-compliant applications. By using OLE, you can insert files from other applications directly into AutoCAD drawings, and you can insert AutoCAD views and AutoCAD objects into other OLE-compliant applications.

Object linking and embedding refers to the two different ways you can insert a file from another application into your drawing. You can insert an OLE object either as a linked object or as an embedded object. A linked object inserts a copy of a file, where the copy references the original source file. A linked OLE object behaves like xrefs in that any modifications made to the source file are reflected in the linked OLE object when the link is updated in your drawing.

In contrast, while an embedded object also inserts a copy of a file into your drawing, it does not maintain a link to the source file. An embedded OLE object behaves like a block inserted from another drawing in that the inserted file exists independent of the source from which it was copied, and it may be edited independently without affecting the source file. More importantly, any edits made to the source file are never reflected in the embedded OLE object. Use linked objects when you want modifications to the source file to be displayed in your drawing; use embedded objects when you want to insert a copy of a file but do not want edits to the source file to be displayed in your drawing.

OLE objects inserted into AutoCAD drawings have certain limitations. For example, you can only insert one page of a word processor document at a time. Also, you can only insert a portion of a spreadsheet file: a limited number of rows and columns amounting to an area approximately 10 inches wide by 13 inches long. Another limitation is that OLE objects cannot be resized if they are rotated in your drawing. Even with these limitations, you will find object linking and embedding a very useful feature.

Note: Windows, not AutoCAD, imposes the limitations of OLE.

Tip: You can use the OLESTARTUP system variable to optimize the quality of plotted OLE objects. The variable controls whether the source application of an inserted OLE object is loaded when the object is plotted. Setting the value to 1 instructs AutoCAD to load the OLE source application when plotting objects. Setting the value to zero instructs AutoCAD not to load the OLE source application when plotting objects.

IMPORTING OBJECTS INTO AUTOCAD USING OLE

You can create compound documents in AutoCAD by linking or embedding objects from other applications. For example, you can insert a table from a spreadsheet application, a set of notes from a word processing application, and a graphic image from a paint program. When you insert the desired objects into your AutoCAD drawing, you create a compound document.

AutoCAD provides several options for linking and embedding objects in drawings, as described in the following sections.

INSERTING OLE OBJECTS FROM WITHIN AUTOCAD

You can insert OLE objects into AutoCAD using the Insert Objects dialog box. This procedure allows you to insert linked or embedded objects from within AutoCAD by executing an AutoCAD command. From the Insert Objects dialog box, you can insert an object from an existing file, or you can create a new OLE object that exists only in the current drawing.

You open the Insert Object dialog box from the Insert menu by choosing OLE Object. Once it is opened, the Insert Object dialog box presents a list of types of objects that it can link or embed, as shown in Figure 27–1.

In the Insert Object dialog box, you select whether you want to create a new OLE object or insert an OLE object from an existing file. The Create New option opens

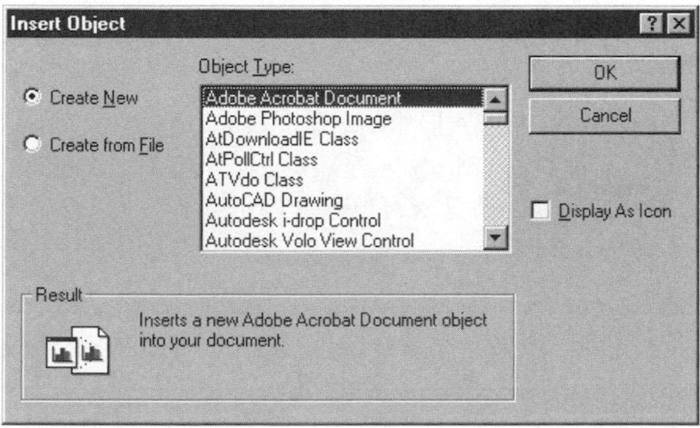

Figure 27–1 *The Insert Object dialog box allows you insert OLE objects from within AutoCAD.*

the selected application so you can create the object. Then, when the object is saved, the selected application closes, and AutoCAD embeds the object in the current drawing.

In contrast, when you choose the Create from File option, the Insert Object dialog box changes its display, allowing you to browse for an object to link or embed, as shown in Figure 27–2. When you select the Link check box, the selected object is inserted into AutoCAD and linked back to the original file.

The Insert Object dialog box provides a straightforward method for inserting OLE objects. Because you have the option of either creating new OLE objects or browsing for existing object files, you can easily insert the needed OLE object into your AutoCAD drawing.

PASTING OLE OBJECTS INTO AUTOCAD

You can insert OLE objects into AutoCAD by pasting them from the Windows Clipboard. This procedure is a very common way to insert OLE objects from one application to another. Using this feature you can copy an object directly from its application to the Clipboard, and then paste the Clipboard's contents into AutoCAD.

You paste objects from the Clipboard using either the Paste command or the Paste Special command. You can access these commands from AutoCAD's Edit menu. You can also access the Paste command from the shortcut menu, which you display by right-clicking in the drawing area. These commands are only available when the Clipboard contains objects.

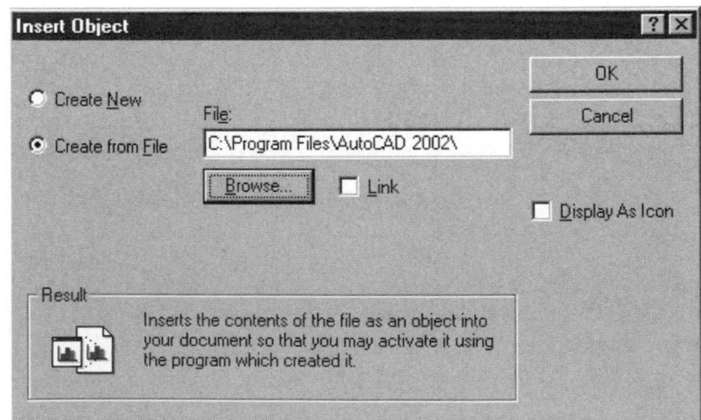

Figure 27–2 *The Insert Object dialog box allows you to browse for existing OLE object files to insert into AutoCAD.*

Note: You can view the contents of the Clipboard using the Clipboard Viewer, and you can also delete the Clipboard's contents. You access the Clipboard Viewer from the Windows Taskbar by choosing Start>Programs>Accessories>Clipboard Viewer. You view the Clipboard's contents by opening the Clipboard window, which appears as an icon in the Clipboard Viewer. To delete the Clipboard's contents, from the Clipboard Viewer's Edit menu, choose Delete.

While both commands paste objects into the current drawing from the Clipboard, they differ in one important way. The Paste command only embeds objects. The Paste Special command allows you to embed objects or insert them as linked objects.

When you choose the Paste command, the object is immediately embedded into AutoCAD. Additionally, the OLE Properties dialog box displays if the Display dialog box when pasting new OLE objects check box is selected. The OLE Properties dialog box allows you to control the size of the OLE object, and is discussed in detail later in this chapter.

When you choose the Paste Special command, AutoCAD displays the Paste Special dialog box. From this dialog box you can choose either the Paste option or the Paste Link option.

When you use the Paste option, the OLE object is embedded into the drawing. The difference between pasting an object from the Paste Special dialog box and pasting it directly from the Edit or shortcut menus is that when you use the Paste Special dialog box, you have more control over the OLE object type you are embedding.

When you choose the Paste option in the Paste Special dialog box, the available object types are displayed in the As list. The object types listed depend on the OLE object you arc pasting from the Clipboard. For example, if the Clipboard contains a Microsoft Word document, you can embed the Clipboard's contents as one of several object types shown in Figure 27–3. The list displays only acceptable types for the particular object. Several object type options are described as follows:

- **Picture (Metafile):** Inserts the contents of the Clipboard into your drawing as a vector-based picture.
- **AutoCAD Entities:** Inserts the contents of the Clipboard into your drawing as circles, arcs, lines, and polylines. Text is inserted as text objects, with each line of text located in a paragraph in the source file converted to individual AutoCAD text objects.
- **Image Entity:** Inserts the contents of the Clipboard into your drawing as an AutoCAD raster image object.
- **Text:** Inserts the contents of the Clipboard into your drawing as an AutoCAD MTEXT object. Any line objects are ignored.

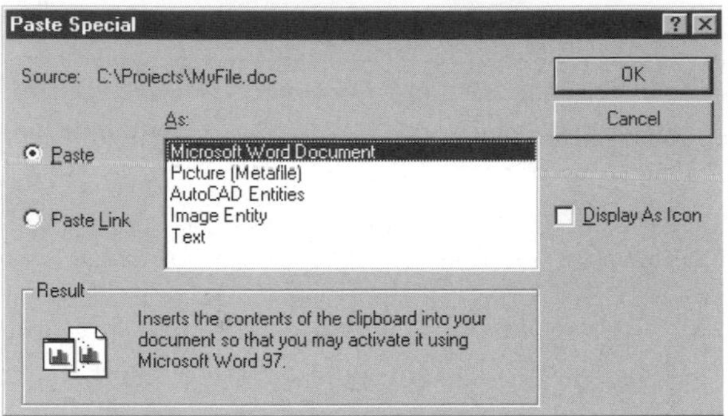

Figure 27–3 *The Paste Special command allows you to select an object's type when it is embedded.*

- **Package:** Inserts the contents of the Clipboard into your drawing as a Windows Package object. A package is an icon that represents embedded or linked information. The information may consist of a complete document, such as a Paint bitmap, or part of a document, such as a spreadsheet cell. You create packages using the Windows Object Packager, which you access from the taskbar by choosing Start>Programs>Accessories>Object Packager.
- **Bitmap Image:** Inserts the contents of the Clipboard into your drawing as a bitmap image object.

When you choose the Paste Link option, you can insert the OLE object as its original object type only. For example, if you choose the Paste Link option to insert a Microsoft Word document, you can insert it as a Microsoft Word document only. This ensures that you can open and edit the source Word document to which the OLE object is linked.

Tip: If you work in a black background in AutoCAD and paste an image from a word processing or spreadsheet application, the pasted image will appear with a white background in the drawing. If this is undesirable, change your background in either program to match the other.

By using the Paste and Paste Special commands, you can embed or link OLE objects in your AutoCAD drawing from the Clipboard. Next you will learn about another method for inserting OLE objects.

USING DRAG & DROP TO INSERT OLE OBJECTS

The Windows operating system provides the ability to drag and drop selected objects from another application into an AutoCAD drawing. When you select objects in an open application, then drag the selected objects into AutoCAD, you in effect cut the objects from the application and embed them into the AutoCAD drawing. To copy the objects from the open application instead of cutting them, hold down the Ctrl key while you drag the objects into the drawing.

Tip: After you select objects in an application, drag them into AutoCAD by right-clicking with your pointing device. AutoCAD displays the shortcut menu and allows you to move the objects (Cut & Paste), copy the objects (Copy & Paste), or link the objects (Paste Special) into the AutoCAD drawing. You can also create a hyperlink to the file or cancel the operation from the shortcut menu.

Additionally, you can drag and drop objects from Windows Explorer. If AutoCAD recognizes the object type, it embeds the object into the drawing. If AutoCAD does not recognize the file type, or if the type of object cannot be inserted as an OLE object, AutoCAD issues an error and cancels the function.

CONTROLLING OLE OBJECT PROPERTIES

AutoCAD provides specific tools for manipulating an OLE object, because common AutoCAD commands typically do not affect OLE objects. For example, you cannot select an OLE object and erase it with the ERASE command, nor can you resize it using the SCALE command. However, by using specific tools designed for manipulating OLE objects, you can control an OLE object's appearance in AutoCAD.

Tip: When you select an OLE object in AutoCAD to display its grips, you can press the Delete key to delete the object.

CONTROLLING OLE OBJECT SIZE

AutoCAD lets you control an OLE object's size in a drawing through the OLE Properties dialog box. The dialog box allows you to control an object's size in one of three ways, as shown in Figure 27–4. You access the dialog box by right-clicking over an OLE object and then selecting Properties from the shortcut menu.

In the Size area, you control an OLE object's size by entering values in the Height and Width fields. If the Lock Aspect Ratio check box is selected in the Scale area, then when one field value is changed, the other is updated automatically, proportionally maintaining the OLE object's aspect ratio size. The units entered in the Height and Width fields are based on the drawing's current units setting. You can also set the OLE object back to its original size by choosing the Reset button.

In the Scale area, you control the OLE object's size by entering a percentage of the object's size in the Height and Width fields. As with the values in the Size area, if the Lock Aspect Ratio check box is selected, when one value in the Height or Width fields is changed, the other value is updated automatically.

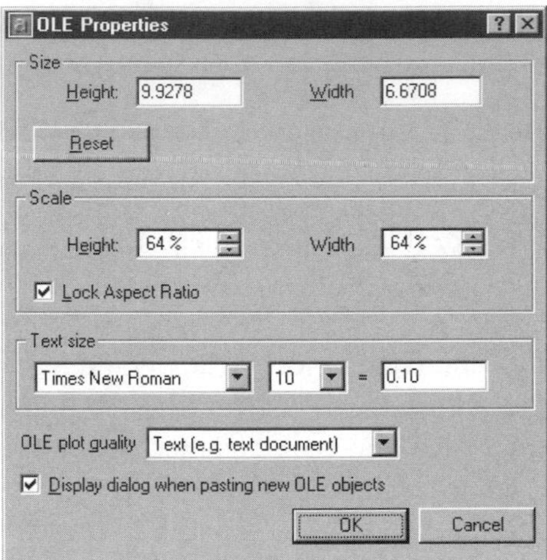

Figure 27–4 *The OLE Properties dialog box lets you control an OLE object's size.*

A third method for controlling an OLE object's size is available in the Text Size area. If the OLE object contains text, you can enter a new text size value to adjust the object's size. The first, or leftmost, field displays the list of font styles in the OLE object, and the second field contains a list of the selected font's sizes. By choosing the desired font style and size in these first two fields, you can control the object's overall size by entering the desired height of the text in the third, or rightmost, field. For example, in Figure 27–4, the OLE object will be resized based on the 10 point Times New Roman font being set to 0.10 drawing units.

Note: It is important to understand that the three areas provided for controlling an OLE object's size work in unison. When one set of values is changed in one area, the values are automatically changed in the other two areas. The values in the three areas cannot be set independent of each other.

In addition to controlling an OLE object's size, the OLE Properties dialog box provides the ability to control the plot quality of an OLE object. From the OLE Plot Quality list, you can choose one of the five plot quality options:

- **Line Art**—Intended for plotting objects such as a spreadsheet
- **Text**—Intended for plotting objects such as a Word document
- **Graphics**—Intended for plotting objects such as a pie chart
- **Photograph**—Intended for plotting objects that are color images
- **High-Quality Photograph**—Intended for plotting objects that are true-color images

The plot quality options are applied specifically to the selected OLE object. Therefore, you can insert a Word document that contains only text and then set the plot quality to Text. Then you can insert a true-color image, and set its plot quality to High-Quality Photograph. By applying the desired plot quality to each OLE object, you can control an object's appearance when it is plotted.

CONTROLLING OLE OBJECTS USING THE SHORTCUT MENU

Once an OLE object is inserted into a drawing, you can control several object properties and perform edits through commands accessed from the shortcut menu. Using these commands, you can delete the OLE object or copy it to the Clipboard. You can determine if the object appears on top of or below other objects in the drawing, and you can control whether or not the object may be selected for editing. The shortcut menu offers these commands and more, providing useful control over OLE objects.

Cutting, Copying, and Clearing OLE Objects

When you right-click over an OLE object, AutoCAD displays the available OLE shortcut commands, as shown in Figure 27–5. The first three commands are described as follows:

- **Cut**—Erases the selected object from the drawing and places a copy in the Clipboard. You can also execute the Cut command by pressing Ctrl+X.
- **Copy**—Leaves the selected object in the drawing and places a copy in the Clipboard. You can also execute the Copy command by pressing Ctrl+C.
- **Clear**—Erases the selected object from the drawing without placing a copy on the Clipboard. You can also execute the Clear command by entering **E** and then pressing Enter.

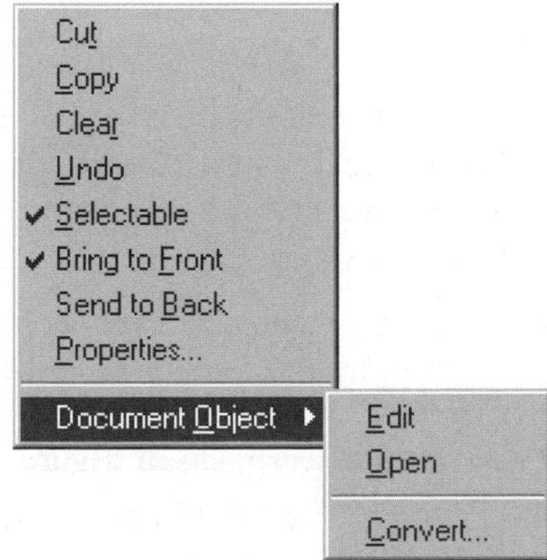

Figure 27–5 *Right-click over an OLE object to display the shortcut menu commands.*

When you use the Cut or Copy commands on an OLE object, the object is placed in the Clipboard in its original object format, not as an AutoCAD object. For example, suppose a Word document resides in your drawing as an OLE object. If you use the Copy command to copy the Word document object from your drawing to the Clipboard, when you paste the object into another application, it will be pasted as a Word document object.

Undoing OLE Object Edits

The next command on the shortcut menu is Undo, which undoes edits made to the OLE object while it is in your drawing. For example, if you move or resize the object in your drawing, you can undo the edit by selecting Undo from the shortcut menu. By selecting the Undo command repeatedly, you can undo a series of edits made to the object.

The Undo command has one important limitation: it does not undo edits made to the OLE object in the object's source application. For example, if you paste a Word document object into your drawing and then edit the document object in Word by adding additional text, when you save your edits and return to your drawing, the edits made in Word cannot be undone in AutoCAD. In other words, the text added to the document in the Word application cannot be undone using the Undo command in AutoCAD.

Note: Do not confuse the Undo command found on the OLE object shortcut menu with AutoCAD's UNDO command. The UNDO command (Ctrl+Z), which is accessed from AutoCAD's Edit menu, does not undo edits, such as moving or resizing, made to an OLE object while it is in your drawing. However, AutoCAD's UNDO command will undo the command used to paste the object into your drawing, thereby removing the entire object from your drawing.

Controlling OLE Object Selectability

The next item on the shortcut menu is the Selectable property, which toggles the selectability of the OLE object. When Selectable is toggled on, a check appears next to the property, indicating that the OLE object may be selected and then moved or resized in your drawing. When the property is toggled off, the check is cleared, indicating that the object cannot be selected.

When you select an OLE object whose Selectable property is toggled on, an object frame and its sizing handles appear around the object. When you place your cursor inside the object frame, the cursor changes to the Move cursor, which is an icon comprised of a cross with four arrows, as shown in Figure 27–6. The Move cursor allows you to drag the object to a new position in your drawing.

The sizing handles are the small, solid squares that appear at the corners and midpoints of the object frame. When you place your cursor over a sizing handle, the

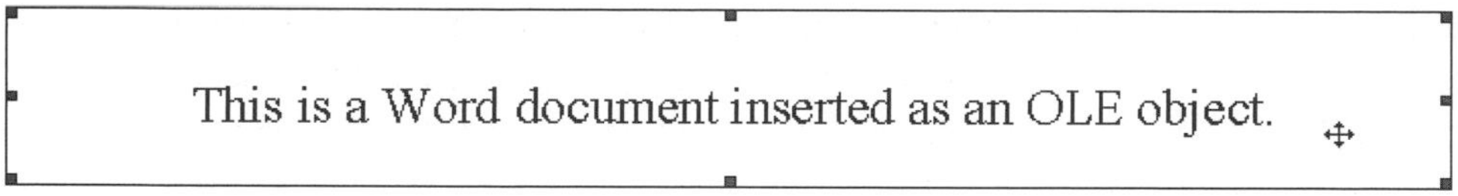

This is a Word document inserted as an OLE object.

Figure 27–6 *AutoCAD lets you drag an OLE object to a new position.*

cursor changes to a double-headed arrow. You can then resize the object by dragging the sizing handle. The sizing handles at the midpoints of the object frame stretch the object, distorting its appearance. The sizing handles at the corners of the object frame scale the object proportionally, maintaining the object's aspect ratio.

Tip: The sizing square color is controlled by the unselected grip color. This can be changed from the Options dialog>Selection> in the Grips area.

The three objects in Figure 27–7, which are OLE objects inserted into a drawing from a Word document, provide an example of how stretching and resizing affects an object. The top object shows how all three word objects appeared in their original size. The middle object is a copy of the top object, which we stretched by dragging its sizing handles at the midpoints on each side of the object frame. Notice that the height of the text is the same as in the original; only the width of the text is changed. The bottom object is also a copy of the top object, which we resized by dragging its sizing handles in the corners of the object frame. Notice that while the text's height and width are larger than the original, their aspect ratio is maintained, and their overall size is correctly proportioned.

When the Selectable property is toggled off, you cannot select the object to display its object frame and sizing handles. This feature is useful for maintaining an object's size and position in your drawing once they are set as desired. By clearing an OLE object's Selectable property, you ensure that the object will not be moved or resized in your drawing.

It is important to note that the Selectable property does not affect the other commands and properties on the shortcut menu. For example, while you cannot drag an OLE object's sizing handles when its Selectable property is cleared, you can resize the object through the OLE Properties dialog box. When you right-click over the OLE object and choose the Properties command from the shortcut menu, you display the OLE Properties dialog box. Any changes made to the object's size in the dialog box modify the object, even though its Selectable property is cleared. This is true for the shortcut menu's other commands, including Cut, Copy, Clear, and Undo.

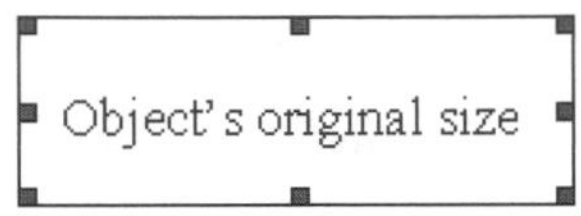

Figure 27–7 *The original object is at the top, the stretched object is in the middle, and the resized object is at the bottom.*

Tip: To duplicate quickly an OLE object inserted in AutoCAD, select the object and then drag it while pressing the Ctrl key.

Controlling OLE Object Display Order

The next two properties are Bring to Front, and Send to Back. These two properties control an object's position in the drawing relative to other objects, and they perform the same function as AutoCAD's Display Order tools. When you right-click your cursor over an OLE object and then select Bring to Front or Send to Back from the shortcut menu, you place the object above or below other objects, as shown in Figure 27–8. These two properties are actually a toggle, and AutoCAD allows you to set only one or the other for each OLE object.

Note: Clicking inside an OLE object always selects it, even if the object is behind other AutoCAD objects.

Tip: To select AutoCAD objects that lie within an OLE object's frame, clear the object's Selectable property.

Editing and Converting OLE Objects

At the bottom of the shortcut menu is the Object menu item. When it is selected, the Object menu item displays a cascading menu that provides access to the OLE object's source application through the Edit and Open commands, allowing you to modify the object. Additionally, you can access the Convert dialog box, which spec-

This OLE Object is in Front.
This OLE Object is in Back.

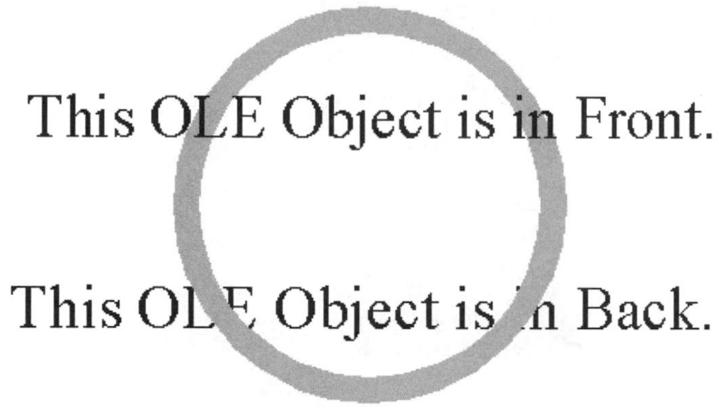

Figure 27–8 *The OLE object at the top has its Bring to Front property set, while the object at the bottom has its Send to Back property set.*

ifies a different source application for the OLE object. From the Object item menu, you can modify an OLE object or convert it to a different object type.

The name displayed for the Object menu item in the shortcut menu changes depending on the type of OLE object selected when the menu is accessed. For example, the shortcut menu in Figure 27–5 indicates that the object is a document object, specifically a Microsoft Word document object. If a Microsoft Excel worksheet is inserted as an OLE object and the shortcut menu is accessed, the Object menu item is listed as Worksheet Object, as shown in Figure 27–9.

The cascading menu's Edit and Open commands launch the object's source application, which displays the object's source file and allows you to make modifications. In AutoCAD, both commands perform the same function. The fact that the two commands perform the same function is due to how AutoCAD interfaces with the Windows operating system when it deals with OLE objects.

In some applications other than AutoCAD, the Edit command only opens a source application file window inside the current application, while the Open command launches the entire source application. For example, Figure 27–10 shows what happens when you access the Edit command in Excel to modify a Word document object inserted into an Excel spreadsheet. Notice that the Word document window is inside the Excel spreadsheet. This allows you to make edits to the Word document from inside Excel without actually launching the Word application. In AutoCAD, however, the Edit command launches the entire source application. Therefore, choosing either the Edit or the Open command in AutoCAD launches the source application, allowing you to make modifications.

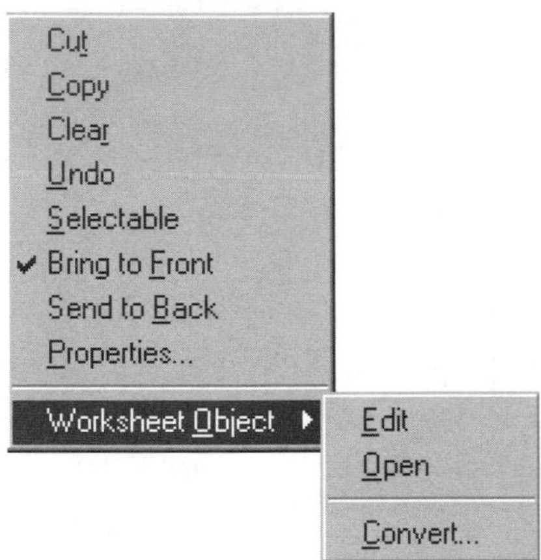

Figure 27–9 *The name of the Object menu item at the bottom of the shortcut menu changes to indicate that a Microsoft Excel Worksheet is selected.*

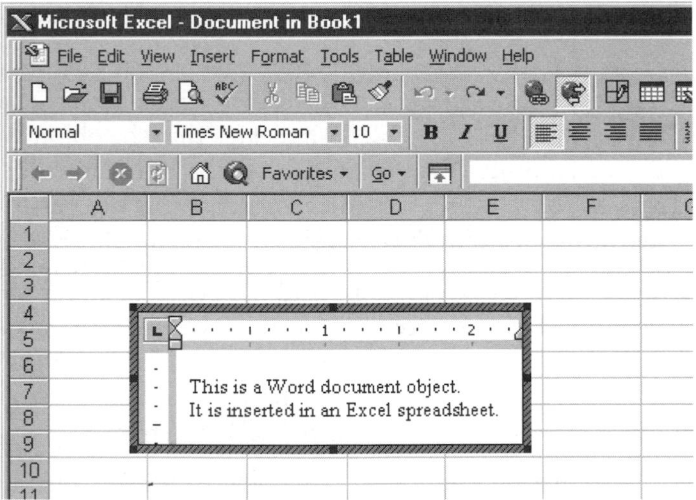

Figure 27–10 *The Edit command allows you to modify a Word document object in Excel.*

Tip: You can execute the object Edit command by double-clicking inside an object's frame.

The Convert command opens the Convert dialog box, which allows you to specify a different source application for an embedded object. When you select the desired source application and then choose OK, the object's source application type is changed to the new application type.

The different object types to which you can convert an object depend on the object selected. For example, when you select an embedded Word document object and then choose the Convert command, you are allowed to convert the document object to a Word picture object, as shown in Figure 27–11. The object types listed are those supported by the source application.

When you convert an object, you can choose from one of two options: Convert To, or Activate As. The Convert To option converts an embedded object to the type specified under Object Type. This means the object is actually converted to the new selected object type. For example, if you converted a Word document object into a Word picture object, and then right-clicked over the object, the shortcut menu would list the object as a Picture object. This means that when you edit the object, you would edit it as a Word picture, not a Word document.

The Activate As option acts similar to the Convert To option, except it only temporarily converts the object during the editing process to the selected object type. Once the editing is complete, the object returns to its original type. For example, you can edit a Word document object, temporarily activating it as a Word picture object. This means that the document object opens as a picture object in the Word application, allowing you to modify the object using Word's picture-editing tools. Once you finish modifying the picture and close and save the file, the modified object is displayed in AutoCAD in its original format, as a Word document object.

If you look at Figure 27–11, you will notice that you can also convert the Word document object to a Word document object. This option is intended to maintain an object in its current type and work in conjunction with the Display As Icon check box. For example, when you choose to convert the Word document object into a Word document object, and then choose the Display As Icon option, the document

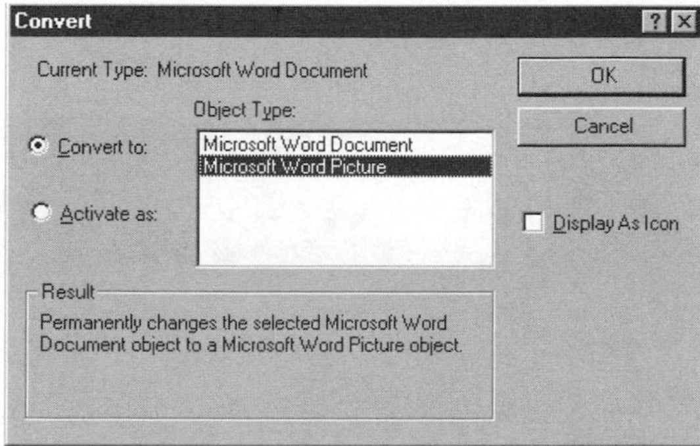

Figure 27–11 *The Convert dialog box allows you to convert the selected object to a different object type.*

object maintains its current type and changes its appearance so it is displayed as a Word document icon, as shown in Figure 27–12. To return the document object back to its original display, convert the Word document object into a Word document object, and then clear the Display As Icon check box.

Tip: Double-clicking on the object icon launches the object's source application and allows you to modify the file.

Note: When you select the Display As Icon option, the Change Icon button is activated. Selecting this button displays the Change Icon dialog box, which allows you to select a new icon to display as the object icon.

The OLE object shortcut menu provides several useful commands and options that allow you to control an OLE object's appearance and its behavior. Next you will learn about two features that allow you to control an OLE object's visibility.

CONTROLLING OLE OBJECT VISIBILITY

When you insert OLE objects into your drawing, you may want to control their visibility. Whether objects are inserted only temporarily or they are used for drawing construction purposes the way you might use construction lines or rays, you may want the visibility of OLE objects turned off in your drawing or in plotted sheets. With object visibility control, you can display reference information to assist you in your work, and you can control when objects are visible in your drawings or whether they appear in plots.

AutoCAD allows you to control the visibility of OLE objects through two methods. The first is simply to insert the OLE object on a layer that you turn off or on, or that you freeze or thaw. The second is to use a special command that allows you to control the visibility of all OLE objects globally. By using these two methods, you can easily control the visibility of OLE objects inserted in your drawing.

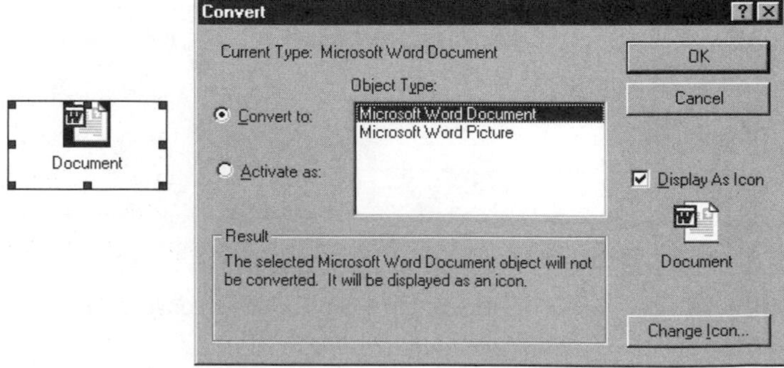

Figure 27–12 *The Convert dialog box allows you to display an OLE object as an icon.*

Controlling OLE Object Layer Properties

Controlling object visibility from the layer on which it is inserted is a very common method for controlling whether or not an object is displayed in your drawing. One of the chief reasons for using layers to organize the objects in your drawing is to control the visibility of groups of objects that reside on a common layer. By inserting OLE objects on their own layers, you can easily control their visibility from the Layer Properties Manager.

Tip: To move or copy an OLE object to a new layer, you must cut or copy the object to the Clipboard, make current the layer to which you want to move or copy the object, and then paste the object.

Note: When you cut or copy an OLE object from an AutoCAD drawing and then paste it back into the drawing, the object's size will revert to its original size, and any modifications to the object will be lost. Therefore, the OLE object's size must be reset to the desired value after it has been pasted back into the drawing. You can resize the OLE object using the Object Properties dialog box.

The Layer Properties Manager allows you to control more than just an OLE object's visibility. Specifically, OLE objects react to the following layer properties:

- **On/Off and Freeze/Thaw**—These properties control an OLE object's visibility, both on the screen and when the object is plotted. Turning off or freezing the layer on which an OLE object resides no longer displays the object in your drawing, either on the screen or when the object is plotted. To restore the object's visibility, turn on or thaw the layer.
- **Lock**—This property prevents the OLE object from being selected. It functions like the OLE object's Selectable property, except it does not allow any type of edits from the OLE object shortcut menu. For example, if an OLE object's Selectable option is turned off, you can still edit the object using commands from the shortcut menu, such as Properties, and Edit or Open. In contrast, when the layer on which an OLE object is inserted is locked, the OLE object shortcut menu cannot be invoked. This property absolutely prevents the OLE object from being edited.
- **Plot**—This property allows an object to remain visible on the screen but prevents it from being plotted. This feature is useful if you need to display an OLE object during an editing session for reference information only, and you do not want the object to appear when you plot drawings. When you turn this property off, you ensure that the OLE object will not appear on plotted drawings.

By using the layer properties discussed in this section, you can control both the appearance and the behavior of an OLE object through the layer on which it resides. In the next section you learn how to control OLE object visibility globally.

Globally Controlling OLE Object Visibility

AutoCAD provides a method of globally controlling OLE object visibility. The OLEHIDE command allows you to determine if OLE objects are visible in a drawing. When you use the OLEHIDE command, you control the visibility of all OLE objects, and you control whether OLE objects are displayed in paper space or in model space.

The OLEHIDE command is actually a system variable whose current setting is stored in your computer's system registry. This means that when you set a value for OLEHIDE, the setting affects all drawings in the current editing session, as well as in future sessions. To control the display of all OLE objects in all drawings, set the desired display value for the OLEHIDE system variable.

Typing **OLEHIDE** at the command prompt allows you to set the current OLEHIDE system variable value. The four possible integer values that you can set are described as follows:

- **0:** Makes all OLE objects visible, both in paper space and in model space
- **1:** Makes OLE objects visible only in paper space
- **2:** Makes OLE objects visible only in model space
- **3:** Makes all OLE objects invisible, both in paper space and in model space

By setting these values, you can control the appearance of all OLE objects, both in paper space and in model space.

Note: If the OLEHIDE system variable is set to 1 or 2 when you insert a new OLE object, AutoCAD automatically changes the OLEHIDE system variable value to allow the new OLE object to be displayed in the current space. This will also cause all OLE objects in the current space to appear.

In the next section you will learn how to work with OLE objects that are linked to their original file.

WORKING WITH LINKED OLE OBJECTS

When you insert an OLE object and link it to its original file, the file can be edited in its source application, and the linked object can be automatically updated. This means that when a linked object is inserted in a drawing and the object's original file is modified, the linked OLE object in AutoCAD is automatically updated to reflect the modifications. This feature is very useful for ensuring that the latest version of an inserted OLE object is displayed in an AutoCAD drawing.

While the ability to update and display the latest version of a linked file automatically is very useful, there will probably be occasions when you do not want the linked object to be updated automatically. For example, if you want to save permanently a set of drawings that represent a 50% completion set, then you do not want OLE objects to be automatically updated when the 50% completion set of drawings is re-opened for reference in the future. You need the ability to control whether or not linked OLE objects are automatically updated.

AutoCAD provides a tool that allows you to control whether or not linked OLE objects are automatically updated. When you choose the OLE Links command from the Edit menu, the Links dialog box is displayed, as shown in Figure 27–13. In the Links dialog box you can choose either the Automatic or Manual Update option, which controls whether linked objects are automatically or manually updated. Additionally, you can restore links lost because the original file cannot be found, and you can associate the link to a different file. You can also break the link connection between the OLE object and the original file, which converts the linked object to an embedded object. In the Links dialog box, you control the link between an OLE object and the file to which it is linked.

In the Links dialog box, if you choose the Manual Update link option, you may update the link by choosing the Update Now button. The Open Source button opens the linked file's source application, allowing you to edit the source file. The Change Source button allows you to locate a missing source file or select a new file. The Break Link button terminates the link between the OLE object and the source file. The link cannot be reestablished; once the link is broken, the OLE object is permanently detached from its original source file.

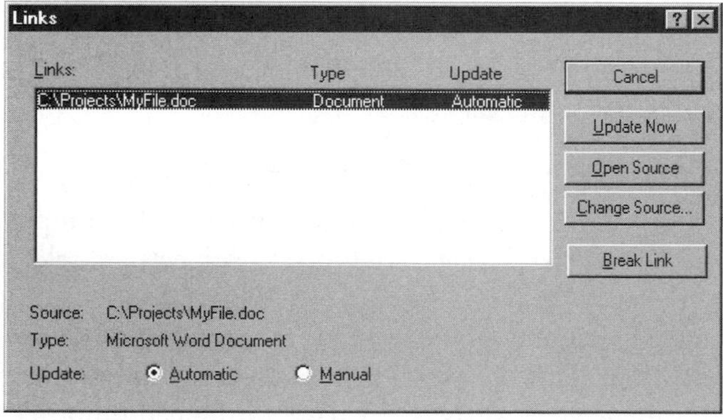

Figure 27–13 *The Links dialog box lets you control the link between an OLE object and its source file.*

EXPORTING AUTOCAD OBJECTS USING OLE

Just as you can insert files from other applications into AutoCAD, you can also insert AutoCAD drawings into other application files. By using certain commands created specifically for AutoCAD drawings and AutoCAD objects, you can insert either linked or embedded AutoCAD files into other application files. Therefore, AutoCAD's OLE features are designed for both inserting files into AutoCAD and for inserting AutoCAD drawings into other applications.

Exporting AutoCAD objects into other application files as OLE objects involves determining if the AutoCAD objects will be linked or embedded. Linked objects are based on a named view in AutoCAD. When the view is updated in the AutoCAD drawing, the link is updated and the modified view appears in the application's file. In contrast, embedded objects are AutoCAD objects that are selected in the drawing and then copied to the Clipboard. Once they are pasted from the Clipboard, the objects are inserted as independent objects with no association to the original AutoCAD objects. Therefore, if the original AutoCAD objects are edited in the drawing from which they were copied, the objects embedded in the application will not be updated.

AutoCAD provides three commands for exporting AutoCAD information into other applications for linking and embedding:

- **Cut**—Executes the CUTCLIP command, which copies AutoCAD objects to the Clipboard, erasing the selected objects from the drawing
- **Copy**—Executes the COPYCLIP command, which copies AutoCAD objects to the Clipboard
- **Copy Link**—Executes the COPYLINK command, which copies the current AutoCAD view to the Clipboard

All three commands are located on AutoCAD's Edit menu.

When you use the Cut or Copy commands, AutoCAD prompts you to select objects if no objects are currently selected. Once the objects are selected, AutoCAD copies the selected objects to the Clipboard. If objects are selected before you execute the commands, the selected objects are immediately copied to the Clipboard, and the commands end. If the Copy Link command is selected, AutoCAD copies all objects in the current view to the Clipboard, without prompting you for object selection. Therefore, the main difference between the Cut and Copy commands and the Copy Link command is that the Cut and Copy commands prompt you to select objects, while the Copy Link command does not.

When the AutoCAD objects are pasted into the target application, an object frame surrounds the objects and represents the drawing's viewport display at the time the objects were copied. This is true for all three commands. Therefore, whether you use the Cut or Copy commands to select objects, or you use the Copy Link command to select objects automatically, the AutoCAD OLE object pasted into the target application includes the visible area displayed in the current viewport.

When you paste an AutoCAD OLE object that was copied using the Copy Link command, AutoCAD creates a named view representing the current viewport display. This is necessary to maintain the link and update the OLE object accurately when the drawing file is modified. Associating the OLE object with a named view causes modifications to AutoCAD objects in the area of the drawing defined by the named view to update their display automatically in the target application.

Tip: You can use the Copy Link command to paste an existing named view to the Clipboard by setting the named view current immediately before you execute the Copy Link command.

Tip: Often the main complaint with importing drawings into other applications is the lineweight control. In the past, it was difficult to have weights that accurately reflected a plotted drawing. However, starting with AutoCAD 2000, the use of lineweights lets you create drawings that display line widths when they are inserted as OLE objects in other applications.

By using AutoCAD's commands for exporting AutoCAD drawing information into target applications, you can create compound documents in other applications using AutoCAD drawings.

SUMMARY

In this chapter you reviewed how to insert OLE objects from within AutoCAD, paste OLE objects into AutoCAD from the Clipboard, and how to use Drag & Drop to insert OLE objects into AutocAD drawings. You learned how to control various OLE object properties, including how to resize an OLE object, and how to control object visibility. You also reviewed how to edit OLE objects, and how the Layer Properties Manager can affect certain OLE object properties. Finally, you learned how to export objects from AutoCAD into other applications, creating AutoCAD OLE objects.

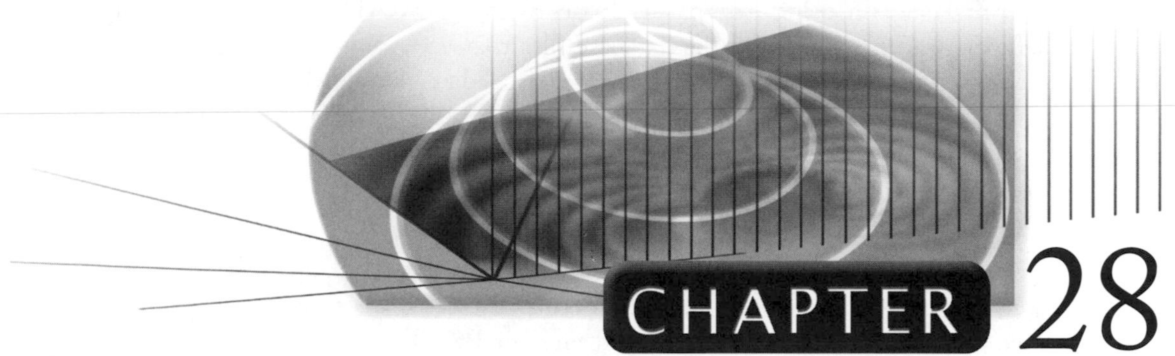

Understanding External Databases

External databases are files that store information. The files are typically composed of tables that look similar to a spreadsheet, with data organized in columns called *fields*, and with each unique set of data stored in rows called *records*. By using external databases to organize and store data, you can create, edit, and retrieve tremendous volumes of data.

AutoCAD 2002 provides tools that let you work with external database files. You can open database tables and view or edit their data. You can link database records to AutoCAD objects such as lines, circles, and polylines. You can insert data into your drawing as labels (text objects) that are automatically updated as data records change. You can run *queries*, which retrieve a subset of records based on certain criteria. By using AutoCAD's tools, you can access and use external database files from entirely within AutoCAD.

This chapter reviews AutoCAD's database tools and explains the steps necessary to work with external database files from directly within AutoCAD. This chapter covers the following subjects:

- Setting up AutoCAD to work with external databases
- Working with database tables
- Working with data and objects
- Using queries

SETTING UP AUTOCAD TO WORK WITH EXTERNAL DATABASES

The first step in using external databases with AutoCAD 2002 drawings entails defining information about the database files that you are using. Before AutoCAD (or any application) can access a database, it needs to know which application created the database (Oracle, Excel, Paradox, etc.), and where the database file is located (its path on your system). When you define the necessary information, AutoCAD can interface with external databases and thereby link records in tables to objects in drawings.

Once you define the information file that lists the database's application and location, you can then access its data from AutoCAD. This is true even if you do not have the database application installed on your PC. This means, for example, that you can access data in an Oracle database even though you do not have Oracle installed on your system. All you need is the database file that was created in Oracle. AutoCAD 2002 is designed to access external database files without using the original database application.

AutoCAD 2002 is designed to access data from the following database applications:

- Microsoft Access 97
- dBase V and III
- Microsoft Excel 97
- Oracle 8.0 and 7.3
- Paradox 7.0
- Microsoft Visual FoxPro 6.0
- SQL Server 7.0 and 6.5

If you are provided database files created by any of these database applications, you can access the data in the files directly from AutoCAD 2002.

Note: Microsoft Excel is not a true database application; therefore, it contains no tables. Consequently, to access Excel data from within AutoCAD, you must first specify at least one named range of cells to act as a database table.

A single Excel file can contain multiple ranges of cells, with AutoCAD treating each range as a unique table.

To define the information that AutoCAD 2002 needs to access external database information, you must do two things. First, you must use Microsoft's ODBC Data Source Administrator to create a **data source**, which lists the database's application type and its location. (You access the ODBC Data Source Administrator through Windows' Control Panel.) Second, you must use Microsoft's OLE DB application to establish a link between AutoCAD and the data source. (You access the OLE DB application from within AutoCAD though the Data Link Properties dialog box.) By using these two features, you can define the information AutoCAD needs to access external databases successfully.

Note: ODBC stands for Open Database Connectivity, which is a standard protocol for accessing information in SQL (Structured Query Language) database servers such as the Microsoft SQL Server.

OLE DB technology is newer than ODBC, and it performs a link-to-database function similar to that of ODBC. OLE DB, however, is designed to address issues encountered when you work with non-relational database files or when you access distributed databases across the Internet, Intranet, and Extranets.

In the following two sections you will use the ODBC Data Source Administrator and the Data Link Properties dialog box to create the information necessary for AutoCAD to access data from an external database file.

CREATING AN ODBC DATA SOURCE FILE

To provide AutoCAD access to data in an external database file, you must first use the ODBC Data Source Administrator to create a data source file. The data source file identifies the database file's application type and the folder in which the database file is located. The data source file is a gateway that AutoCAD uses to access a database file. By using the ODBC Data Source Administrator, you can create a data source file that lets AutoCAD connect directly to a database file.

Note: If you work with Microsoft Access, Oracle, or Microsoft SQL Server database files, you can bypass setting up an ODBC data source and, instead, use the OLE DB direct drivers available on your system. Using the direct drivers, you access the database files through an OLE DB (.udl) configuration file, which you create using AutoCAD's Data Link Properties dialog box.

For more information, refer to AutoCAD 2002's *Driver and Peripheral Guide*, in the section titled Configure External Databases. In this section's Procedures folder you can access the information necessary to set up a direct OLE DB configuration file. The guide is located in the acad_dpg.chm file, which is stored in the AutoCAD 2002/Help folder.

In the following exercise you will use the ODBC Data Source Administrator to create a data source for dBase files.

EXERCISE: USING THE ODBC DATA SOURCE ADMINISTRATOR

1. Create a new folder called DB Files on your PC.
2. Copy the Manholes.dbf and Pipes.dbf files found on the accompanying CD to the DB Files folder. After you copy the dBase files, right-click on the *.dbf files, choose Properties, and then clear the Read-only attribute.

   Next you will create a data source file using the two *.dbf files.

3. From the Windows taskbar, choose Start>Settings>Control Panel. The Control Panel folder opens.
4. From Control Panel, double-click on the Data Sources (ODBC) icon. The ODBC Data Source Administrator is displayed.
5. In the ODBC Data Source Administrator, choose the User DSN tab.

   An ODBC data source stores information on how AutoCAD connects to database files. There are three different methods for defining a data source:

   - Use the User DSN folder to create a data source that is visible only to you, and that can only be accessed from the computer on which the data source is created.
   - Use the System DSN folder to create a data source that is visible to all users who have access rights to the computer on which the data source is created.
   - Use the File DSN folder to create a data source that can be shared with other users who have the same ODBC drivers installed on their computer systems.

6. Choose Add. The Create New Data Source dialog box is displayed.
7. From the list of available database drivers, choose Microsoft dBase Driver (*.dbf), as shown in Figure 28–1.
8. Choose Finish. The ODBC dBase Setup dialog box is displayed.

   By selecting the Microsoft dBase Driver in step 7, you indicated the database file's application type (dBase). Next you will identify the database file's location (its path). Through the ODBC dBase Setup dialog box, you locate the database

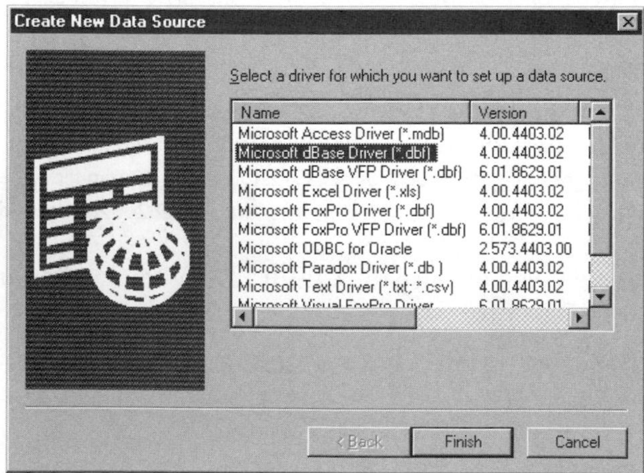

Figure 28–1 *The Create New Data Source dialog box identifies the driver to use to access database files.*

file you wish to access, and you also assign a name and description that makes it easy to identify the new data source from within AutoCAD.

9. In the Data Source Name text box, type **StormDrains**.
10. In the Description text box, type **Storm Drain Tables**.
11. In the Database area, choose dBase 5.0 from the Version list. This is the appropriate version to select because the database files you copied to the DB Files folder are dBase 5.0 files.
12. In the Database area, clear the Use Current Directory check box. The Select Directory button is activated, as shown in Figure 28–2.
13. In the Database area, choose the Select Directory button. The Select Directory dialog box is displayed.
14. In the Select Directory dialog box, browse to your DB Files directory, as shown in Figure 28–3.

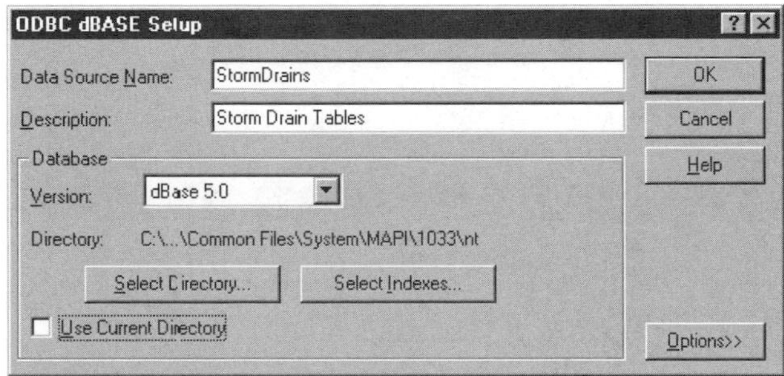

Figure 28–2 *The ODBC dBase Setup dialog box identifies the location of the dBase database files and assigns a data source name.*

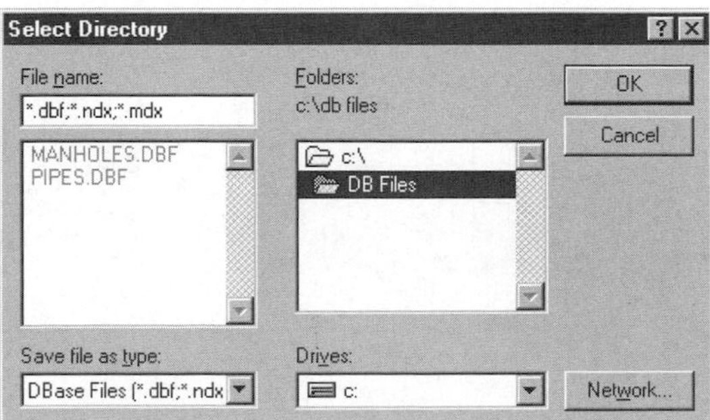

Figure 28–3 *The Select Directory dialog box allows you to locate the folder that contains the database files you use in AutoCAD.*

15. Choose OK to dismiss the Select Directory dialog box, and then choose OK to dismiss the ODBC dBase Setup dialog box. The ODBC Data Source Administrator displays the new StormDrains data source, as shown in Figure 28–4.
16. Choose OK. The data source is saved and can now be accessed from AutoCAD.
17. You may close the Control Panel folder.

Using the ODBC Data Source Administrator, you can create multiple data source files, with each data source file providing the information necessary for AutoCAD to access database files. When you define the data source files, you tell AutoCAD the type of database application that created the files, and where the files are located on your PC. By using the ODBC Data Source Administrator to create a data source file, you can access data stored in database files from within AutoCAD.

Note: The steps used in the preceding exercise are those necessary to create data source files for accessing dBase files. The steps for accessing database files created in other applications vary slightly.

For more information, refer to AutoCAD 2002's *Driver and Peripheral Guide*, in the section titled Configure External Databases. The guide is located in the acad_dpg.chm file, which is stored in the AutoCAD 2002/Help folder.

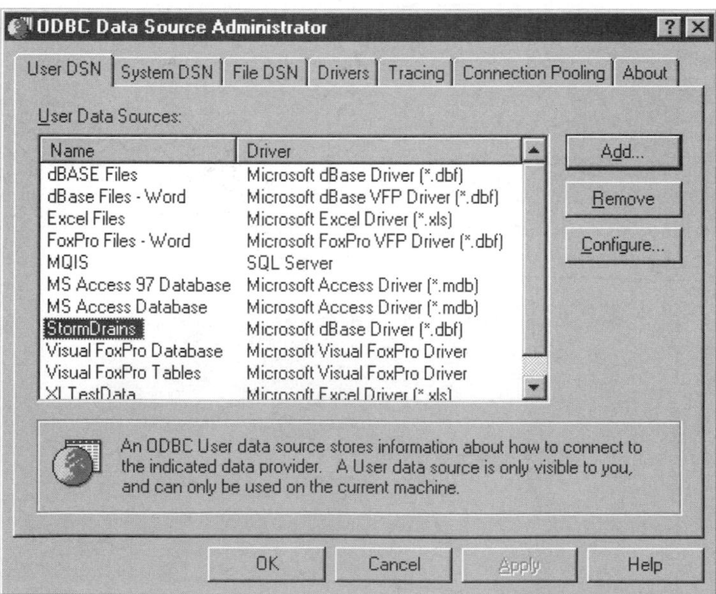

Figure 28–4 *The StormDrains data source has been created and can now be accessed from AutoCAD.*

In the next section you will use the data source file you just created to define an OLE DB configuration (.udl) file, which is the final component necessary for AutoCAD to access database files.

CREATING AN OLE DB CONFIGURATION FILE

An OLE DB configuration file contains information that AutoCAD uses to access data in database files. The configuration file is where you add information that identifies the ODBC data source file name, that lets you password-protect the connection, and that defines the location of the source database files. By creating an OLE DB configuration file, you give AutoCAD the information it needs to access database files.

In the following exercise you will create an OLE DB configuration file.

EXERCISE: CREATING AN OLE DB CONFIGURATION FILE

1. From the AutoCAD Tools menu, choose dbConnect. AutoCAD loads the dbConnect pull-down menu, inserting it into AutoCAD's menu bar, and then AutoCAD loads the dbConnect Manager window.

Note: The dbConnect command located on the Tools menu is a toggle. Once selected, AutoCAD places a check mark next to the dbConnect command. To remove the dbConnect menu and the dbConnect Manager, choose dbConnect from the Tools menu to clear the check mark.

2. From the dbConnect pull-down menu, choose Data Sources>Configure. AutoCAD displays the Configure a Data Source dialog box.
3. In the Data Source Name text box, enter **Storm_Drain**, as shown in Figure 28–5. This is the name that you assign to the OLE DB (.udl) configuration file and that will appear in the dbConnect Manager.

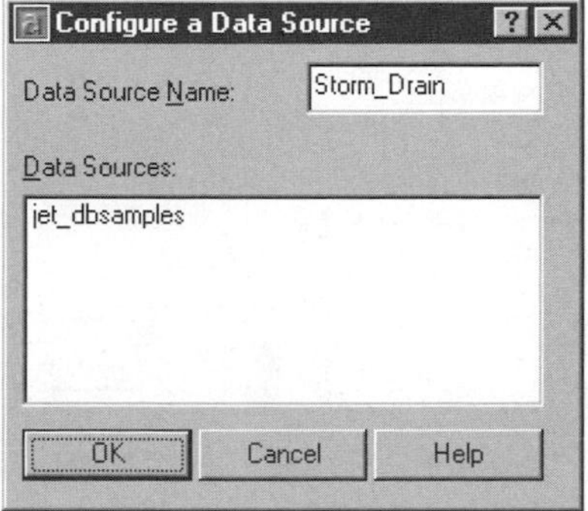

Figure 28–5 *AutoCAD adds the data source name to the dbConnect Manager, which lets you connect to a database.*

4. Choose OK. The Data Link Properties dialog box is displayed.
5. In the Provider folder, select Microsoft OLE DB Provider for ODBC Drivers, as shown in Figure 28–6, and then choose Next. The Connection folder is displayed.
6. In the Connection folder, under step 1, choose the Use Data Source Name radio button.
7. From the Use Data Source Name list, choose the StormDrains data source name.

Note: The StormDrains data source was created in the exercise located in the previous section, Creating an ODBC Data Source File.

Tip: If the StormDrains data source name does not appear in the Use Data Source Name list, choose the Refresh button to update the list.

8. Under step 2, select the Blank Password check box.
9. Under step 3, choose the DB Files catalog from the list, as shown in Figure 28–7.

Note: The DB Files catalog is the location (path) of the dBase files. The location was defined when the StormDrains data source was created through the exercise in the previous section, Creating an ODBC Data Source File.

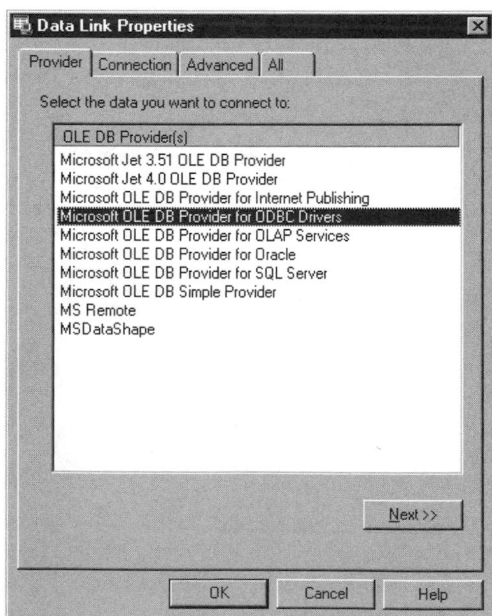

Figure 28–6 *In the Provider folder, select the Microsoft OLE DB Provider for ODBC Drivers.*

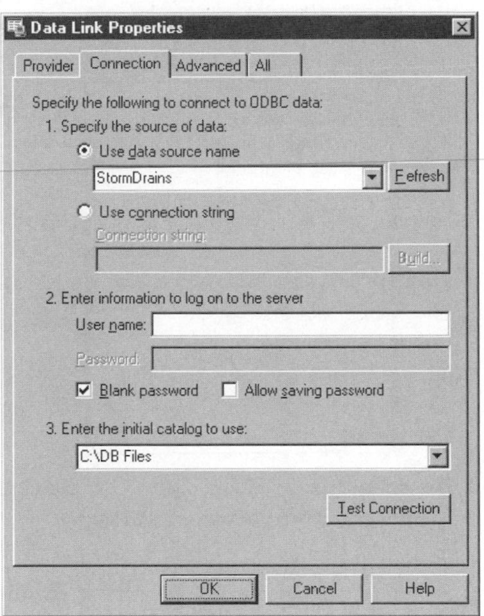

Figure 28–7 *The Connection folder lets you identify the data source name, password-protect access to the database files, and locate the folder, or catalog, that holds the database files.*

Note: A catalog in a database refers, in essence, to the folder that contains the database files.

10. Choose the Test Connection button. AutoCAD performs a test to ensure that it is successfully connected to the database files. If it is, AutoCAD displays a message noting that the test connection succeeded.

    If the test connection is successful, choose OK to dismiss the dialog box. If the text connection fails, you must check to ensure that you chose the correct data source name and the correct catalog folder.

11. Choose OK to dismiss the Data Link Properties dialog box.

    Once the Data Link Properties dialog box is dismissed, the Storm_Drain OLE DB configuration file appears in the dbConnect Manager, as shown in Figure 28–8. This means that you successfully configured the database, and AutoCAD can now access the two tables located in the DB Files folder.

12. You may close the drawing without saving your changes.

Once you successfully configure the database for use with AutoCAD, a configuration file with the extension .udl is created. This configuration file contains the information AutoCAD needs to access the configured database.

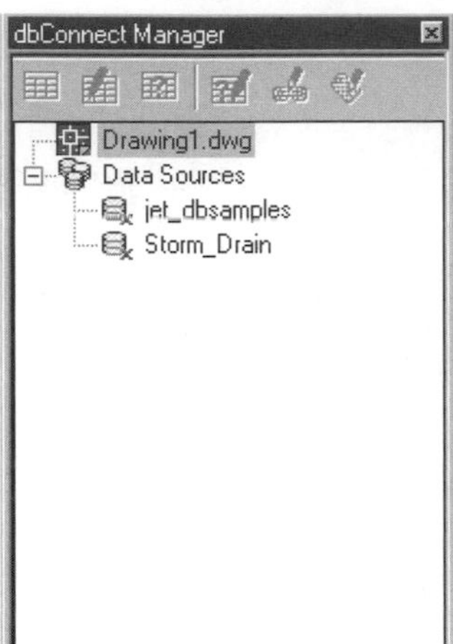

Figure 28–8 *The newly created Storm_Drain OLE DB configuration file appears in the dbConnect Manager.*

Tip: By default, AutoCAD stores the .udl files that it creates in the Data Links folder, which is located in the AutoCAD 2002 folder. You can specify a different location for .udl files in the Options dialog box.

Note: To remove an OLE DB configuration file from AutoCAD's dbConnect Manager, you must delete its .udl file from the Data Links folder.

Note: There are some limitations with using AutoCAD's database connectivity feature, because AutoCAD uses a limited subset of the ODBC and OLE DB database configuration utilities.

For more information, refer to AutoCAD 2002's *Driver and Peripheral Guide*, in the section titled Configure External Databases, under the heading, Known Limitations of Configuring External Databases. The guide is located in the acad_dpg.chm file, which is stored in the AutoCAD 2002/Help folder.

In the preceding two sections, you created a data source for a dBase database and configured an OLE DB file to allow AutoCAD to access the dBase files stored in the DB Files folder. Next you will learn how to access the data in the database files.

WORKING WITH DATABASE TABLES

AutoCAD provides tools for working with external database files. By using these tools, you can view database tables and edit their data. When you view tables, you can move, resize, and hide columns. You can specify a sort order for the table's records. You can edit or delete a record, and you can add a new record to the table. You can search the table for particular values and then replace those values once they are found. By using AutoCAD's database table tools, you can control the display of and edit table data.

ACCESSING TABLES FROM THE DBCONNECT MANAGER

The dbConnect Manager provides a simple, visual way to review drawings linked to database files, and the data sources to which they are linked. You can display links, labels, and queries defined in drawings. You can also display the tables associated with data sources. Through the dbConnect Manager, you can easily view the tables available on your system, and the relationship between the tables and the drawings to which they are linked.

The dbConnect Manager lets you connect to data sources. Once you are connected, you can select a table in the data source and view or edit its data. You can also create a new link template or a new label template—which is used to link objects in drawings to data in tables—and then label the objects using data extracted from the tables. You can also define a new query or execute an existing one. By using the dbConnect Manager, you can perform several useful functions on tables and their data.

To view the dbConnect Manager you can either select dbConnect from the Tools menu or type **DBCONNECT** at the command prompt. Once the command is executed, AutoCAD loads the dbConnect pull-down menu and the dbConnect Manager. The Manager is a dockable, resizable window that can float over your AutoCAD drawing windows. To close the Manager, either execute the dbConnect command again or choose the Exit button (the small "X" button in the upper-right corner of the dbConnect Manager window). Once the Manager is closed, both it and the dbConnect pull-down menu are removed.

In the following exercise you will use the dbConnect Manager to connect to a data source and then view one of its tables.

EXERCISE: USING THE DBCONNECT MANAGER

1. From the AutoCAD Tools menu, choose dbConnect. AutoCAD loads the dbConnect pull-down menu and the dbConnect Manager.

Note: The dbConnect menu and the dbConnect Manager are already loaded if you are continuing from the previous section's exercise.

2. Under Data Sources, right-click on jet_dbsamples, and then choose Connect from the shortcut menu. AutoCAD connects to the jet_dbsamples data source and displays the names of its tables, as shown in Figure 28–9.

   The jet_dbsamples data source is an OLE DB (.udl) configuration file that is automatically installed with AutoCAD 2002. The file is located in the AutoCAD 2002/Data Links folder.

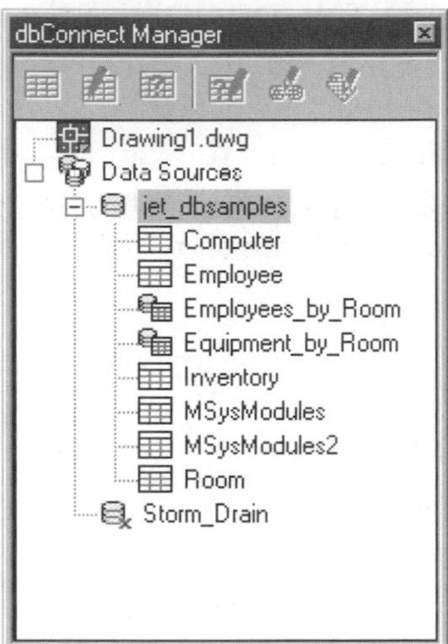

Figure 28–9 *When you connect to a data source, the dbConnect Manager displays the names of its tables.*

Note: If you did not install AutoCAD in its default directory, you may need to update the jet_dbsamples.udl configuration file before you work with these tables. Updating the file is only necessary if AutoCAD cannot locate the files on your system.

3. Under the jet_dbsamples data source, choose the Employee table. The table name is highlighted, and several tool buttons are activated on the dbConnect Manager, as shown in Figure 28–11.
4. Choose the View Table button. The Data View window displays the table in read-only mode, as shown in Figure 28–10.
5. Leave the Data View window open to continue with the next exercise.

The dbConnect Manager provides an easy way to display information about drawing links, labels, and queries, and about data sources available on your system. From this window you can also access table data and define queries. Using the dbConnect Manager, you can quickly perform common database tasks.

In the next section you will use the Data View window to view data in a table.

VIEWING TABLE DATA WITH DATA VIEW

AutoCAD's Data View window allows you to manipulate its view to display data in a table the way you want. For example, you can hide and unhide columns. You can move columns to new positions in the table, changing their order. You can also sort the data in columns, in either ascending or descending order. You can even control the appearance of the text displayed in the table by changing its font or font size. Through the Data View window, you can manipulate the display of a table to make viewing data easier.

Note: When you choose the View Table button, AutoCAD also displays the Data View menu, which appears in AutoCAD's menu bar.

Data View - Employee (Drawing1.dwg) - Read Only

-- New Link Template -- | -- New Label Template --

Emp_Id	Last_Name	First_Name	Gender	Title	Department	Room
1000	Torbati	Yolanda	F	Programmer	Eng neering	6044
1001	Kleinn	Joel	M	Programmer	Eng neering	6058
1002	Ginsburg	Laura	F	President	Corporate	6050
1003	Cox	Jennifer	F	Programmer	Eng neering	6056
1005	Ziada	Mauri	M	Product Designer	Eng neering	6055
1006	Keyser	Cara	F	Account Executive	Sales	6106
1063	Ford	Janice	F	Accountant	Accounting	6020
1010	Smith	Roxie	M	Programmer	Eng neering	6054
1011	Nelson	Robert	M	Programmer	Eng neering	6042
1012	Sachsen	Lars	M	Support Technician	Product Support	6104
1013	Shannon	Don	M	Product Designer	Eng neering	6053
1016	Miro	Terri	F	Network Administrator	IS	6074
1017	Lovett	Greg	M	Programmer	Eng neering	6060
1018	Larson	Steve	M	Programmer	Eng neering	6052
1019	Haque	Cintra	F	Sales Representative	Sales	6124
1020	Sampson	Heather	F	Marketing Representative	Marketing	6076
1023	Gupta	Rebecca	F	Programmer	Eng neering	6051
1024	Feaster	John	M	Vice President	Sales	6153

Record 1

Figure 28–10 *The Data View window can be accessed from the dbConnect Manager.*

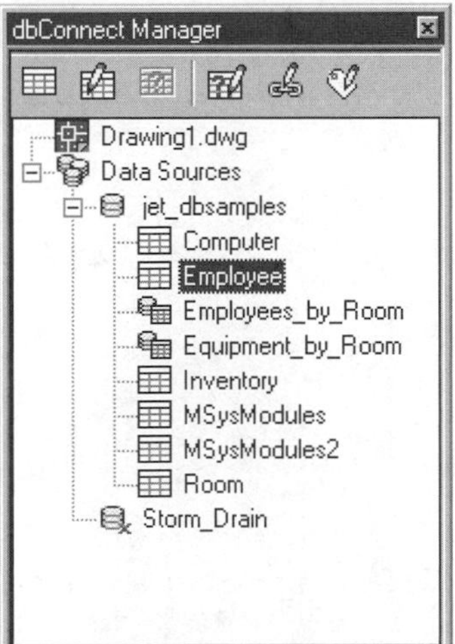

Figure 28–11 *When you select a table under a data source, the tool buttons are activated on the dbConnect Manager.*

In the following exercise you will learn about controlling the appearance of data in the Data View window.

EXERCISE: MANIPULATING THE DATA VIEW WINDOW

1. Continue from the previous exercise and choose the Title column's header. You may need to pan to the right in order to see the Title column.

   The column header is actually a button that displays the column's name at the top of the column. When you choose the button, AutoCAD highlights the entire column.

2. Drag the column to the left side of the Gender column. When a red line appears between the First_Name and Gender columns, release your mouse button to drop the column. AutoCAD places the Title column to the left of the Gender column, as shown in Figure 28–12.

   You can resize a column as desired by clicking and dragging the line between header buttons. You can also automatically resize a column by double-clicking on the line between two columns.

   AutoCAD lets you hide columns in the Data View window, and you can then unhide hidden columns. Next you will hide and then unhide the First_Name column.

3. Choose the First_Name column's header. The column is highlighted.

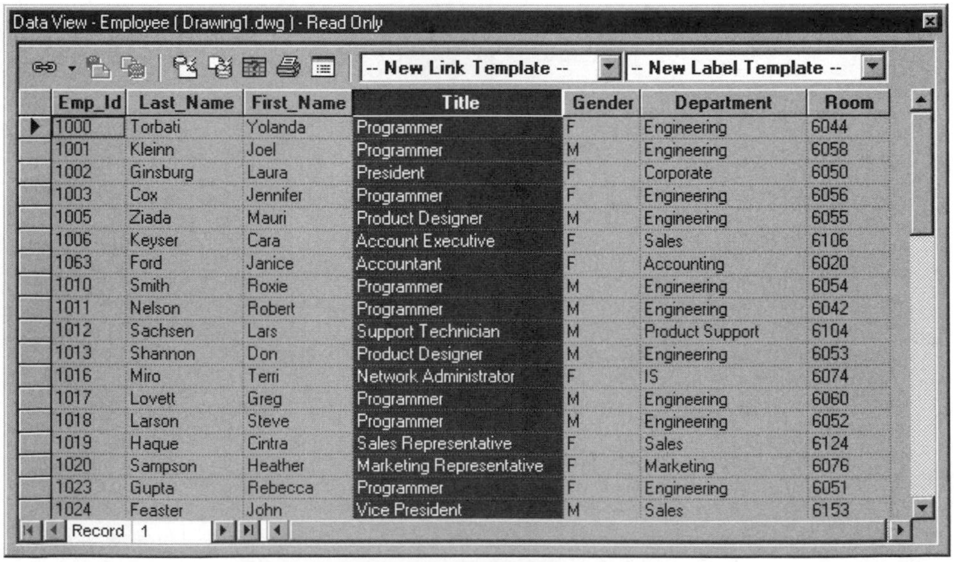

Data View - Employee (Drawing1.dwg) - Read Only

-- New Link Template -- | -- New Label Template --

Emp_Id	Last_Name	First_Name	Title	Gender	Department	Room
1000	Torbati	Yolanda	Programmer	F	Engineering	6044
1001	Kleinn	Joel	Programmer	M	Engineering	6058
1002	Ginsburg	Laura	President	F	Corporate	6050
1003	Cox	Jennifer	Programmer	F	Engineering	6056
1005	Ziada	Mauri	Product Designer	M	Engineering	6055
1006	Keyser	Cara	Account Executive	F	Sales	6106
1063	Ford	Janice	Accountant	F	Accounting	6020
1010	Smith	Roxie	Programmer	M	Engineering	6054
1011	Nelson	Robert	Programmer	M	Engineering	6042
1012	Sachsen	Lars	Support Technician	M	Product Support	6104
1013	Shannon	Don	Product Designer	M	Engineering	6053
1016	Miro	Terri	Network Administrator	F	IS	6074
1017	Lovett	Greg	Programmer	M	Engineering	6060
1018	Larson	Steve	Programmer	M	Engineering	6052
1019	Haque	Cintra	Sales Representative	F	Sales	6124
1020	Sampson	Heather	Marketing Representative	F	Marketing	6076
1023	Gupta	Rebecca	Programmer	F	Engineering	6051
1024	Feaster	John	Vice President	M	Sales	6153

Record 1

Figure 28–12 *You can highlight and move a column by picking and dragging its header, which is the button that displays the column's name.*

4. Right-click on the First_Name column's header and then choose Hide. The First_Name column is hidden, as shown in Figure 28–13.

Tip: By holding down the Ctrl key when you choose column headers, you can select and then hide multiple columns at the same time.

5. Right-click on the Emp_ID column and then choose Unhide All. The First_Name column is redisplayed.

   The Data View window allows you to sort the data in columns. The Data View window can sort both numerically and alphabetically, and it allows you to sort data in ascending or descending order. You can also execute sorts on up to five columns at one time. Next you will sort two different columns at once.

6. Right-click on the Emp_ID column's header button and then choose Sort. The Sort dialog box is displayed.
7. In the Sort By area, choose Emp_ID from the drop-down list. Make sure the Ascending radio button is selected.
8. In the Then By area immediately below the Sort By area, choose Department from the drop-down list, as shown in Figure 28–14. Make sure the Ascending radio button is selected.
9. Choose OK. AutoCAD executes the sort and redisplays the data based on the sort results.

Data View - Employee (Drawing1.dwg) - Read Only

-- New Link Template -- | -- New Label Template --

Emp_Id	Last_Name	Title	Gender	Department	Room
1000	Torbati	Programmer	F	Engineering	6044
1001	Kleinn	Programmer	M	Engineering	6058
1002	Ginsburg	President	F	Corporate	6050
1003	Cox	Programmer	F	Engineering	6056
1005	Ziada	Product Designer	M	Engineering	6055
1006	Keyser	Account Executive	F	Sales	6106
1063	Ford	Accountant	F	Accounting	6020
1010	Smith	Programmer	M	Engineering	6054
1011	Nelson	Programmer	M	Engineering	6042
1012	Sachsen	Support Technician	M	Product Support	6104
1013	Shannon	Product Designer	M	Engineering	6053
1016	Miro	Network Administrator	F	IS	6074
1017	Lovett	Programmer	M	Engineering	6060
1018	Larson	Programmer	M	Engineering	6052
1019	Haque	Sales Representative	F	Sales	6124
1020	Sampson	Marketing Representative	F	Marketing	6076
1023	Gupta	Programmer	F	Engineering	6051
1024	Feaster	Vice President	M	Sales	6153

Record 1

Figure 28–13 *The First_Name column is hidden from view.*

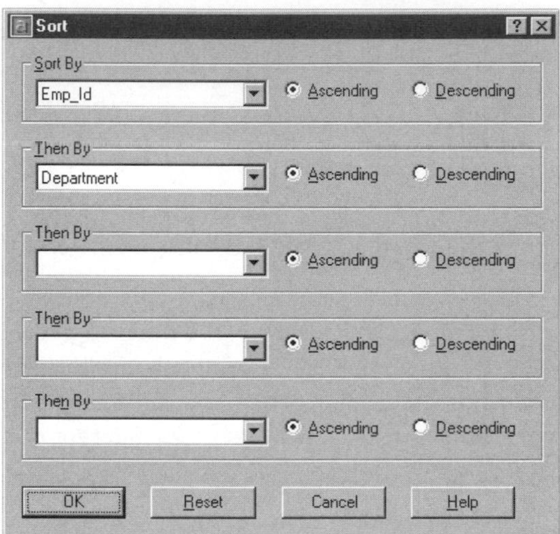

Figure 28–14 *The Sort dialog box allows you to create a sort of up to five levels.*

10. Close the Data View window and the dbConnect Manager by choosing the Exit button, which is the small button with an "X" in it located in the upper-right corner of each window.
11. You may now exit AutoCAD without saving your changes.

As you work with large database files, some tables you display in the Data View window will contain many columns. To view these columns you will need to drag the slide bar at the bottom of the window left or right. The downside to dragging the slide bar to view columns outside the window's current display is that columns you are currently viewing will be moved out of view as you move the slide bar.

The Data View window allows you to freeze a column's position. This means that when you move the slide bar to view other columns, the frozen columns do not move from view, and their positions remain fixed. This feature allows you to drag the slide bar to view columns outside the current display, while always keeping the frozen columns displayed.

To freeze a column, right-click on the column's header button and then choose Freeze. To unfreeze a column, right-click on any column's header button and then choose Unfreeze All.

The Data View window also allows you to control the alignment of text in columns and change the font and font size of the text used to display a table's data. To change the alignment of a column, right-click on the column's header button, choose Align, and then choose the desired alignment from the fly-out menu, as shown in Figure 28–15.

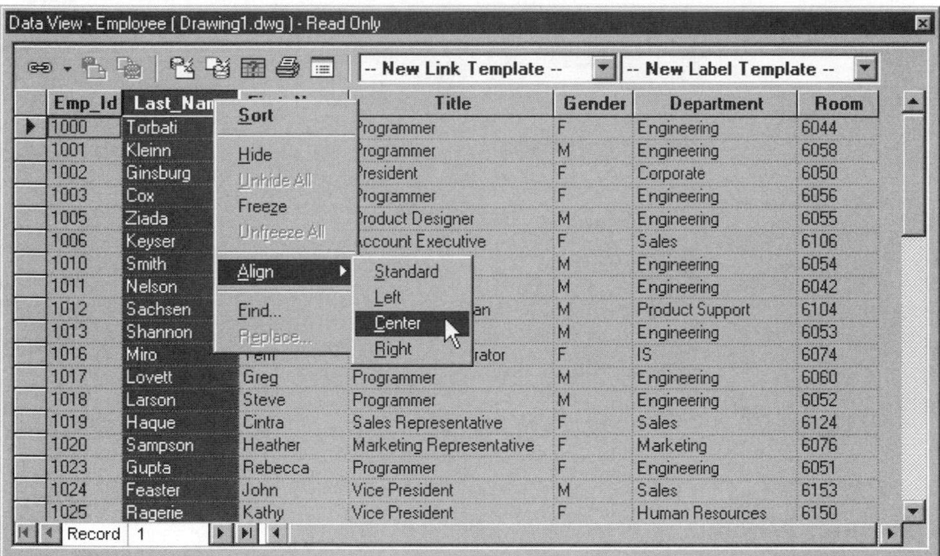

Figure 28–15 *Right-click on a column's header to control its text justification.*

To control the font properties of text that appears in the Data View window, from the Data View pull-down menu, choose Format to display the Format dialog box, as shown in Figure 28–16. In the Format dialog box you can modify font properties such as style and size.

The Data View window also allows you to control its appearance. You can hide columns, change the position of columns, or lock columns by freezing them. You can also sort the table's data using the Sort dialog box. In addition to these features, you can control the appearance of the text displayed in the columns and rows; you can change their justification and their font size. By manipulating its display, you can modify the Data View window to display the data the way you want.

EDITING TABLE DATA USING THE DATA VIEW WINDOW

AutoCAD's Data View window has two modes. The first is read-only mode, and the second is edit mode. The read-only mode allows you to view data and manipulate how it is displayed in the Data View window. The edit mode also allows you to manipulate the window's appearance, but more importantly, it allows you to manipulate the table and its data. For example, in edit mode, you can edit, delete, and add new records to a table. You can also search the table for certain values, and then replace those values with new ones. By using Data View's edit mode, you can view, edit, and explore the data in database tables.

In the following exercise you will learn about the various features of the Data View window's edit mode.

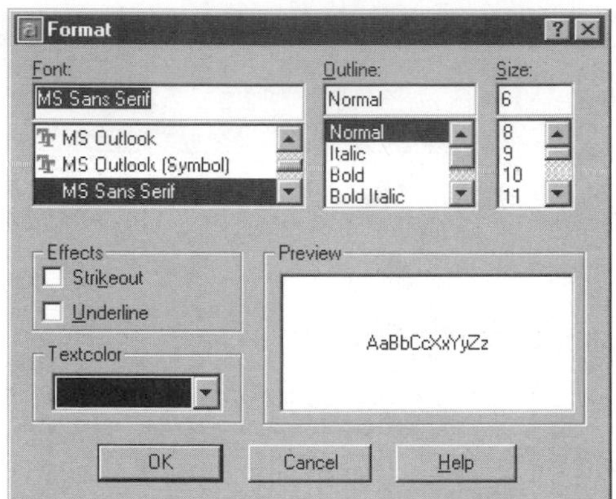

Figure 28–16 *The Format dialog box allows you to change the Data View window's font properties.*

EXERCISE: USING DATA VIEW'S EDIT MODE

1. If you have not done so in a previous exercise, create a new folder called DB Files on your PC.
2. Copy the StormDrain.mdb file found on the accompanying CD to the DB Files folder. Once the file is copied into the folder, right-click on the *.mdb file, choose Properties, and then clear the Read-only attribute.

   The StormDrain.mdb file is a Microsoft Access 2000 database. As was noted previously in this chapter, in the section titled Creating an ODBC Data Source File, it is not necessary to create an ODBC data source file to connect to Microsoft Access databases. Instead, you can use the OLE DB direct drivers available on your system to create an Access OLE DB (.udl) configuration file from within AutoCAD.

   Next you will create an Access OLE DB (.udl) configuration file for the StormDrain.mdb database.
3. From the dbConnect pull-down menu, choose Data Sources>Configure. AutoCAD displays the Configure a Data Source dialog box.
4. In the Data Source Name text box, enter **StormDrains**. This is the name you assign to the Access OLE DB (.udl) configuration file, and it is the name that will appear in the dbConnect Manager.
5. Choose OK. The Data Link Properties dialog box is displayed.
6. In the Provider folder, select Microsoft Jet 4.0 OLE DB Provider and then choose Next. The Connection folder is displayed.
7. In the Connection folder, under step 1, choose the "..." button and then browse to the DB Files folder and open StormDrain.mdb.

8. Under step 2, select the Blank Password check box, as shown in Figure 28–17. Choose the Test Connection button to confirm that AutoCAD is connected to the database files.

Note: If the test connection is successful, choose OK to dismiss the dialog box. If the text connection fails, you must check to ensure that you chose the correct Microsoft Access database.

9. Choose OK to dismiss the Data Link Properties dialog box.

   Once the Data Link Properties dialog box is dismissed, the StormDrains OLE DB configuration file appears in the dbConnect Manager.

10. Right-click on the StormDrains data source and then choose Connect. AutoCAD connects to the data source and displays its tables.

11. Choose the Pipes table and then choose the Edit Table button. AutoCAD opens the Data View window in edit mode.

    In the Data View window you can edit individual cells in the table. Next you will edit the DIA value for one of the records.

12. In the DIA field for record 1008, replace value 24 with **36**, and then press Enter. AutoCAD replaces the old cell value with the new one, as shown in Figure 28–18.

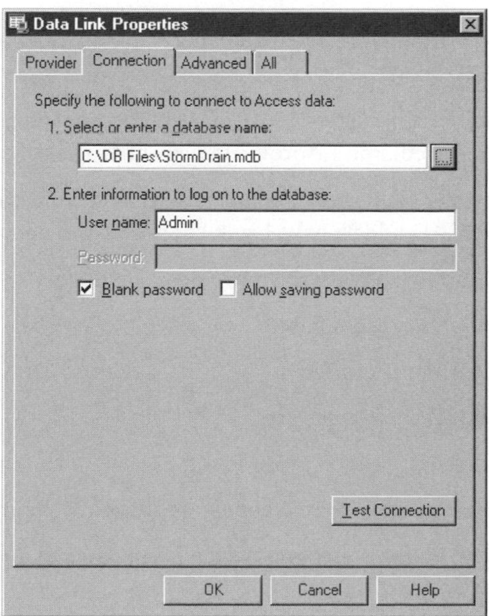

Figure 28–17 *The Connection folder lets you identify the Microsoft Access database file.*

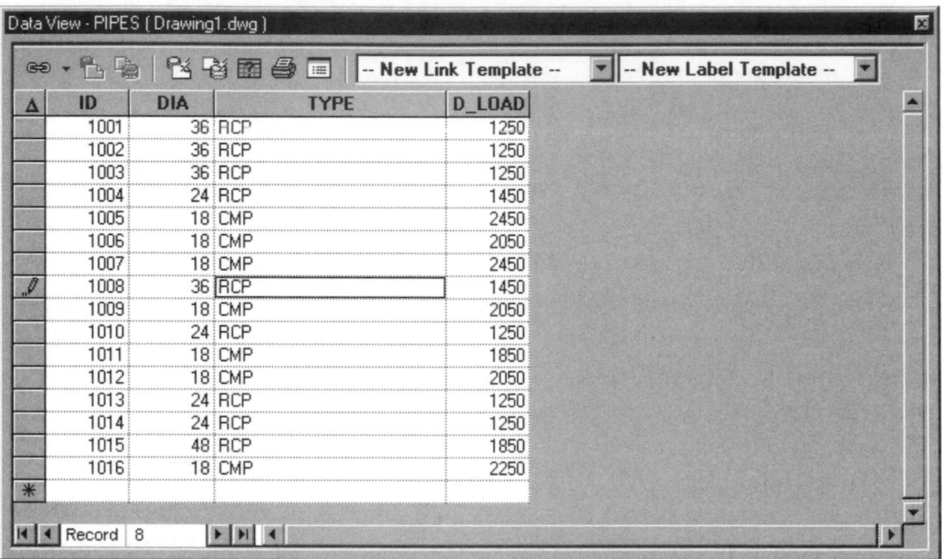

ID	DIA	TYPE	D_LOAD
1001	36	RCP	1250
1002	36	RCP	1250
1003	36	RCP	1250
1004	24	RCP	1450
1005	18	CMP	2450
1006	18	CMP	2050
1007	18	CMP	2450
1008	36	RCP	1450
1009	18	CMP	2050
1010	24	RCP	1250
1011	18	CMP	1850
1012	18	CMP	2050
1013	24	RCP	1250
1014	24	RCP	1250
1015	48	RCP	1850
1016	18	CMP	2250

Figure 28–18 *The DIA field value for record 1008 is changed to 36.*

Note: Once you enter a new value, AutoCAD instantly updates the table. Edits cannot be undone.

The Data View window lets you add new records to an existing table. Next you will add a new record.

13. Select the header for record 1010. (The record header is the small button to the left of the ID column.) AutoCAD highlights the record.
14. Right-click on the highlighted record header and then choose Add New Record. AutoCAD adds a new, blank record at the end of the records list.
15. Enter values for each cell in the new record as follows:

    ID: Type **1017**, then press Tab.

    DIA: Type **24**, then press Tab.

    TYPE: Type **RCP**, then press Tab.

    D_LOAD: Type **1450**, then press Enter.

    The Data View window should appear as shown in Figure 28–19.

    The Data View window lets you delete records from an existing table. Next you will delete a record.
16. Select the record header 1007. AutoCAD highlights the record.

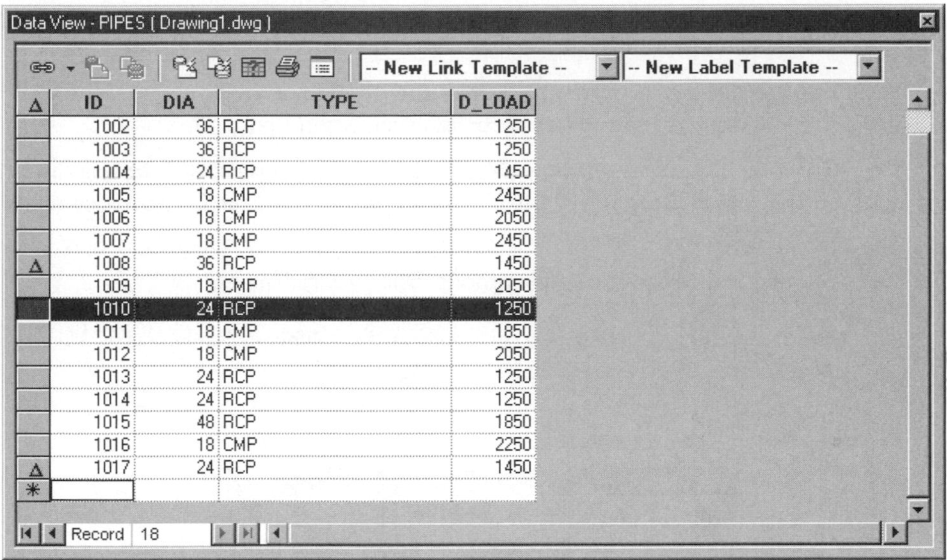

ID	DIA	TYPE	D_LOAD
1002	36	RCP	1250
1003	36	RCP	1250
1004	24	RCP	1450
1005	18	CMP	2450
1006	18	CMP	2050
1007	18	CMP	2450
1008	36	RCP	1450
1009	18	CMP	2050
1010	24	RCP	1250
1011	18	CMP	1850
1012	18	CMP	2050
1013	24	RCP	1250
1014	24	RCP	1250
1015	48	RCP	1850
1016	18	CMP	2250
1017	24	RCP	1450

Figure 28–19 *The new record 1017 is added to the existing table.*

17. Right-click on the highlighted record header and then choose Delete Record. AutoCAD displays a window asking you to verify that you want to delete the selected row.
18. Choose Yes. AutoCAD deletes the selected record.

Tip: By holding down the Ctrl key when you choose record headers, you can select multiple records to delete at the same time.

The Data View window allows you to find specified values in selected columns. When you choose a cell in a column, then right-click, and then choose Find from the shortcut menu, AutoCAD searches the selected column and highlights the cell in which it finds the specified value. Additionally, you can replace the specified value with a new value.

Next you will search for values in the TYPE column and replace them with new values.

19. Select a cell in the TYPE field. AutoCAD highlights the cell.
20. Right-click on the cell and then choose Replace from the shortcut menu. AutoCAD displays the Replace dialog box.
21. In the Find What text box, type **CMP**.
22. In the Replace With text box, type **RCP**, as shown in Figure 28–20.

23. Choose Replace All. AutoCAD replaces all instances of the CMP value in the TYPE column with RCP.
24. Choose the Cancel button to dismiss the Replace dialog box. The Data View window displays the updated records, as shown in Figure 28–21.
25. Close the Data View window and the dbConnect Manager by choosing the Exit button, which is the small button with an "X" in it located in the upper-right corner of each window.
26. You may now exit AutoCAD without saving your changes.

Note: AutoCAD searches only for values in the column indicated. AutoCAD cannot globally search all columns for values in a table.

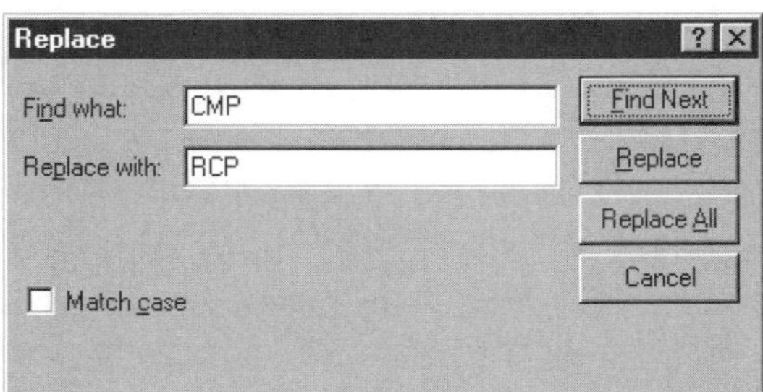

Figure 28–20 *The CMP value will be replaced by the RCP value.*

Data View - PIPES (Drawing1.dwg)

-- New Link Template -- | -- New Label Template --

ID	DIA	TYPE	D_LOAD
1001	36	RCP	1250
1002	36	RCP	1250
1003	36	RCP	1250
1004	24	RCP	1450
1005	18	RCP	2450
1006	18	RCP	2050
1008	36	RCP	1450
1009	18	RCP	2050
1010	24	RCP	1250
1011	18	RCP	1850
1012	18	RCP	2050
1013	24	RCP	1250
1014	24	RCP	1250
1015	48	RCP	1850
1016	18	RCP	2250
1017	24	RCP	1450

Record 3

Figure 28–21 *The Data View window displays the table's updated records.*

Note: In this exercise, modifications made to the table instantly updated the original table. Other databases do not instantly update the original table; they allow you to choose whether you want to accept or reject your modifications before updating the original table.

When you work with databases that allow you to accept or reject your modifications, you are prompted either to commit modifications or to restore the table's original values. You indicate your choice by right-clicking on the grid header and then choosing the desired option from the shortcut menu. The grid header is the small button in the upper-left corner of the column and record headers.

The Data View window's edit mode allows you to edit an existing table. You can change cell values individually, or you can search and replace specified values instantly in an entire column. You can add new records to the table or delete existing ones. By using the Data View window's edit mode, you can easily modify a table's data.

WORKING WITH DATA AND OBJECTS

AutoCAD allows you to create a link between records in a database table and objects in a drawing. Once you have established a link between an object and a record, you can use the link either to locate objects by selecting the records to which they are attached, or to locate records in a table by selecting objects in a drawing. Additionally, you can extract the data values stored in a table and then automatically insert them as a text label that is attached to its linked object. By using AutoCAD's ability to link data to objects, you can quickly locate linked objects and records, and you can easily label objects by extracting the data to which they are attached.

LINKING DATA TO AUTOCAD OBJECTS

Up to this point you have learned how to use AutoCAD to connect to a database. By defining an ODBC data source file and then creating an OLE DB (.udl) configuration file, you can access a variety of databases. Through the data source and .udl files, you establish connections that let you view and edit databases from within AutoCAD. While the ability to connect to databases is useful, there is more to working with databases from within AutoCAD than just viewing and editing. The real power of connecting to databases from within AutoCAD lies in the ability to link objects in drawings to records in databases.

You may have noticed in the previous exercises in this chapter that while you connected AutoCAD to databases, the connections occurred independent of the current drawing. The fact is that while you followed the previous exercises, you were simply connecting a database to the AutoCAD application but not to the AutoCAD drawing. Therefore, after you connect AutoCAD to a database, the next step is to link records in the database to objects in a drawing.

To link records in a database to objects in a drawing, you must first define a link template. Using a link template, AutoCAD lets you select a record in a table and link it to a graphic object in a drawing. So, by first connecting to a database through a data source and .udl files, and then by defining a link template, you can create links between the graphic objects in a drawing and the records stored in database tables.

Creating a Link Template

A link template identifies the field (column) in a database's table to use to link a record to an object. Once you identify the field in a table, you can then establish a link between a specific record's value and an AutoCAD object. For example, in the previous section, you used AutoCAD to add a new record to an existing table in an Access database. The record included the unique ID value 1017. To link record 1017 to an object in a drawing, you must first create a link template that tells AutoCAD that the ID field contains the unique identifiers to use when it links a record to an object. So, after you connect a database to the AutoCAD application through the data source and .udl files, you then create a link template that tells AutoCAD which field's data to use to link table records in a database to objects in a drawing.

In the following exercise you will create a link template.

EXERCISE: CREATING A LINK TEMPLATE

1. Copy the 28DWG01.DWG file found on the accompanying CD to the DB Files folder. The DB Files folder was created in a previous exercise in this chapter, in the section titled Creating an ODBC Data Source File.

   After you copy the drawing file, right-click on the file, choose Properties, and then clear the Read-only attribute.
2. Open the drawing 28DWG01.DWG located in the DB Files folder. The drawing contains a series of polylines and circles, which represent storm drain lines and manholes.
3. From the Tools menu, choose dbConnect. AutoCAD displays the dbConnect pull-down menu and the dbConnect Manager.
4. In the dbConnect Manager, right-click on the StormDrains data source and then choose Connect. AutoCAD connects to the data source and displays its tables.
5. From the dbConnect pull-down menu, choose Templates>New Link Template. AutoCAD displays the Select Data Object dialog box.
6. In the Select Data Object dialog box, choose the PIPES table, as shown in Figure 28–22, and then choose Continue. AutoCAD displays the New Link Template dialog box.
7. In the New Link Template dialog box, in the New Link Template Name text box, type **PipesLink** as shown in Figure 28–23, and then choose Continue. AutoCAD displays the Link Template dialog box.
8. In the Link Template dialog box, in the Key Fields list, choose the check box next to the ID key field, as shown in Figure 28–24, and then choose OK.

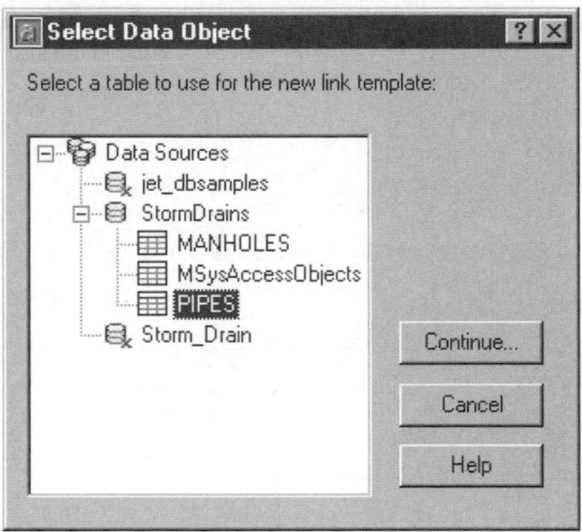

Figure 28–22 *The Select Data Object dialog box identifies the table for which you will create a link template.*

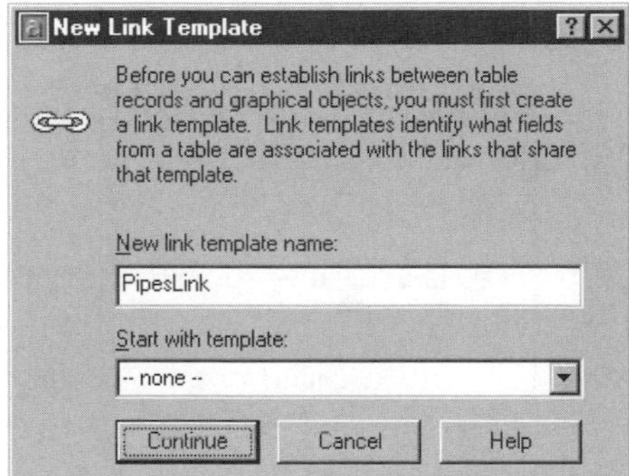

Figure 28–23 *The New Link Template dialog box lets you define the name for the new link template.*

AutoCAD defines the new link template for the 28DWG01.DWG drawing, as is indicated by the PipesLink connection displayed in the dbConnect Manager shown in Figure 28–25.

9. Save your changes, and leave the drawing open for use in the next exercise.

Note: Notice in Figure 28–25 that the PipesLink link template appears as an object under the 28DWG01.DWG drawing. As this hierarchy implies, the PipesLink link template is saved within the 28DWG01.DWG drawing. In contrast, the Data Sources shown in Figure 28–25 are independent of the 28DWG01.DWG drawing.

When you work with databases from within AutoCAD, creating link templates is a necessary step that lets AutoCAD link records in a database to objects in drawings.

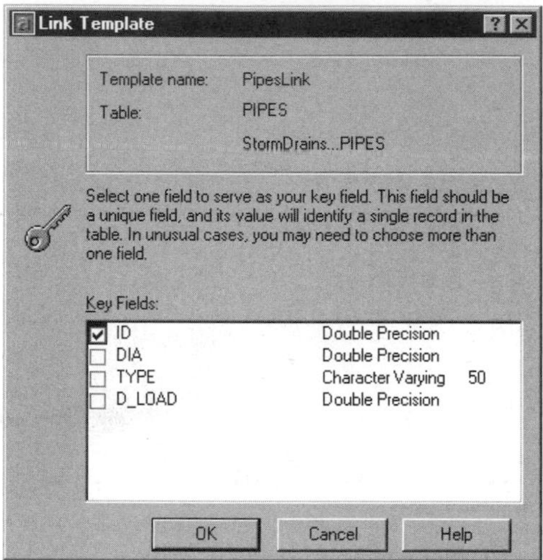

Figure 28–24 *The Link Template dialog box let you identify the key field for the new link template.*

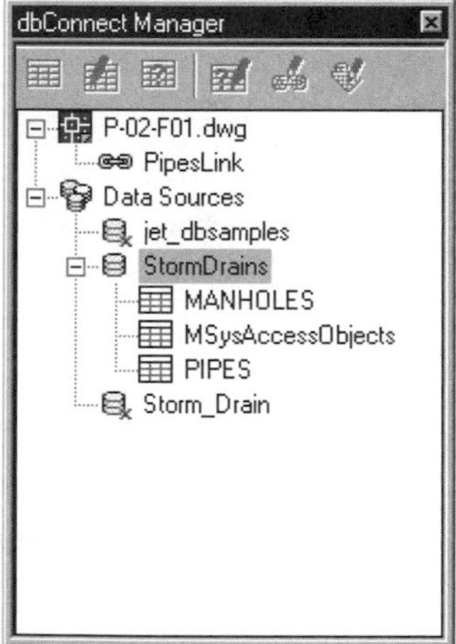

Figure 28–25 *The PipesLink link template is defined for the 28DWG01.DWG drawing and displayed in the dbConnect Manager.*

When you create link templates that are stored within drawings, you identify the field in a table that AutoCAD uses to link records to objects.

Next you will use the link template that you just created to link records in the PIPES table to polyline objects in a drawing.

Creating Links

Once you define a link template for a drawing, you can then link records in a table to objects in a drawing. You can link multiple records to a single object, or you can link a single record to multiple objects. Once a link is created, you can select an object to highlight the record to which it is attached, or you can select a record to highlight the object to which it is linked. Once a link template is defined for a drawing, you can link records to objects and then view their link associations.

In the following exercise you will create links between records in a table and objects in a drawing.

EXERCISE: LINKING RECORDS TO OBJECTS

1. Continue from the previous exercise.
2. In the dbConnect Manager, choose the PIPES table, and then choose the View Table button. AutoCAD displays the Data View window in read-only mode.
3. In the Data View window, from the New Link Template list, choose the PipesLink template.
4. Choose the record header for record ID 1001. AutoCAD highlights the record.
5. From the Data View pull-down menu (located in the AutoCAD menu bar), if the option is not already checked, choose Link and Label Settings>Create Links. AutoCAD sets the link mode to "create links."
6. From the Data View menu, choose Link!. AutoCAD prompts you to select objects to which it should link the highlighted record.
7. Choose the polyline that connects circle 1 and circle3, and then press Enter to end object selection. AutoCAD links the highlighted record to the selected polyline and changes the highlighted record's color to yellow.
8. In the Data View window, choose the record header for record ID 1004. AutoCAD highlights the record.
9. From the Data View menu, choose Link!. AutoCAD prompts you to select objects to which it should link the highlighted record.
10. Choose the polyline that connects circle 2 and circle 4, then choose the polyline that connects circle 4 and circle 5, and then press Enter to end object selection. AutoCAD links the highlighted record to the selected polylines and changes the highlighted record's color to yellow, as shown in Figure 28–26.

    Once links are created between records and objects, you can view the links by either selecting records to highlight the objects to which they are attached, or by selecting objects to highlight the records to which they are linked.

    Next you will use the link associations you just created to highlight records and objects.

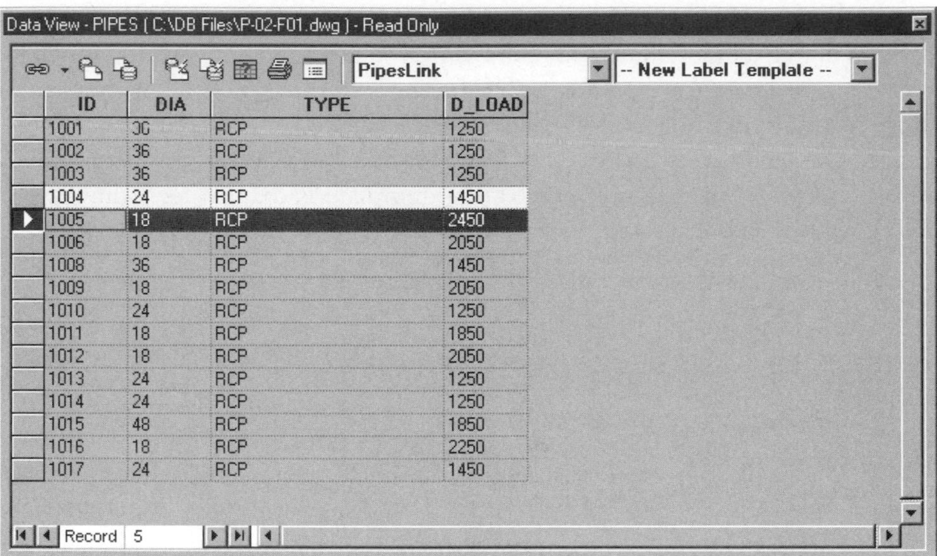

ID	DIA	TYPE	D_LOAD
1001	36	RCP	1250
1002	36	RCP	1250
1003	36	RCP	1250
1004	24	RCP	1450
1005	18	RCP	2450
1006	18	RCP	2050
1008	36	RCP	1450
1009	18	RCP	2050
1010	24	RCP	1250
1011	18	RCP	1850
1012	18	RCP	2050
1013	24	RCP	1250
1014	24	RCP	1250
1015	48	RCP	1850
1016	18	RCP	2250
1017	24	RCP	1450

Figure 28–26 *In the Data View window, record 1004 is highlighted yellow after it is linked to an object in a drawing.*

11. In the Data View window, choose the record header for record ID 1001. AutoCAD highlights the record.
12. From the Data View menu, choose View Linked Objects. AutoCAD highlights the object to which the record is linked, as shown in Figure 28–27.

    Next you will use an object to highlight a record.
13. Press Esc to deselect the highlighted polyline object.
14. Select the polyline that connects circles 4 and 5. AutoCAD highlights the polyline.
15. From the Data View menu, choose View Linked Records. AutoCAD displays only the record to which the object is linked in the Data View window, as shown in Figure 28–28.
16. Save your changes and then close the drawing.

Once you establish links between objects and records, you can use the links to locate and highlight linked objects. You can select objects and locate the records to which they are linked, or you can choose records to locate and select the objects to which they are attached. By creating links between records and objects, you add intelligence to your drawing that you can use to locate and select records and objects.

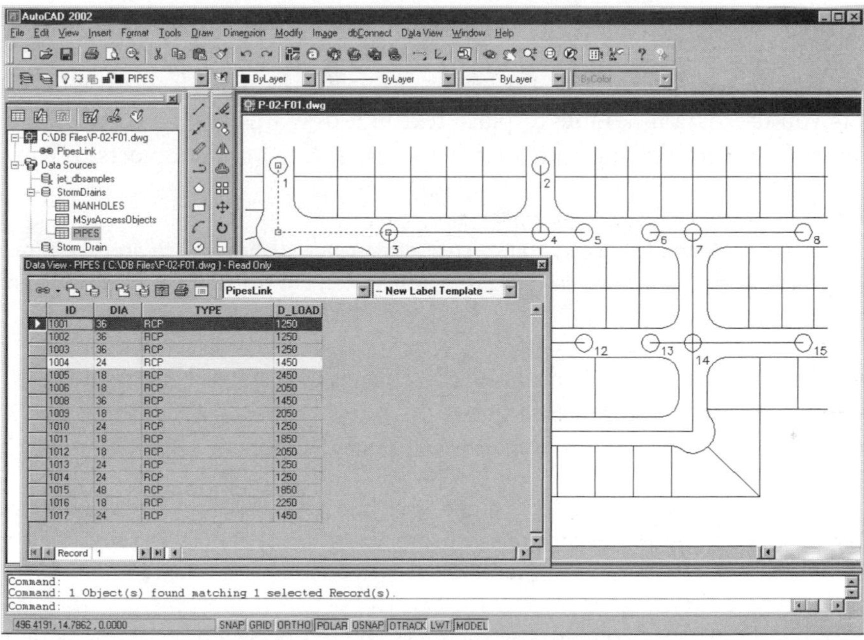

Figure 28–27 *The selected record is used to highlight the object to which it is linked.*

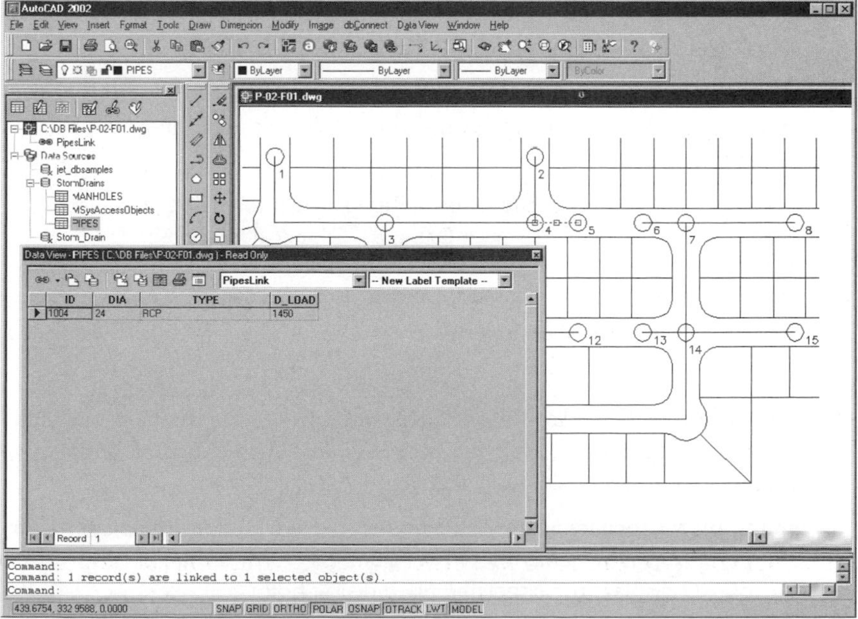

Figure 28–28 *The selected object is used to locate and display the record to which it is linked.*

LABELING OBJECTS WITH DATA FROM TABLES

Once you have established links between records in a table and objects in a drawing, you can use those links to label the objects. AutoCAD provides a labeling feature that lets you use data in a table to place text in a drawing. Using AutoCAD's label feature, you can extract data from a table and use it to label the objects to which the data is linked.

Note: AutoCAD allows you to create freestanding labels, which are labels (text) that are inserted into a drawing using data from tables, but that are not attached to an object in a drawing.

Creating labels from data linked to objects requires two steps. First, you must create a label template. Second, you must insert the label. By performing these two simple steps, you can insert a label into a drawing and attach it to an object in a drawing.

Note: The following exercise uses a drawing that was modified in previous exercises. To perform the steps in the following exercise, you must use the drawing modified in the two exercises located in this chapter, in the section titled Linking Data to AutoCAD Objects.

In the following exercise you will use the links created between records in a table and objects in a drawing to label the objects.

EXERCISE: LABELING OBJECTS LINKED TO RECORDS

1. Open the 28DWG01.DWG file located in the DB Files folder.
2. If the dbConnect Manager is not displayed, from the Tools menu, choose dbConnect. AutoCAD displays the dbConnect Manager and the dbConnect pull-down menu.
3. From the dbConnect menu, choose Templates>New Label Template. AutoCAD displays the Select a Database Object dialog box.

Note: A label template requires a link template. You can only create a label template after you have created a link template.

4. In the Select a Database Object dialog box, make sure the PipesLink link template is selected as shown in Figure 28–29, and then choose Continue. AutoCAD displays the New Label Template dialog box.
5. In the New Label Template dialog box, in the New Label Template Name text box, type **PipesLabel** as shown in Figure 28–30, and then choose Continue. AutoCAD displays the Label Template dialog box.
6. In the Label Template dialog box, choose the Label Fields tab.

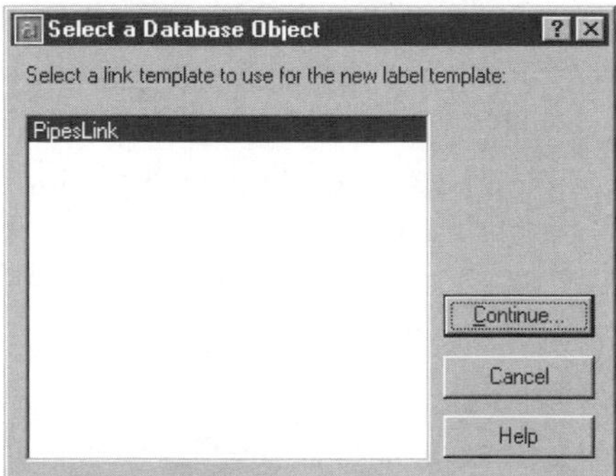

Figure 28–29 *The Select a Database Object dialog box is where you identify the link template to use with the label template.*

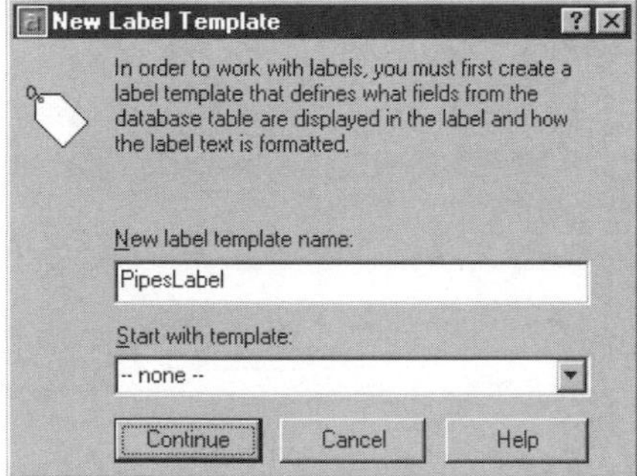

Figure 28–30 *The New Label Template dialog box lets you enter a name for the new label template.*

7. In the Label Fields folder, from the Field list, choose DIA and then choose Add. AutoCAD adds the DIA field to the text window.
8. In the text window after #(DIA), type a double quote symbol (").
9. In the Label Template dialog box, choose the Properties tab.
10. In the Properties folder, from the Justification list, choose Middle Left ML. The mtext object's text insertion point is set to middle-left.

Note: The Label Template dialog box is actually the Mtext dialog box, and the label is inserted as an mtext object.

Note: The field value displayed in the text area represents the column from which AutoCAD will extract the value of the record that is linked to the selected AutoCAD object. You can add additional text, such as a prefix or suffix, to the field value displayed in the text window.

11. In the Label Template dialog box, choose the Character tab.
12. In the Character folder, right-click in the text window and then choose Select All. The text is selected in the window.
13. In the Font Height list, type **10.0** as shown in Figure 28–31, and then choose OK. AutoCAD defines the new label template for the 28DWG01.DWG drawing, as is indicated by the PipesLabel symbol displayed in the dbConnect Manager shown in Figure 28–32.

    Next you will insert a label.
14. In the dbConnect Manager, choose the PIPES table, and then choose the View Table button. AutoCAD displays the Data View window in read-only mode.
15. From the New Link Template list, make sure the PipesLink link template is selected.
16. From the New Label Template list, make sure the PipesLabel label template is selected.

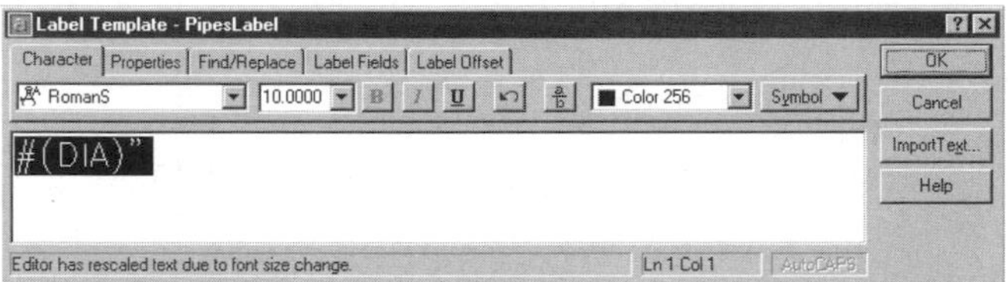

Figure 28–31 *The field from which AutoCAD should extract the table data is defined, and a quote symbol is added as a suffix to include with the label.*

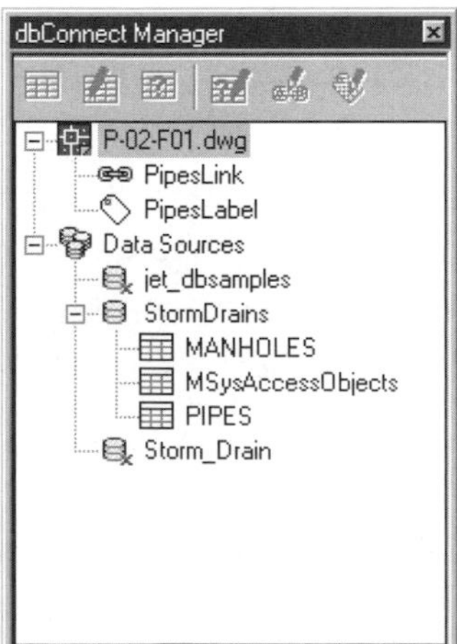

Figure 28–32 *The label template is defined for the 28DWG01.DWG drawing.*

17. In the Data View window, select the record header for record ID 1003.
18. From the Data View pull-down menu, choose Link and Label Settings>Create Attached Labels. AutoCAD switches to "create attached labels" mode.
19. From the Data View menu, choose Link!, then select the polyline that connects to circles 3 and 4, and then press Enter. AutoCAD links the record to the selected polyline object and then inserts the label at the midpoint of the polyline, as shown in Figure 28–33.
20. Save your changes and then close the drawing.

Tip: After you insert the label, you can move the label above the polyline for easier viewing.

AutoCAD creates a link between the selected record and the selected AutoCAD object, and then it extracts the cell value from the field indicated by the label template. The label's insertion point is controlled in the Label Template dialog box's Label Offset folder, and it may be modified. Additionally, you can edit label templates by choosing Templates>Edit Label Template from the dbConnect menu.

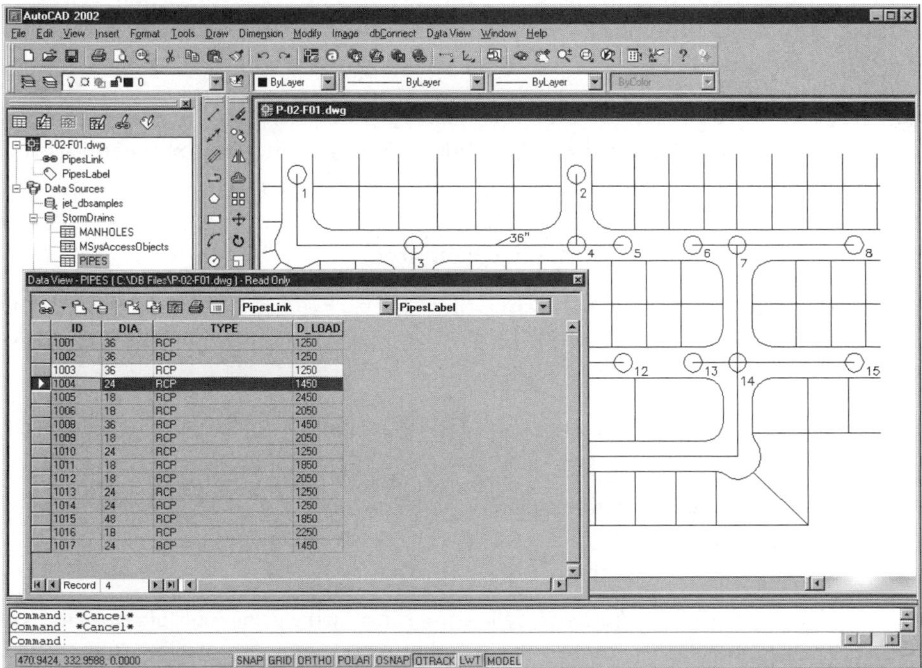

Figure 28–33 *The selected record is used to label the polyline.*

Tip: You can control the layer on which the label is inserted by switching to the desired layer before you insert the label.

By using AutoCAD's labeling feature, you can easily label objects in drawings by extracting values from data in tables. Using this technique, you can automate the process of labeling objects, and you can insert text values accurately by extracting the text values directly from the linked database table. AutoCAD's labeling feature makes adding text to drawings easier. In the next section you will use AutoCAD's querying features to find data quickly.

USING QUERIES

Databases can contain enormous amounts of data. They can consist of dozens of tables, with each table containing hundreds of records, and each record consisting of many fields. The amount of data in a database can be overwhelming.

The goal of any database is to provide a place to organize and store large amounts of information, which you can then query for subsets of data. A query consists of search criteria that you specify. Once the search criteria is defined, the query is run, and the query searches the entire database for data that matches the specified criteria. If matching data is found, the query returns only those records that contain the matching data. By defining and running queries, you can quickly extract the particular set of data you need.

AutoCAD allows you to create and run queries. You can create queries that search through a table for specified values. You can also define queries that search through the objects in a drawing, returning a selection set of records that meet the query criteria. By using AutoCAD's Query Editor, you can build and run queries that search through a drawing or a table for the data you need.

You create queries in AutoCAD using the Query Editor, which consists of four folders, as shown in Figure 28–34. The folders are designed for building queries, and their purposes are described as follows:

- **Quick Query**—Lets you define and run a query using basic operators such as Is Equal To or Is Greater Than.
- **Range Query**—Lets you define and run a query based on a range of values. For example, you can query for all objects whose field value is greater than or equal to 18 and less than or equal to 36.
- **Query Builder**—Lets you define and run a query using multiple operators and ranges, and allows you to use parentheses to group the criteria. Additionally, you can use Boolean operators such as AND and OR to refine your query further. This folder represents AutoCAD's primary query builder.

- **SQL Query**—Lets you define and run a query by creating SQL statements that conform to Microsoft's implementation of the SQL 92 protocol. This folder allows you to build free-form SQL queries, queries that perform relational operations on multiple database tables using the SQL join operator.

By using AutoCAD's Query Editor, you can construct a variety of queries that range from simple to complex.

In the following section you will use the Query Editor to define and run a simple query.

QUERYING OBJECTS

The simplest way to create a query is to use the Query Editor's Quick Query feature. The Quick Query feature allows you to define a basic query that can find data using simple comparison operators such as Is Equal To or Is Greater Than. By using the Quick Query feature, you can easily define and run a query that searches your data and returns the values you need.

Note: To perform the steps in the following exercise, you must use the drawing, that was modified in the exercises located in the section titled Working with Data and Objects in this chapter.

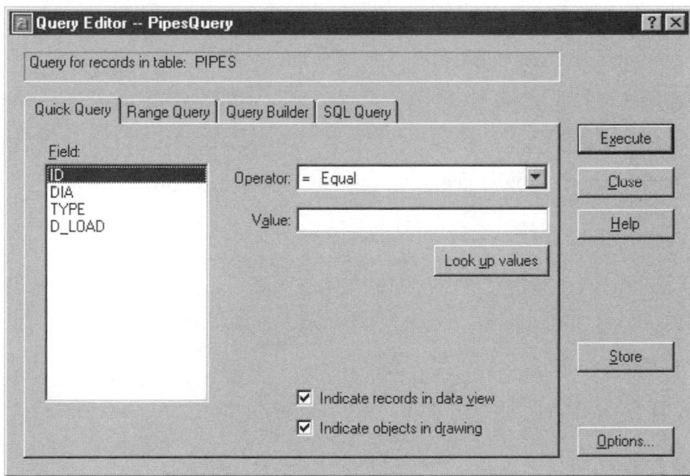

Figure 28–34 *The Query Editor lets you build queries that range from simple to complex.*

In the following exercise you will use the Quick Query feature to build and run a query that searches for pipes with a diameter equal to 36 inches.

EXERCISE: CREATING AND RUNNING A QUERY

1. Open the 28DWG01.DWG file located in the DB Files folder.
2. From the Tools menu, choose dbConnect. AutoCAD displays the dbConnect Manager and the dbConnect pull-down menu.
3. In the dbConnect Manager, right-click on the StormDrains data source, and then choose Connect. AutoCAD connects to the data source and displays its tables.
4. From the dbConnect menu, choose Queries>New Query on an External Table. AutoCAD displays the Select Data Object dialog box.
5. In the Select Data Object dialog box, select the PIPES table as shown in Figure 28–35, and then choose Continue. AutoCAD displays the New Query dialog box.
6. In the New Query dialog box, in the New Query Name text box, type **PipesQuery** as shown in Figure 28–36, and then choose Continue. AutoCAD displays the Query Editor.
7. In the Query Editor, select the Quick Query tab.
8. In the Quick Query folder, in the Field list, choose DIA.
9. From the Operator list, choose = Equal.
10. In the Value text box, type **36**.
11. Make sure that the Indicate Records in Data View and Indicate Objects in Drawing check boxes are selected as shown in Figure 28–37.

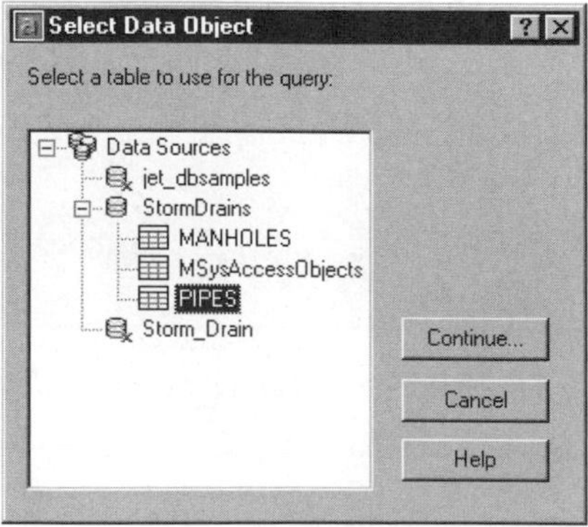

Figure 28–35 *In the Select Data Object dialog box, you choose the table you wish to query.*

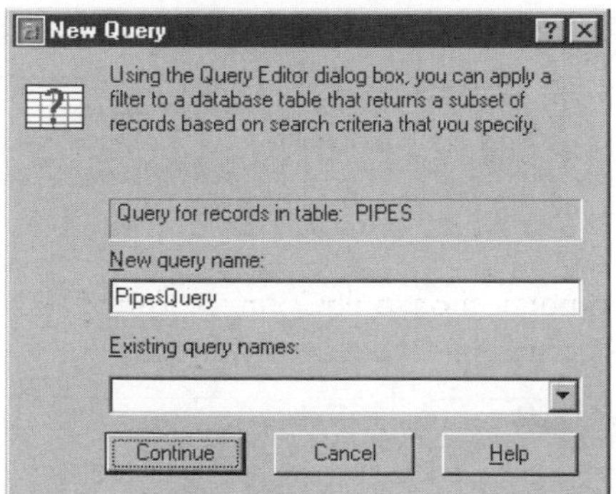

Figure 28–36 *In the New Query dialog box, you define the name for the new query.*

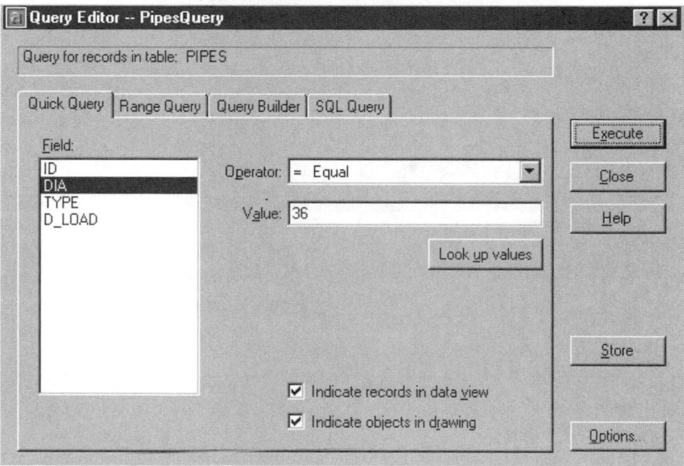

Figure 28–37 *The quick query is defined and ready to run.*

12. Choose Execute. AutoCAD executes the query, returns the subset records that match the query criteria, and displays the records in the Data View window, as shown in Figure 28–38.
13. You may exit AutoCAD without saving your changes.

With the query's selection set returned and displayed in the Data View window, you could highlight all the records in the Data View window and then choose the View Linked Objects in Drawing button to have AutoCAD highlight the selection set's linked objects in the drawing, thereby visually locating all pipes with a diameter equal to 36 inches.

AutoCAD's Query Editor allows you to define and run a query. You can quickly define a query using the Quick Query folder, or you can create more complex queries using the Range Query, Query Builder, or SQL Query folders. By using the Query Editor, you can easily define and execute a query that returns the data you need.

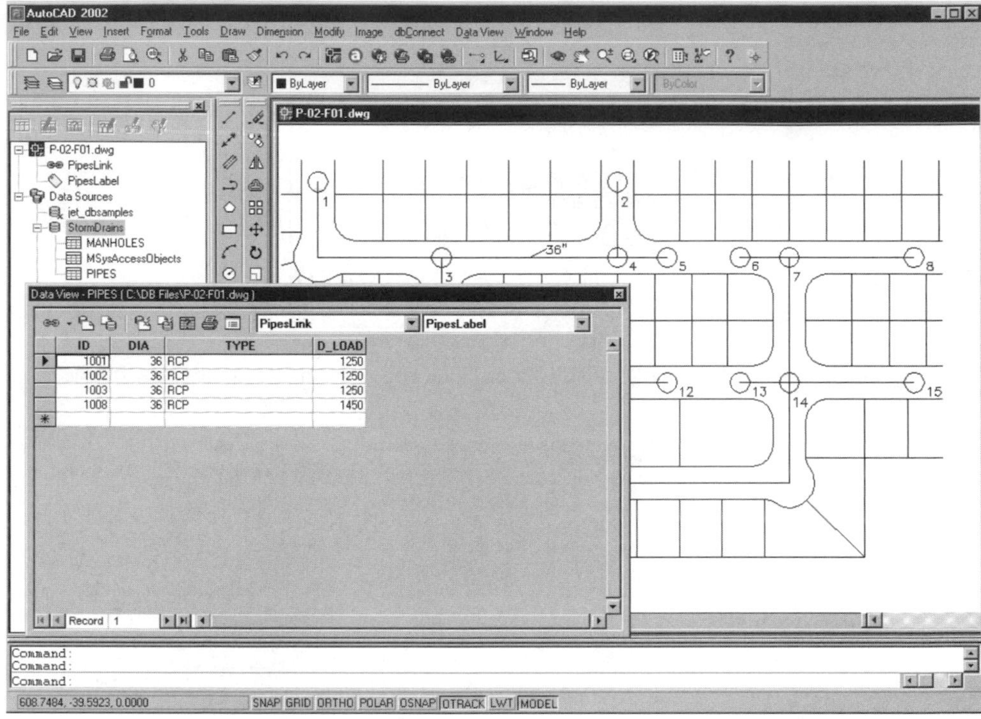

Figure 28–38 *The quick query returns the subset of records that match the query's criteria.*

SUMMARY

In this chapter you reviewed how to use the ODBC Data Source Administrator provided by the Windows operating system to create a data source for a database file. You learned how to use the Data Link Properties dialog box to create an OLE DB (.udl) configuration file, which lets AutoCAD connect to an external database. You used the dbConnect Manager to access external database tables, and you used the Data View window to view and edit the information in tables. You learned how to link data to AutoCAD objects, and how to label objects with data extracted from external database tables. Finally, you learned how to execute queries and query for objects in either AutoCAD drawings or tables.

By using the techniques discussed in this chapter, you can associate external database files with your AutoCAD drawings and use the tools provided by AutoCAD to work with the data stored in database tables.

INDEX

SYMBOLS

NUMERICS

A

C

D

E

G

H

I

J

K

L

M

O

P

Q

R

S

X

Y

Z